Commercial Drafting and Detailing

Commercial Drafting and Detailing

Second Edition

ALAN JEFFERIS

Clackamas Community College, Oregon City, OR

KENNETH D. SMITH, A.I.A.

THOMSON LEARNING

Australia Canada Mexico Singapore Spain United Kingdom United States

Commercial Drafting and Detailing
by Alan Jefferis & Kenneth D. Smith, AIA

Business Unit Director:
Alar Elken

Acquisitions Editor:
James DeVoe

Executive Editor:
Sandy Clark

Development Editor:
John Fisher

Editorial Assistant:
Jasmine Hartman

Executive Marketing Manager:
Maura Theriault

Marketing Coordinator:
Karen Smith

Production Manager:
Larry Main

Production Editor:
Stacy Masucci

Art/Design Coordinator:
Mary Beth Vought

Cover:
Photo of the Bryan Street Garage, Savannah, GA, courtesy of Carl Walker, Inc. and Hansen Architects

Library of Congress Cataloging-in-Publication Data

Jefferis, Alan
  Commercial drafting and detailing / Alan Jefferis, Kenneth D. Smith -- 2nd ed.
   p. cm.
  Includes index
  ISBN 0–7668–3886–2
  1. Architectural drawing. 2. Architectural drawing--Detailing. 3. Commercial buildings--Designs and plans. 4. Structural drawing. 5. Industrial buildings--Details--Drawings I. Smith, Kenneth D. II. Title

NA2708 .J445 2001
720'.28'4--dc21
                                            20010117350

NOTICE TO THE READER

Contents

SECTION 4 PREPARING STRUCTURAL DRAWINGS

Preface

COMMERCIAL DRAFTING AND DESIGN is a practical, comprehensive textbook that is intended to introduce students to the development of architectural and structural drawings required to develop a commercial structure. Students are expected to have previously completed either a basic drawing class or a class dealing with residential drafting as well as have a working knowledge of AutoCAD. Students will build on their knowledge gained through residential drafting and design as common materials and construction methods for commercial structures are explored. Throughout the text, students will be exposed to the work that the engineer or architect will do to develop the project. This book is in no way an attempt to transform drafters into junior engineers. The work of the engineer and the architect are presented to help develop an understanding of the drafter's role in the overall process of developing a set of plans.

LAYOUT

This text has two major divisions. Chapters 1 through 12 are design to introduce the drafting student to the materials and construction methods of commercial construction. This portion of the book is itself divided into two major sections. Each chapter will provide quizzes to test understanding, as well as practical drawing problems to apply your knowledge. Many of the drawing problems are intended to force the student to make use of Sweets's catalogs, vendor Web sites, or local suppliers to prepare the student for the research that is expected in most professional offices.

Section I will introduce students to the major codes that they will encounter in an engineers or architects office. The content of the entire book is based on the 2000 edition of the International Building Code. Students are also introduced to ADA standards that effect commercial structures. Although this is information that the project designer will be responsible for, we feel it is critical for the drafter to understand the codes that shape design.

Section II deals with the materials that shape the construction process. Wood, timber, steel, concrete block, and poured concrete construction are addressed. Each chapter could serve as the core for a class dealing with each material or as an overview of the entire process.

The second half of this text is intended to be a practical approach to the development to the architectural and structural drawings. The key elements of each drawing that comprises the architectural and structural drawings will be introduced in Chapters 13 through 23. In Chapter 14, projects will be introduced, with each one reflecting one of the major building materials. These projects include the following:

- Wood: Multi-family complex
- Concrete block\truss roof: retail sales
- Concrete block: warehouse with mezzanine, panelized roof
- Pre-cast concrete: warehouse with mezzanine, truss roof

These projects are intended to be used throughout the remaining chapters so that students will develop drawings that relate to one structure. Portions of each of the projects contained in Chapters 15 to 23 can be accessed from http://www.delmar.com/resources/ocl.html and used as a base from which to complete the assignment. The drawings are a rough draft only and need to be updated to common standards. Use appropriate symbols, linetypes, dimensioning methods, and notations to complete the drawing.

Note: The drawings contained at the end of the chapters are to be used as a guide only. As you progress through them, you will find that some portions of the drawings do not match things that have been drawn on previous drawings. Each project has errors that will need to be solved. The errors are placed to force you to think in addition to draw. Most of the errors are so obvious that you will have no trouble finding them. If you think you have found an error, do not make changes to the drawings until you have discussed the problem and possible solutions with your engineer (your instructor).

ACKNOWLEDGEMENTS

The following architects and engineers have reviewed portions of the manuscript and provided the illustrations that are provided throughout the text:

G. Williamson Archer A.I.A., Archer & Archer P.A.

Scott Beck, Scott R. Beck, Architect

Bill Berry, Berry-Nordling Engineers, Inc.

Laura Bourland, Halliday Associates

LeRoy Cook, Architectural Instructor, Clackamas Community College

Charles J. Conlee, P.E., Conlee Engineers, Inc.

Chris DiLoreto, DiLoreto Architects, LLC.

Russ Hanson, H.D.N. Architects, A.I.A.

Havlin G. Kemp P.E., Van Domelen/Looijenga/McGarrigle/Knauf Consulting Engineers

Tom Kuhns, Michael and Kuhns Architects, P.C.

Ron Lee, A.I.A., Architects Barrentine, Bates & Lee, A.I.A.

Ned Peck, Peck, Smiley, Ettlin Architects

David Rogencamp, KPFF Consulting Engineers

Sherry Bryan-Hagge, Memphis State University, Memphis, TN

Jeffrey Callahan, University of Alaska, Ankorage, AK

Nadine Deiterman-Greni, Western Dakota Techinical Institute

Steven Dinin, Finger Lakes Community College, Canandaigua, NY

Debra Dorr, Phoenix College, Phoenix, AZ

Dan Masters, Southeast Community College, Lincoln, NE

Jesse Mireles, Phoenix College, Phoenix, AZ

Scott Rathburn, Triangle Institute of Technology, Erie, PA

The following Associations and companies have also made major contributions to this text and would be excellent references for students who desire further information.

Robert G. Wiedyke
Director of Publications Services
American Concrete Institute
P.O. Box 19150
Detroit MI 48219
American Institute of Architects Press
1735 New York Avenue NW
Washington, DC 20006-5292
American Institute of Steel Construction
1 East Wacker Drive, Suite 3100
Chicago, IL 60601

Bruce D. Pooley P.E.
Director of Technical Services
American Institute of Timber Construction
7012 South Revere Parkway
Suite 140
Englewood, CO 80112

Mike Shultz
American Plywood Association
P.O. Box 11700
Tacoma, WA 98411-0700

American Society for Concrete Construction
1902 Techny Court
Northbrook, IL 60062

Americans with Disabilities Act
Federal Register Vol 56, No 144, July 26, 1991

Bethlehem Steel Corporation
S.G.O. Building
Structural Products Division
701 E. Third Street.
Bethlehem, PA 18016

Pamela Allsebrook
Publicity Manager
California Redwood Association
405 Enfrente Drive, Suite 200
Novato, CA 94949

Concrete Masonry Association of California and Nevada
6060 Sunrise Vista DriveSuite 1875
Citrus Heights, CA 95610

Steven E. Ellingson P.E.
Director of Marketing
Concrete Reinforcing Steel Institute
933 N. Plum Grove Rd.
Schaumburg, IL 60173-4758

Construction Metrication Council
National Institute of Building Sciences
1201 L Street, N.W. Suite 400
Washington, DC 20005

Construction Specifications Institute
601 Madison Street
Alexandria, VA 22314-1791

Kathy Ziprik
Public Relations Manager
Georgia Pacific Corporation
133 Peachtree Street, N.E.
Atlanta, GA 30303

Richard P. Kuchnicki
Executive Vice President
International Code Council
5203 Leesburg Pike, Suite 708
Falls Church, Va. 22041-3401

Anna Lopez & Mark Johnson
International Conference of Building Officials
5360 Workman Mill Road
Whittier, CA 90601-2298

Terry C. Peterson
Marketing Manager
MERO Structures
N112 W18810 Mequon Rd.
Germantown, WI 53022

Patsy Harms
Portland Cement Association
54200 Old Orchard Road
Skokie, Il. 60077-1083

Post-Tensioning Institute
1717 W. Northern Avenue
Suite 218
Phoenix, Arizona 85021

Robert C. Guzikowski
Simpson Strong-tie Connectors
4637 Chabot Dr. Ste 200
Pleasanton, Ca 94588

Richard Wallace
Media Director
Southern Forest Products Association
P.O. Box 641700
Kenner Louisiana 70064

R. Donald Murphy
Managing Director
Steel Joist Institute
1205 48th Avenue North
Suite A
Myrtle Beach, S.C. 29577-5424

Trus Joist Corporation
P.O. Box 60
Boise, Idaho 83707

U.S.G. Steel Framing Systems
101 South Wacker Drive
Chicago, Il. 60606-4385

Western Wood Products Assoc.
1500 Yeon Building
Portland, Or. 97204

SECTION 1

Drafting & Design Considerations of Structures

CHAPTER 1

Professional Careers and Commercial Drafting

To many in the construction field, commercial drafting is the development of the construction drawings for any non-residential structure. For most professionals involved in the design process, the general area of commercial drafting and construction is broken into the areas of public, industrial, institutional, and commercial projects. Examples of each include

- **Public structures**—buildings or portions of structures used for assembly, education, or civic administration such as schools, churches, stadiums, post offices, and libraries
- **Industrial structures**—buildings or portions of structures used for manufacturing, assembly, or storage such as factories, warehouses, and businesses involved in hazardous, toxic, or unstable materials
- **Institutional structures**—buildings or portions of structures used by occupants with limitations caused by health, age, correctional purposes, or in which the mobility is restricted such as residential care facilities, medical facilities where 24-hour care is provided, day care facilities jails, and prisons
- **Commercial structures**—buildings or portions of structures used for sales, business and professional or service transactions, including office buildings and eating and drinking establishments

This text will look at commercial drafting as it relates to all of these fields. Your role in the development of commercial drawings will vary depending on the size and structure of the firm where you work. Some design firms are very specialized and work only in one field, such as educational or institutional facilities. Some firms have several different divisions within the company for designing various areas of occupancy. Chapter 4 will introduce specific areas of construction based on the occupancy of the structure.

No matter how you define commercial construction, commercial drafting offers far more career options and challenges than residential drafting. Commercial projects are typically much larger and are completed by a team rather than an individual. In addition to the design team, engineers, architects, and drafters from several different firms in a variety of construction fields complete the project.

The field of architecture has always offered many career opportunities. The acceptance of computers and the use of AutoCAD® have revolutionized many offices and opened an almost unlimited potential for design. This chapter will explore some of the main areas of employment in the field of architecture and explain how

CAD has affected each area. The four major areas of employment in the architectural field are that of drafter, designer, architect, and engineer.

DRAFTER

A drafter is typically a person who draws the designs that originate with another person. The drafter's main responsibility is to take a drawing similar to Figure 1-1 and fill in the missing material, using acceptable office standards so that it resembles the drawing in Figure 1-2. The experience and education of the drafter affect the actual job assignments.

The Beginning Drafter

The beginning or junior drafter is typically preoccupied with making corrections, running prints, and completing simple drawings while confidence is gained in the standard office procedure. In some areas of the country, it is possible to obtain a beginning drafting job as a high school student. Because of the complexities in commercial drafting, most offices require college drafting experience. A new drafter might be limited to making corrections to existing drawings. An understanding of AutoCAD's basic command structure and knowledge of basic construction techniques are essential for advancement. One of the best ways to gain an understanding of typical construction practice is to spend time at construction sites. Being able to follow a project through the various stages of construction will greatly aid a beginning drafter in gaining an understanding of the information that is entered at the keyboard.

In addition to improving CAD drafting skills, most employers expect academic skills to increase before a beginning drafter is assigned to more challenging projects. Although most office procedures for a junior drafter involve only basic math skills, an understanding of algebra, geometry and trigonometry will help in career advancement.

Other necessities for advancement are confidence, reliability, and the ability to work in a team. As the beginning drafter gains confidence and an understanding of the types of drawings being completed, the supervisor will be able to provide sketches with less detail and rely on the drafter to research a solution based on the governing code and similar projects that have been completed in the office. Reliability within an office is measured by the maintenance of good attendance patterns and the production of drawings as scheduled. It is important that a drafter be able to accurately estimate the

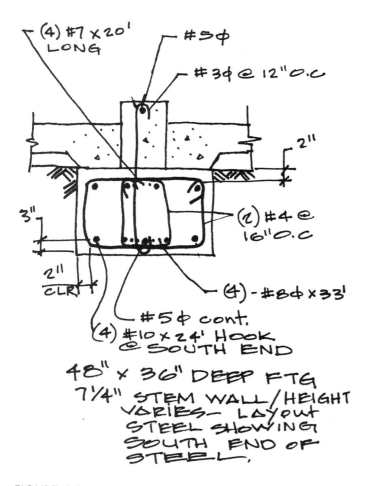

FIGURE 1-1 ■ The project manager will typically provide a sketch for the drafter to use as a guide.

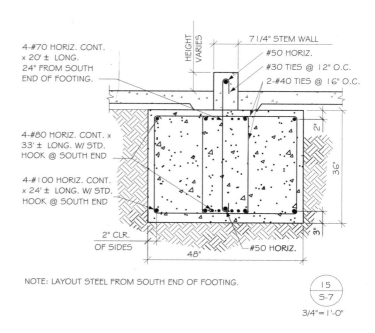

FIGURE 1-2 ■ The drafter's main responsibility is to complete drawings using acceptable office standards.

time required to complete each project. You can learn this skill while working on school projects by estimating the amount of time required to complete the project prior to starting the drawing. In the planning stage, break the project into components and estimate how long each component will take. When the drawing is complete, review your estimates. Although most firms would prefer a drafter who can quickly complete a project, speed is no substitute for accuracy. Push yourself as a student to meet a self-imposed time deadline while maintaining quality and accuracy. Reliability is also important because the drawings for a structure are completed by a team. The ability to get along with others, to complete assigned projects in a timely manner, and to coordinate different parts of a project with others will greatly affect how fast a drafter will advance.

Don't be discouraged as you consider the typical entry-level drafting position. This is the type of position you might consider during your first year in school, not as a career. Although you might aspire to design the eighth, ninth, and tenth wonders of the world, an entry-level position can offer valuable insight into the world of architecture and supplement your academic classwork.

Senior Drafter

A senior drafter, often referred to as a project manager or a job captain, is part of a design team in most mid-sized offices. This person is responsible for supervising several junior drafters. The senior drafter is expected to be familiar with the governing building codes that govern the project and to maintain legal building standards. In small firms the senior drafter is typically in charge of producing the working drawings and is often expected to perform simple beam calculations and preliminary wind and seismic studies. Other job duties might include assigning projects for the drawing team, site visitations, and conferences with municipal building officials. A senior drafter is often instrumental in assigning file and layer names, linetype, and other CAD drawing standards and requirements to be used throughout the project.

In many offices that draw multifamily and light commercial projects, senior drafters are part of a larger design team comprised of other drafters and supervised by an architect or engineer. The drafter is expected to select material from vendor catalogs such as *Sweet's Catalogs* and to apply basic information from *Architectural Graphics* or *Time Saver Standards* to specific items within each project.

Educational Requirements

The education required to be a drafter could range from a year of high school drafting classes to a degree from an accredited junior college or technical school. Basic writing, math, art, and drafting classes will enhance job opportunities. Computer classes such as keyboarding, DOS, and LISP will also be helpful. Classes offering a complete mastery of AutoCAD's 2D commands are a must and update classes in the most recent release of AutoCAD are often required. Once you are established within the firm, continuing education that will improve job performance is often a company benefit.

Employment Opportunities

Firms of all sizes hire drafters to help complete drawing projects. Many opportunities exist for converting hand-drawn details and drawings to AutoCAD. In addition to jobs in architectural and engineering offices, many drafters are employed by suppliers of architectural equipment such as heating, air conditioning, electrical, and plumbing companies. Many construction subcontractors hire drafters to do their shop drawings. Firms that manufacture the actual construction components such as trusses, steel, and concrete beams also employ drafters to produce the needed drawings to manufacture their components. Many firms hire drafters for their CAD and construction knowledge to be sales representatives. Utility companies such as cable television, telephone, water, and gas companies all employ large numbers of CAD drafters to draw maps of individual routes of service. City, county, state, and federal agencies also hire large numbers of CAD drafters to update municipal records and projects such as parcel and subdivision maps, sewer layouts, and roadway projects. Figure 1-3 shows an example of a sewer drawing prepared by a drafter working for a city sanitation department. Information about drafting opportunities in related construction trades can be obtained from the American Design Drafting Association. This organization contains members who work in areas such as heating and air conditioning, plumbing, and electrical design. Contact the American Design Drafting Association for further information:

American Design Drafting Association
P.O. Box 11937
Columbia, SC 29211
803-771-0008

DESIGNER

In some states a designer is a drafter who, by experience or state certification, is allowed to design multifamily and small commercial projects. The size of the project is dictated by the state. State laws vary on the types of buildings a designer may assume responsibility for without the supervision of an architect or engineer. In some areas of the country, anyone can decide that they are a building designer. In most states a designer working with commercial projects is required to work under the direct supervision of a licensed architect or engineer.

Because of increasing liability problems, many states now require any person using the title *designer* to be licensed. The American Institute of Building Design (AIBD) now oversees designers in forty states throughout the United States. Contact the AIBD for further information:

American Institute of Building Design
P.O. Box 1148
Pacific Palisades, CA 64057

ARCHITECT

An architect is a person who is licensed by individual states to design and supervise the construction of structures. Although a few work full time in the design or supervision of residential projects, most work in multifamily and commercial design because of the design challenges, job security, pay, and fringe benefits. Figure 1-4 shows a project designed by an architect. An architect is responsible for the design of a structure, from the preliminary drawings through the construction process, and must be able to coordinate the desires and limitations of the client, the financial restraints of the project, the physical setting of the project, and the structural elements of the material to be used.

Use of the title *architect* is legally restricted by each state. In order to be a licensed architect, a person must obtain a master's degree from an accredited school of architecture, complete three years of work experience for a licensed architect, and then pass the state test to demonstrate competency in several areas related to architectural practice. In some states an architect's license can also be obtained by gaining practical experience under the direct supervision of an architect and then passing the written exams. The length of time varies with each state, but typically five to eight years of work experience under the direct supervision of an architect are required before the state examinations may be taken.

Educational Requirements

Because of the wide range of areas that an architect must be competent in, the educational requirements must be diversified. In addition to the skills recommended for drafters, classes in the fine arts, history, and the humanities will help provide an understanding of how structures relate to the cultures that will occupy the structure. Training in math and science will provide the basis for the structural design. Sociology and psychology classes will help provide an understanding of how individuals and groups will interact with the structure and various parts of the building.

Employment Opportunities

An architect performs the tasks of many professionals, including designer, artist, project manager, and construction supervisor. Common positions listed by the American Institute of Architects (AIA) that architects fill include

Technical Staff—CAD operators, drafters, consulting engineers such as mechanical, electrical, and structural engineers, landscape architects, interior designers.

Intern—unlicensed architectural graduate with less than three years experience. Common responsibilities include developing design and technical solutions under the supervision of an architect.

Architect I—licensed architect with 3 to 5 years of experience. The job description typically includes responsibility for a specific portion of a project within the parameters set by a supervisor.

Architect II—licensed architect with 6 to 8 years of experi-

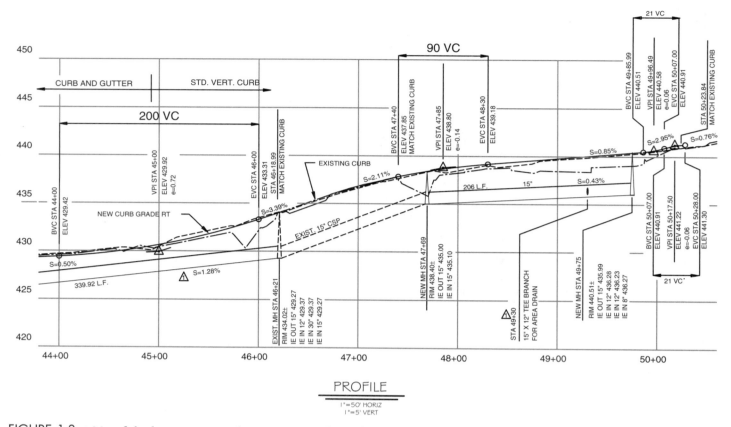

FIGURE 1-3 ■ Many federal, state, county and city agencies employ drafters to complete required drawings. This profile drawing was completed by a drafter employed by a city sanitation department. *Courtesy City of Oregon City.*

ence. The job description typically includes responsibility for the daily design and technical development of a project.

Architect III—licensed architect with 8 to 10 years of experience. The job description typically includes responsibility for the management of major projects.

Manager—licensed architect with more than 10 years of experience. The job description includes management of several projects, project teams, client contacts, project scheduling, and budgeting.

Associate—senior management architect, but not an owner in the firm. This person is responsible for major departments and their functions.

Principal—owner or partner in an architectural firm.

The work of an architect is not limited to the design of structures. Architects are often employed by local and state governments to work as urban planners, planning whole areas rather than a single structure. Some architects, such as interior planners, lighting specialists, or audio experts, specialize in specific areas of design. Contact the AIA for more information about the role of an architect:

The American Institute of Architects
1735 New York Avenue, NW
Washington, DC 20006

Write to the these organizations for related information about specialties in the field of architecture:

American Society of Landscape Architects, Inc.
636 Eye Street, NW
Washington, DC 20001-3736
202-898-2444

American Institute of Interior Design
730 Fifth Avenue
New York, NY 10019

ENGINEER

The title *engineer* encompasses a wide variety of professions. Structural, electrical, mechanical, and civil engineers are typically involved in the design of a structure. *Structural engineers* are licensed professionals who typically specialize in the design of the structural skeleton of a structure. Figure 1-5 shows the framework of a structure that is the result of an engineer's calculations and drawings. *Electrical engineers* work with architects and structural engineers and are responsible for the design of lighting, power supply, and communications systems. This would include supervision of the design and installation of specific lighting fixtures, telephone services, and requirements for computer networking.

FIGURE 1-4 ■ The entry way to the Interfirst II project by S.O.M Architects of Houston Texas is representative of the wide range of projects architects design and draw. *Courtesy MERO Structures, Inc.*

Mechanical engineers are responsible for the sizing and layout of the heating, ventilation, and air conditioning systems (HVAC) and details similar to those shown in Figure 1-6. This involves working with the project architect, who determines the number of occupants and the heating and cooling loads, and designing the route that conditioned air will take through the building. The mechanical engineer also plans the size and locations of plumbing lines for fresh and waste water as well as lines for other liquids or gases that might be required for specific occupancies.

Civil engineers are responsible for the design and supervision of a wide variety of construction projects such as subdivision plans, as well as grading and utility plans for construction projects such as buildings, highways, bridges, sanitation facilities, and water treatment plants. See Figure 1-7. Civil engineers are often directly employed by construction companies to oversee the construction of large projects and to verify that the specifications of the design architects and engineers have been carried out.

Educational Requirements

Success in any of the engineering fields requires a high degree of proficiency in math and science. Similar to the requirements for

becoming an architect, engineers are required to complete five years of education at an approved college or university, followed by successful completion of a state-administered examination. Certification can also be accomplished by training under a licensed engineer and then successfully completing the examination. Contact these organizations for additional information about the field of engineering:

American Society of Civil Engineers (ASCE)
1801 Alexander Bell Drive
Reston, VA 20191-4400
800-548-2723

American Consulting Engineers Council (ACEC)
1015 15th Street, NW, Suite 802
Washington, DC 20005
202-347-7474

American Society of Heating, Refrigerating, and Air
    Conditioning Engineers, Inc. (ASHRAE)
1791 Tullie Circle NE
Atlanta, GA 30329-2305
800-527-4723

Illuminating Engineering Society of North America
    (IESNA)
120 Wall Street, 17th Floor
New York, NY 10005
212-248-5000

RELATED FIELDS

In addition to the drawing and design aspects of construction drawings, a drafting background can lead to several other careers such as illustrator, model maker, specification writer, and inspector. A drafting background can also be useful in careers such as sales representative, contractor, and estimator. Although these jobs do not require drafting as a job requirement, a knowledge of drafting fundamentals will help in the performance of job requirements.

Illustrator

Commercial drawing requires the use of a rendering such as that seen in Figure 1-8. By combining artistic talent with a basic understanding of architectural principles, an illustrator is able to produce drawings that realistically show a proposed structure. Traditionally, graphite, ink, and watercolor have been the main tools of an illustrator. Many firms are now adopting the use of AutoCAD's solid modeling programs and third-party software to create computer-generated renderings similar to the building in Figure 1-9. Drawings created in 3D allow the viewpoint to be altered, providing a realistic look of the structure from any vantage point. Illustrators also use *virtual reality* to represent how a structure will blend with its environment and to allow clients to walk through proposed buildings and see how components can be arranged. Some program developers are predicting that the traditional 2D drawings of the construction industry could be replaced entirely by 3D color drawings and models.

FIGURE 1-5 ■ The steel skeleton for a structure can be seen before the nonstructural materials are applied. Engineers are responsible for the design of the structural skeleton. *Courtesy Janice Jefferis.*

FIGURE 1-7 ■ Civil Engineers are responsible for the design of a variety of construction projects including grading and utility plans. *Courtesy Bethlehem Steel Corporation.*

Model Maker

In addition to presentation drawings, many architectural firms use models of building projects to help convey design concepts. Models similar to the one in Figure 1-10 are often used for public presentations to groups such as zoning commissions, citizens' advisory committees, and other regulatory commissions. Model makers need basic drafting skills to interpret the working drawings so that the model can be an accurate representation of the proposed structure. Model makers can be employed within a large architectural firm or by a company that specializes in modeling.

Specifications Writer

Most architectural and engineering firms have an internal staff that specializes in written documentation. The written specifications

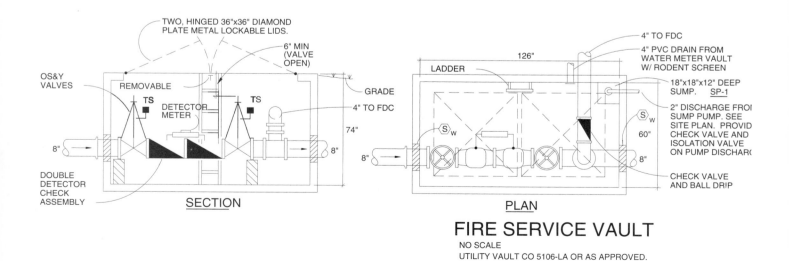

FIGURE 1-6 ■ Mechanical engineers are responsible for the design of the HVAC system. *Courtesy Manfull Curtis Consulting Engineers.*

that accompany the working drawings, environmental impact reports, and addendums to the permit forms are all written by a specifications writer. An understanding of basic drafting principles, a mastery of grammar and basic English skills, keyboarding, word processing, and spreadsheet skills are all essential for a specifications writer. A thorough understanding of the Construction Specifications Institute's (CSI) chapter format, which will be introduced in Chapter 13, is also required.

Inspector

Municipalities require that the plans and the construction process be inspected to ensure that the required codes for public safety have been met. A plans examiner must be licensed by the state to certify a thorough understanding of the construction process and common engineering practices. Most states require a degree in architecture or engineering for a state or city inspector or plans examiner.

Most architectural and engineering firms also require job inspection as part of their design process to ensure that contractors have rigidly followed the plans and specifications. This also ensures that conflicts between the numerous firms involved in the construction process can be resolved to the benefit of the client. An inspector or job supervisor is not required to be a licensed architect or engineer, but a thorough understanding of the drawings and the construction process is essential.

CAD INFLUENCES IN ARCHITECTURAL AND ENGINEERING OFFICES

The use of computers in the construction industry has had a substantial impact on the design and documentation of a structure. Given the magnitude of the project size and the number of professionals to be coordinated, the use of computers has greatly aided the management and production of architectural projects. This text assumes you have mastered basic CAD skills and will attempt to help you understand how the skills taught in those classes relate to construction drawings.

Working Drawings

Computers have greatly aided in the drawing and coordination of the working drawings of a structure. Figure 1-11 shows an example of a CAD drawing completed by a drafter. Chapter 3 will review basic CAD drawing skills that can be used to complete the required drawings.

Mathematical Calculations

Third-party software programs that work within AutoCAD are available to provide many of the mathematical calculations for structural analysis. Programs have been available for several years to design wood, steel, and concrete beams and supports. Computer

FIGURE 1-8 ■ Drawings are often prepared by an architectural illustrator to help convey design concepts. *Courtesy Dean Smith, Kenneth D. Smith Architect & Associates, Inc.*

models of a structure can also be created to analyze how specific features of a structure, and the entire structure as a whole, will react under various loading conditions. These studies can be used to determine how a structure will respond to wind pressure, floodwater, or an earthquake of a specific magnitude. Software programs

FIGURE 1-9 ■ Many architectural illustrators generate renderings using AutoCAD or third-party software.

FIGURE 1-10 ■ Models are used for public displays to convey design ideas. *Courtesy KOIN Center, Olympia & York Properties (Oregon), Inc.*

are also available for designing static and fatigue calculations to predict repair and maintenance cost.

Third-party software is available to provide computer analysis of heat loss. Many programs will also show the average outdoor temperature and the average amount of energy needed to reach specific temperatures. Cost comparisons can also be determined based on predicted weather patterns, cost of a specific fuel, and predicted efficiency of a heating unit.

Three-Dimensional Drawings

Software developments by AutoCAD and third-party developers provide the ability to draw three-dimensional drawings similar to Figure 1-12. These drawings can be used to help develop the working drawings as well as the artistic drawings for preliminary planning. In addition to 3D and wire-frame drawings, some of the biggest advances in the architectural field include the uses of solid models. Once drawn, the structure can be rotated so that the client can view the structure at any angle and to reflect the shade that would be generated at specific times of the day, at any specified time of the year.

Three-dimensional models can also be used to provide the client a walk-through of the structure before the working drawings are even started. These models can save huge amounts of time by providing insight to placement of heating ducts, electrical conduits, and various piping runs. Problems in placement can be identified in 3D drawings that may not be easily identified in two-dimensional drawings.

ADDITIONAL READING

The following Web sites can be used as a resource to help you keep current with changes in professional design careers.

www.acec.org	American Consulting Engineers Council
www.aiaonline.com	American Institute of Architects
www.aibdnat@aol.com	American Institute of Building Designers
www.adda.org	American Design Drafters Association
www.asce.org	American Society of Civil Engineers
www.ashrae.org	ASHRAE—American Society of Heating, Refrigerating, and Air-Conditioning Engineers.
www.asla.org	American Society of Landscape Architects
www.thebluebook.com	The Construction Information Network
www.iesna.org	Illuminating Engineering Society of North America
www.nspe.org	NSPE—The National Society of Professional Engineers

REVIEW ACTIVITIES

1-1. Contact your state's branch of the AIA and list the requirements for becoming an architect in your state.

1-2. Visit the Information for Consumers section of the AIA Web site and list the three areas of information listed under Commercial.

1-3. List the benefits of using an architect that are described in the Institutional area of the AIA Web site.

1-4. Visit the National Society of Professional Engineers Web site and briefly describe three current news items that are listed.

1-5. Use a search engine to list and describe a minimum of five different web sites for engineers.

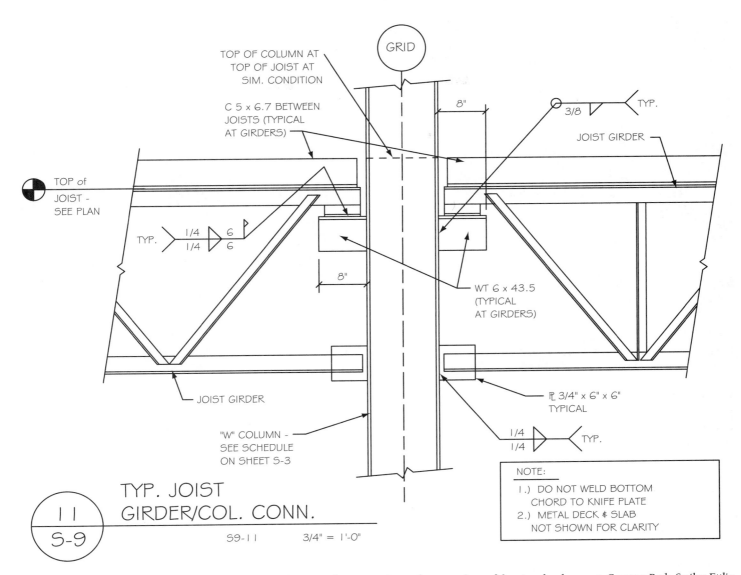

FIGURE 1-11 ■ Computers are used to aid drafters, architects and engineers in nearly every phase of drawing development. *Courtesy Peck, Smiley, Ettlin Architects.*

CHAPTER 1

Professional Careers and Commercial Drafting

CHAPTER QUIZ

Answer the questions on a separate sheet of paper. Print the chapter title, the question number, and a short complete statement for each question.

Question 1-1 List the main job responsibilities of a senior drafter.

Question 1-2 List several skills or assets needed to become a senior drafter.

Question 1-3 List and describe nine different careers in which drafting could be helpful.

Question 1-4 List three different duties a junior drafter might perform.

Question 1-5 Why would a student in an architecture program need to take classes in social science?

FIGURE 1-12 ■ 3D drawings can be drawn and rendered using AutoCAD and third party software. *Courtesy Ginger Smith, Kenneth D. Smith Architect & Associates, Inc.*

Question 1-6 List the addresses of five different architectural or engineering associations in your area.

Question 1-7 Why are engineering students required to have such a strong understanding of math?

Question 1-8 List five computer skills that will be useful in construction drawings.

Question 1-9 List five areas of construction and the impact that CAD has had on these areas.

Question 1-10 Contact one member of the AIBD, AIA, or ASCE in your area and determine specific requirements for education, employment, and advancement in your area.

CHAPTER 2

The Drafter's Role in Office Practice and Procedure

The office that you work in, the size of the staff, and the type of project being drawn will all affect the role of the drafter. This chapter explores common roles of a drafter and common types of drawings included in a set of plans. Because most graduates of two-year technical and junior college programs work for architectural and engineering firms, these will be the only types of office work explored in this text.

COMMON OFFICE PRACTICE

The design of a structure starts with the selection of the design team. The client typically selects the architectural firm based on several common reasons. The reputation of the design firm is typically of utmost importance. The relationship with previous clients and the ability to meet the design requirements in a timely and economical manner are key elements in any designer's reputation. Previous experience with projects similar to the structure to be designed is also an important consideration. Although many large firms work with many types of structures and materials, some smaller architectural firms are very specialized in the type of structures they work with. For instance, some firms work primarily with educational structures. Some firms work only with specific materials such as concrete tilt-up structures or with rigid steel frames. Recommendations from a realtor, builder, or previous client, as well as marketing also play an important part in the selection of an architectural firm.

Once selected, the architectural team can provide valuable assistance to the client in the selection of the construction site and in the refinement of the design criteria. With an understanding of the basic design requirements, analysis of the potential site based on access, climate, and zoning laws can be started. Chapters 4 and 5 will provide further information on how building codes affect the design process. Other areas such as the shape of the potential job site, availability of utilities, topography, soil conditions, drainage patterns, and proposed building materials also must be considered in the initial stages of design. With a specific site purchased, design studies can be done to determine the best possible merging of the client's desires with the limitations of the site and the financial constraints of the project.

The Role of the Drafter

The experience of the drafter, the number of drafters assigned to a project, and the complexity of the project will dictate when the drafter will become involved in the project. Typically a drafter will become involved in the completion of the preliminary drawings that make up the design presentation. This would include work on the floor plan, elevations, and sections so that the key elements of the design can be presented to the client. Once approved, the preliminary drawings can be used as the base for the working drawings. These drawings will become the core of the plans that will be given to the structural engineering firm. A structural engineering firm typically works in conjunction with the architectural firm to analyze the forces that will be imposed on the structure. Chapter 12 will introduce common structural considerations affecting the design of a structure.

Depending on the size and complexity of the project, the structural engineer might provide calculations and details for the architectural firm to use to complete the project or might provide a complete set of structural drawings to accompany the architectural drawings. In either case, drafters will be used to ensure that all of the mathematical calculations of the engineer have been incorporated into the drawings. Drafters will also be used to incorporate information supplied by several other firms including mechanical, electrical, plumbing, lighting, and interior design firms.

TYPES OF DRAWINGS

Because of their simplicity, residential drawings are typically the first construction drawings that students are introduced to. Most plans consist of a site plan, floor plan, elevations, foundation plan, and sections and construction details. Commercial drawings often include over 100 sheets of 24" x 36" or 30" x 42" drawings. Because of the size and scope of most commercial projects, plans are arranged into seven major groups. Common drawings groups include site, architectural, structural, mechanical, plumbing, electrical, and fixture drawings. These drawings might be prepared by separate firms that are coordinated by the architectural firm, or they might be prepared by the architectural team under the supervision of various architectural and engineering firms. Regardless of the source of origination, drafters will be employed to develop each type of drawing. The first page of most commercial plans is a title page, which contains a table of contents similar to the one seen in Figure 2-1, as well as a list of each consulting firm responsible for some aspect of the plans. A list of abbreviations, project data, a location map, and general notes are also placed on the title page.

Site Drawings

Drawings that are related to the construction site typically follow the title page. Each drawing is typically numbered in successive

INDEX OF DRAWINGS

CIVIL

U1 — SITE UTILITIES
U2 — EROSION CONTROL PLAN & DETAILS
U3 — PUBLIC UTILITY PLAN
U4 — UTILITY DETAILS

LANDSCAPE

L1 — IRRIGATION SYSTEM PLAN & LEGEND
L2 — IRRIGATION SYSTEM DETAILS & NOTES
L3 — PLANTING PLAN, DETAILS & LEGEND

ARCHITECTURAL

T1 — TITLE SHEET, SITE DEMOLITION & SURVEY
A1 — SITE PLAN & DETAILS
A2 — GRADING PLAN & DETAILS
A3 — FIRST FLOOR PLAN, SCHEDULES & DETAILS
A4 — SECOND FLOOR PLAN, SCHEDULES & DETAILS
A5 — ENLARGED PLANS, INTERIOR ELEV. & DETAILS
A6 — EXTERIOR ELEVATIONS & DETAILS
A7 — BUILDING & WALL SECTIONS & DETAILS
A8 — ROOF PLAN & DETAILS
A9 — DETAILS
A10 — REFLECTED CEILING PLAN & DETAILS

STRUCTURAL

S1 — FOUNDATION PLAN
S2 — FOUNDATION DETAILS
S3 — SECOND FLOOR FRAMING PLAN
S4 — ROOF FRAMING PLAN
S5 — DETAILS
S6 — DETAILS

MECHANICAL

M1 — FIRST FLOOR PLUMBING PLAN, LEGENDS
M2 — SECOND FLOOR PLUMBING PLAN, NOTES
M3 — DETAILS, SCHEDULES
M4 — FIRST FLOOR H.V.A.C PLAN, LEGENDS
M5 — SECOND FLOOR H.V.A.C. PLAN
M6 — ROOF MOUNTED H.V.A.C. EQUIP. PLAN
M7 — H.V.A.C. DETAILS, SCHEDULES

ELECTRICAL

E1 — NOTES, LEGEND, RISER
E2 — FIRST FLOOR LIGHTING PLAN
E3 — SECOND FLOOR LIGHTING PLAN
E4 — FIRST FLOOR POWER PLAN
E5 — SECOND FLOOR POWER PLAN
E6 — ROOF MOUNTED EQUIP. POWER PLAN
E7 — DETAILS, SCHEDULES
E8 — FIRST FLOOR COMMUNICATIONS PLAN, LEGEND
E9 — SECOND FLOOR COMMUNICATIONS PLAN
E10 — H.V.A.C. DETAILS, SCHEDULES

FOOD SERVICE

FS1 — EQUIPMENT PLAN, SCHEDULES
FS2 — PLUMBING PLAN, MECHANICAL PLAN
FS3 — EQUIPMENT, DETAILS,

FIGURE 2-1 ■ The first sheet of each project typically contains a table of contents to help print readers use the drawings effectively. *Courtesy Architects Barrentine, Bates & Lee, A.I.A.*

order starting with **L-1** (for **Land**) or **C-1** (for **Civil**). Figure 2-2 shows an example of a commercial site plan. Depending on the complexity of the project and the site, the site drawings might include an existing site plan, a proposed site plan, a grading plan, a utility plan, an irrigation or sprinkler plan, and a landscape plan. Each of these drawings will be introduced in Chapter 14.

Architectural Drawings

The architectural drawings are the drawings that describe the size and shape of a structure. They are prepared by or under the direct supervision of an architect. Common architectural drawings for a commercial project include floor plans, enlarged floor plans, elevations, wall sections, roof plan, reflected ceiling plan, interior elevations, finish schedules, and interior details. Each type of architectural drawing will be introduced in Section III.

Architectural drawings can be numbered using one of two formats. Some offices place each page numbered in successive order starting with **A-1**. This method presents problems in detail numbering and planning if a project is revised and a page must be added to the set. Most firms use the numbering system recommended by the *AIA Handbook of Professional Practice*, where page numbers contain a decimal number such as **A1.01**, **A2.01**, **A2.02**, or **A3.01**. All of the pages numbered A1.00 are related to the site. Pages numbered **A2.00** are related to the floor plan, and pages numbered **A3.00** would be related to elevation. A four-story building will require four different floor plans, with pages numbered **A2.1**, **A2.2**, **A2.3** and **A2.4**. Recommended page numberings:

A0.01	Index, symbols, abbreviations, notes, location map
A1.01	Demolition, site plan, temporary work
A2.01	Plans, room material schedule, door schedule, key drawings
A3.01	Sections, elevations
A4.01	Detailed floor plans
A5.01	Interior elevations
A6.01	Reflected ceiling plans
A7.01	Vertical circulation—stairs, elevator and escalator drawings
A8.01	Exterior details
A9.01	Interior details

Chapter 3 presents guidelines for organizing projects.

Structural Drawings

As their name implies, these are the drawings used to construct the skeleton of the structure. These drawings are prepared by engineers or drafters working directly under the supervision of an engineer. Structural drawings will be discussed in Section V and are numbered in successive order starting with **S-1**. These drawings include framing plans, foundation plan, and related details. Figure 2-3 shows an example of a framing plan by an engineering firm.

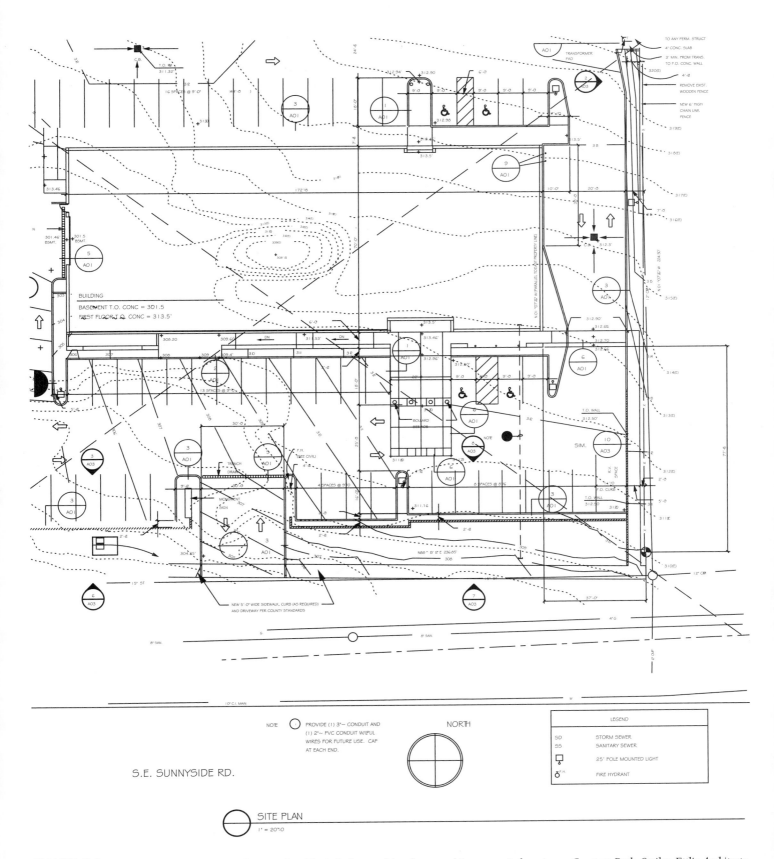

FIGURE 2-2 ■ ■ A commercial site plan can be completed by a drafter working for an architect or a civil engineer. *Courtesy Peck, Smiley, Ettlin Architects.*

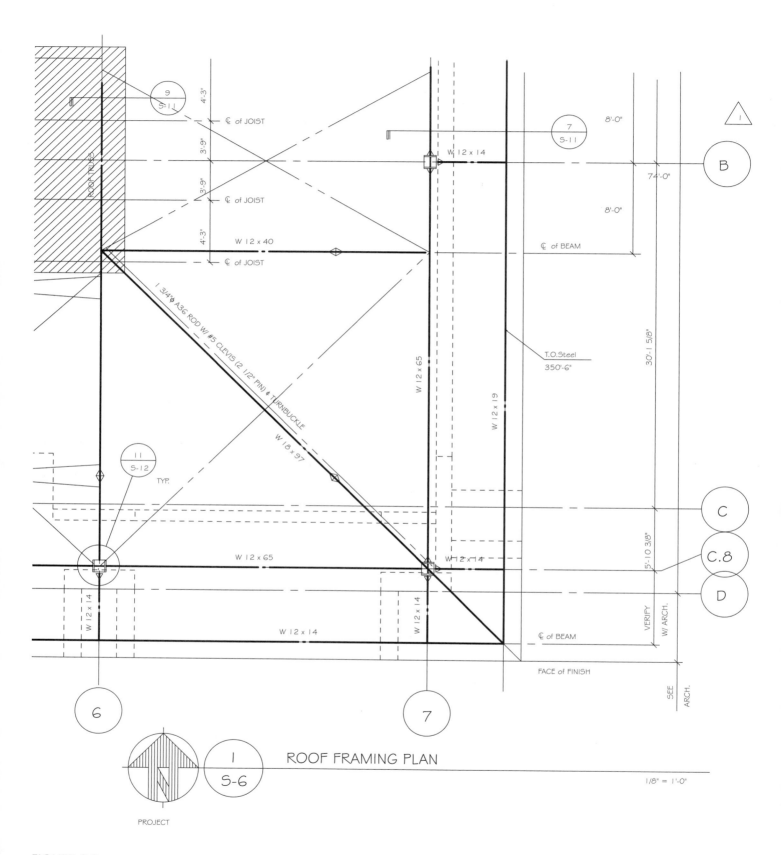

FIGURE 2-3 ■ The structural drawings show how to construct the skeleton of a structure. *Courtesy Van Domelen/Looijenga/McGarrigle/Knauf Consulting Engineers.*

Cabinet and Fixture Drawings

On structures with few cabinets, the interior details and elevations similar to Figure 2-4 are part of the architectural drawings. On structures such as a restaurant or a medical facility that will contain a magnitude of specialty cabinet or trim drawings, these drawings will be in their own section and be in successive order starting with **F-1** (for **Fixtures**). This section will also include details covering interior trim and specialized equipment, as well as cabinets. Figure 2-5 shows an example of interior drawings for a commercial kitchen.

Electrical Drawings

The electrical drawings include the electrical plans, lighting plans, equipment plan, and related schedules and details needed to completely specify the electrical requirements of the structure. For simple projects, drafters working for the architectural firm might complete the electrical drawings. Electrical drawings are usually prepared by drafters working under the direct supervision of a licensed electrical engineer. These drawings will be successively arranged starting with drawing **E-1**. The preparation of electrical drawings and details for commercial plans will not be discussed in this text.

Mechanical Drawings

The drawings that are used to show the movement of air throughout the structure comprise the mechanical drawing portion of a project. These drawings will be successively arranged starting with **M-1**. Several schedules and details are also typically very instru-mental to these drawings. The preparation of mechanical drawings and details will not be discussed in this text.

Plumbing Drawings

The plumbing drawings are used to show how fresh water and waste water will be routed throughout the structure. Plumbing drawings, schedules and details are successively arranged starting with page **P-1**. The preparation of plumbing drawings and details will not be discussed in this text.

WORKING WITH CALCULATIONS

With so many different types of drawings and so many different firms contributing to the structure, an inexperienced drafter could easily get lost in the developmental process of a structure. To ensure that all required information is incorporated in the drawings, architects and engineers typically provide a set of calculations and sketches for the drawing team to use in order to complete the plans. The calculations for the structural drawings are usually required to be signed by an architect or engineer and must be submitted to the building department with the completed drawings in order for a a building permit to be obtained.

Figure 2-6 shows a portion of one page of an engineer's calculations. Typically, calculations are divided into three sections that include a problem statement, mathematical solution, and the material to be used to meet the imposed stress. In Figure 2-6, the engineer is determining the size of two beams and how they will be connected. Although it is important that the drafter understand

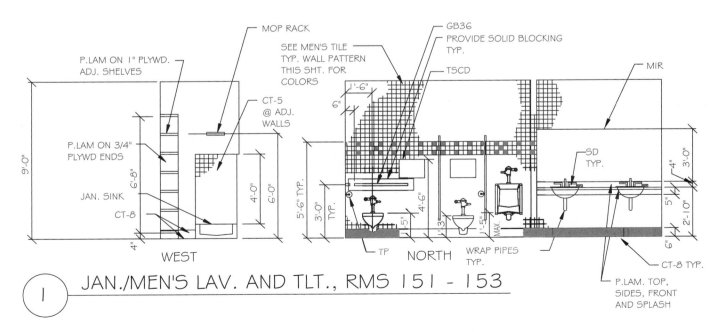

FIGURE 2-4 ■ Cabinet drawings and interior elevations are drawn by the architectural team. *Courtesy Michael & Kuhns Architects, P. C.*

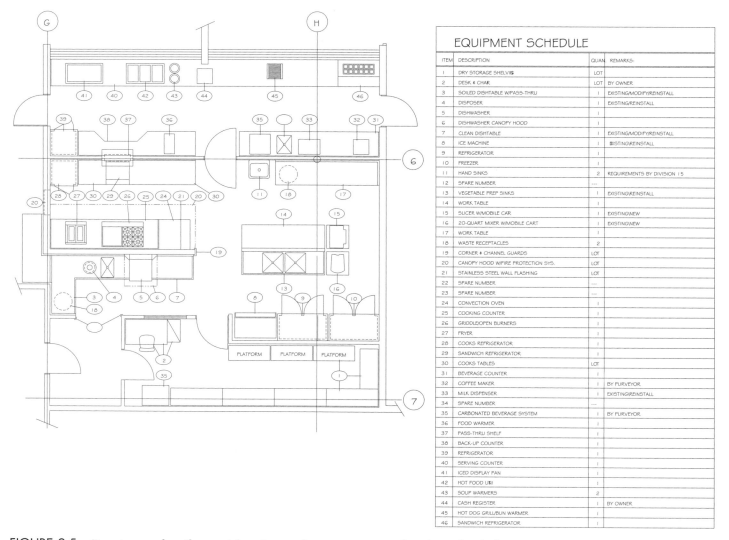

ITEM	DESCRIPTION	QUAN.	REMARKS:
1	DRY STORAGE SHELVING	LOT	
2	DESK & CHAIR	LOT	BY OWNER
3	SOILED DISHTABLE W/PASS-THRU	1	EXISTING/MODIFY/REINSTALL
4	DISPOSER	1	EXISTING/REINSTALL
5	DISHWASHER	1	
6	DISHWASHER CANOPY HOOD	1	
7	CLEAN DISHTABLE	1	EXISTING/MODIFY/REINSTALL
8	ICE MACHINE	1	EXISTING/REINSTALL
9	REFRIGERATOR	1	
10	FREEZER	1	
11	HAND SINKS	2	REQUIREMENTS BY DIVISION 15
12	SPARE NUMBER	---	
13	VEGETABLE PREP SINKS	1	EXISTING/REINSTALL
14	WORK TABLE	1	
15	SLICER W/MOBILE CAR	1	EXISTING/NEW
16	20-QUART MIXER W/MOBILE CART	1	EXISTING/NEW
17	WORK TABLE	1	
18	WASTE RECEPTACLES	2	
19	CORNER & CHANNEL GUARDS	LOT	
20	CANOPY HOOD W/FIRE PROTECTION SYS.	LOT	
21	STAINLESS STEEL WALL FLASHING	LOT	
22	SPARE NUMBER	---	
23	SPARE NUMBER	---	
24	CONVECTION OVEN	1	
25	COOKING COUNTER	1	
26	GRIDDLE/OPEN BURNERS	1	
27	FRYER	1	
28	COOKS REFRIGERATOR	1	
29	SANDWICH REFRIGERATOR	1	
30	COOKS TABLES	LOT	
31	BEVERAGE COUNTER	1	
32	COFFEE MAKER	1	BY PURVEYOR
33	MILK DISPENSER	1	EXISTING/REINSTALL
34	SPARE NUMBER	---	
35	CARBONATED BEVERAGE SYSTEM	1	BY PURVEYOR
36	FOOD WARMER	1	
37	PASS-THRU SHELF	1	
38	BACK-UP COUNTER	1	
39	REFRIGERATOR	1	
40	SERVING COUNTER	1	
41	ICED DISPLAY PAN	1	
42	HOT FOOD URN	1	
43	SOUP WARMERS	2	
44	CASH REGISTER	1	BY OWNER
45	HOT DOG GRILL/BUN WARMER	1	
46	SANDWICH REFRIGERATOR	1	

FIGURE 2-5 ■ Drawings to describe special equipment in a structure are often drawn by drafters employed by a consulting firm. *Courtesy Halliday Associates.*

basic structural principles to advance in the field, the drafter will not be expected to do structural calculations. Sections III and V will introduce concepts the drafter should be familiar with to complete structural drawings. The drafter's role is to be sure that the third portion of the calculations—the solution—is correctly placed on the drawings and to draw the connecting detail.

Although the layout of the calculations can take many forms, usually the loads are determined from the top of the structure to the foundation. This allows loads to be accumulated as formulas are completed for the upper levels and used to solve problems at lower levels. Figure 2-7 shows the framing detail that was created to comply with the engineer's design. Notice that several details in Figure 2-7 were not specified by the engineer's sketch. As a drafter gains experience and knowledge of the construction industry, the drafter is expected to be able to complete the detail. Depending on the engineer and the experience of the drafting team, the calculations may or may not contain sketches. When an engineer is specifying a common construction method to a skilled team, usually a sketch

will not be provided. Chapter 3 provides an introduction to drawing common construction details.

USING VENDOR CATALOGS

Closely related to the use of the engineers's calculations is the use of vendor catalogs. Common sources of building materials include the Internet, *Sweet's Catalogs*, *Architects' First Source for Products*, and the local yellow pages of your phone book. The World Wide Web is an excellent source for obtaining materials related to every phase of construction. *Sweet's* is a collection of vendor catalogs that are available in book, CD-ROM and Internet formats. Published yearly, *Sweet's* has become a staple of every architectural firm as a convenient source for construction materials and supplies. Information is listed using the numbering system of the Construction Specifications Institute's *Masterformat*, which will be

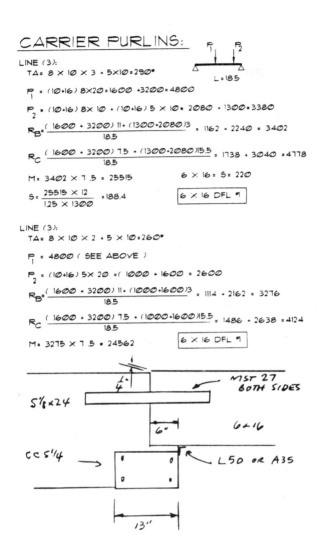

CARRIER PURLINS:

LINE (3):
TA= 8 X 10 X 3 • 5X10=290#

P_1 = (10•16) 8X20=1600 +3200=4800

P_2 = (10•16) 8 X 10 + (10•16) 5 X 10 = 2080 + 1300=3380

$R_B = \dfrac{(1600 + 3200)\,11 + (1300+2080)3}{18.5}$ = 1162 + 2240 = 3402

$R_C = \dfrac{(1600 + 3200)\,7.5 + (1300+2080)15.5}{18.5}$ = 1738 + 3040 = 4778

M = 3402 X 7.5 = 25515   6 X 16 = S = 220

$S = \dfrac{25515 \times 12}{1.25 \times 13.00}$ = 188.4   [6 X 16 DFL #]

LINE (3):
TA= 8 X 10 X 2 • 5 X 10=260#

P_1 = 4800 (SEE ABOVE)

P_2 = (10•16) 5X 20 =(1000 + 1600 = 2600

$R_B = \dfrac{(1600 + 3200)\,11 + (1000+1600)3}{18.5}$ = 1114 + 2162 = 3276

$R_C = \dfrac{(1600 + 3200)\,7.5 + (1000•1600)15.5}{18.5}$ = 1486 + 2638 = 4124

M = 3275 X 7.5 = 24562   [6 X 16 DFL #]

5⅞ x 24   MST 27 BOTH SIDES

6"   6 + 16

CC 5¼   L50 OR A35

13"

FIGURE 2-6 ■ The engineer's calculations provide the mathematical proof that components in the structure will resist all loads and stresses. *Courtesy StructureForm Masters, Inc.*

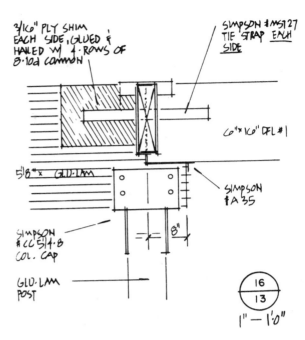

3/16" PLY SHIM EACH SIDE, GLUED & NAILED W/ 4 ROWS OF 8-10d COMMON

SIMPSON #MST27 TIE STRAP EACH SIDE

6"x 16" DFL #1

5⅛" x GLU-LAM

SIMPSON #A35

SIMPSON #CC 5¼-8 COL. CAP

GLU-LAM POST

16 / 13   1" — 1'0"

FIGURE 2-7 ■ A framing detail based on the calculations shown in Figure 210. *Courtesy StructureForm Masters, Inc.*

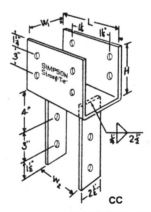

CC

MODEL NO.	DIMENSIONS				FASTENERS		ALLOWABLE LOADS		
	W₁	W₂	L	H	BEAM	POST	UPLIFT²	DOWN¹	
								CC	ECC
CC3¼-4	3¼	3⅝	11	6½	4-5/8 MB	2-5/8 MB	2795	15470	4920
CC3¼-6	3¼	5½	11	6½	4-5/8 MB	2-5/8 MB	2795	15470	7735
CC44	3⅝	3⅝	7	4	2-5/8 MB	2-5/8 MB	2430	9430	4715
CC46	3⅝	5½	11	6½	4-5/8 MB	2-5/8 MB	3005	14820	7410
CC5¼-6	5¼	5½	13	8	4-3/4 MB	2-3/4 MB	9440	29980	12615
CC5¼-8	5¼	7½	13	8	4-3/4 MB	2-3/4 MB	9440	29980	17295
CC64	5½	3⅝	11	6½	4-5/8 MB	2-5/8 MB	3750	23290	7410
CC66	5½	5½	11	6½	4-5/8 MB	2-5/8 MB	3750	23290	11645
CC68	5½	7½	11	6½	4-5/8 MB	2-5/8 MB	3750	23290	15880
CC76	6⅞	5½	13	8	4-3/4 MB	2-3/4 MB	9440	40220	16705
CC77	6⅞	6⅞	13	8	4-3/4 MB	2-3/4 MB	9440	40220	20500
CC78	6⅞	7½	13	8	4-3/4 MB	2-3/4 MB	9440	40220	22780
CC86	7½	5½	13	8	4-3/4 MB	2-3/4 MP	9440	37540	15880
CC88	7½	7½	13	8	4-3/4 MB	2-3/4 MB	9440	37540	21655
CC96	8⅞	5½	13	8	4-3/4 MB	2-3/4 MB	9025	51190	21655
CC98	8⅞	7½	13	8	4-3/4 MB	2-3/4 MB	9025	51190	29530
CC106	9½	5½	13	8	4-3/4 MB	2-3/4 MB	8745	47550	20115

FIGURE 2-8 ■ Drafters are often required to consult vendor catalogs to find needed information to complete drawings. *Courtesy Simpson Strong-Tie Co., Inc.*

discussed in Chapter 13. *Architects' First Source* is an index of materials with a short listing of the product descriptions and performance features. Book and Web site options are available. The yellow pages of your area phone book are also an excellent source of local suppliers. Most firms will gladly supply written specifications or construction details on disk or CD-ROM to help ensure the use of their products.

As seen in Figure 2-7, the engineer has specified a specific column cap to be used. Notice that no information is supplied about the cap other than the manufacture and the model number. The drafter is expected to be able to research a specific product and completely specify the product on the working drawings. Figure 2-8 shows a sample listing from the Simpson Strong-Tie brochure from the *Sweet's Catalogs*. The sizes and bolting for the specific connector can now be determined.

WORKING WITH CODES

In addition to calculations and vender catalogs, a drafter will also be required to have a working knowledge of the building code that governs the municipality where the structure will be built. Chapter 4 introduces basic code considerations for design and construction.

Building codes deal with two general categories in the design of a structure. To ensure public safety, most aspects of a code relate directly to either fire protection or the ability of a material or portion of a structure to resist collapsing under the loads that will be resisted. To resist fire, building codes deal with three general areas. These include the tendency of a material to fail by combustion, the tendency of a material to lose its strength once a fire has broken out, and methods of containment of a fire. You will be introduced to basic code requirements as you progress throughout this text.

The second major concern of the building code is the ability of the materials and the structure as a whole to resist loads. Building codes typically use either the working stress method or the factored load method to assure the safety of a structure. Each method deals with the resistance of loads from stress on materials and the predetermined ability of a material to resist stress. Chapter 12 will provide an introduction to forces that affect the design of a structure. Subsequent chapters will deal with how these stresses are resisted throughout a structure.

Although the architect or engineer will complete the majority of work with codes, it is important for the drafter to have a basic understanding of the code. The architect will determine the size, height, and type of materials to be used in construction, based on a thorough knowledge of the code. The drafter will need an understanding of the code to understand many of the components that will be added to a drawing. This will require knowledge of basic code layout and, specifically, an understanding of the chapters dealing with basic building requirements based on a specific type of construction.

DRAWING LAYOUT

The drawing layout must be one of the first things considered in planning a project. General office practice is to plan the drawing requirements and placement prior to starting the working drawings. The general order of a set of drawings was introduced earlier in this chapter. As the drawing layout is being planned, the placement on the page should be carefully considered. Great care must be taken, however, when several drawings will be placed on the same sheet. The architect, engineer, or project coordinator will typically provide new drafters with a sketch of the intended page layout along with individual sketches of each drawing to be placed on the page. These sketches of the proposed project are similar to an outline that is developed for a written project. The sketches provide a tentative layout for the placement of each plan, elevation, section, schedule, and detail that is anticipated.

When all of the drawings are complete and printed, they will be assembled into a set and bound along the left hand margin. The larger the sheets used, the more likely that drawings, notes or parts of drawings placed on the left side of the sheet will not be seen.

Whenever only one drawing is to be placed on a sheet, it should be placed near the right margin, leaving any blank space on the left or binding side. This same guideline should be used if smaller drawings, such as details, are to be placed with a large drawing. Placing the details on the left could cost the print reader needless time in skimming the drawings because the details might not be seen. With the details placed on the right side, as seen in Figure 2-9, they can be clearly seen while the reader thumbs through the drawing. It is true that the print reader should be careful when examining a set of drawings; however, anything the drafter can do to improve communication should be done.

Another consideration in placing details is to place them as close as possible to where they occur, but still being careful that small drawings will not be lost. The layout in Figure 2-10 places the detail close to its source and in a location where it can't be hidden by other pages. Wherever possible on symmetrical drawings, place the detail symbol where the right side can be seen.

In positioning drawings together, it is important that the limits of each drawing not interfere with its neighboring drawing. Lettering from one drawing should not be intertwined with a neighboring drawing. Some offices share text between two drawings, as seen in Figure 2-11. Text that relates only to the drawing on the right should be separated to avoid confusion. Other offices use a layout similar to Figure 2-12, which places details in neat rows with no shared text. Another option is to use a grid system similar to Figure 2-13. All boxes do not need to be the same size, but the box edges must align. Care must be taken to number the detail grids in an orderly manner. Order can vary, with the numbering of details starting in any one of the corners, but the same order of progression should be used throughout the project. Text assigned to details symbols will be further discussed as the drawings are being coordinated.

Keep in mind that there is no one correct way to arrange details. When entire sheets will be filled with details, it is important to have a logical order to their arrangement. Figure 2-14 shows a section of a brick wall. Notice that three details are required to completely explain this component. These details have been placed in an order that matches their arrangement within the section.

Another consideration in placing details throughout the drawing set is to place them in groups according to the labor force that will do the work. Roof drawings should not be mixed with concrete drawings because two completely different work crews will complete those jobs. If possible, place all concrete details with the foundation plan, or on a separate sheet following the foundation, and all of the roof details with the roof plan or very close to it.

DRAWING COORDINATION

Completion of the last drawing is no cause to celebrate. All working drawings must now be coordinated. The architectural office will arrange all of the architectural drawings and each consulting engineer will be responsible for the coordination of their drawings. Consulting firms will provide disk or paper copies ready to be added to the project. *Drawing coordination* involves the placing of

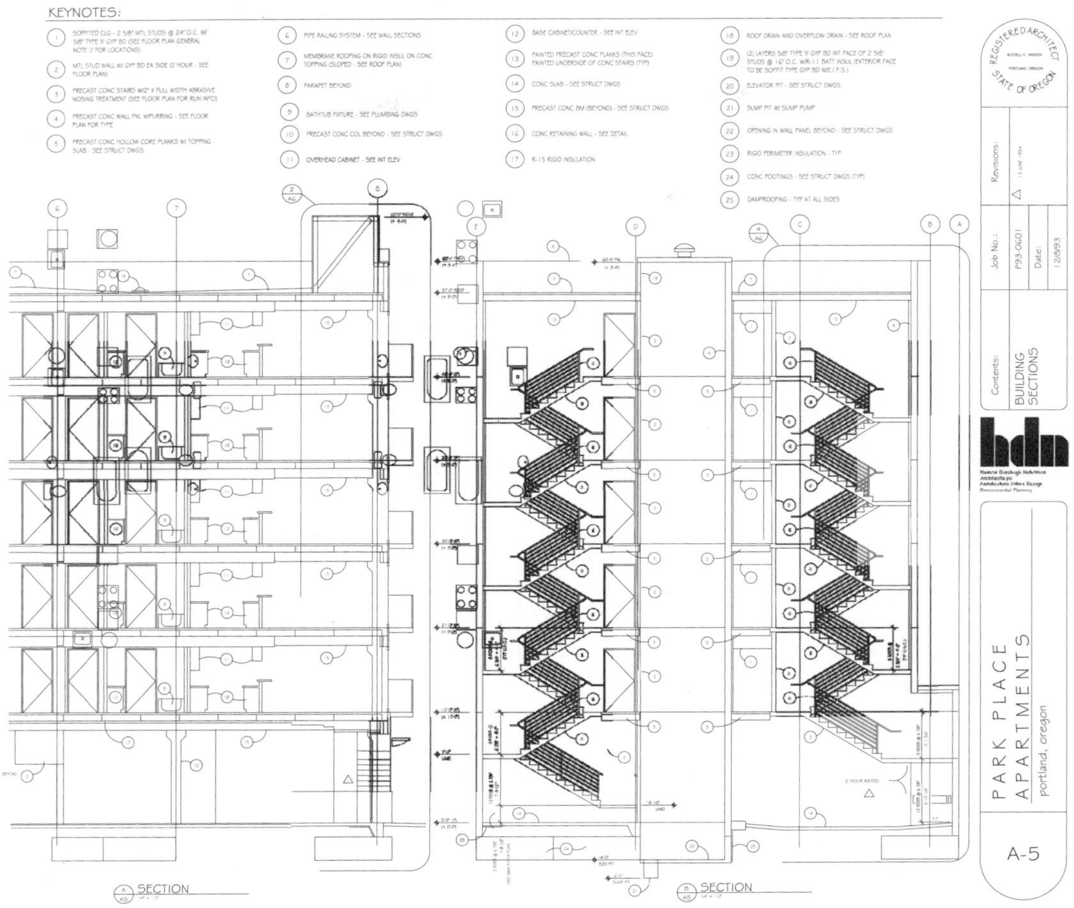

FIGURE 2-9 ■ The smaller the detail, the more important that it be placed on the right side of the page to avoid being ignored as pages are skimmed. *Courtesy H.D.N. Architects A.I.A.*

each page in its final order within the drawing set and assigning numbers to each page and detail. Job coordination is typically done working with a paper copy of each sheet. Pages are usually arranged in the order presented earlier in this chapter. Although this order will vary for each office and from job to job, a typical order for the architectural drawings might include the following:

T-1   Title sheet

A-1   Site plan and details

A-2   First floor plan, schedules

A-3   Second floor plan, details

A-4   Enlarged floor plans, interior elevations, details

A-5   Exterior elevations

A-6   Exterior elevation, details

A-7   Roof plan, details

A-8   Building sections

A-9   Reflected ceiling plan details

A-10 Details

A-11 Details

Assigning Page Numbers

A space in the lower right corner of the title block is generally reserved for the display of the page number. Two methods of page numbering have been introduced. As long as the drawings communicate effectively, they can be arranged in either order, based on office procedure. Generally offices try and maintain a similar order for each of their jobs so that bidders and construction crews will have a sense of familiarity with their plans. Sheets that contain abbreviations or symbols should be placed very early in the drawing set.

Assigning Detail References

Once the page numbers have been assigned, details on each specific page can be numbered. Each detail should have a detail symbol that shows the detail number over the page number. A title is often placed near the detail symbol similar to the drawing shown in Figure 2-14. The text used for the title and the detail numbers is usually between 1/8" and 1/4" high. A uniform height should be used throughout the entire project. Pages, details, and scales are usually placed in 1/8" high text. One of two methods is typically used to assign detail numbers.

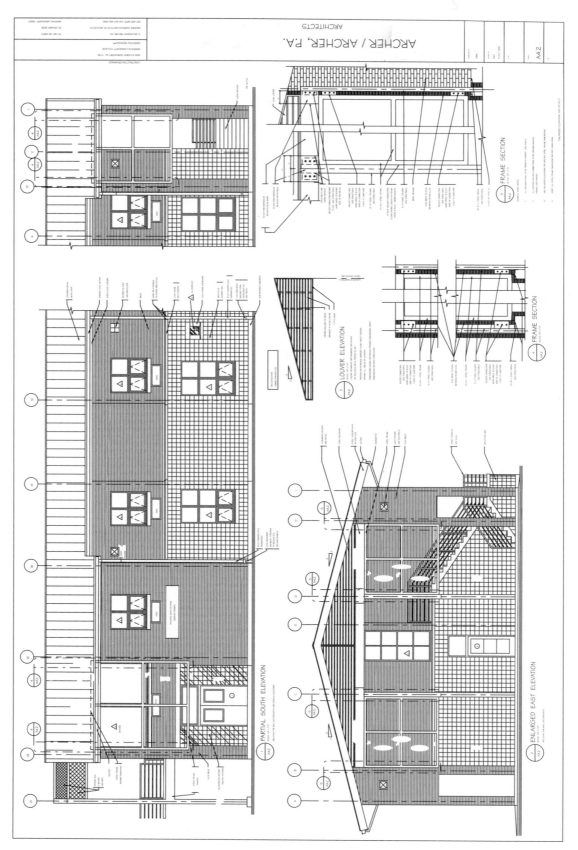

FIGURE 2-10 ■ Details should be placed as close as possible to the drawing that they reference. *Courtesy G. Williamson Archer A.I.A., Archer & Archer P.A.*

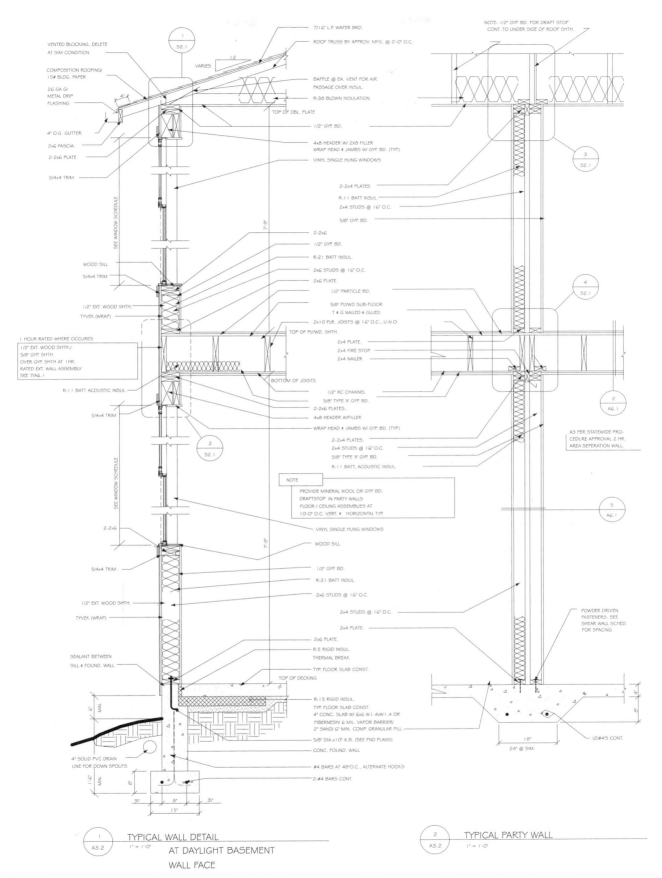

FIGURE 2-11 ■ Text can be shared between two similar details to save space and drawing time. *Courtesy Scott R. Beck, Architect.*

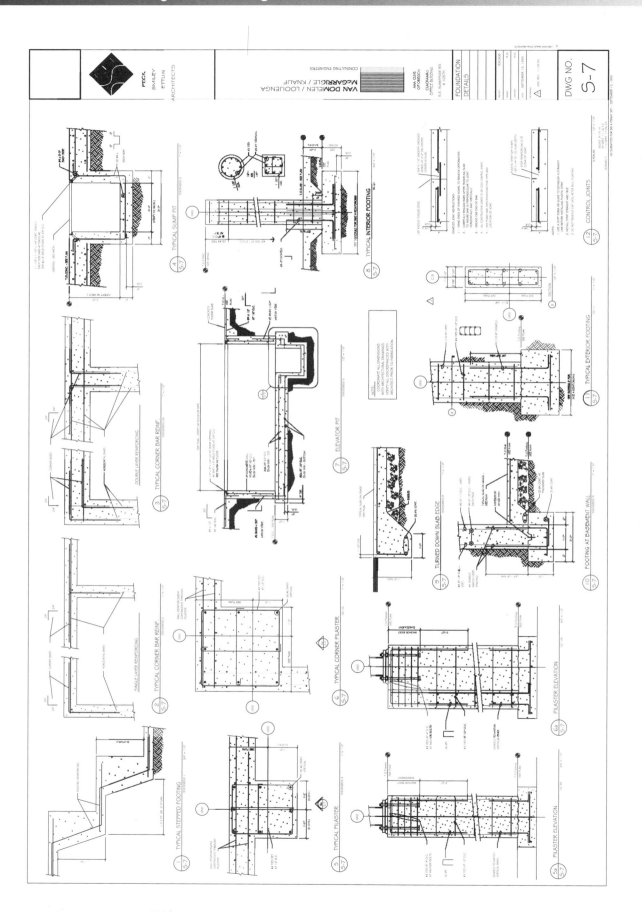

FIGURE 2-12 ■ Careful placement of details and other small drawings is critical to good page clarity. *Courtesy Van Domelen/Looijenga/McGarrigle/Knauf Consulting Engineers.*

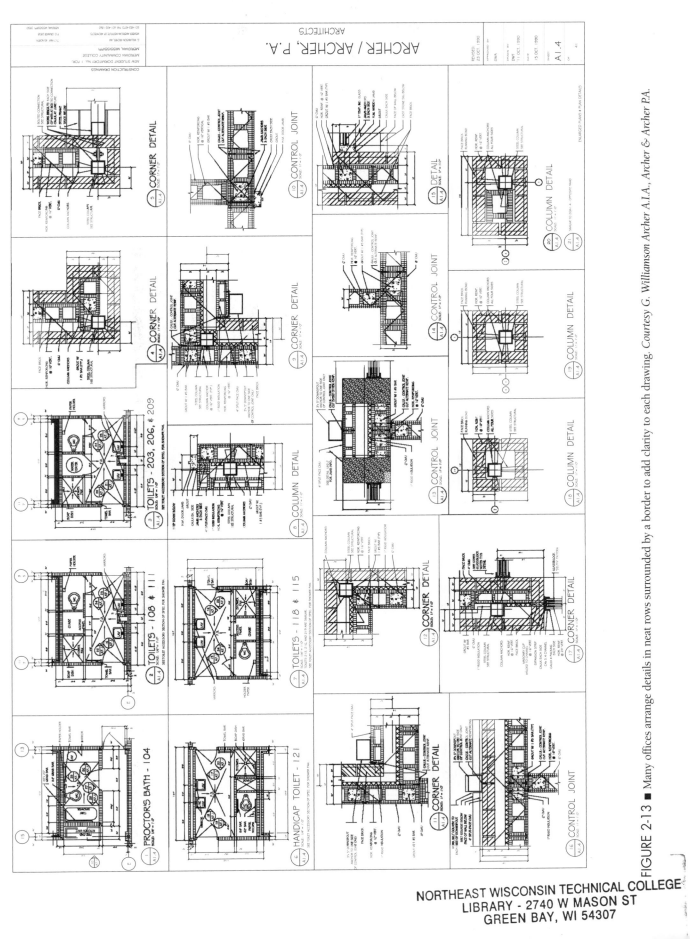

FIGURE 2-13 ■ Many offices arrange details in neat rows surrounded by a border to add clarity to each drawing. *Courtesy G. Williamson Archer A.I.A., Archer & Archer P.A.*

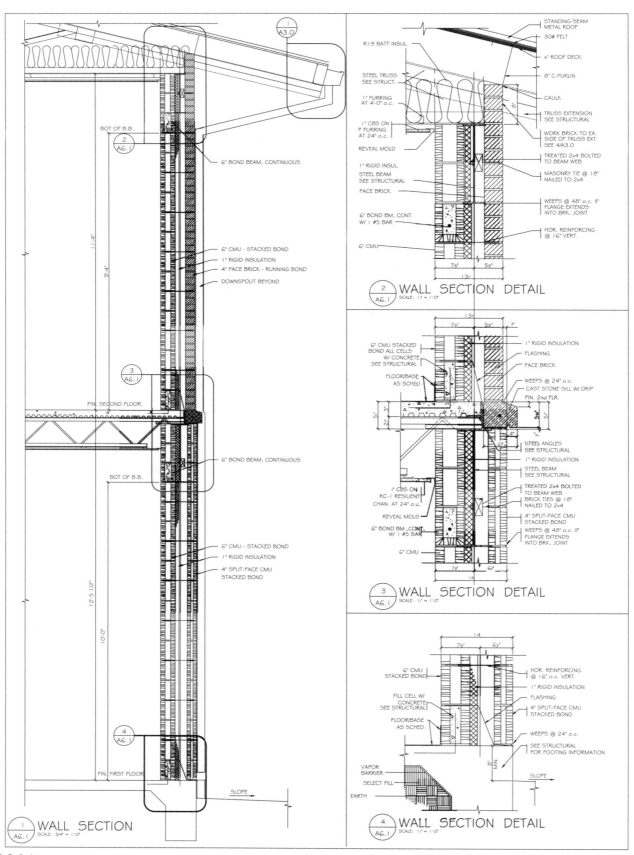

FIGURE 2-14 ■ Placing details near the source of the reference and using a logical order of presentation can aid the print reader. *Courtesy G. Williamson Archer A.I.A., Archer & Archer P.A.*

Some offices assign consecutive numbers for each detail on a specific page, but begin with 1 for every page containing a detail, as shown in Figures 2-12 and 2-13.

■ Numbering begins with page 1 and every detail is numbered consecutively from start to finish. This method eliminates several details with a detail number of 1.

Once every detail in the project has been assigned a detail and page number, each reference to the detail must be represented on the plan, elevation, or section where it occurs. Junior drafters are often given a check print with all numbers assigned to be filled in. Experienced drafters are generally expected to coordinate a project without the benefit of a check print. This is best done by working with a paper copy of the drawings. Although the detail does not have to be referenced to every occurrence on a plan view, it should be marked so that there is no confusion about where a particular detail occurs. Referencing every occurrence of the detail would often decrease drawing clarity and take unnecessary drawing time. The print reader is required to make intelligent decisions regarding where every occurrence of the detail will take place.

REVISIONS

No matter how carefully a set of drawings is prepared, portions of the drawing set might need to be revised. Common causes for revisions:

■ Changes in building codes after the drawings were completed but before the building permits issued

■ Changing owner or tenant requirements

■ Changes occurring at the job site

■ Errors by the architectural team

To make all users of the plans aware of the change, an addendum is issued. An *addendum* is a written notification of the changes, along with a drawing showing the new design requirements. Figure 2-15 shows an example of a revised framing plan. Notice that a portion of the drawing is encircled by a squiggly line to draw attention to the changes. A number inside a triangle is also used to draw attention to the changes that have been made from earlier prints. The number relates to a note that is then placed somewhere on the page to explain the required changes. Many offices have a specific portion of the title block or the drawing area reserved for revision references or explanations.

METRIC AND THE CONSTRUCTION INDUSTRY

Federal law mandated in 1988 that the metric system will be the preferred system of measurement for the United States. Federal agencies involved in construction agreed to the use of metric in the design of all federal construction projects as of January 1994. Although most firms are still working with traditional units of measurement, metric measurement will become important to your drafting future. Although the construction industry does not have one uniform standard, metric guidelines expressed throughout this text are based on the recommendations of the *Metric Guide for Federal Construction*, the *International Building Code* and the Construction Metrication Council.

Basic Metric Units

Although many in the construction industry are not familiar with metric units (SI), the system is logical and easy to use. Six base units of metric measurement are typically used in the construction industry, as Table 2-1 shows.

TABLE 2-1 ■ Base Units of Metric Measurement in the Construction Industry

Quality	Unit	Symbol
length	meter	m
mass (weight)	kilogram	kg
time	second	s
electric current	ampere	A
temperature	kelvin	K
luminous intensity	candela	cd

The numerical base of the metric system is 10, with all functions either a multiple or decimal fraction of 10, as shown in Table 2-2:

TABLE 2-2 ■ Common Metric System Units

Multiplication Factor		Prefix	Symbol
$1000000000000000000 = 10^{18}$		exa	E
$1000000000000000 = 10^{15}$		peta	P
$1000000000000 = 10^{12}$		tera	T
$1000000000 = 10^{9}$		giga	G
$1000000 = 10^{6}$		mega	M
$1000 = 10^{3}$		kilo	k
$100 = 10^{2}$		hecto	h
$10 = 10^{1}$		deka	da
$0.1 = 10^{-1}$		deci	d

TABLE 2-2 ■ Common Metric System Units (continued)

Multiplication Factor		Prefix	Symbol
$0.01 = 10^{-2}$		centi	c
$0.001 = 10^{-3}$		milli	m
$0.000001 = 10^{-6}$		micro	m
$0.000000001 = 10^{-9}$		nano	n
$0.000000000001 = 10^{-12}$		pica	p
$0.000000000000001 = 10^{-15}$		femto	f
$0.000000000000000001 = 10^{-18}$		atto	a

Each linear unit is smaller or larger than its predecessor by a factor of 10, each square unit by a factor of 100, and each volume by a factor of 1000. Units for expressing linear, area, and volume measurements can be seen in the following lists:

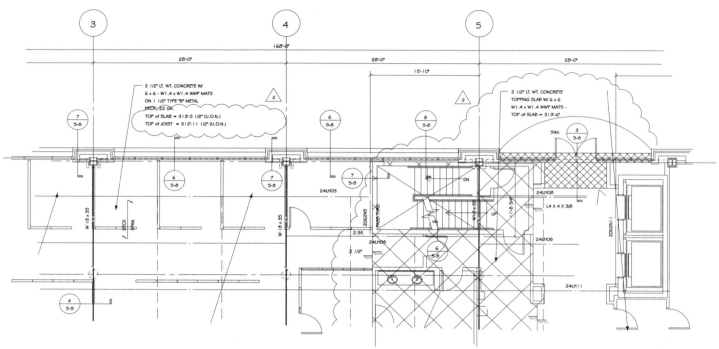

FIGURE 2-15 ■ Revisions are often enclosed by a squiggly line to draw attention to the change. The 2 inside the triangle refers to a notation in the title block. In this case, floor elevations were added in two places on the drawing since the last printing. *Courtesy Van Domelen/Looijenga/McGarrigle/Knauf Consulting Engineers.*

Linear Units × 10

basic   = 1 mm
10 mm = 1 cm
10 cm = 1 dm
10 dm = 1 m
10 m   = 1 dam
10 dam = 1 hm
10 hm = 1 km

Units of Area × 100

basic   = 1 mm²
100 mm² = 1 cm²
100 cm²   = 1 dm²
100 dm²   = 1 m²
100 m²   = 1 dam²
100 dam² = 1 hm²
100 hm² = 1 km²

Units of Volume × 1,000

basic   = 1 mm³
1000 mm³ = 1 cm³
1000 cm³   = 1 dm³
1000 dm³   = 1 m³
1000 m³   = 1 dam³
1000 dam³ = 1 hm³
1000 hm³ = 1 km³

The basic module in construction will be based on 100 mm (about 4"). The preferred order for submodules and their approximate inch equivalent are 50 mm (2"), 25 mm (1"), 20 mm (3/4"), 10 mm (3/8"), and 5 mm (3/16"). The preferred order for multimodules and their approximate inch/foot equivalents are 300 mm (about 12"), 600 mm (2'), 1200 mm (4'), 3000 mm (10'), and 6000 mm (20').

Metric Conversion Factors

The International Building Code (IBC), many construction suppliers, and this text feature dual units where measurements are specified. Conversion from imperial units to metric units can be done by hard or soft conversions. *Hard conversions* are made by using a mathematical formula to change a value of one system (for example, 1") to the equivalent value in another system (for example, 25.4 mm). A 6 × 12 beam using hard conversion methods would now be a 152 × 305 mm (6 × 25.4 and 12 × 25.4). *Soft conversions* change a value from one system (for example, 1") to a rounded value in another system (for example, 25 mm). A 6 × 12 using soft conversion would be a 150 × 300 mm. Many construction products can be soft converted and still be reliable in metric form. Hard conversions will be used throughout this text for all references to minimum standards listed in the IBC. All other references to metric sizes will be soft converted. Common conversion factors from imperial to metric can be seen in Table 2-3:

TABLE 2-3 ■ Conversion Factors

Length		Multiply By
1 mile	km	1.609344
1 yd	m	0.9144
1 ft	m	0.3048
	mm	304.8
1 in	mm	25.4

Area		Multiply By
1 mile^2	km^2	2.589998
1 acre	ha	0.4046873
	m^2	4046.873
1 yd^2	m^2	0.8361274
1 ft^2	m^2	0.09290304
1 in^2	mm^2	645.16

Volume, Modules of Section		Multiply By
1 acre ft	m^3	1233.489
1 yd^3	m^3	0.764 5549
100 board ft	m^3	0.2359737
1 ft^3	m^3	0.02831685
	L(dm^3)	28.3168
1 in^3	mm^3	16387.06
1 barrel (42 US gal)	m^3	0.1589873

Fluid Capacity		Multiply By
1 U.S. gal	L	3.785412
	m^3	0.003785412
1 U.S. qt	mL	946.3529
1 U.S. pt	mL	473.1765
1 fl oz	mL	29.5735

Plane Angle		Multiply By
1°	rad	0.01745329
	mrad	17.45329
1'(minute)	urad	290.8882
1"(second)	urad	4.848137

Velocity, Speed		Multiply By
1 ft/s	m/s	0.3048
1 mile/h	km/h	1.609344
	m/s	0.47704

Volume—Rate of Flow		Multiply By
1 ft^3/s	m^3/s	0.02831685
1 ft^3/min	L/s	0.4719474
1 gal/min	L/s	0.0630902
	m^3	0.0038
1 gal/h	mL/s	1.05150
1 million gal/d	L/s	43.8126
1 acre ft/s	m^3/s	1,233.49

Mass		Multiply By
1 ton(short)	metric ton	0.907185
	kg	907.1847
1 lb	kg	0.4535924
1 oz	g	28.34952

Mass per Unit Area		Multiply By
1 lb/ft^2	kg/m^2	4.882428
1 oz/yd^2	g/m^2	33.90575
1 oz/ft^2	g/m^2	305.1517

Density (Mass per Unit Volume)		Multiply By
1 lb/ft^3	kg/m^3	16.01846
1 lb/yd^3	kg/m^3	0.5932764
1 ton/yd^3	t/m^3	1.186553

Force		Multiply By
1 tonf (ton-force)	kN	8.89644
1 kip (1000 lbf)	kN	4.44822
1 lbf pound-force	N	4.44822

Moment of Force, Torque		Multiply By
1 lbf ft	N·m	1.355818
1 lbf in	N·m	0.1129848
1 ton ft	nK·m	2.71164
1 kip ft	kN·m	1.35582

Force per Unit Length		Multiply By
1 lbf/ft	N/m	14.5939
1 lbf/in	N/m	175.1268
1 tonf/ft	kN/m	25.1878

Pressure, Stress, Modulas of Elasticity (force per unit area) (1 Pa = 1 N/m^2)		Multiply By
1 tonf/in^2	MPa	13.7895
1 tonf/ft^2	kPa	95.7605
1 kip/in^2	MPa	6.894757
1 lbf/in^2	kPa	6.894757
1 lbf/ft^2	Pa	47.8803
Atmosphere	kPa	101.3250

Care must be taken in rounding numbers so that unnecessary accuracy is not specified. Remember that it is easiest for field personnel to measure in 10 mm or 5 mm increments. Generally any dimension over a few inches long can be rounded to the nearest 5 mm (1/5") and anything over a few feet long can be rounded to the nearest 10 mm (2/5"). Dimensions between 10' and 50' can be rounded to the nearest 100 mm, and dimensions over 100' can be rounded to the nearest meter.

Uncommon but Useful Conversions

Time between slipping on a peel and smacking the pavement:	1 Bananosecond
Two monograms:	1 Diagram
Ratio of an igloo's circumference to its diameter:	Eskimo Pi
1 kilogram of figs:	1 Fig Newton
Basic unit of laryngitis:	1 Hoarsepower
1,000 aches:	1 Kilohurtz
Time it takes to sail 220 yards at 1 nautical mph:	Knot-furlong
365.25 day of drinking less filling low-calorie beer:	1 Lite year
1,000 cubic centimeters of wet socks:	1 Literhosen
1,000,000 bicycles:	2 Megacycles
1,000,000 microphones:	1 Megaphone
1/1,000,000 of a fish:	1 Microfish
1/1,000,000 of a mouthwash:	1 Microscope
453.6 graham crackers:	1 Pound cake
Half of a large intestine:	1 Semicolon

Metric Paper and Scale Sizes

Because of the abundance of preprinted paper, many professional firms might not convert to metric paper sizes. Metric projects can be plotted on standard drawing paper. The five standard sizes of metric drawing material:

A0 1189 × 841 mm (46.8 × 33.1 in.)
A1 841 × 594 mm (33.1 × 23.4 in.)
A2 594 × 420 mm (23.4 × 16.5 in.)
A3 420 × 297 mm (16.5 × 11.7 in.)
A4 297 × 210 mm (11.7 × 8.3 in.)

When drawings are to be produced in metric, an appropriate scale should be used. Metric scales are true ratios and are the same for both architectural and engineering drawings. A conversion of common architectural and engineering scales to metric is shown in Table 2-4:

TABLE 2-4 ■ Scale Conversions to Metric

Inch/Foot Scale	Ratio
Full size	1:1
Half size	1:2
4" = 1'–0"	1:3
3" = 1'–0"	1:4
1 1/2" = 1'–0"	1:8
1" = 1'–0"	1:12
3/4" = 1'–0"	1:16
1/2" = 1'–0"	1:24
3/8" = 1'–0"	1:32
1/4" = 1'–0"	1:48
3/16" = 1'–0"	1:64
1/8" = 1'–0"	1:96
1" = 10'–0"	1:120
3/32" = 1'–0"	1:128
1/16" = 1'–0"	1:192
1" = 20'–0"	1:240
1" = 30'–0"	1:360
1/32" = 1'–0"	1:384
1" = 40'–0"	1:480

TABLE 2-4 ■ Scale Conversions to Metric (continued)

Inch/Foot Scale	Ratio
1" = 60'–0"	1:720
1" = 80'–0"	1:1000

Many of the scales traditionally used in architecture cannot be found on a metric scale. Although this does not affect CAD drafters, scales should be established which the print reader can easily work with. With the print reader in mind, preferred metric drawing scales and their approximate inch/foot equivalent are shown in Table 2-5:

TABLE 2-5 ■ Preferred Metric Scales

Ratio Inch/foot	equivalent
1:1	Same as full size
1:2	Same as half size
1:5	Close to 3" = 1'–0"
1:10	Between 1" = 1'–0" and 1 1/2" = 1'–0"
1:20	Between 1/2" = 1'–0" and 3/4" = 1'–0"
1:50	Close to 1/4" = 1'–0"
1:100	Close to 1/8" = 1'–0"
1:200	Close to 1/16" = 1'–0"
1:500	Close to 1" = 40'–0"
1:1000	Close to 1" = 80'–0"

Expressing Metric Units on Drawings

Units on a drawing should be expressed in feet and inches or in the metric equivalent, but not as dual units. Metric dimensions on most drawings should be represented as millimeters. The mm symbol does not need to be placed after the specified size. Large dimensions on site plans or other civil drawings can be expressed as meters and should be followed by the m symbol so that no confusion results. When expressing sizes in notation, mm should be used. Plywood thickness would be noted as

12.7 MM STD. GRADE PLY ROOF SHEATHING

Chapter 13 covers other guidelines for expressing written metric measurements. Chapters 6 through 10 will discuss the effects of metric conversion on specific building materials.

ADDITIONAL READING

The following Web sites can be used as a resource to help you keep current with changes in professional design careers.

ADDRESS	COMPANY/ORGANIZATION
Major research tools:	
www.afsonl.com	Architects' First Source for Products
www.sweets.com	Sweet's Building Product Information
Major testing labs, quality assurance, and inspection agencies:	
www.apawood.org	APA – The Engineered Wood Association
www.etl.go.jp	ETL Testing Services
www.fmglobal.com/researchstandard_testing/research.html	
	Factory Mutual Research Corporation
www.itsglobal.com	Intertek Testing Services
www.ul.com	Underwriters Laboratories Inc.
www.wwpa.org	Western Wood Products Association
Other United States and Federal Agencies Related to Construction:	
www.access-board.gov	*The Access Board*—The U.S. Architectural and Transportation Barriers Compliance Board
www.eren.doe.gov	U.S. Department of Energy's Energy Efficiency and Renewable Energy Network

www.epa.gov	*EPA*—The United States Environmental Protection Agency
www.nara.gov/fedreg	*Federal Register*—The Government Printing Office
www.thomas.loc.gov	Thomas—Federal legislative information, status of bills, Congressional Record, and related information
www.nist.gov	*NIST*—National Institute of Standards and Technology

Other agencies and their web sites that impact building codes in your area:

www.ansi.org	*ANSI*—American National Standards Institute
www.astm.org	*ASTM*—American Society for Testing and Materials
www.nfpa.org	*NFPA* —National Fire Protection Association

In addition to these regulatory agencies that influence construction, several related organizations are often used by architects and engineers to ensure quality materials. These agencies and their Web sites include:

www.iafconline.org	*IAFC*—International Association of Fire Chiefs
www.iec.ch	*IEC*—International Electrotechnical Commission
www.ieee.org	*IEEE*—The Institute of Electrical and Electronics Engineers, Inc.
www.iso.ch	*ISO*—The International Organization for Standardization
www.ncma.org	*NCMA*—The National Concrete Masonry Association

CHAPTER 2

The Drafter's Role in Office Practice and Procedure

CHAPTER QUIZ

Answer the questions on a separate sheet of paper. Print the chapter title, the question number, and a short complete statement for each question.

Question 2-1 List the three major portions of a set of calculations.

Question 2-2 How is information arranged in the *Sweet's Catalogs*?

Question 2-3 List the four main national building codes.

Question 2-4 Where should small details be located? Explain your answer.

Question 2-5 Explain the guidelines for labeling details.

Question 2-6 Describe methods of arranging detail pages.

Question 2-7 What does drawing coordination entail?

Question 2-8 List and describe two common methods of assigning page numbers.

Question 2-9 List two methods of describing revisions.

Question 2-10 What is an addendum?

Use Figures 2-7 and 2-8 to answer the following questions:

Question 2-11 List the quantity and size of bolts to be used to connect a CC68 to a 6 × 8 column.

Question 2-12 How many bolts will be needed to connect a CCO68 to a 6 × 12 beam?

Question 2-13 How much uplift can a CC88 resist?

Question 2-14 When would a CCO cap be useful and how is it attached?

Question 2-15 How many pounds can an ECC106 support?

Use Figure 2-6 to answer the following questions:

Question 2-16 L represents the length of the beam. What length of beam will be used?

Question 2-17 The letter P is used to represent the loads on the beam. What loads will this beam support?

Question 2-18 List the size of the beams to be supported.

Question 2-19 List three connectors that will be used at this beam intersection.

Question 2-20 List the size, quantity and location of the bolts to be used with the column cap.

Question 2-21 Why would the construction industry be interested in converting to metric measurement?

Question 2-22 What are the basic units of length used with the metric system?

Question 2-23 List and describe two methods of converting numbers to metric.

Question 2-24 List examples of each conversion methods for a 4 × 12 beam.

Question 2-25 What scale should be used to draw a floor plan so that it would resemble 1/4" = 1'–0"?

Question 2-26 In addition to altering a drawing, explain how a revision is noted on a drawing.

Question 2-27 List the exact ratio and preferred scale for the following scales.

Full size

3" = 1'–0"

1 1/2" = 1'–0"

1" = 1'–0"

3/4" = 1'–0"

1/2" = 1'–0"

3/8" = 1'–0"

1/4" = 1'–0"

3/16" = 1'–0"

1/8" = 1'–0"

3/32" = 1'–0"

1/16" = 1'–0"

1/32" = 1'–0"

1" = 10'–0"

1" = 20'–0"

1" = 30'–0"

1"=40'–0"

1" = 60'–0"

Question 2-28 Convert the following dimensions using hard conversions.

a) 1/2"   f) 12"

b) 3/4"   g) 16"

c) 2"   h) 2'–6"

d) 4"   i) 4'–0"

e) 9"   j) 10'–0"

Question 2-29 Convert the following dimensions using soft conversions.

a) 1/2"   f) 12"

b) 3/4"   g) 16"

c) 2"   h) 2'–6"

d) 4"   i) 4'–0"

e) 9"   j) 10'-0"

CHAPTER 3

Applying AutoCAD Tools to Commercial Drawings

This chapter will provide an overview and practical application of the computer skills necessary to produce a set of drawings. It is assumed that you have already completed an entry-level class to CAD drafting prior to working through this text and that you are working with AutoCAD 2000 (or newer). This chapter will help you utilize your CAD skills in the preparation of commercial drawings. Consideration will be given to managing folders and files, efficient use of drawing templates, assembling drawings for plotting at multiple scales, working with multiple documents, using DesignCenter, and applying the metric system to construction drawings

MANAGING FOLDERS AND FILES

The method used to organize drawing files will vary depending on existing office practice and the size of the structure. This chapter is intended to provide you with a method of organizing your drawing files while in school and to give you an understanding of some of the common methods that offices use to organize their computer files. Consideration will be given to common drawing storage methods, the naming of drawing folders and files, methods of storing drawing files, and methods of maintaining drawing files. It is important to remember that when you leave school and enter an office, it is rarely the drafter's role to come into an office and develop a new filing system. It will be your job as a drafter to learn the existing system and make your work conform to your employer's standards.

Storage Locations for Drawing Files

If you are a network user, your instructor or network administrator must provide you with a user name or account number before you can access or store information on a network. Once you have access, you will be given a folder that is located on the network server. Network servers should not be used for long-term storage, but they do provide an excellent storage location for active drawing projects. Many design firms place active drawing files in folders that can be accessed by anyone on the network, ensuring that each member of the design team is working on the most current file without having to pass a diskette between team members. Networks can also be configured to allow access over the Internet to consulting firms. A designer working for the structural engineer can be allowed access to the floor plan created by the architectural

firm, ensuring the electronic transfer of the most up-to-date drawing files between offices.

Your school or office administrator will determine where you will save your drawing files. Most users save their projects at about 15-minute intervals on the hard drive. At the end of a drawing session, the file is saved in your folder on the network and to a diskette. Diskettes and fixed drives are often used for the day-to-day storage of drawing files because of their access speed. Once the drawing session or project is completed, it can be stored on a CD, Zip or Jaz disk or on a tape cartridge where it is less subject to damage.

Naming Folders

You were exposed to creating folders in an introductory computer class, but their importance can't be overlooked as you work on large commercial projects. An efficient CAD drafter will create folders using Windows Explorer to aid filing. It is also wise to divide your folder into subfolders. This might include subfolders based on specific classes you're enrolled in, different types of drawing, or different projects. Many small offices keep work for each client in separate folders or disks with labels based on the client name. Drawings are then saved by contents such as floor, foundation, elevation, sections, specs, or site. Some offices assign a combination of numbers and letters to name each project. Numbers are usually assigned to represent the year the project is started, as well as a job number with letters representing the type of drawing. For example, 0153fl would represent the floor plan for the fifty-third project started in 2001. You're honing your CAD and commercial drafting skills at a time when the construction industry is becoming increasingly connected by the Internet. Plans that were transported between offices by messenger are now shared between offices electronically. Because of the need by so many firms to work with a set of drawings, an efficient layer system is essential. Most architectural and engineering offices use file names based on the AIA or Construction Specifications Institute (CSI) guidelines. The AIA guidelines provide a uniform file naming system that can be recognized by the various consultants that work on each project. Architectural firms often use two variations of AIA guidelines for naming files.

One common practice is to name drawing files by the page number it will occupy in the drawing set based on AIA page numbering guidelines. These numbers include:

Common Page Numbers and File Names

A0.01	Index, symbols, abbreviations, notes, and location maps
A1.01	Demolition, site plans, and temporary work
A2.01	Plans, and schedules such as room material, door, or windows and keyed drawings
A3.01	Sections and exterior elevations
A4.01	Detailed, large-scale floor plans
A5.01	Interior elevations
A6.01	Reflected ceiling plans
A7.01	Vertical circulation drawings such as stair, elevator, and escalator drawings
A8.01	Exterior details
A9.01	Interior details

When this system is followed, a drawing file name of A4.01 would represent a detailed floor plan. If plans for a five-level building were to be drawn, each floor could be saved as a separate drawing file using drawing names such as

A2.01	level 1
A2.02	level 2
A2.03	level 3
A2.04	level 4
A2.05	level 5

A second method of naming files is based on a letter that represents the discipline of the drawing originator, and a two-letter code that represents drawing types common to all disciplines. A number that represents the sheet sequence number is sometimes added to the filename. Under this system, drawing files would have names such as A–FP01 or S–DT13. Some of the most common discipline and drawing codes are shown in Table 3-1:

TABLE 3-1 ■ Common Discipline and Drawing Codes

Discipline Codes		Drawing Codes	
A	Architectural	FP	Floor Plan
M	Mechanical	DT	Detail
C	Civil	SP	Site Plan
P	Plumbing	SH	Schedules
E	Electrical	DP	Demolition Plan
Q	Equipment	3D	Isometric/3D drawings
F	Fire Protection	QP	Equipment Plan
S	Structural	DG	Diagrams
I	Interiors	XP	Existing Plan
T	Telecommunications	EL	Elevation
L	Landscape	SC	Section

CSI Guidelines for Naming Files

Another popular method of naming drawing files is to use the guidelines provided by CSI. These guidelines recommend titles that start with a letter to represent a discipline code, a sheet designator, and a page number. Page numbers in filenames are designated sequentially from 01 through 99. Discipline codes match those used by the AIA. Using the CSI format, a file name such as A-501, **A** is a drawing by the architectural team, **5** represents details, and **01** represents sheet 1 of the detail section.

Sheet Type Designators

0	General symbols, abbreviations, notes, and location maps
1	Plans, and schedules such as room material, door, or windows and keyed drawings
2	Exterior elevations
3	Sections
4	Detailed, large-scale floor plans
5	Details
6	Schedules and diagrams
7	User defined
8	User defined
9	3D views such as isometrics, perspectives, and photographs

Project Storage Methods

Before you begin a project, you also need to develop a method of saving drawing files. Common methods of saving drawing projects, in their order of usefulness and power include single-drawing files, layered files, referenced files and layouts.

Single Drawing Files

As you completed basic computer classes, you probably saved each drawing project as a separate drawing (.dwg) file. This method has advantages as you work on larger drawing projects, but it also has its limitations. The main advantage of storing one drawing per file is ease of plotting. Individual drawings can be saved using the project name and contents such as \ARTIMIZ\SITE. The site plan can be easily retrieved and plotted. The drawbacks to with this storage system are the amount of storage space required and the difficulty in making revisions. If the client changes the overall size of the project, the correction would need to be reflected in the \ARTIMIZ\SITE, \ARTIMIZ\GRADE, and \ARTIMIZ\LANDSCAP drawing files. This would triple drafting time, cutting into company profits and your break time.

Layered Drawing Files

One way to reduce the problems created with single drawing files is to store related drawings in one drawing file. For plan views, this might include the floor, framing, electrical, heating, and plumbing drawings. To be effective, these drawings must be stacked one above the other. Common elements will be drawn only once, rather than repeated for each drawing, effectively reducing the file size. Drawings can be separated using the **LAYER** command. Prefixes for layer names can be assigned to easily distinguish the drawing contents and make drawing assembly for plotting easier. Figures 3-1a and 3-1b show the floor and framing plans, which were stored in one drawing file and separated by the use of layers.

Naming Layers

Although every drawing file should make extensive use of layering, storing all related drawings in one file can only be done if each major group of components is placed on a separate layer. Layers should be given names that describe both the base drawing and the contents of the layer to make plotting easier. Prefixes such as flor, hvac, plumb, and elect will help identify the floor, plumbing, and electrical drawings. Titles such as flordim, hvacanno, or plumbsymb will clearly describe the contents. Information that will be required on all drawings—such as the walls, windows, doors, and appliances—can be stored on a layer named base. To plot the framing plan, the base and all framing layers would be set to **THAW,** with all layers related to other drawings set to **FREEZE**. Further guidelines on naming layers regardless of the filing method will be given later in this chapter.

Viewing Complex Drawings

One of the disadvantages of using files containing multiple files is that because they contain so much information, moving through the drawings using the **ZOOM** command can be slow. You can overcome this by using the **PARTIALOAD** or **PARTIALOPEN** command, the dynamic option of the **ZOOM** command or by using the **VIEW** command. When you work with large files, you can use the **PARTIALOAD** and **PARTIALOPEN** command to allow a specific portion of the drawing to be accessed. Dynamic **ZOOM** provides quick access to a specific area of the drawing and increases efficiency. The **VIEW** command allows specific areas of the drawing to be named for quick reference. On a large drawing, specific areas can be named, saved and quickly retrieved and magnified, saving valuable drafting time. Figure 3-2a shows a sheet of details and Figure 3-2b shows the use of the **VIEW** command to display a zoom of view DTL-Z3.

External Referenced Drawings

The **XREF** command is an effective method for working with related drawings. Floor-related drawings provide an excellent example of drawings that can be referenced. A drafter working for the architectural team can draw information in a base drawing that will be reflected on all other plan drawings. Copies of this drawing file can be given electronically to other firms that will develop related drawings, such as the HVAC, plumbing and electrical consultants. This will allow each firm to have a current drawing file as a base, while each firm progresses with its work. A drafter working for the mechanical engineer can add information to the base drawing, which is kept by the originating office and updated as needed. Any time a drawing that has a referenced drawing attached to it is accessed, the most current version of the base drawing is provided.

External referencing is executed by using the **XREF** command. A referenced drawing is similar to a wblock in that it is displayed each time a master drawing is accessed, but the material on the referenced drawing is not stored as part of the drawing file. It is brought into other drawing files for viewing, but it does not become part of the current drawing base. Only the name of the referenced drawing and a small amount of information are stored in the new drawing base. Attaching drawings using **XREF** also helps when sheets containing details with multiple scales must be plotted. In addition to making the assembly of drawings easy, this method ensures that stock details are current. If a stock detail must be changed to reflect new code requirements, the base detail can be corrected and all drawings that contain the referenced drawing will be automatically updated.

Layouts

A layout is a paper space tool that consists of one or more viewports to aid plotting. Layouts allow the ease of plotting associated with single drawing files and the benefits of referenced drawings to be used in displaying multiple drawings for display and plotting at varied scales. To understand the process of viewing a drawing in a layout, visualize a sheet of 24" × 36" vellum with a title block and border printed on the sheet. Imagine a hole, the viewport, cut in the vellum that allows you to look through the paper and see the floor plan on another sheet of vellum. This is a simplified version of what is required to display a drawing for plotting. Figure 3-3 shows the theory of displaying a drawing in model space in a layout created in paper space. Now imagine the floor plan shown is 90'–0" wide in model space. If you were to hold a sheet of D size vellum in front of the plan, the paper would be minute. To make the floor plan fit inside the viewport on the paper, you're going to have to hold the paper a great distance away from the floor plan until the drawing is small enough to be seen through the hole. AutoCAD will figure the distance as drawings are placed in the viewport. By entering the desired scale factor for plotting, you can reduce the floor plan to fit inside the viewport and maintain a scale typically used in the construction trade. Multiple viewports can be placed in a single layout, and multiple layouts can be created within a single drawing file. Multiple viewports will be discussed later in this chapter.

WORKING WITH LAYOUTS

Although the purpose of this chapter is not to teach specific computer skills, the use of layouts is very important to commercial drafting, because you need to be comfortable moving from model space to layouts. You will be creating drawings in model space. When your drawing is complete, use the following steps to view a drawing in a paper space layout:

1. Select the Architectural tab.
2. Select the **PAPER** button at the bottom of the screen to enter into model space in the viewport.
3. Use the All or Extents option of **ZOOM** to view the drawing.
4. Use **PAN** to center the drawing in the layout.
5. Select the **MODEL** button at the bottom of the screen to return to paper space.

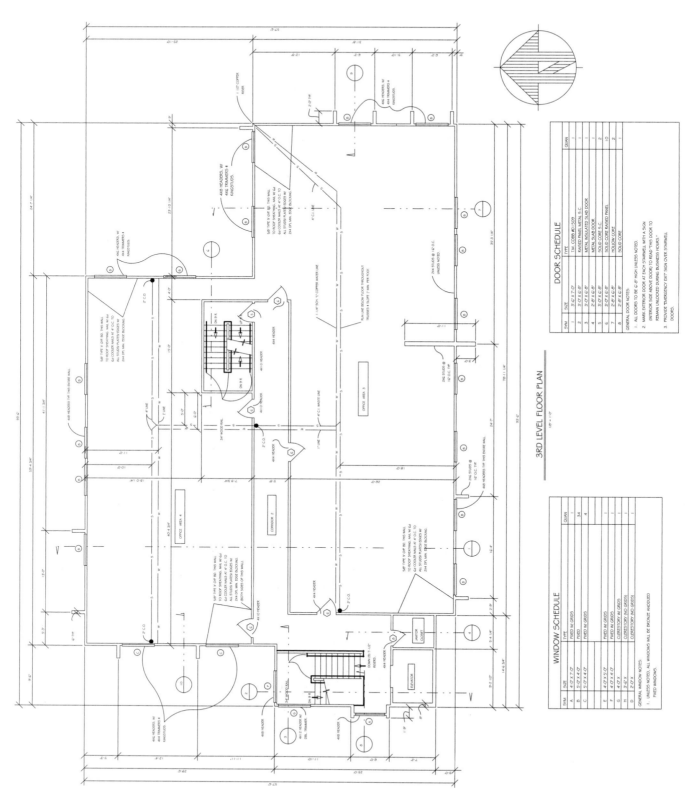

FIGURE 3-1A ■ Layered drawing files allow storage of several drawings in one file. *Courtesy StructureForm Masters, Inc.*

3RD LEVEL FLOOR PLAN

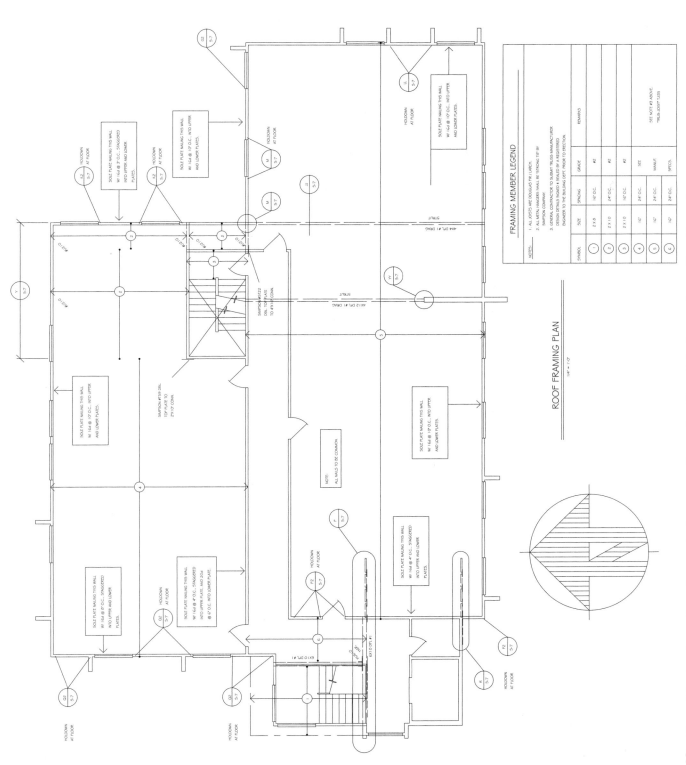

FIGURE 3-1B ■ The roof framing plan can be easily created using common features from Figure 3-1a. *Courtesy StructureForm Masters, Inc.*

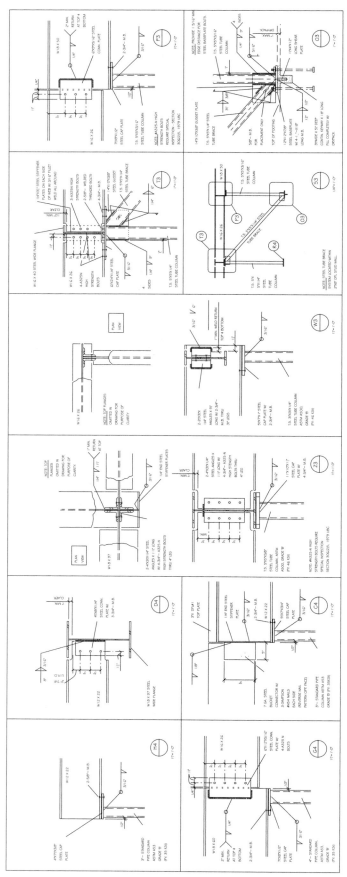

FIGURE 3-2A ■ A display of an entire drawing is often too small to be read. *Courtesy StructureForm Masters, Inc.*

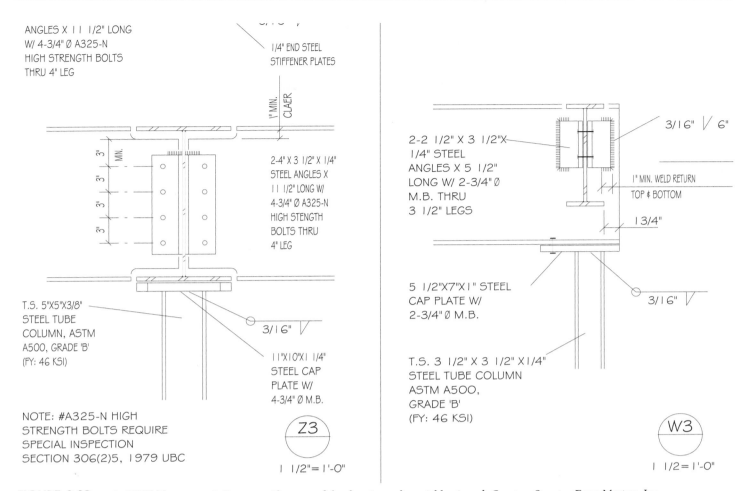

ANGLES X 1 1 1/2" LONG
W/ 4-3/4" Ø A325-N
HIGH STRENGTH BOLTS
THRU 4" LEG

1/4" END STEEL
STIFFENER PLATES

1" MIN. CLAER

2-4" X 3 1/2" X 1/4"
STEEL ANGLES X
11 1/2" LONG W/
4-3/4" Ø A325-N
HIGH STENGTH
BOLTS THRU
4" LEG

T.S. 5"X5"X3/8"
STEEL TUBE
COLUMN, ASTM
A500, GRADE 'B'
(FY: 46 KSI)

3/16"

11"X10"X1 1/4"
STEEL CAP
PLATE W/
4-3/4" Ø M.B.

NOTE: #A325-N HIGH
STRENGTH BOLTS REQUIRE
SPECIAL INSPECTION
SECTION 306(2)5, 1979 UBC

Z3

1 1/2"= 1'-0"

2-2 1/2" X 3 1/2"X
1/4" STEEL
ANGLES X 5 1/2"
LONG W/ 2-3/4" Ø
M.B. THRU
3 1/2" LEGS

3/16" 6"

1" MIN. WELD RETURN
TOP & BOTTOM

1 3/4"

5 1/2"X7"X1" STEEL
CAP PLATE W/
2-3/4" Ø M.B.

3/16"

T.S. 3 1/2" X 3 1/2" X1/4"
STEEL TUBE COLUMN
ASTM A500,
GRADE 'B'
(FY: 46 KSI)

W3

1 1/2= 1'-0"

FIGURE 3-2B ■ The **VIEW** command allows specific areas of the drawing to be quickly viewed. *Courtesy StructureForm Masters, Inc.*

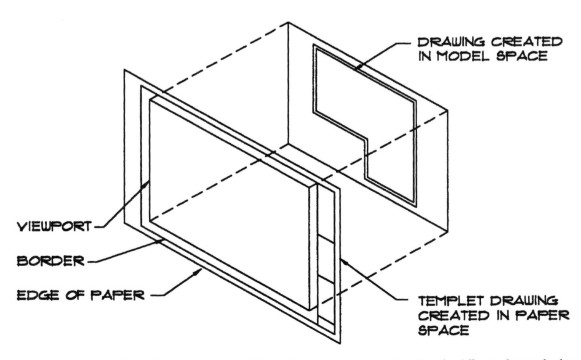

DRAWING CREATED
IN MODEL SPACE

VIEWPORT

BORDER

EDGE OF PAPER

TEMPLET DRAWING
CREATED IN PAPER
SPACE

FIGURE 3-3 ■ A viewport is provided in each drawing layout. Additional layouts can be created to allow for different plotting displays.

Creating New Layouts

In addition to the use of layouts for plotting, several layouts can be created for the same drawing so that different parameters can be emphasized in the plotting process. Different scales and paper sizes can be assigned to different layouts, or different plotting devices can be assigned to the layouts. Layouts can be assigned a unique name, and multiple paper space viewports can be assigned to one or more of the layouts. Select **Create Layout from Wizards** from the **Tools** pull-down menu to create a new layout. The layout wizard contains a series of pages that walk you through the process of creating a new layout. The steps necessary to create a new layout with a wizard are as follows:

1. Select **Create Layout from Wizard** from the **Tools** pull-down menu.
2.. Provide a name for the layout being created.
3. Select the plotting device.
4. Select the paper size and the drawing units.
5. Select the orientation.
6. Select the title block to be used.
7. Select the number of viewports and the scale to be used in the viewport.
8. Select the corners of the viewport.
9. Click the **Finish** button to return to the drawing area.

Cleaning Drawing Files

No matter what method is used to store drawing files, care should be taken to keep the files free of useless material. As you draw on a computer screen, you need to think of how you work with pencil and paper. Notes, file folders, and unfinished drawings tend to accumulate on and around your desk until it is buried. The same can happen to a computer file. AutoCAD saves its scratch paper and notes from past files unless they are removed. These files can be removed using either the **WBLOCK*** or the **PURGE** command.

WBLOCK*

The **WBLOCK** asterisk option will save your drawing as a block, but each object is saved individually rather than as one object. Information about commands that might have been tried and then erased or deleted is removed from the drawing base.

PURGE

The **PURGE** command can be a useful tool within a drawing file to eliminate unused items and minimize the amount of storage space used. Objects that have been named during a drawing session such as blocks, dimstyles, layers, linetypes, shapes, and text styles are examples of items that can be purged from a .dwg file. **PURGE** can be used, for instance, if a template drawing was used to start a drawing file. Once the new drawing is completed, unused information on the template drawing can be eliminated, reducing the file size.

DRAWING TEMPLATES

Once the decision has been made about how drawing projects will be organized, consideration should be given to how individual drawing files can be developed. One of the best methods of creating a drawing is by using a drawing template. A *template drawing* contains all of the typical settings, layers, linetypes, line weights, blocks, text styles, and dimension styles normally required for a specific type of drawing. The template will also include a border, title block with job specific attributes, and the architect's or engineer's stamp. Although all of this material should have been covered in your first AutoCAD class, it is important that you are able to relate all of this information to a construction project. As templates are developed for commercial drawings, the title block should include

■ Company name and address

■ Project name and address

■ Sheet title and page number

■ Project number

■ A space for the licensing stamp

■ A list of revisions

A template drawing is usually geared to plotting on a specific size such as D (24" × 36") or E (36" × 48") paper. Common metric paper sizes will be discussed at the end of this chapter. Many offices have template drawings for plan views, typically drawn at 1/4" = 1'–0"; site drawings drawn at an engineering scale of 1" = 10' and detail sheets. Templates for detail sheets are often specific to particular types of construction, such as standard laminated and I-joist details, concrete block details, and concrete tilt-up details.

Units of Measurement

The units of measurement should be included in a template drawing. Architectural and engineering are the units of measurement normally associated with construction drawings. With the exception of large-scale site drawings that are drawn using the engineering units, most architectural and structural drawings use architectural units. This would produce numbers that are expressed as 125'–6 3/4". Occasionally offices use engineering units that express numbers in decimal fractions, such as 75'–3.5". The unit of measurements selected can be changed throughout a drawing session, but they should be selected for a template drawing based on the office standard for a particular type of drawing.

Angle Measurement

Decimal degrees expressed as 28.30° and surveyor's units expressed as N30°30'15"E are the two most common methods of angular measurement used with construction drawings. More important than the method of expressing the measurement is adjusting the starting point that angles are measured from. Standard AutoCAD setup allows for 0° to be measured from the 3 o'clock position. This is not appropriate for site-related drawings. Common office practice for adjusting the direction of north is to draw the site plan while assuming north is at the top of a page. This may result in property lines that are not parallel to the border of the paper. Once drawn, however, the site plan and the north arrow can be rotated so that the longest property line is parallel to a border or paper edge.

Drawing Limits

The limits for a drawing template should be set for both the model and layout views as the template is created. Buildings drawn in model space are drawn at full scale. Finished drawings will be plotted to a reduced scale for easy reference by various users. This is done by the use of the **LIMITS** command. Adjusting the drawing limits is equal to selecting the paper size to use for a manual drawing. It is important to remember that a drawing can be started with the limits set to any convenient size, and that the model space limits can be adjusted at any point throughout the drawing session.

When the limits for a layout in a template drawing are established, the limits should be set based on the size of the paper that will be used for plotting. Determine the limits by multiplying the scale to be used by the paper size to be used for plotting. If a floor plan is to be drawn at a scale of 1/4"=1'–0" on D-sized paper, multiply 4 X 36 (4=the number of feet per inch X paper size) and 4 X 24 to determine the drawing limits. This will result in limits of 96' X 144'. The viewing screen can be larger than these limits to provide drawing space around the drawing, but only information within the specified limits will be plotted to scale for the finished drawing. Figure 3-4 shows a listing of common architectural and engineering scales and the resulting limits for establishing a template drawing.

Layers

Although layers are used on all drawings, they are particularly important to a template drawing. By placing layers in a template drawing, you can create, save, and reuse layers for all similar drawings. Layering can also be especially useful in assigning particular linetypes and colors to specific layers of the template. Because of their importance to a template, layer titles must be easily recognized by each member of the design team. Possible layer prefixes for simple drawings were given earlier in this chapter as drawing setup was considered. On drawing projects involving multiple design firms, a uniform layer and drawing filing naming system will aid efficiency.

AIA Standardized Layer Names

Although there is no standard method of naming and using layers on construction drawings, the AIA CAD Layer Guidelines provide the basis of layer management for many architectural and engineering offices. The AIA recommends a name comprised of a discipline code and a major group name for layers. This is similar to the guidelines for naming drawing files. A minor group name and a status name can also be added to the layer name, depending on the complexity of the project. The guidelines can be used to label both layers and drawing files with components of file names also used in the layer names. Because name components are common to both file and layer names, care must be taken to avoid creating project-specific references within a layer name. Project designations should not be included in folder names. Use the guidelines presented earlier in this chapter for naming project folders, and the following guidelines for naming files and layers. Figure 3-5 shows examples of the guidelines and how they can be applied to file and layer names.

DRAWING SCALE FACTORS

ARCHITECTURAL SCALE FACTORS				ENGINEERING SCALE FACTORS			
SCALE	LIMITS		TEXT SCALE	SCALE	LIMITS		TEXT SCALE
	18 X 24	24 X 36			18 X 24	24 X 36	
1" = 1'-0"	18 X 24	24 X 36	12	1"= 1'-0"	18 X 24	24 X 36	12
3/4" = 1'-0"	24 X 32	32 X 48	16	1"= 10'-0"	180 X 240	240 X 360	120
1/2" = 1'-0"	36 X 48	48 X 72	24	1"= 100'-0"	1800 X 2400	2400 X 3600	1200
3/8" = 1'-0"	48 X 64	64 X 96	32	1"= 20'-0"	360 X 480	480 X 720	240
1/4" = 1'-0"	72 X 96	96 X 144	48	1"= 200'-0"	3600 X 4800	4800 X 7200	2400
3/16" = 1'-0"	96 X 128	128 X 192	64	1"= 30'-0"	540 X 864	864 X 1080	360
1/8" = 1'-0"	144 X 192	192 X 288	96	1"= 40'-0"	720 X 960	960 X 1440	480
3/32" = 1'-0"	192 X 256	256 X 384	128	1"= 50'-0"	900 X 1200	1200 X 1800	600
1/16" = 1'-0"	288 X 384	384 X 576	192	1"= 60'-0"	1080 X 1440	1440 X 2160	720

NOTE: ALL LIMIT SIZES ARE GIVEN IN FEET.

FIGURE 3-4 ■ Common architectural and engineering scale factors. Although AutoCAD will automatically set the scale factor as drawings are inserted into a viewport, knowing the drawing limits will aid in establishing basic drawing content.

MODEL FILE NAME FORMAT

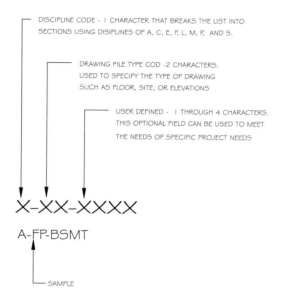

DISCIPLINE CODE - I CHARACTER THAT BREAKS THE LIST INTO SECTIONS USING DISIPLINES OF A, C, E, F, L, M, P, AND S.

DRAWING FILE TYPE COD -2 CHARACTERS. USED TO SPECIFY THE TYPE OF DRAWING SUCH AS FLOOR, SITE, OR ELEVATIONS

USER DEFINED - I THROUGH 4 CHARACTERS. THIS OPTIONAL FIELD CAN BE USED TO MEET THE NEEDS OF SPECIFIC PROJECT NEEDS

X-XX-XXXX

A-FP-BSMT

SAMPLE

SHEET FILE NAME FORMAT

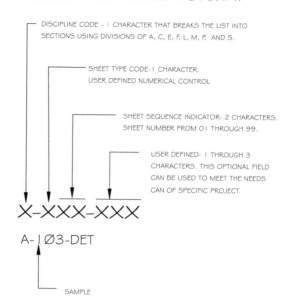

DISCIPLINE CODE - I CHARACTER THAT BREAKS THE LIST INTO SECTIONS USING DIVISIONS OF A, C, E, F, L, M, P, AND S.

SHEET TYPE CODE-I CHARACTER. USER DEFINED NUMERICAL CONTROL

SHEET SEQUENCE INDICATOR- 2 CHARACTERS. SHEET NUMBER FROM 01 THROUGH 99.

USER DEFINED- I THROUGH 3 CHARACTERS. THIS OPTIONAL FIELD CAN BE USED TO MEET THE NEEDS CAN OF SPECIFIC PROJECT.

X-XXX-XXX

A-IØ3-DET

SAMPLE

LAYER NAME FORMAT

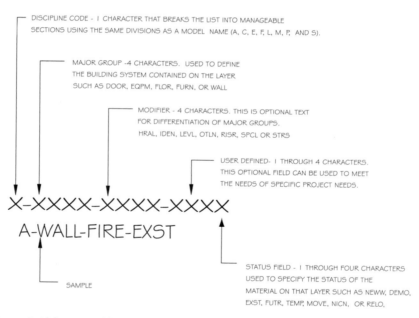

DISCIPLINE CODE - I CHARACTER THAT BREAKS THE LIST INTO MANAGEABLE SECTIONS USING THE SAME DIVISIONS AS A MODEL NAME (A, C, E, F, L, M, P, AND S).

MAJOR GROUP -4 CHARACTERS. USED TO DEFINE THE BUILDING SYSTEM CONTAINED ON THE LAYER SUCH AS DOOR, EQPM, FLOR, FURN, OR WALL

MODIFIER - 4 CHARACTERS. THIS IS OPTIONAL TEXT FOR DIFFERENTIATION OF MAJOR GROUPS. HRAL, IDEN, LEVL, OTLN, RISR, SPCL OR STRS

USER DEFINED- I THROUGH 4 CHARACTERS. THIS OPTIONAL FIELD CAN BE USED TO MEET THE NEEDS OF SPECIFIC PROJECT NEEDS.

X-XXXX-XXXX-XXXX

A-WALL-FIRE-EXST

SAMPLE

STATUS FIELD - I THROUGH FOUR CHARACTERS USED TO SPECIFY THE STATUS OF THE MATERIAL ON THAT LAYER SUCH AS NEWW, DEMO, EXST, FUTR, TEMP, MOVE, NICN, OR RELO.

FIGURE 3-5 ■ The AIA CAD Layer Guidelines use abbreviations to name drawing files and layers. By using a uniform system to name files and layers, architects, engineers, and consultants working for several different firms can easily produce a coordinated set of drawings.

Guidelines for Naming Layers

The method for naming layers is similar to naming files presented earlier. A layer name can be composed of the discipline and major group name. A minor name and status code can be added to the sequence. Figure 3-6 shows some alternatives for using components with layer names. The discipline code is the same as the one- or two-character code that is used for model and sheet file names.

Major Group Name—The major group code is a four-character code that identifies a building component specific to the defined layer. Codes such as ANNO (annotation), EQUIP (equipment), FLOR (floor), GLAZ (glazing), and WALL (walls) are examples of major group codes that are associated with the architectural layers. A sample listing of layer names can be seen in Figure 3-7. A complete listing of major codes for each group can be found in the AIA Layer Guidelines.

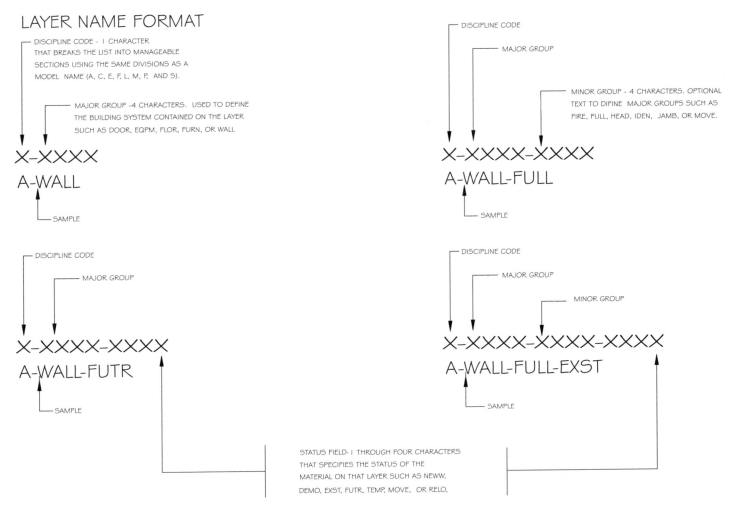

FIGURE 3-6 ■ Layer names are described by four categories when based on the AIA CAD Layer Guidelines.

Minor Group Name—The minor group code is an optional four-letter code that can be used to define subgroups to the major group. The code A–FLOR (architectural–floor) might include minor group codes for OTLN (outline), LEVL (level changes), STRS (stair treads or escalators), EVTR (elevators), or PFIX (plumbing fixtures). A layer name of A–FLOR–IDEN would contain room names, numbers, and other related titles or tags. Sample listings can be found in Figure 3-7. A sample listing of minor group codes specific to each discipline is listed in the AIA Layer Guidelines.

Status Code—The status code is an optional four-letter code that can be used to define a subgroup of either a major or minor group (see Figure 3-7). The code is used to specify the phase of construction. The layer names for the walls of a floor plan (A–WALL–FULL) could be further described using the status code of NEWW (new work), EXST (existing walls), or DEMO (demolition). A wall shown on the floor plan that is to be moved would be displayed on the A–WALL–FULL–MOVE layer.

Linetypes

A linetype is a specific pattern of line segments of varied lengths and widths. Varied line types have traditionally been used throughout construction drawings to represent different materials and to contrast different components. Figure 3-8 shows examples of common linetypes used in commercial drawings. Assigning varied linetypes to a specific layer on a template drawing can save a great deal of drafting time. The exact type of line to be used will vary based on the type of drawing. Figure 3-9 shows a floor plan created using several varied types and widths of line types. Sections III and IV discuss specific line widths and types that relate to specific drawings.

In addition to linetypes being created and set based on a specific layer, the scale factor for linetypes should also be set for a template drawing. The scale can be adjusted using the **LTSCALE** command. To display linetypes for a drawing in a paper space layout, **PSLTSCALE** must be set to 0. Common line scales can be seen in Figure 3-4. The line scale is determined by the size the drawing will be plotted at. Using an **LTSCALE** of 2 will make the line pattern twice as big as the line pattern created by a scale of 1.

ARCHITECTURAL LAYER NAMES

ANNOTATION

A-ANNO-DIMS	DIMENSIONS
A-ANNO-KEYN	KEY NOTES
A-ANNO-LEGN	LEGENDS & SCHEDULES
A-ANNO-NOTE	NOTES
A-ANNO-NPLT	NON-PLOTTING INFORMATION
A-ANNO-NRTH	NORTH ARROW
A-ANNO-REVS	REVISIONS
A-ANNO-REDL	REDLINE
A-ANNO-SYMB	SYMBOLS
A-ANNO-TEXT	TEXT
A-ANNO-TTLB	BORDER & TITLE BLOCK

FLOOR PLAN

A-FLOR-EVTR	ELEVATOR CAR & EQUIPMENT
A-FLOOR-FIXT	FLOOR INFORMATION
A-FLOR-FIXD	FIXED EQUIPMENT
A-FLOR-HRAL	HANDRAILS
A-FLOR-NICN	EQUIPMENT NOT IN CONTRACT
A-FLOR-OVHD	OVERHEAD ITEMS (SKYLIGHTS)
A-FLOR-PFIX	PLUMBING FIXTURES
A-FLOR-STRS	STAIR TREADS
A-FLOR-TPIN	TOILET PARTITIONS

ROOF PLAN

A-ROOF	ROOF
A-ROOF-OTLN	ROOF OUTLINE
A-ROOF-LEVL	ROOF LEVEL CHANGES
A-ROOF-PATT	ROOF PATTERNS

ELEVATIONS

A-ELEV	ELEVATIONS
A-ELEV-FNSH	FINISHES & TRIM
A-ELEV-IDEN	COMPONENT IDENT. NUMBERS
A-ELEV-OTLN	BUILDING OUTLINES
A-ELEV-PATT	TEXTURES & HATCH PATTERNS

SECTIONS

A-SECT	SECTIONS
A-SECT-IDEN	COMPONENT IDENT. NUMBERS
A-SECT-MBND	MATERIAL BEYOND SECTION CUT
A-SECT-MCUT	MATERIAL CUT BY SECTION PLANE
A-SECT-PATT	TEXTURES AND HATCH PATTERNS

CIVIL LAYER NAMES

ELEVATIONS

C-BLDG	PROPOSED BUILDING FOOTPRINT
C-PKNG	PARKING LOTS
C-PKNG-CARS	AUTOMOBILES
C-PKNG-DRAN	DRAINAGE INDICATORS
C-PROP	PROP. LINES & BENCHMARKS
C-PROP-BRNG	BEARINGS & DISTANCE LABELS
C-PROP CONS	CONSTRUCTION CONTROLS
C-PROP-ESMT	EASEMENTS, RIGHTS- OF -WAY
C-ROAD	ROADWAYS
C-ROAD-CNTR	CENTER LINES
C-ROAD-CURB	CURBS
C-SSWR	SANITARY SEWERS
C-STRM	STORM DRAIN / CATCH BASINS
C-STRM-UNDR	UNDERGROUND DRAIN. PIPES
C-TOPO	PROPOSED CONTOURS & ELEV..
C-TOPO-BORE	TEST BORINGS
C-TOPO-RTWL	RETAINING WALL

STRUCTURAL LAYER NAMES

GENERAL LAYERS

S-ABLT	ANCHOR BOLTS
S-BEAM	BEAMS
S-COLS	COLUMNS
S-DECK	STRUCTURAL FLOOR DECKING
S-GRID	COLUMN GRIDS
S-GRID-DIMS	COLUMN GRID DIMENSIONS
S-GRID-EXTR	COLUMN GRIDS OUTSIDE OF BUILD.
S-GRID-IDEN	COLUMN GRID TAGS
S-JOIS	JOIST
S-METL	MISCELLANEOUS METAL
S-WALL	STRUCT. BEARING / SHEAR WALLS

FOUNDATION /SLAB PLAN LAYERS

S-FNDN	FOUNDATION
S-FNDN-PILE	PILES / DRILLED PIERS
S-FNDN-RBAR	FOUNDATION REINFORCING
S-SLAB	SLAB
S-SLAB-EDGE	EDGE OF SLAB
S-SLAB-REBAR	REINFORCING
S-SLAB-JOIN	SLAB CONTROL JOINTS

FIGURE 3-7 ■ Sample layer names based on the AIA CAD Layer Names use a discipline code, major group code, minor group code, and status group format.

For a drawing plotted at 1/4" = 1'–0" a dashed line that should be 1/8" when plotted would need to be 3" long when drawn at full scale. This was determined by multiplying .125" × 24 (the **LTSCALE**). The scale factor can be determined by using half of the scale factor. The drawing scale factor is always the reciprocal of the drawing scale. Using a drawing scale of 1/4" = 1'–0" would equal .25"=12". Dividing 12 by .25 produces a scale factor of 48. Keep in mind that when the **LTSCALE** is adjusted, it will affect all linetypes within the drawing. This is typically not a problem unless you are trying to create one unique line and **CTLTSCALE** can be used.

Lineweight

Varied line widths will help you and the print reader keep track of various components of a complex drawing. As you assign lineweight to template drawings, remember that you're using thick and thin lines to add contrast, not represent thickness of an object. Lineweights can be assigned to a layer by selecting **Lineweights** from the **Format** pull-down menu, or by entering **LWEIGHT** at the command prompt. Each method will display the **Lineweight Settings** dialog box. Once you've made some basic settings, lineweights can be altered using the **Lineweight Control** menu

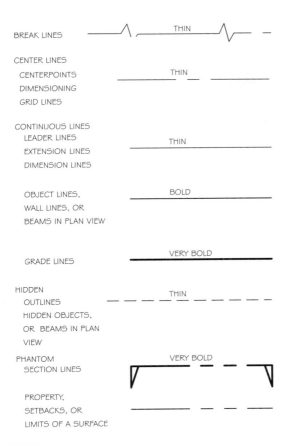

BREAK LINES	THIN	
CENTER LINES CENTERPOINTS DIMENSIONING GRID LINES	THIN	
CONTINUOUS LINES LEADER LINES EXTENSION LINES DIMENSION LINES	THIN	
OBJECT LINES, WALL LINES, OR BEAMS IN PLAN VIEW	BOLD	
GRADE LINES	VERY BOLD	
HIDDEN OUTLINES HIDDEN OBJECTS, OR BEAMS IN PLAN VIEW	THIN	
PHANTOM SECTION LINES	VERY BOLD	
PROPERTY, SETBACKS, OR LIMITS OF A SURFACE		

FIGURE 3-8 ■ Common linetypes found throughout commercial drawings.

from the **Object Properties** toolbar. Common lineweights and their widths in millimeters and inches, point size and pen size can be seen in Figure 3-10.

Assigning Lineweights

The **Linetype Settings** dialog box can be used to set the units for measuring line width, the default lineweight, the display scale, and as a display of the current settings. When the dialog box is first displayed, the **Display Lineweights** box is OFF. In its current state, if varied lineweights are used, the width will be reflected only on a print and not in the drawing display. Select the **Display Lineweight** box to make it active, and varied lineweights will be displayed on the screen. To assign a lineweight, scroll down the list of lineweights, find the desired line weight, and select the **OK** button. Once the unit and display settings have been set, lineweights can be altered by selecting **Lineweight Control** from the **Object Properties** toolbar.

Displaying Lineweights

In addition to activating the **Display Lineweight** in the **Lineweight Settings** dialog box, you also must consider the difference in lineweight display between model space and paper space. In model space, lineweights are displayed in relation to pixels. Lineweights in paper space are displayed in the exact plotting width (see Table 3-2). For work in model space, lineweights should be selected that are proportional to their pixel value and not their real world width.

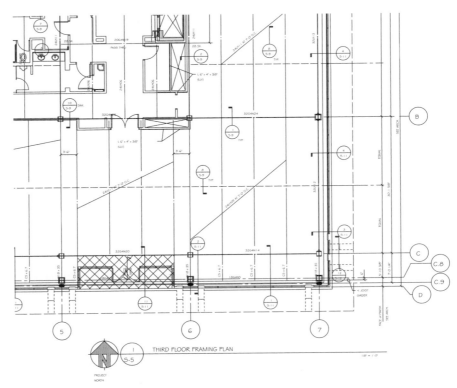

FIGURE 3-9 ■ Examples of varied line types and weights used in commercial drawings. *Courtesy Peck, Smiley, Ettlin Architects.*

LINEWEIGHT	MILLIMETERS	INCHES	POINTS	PEN SIZE
————	.25	.010	3/4 pt.	OOO
————	.30	.012		OO
————	.35	.014	I pt.	O
————	.50	.020		I
————	.53	.021	1-1/2pt.	
————	.60	.024		2
————	.80	.031		3
————	1.00	.039		3-1/2
————	1.20	.047		4

FIGURE 3-10 ■ Common lineweights and their width in metric, inches, point size, and pen size.

TABLE 3-2 ■ Lineweights for Use in Paper Space

mm	In.	Description	Use
.25	.010	very, very, thin	Very detailed areas
.30	.012	very thin	Window symbols and door swings
.35	.014	thin	Cabinets and plumbing fixtures
.50	.020	moderate	Doors
.60	.024	thick	Outline of major components, such as walls
.80	.031	very thick	Underlining room titles
1.2	.047	very, very thick	Borders

Blocks and Wblocks

A block is a group of objects that are treated as one unit. Because so much of construction drawings are comprised of symbols, these symbols can be drawn once and stored for future use. Storing blocks in a template drawing places the symbol where it is frequently used. Figure 3-11 shows an example of block drawing of

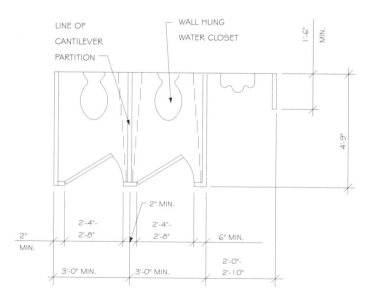

FIGURE 3-11 ■ Blocks and wblocks can be used to reproduce common features and symbols.

toilet partitions that could be attached to a floor plan template. Because this is such a common fixture to a floor plan, the block could be stored on the A-flor-plmb layer, frozen, and then thawed as the plan is drawn.

A wblock is similar to a block but it has a wider range of uses. Blocks can only be used in the drawing they are stored in. Wblocks can be stored as a separate .dwg file and inserted into another drawing using the **INSERT** command. The entire template drawing can be stored as a wblock and inserted into another drawing file, or another wblock can be inserted into the template and then stored with a new file name. Figure 3-12 shows other common symbols that are used throughout commercial drawings that could be stored as wblocks. Blocks and wblocks will be discussed as the DesignCenter is explored and in Sections III and IV, when components of architectural and structural drawings are introduced.

Text

Specific text styles, fonts, styles as well as general and local notes should be stored in drawing templates. Large bodies of text can be stored on the template drawing using the **QTEXT** mode to reduce regen times.

Text Styles and Fonts

Text style refers to specific characteristics of text, such as the height, width, and letter shape or font. The **STYLE** command and the **Text Style** dialog box can be used to store different combinations of text styles for quick usage. As styles are created for the template drawing, easily recognizable titles should be given. Names including the text font and usage are helpful. Much of the lettering in this text was created using a font called archstyl, which is created by a third party vendor for use with AutoCAD. Styles such as archtitl, archtxt, and italitxt can be created in the template drawing. These styles are often used for creating styles for titles and text using the font archstyl and a text style using the font italic for drawing text. If all text throughout a drawing is the same size, the height variable of the **STYLE** command can be set so that the variable will not have to be set each time the font is used. The height of text is best controlled with the height option of the **FONT** command. The **STYLE** command also allows text to be placed on a slant and allows the width to be altered, as well as the position of the text to horizontal.

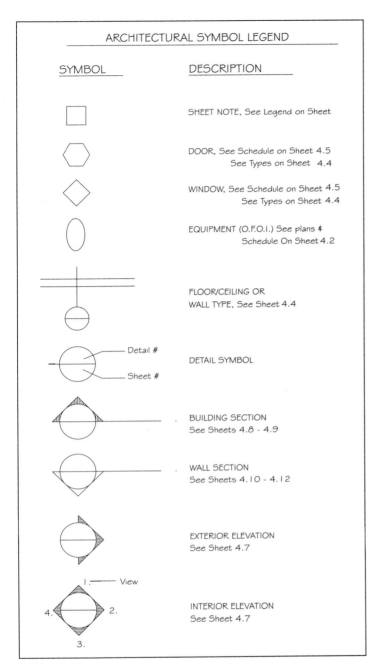

FIGURE 3-12 ■ Common symbols found on commercial drawings can be saved as wblocks.

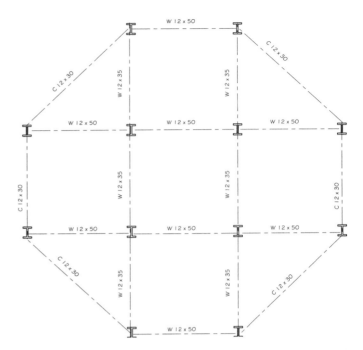

FIGURE 3-13 ■ Common methods of text placement in relation to drawing objects.

Text Placement

Text placement refers to the location of text relative to the drawing. The drafter should be careful about the orientation that is used to place the text. Text should be placed parallel to the bottom or the right edge of the page. When describing structural material shown in plan view, text is ideally placed parallel to the member being described. Figure 3-13 shows typical methods of placing text on plan views. On plan views such as Figure 3-14, text is placed within the drawing, but arranged so that it does not interfere with any part of the drawing. Text is generally placed within a few inches of the object being described. A leader line is used to connect the text to the drawing. On details, text can be placed within the detail if large open spaces are part of the drawing. It is preferable to keep text out of the drawing. Text should be aligned to enhance clarity and can be aligned either left or right. Figure 3-15 shows an example of a detail using good text placement.

General Notes

Much of the text used to describe a drawing can be standardized and placed in a wblock or template drawing. Figure 3-16 shows an example of standard notes that an office includes for all site plans. These notes can be typed once and saved as a wblock. The wblock can then be *nested* in the template drawing on the C-site-anno-genn layer. The layer can be frozen, thawed when needed, and then edited using the **DDEDIT** command to make minor changes based on specific requirements of the job. Notice in Figure 3-16 that several of the notes and all the schedules include areas labeled *xxx*. These are notes that have been set up as blocks and stored with attributes that can be altered for each usage.

The term *font* refers to the design used to represent the shape of each letter. Many third-party vendors supply text libraries that can be used in AutoCAD featuring fonts that resemble handwritten printing. Architectural and engineering offices often use different fonts throughout a drawing to distinguish between titles and text, or to highlight specific information. Whenever text is being selected for a template drawing, it is important to select fonts that are not difficult to read and fonts that can be quickly reproduced by a plotter. The complexity of a particular font also affects the amount of storage space required, the speed of regens, and the plotter speed.

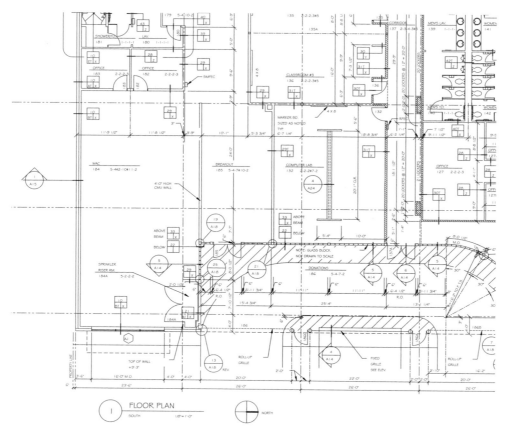

FIGURE 3-14 ■ Text must be placed within the drawings so that it can be easily read. *Courtesy Michael & Kuhns Architects, P.C.*

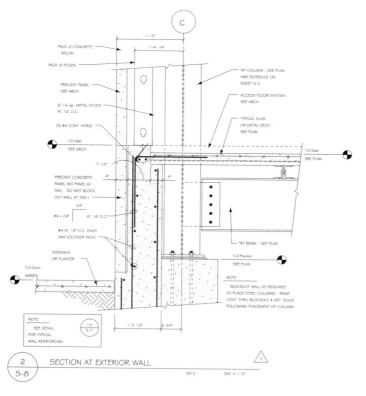

FIGURE 3-15 ■ Text placed neatly around the drawing adds to clarity. *Courtesy Van Domelen/Looijenga/McGarrigle/Knauf Consulting Engineers.*

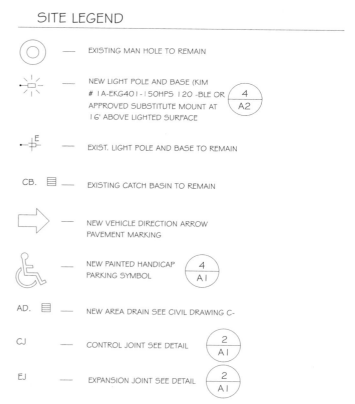

SITE LEGEND

EXISTING MAN HOLE TO REMAIN

NEW LIGHT POLE AND BASE (KIM # 1A-EKG401-150HPS 120 -BLE OR APPROVED SUBSTITUTE MOUNT AT 16' ABOVE LIGHTED SURFACE

EXIST. LIGHT POLE AND BASE TO REMAIN

CB. EXISTING CATCH BASIN TO REMAIN

NEW VEHICLE DIRECTION ARROW PAVEMENT MARKING

NEW PAINTED HANDICAP PARKING SYMBOL

AD. NEW AREA DRAIN SEE CIVIL DRAWING C-

CJ CONTROL JOINT SEE DETAIL

EJ EXPANSION JOINT SEE DETAIL

FIGURE 3-16 ■ Standard general notes can be saved as a wblock or as part of a template drawing to save time.

Local Notes

Many drawings, such as sections, contain basically the same notes. Although sections for a tilt-up structure will be radically different from a section for a steel rigid frame, a template sheet can be developed for each type of construction. The actual drawing might look different, but the notes to specify standard materials are similar for most buildings using the same type of construction. Local notes can also be placed in the template drawing as a block, as seen in Figure 3-17. These notes can be thawed, moved into the needed position and edited, as seen in Figure 3-18. This can greatly reduce drafting time and increase drafting efficiency. The **MIRRTEXT** command should also be adjusted when you work with sections or other drawings that are created using the **MIRROR** command. A **MIRRTEXT** variable setting of 0 will flip a drawing while leaving the text readable.

Dimension Variables

One of the best features of a template drawing is the ability to store dimension variables so that they do not have to be set with each new drawing. Dimension variables can be combined and saved as a style and then altered using the **Dimension Style Manager**. The manager can be used to control the lines and arrow type, text height, color, and placement, the relationship of the text to the dimension lines, and the units used for the dimensions.

1/2" CD APA 32/16 PLY. ROOF SHTG. INT. GR. W/ EXT. GLUE. LAY FACE GRAIN PERP. TO JOISTS & STAGGER JOINTS. USE: 8d COM. NAILS @ 6" O.C. BOUNDARY & EDGES, 8d COM. NAILS @ 12" O.C. FIELD. SEE ROOF-FRAMING PLANS FOR SPECIAL NAILING.

'TRUS-JOIST' 16" TJI 35" @ 24" O.C.

MINERAL CAP SHEET OVER 2 LAYERS OF ASB. FELT AS PER. JOHN MANSVILLE SPEC. #406

PROVIDE 1/2" GYP. BD. DRAFT STOPS FROM CEILING TO PLY ROOF SHTG. FOR EVERY 3,000 SQ. FT. MAX.

11/2" 100#/FT LT. WT. CONC. OVER 15# ASPH. SATURATED FELT W/ 6"X6" W-4"X4" WWM IN SLAB OVER PLY. SHTG.

3/4" CD APA 42/20 T&G PLY FLR. SHTG., INT. GRD. W/ EXT. GLUE. LAY FACE GRAIN PERP. TO JOISTS & STAGGER JOINTS. USE: 10d COM. NAILS @ 6" O.C. BOUNDARY & EDGES, 10d COM. NAILS O.C. FIELD. SEE FLOOR FRAMING PLANS FOR SPECIAL NAILING.

LINE OF SUSP. CEILING (CLASS 'A')

11/2" 100#/FT LT. WT. CONC.

3/4" CD APA 42/20 T&G PLY FLOOR SHEATHING.

R-25 RATED INSULATION

PROVIDE 1/2" GYP. BD. DRAFT STOPS FROM CEILING TO PLY FLOOR SHEATHING. FOR EVERY 1,000 SQ. FT. MAX.

2-2"X4" DFL TOP PLATES LAP 48" MIN. W/ 10-16d COM.

5/8" TYPE "X" GYP. BD. 2"X4" STUDS @ 16" O.C. UNLESS NOTED

2-2"X6" DFL TOP PLATES LAP 48" MIN. W/ 12-16d COM.

R-21 RATED INSULATION

2"X6" SOLE PLATES.

R-38 RATED INSULATION

FIGURE 3-17 ■ Common local notes can also be saved as a wblock or nested in a template drawing.

The **Lines and Arrows** tab of the **Dimension Style Manage** should be used to establish the dimension line criteria of the template drawing. This would include setting the color, usage, lineweight, offset and extension of the extension and dimension lines. This tab also controls the choice of line terminators to be used for the template. Figure 3-19 shows the options for usage of extension lines. Figure 3-20 shows several common dimension line terminators that are common to architectural templates. Figure 3-21 shows common alternatives for placing dimension text using the **Dimension Style Manager**.

ASSEMBLING DRAWINGS AT MULTIPLE SCALES

One of the most valuable computer skills needed to complete commercial drawing projects is the ability to combine drawings to be plotted at multiple scales. Because of its importance, a brief review will be provided. Drawing sheets can be assembled for plotting using layouts and viewports. Each time you enter a layout, you're working with a floating viewport. The viewport allows you to look through the paper and see the drawing created in model space and to display the drawing in the layout for plotting. An unlimited amount of viewports can be created in paper space, but only 64 viewports can be visible at once. Because floating viewports are considered objects, drawing objects displayed in the viewport cannot be edited. For objects to be edited within the viewport, model space must be restored. You can toggle between model space and a floating viewport by choosing either the Model or Layout tab, depending on which one is inactive. Double clicking in a floating viewport will also switch the display to model space.

Note: Remember, while editing in model space in a layout, you risk changing the drawing setup. This should only be used by experienced AutoCAD users or when in the initial setup of the view inside of the viewport.

Most professionals work in model space and arrange the final drawing for output in paper space. While working in a layout, with the viewport in paper space, any material added to the drawing will be added to the layout, but not shown in the model display. This will prove useful as the final plotting layout is constructed.

Creating a Single Floating Viewport

Floating viewports provide an excellent means of assembling multi-scaled drawings for plotting. Figure 3-22 shows a sheet of details assembled for plotting using multiple viewports. Construction projects typically comprise drawings that are drawn at a variety of scales. These details range in scale from 3/8" = 1'–0" to 1 1/2" = 1'–0". This sheet of details cannot be easily assembled without the use of multiple viewports. The process for creating multiple viewports will be similar to the creation of one viewport. The following example will demonstrate the steps used to prepare the section and three details shown in Figure 3-23 for plotting. The section will be displayed at a scale of 1/4" = 1'–0", two of the details will be displayed at a scale of 3/4"=1'–0", and the stair section will be displayed at a scale of 1/2" = 1'–0". The following sections will walk you through the process of creating multiple viewports in a

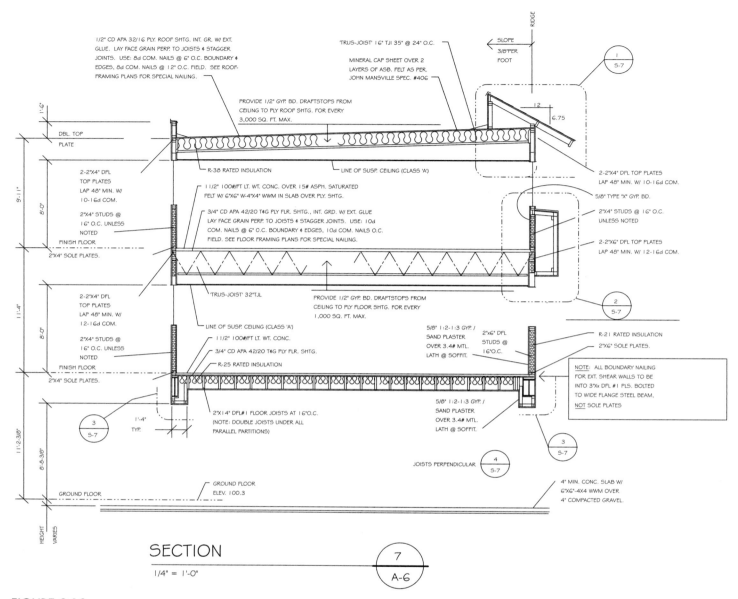

1/2" CD APA 32/16 PLY. ROOF SHTG. INT. GR. W/ EXT. GLUE. LAY FACE GRAIN PERP. TO JOISTS & STAGGER JOINTS. USE: 8d COM. NAILS @ 6" O.C. BOUNDARY & EDGES, 8d COM. NAILS @ 12" O.C. FIELD. SEE ROOF-FRAMING PLANS FOR SPECIAL NAILING.

TRUS-JOIST' 16" TJI 35" @ 24" O.C.

MINERAL CAP SHEET OVER 2 LAYERS OF ASB. FELT AS PER. JOHN MANSVILLE SPEC. #406

SLOPE
3/8"PER
FOOT

RIDGE

PROVIDE 1/2" GYP. BD. DRAFTSTOPS FROM CEILING TO PLY ROOF SHTG. FOR EVERY 3,000 SQ. FT. MAX.

12
6.75

1
S-7

1'-6"

DBL. TOP PLATE

R-38 RATED INSULATION

LINE OF SUSP. CEILING (CLASS 'A')

2-2"X4" DFL TOP PLATES LAP 48" MIN. W/ 10-16d COM.

5/8" TYPE "X" GYP. BD.

2-2"X4" DFL TOP PLATES LAP 48" MIN. W/ 10-16d COM.

9'-11"

8'-0"

2-2"X4" DFL TOP PLATES LAP 48" MIN. W/ 10-16d COM.

2"X4" STUDS @ 16" O.C. UNLESS NOTED

FINISH FLOOR

2"X4" SOLE PLATES.

1 1/2" 100#/FT LT. WT. CONC. OVER 15# ASPH. SATURATED FELT W/ 6"X6" W-4"X4" WWM IN SLAB OVER PLY. SHTG.

3/4" CD APA 42/20 T&G PLY FLR. SHTG., INT. GRD. W/ EXT. GLUE LAY FACE GRAIN PERP. TO JOISTS & STAGGER JOINTS. USE: 10d COM. NAILS @ 6" O.C. BOUNDARY & EDGES, 10d COM. NAILS O.C. FIELD. SEE FLOOR FRAMING PLANS FOR SPECIAL NAILING.

2"X4" STUDS @ 16" O.C. UNLESS NOTED

2-2"X6" DFL TOP PLATES LAP 48" MIN. W/ 12-16d COM.

11'-4"

2-2"X4" DFL TOP PLATES LAP 48" MIN. W/ 12-16d COM.

2"X4" STUDS @ 16" O.C. UNLESS NOTED

FINISH FLOOR

2"X4" SOLE PLATES.

TRUS-JOIST' 32"TJL

LINE OF SUSP. CEILING (CLASS 'A')

1 1/2" 100#/FT LT. WT. CONC.

3/4" CD APA 42/20 T&G PLY FLR. SHTG.

R-25 RATED INSULATION

PROVIDE 1/2" GYP. BD. DRAFTSTOPS FROM CEILING TO PLY FLOOR SHTG. FOR EVERY 1,000 SQ. FT. MAX.

5/8" 1:2-1:3 GYP. / SAND PLASTER OVER 3.4# MTL. LATH @ SOFFIT.

2"X6" DFL STUDS @ 16"O.C.

2
S-7

R-21 RATED INSULATION

2"X6" SOLE PLATES.

NOTE: ALL BOUNDARY NAILING FOR EXT. SHEAR WALLS TO BE INTO 3"x DFL #1 PLS. BOLTED TO WIDE FLANGE STEEL BEAM, NOT SOLE PLATES

8'-0"

2"X14" DFL#1 FLOOR JOISTS AT 16"O.C. (NOTE: DOUBLE JOISTS UNDER ALL PARALLEL PARTITIONS)

5/8" 1:2-1:3 GYP. / SAND PLASTER OVER 3.4# MTL. LATH @ SOFFIT.

JOISTS PERPENDICULAR

3
S-7

1'-4"
TYP.

3
S-7

4
S-7

11'-2-3/8"

8'-6-3/8"

GROUND FLOOR

GROUND FLOOR ELEV. 100.3

4" MIN. CONC. SLAB W/ 6"X6"-4X4 WWM OVER 4" COMPACTED GRAVEL.

HEIGHT VARIES

SECTION
1/4" = 1'-0"

7
A-6

FIGURE 3-18 ■ Once a view is drawn, required local notes stored in blocks can be exploded and moved into the required position.

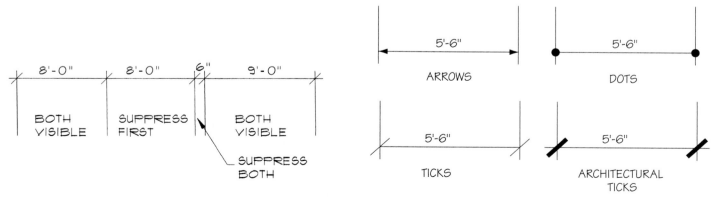

8'-0"   8'-0"   6"   9'-0"

BOTH VISIBLE | SUPPRESS FIRST | BOTH VISIBLE

SUPPRESS BOTH

FIGURE 3-19 ■ Practical applications for suppressing extension lines.

5'-6"
ARROWS

5'-6"
DOTS

5'-6"
TICKS

5'-6"
ARCHITECTURAL TICKS

FIGURE 3-20 ■ Variations of dimension line terminators available in AutoCAD.

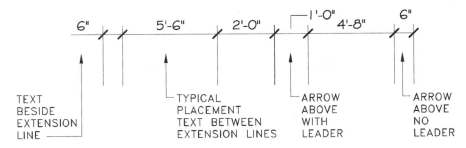

FIGURE 3-21 ■ Common placement of dimension text.

template and inserting multiple drawings. For this discussion, the **INSERT** command will be used.

Displaying Model Space Objects in the Viewport

The easiest method to display multiple drawings at multiple scales is by inserting each of the drawings to be plotted into the template in model space. With the details arranged in one drawing, similar to Figure 3-23, switch to the desired layout. Use the following steps to display the drawings in multiple viewports:

1. In the layout, click the **PAPER** button on the status bar to toggle the drawing to model space, use the **ZOOM** All option to enlarge the area of model space to be displayed in the existing viewport. Each of the four drawings will be displayed, but each will be at an unknown scale.
2. Select what will be the largest drawing to work with first. In our example, the viewport for displaying the section will be adjusted first.
3. Activate the **Viewport** toolbar.
4. Set the scale of the viewport to the appropriate value for displaying the section using the **Viewport Scale Control** menu on the **Viewport** toolbar. For this example a scale of 1/4" = 1'–0" was used.
5. Click the **MODEL** button on the status bar to return to paper space in your current layout.
6. New viewports will need to be created to display the other details at different scale factors. Refer to the section "Creating Additional Viewports." Once all of the drawings to be plotted have been displayed in a viewport with the proper scale factor you are ready to plot.

Adjusting the Existing Viewport

Once the desired drawing is inserted into the viewport, the size of the viewport can be altered if the entire drawing can't be seen. To alter the viewport size, move the cursor to touch the viewport and activate the viewport grips. Select one of the grips to make it hot, and then use the hot grip to drag the window to the desired size. The Section has now been inserted into the template and is ready to be plotted at a scale of 1/4" = 1'–0". The display would resemble Figure 3-24.

Creating Additional Viewports

Use the following steps to prepare additional viewports:

1. Set the current layer to Viewport.
2. Select **Single Viewport** from the **Viewport** toolbar.
3. Select the corners for the new viewport. Although you should try to size the viewport accurately, it can be stretched to enlarge or reduce its size once the scale has been set.

The results of this command can be seen in Figure 3-25.

Altering the Second Viewport

As a new viewport is created, the display from the existing viewport will be displayed in the new viewport as well. Use the following steps to alter the display.

1. Click the **PAPER** button on the status bar to toggle the drawings to model space.
2. Make the new viewport the current viewport by placing the cursor in the viewport and single clicking.
3. Use the **ZOOM ALL** option to display all of the model contents in the second viewport.
4. Set the viewport scale to the appropriate value for displaying the section. For this example, a scale of 3/4" = 1'–0" will be used to display the wall section.
5. Use the **PAN** command to center the wall section in the viewport.
6. Click the **MODEL** button on the status bar to toggle the drawings to paper space.
7. Select the edge of the viewport to display the viewport grips.
8. Make a grip hot and shrink the viewport so that only the wall section is shown.
9. Use the **MOVE** command to move the viewport to the desired position in the template.
10. Save the drawing for plotting.

The drawing should now resemble Figure 3-26. Once the viewport for the wall section been adjusted, additional viewports can be created. Once all of the desired details have been added to the lay-

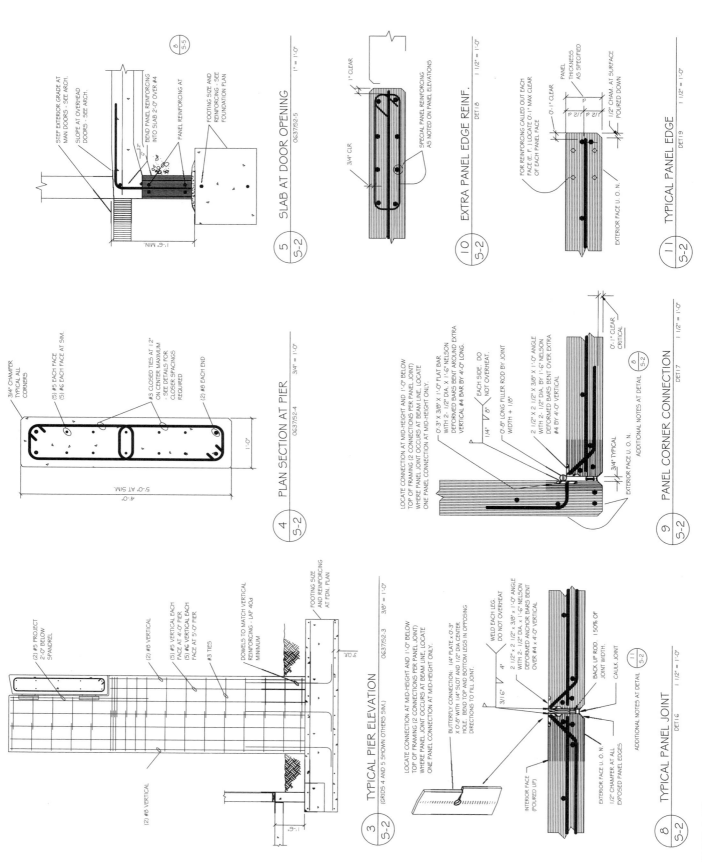

FIGURE 3-22 ■ Details drawn at varied scales can be assembled on one sheet for plotting. *Courtesy Van Domelen/Looijenga/McGarrigle/Knauf Consulting Engineers.*

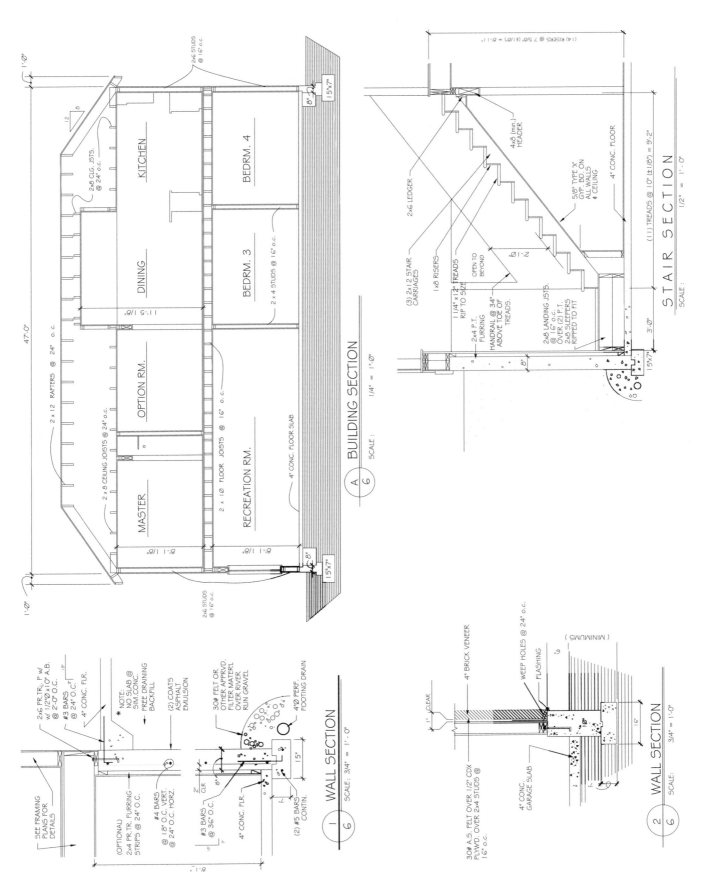

FIGURE 3-23 ■ Layout drawings are started by inserting all required drawings into the template in model space.

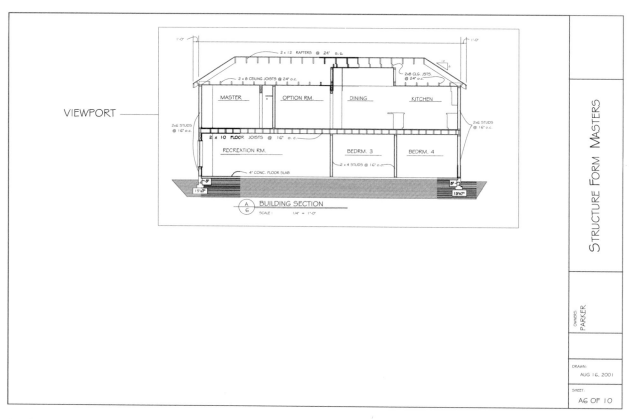

FIGURE 3-24 ■ The appropriate scale for plotting can be set as the drawing is inserted in the viewport. The size of the viewport can be adjusted if the entire drawing can't be seen.

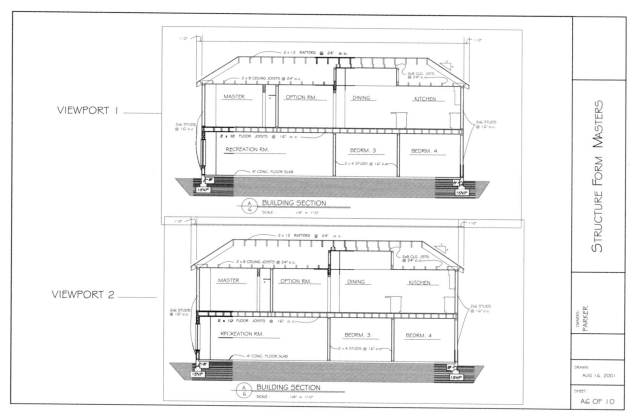

FIGURE 3-25 ■ If a new viewport is created, the display from the original viewport will be displayed in the new viewport.

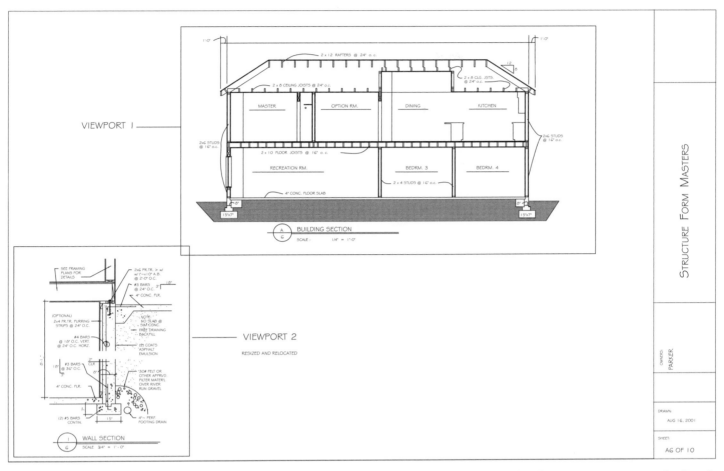

FIGURE 3-26 ■ Use the **PAN** command to move the drawings so that the desired information is centered in the viewport. Once centered, adjust the size of the viewport so that only the desired material is displayed.

out, the viewport outlines can be frozen. Freezing the viewports will eliminate the line of the viewport being produced as the drawing is reproduced. The finished drawing, with the viewports frozen, will resemble Figure 3-27.

WORKING IN A MULTI-DOCUMENT ENVIRONMENT

An alternative to using template drawings to store frequently used styles, settings, and objects is to move objects and information between two or more drawing files. Just as you can have several programs open on your desktop at once and rapidly switch from one program to another, AutoCAD allows you to have several drawings open at the same time. Objects or drawing properties can be moved between drawings using cut/copy and paste, object drag and drop, the Property Painter, and concurrent command execution. Important considerations to remember when having multiple drawings open include the following:

■ Any one of the open drawings can be made active.

■ Any one of the open drawings can be maximized or mini-

mized to ease viewing, or all can be left open and you can switch from drawing to drawing as needed.

■ You can start a command in one drawing, switch to another drawing and perform a different command, and then return to the original drawing and complete the command in progress. As you return to the original drawing, single-click anywhere in the drawing area, and the original command will be continued.

■ Each drawing is saved as a separate drawing file, independent of the other open files. You can close a file and not affect the contents of the remaining open files.

Using the Cut, Copy, and Paste Commands

AutoCAD allows drawings objects and properties to be transferred directly from one drawing to another using the **Cut**, **Copy**, and **Paste** commands. Each command can be performed using the **Edit** menu, by keyboard, or by shortcut menu. **Cut** will remove objects from a drawing and move them to the Clipboard. Once cut, the object can be placed in a new drawing with **Paste**. **Copy** allows an object to be reproduced in another location. The object

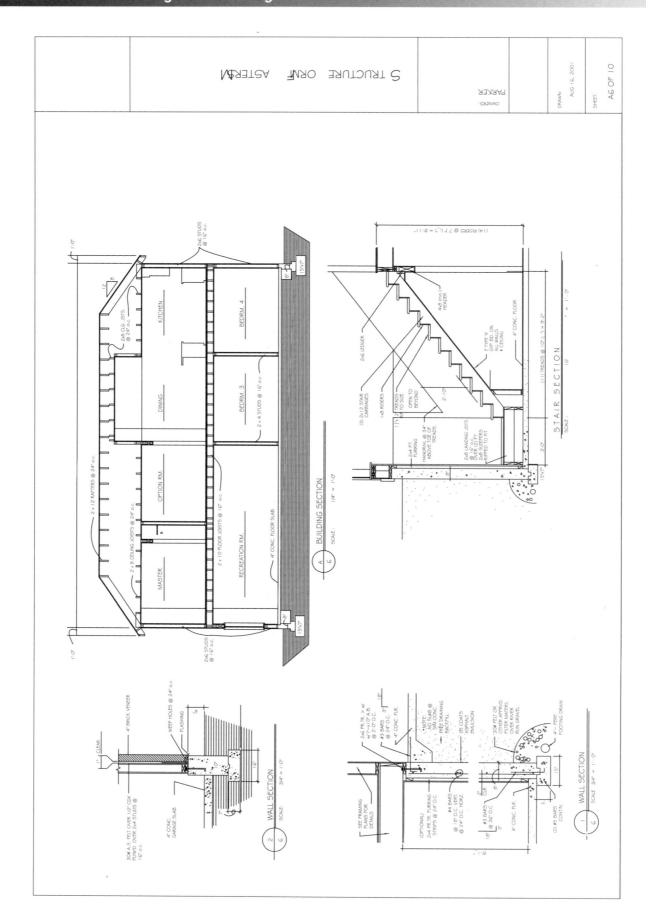

FIGURE 3-27 ■ The finished drawing, with the *Viewports* layer frozen or set as a non-plotting layer, is ready for plotting.

to be copied is reproduced in its new location with the **Paste** command. As the object is copied to the new drawing, so are any new layers, linetypes, or other properties associated the objects.

Copy With Base Point—The **Copy with Base Point** option of the shortcut menu is similar to the **Copy** option. With this option, before you select objects to copy, you'll be prompted to provide a base point. This option works well when objects need to be inserted accurately.

Paste as Block—The **Paste as Block** option can be used to paste a block into a different drawing. It functions similarly to the **Paste** command. Select the objects to be copied in the original drawing and select the **Copy** command. Selected objects do not need to be a block, because the command will turn them into a block. Next activate the new drawing and right-click to select the **Paste as Block** option. The copied objects are now a block.

Paste to Original Coordinates—This option can be used to copy an object in one drawing to the exact same location in another drawing. The command could be useful in copying an object from one apartment unit to the same location in another unit. The command is active only when the Clipboard contains AutoCAD data from a drawing other than the current drawing. Selecting an object to be copied starts the command sequence, and it is completed by making another drawing active and then selecting the **Paste to Original Coordinates** option.

Dragging Objects

AutoCAD allows objects to be moved between drawings using a drag and drop sequence. Objects can be

■ Moved or copied to a new location in the same drawing

■ Moved or copied to another drawing in the same drawing session

■ Copied to a drawing in another session

■ Copied to another application

Dragging can be completed using either the left or right mouse buttons. Slightly different results and procedures will be required using each method.

Selecting Objects to Drag with Left-Click—Dragging using the left mouse button begins by selecting the objects to be moved or copied. Once the set is selected, move the cursor over one of the selected objects and press the left mouse button. Moving the cursor while the left button is held down will show the effect of moving the selected objects. Release the mouse button when the objects are in the desired location.

Selecting Objects to Drag with Right-Click—Dragging using the right mouse button begins by selecting the objects to be moved or copied. Once the set is selected, move the cursor over one of the selected objects and press the right mouse button. Moving the cursor while the button is held down will show the effect of moving the select-

ed objects. Release the mouse button when the objects are in the desired location. As the right button is released, a menu will be displayed. The menu will depend on what type of drawing you're working in.

Using the Property Painter

The **Property Painter** can also be used to transfer properties from an active drawing to other drawings. The process to change the layer and linetype of objects in a drawing to match another drawing can be completed using the following process:

1. Select the source object using any selection method.

2. Select **Match Properties** from the **Standard** toolbar.

3. Move to the target drawing and make it the current drawing. The **Match Properties** cursor will be displayed by the selection box.

4. Select the target objects.

The command can be continued within the current drawing, or another drawing can be made current, and the properties of the original object can be transferred to additional objects.

BUILDING DRAWINGS USING THE DESIGNCENTER

The AutoCAD DesignCenter can be used to locate, organize, and customize drawing information. The DesignCenter can be used to transfer data such as blocks, layers, and external references directly from one drawing to another without using the Clipboard. Drawing content can be moved from other open files in AutoCAD or from files stored on your machine, on a network drive, or from an Internet site. Access the DesignCenter by selecting the **AutoCAD DesignCenter** icon from the **Standard** toolbar.

Features of the DesignCenter

The DesignCenter provides direct access to any drawing file that you are permitted to use. Files stored on diskettes, the hard drive, network, or the Internet can be accessed and reused as a library for retrieving or coping blocks, referenced drawings, layouts, layers, linetypes, text styles or dimension styles. The DesignCenter enables you to do the following:

■ Create a Favorites folder to create a shortcut for frequently accessed drawings, folders, or Internet locations.

■ View, attach, insert, or copy and paste layer and block definitions of any file you have access to into the current drawing. This feature could be used to update a drawing supplied by a subcontractor to match current standards.

■ Open drawing files by dragging the file from the tree view into the current drawing area.

■ Find drawing contents stored in any folder, drive, or network, or from any Internet site that you have access to, based on the quality of the source. Search tools such as a name or date can be used to locate materials to be dragged into the current drawing file.

■ Use the DesignCenter palette to view aspects of the object, block, or aspect of the drawing content.

Adding Content to Drawings Using the DesignCenter

One of the best uses for the DesignCenter is for adding content from the palette into a new or existing drawing. Information can be selected from the palette or from **Find** and dragged directly into a drawing without opening the drawing containing the original. DesignCenter can also be used to attach external referenced drawings and to copy layers between drawings.

Inserting Blocks—The DesignCenter provides two methods for inserting blocks into a drawing. Blocks can be inserted using the default scale and rotation of the block, or the block parameters can be altered as the block is dragged into the new drawing.

Attaching Referenced Drawings—The DesignCenter can be used to attach a referenced drawing using similar steps used to attach a block and provide the parameters.

Working with Layers—The DesignCenter can be used to copy layers from one drawing to another. Typically, a template drawing containing stock layers can be used when creating new drawing files. This option of the DesignCenter is useful when you're working with drawings created by a consulting firm and the drawing needs to conform to office standards. DesignCenter can be used to drag layers from a template, or any other drawing, into the new

drawing and ensure drawing consistency.

Accessing Favorite Contents—As with most programs, the DesignCenter allows frequently used sources to be stored for easy retrieval. The final section of this chapter will explore methods of adding shortcuts, displaying, and organizing your favorite drawing tools.

ADDITIONAL READING

The following Web sites can be used as a resource to help you keep your CAD skills current.

ADDRESS	COMPANY/ORGANIZATION
www.beamchek.com	AC Software, Inc.
www.ansys.com	Ansys, Inc.
www.autodesk.com	Autodesk
www.chiefarch.org	Chief Architect
www.eaglepoint.com	Eagle Point Software Incorporated
www.rasterex.com	Rasterex
www.graphisoft.com	Graphisoft
www.ketiv.com	Ketiv Technologies, Inc.
www.pointa.autodesk.com	Autodesk Point A
www.strucalc.com	StruCalc 5.0 developed by Cascade Consulting Associates
www.softplan.com	SoftPlan Systems, Inc.
www.thebluebook.com	The Construction Information Network
www.miscrosoft.com/office/visio	
	Microsoft Visio

CHAPTER 3 *Applying AutoCAD Tools to Commercial Drawings*

CHAPTER QUIZ

Answer the questions on a separate sheet of paper. Print the chapter title, question number, and a short complete statement for each question.

Question 3-1 What is the advantage of using folders when working on the hard disk?

Question 3-2 List two benefits of storing related drawings in a single drawing file.

Question 3-3 List six items typically included in a title block.

Question 3-4 What are the disadvantages of storing related drawings in a single drawing file?

Question 3-5 What is a template drawing?

Question 3-6 List the two types of measurement units typically associated with construction drawings.

Question 3-7 Describe the differences between **ON/OFF** and **THAW/FREEZE** as they relate to the display of layers.

Question 3-8 How can external referencing be used within a set of construction drawings?

Question 3-9 What are the benefits of the **PURGE** command, and when can it be used?

Question 3-10 List the eight major groups of layer names recommended by the AIA.

Question 3-11 Describe guidelines for naming minor groups for long and short formats.

Question 3-12 Describe the difference between a block and a wblock.

Question 3-13 List two different types of notes found on construction drawings and describe their purpose.

Question 3-14 List options for places where drawings can be saved and explain any advantages or disadvantages of each.

Question 3-15 List an alternative to the AIA for naming layers.

Question 3-16 Enter AutoCAD and research the difference between **PARTIALOAD** and **PARTIALOPEN**.

Question 3-17 Other than plotting, what advantages do layouts provide?

Question 3-18 You've adjusted the width for line weight, but your drawing does not reflect the change. Why not?

Question 3-19 What is the difference between a layout and a viewport?

Question 3-20 You've created a second viewport, but the contents of the first viewport are displayed. How can this be corrected?

Drawing Problems

For all three of the problems below, create a title block using your school name as the company name along with other recommended contents. Use a 1/2" wide margin on the top, right, and bottom of the page. Use a 1 1/2" margin on the left side. Establish measurement units, and angle measurements suitable for an architectural drawing.

Problem 3-1 Create a drawing template for a drawing using 1/4" = 1'–0" using D-size material.

Problem 3-2 Create a template drawing for site plans drawn at a scale of 1" = 10'–0" using D size material.

Problem 3-3 Create a template drawing for architectural drawings using metric features. Assume a scale of 1:50 will be used.

CHAPTER

Introduction to National Building Codes

Most construction is governed by building codes. Rather than writing their own codes, most cities adopt model building codes. The first and most important purpose of a building code is to protect human life. The building code does this by regulating the building's structural integrity, exit patterns, height, fire resistance, total size, and number of occupants.

Formerly there were three model building codes in use in the United States. They were the Uniform Building Code (UBC), the National Building Code (BOCA), and Standard Building Code (SBC). These three codes have become one, the International Building Code. This code was developed by the members of the three code groups listed above. They combined their extensive knowledge to produce a comprehensive and well researched model code to be used by the nation and beyond. This is basically a performance-based code that describes the desired results, not the exact method used to achieve those results.

The International Building Code is part of a family of codes and books that includes plumbing, mechanical, and fire. There are also handbooks, interpretive manuals, and learning aids available for each model code. The interpretive manuals are very useful in helping achieve an understanding of the intent of the code.

The building code has an influence over every aspect of the project. Its influence extends from the very beginning of design, by controlling the location, size, and construction type of the building, through engineering and the working drawing phase. Familiarity with the code and an understanding of the intent of its sections are important attributes for any member of the team putting a project together.

The code book should be used as a drafting tool just as a keyboard and mouse are. It should be at your side at the drafting table and referred to frequently. It should be marked in and tabbed so that you can easily find information such as frequently needed tables and formulas. If possible, purchase the loose-leaf version of the code so that it will be easier to add updates and local amendments, as well as your own comments, formulas, and notes. The IBC Handbook, which explains nonstructural aspects of the code, is another excellent source of information.

SUBJECTS COVERED BY THE MODEL CODES

The IBC is broken down into 35 chapters and 10 Appendices as listed below.

1. ADMINISTRATION
2. DEFINITIONS
3. USE AND OCCUPANCY CLASSIFICATION
4. SPECIAL DETAILED REQUIREMENTS BASED ON USE AND OCCUPANCY
5. GENERAL BUILDING HEIGHTS AND AREAS
6. TYPES OF CONSTRUCTION
7. FIRE-RESISTANT-RATED CONSTRUCTION
8. INTERIOR FINISHES
9. FIRE PROTECTION SYSTEMS
10. MEANS OF EGRESS
11. ACCESSIBILITY
12. INTERIOR ENVIRONMENT
13. ENERGY EFFICIENCY
14. EXTERIOR WALLS
15. ROOF ASSEMBLIES AND ROOFTOP STRUCTURES
16. STRUCTURAL DESIGN
17. STRUCTURAL TESTS AND SPECIAL INSPECTIONS
18. SOIL AND FOUNDATIONS
19. CONCRETE
20. ALUMINUM
21. MASONRY
22. STEEL
23. WOOD
24. GLASS AND GLAZING
25. GYPSUM BOARD AND PLASTER
26. PLASTIC
27. ELECTRIC
28. MECHANICAL SYSTEMS
29. PLUMBING SYSTEMS
30. ELEVATORS AND CONVEYING SYSTEMS
31. SPECIAL CONSTRUCTION
32. ENCROACHMENT INTO THE PUBLIC RIGHT-OF-WAY
33. SAFEGUARDS DURING CONSTRUCTION
34. EXISTING STRUCTURES
35. REFERENCED STANDARDS

APPENDIX

A. EMPLOYEE QUALIFICATIONS
B. BOARD OF APPEALS

C. GROUP U—AGRICULTURE BUILDINGS

D. FIRE DISTRICTS

E. SUPPLEMENTARY ACCESSIBILITY REQUIREMENTS

F. RODENT PROOFING

G. FLOOD RESISTANT CONSTRUCTION

H. SIGNS

I. PATIO COVERS

J. SUPPLEMENTARY ACCESSIBILITY REQUIREMENTS FOR QUALIFIED HISTORIC BUILDINGS AND FACILITIES

General Construction Requirements

To become familiar with the code, start by reading the chapters that deal with general building limitations of the code. Chapter 6 of the code discusses types of construction as classified by the model codes. Construction types are numbered from I to V, with I being the most fire resistive and V being the least fire resistive. Each construction type has restrictions on the structural frame, exterior walls, interior walls, openings in walls, stair construction and roof construction. Construction type will be used later in this chapter to calculate allowable building size. Figure 4-1 lists the fire resistance rating for various building components, for each construction type.

Safety Requirements

Once you feel comfortable with general guidelines that govern placement and types of structures that are allowed in a specific area, read Chapters 7 through 12 of the code:

■ Chapter 7 deals with the materials required to obtain a specific fire rating and will be covered in Chapter 12 of this text.

TABLE 601
FIRE-RESISTANCE RATING REQUIREMENTS FOR BUILDING ELEMENTS (hours)

BUILDING ELEMENT	TYPE I		TYPE II		TYPE III		TYPE IV	TYPE V	
	A	B	A[d]	B	A[d]	B	HT	A[d]	B
Structural frame[a] Including columns, girders, trusses	3[b]	2[b]	1	0	1	0	HT	1	0
Bearing walls Exterior[f] Interior	3 3[b]	2 2[b]	1 1	0 0	2 1	2 0	2 1/HT	1 1	0 0
Nonbearing walls and partitions Exterior Interior[e]	See Table 602 See Section 602								
Floor construction Including supporting beams and joists	2	2	1	0	1	0	HT	1	0
Roof construction Including supporting beams and joists	$1^{1/2c}$	1[c]	1[c]	0[c]	1[c]	0	HT	1[c]	0

For SI: 1 foot = 304.8 mm.

a.   The structural frame shall be considered to be the columns and the girders, beams, trusses and spandrels having direct connections to the columns and bracing members designed to carry gravity loads. The members of floor or roof panels which have no connection to the columns shall be considered secondary members and not a part of the structural frame.

b.   Roof supports: Fire-resistance ratings of structural frame and bearing walls are permitted to be reduced by 1 hour where supporting a roof only.

c.   1. Except in Factory-Industrial (F-I), Hazardous (H), Mercantile (M) and Moderate Hazard Storage (S-1) occupancies, fire protection of structural members shall not be required, including protection of roof framing and decking where every part of the roof construction is 20 feet or more above any floor immediately below. Fire-retardant-treated wood members shall be allowed to be used for such unprotected members.
       2. In all occupancies, heavy timber shall be allowed where a 1-hour or less fire-resistance rating is required.
       3. In Type I and Type II construction, fire-retardant-treated wood shall be allowed in buildings not over two stories including girders and trusses as part of the roof construction.

d.   An approved automatic sprinkler system in accordance with Section 903.3.1.1 shall be allowed to be substituted for 1-hour fire-resistance-rated construction, provided such system is not otherwise required by other provisions of the code or used for an allowable area increase in accordance with Section 506.3 or an allowable height increase in accordance with Section 504.2. The 1-hour substitution for the fire resistance of exterior walls shall not be permitted.

e.   For interior nonbearing partitions in Type IV construction, also see Section 602.4.6.

f.   Not less than the fire-resistance rating based on fire separation distance (see Table 602.)

FIGURE 4-1 ■ The fire resistance of structural elements must be considered as the project is designed. The architect will determine each element using this table. Reproduced from the 2000 edition of the International Building Code, copyright © 2000, with the permission of the publisher, the *International Conference of Building Officials*, under license from the International Code Council. The 2000 IBC is a copyrighted work of the International Code Council.

■ Chapter 8 primarily deals with flame spread and smoke classification of interior finishes such as wallpaper and carpet.

■ Chapter 9 contains information about fire suppression and detection systems, and smoke control and venting. Plans for fire sprinkler systems and fire detection systems are usually prepared by consultants specializing in this field. Wet and dry stand pipes are frequently located by the project architect. *Stand pipes* are systems that fire hoses connect to. Dry stand pipes are dry until a fire truck connects to them and pumps water through them. The code specifies when these systems are required and gives criteria for locating them.

■ Chapter 10 addresses exits and is one of the most important chapters for a space planner to understand. Although the drafter is not usually responsible for the placement of walls, a knowledge of code requirements will aid in understanding the building design. Placement of hallways and exits will be discussed later in this chapter.

■ Chapter 11 is an extension of ADA (Americans with Disabilities Act) requirements discussed in Chapter 5 of this book.

■ Chapter 12 contains minimum requirements for light and ventilation, both natural, and artificial or mechanical. It also discusses minimum room sizes and ceiling heights.

Building Systems

Several chapters in each code deal with roof, wall, and floor systems. The primary focus of Chapter 15 of the code is roof coverings. Minimum roof slope, and maximum exposure to the weather is discussed with tables similar to Figure 4-2. The fire classification of the roof as shown in Figure 4-3, is also discussed as well as roof flashing and roof drainage.

Chapter 16 list loads and loading conditions that a building must be designed to resist. The chapter is primarily intended for the engineer, but knowledge of the loads acting on each portion of the structure will add to the drafter's understanding of the project. Figure 4-4 shows required live loads for which different types of occupancies must be designed. Chapter 12 of this text will explore how the engineer makes use of this information in designing the structure. Snow loads, wind loads and earth quake loads will be discussed later in this chapter as you are introduced to how the architect begins the preliminary design of a structure.

Chapter 18 of the code is about foundations and retaining walls. Foundation investigations, which are usually done by a soils lab, are discussed. Requirements for standard foundations, such as minimum thickness of foundation walls (see Figure 4-5) and allowable foundation loads are also addressed. A substantial portion of this chapter covers the design of pile foundations. Chapters 6 and 23 of this text will explain how each of these systems affects a drafter.

TABLE 1507.9.7
WOOD SHAKE WEATHER EXPOSURE AND ROOF SLOPE

ROOFING MATERIAL	LENGTH (inches)	GRADE	EXPOSURE (inches) 4:12 PITCH OR STEEPER
Shakes of naturally durable wood	18	No. 1	7.5
	24	No. 1	10[a]
Preservative treated taper sawn shakes of Southern Yellow Pine	18	No. 1	7.5
	24	No. 1	10
	18	No. 2	5.5
	24	No. 2	7.5
Taper-sawn shakes of naturally durable wood	18	No. 1	7.5
	24	No. 1	10
	18	No. 2	5.5
	24	No. 2	7.5

For SI: 1 inch = 25.4 mm. a. For 24-inch by 0.375-inch handsplit shakes, the maximum exposure is 7.5 inches.

FIGURE 4-2 ■ The project designer must consider each major building system as it relates to the code. The weather exposure and roof pitch table will guide the design and selection of roof materials. Reproduced from the 2000 edition of the International Building Code, copyright © 2000, with the permission of the publisher, the *International Conference of Building Officials*, under license from the International Code Council. The 2000 IBC is a copyrighted work of the International Code Council.

Building Materials

Once you feel at ease with the guidelines of building systems, work through the chapters that deal with specific types of materials. Wood is one of the most common materials of light construction

TABLE 1505.1[a,b]
MINIMUM ROOF COVERING CLASSIFICATION
FOR TYPES OF CONSTRUCTION

IA	IB	IIA	IIB	IIIA	IIIB	IV	VA	VB
B	B	B	C[c]	B	C[c]	B	B	C[c]

For SI: 1 foot = 304.8 mm, 1 square foot = 0.0929 m^2.
a. Unless otherwise required in accordance with the *Urban Wildland Interface Code* or due to the location of the building within a fire district in accordance with Appendix D.
b. Nonclassified roof coverings shall be permitted on buildings of Group R-3 as applicable in Section 101.2 and U occupancies, where there is a minimum fire-separation distance of 6 feet measured from the leading edge of the roof.
c. Buildings that are not more than two stories in height and having not more than 6,000 square feet of projected roof area and where there is a minimum 10-foot fire-separation distance from the leading edge of the roof to a lot line on all sides of the building, except for street fronts or public ways, shall be permitted to have roofs of No. 1 cedar or redwood shakes and No. 1 shingles constructed in accordance with Section 1505.6.

FIGURE 4-3 ■ The fire safety of materials must be considered as the structure is designed using appropriate tables. Reproduced from the 2000 edition of the International Building Code, copyright © 2000, with the permission of the publisher, the *International Conference of Building Officials*, under license from the International Code Council. The 2000 IBC is a copyrighted work of the International Code Council.

TABLE 1607.1
MINIMUM UNIFORMLY DISTRIBUTED LIVE LOADS AND MINIMUM CONCENTRATED LIVE LOADS[g]

OCCUPANCY OR USE	UNIFORM (psf)	CONCENTRATED (lbs.)
1. Apartments (see residential)		
2. Access floor systems		
Office use	50	2,000
Computer use	100	2,000
3. Armories and drill rooms	150	—
4. Assembly areas and theaters		—
Fixed seats (fastened to floor)	60	
Lobbies	100	
Movable seats	100	
Stages and platforms	125	
Follow spot, projection and control rooms	50	
Catwalks	40	
5. Balconies (exterior)	100	—
On one- and two-family residences only, and not exceeding 100 ft.[2]	60	
6. Decks	Same as occupancy served[h]	
7. Bowling alleys	75	—
8. Cornices	60	—
9. Corridors, except as otherwise indicated	100	—
10. Dance halls and ballrooms	100	—
11. Dining rooms and restaurants	100	—
12. Dwellings (see residential)	—	—
13. Elevator machine room grating (on area of 4 in.[2])	—	300
14. Finish light floor plate construction (on area of 1 in.[2])	—	200
15. Fire escapes	100	—
On single-family dwellings only	40	
16. Garages (passenger cars only)	50	Note a
Trucks and buses	See Section 1607.6	
17. Grandstands (see stadium and arena bleachers)	—	—
18. Gymnasiums, main floors and balconies	100	—
19. Handrails, guards and grab bars	See Section 1607.7	
20. Hospitals		
Operating rooms, laboratories	60	1,000
Private rooms	40	1,000
Wards	40	1,000
Corridors above first floor	80	1,000
21. Hotels (see residential)	—	—
22. Libraries		
Reading rooms	60	1,000
Stack rooms	150[b]	1,000
Corridors above first floor	80	1,000
23. Manufacturing		
Light	125	2,000
Heavy	250	3,000
24. Marquees and canopies	75	—

OCCUPANCY OR USE	UNIFORM (psf)	CONCENTRATED (lbs.)
25. Office buildings		
File and computer rooms shall be designed for heavier loads based on anticipated occupancy		
Lobbies and first floor corridors	100	2,000
Offices	50	2,000
Corridors above first floor	80	2,000
26. Penal Institutions		
Cell blocks	40	—
Corridors	100	
27. Residential		—
Group R-3 as applicable in Section 101.2		
Uninhabitable attics without storage	10	
Uninhabitable attics with storage	20	
Habitable attics and sleeping areas	30	
All other areas except balconies and decks	40	
Hotels and multifamily dwellings		
Private rooms	40	
Public rooms and corridors serving them	100	
28. Reviewing stands, grandstands and bleachers	100[c]	—
29. Roofs	See Section 1607.11	
30. Schools		
Classrooms	40	1,000
Corridors above first floor	80	1,000
First floor corridors	100	1,000
31. Scuttles, skylight ribs, and accessible ceilings	—	200
32. Sidewalks, vehicular driveways and yards, subject to trucking	250[d]	8,000[e]
33. Skating rinks	100	—
34. Stadiums and arenas		
Bleachers	100[c]	
Fixed seats (fastened to floor)	60[c]	
35. Stairs and exits	100	Note f
One- and two-family dwellings	40	
All other	100	
36. Storage warehouses (shall be designed for heavier loads if required for anticipated storage)		
Light	125	
Heavy	250	
37. Stores		
Retail		
First floor	100	1,000
Upper floors	75	1,000
Wholesale, all floors	125	1,000
38. Vehicle barriers	See Section 1607.7	
39. Walkways and elevated platforms (other than exitways)	60	—
40. Yards and terraces, pedestrians	100	

FIGURE 4-4 ■ How the structure will be used will determine the loads that must be designed for by both the architectural and engineering team. Reproduced from the 2000 edition of the International Building Code, copyright © 2000, with the permission of the publisher, the *International Conference of Building Officials*, under license from the International Code Council. The 2000 IBC is a copyrighted work of the International Code Council.

TABLE 1805.5(2)
8-INCH REINFORCED CONCRETE AND MASONRY FOUNDATION WALLS WHERE d ≥ 5 INCHES[a, b, c]

WALL HEIGHT (feet)	HEIGHT OF UNBALANCED BACKFILL (feet)	VERTICAL REINFORCEMENT		
		Soil classes and lateral soil load[a] (psf per foot below natural grade)		
		GW, GP, SW and SP Soils 30	GM, GC, SM, SM-SC and ML Soils 45	SC, MH, ML-CL and Inorganic CL Soils 60
7	4 (or less)	#4 at 48" o.c.	#4 at 48" o.c.	#4 at 48" o.c.
	5	#4 at 48" o.c.	#4 at 48" o.c.	#4 at 40" o.c.
	6	#4 at 48" o.c.	#5 at 48" o.c.	#5 at 40" o.c.
	7	#4 at 40" o.c.	#5 at 40" o.c.	#6 at 48" o.c.
8	4 (or less)	#4 at 48" o.c.	#4 at 48" o.c.	#4 at 48" o.c.
	5	#4 at 48" o.c.	#4 at 48" o.c.	#4 at 40" o.c.
	6	#4 at 48" o.c.	#5 at 48" o.c.	#5 at 40" o.c.
	7	#5 at 48" o.c.	#6 at 48" o.c.	#6 at 40" o.c.
	8	#5 at 40" o.c.	#6 at 40" o.c.	#7 at 40" o.c.
9	4 (or less)	#4 at 48" o.c.	#4 at 48" o.c.	#4 at 48" o.c.
	5	#4 at 48" o.c.	#4 at 48" o.c.	#5 at 48" o.c.
	6	#4 at 48" o.c.	#5 at 48" o.c.	#6 at 48" o.c.
	7	#5 at 48" o.c.	#6 at 48" o.c.	#7 at 48" o.c.
	8	#5 at 40" o.c.	#7 at 48" o.c.	#8 at 48" o.c.
	9	#6 at 40" o.c.	#8 at 48" o.c.	#8 at 32" o.c.

For SI: 1 inch = 25.4 mm, 1 foot = 304.8 mm, 1 pound per square foot per foot = 0.157 kPa/m.

a. For design lateral soil loads, see Section 1610. Soil classes are in accordance with the Unified Soil Classification System and design lateral soil loads are for moist soil conditions without hydrostatic pressure.

b. Provisions for this table are based on construction requirements specified in Section 1805.5.2.

c. For alternative reinforcement, see Section 1805.5.3.

FIGURE 4-5 ■ Building codes specify minimum standards that must be considered in the design process. The table for the thickness of foundation walls will play an important role in the design of the foundation. Reproduced from the 2000 edition of the International Building Code, copyright © 2000, with the permission of the publisher, the *International Conference of Building Officials*, under license from the International Code Council. The 2000 IBC is a copyrighted work of the International Code Council.

and is covered in Chapter 23. Although you may be familiar with parts of these chapters from experience with residential construction, commercial construction will require a greater depth of understanding. This chapter has a lot of technical design information for the engineer, but it also has many tables and standard construction provisions that can be used by the drafter. Chapters 7 and 8 of this text will explore wood and timber construction methods.

Chapter 22 is a very technical chapter intended for the engineer involved with the structural steel. Chapter 19 of the code deals with concrete, and it is also very complex material. That chapter sets the rules for its use and discusses highly technical subjects such as concrete mixes and seismic performance of concrete. Minimum required slab thickness, formwork, and durability are also addressed. One of the most important pieces of information for the drafter is shown in Figure 4-6. This is a table showing the minimum requirements for covering steel reinforcing with concrete. This information is critical to developing proper details for reinforced concrete construction. Chapters 9 and 11 of this text will introduce the drafter to materials and construction methods generally associated with steel and concrete construction.

Structural design of masonry construction is addressed in Chapter 24 of the IBC. Required reinforcing, grout and mortar strength, and construction with glass block and screen block are

examined. Each of these subjects will be discussed in Chapter 10 of this text.

Chapter 24 of the code deals with structural requirements for glass. It covers vertical and sloped glazing as well as unusual uses for glass such as glass handrails and racquetball courts. Much of the information is presented in tables similar to Figure 4-7 (Minimum glass thickness).

USING THE CODE

To effectively use the code during the initial design phase, you must determine five basic classifications of the building:

■ Occupancy group

■ Location on the site

■ Floor area

■ Height or number of stories

■ Construction type

There is no set order for determining these classifications when designing a building. The starting point depends on what is already known about the project.

TABLE 1907.7.1
MINIMUM CONCRETE COVER

CONCRETE EXPOSURE	MINIMUM COVER inches
1. Concrete cast against and permanently exposed to earth	3
2. Concrete exposed to earth or weather No. 6 through No. 18 bar No. 5 bar, W31 or D31 wire, and smaller	2 $1^1/_2$
3. Concrete not exposed to weather or in contact with ground Slabs, walls, joists: No. 14 and No. 18 bars No. 11 bar and smaller Beams, columns: Primary reinforcement, ties, stirrups, spirals Shells, folded plate members: No. 6 bar and larger No. 5 bar, W31 or D31 wire, and smaller	$1^1/_2$ $^3/_4$ $1^1/_2$ $^3/_4$ $^1/_2$

For SI: 1 inch = 25.4 mm.

FIGURE 4-6 ■ Drafters must complete details that must meet or exceed minimum concrete cover for steel reinforcing. Reproduced from the 2000 edition of the International Building Code, copyright © 2000, with the permission of the publisher, the *International Conference of Building Officials*, under license from the International Code Council. The 2000 IBC is a copyrighted work of the International Code Council.

Occupancy Classification

A good way to begin checking a project for code compliance is to determine what occupancy classification the building will fall into. Basic occupancy classifications are listed in Chapter 3 of the code. The IBC lumps various uses into categories (classifications), for example:

A for Assembly uses such as theaters and churches

B for non hazardous Business such as an office building or buildings for services uses

E for Educational buildings

F for Factories, industrial

H for Hazardous uses such as areas where large volumes of flammable liquids or highly toxic materials are stored or used

I for Institutional uses such as hospitals and jails

M for Mercantile uses such as department stores and markets

R for single or multifamily Residential

S for Storage of moderate- and low-hazard materials

U for Utility and miscellaneous

These uses are further broken down into subcategories. A partial list of these subcategories is shown in Figure 4-8).

TABLE 2404.1
c_1 FACTORS FOR VERTICAL AND SLOPED GLASS[a]
[For use with Figures 2404(1) through 2404(12)]

GLASS TYPE	FACTOR
Single Glass	
Regular (annealed)	1.0
Heat Strengthened	2.0
Fully Tempered	4.0
Wired	0.50
Patterned[c]	1.0
Sandblasted[d]	0.50
Laminated - regular plies[e]	0.7/0.90[f]
Laminated - heat-strengthened plies[e]	1.5/1.8[f]
Laminated - fully tempered plies[e]	3.0/3.6[f]
Insulating Glass[b]	
Regular (annealed)	1.8
Heat Strengthened	3.6
Fully Tempered	7.2
Laminated - regular plies[e]	1.4/1.6[f]
Laminated - heat-strengthened plies[e]	2.7/3.2[f]
Laminated - fully tempered plies[e]	5.4/6.5[f]

a. Either Table 2404.1 or 2404.2 shall be appropriate for sloped glass depending on whether the snow or wind load is dominant. See Section 2404.2. For glass types (vertical or sloped) not included in the tables, refer to ASTM E 1300 for guidance.

b. Values apply for insulating glass with identical panes.

c. The value for patterned glass is based on the thinnest part of the pattern; interpolation between graphs is permitted.

d. The value for sandblasted glass is for moderate levels of sandblasting.

e. Values for laminated glass are based on the total thickness of the glass and apply for glass with two equal glass ply thicknesses.

f. The lower value applies if, for any laminated glass pane, either the ratio of the long to short dimension is greater than 2.0 or the lesser dimension divided by the thickness of the pane is 150 or less; the higher value applies in all other cases.

FIGURE 4-7A ■ Minimum glass thickness. Reproduced from the 2000 edition of the International Building Code, copyright © 2000, with the permission of the publisher, the *International Conference of Building Officials*, under license from the International Code Council. The 2000 IBC is a copyrighted work of the International Code Council.

Location on the Property

Zoning ordinances will have a significant impact on where the building is located on the property. Where the building is located on the lot will also have an impact on the construction of the building. The location of the building in relationship to other buildings or the property lines is very important in regard to fire safety and

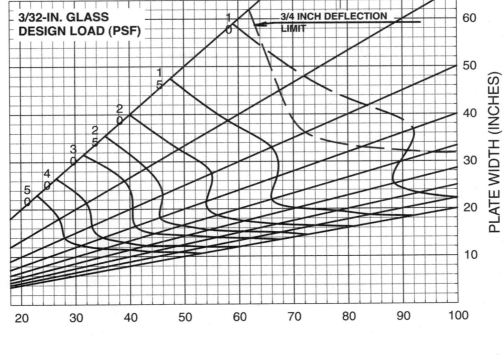

PLATE LENGTH (inches)

For SI: 1 inch = 25.4 mm, 1 pound per square foot = 0.0479 kPa.

FIGURE 2404(1)[a, b, c, d, e, f]
MAXIMUM ALLOWABLE LOAD FOR VERTICAL AND SLOPED
RECTANGULAR GLASS SUPPORTED ON ALL SIDES

Notes:
  a. In each graph, the vertical axis is the lesser dimension; the horizontal axis is the greater dimension.
  b. The diagonal number on each graph shows the equivalent design load in psf.
  c. The dashed lines indicate glass that has deflection in excess of 3/4 inch.
  d. Interpolation between lines is permitted. Extrapolation is not allowed.
  e. For laminated glass, the applicable glass thickness is the total glass thickness.
  f. For insulating glass panes, the applicable glass thickness is the thickness of one pane.

FIGURE 4-7B ■ Minimum glass thickness. Reproduced from the 2000 edition of the International Building Code, copyright © 2000, with the permission of the publisher, the *International Conference of Building Officials*, under license from the International Code Council. The 2000 IBC is a copyrighted work of the International Code Council.

will affect the requirements for the fire resistance of exterior walls and the openings in those walls. The closer two building are located together, the more likely it is that a fire in one will damage the other. Larger space between buildings will also provide greater access for fire fighting equipment. The code has provisions for increasing the allowable size of a building when large yards are provided. For the purpose of area increases, a yard includes adjacent public streets, alleys, etc. Yards on adjacent lots do not count. Figure 4-9 (Sec. 506 of IBC) contains the formulas for calculating the positive impact of large yards around the building on the allowable size of the building. With the increase in the size of yards, fire safety is increased and therefore the code allows larger buildings to be built.

It might be beneficial to increase the size of a yard to avoid required fire-resistive construction. Columns 3 and 4 of Figure 4-10 (Table 602 of the IBC) call for fire-resistive walls when yards become less than a given size.

Allowable Floor Area, Heights and Number of Stories, Type of Construction

Allowable floor area, height and number of stories, and construction type are three factors that need to be worked out together. After the building has been laid out on the site, you will usually have a building footprint determined. The number of stories need-

SECTION 303
ASSEMBLY GROUP A

303.1 Assembly Group A. Assembly Group A occupancy includes, among others, the use of a building or structure, or a portion thereof, for the gathering together of persons for purposes such as civic, social or religious functions, recreation, food or drink consumption or awaiting transportation. A room or space used for assembly purposes by less than 50 persons and accessory to another occupancy shall be included as a part of that occupancy. Assembly occupancies shall include the following:

A-1   Assembly uses, usually with fixed seating, intended for the production and viewing of the performing arts or motion pictures including, but not limited to:

> Motion picture theaters
> Television and radio studios admitting an audience
> Theaters

A-2   Assembly uses intended for food and/or drink consumption including, but not limited to:

> Banquet halls
> Night clubs
> Restaurants
> Taverns and bars

A-3   Assembly uses intended for worship, recreation or amusement and other assembly uses not classified elsewhere in Group A, including, but not limited to:

> Amusement arcades
> Art galleries
> Auditoriums
> Bowling alleys
> Churches
> Community halls
> Courtrooms
> Dance halls
> Exhibition halls
> Funeral parlors
> Gymnasiums
> Indoor swimming pools
> Indoor tennis courts
> Lecture halls
> Libraries
> Museums
> Passenger stations (waiting area)
> Pool and billiard parlors

A-4   Assembly uses intended for viewing of indoor sporting events and activities with spectator seating, including, but not limited to:

> Arenas

> Skating rinks
> Swimming pools
> Tennis courts

A-5   Assembly uses intended for participation in or viewing outdoor activities including, but not limited to:

> Amusement park structures
> Bleachers
> Grandstands
> Stadiums

SECTION 304
BUSINESS GROUP B

304.1 Business Group B. Business Group B occupancy includes, among others, the use of a building or structure, or a portion thereof, for office, professional or service-type transactions, including storage of records and accounts. Business occupancies shall include, but not be limited to, the following:

> Airport traffic control towers
> Animal hospitals, kennels and pounds
> Banks
> Barber and beauty shops
> Car wash
> Civic administration
> Clinic—outpatient
> Dry cleaning and laundries; pick-up and delivery stations and self-service
> Educational occupancies above the 12th grade
> Electronic data processing
> Fire and police stations
> Laboratories; testing and research
> Motor vehicle showrooms
> Post offices
> Print shops
> Professional services (architects, attorneys, dentists, physicians, engineers, etc.)
> Radio and television stations
> Telephone exchanges

SECTION 305
EDUCATIONAL GROUP E

305.1 Educational Group E. Educational Group E occupancy includes, among others, the use of a building or structure, or a portion thereof, by six or more persons at any one time for educational purposes through the 12th grade.

305.2 Day care. The use of a building or structure, or portion thereof, for educational, supervision or personal care services for more than five children older than $2^1/_2$ years of age, shall be classified as a Group E occupancy.

FIGURE 4-8 ■ How a structure will be used determines its occupancy classification. The project designer uses this section of the code to determine the occupancy classification. Reproduced from the 2000 edition of the International Building Code, copyright © 2000, with the permission of the publisher, the *International Conference of Building Officials*, under license from the International Code Council. The 2000 IBC is a copyrighted work of the International Code Council.

SECTION 506
AREA MODIFICATIONS

506.1 General. The areas limited by Table 503 shall be permitted to be increased due to frontage (I_f) and automatic sprinkler system protection (I_s) in accordance with the following:

$$A_a = A_t + \left[\frac{A_t I_f}{100}\right] + \left[\frac{A_t I_s}{100}\right] \qquad \textbf{(Equation 5-1)}$$

where:

A_a = Allowable area per floor (square feet).
A_t = Tabular area per floor in accordance with Table 503 (square feet).
I_f = Area increase due to frontage (percent) as calculated in accordance with Section 506.2.
I_s = Area increase due to sprinkler protection (percent) as calculated in accordance with Section 506.3.

506.1.1 Basements. A single basement need not be included in the total allowable area provided such basement does not exceed the area permitted for a one-story building.

506.2 Frontage increase. Every building shall adjoin or have access to a public way to receive an area increase for frontage. Where a building has more than 25 percent of its perimeter on a public way or open space having a minimum width of 20 feet (6096 mm), the frontage increase shall be determined in accordance with the following:

$$I_f = 100 \left[\frac{F}{P} - 0.25\right] \frac{W}{30} \qquad \textbf{(Equation 5-2)}$$

where:

I_f = Area increase due to frontage (percent).
F = Building perimeter which fronts on a public way or open space having 20 feet (6096 mm) open minimum width.
P = Perimeter of entire building.
W = Minimum width of public way or open space.

506.2.1 Width limits. W must be at least 20 feet (6096 mm) and the quantity W divided by 30 shall not exceed 1.0 except that for buildings which are permitted to be unlimited in area by Section 503.1.2, Section 507 or Section 508, the quantity W divided by 30 shall not exceed 2.0.

506.2.2 Open space limits. Such open space shall be either on the same lot or dedicated for public use and shall be accessed from a street or approved fire lane.

506.3 Automatic sprinkler system increase. Where a building is protected throughout with an approved automatic sprinkler system in accordance with Section 903.3.1.1, the area limitation in Table 503 is permitted to be increased by 200 percent (I_s = 200 percent) for multi-story buildings and 300 percent (I_s = 300 percent) for single-story buildings.

Exception: Group H-1, H-2 or H-3.

FIGURE 4-9 ■ The size of a structure can often be increased because of the yard size or fire sprinklers. Reproduced from the 2000 edition of the International Building Code, copyright © 2000, with the permission of the publisher, the *International Conference of Building Officials*, under license from the International Code Council. The 2000 IBC is a copyrighted work of the International Code Council.

ed to achieve the required size of the building for the project can then be determined.

The next challenge is to find an allowable way to construct the building. The code will classify a building according to the materials it is built with. Buildings made of wood frame construction are not allowed to be as large or as tall as buildings built of concrete and steel, because wood is not as stable and permanent as steel and concrete. Buildings that have their structural members protected from fire are allowed to be larger and taller than buildings without such fire protection. Referring to Figure 4-1, we see that Construction Types with an 'A' following are required to be more fire resistant than those with a 'B.' Type III-A is required to be more fire resistive than Type III-B. The Code has tables that specify allowable areas for a variety of construction types. These values will be different for each occupancy classification. Figure 4-11 shows the square footage allowed by the IBC for each Occupancy Group and each Construction Type. According to that table, a 'B' occupancy of

Type III-A construction has a basic allowable area of 28,500 square feet. If the 'B' occupancy is built of Type III-B construction its Basic allowable area is only 19,000 square feet. Why would anyone build with Type III-B construction instead of Type III-A construction? Because Type III-B is less expensive to build.

Notice that the allowable building sizes listed above were referred to as "Basic allowable area." These values are a starting point, a basic area. You will be allowed to construct a larger building based on the size of the yard around the building and if fire sprinklers are added, as described in Figure 4-9.

APPLYING THE CODE

The information that has just been described can be applied to the site plan in Figure 4-12. The building is to be a professional office building, two stories high, with 30,000 sq ft on each floor. Section 304 of the IBC (Figure 4-8) lists office buildings as a B Occupancy.

TABLE 602
FIRE-RESISTANCE RATING REQUIREMENTS FOR EXTERIOR WALLS BASED ON FIRE SEPARATION DISTANCE[a]

FIRE SEPARATION DISTANCE (feet)	TYPE OF CONSTRUCTION	GROUP H	GROUP F-1, M, S-1	GROUP A,B,E, F-2, I, R[b], S-2, U
< 5[c]	All	3	2	1
≥ 5 < 10	I-A Others	3 2	2 1	1 1
≥ 10 < 30	I-A, I-B II-B, V-B Others	2 1 1	1 0 1	1 0 1
≥ 30	All	0	0	0

For SI: 1 foot = 304.8 mm.

a. Load-bearing exterior walls shall also comply with the fire-resistance rating requirements of Table 601.
b. Group R-3 and Group U when used as accessory to Group R-3, as applicable in Section 101.2 shall not be required to have a fire-resistance rating where fire separation distance is 3 feet or more.
c. See Section 503.2 for party walls.

FIGURE 4-10 ■ The required fire resistance of exterior walls is determined by the occupancy classification and the fire separation distance. Reproduced from the 2000 edition of the International Building Code, copyright © 2000, with the permission of the publisher, the *International Conference of Building Officials*, under license from the International Code Council. The 2000 IBC is a copyrighted work of the International Code Council.

Next, determine how to achieve 30,000 sq ft per floor. Referring to Figure 4-11 (Table 503) under B you will see that only type I-A & I-B, type II-A, and type IV construction have a large enough basic allowable area for this building. These types of construction are relatively expensive, therefore less costly answers should be explored. As mentioned earlier, you can increase the basic allowable area by installing fire sprinklers and by having large yards around the building.

If Section 506 of the IBC (Figure 4-9) is explored, you will see that the building size can be increased if is has enough of its walls fronting on yards larger than 20' (6100 mm). The building in Figure 4-12 has 452' of wall facing a yard larger than 20' (212' of south wall +150' of east wall + 90' of west wall). In the second formula shown in Figure 4-9, 452 will be *F*. The building has a total of 700' of exterior wall (perimeter). In the formula, 700 will be *P*. Next, the smallest of the three yards that are being used for the area increase needs to be determined. The full width of the street is allowed to be used, so it is rather obvious that the west yard is the smallest. In the second formula of Figure 4-9, 36 will be *W*.

By doing the calculations (see Figure 4-13) we see that the Basic Allowable area can be increased by 48%.

Now go back to Table 503 (Figure 4-11) and look at the available options. By multiplying the allowable area of a type V-B building by 1.48, a 13,320 sq ft area is allowed (9000 × 1.48); with type V, 26,640 sq feet are allowed. Neither of the numbers is large enough. Using the same process with III-B and II-B results in 28,120and 34,040 respectively. Type II-B is large enough. Type III-A is also large enough at 42,180 sq ft. There are now six construction types to choose from.

There is another option for increasing the allowable area of the building and this is to install a fire sprinkler system. Using the first formula in figure 4-9 will give the maximum allowable area after considering yard increases and fire sprinkler increases. This formula looks rather complicated at first, but it really is not that difficult to use. Figure 4-14 illustrates the calculation for the least restrictive construction method, Type V-B construction. Because this example results in an area greater than is needed, we do not need to work out the allowable area for any other construction type. Using a spreadsheet to solve this formula would allow you to explore several solutions in a very short period of time. A spreadsheet has been provided on the Internet. Try using it with Figure 4-12. If you do not use the west yard, you will actually get a larger allowable area.

Reading Table 602 (Figure 4-10) for a B occupancy classification, it can be seen that walls facing a yard smaller than 10' (6100 mm) will have to be built of fire-resistive construction no matter what construction type the building is. This means that the north wall of the building will have to be fire resistive. A portion of the west wall will have to be fire resistive for many construction types but not for Type V-B. Chapter 12 of this text will introduce methods to achieve specific fire ratings. The east wall of the building will not have to be fire resistive (unless a fire rated construction type is chosen for the entire building) even though it is only 8' (3050 mm) from the property line, because one half of the street can be counted as part of the yard.

Reading the last column of Table 602 (Figure 4-10) it can be seen that the structure is allowed to have openings in the north wall, but they must be fire resistive (protected). Again, unprotect-

TABLE 503
ALLOWABLE HEIGHT AND BUILDING AREAS
Height limitations shown as stories and feet above grade plane.
Area limitations as determined by the definition of "Area, building", per floor.

GROUP	Hgt(S)	TYPE I A	TYPE I B	TYPE II A	TYPE II B	TYPE III A	TYPE III B	TYPE IV HT	TYPE V A	TYPE V B
HGT(ft)		UL	160	65	55	65	55	65	50	40
A-1	S	UL	5	3	2	3	2	3	2	1
	A	UL	UL	15,500	8,500	14,000	8,500	15,000	11,500	5,500
A-2	S	UL	11	3	2	3	2	3	2	1
	A	UL	UL	15,500	9,500	14,000	9,500	15,000	11,500	6,000
A-3	S	UL	11	3	2	3	2	3	2	1
	A	UL	UL	15,500	9,500	14,000	9,500	15,000	11,500	6,000
A-4	S	UL	11	3	2	3	2	3	2	1
	A	UL	UL	15,500	9,500	14,000	9,500	15,000	11,500	6,000
A-5	S	UL	UL	UL	UL	UL	UL	UL	UL	UL
	A	UL	UL	UL	UL	UL	UL	UL	UL	UL
B	S	UL	11	5	4	5	4	5	3	2
	A	UL	UL	37,500	23,000	28,500	19,000	36,000	18,000	9,000
E	S	UL	5	3	2	3	2	3	1	1
	A	UL	UL	26,500	14,500	23,500	14,500	25,500	18,500	9,500
F-1	S	UL	11	4	2	3	2	4	2	1
	A	UL	UL	25,000	15,500	19,000	12,000	33,500	14,000	8,500
F-2	S	UL	11	5	3	4	3	5	3	2
	A	UL	UL	37,500	23,000	28,500	18,000	50,500	21,000	13,000
H-1	S	1	1	1	1	1	1	1	1	NP
	A	21,000	16,500	11,000	7,000	9,500	7,000	10,500	7,500	NP
H-2	S	UL	3	2	1	2	1	2	1	1
	A	21,000	16,500	11,000	7,000	9,500	7,000	10,500	7,500	3,000
H-3	S	UL	6	4	2	4	2	4	2	1
	A	UL	60,000	26,500	14,000	17,500	13,000	25,500	10,000	5,000
H-4	S	UL	7	5	3	5	3	5	3	2
	A	IL	UL	37,500	17,500	28,500	17,500	36,000	18,000	6,500
H-5	S	3	3	3	3	3	3	3	3	2
	A	UL	UL	37,500	23,000	28,500	19,000	36,000	18,000	9,000
I-1	S	UL	9	4	3	4	3	4	3	2
	A	UL	55,000	19,000	10,000	16,500	10,000	18,000	10,500	4,500
I-2	S	UL	4	2	1	1	NP	1	1	NP
	A	UL	UL	15,000	11,000	12,000	NP	12,000	9,500	NP
I-3	S	UL	4	2	1	2	1	2	2	1
	A	UL	UL	15,000	10,000	10,500	7,500	12,000	7,500	5,000
I-4	S	UL	5	3	2	3	2	3	1	1
	A	UL	60,500	26,500	13,000	23,500	13,000	25,500	18,500	9,000

FIGURE 4-11 ■ The basic size and height of a structure is affected by the fire resistance of the materials used. The architect must consider the materials of construction very early in the design process. Reproduced from the 2000 edition of the International Building Code, copyright © 2000, with the permission of the publisher, the *International Conference of Building Officials*, under license from the International Code Council. The 2000 IBC is a copyrighted work of the International Code Council.

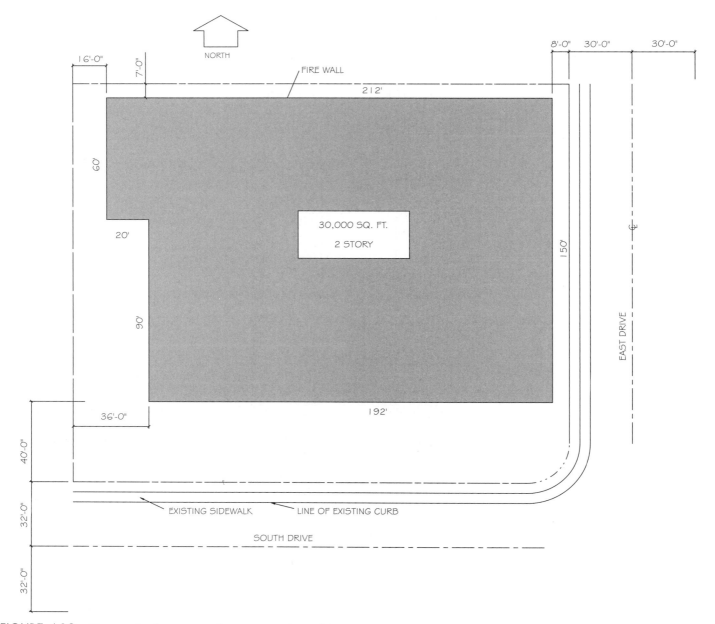

FIGURE 4-12 ■ The site plan for a proposed structure can be useful in visualizing required construction requirements.

ed openings are allowed on the east side of the building because one half of the street may be used as a part of the required yard.

MIXED OCCUPANCY

Buildings often contain more than one occupancy classification. This complicates their design in several ways. If a 5000 sq ft retail store is to be provided on the lower floor of the office building just examined, you will have a building with a B and an M occupancy. Referring to Figure 4-15 (Table 302.3.3 of the IBC) you see that a 2-hour fire separation between the two occupancies is required. This means that the walls surrounding the retail store and the floor

ceiling assembly over the retail store will have to be constructed to resist a fire for at least two hours. There are descriptions of various fire-resistive constructions in the codes and in other sources. Chapter 12 of this text further explores options for creating fire assemblies. If the entire building is built to the requirements of the more restrictive occupancy, M, the area separation wall is not required.

Determining Mixed Occupancy Size

One of the other aspects affected by mixed occupancy is the allowable area of the building. The building's allowable size is not as

$$I_f = 100[F/P - .25] W/30$$

$$F = 452$$
$$P = 724$$
$$W = 36$$

Divide 724 into 452 first

$$I_f = 100[452/724 - .25] 36/30$$

Divide 36 by 30 second

The answer rounded to 2 decimal places is close enough

$$I_f = 100[.62 - .25]1.2$$

Work inside the brackets first

$$I_f = 100[.37]1.2$$

100 times .37 times 1.2

$$I_f = 44.4$$

FIGURE 4-13 ■ The Basic Allowable Area of a building may be increased if unoccupied yards of greater than 20' are provided around the building. The formula and the percentage of increase for the building in Figure 4-12 are shown.

$$A_a = A_t + [A_t I_f /100] + [A_t I_s /100]$$

$A_t = 9000$ from Figure 4-11 for a B occupancy of Type V-B construction
$I_f = .37$ from Figure 4-13
$I_s = 200$ from Figure 4-9 506.3 for a multi-story building

Work inside the brackets first

$$A_a = 9000 + [9000(37)/100] + [9000(200)/100]$$

$$A_a = 9000 + [3330] + [18000]$$

$$A_a = 30330$$

FIGURE 4-14 ■ The Basic Allowable Area of a building can also increase if fire sprinklers are installed. The formula and the Allowable Area for the building in Figure 4-12 are shown.

apparent as with a single occupancy type. If a 5000 sq ft art gallery is include in our office building shown in Figure 4-12 (instead of the retail sales area), there will be an A-3 occupancy (see Figure 4-8) and a B occupancy. Each of these occupancies has a different Basic Allowable Area for every construction type. The allowable area for each of the occupancy classifications needs to be calculated. The sum of the ratios of the actual areas to the allowable areas may not exceed 1.

Looking over the formula in Figure 4-14, the only variable that changes is At, the Tabular area or Basic Allowable Area. To check the building to see if Type V-B construction still works, rework the formula using an At of 6000 from Figure 4-11 for an A-3 occupan-

cy. The result will be an allowable area of 20,695 for type V-B construction.

The formula for calculating the sum of the ratios of the actual areas to the allowable areas is shown in Figure 4-16. The actual area for the B occupancy for 25,000 sq ft (30,000 – 5000) is divided by the allowable area of 30,330 equals .824. The actual area of the A-3 occupancy of 5000 divided by the allowable area of 20,695 equals .241. The sum of .824 plus .241 is greater than 1 (see Figure 4-16), so Type V-B construction is not acceptable. A higher classification of construction will need to be used. Going through the same steps for Type III-B construction proves that it will work, as shown in Figure 4-16.

EXITS

After the basic exterior shape of the building has been determined, you will begin to fill it with rooms. One of the most important steps in this process is to determine the exit path. The number and size of the exits will be determined in most cases by the number of occupants in the building and the occupancy classification.

The number of occupants in the building is not necessarily determined by the number of people you say will be in the various rooms. It is determined by a ratio of the room area to the code specified area per person, as shown in Figure 4-17 (Table 1003.2.2.2 of IBC), or by the number of occupants that are to use the room, whichever is greater. In some cases, the number of fixed seats in the room determines the number of occupants. Theaters and churches are examples of spaces where the number of occupants is determined by the number of seats. In most other spaces, the area of the room is divided by the factor shown in column 2 of Table 1003.2.2.2 (Figure 4-17) to determine the number of occupants.

After determining the number of occupants in the building, you can determine the number and width of the required exits. Figure 4-18 lists the IBC requirements for two exits under most conditions. Three exits are required if the building has 501 to 1000 occupants and four exits are required for 1001 and more occupants.

The location of the exits is also an important factor. If more than one exit is required, the code requires at least two of them to be separated by one-half of the diagonal of the area served, as shown in Figure 4-19. Dead-end corridors over 20' (6100 mm) in length in an unsprinklered building for most occupancies (see Figure 4-19) are not allowed. The length of a dead-end corridor can be extended to 50' (15 240 mm) for some occupancies if the building is fully sprinklered. If the exits are not reasonably separated, there is a chance that a fire might block both exits, denying people an escape route. If the occupants of a building run down a long dead-end corridor, they will lose valuable time retracing their steps to get out of the building. If the exits are wide enough, people will be able to flow through them rapidly in an emergency.

Determining Exits

A single-story unsprinklered building with 35,000 sq ft of office space and a 900 sq ft conference rooms will be considered. Figure 4-17 states that for the purpose of calculating exits, you are to use one occupant for every 100 sq ft for business uses and one occu-

TABLE 302.3.3
REQUIRED SEPARATION OF OCCUPANCIES (HOURS)[a]

USE	A-1	A-2	A-3	A-4	A-5	B[b]	E[e]	F-1	F-2	H-1	H-2	H-3	H-4	H-5	I-1	I-2	I-3	I-4	M[b]	R-1	R-2	R-3,R-4	S-1	S-2[c]	U
A-1	2	2	2	2	2	2	2	3	2	NP	4	3	2	4	2	2	2	2	2	2	2	2	3	2	1
A-2[h]	—	2	2	2	2	2	2	3	2	NP	4	3	2	4	2	2	2	2	2	2	2	2	3	2	1
A-3[d,f]	—	—	2	2	2	2	2	3	2	NP	4	3	2	4	2	2	2	2	2	2	2	2	3	2	1
A-4	—	—	—	2	2	2	2	3	2	NP	4	3	2	4	2	2	2	2	2	2	2	2	3	2	1
A-5	—	—	—	—	2	2	2	3	2	NP	4	3	2	4	2	2	2	2	2	2	2	2	3	2	1
B[b]	—	—	—	—	—	2	2	3	2	NP	2	1	1	1	2	2	2	2	2	2	2	2	3	2	1
E	—	—	—	—	—	—	2	3	2	NP	4	3	2	3	2	2	2	2	2	2	2	2	3	2	1
F-1	—	—	—	—	—	—	—	3	3	NP	2	1	1	1	3	3	3	3	3	3	3	3	3	3	3
F-2	—	—	—	—	—	—	—	—	2	NP	2	1	1	1	2	2	2	2	2	2	2	2	3	2	1
H-1	—	—	—	—	—	—	—	—	—	4	NP	NP	NP	NP	NP	NP	NP	NP	NP	NP	NP	NP	NP	NP	NP
H-2	—	—	—	—	—	—	—	—	—	—	4	1	2	2	4	4	4	4	2	4	4	4	2	2	1
H-3	—	—	—	—	—	—	—	—	—	—	—	3	1	1	4	3	3	3	1	3	3	3	1	1	1
H-4	—	—	—	—	—	—	—	—	—	—	—	—	2	1	4	4	4	4	1	4	4	4	1	1	1
H-5	—	—	—	—	—	—	—	—	—	—	—	—	—	2	4	4	4	3	1	4	4	4	1	1	3
I-1	—	—	—	—	—	—	—	—	—	—	—	—	—	—	2	2	2	2	2	2	2	2	4	3	2
I-2	—	—	—	—	—	—	—	—	—	—	—	—	—	—	—	2	2	2	2	2	2	2	3	2	1
I-3	—	—	—	—	—	—	—	—	—	—	—	—	—	—	—	—	2	2	2	2	2	2	3	2	1
I-4	—	—	—	—	—	—	—	—	—	—	—	—	—	—	—	—	—	2	2	2	2	2	3	2	1
M[b]	—	—	—	—	—	—	—	—	—	—	—	—	—	—	—	—	—	—	2	2	2	2	3	2	1
R-1	—	—	—	—	—	—	—	—	—	—	—	—	—	—	—	—	—	—	—	2	2	2	3	2	1
R-2	—	—	—	—	—	—	—	—	—	—	—	—	—	—	—	—	—	—	—	—	2	2	3	2	1
R-3, R-4	—	—	—	—	—	—	—	—	—	—	—	—	—	—	—	—	—	—	—	—	—	2	3	2[g]	1[g]
S-1	—	—	—	—	—	—	—	—	—	—	—	—	—	—	—	—	—	—	—	—	—	—	3	3	3
S-2[c]	—	—	—	—	—	—	—	—	—	—	—	—	—	—	—	—	—	—	—	—	—	—	—	2	1
U	—	—	—	—	—	—	—	—	—	—	—	—	—	—	—	—	—	—	—	—	—	—	—	—	1

For SI: 1 square foot = 0.0929 m².
NP = Not permitted.
a. See Exception 1 to Section 302.3.3 for reductions permitted.
b. Occupancy separation need not be provided for incidental storage areas within Groups B and M if the:
  1. Area is less than 10 percent of the floor area, or
  2. Area is provided with an automatic fire-extinguishing system and is less than 3,000 square feet, or
  3. Area is less than 1,000 square feet.
c. Areas used only for private or pleasure vehicles may reduce separation by 1 hour.
d. Accessory assembly areas are not considered separate occupancies if the floor area is 750 square feet or less.
e. Assembly uses accessory to Group E are not considered separate occupancies
f. Accessory religious educational rooms and religious auditoriums with occupant loads of less than 100 are not considered separate occupancies.
g. See exception to Section 302.3.3.
h. Commercial kitchens need not be separated from the restaurant seating areas that they serve.

FIGURE 4-15 ■ When there is going to be more than one type of Occupancy Classification, the required separation of common walls and floor systems must be determined. Reproduced from the 2000 edition of the International Building Code, copyright © 2000, with the permission of the publisher, the *International Conference of Building Officials*, under license from the International Code Council. The 2000 IBC is a copyrighted work of the International Code Council.

Mixed occupany

$$\frac{\text{actual area 1}}{\text{allowable area 1}} + \frac{\text{actual area 2}}{\text{allowable area 2}} + \frac{\text{actual area 3}}{\text{allowable area 1}} \leq 1$$

check for Figure 4–12 with a 5000 sq ft art gallery

Type V-B construction

$$\frac{25,000}{30,330} + \frac{5,000}{20,695} = .824 + .241 = 1.065 \quad \text{this is greater than 1 and is not OK}$$

Type III-B construction

$$\frac{25,000}{65,534} + \frac{5,000}{32,767} = .824 + .153 = .534 \leq \text{OK}$$

FIGURE 4-16 ■ The formula for computing the allowable area of a building containing more than one Occupancy Classification is shown along with a sample calculation.

pant for every 15 sq ft of an assembly area with tables and chairs. Dividing 35,000 by 100 results in 350 occupants for the office area, and dividing 900 by 15 results in 60 occupants for a total of 410 occupants. Per Table 1003.2.3 of the IBC, the code requires 0.2" of exit width for every occupant. Multiplying 0.2" by 410 equals 82" (6'–10") of required exit. This building requires only two exits but two 3'–0" doors will not provide the proper exit width. The code also states that the required width shall be divided approximately equally among the separate exits. This means that if you only provide two exits, a 3'–0" door at one of them and a 4'–0" door or a pair of 3'–0" doors at the other might not be acceptable. Because the word "approximately" has been used in the code, it would be advisable to check with the local building official for his interpretation on this matter.

There are minimum requirements in the code for hall and door widths. Three-foot-wide exit doors are required for most condi-

tions. There are cases where corridors as narrow as 36" are acceptable but most conditions required 44" as a minimum. Stairways have required minimum widths and a required minimum width per occupant.

THE CODE AND COMPUTERS

As you can see, using the code can be a lot of work. There is a lot of flipping back and forth between chapters. A thorough code check is important to the safety of the building's occupants. As your familiarity of the code increases, its use will become easier. You will remember what areas of the code to check. It is important that you check the code thoroughly, rather than try to commit it completely to memory, because the code is revised every three years. Your memory may dredge up an outdated version of the code section.

There are computer-based code-checking systems available for each of the model building codes. These range from a simple computerized checklist to a comprehensive analyst program. These programs can greatly speed up code checking and increase your confidence in the adequacy of your design.

ADDITIONAL READING

The following Web sites can be used as a resource to help you keep current with changes in building codes. All four sites are sources for the 2000 IBC, as well as the other code listed with their names. They are sources for interpretive manuals and training.

ADDRESS	COMPANY/ORGANIZATION
www.intlcode.org	International Code Council
www.bocai.org	*BOCA*— Building Officials and Code Administrators International.
www.sbcci.org	Southern Building Code Congress International, Inc. SBC (Standard Building Code)
www.icbo.org	ICBO Code Central (International Conference of Building Officials).

TABLE 1003.2.2.2
MAXIMUM FLOOR AREA ALLOWANCES PER OCCUPANT

OCCUPANCY	FLOOR AREA IN SQ. FT. PER OCCUPANT
Agricultural building	300 gross
Aircraft hangars	500 gross

Airport terminal
    Concourse
    Waiting areas
    Baggage claim
    Baggage handling
	
	100 gross
	15 gross
	20 gross
	300 gross

Assembly Gaming floors (keno, slots, etc.)	 11 gross
Assembly with fixed seats	See 1003.2.2.9

Assembly without fixed seats
    Concentrated (chairs only—not fixed)
    Standing space
    Unconcentrated (tables and chairs)
	
	7 net
	5 net
	15 net

Bowling centers, allow 5 persons for each lane including 15 feet of runway, and for additional areas	 7 net
Business areas	100 gross
Courtrooms—other than fixed seating areas	40 net
Dormitories	50 gross

(continued)

FIGURE 4-17 ■ The number of occupants in a space is usually calculated on a square footage basis when fixed seats are not installed. This table lists the maximum number of square feet that you can assume one person occupies for various uses.

TABLE 1004.2.1
SPACES WITH ONE MEANS OF EGRESS

OCCUPANCY	MAXIMUM OCCUPANT LOAD
A, B, E, F, M, U	50
H-1, H-2, H-3	3
H-4, H-5, I-1, I-3, I-4, R	10
S	30

FIGURE 4-18 ■ If a large number of people occupy a space, more than one exit will be required. Reproduced from the 2000 edition of the International Building Code, copyright © 2000, with the permission of the publisher, the *International Conference of Building Officials*, under license from the International Code Council. The 2000 IBC is a copyrighted work of the International Code Council.

TABLE 1003.2.2.2—continued
MAXIMUM FLOOR AREA ALLOWANCES PER OCCUPANT

OCCUPANCY	FLOOR AREA IN SQ. FT. PER OCCUPANT

Educational
    Classroom area
    Shops and other vocational room areas
	
	20 net
	50 net

Exercise rooms	50 gross
H-5 Fabrication and manufacturing areas	200 gross
Industrial areas	100 gross

Institutional areas
    Inpatient treatment areas
    Outpatient areas
    Sleeping areas
	
	240 gross
	100 gross
	120 gross

Kitchens, commercial	200 gross

Library
    Reading rooms
    Stack area
	
	50 net
	100 gross

Locker rooms	50 gross

Mercantile
    Basement and grade floor areas
    Areas on other floors
    Storage, stock, shipping areas
	
	30 gross
	60 gross
	300 gross

Parking garages	200 gross
Residential	200 gross

Skating rinks, swimming pools
    Rink and pool
    Decks
	
	50 gross
	15 gross

Stages and platforms	15 net
Accessory storage areas, mechanical equipment room	300 gross
Warehouses	500 gross

For SI: 1 square foot = 0.0929 m².

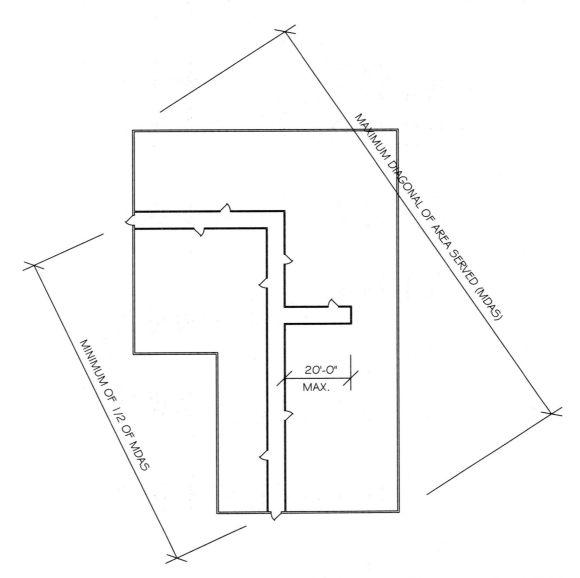

FIGURE 4-19 ■ If more than one exit from a building is required, two of them will need to be seperated by a minimum of one-half of the area served. The code limits the length of dead-end corridors when two or more exits are required. They usually cannot exceed 20' (6100 mm) in length for a building without fire sprinklers and 50' in a building with fire sprinklers.

CHAPTER 4

Introduction to National Building Codes

CHAPTER QUIZ

Answer the questions on a separate sheet of paper. Print the chapter title, question number, and a short complete statement for each question. Answer the questions according to the code that governs your area. If math is required to answer a question, show the work.

Question 4-1 What occupancy classification would a 10-unit apartment building be?

Question 4-2 What occupancy classification would a clothing store be?

Question 4-3 What fire separation is required between a B and an H-3 occupancy using the IBC? Between an A-2 and an E?

Question 4-4 What is the basic allowable area of an H-4 occupancy in a type V building? In a type III building?

Question 4-5 What is the most important purpose of the model code?

Question 4-6 Using the drawing shown below (Figure 4-20) and the IBC, what is the maximum size allowed for the following one-story building without a fire sprinkler system? Assume an occupancy of E-1, building type II-A.

Question 4-7 What is the maximum allowable size for the building in question 4-6 if the building is provided with a fire sprinkler system and is of type II-B construction?

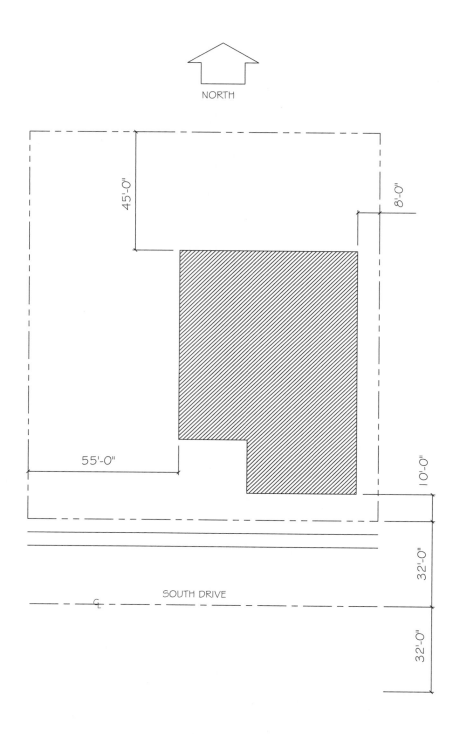

NORTH

45'-0"

8'-0"

55'-0"

10'-0"

32'-0"

SOUTH DRIVE

32'-0"

FIGURE 4-20

CHAPTER 5

Access Requirements for People with Disabilities

The Americans with Disabilities Act, otherwise known as the ADA, is a federal law requiring all new public accommodations and commercial facilities to be accessible and usable by people with disabilities. This landmark law was driven by the fact that medical and scientific technology has allowed people to survive previously fatal conditions and live on with significant disabilities. The ADA gives these citizens a better chance to live productive and fulfilling lives.

This law has had a significant impact on building and site design. It does not guarantee the disabled access to every square foot of every new building, but it does greatly increase the building space they can utilize in a comfortable and convenient manner. Access is required to the ground floor of almost every nonresidential building, and it is required to all floors of most public and medical buildings and buildings over three stories in height. Make note that public buildings include retail stores, restaurants, theaters and other privately owned buildings that the general public utilizes. Public buildings also include public-owned facilities such as libraries and courthouses.

The ADA addresses both new and existing structures. Some existing public facilities are required to remove some barriers now. Other facilities are required to remove barriers when they remodel or do an addition. In both cases the term *readily achievable* is used, which means that it would be easy to do, not creating an undue hardship. The ADA provides several alternative solutions for accessible facilities in existing buildings. These alternative solutions and the term *readily achievable* do not apply to new construction. This chapter is intended to address the requirements of new construction.

ADA ACCESSIBILITY GUIDELINES FOR BUILDINGS AND FACILITIES

The ADA addresses the requirements of people with disabilities. Many of the requirements are to accommodate people in wheelchairs, but the ADA also addresses the problems of the elderly, the blind, the hearing impaired, and those with other forms of physical disabilities. It is important to realize that the ADA may not be the only set of regulations governing access. Several states had access laws prior to the ADA that might still be in effect with requirements that exceed those of the ADA. The requirements of the ADA do not necessarily provide the ideal or optimum conditions for people with disabilities, only the minimum acceptable requirements.

Meeting ADA requirements does not excuse the designer from meeting the requirements of any other governing building code.

The primary concerns addressed by the ADA are access to the building, the building spaces it contains, and the use of facilities such as rest rooms, drinking fountains, and telephones. The *Accessibility Guidelines For Buildings And Facilities* are available at www.access-board.gov.

PARKING REQUIREMENTS

Special parking spaces must be provided for people with disabilities. This parking must be relatively flat (2% maximum slope) and must be located as close as possible to the entrance of the building. There must be an accessible walkway from the parking space to the building. The parking spaces must be marked with the international symbol of accessibility shown in Figure 5-1.

Parking spaces must be of a size adequate to accommodate the special needs of people with disabilities. At least one space needs to be 8' (2400 mm) wide with an 8' (2400 mm) wide loading space to accommodate a van with a wheelchair lift, as shown at A in Figure 5-2. Other parking spaces for the disabled need only a 5' (1525 mm) wide loading space to allow a wheelchair to be brought alongside a car. Figure 5-3 shows how these minimum requirements relate to each other. An alternative to these two sizes is to provide universal parking spaces, as shown at B in Figure 5-2. The minimum vertical clearance for parking spaces for the disabled is 96" (2440 mm). If there is a curb at the parking space, there will be a need for a curb ramp, as shown in Figure 5-4.

The drafter will have to show the proper size and number of parking spaces on the site plan based on the criteria listed in Figure 5-5. Parking details will normally be stock details stored as wblocks or contained on a standard sheet with other ADA details. Site grades can influence the location of accessible parking spaces. The space must be on an accessible route of travel and meet rigid requirements for slope in both directions. The required slope of the parking space and the accessible path of travel will need to be coordinated with the civil engineer for the project.

METHODS OF EGRESS

Many of the ADA requirements deal with access to and throughout a structure. Major areas to be considered by the design team include walks and hallways, ramps, stairs, elevators, lifts, areas of rescue assistance, and doors.

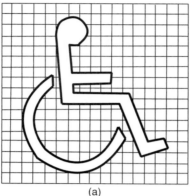

(a)
Proportions
International Symbol of Accessibility

(b)
Display Conditions
International Symbol of Accessibility

(c)
International TDD Symbol

(d)
International Symbol of Access for Hearing Loss
International Symbols

FIGURE 5-1 ■ All parking spaces for the disabled must be marked with the international symbol for accessibility. Other symbols used to designate special provisions include the TDD (telecommunication device for the deaf) and hearing loss symbols. *Courtesy Federal Register/Vol.56 No.144/July 26, 1991.*

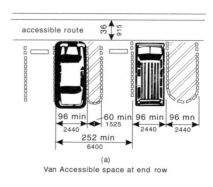

(a)
Van Accessible space at end row

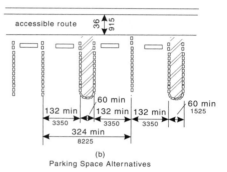

(b)
Parking Space Alternatives

FIGURE 5-2 ■ Van parking and alternatives. *Courtesy Uniform Federal Accessibility Standards.*

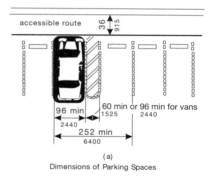

(a)
Dimensions of Parking Spaces

FIGURE 5-3 ■ Car parking minimum sizes. *Courtesy Uniform Federal Accessibility Standards.*

Walks and Hallways

There must be an accessible route from public transportation, parking spaces, loading zones, and the public sidewalk to the building. The route shall lead to a primary building entrance, not a service entrance, and must be at least 36" (915 mm) wide. A wider route of 48" (1220 mm) is recommended so that someone walking can pass a wheelchair, as shown in Figure 5-6. Turns in the walk or hall must be able to accommodate a wheelchair, as shown at A and B in Figure 5-7. If the route is less than 60" (1525 mm) wide, there must be a 60 × 60" (1525 × 1525 mm) minimum passing space located no further than 200' (60 960 mm) apart. The route must have a minimum vertical clearance of 80" (2030 mm). The surface of the route must be slip resistant and must also be a reasonably firm sur-

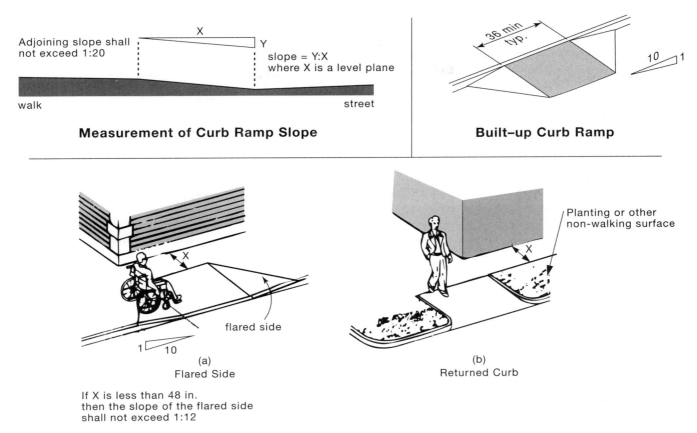

Measurement of Curb Ramp Slope

Built–up Curb Ramp

Side of Curb Ramps

FIGURE 5-4 ■ Curb construction in an accessible route. *Courtesy Uniform Federal Accessibility Standards.*

Total Parking in Lot	Required Minimum Number of Accessible Spaces
1 to 25	1
26 to 50	2
51 to 75	3
76 to 100	4
101 to 150	5
151 to 200	6
201 to 300	7
301 to 400	8
401 to 500	
501 to 1000	2 percent of total
1001 to over	20 plus 1 for each 100 sq. ft. over 1,000

FIGURE 5-5 ■ Required minimum number of handicapped spaces is based on the total parking. *Courtesy Uniform Federal Accessibility Standards.*

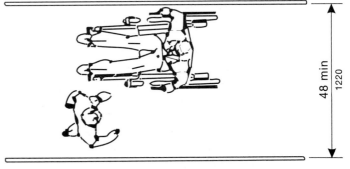

FIGURE 5-6 ■ Recommended width will allow passage of a wheelchair and a pedestrian. *Courtesy Uniform Federal Accessibility Standards.*

face (no plush carpet or pad, etc.). Projections into the access route could restrict the passage of wheelchairs and could be hazardous to the visually impaired. Therefore there are limitations on projections into the access route, as shown in Figure 5-8.

Changes in level along the accessible route of up to 1/2" (13 mm) shall be per C and D in Figure 5-7. Changes in elevation of more than 1/2" (13 mm) must be by curb ramp, elevator, or lift. If the walk slopes more than 5% (1 in 20) in the direction of travel, it is considered a ramp and shall comply with the requirements of a ramp. The cross slope of a travel route may not slope more than 1 in 50, or 2%.

The drafter's job is to draw the walks to their proper width and destination and to assure that there is proper maneuvering space

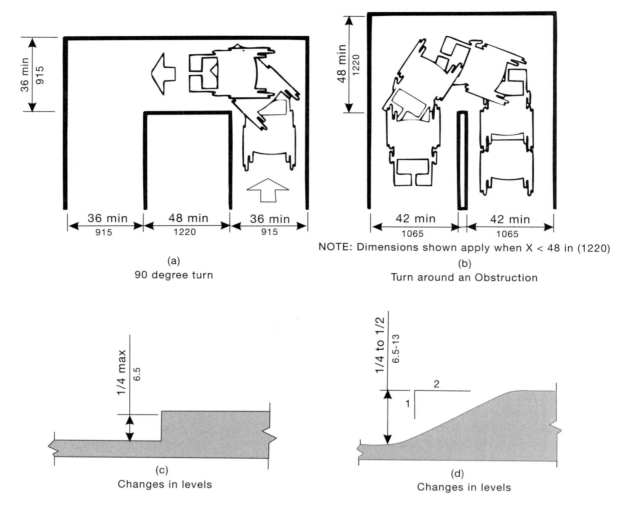

(a)
90 degree turn

NOTE: Dimensions shown apply when X < 48 in (1220)

(b)
Turn around an Obstruction

(c)
Changes in levels

(d)
Changes in levels

FIGURE 5-7 ■ Creating turns in a ramp or hallway must be carefully planned. *Courtesy Uniform Federal Accessibility Standards.*

for turns. Slope of walks, both in the direction of travel and cross fall, must be coordinated with the civil engineer. If ramps or stairs are required, they need to be properly detailed to meet the requirements outlined in the following sections. Most offices will have standard details, reflecting the requirements of the ADA, that can be referenced to cover many of the conditions that occur on a site or within the accessible route in the building.

Ramps

Curb ramps, as the name implies, are intended to make the transition through the change in level caused by a normal curb. These ramps shall be a minimum of 36" (915 mm) wide and shall be sloped as shown in Figure 5-4. Curb ramps were required to have a detectable warning at their top edge, as shown in Figure 5-9. This requirement has been temporarily suspended but may be reinstat-

ed in the future. Curb ramps do not require handrails or guardrails.

Other ramps are usually longer than curb ramps. Like a walk, ramps are required to be a minimum of 36" (915 mm) wide. It is desirable to make the ramps wider, especially if this is also the primary access for people who are not disabled. Any access route having a slope of 1 in 20 or greater is considered to be a ramp. A ramp should be as shallow a slope as possible and may not exceed a slope of 1 in 12. The maximum change in height between landings is 30" (760 mm). Level landings are required at each end of a ramp. They must be at least as wide as the ramp and be 60" (1525 mm) long in the direction of travel (see Figure 5-10). If the ramp changes direction, the landing must be 60" × 60" (1525 × 1525 mm) minimum. If there is a doorway on the landing, all requirements for maneuvering at a door must be met. The requirements for maneuvering at doors is discussed later in this chapter.

Guardrails must be provided on the open side of a ramp, as shown in Figure 5-11. Unless adjacent to a seating area, ramps ris-

CHAPTER 5 — Access Requirements for People with Disabilities ■ 83

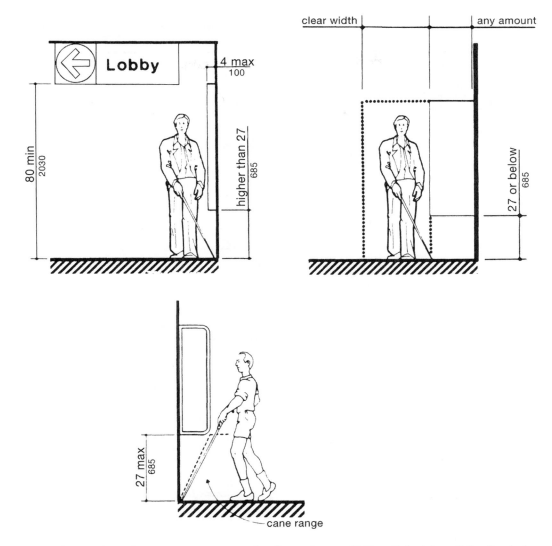

FIGURE 5-8 ■ Limitations for any object that projects into a required walkway. *Courtesy Uniform Federal Accessibility Standards.*
continued

ing more than 6" (150 mm) require a handrail on both sides. The handrail shall be constructed per Figure 5-12, projecting 1 1/2" (38 mm) clear of the wall and extending 12" (300 mm) beyond the ends of the ramp, except for a continuous inside handrail on a switchback or dogleg ramp as shown in Figure 5-13. The handrail shall be continuous and be mounted 34" to 38" (860 to 960 mm) above the ramp surface.

Required ramps frequently fail to receive proper attention during the design stage. Total rise must be considered and, if the ramp must rise over 30" (760 mm), a landing must be inserted. Landings at changes in direction must be properly sized. Ramps and landings take up a surprising amount of room. It is important to analyze the total amount of room needed early in the drawing process so that any deficiencies can be corrected with minimum effort.

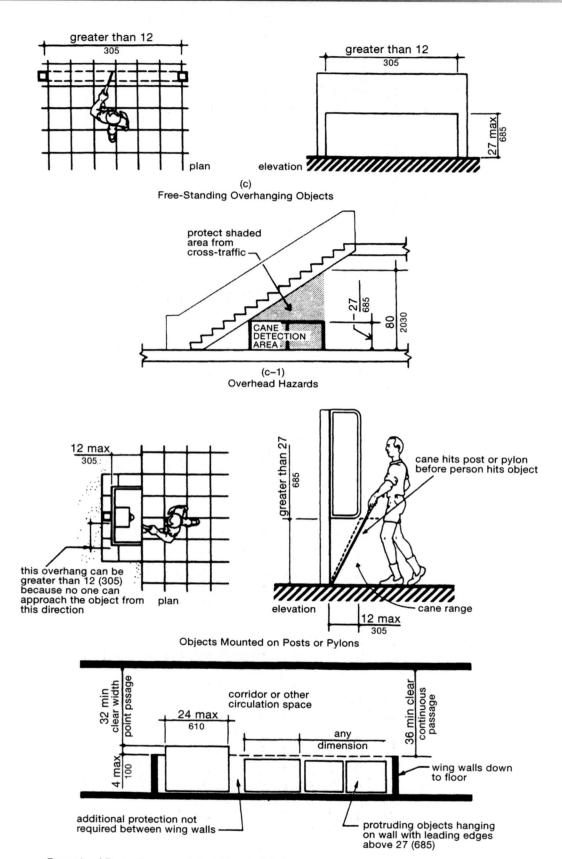

(c)
Free-Standing Overhanging Objects

(c–1)
Overhead Hazards

Objects Mounted on Posts or Pylons

Example of Protection around Wall-Mounted Objects and Measurements of Clear Widths

FIGURE 5-8 ■ *continued*

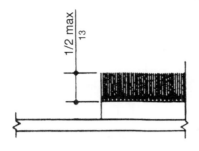

Carpet Pile Thickness

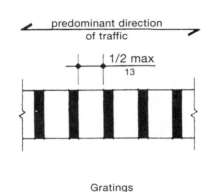

Gratings

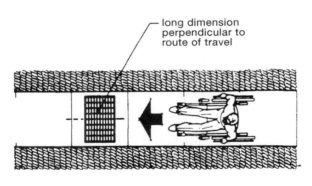

Grating Orientation

FIGURE 5-8 ■ *continued*

Stairs

Stairways serving a floor that is not otherwise accessible to people with disabilities shall meet the following conditions in addition to the requirements of the building code. Treads shall be a uniform width not less than 11" (280 mm) and risers shall be a uniform height. Note that the allowable height of the risers is not called out.

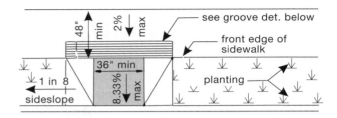

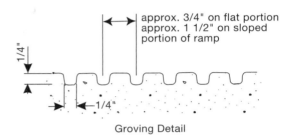

Groving Detail

Warning Strip

FIGURE 5-9 ■ Ramps must have a detectable warning at their top edge. This Provision has been suspended until July 26, 2001. *Courtesy Uniform Federal Accessibility Standards.*

This is a case where the building code would control. When nosings are provided, they shall be constructed as shown in Figure 5-14. Stairways shall not have open risers. Open risers can be confusing to people with poor sight and can be a problem for cane use by the blind. Handrails shall be provided, measuring 34" to 38" (865 to 965 mm) high on both sides of a stair, and shall be constructed as shown in Figures 5-12 and 5-13. Handrails shall be continuous and uninterrupted by newel post or other obstructions. For additional information on stairs, see Chapter 20.

Elevators

An elevator is usually the easiest way to provide access to upper floors when access is required above the ground floor. If an elevator is installed, it must meet the ADA requirements whether the elevator was required or not. Every floor that the elevator serves is accessible and will need to meet access requirements whether they were otherwise required or not. The elevator must be automatic and have proper call buttons, car control, illumination, door timing, and reopening as shown in Figure 5-15. The elevator floor is a part of the access route so it must be similar to the floor described in walks and hallways. The shape and size of the elevator must be such that it is usable by a wheelchair, as shown in Figure 5-16. Much of the drafting for elevators is done by the elevator manufacturer or supplier. The drafter's job with respect to the ADA will normally be limited to details concerning interior finishes of the elevator.

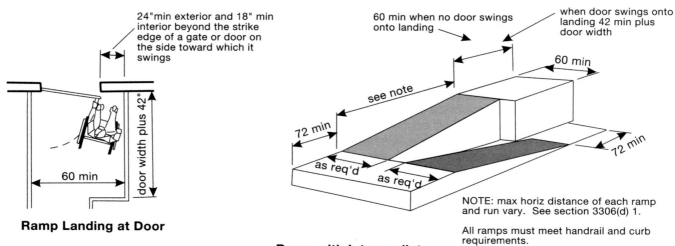

24"min exterior and 18" min interior beyond the strike edge of a gate or door on the side toward which it swings

door width plus 42"

60 min

Ramp Landing at Door

60 min when no door swings onto landing

when door swings onto landing 42 min plus door width

60 min

see note

72 min

72 min

as req'd

as req'd

NOTE: max horiz distance of each ramp and run vary. See section 3306(d) 1.

All ramps must meet handrail and curb requirements.

Ramp with Intermediate Switch-Back Platform

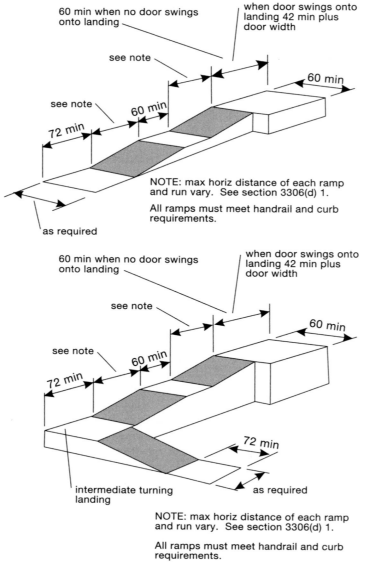

60 min when no door swings onto landing

when door swings onto landing 42 min plus door width

see note

see note

60 min

60 min

72 min

NOTE: max horiz distance of each ramp and run vary. See section 3306(d) 1.

All ramps must meet handrail and curb requirements.

as required

60 min when no door swings onto landing

when door swings onto landing 42 min plus door width

see note

60 min

see note

60 min

72 min

72 min

intermediate turning landing

as required

NOTE: max horiz distance of each ramp and run vary. See section 3306(d) 1.

All ramps must meet handrail and curb requirements.

FIGURE 5-10 ■ Changes in height for ramps. *Courtesy Uniform Federal Accessibility Standards.*

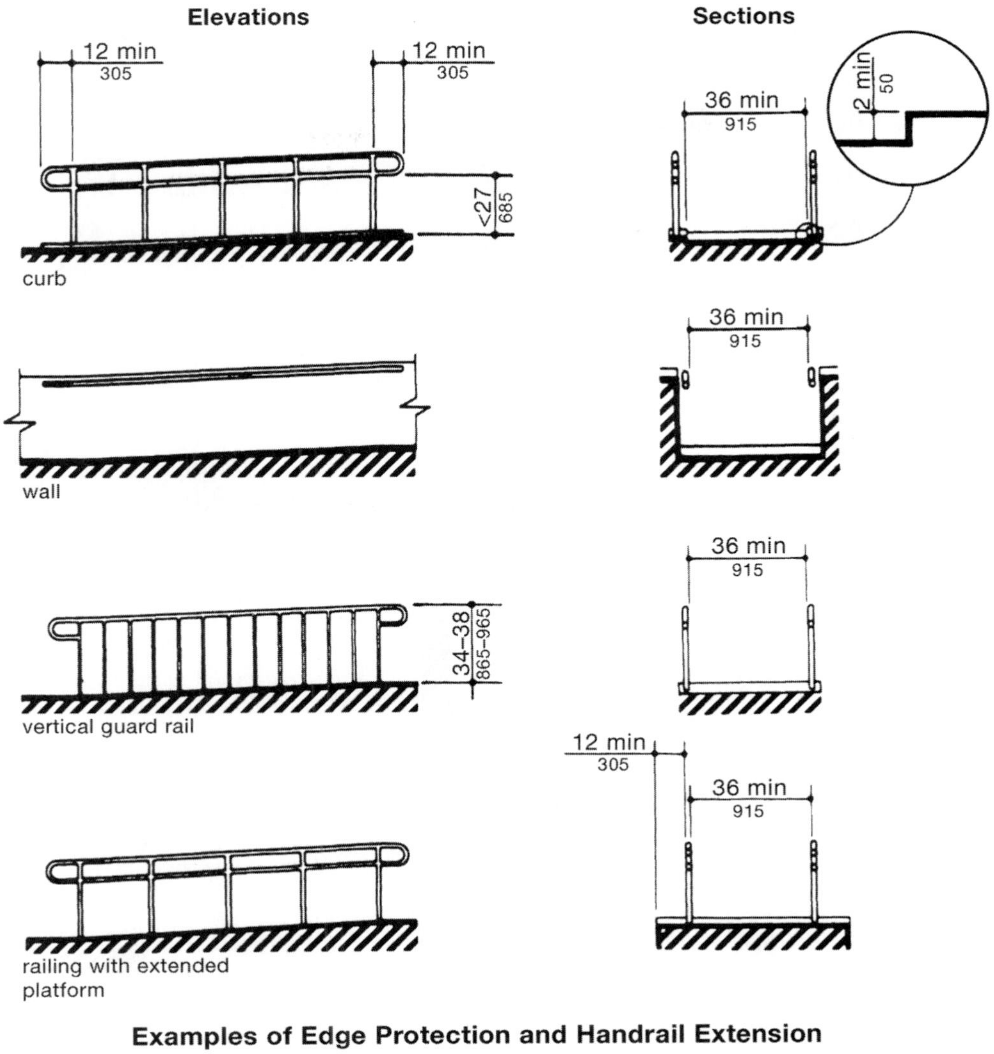

Elevations   **Sections**

12 min / 305   12 min / 305

36 min / 915   2 min / 50

<27 / 685

curb

wall

36 min / 915

vertical guard rail   34-38 / 865-965

36 min / 915

12 min / 305   36 min / 915

railing with extended
platform

Examples of Edge Protection and Handrail Extension

FIGURE 5-11 ■ Guardrail construction. *Courtesy Uniform Federal Accessibility Standards.*

Lifts

Platform or wheelchair lifts may be used in place of a ramp or elevator. If they are used, they must allow for unassisted entry, operation, and exit. They must also meet elevator and escalator safety requirements.

Areas of Rescue Assistance

Elevators do not provide a good means of exit from a building in an emergency situation such as a fire. This presents a substantial exit problem when an elevator is the means of access to the building. People with disabilities have been given a way of getting into the building but no way of getting out. The solution to this problem is an area of rescue assistance where people with disabilities can wait safely until rescue crews come to their aid. This area must be on or near the exit route so that the rescue crew has quick and easy access to its occupants. The space needs to be out of the normal exit flow so that its occupants do not interfere with the exiting

of others and must provide a smoke-free and fire-safe environment. There must be a two-way communication system between the rescue area and the main entrance to the building area so that rescue workers can verify that someone needs their assistance or the people with disabilities can cry for help.

The area of rescue assistance must provide a minimum of two 30 × 48" (760 × 1200 mm) areas. These areas can be located on a stairway landing with a smoke-proof enclosure, on an exterior exit balcony located adjacent to an exit stairway, or in a one-hour fire-rated corridor immediately adjacent to an exit enclosure. They can be located in an elevator lobby that is separated from the balance of the building by two-hour fire-resistive construction when the elevator shaft is properly pressurized. They can also be located on a stairway landing within an exit enclosure that is vented to the exterior and separated from the rest of the building by a one- hour door. There must be at least one 30 × 48" (760 × 1200 mm) rescue location for each 200 occupants of the building. The stairway serving the rescue area must have 48" (1220 mm) minimum between handrails to allow room for someone to be carried down

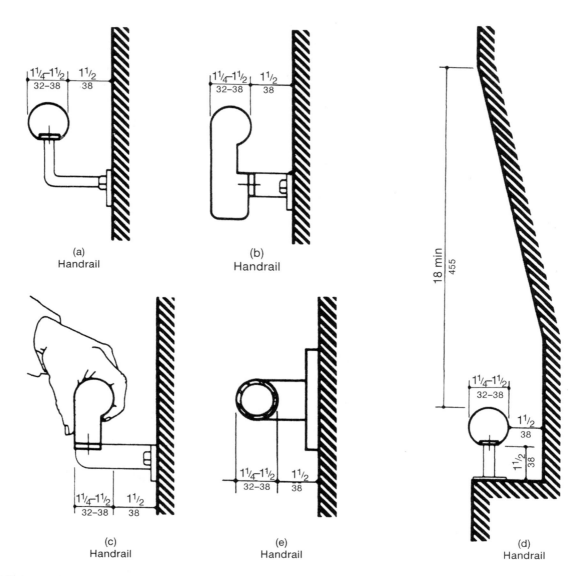

FIGURE 5-12 ■ Handrail construction. *Courtesy Uniform Federal Accessibility Standards.*

them. An area of rescue assistance is not required in buildings having a supervised automatic sprinkler system.

The location of the areas of rescue assistance is determined by the project designer. It is the drafter's job to detail the areas and coordinate with the proper consultants. Fire and smoke protection must be properly detailed. Mechanical ducts must be properly dampered, and electrical devices such as communication systems must be detailed throughout the drawings.

Doors

People with disabilities must have access to all areas on floors that are required to be accessible. At least one door to each space must meet the following requirements: Doors must provide a minimum of 32" (810 mm) clear in the 90° position as shown in Figure 5-17. This normally requires a 36" (915 mm) door. They must open with one hand and without requiring a tight grip, tight pinching, or

twisting of the wrist. This normally requires lever-operated door handles, push-type mechanisms, or U-shaped handles. Note that this requirement also applies to sliding and pocket doors. The hardware may not be mounted more than 48" (1220 mm) above the floor. Interior hinged doors, sliding doors, and folding doors must open with a maximum force of 5 lbs. (2.3 kg). This is important with regard to automatic closing devices. It is recommended that doors have a 10" (255 mm) high kick plate on the bottom.

Maneuvering space at doors shall be as shown in Figure 5-18. The required width in front of doors can have an influence on hall widths. Doors also have a required clear width on the swing side. This can affect widths of halls and alcoves when doors are located at their ends. Assuring that these clearances are provided is often the responsibility of the drafter. If a door is placed in a manner that does not allow for the required clearances, the drafter should point this out to the designer so that the proper adjustments to the design may be made.

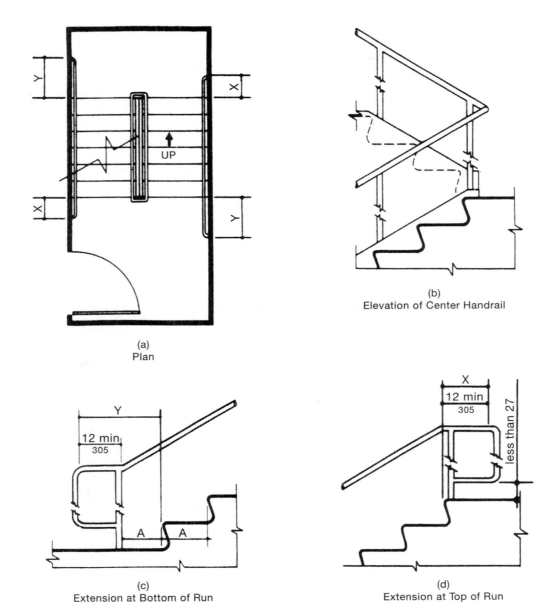

(a)
Plan

(b)
Elevation of Center Handrail

(c)
Extension at Bottom of Run

(d)
Extension at Top of Run

NOTE:
X is the 12 in minimum handrail extension required at
each top riser.

Y is the minimum handrail extension of 12 in plus the
width of one tread that is required at each bottom riser.

Stair Handrails

FIGURE 5-13 ■ Stair handrail construction and placement. *Courtesy Uniform Federal Accessibility Standards.*

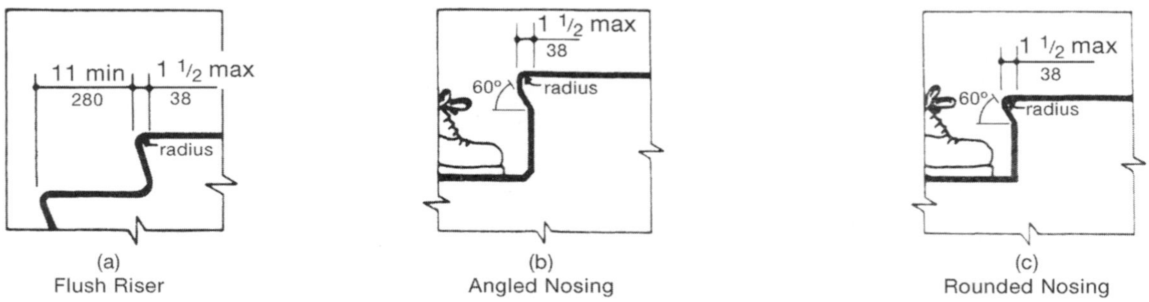

(a)
Flush Riser

(b)
Angled Nosing

(c)
Rounded Nosing

Usable Tread Width and Examples of Acceptable Nosings

FIGURE 5-14 ■ Usable tread widths. *Courtesy Uniform Federal Accessibility Standards.*

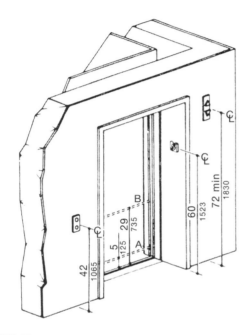

NOTE: The automatic door reopening device is activated if an object passes through either line A or line B. Line A and line B represent the vertical locations of the door reopening device not requiring contact.

Hoistway and Elevator Entrances

FIGURE 5-15 ■ Minimum entrance requirements for an elevator. *Courtesy Uniform Federal Accessibility Standards.*

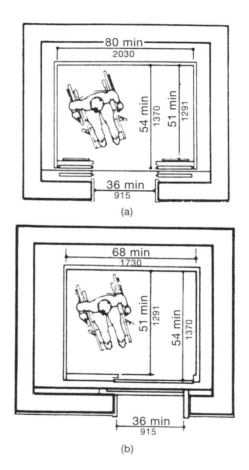

(a)

(b)

Minimum Dimensions of Elevator Cars

FIGURE 5-16 ■ Minimum dimensions of elevator cars must be considered in the initial building design. *Courtesy Uniform Federal Accessibility Standards.*

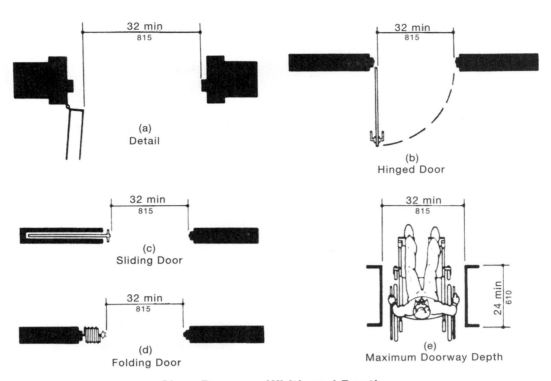

Clear Doorway Width and Depth

FIGURE 5-17 ■ ■ Minimum clear doorway width and depth. *Courtesy Uniform Federal Accessibility Standards.*

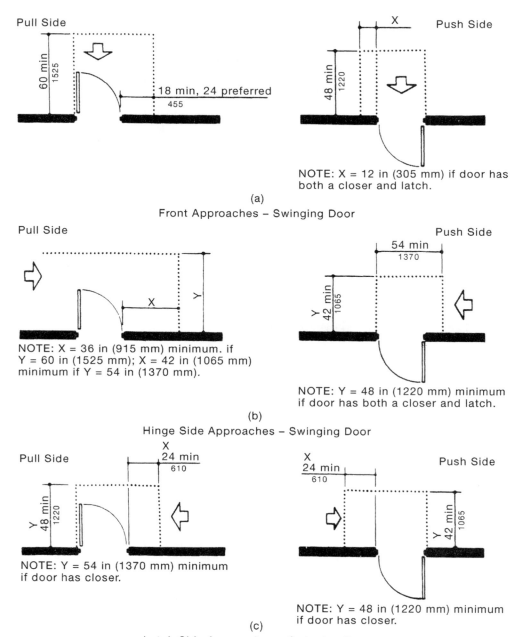

(a)
Front Approaches – Swinging Door

NOTE: X = 12 in (305 mm) if door has both a closer and latch.

(b)
Hinge Side Approaches – Swinging Door

NOTE: X = 36 in (915 mm) minimum. if Y = 60 in (1525 mm); X = 42 in (1065 mm) minimum if Y = 54 in (1370 mm).

NOTE: Y = 48 in (1220 mm) minimum if door has both a closer and latch.

(c)
Latch Side Approaches – Swinging Door

NOTE: Y = 54 in (1370 mm) minimum if door has closer.

NOTE: Y = 48 in (1220 mm) minimum if door has closer.

NOTE: All doors in alcoves shall comply with the clearances for front approaches

Maneuvering Clearances at Doors

FIGURE 5-18A ■ The area near a door must be considered to allow for easy maneuvering. *Courtesy Uniform Federal Accessibility Standards.*

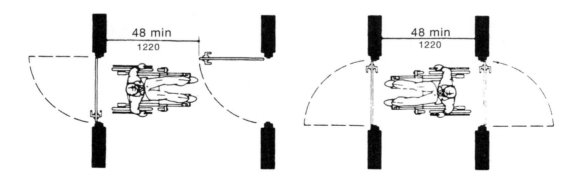

48 min
1220

(d)
Front Approach – Sliding Doors
and Folding Doors

54 min
1370

42 min
1065

(e)
Slide Side Approach – Sliding Doors
and Folding Doors

X
24 min
610

42 min
1065

(f)
Latch Side Approach – Sliding Doors and Folding Doors

NOTE: All doors in alcoves shall comply with the clearances for front approaches

Maneuvering Clearances at Doors

48 min
1220

48 min
1220

Two Hinged Doors in Series

FIGURE 5-18B ■ The area near a door must be considered to allow for easy maneuvering. *Courtesy Uniform Federal Accessibility Standards.*

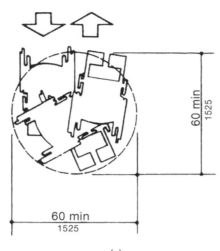

(a)
60 in (1525 mm) – Diameter Space

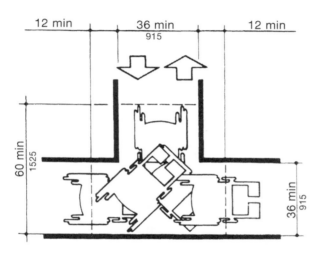

(b)
T-Shaped Space for 180° Turns

FIGURE 5-19 ■ Minimum acceptable turning space. *Courtesy Uniform Federal Accessibility Standards.*

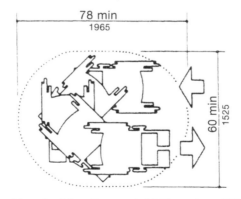

Space Needed for Smooth U-Turn in a Wheelchair

FIGURE 5-20 ■ Preferred turning space. *Courtesy Uniform Federal Accessibility Standards.*

RESTROOMS AND BATHING FACILITIES

The ADA requires at least one restroom to be accessible to people with disabilities for each sex, per accessible floor. In some cases a unisex restroom can satisfy this requirement. The door to the restrooms must meet all door requirements shown in Figure 5-18, including the distance between doors at the vestibule. All of the fixtures intended for people with disabilities shall be on an accessible route. This means that there must be adequate access to the fixtures, and the restroom must have room for a wheelchair to maneuver. Figure 5-19 shows the minimum acceptable maneuvering space. A preferred maneuvering configuration is shown in Figure 5-20.

Toilet Stalls

At least one toilet stall in each restroom shall meet the requirements shown at A in Figure 5-21. These stalls may be reversed to allow for either a left- or right-handed approach. Alternate stall placement (at B in Figure 5-21) is not allowed in new construction. If a restroom contains six or more stalls, in addition to the standard accessible stall, one of these stalls must be at least 36" (915 mm) wide with an outword swinging door. The front and one side partition of the stall must be at least 9" (230 mm) above the floor unless the stall is longer than 60" (1525 mm). A single-person restroom with a toilet and a sink without a stall shall meet the requirements shown in Figure 5-22. The drafter should detail door swings to stalls to maintain required clearances.

The toilet or water closet shall be 17" to 19" (430 to 485 mm) to the top of the toilet seat. Stalls for wall-mounted toilets can be 3" (75 mm) shorter than those for floor-mounted toilets. Grab bars, shown in Figure 5-21, shall meet the requirements illustrated at E in Figure 5-12. The toilet paper dispenser shall be 19" to 36" (485 to 915 mm) above the floor.

Urinals

If urinals are provided, at least one shall meet the following requirements: It shall be stall-type or wall-hung with an elongated rim at a maximum of 17" (430 mm) above the floor. A clear floor space of 30 × 48" (815 × 1220 mm) shall be provided in front of the urinals. This clear space can overlap the accessible route. Partitions that do not project beyond the rim can be installed with 29" (740 mm) minimum clearance.

Lavatories and Mirrors

If lavatories and mirrors are provided, at least one shall meet the following requirements: Lavatories and mirrors shall conform to the requirements shown in Figure 5-23. The clear space can overlap the accessible route. Hot water and drain pipes under lavatories shall be insulated or otherwise configured to protect against contact. There shall be no sharp or abrasive surfaces under the lavatories. Faucets shall be operable with one hand and without requir-

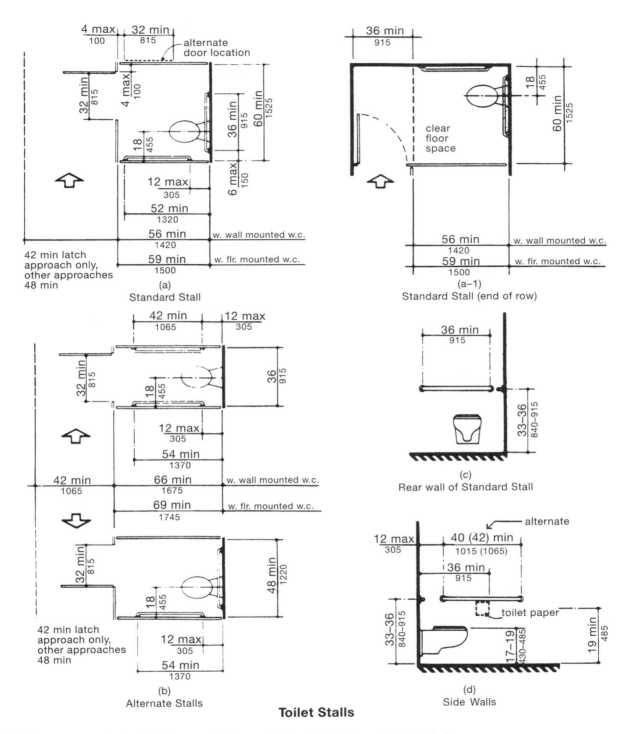

FIGURE 5-21 ■ Design standards for bathroom stalls. *Courtesy Uniform Federal Accessibility Standards.*

ing a tight grip, tight pinching, or twisting of the wrist. They shall be operable with a maximum force of 5 lbs (2.3 kg). Lever-operated, push-type and electrically operated faucets are examples of acceptable designs. If self-closing valves are used, the faucet shall remain open for at least 10 seconds.

Bathtubs

If bathtubs are provided, at least one shall meet the following requirements: There are requirements to provide seats, grab bars and clear space in front of the tub, as shown in Figure 5-24. Doors

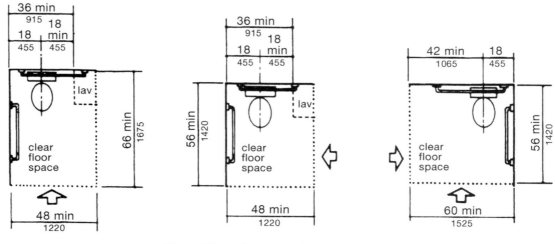

Clear Floor Space at Water Closets

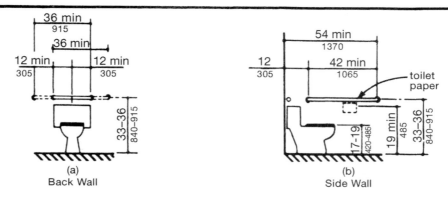

Grab Bars at Water Closets

FIGURE 5-22 ■ Design standards for a single toilet stall. *Courtesy Uniform Federal Accessibility Standards.*

shall not swing into the required clear space. Grab bars shall be per E in Figure 5-12. Faucets shall have controls as described above for lavatories and shall be located as shown in Figure 5-24. The seat must be securely mounted. See "Handrails, Grab Bars, and Tub and Shower Seats" later in this chapter for the structural requirements of the seat and grab bars. Tub enclosures may not obstruct the controls or transfer onto the seat. Enclosures shall not have tracks mounted on the rim of the tub.

Shower Stalls

If shower stalls are provided, at least one shall meet the following requirements: Shower stalls for hotels, motels and other transient lodging facilities shall be provided as shown in Figure 5-25. All other accessible showers shall be the size and shape shown in Figure 5-26. Doors shall not swing into the required clear space. Seats shall be located between 17" and 19" (430 and 485 mm) above the floor and shall be per Figure 5-27. When a seat is provided for in the 30 × 60" (760 × 1525 mm) shower configuration, it shall fold down and be located on the wall adjacent to the con-

trols. This will allow a wheelchair to roll into the shower or a person to use the built-in seat and be able to reach the controls. See Figure 5-24 and E in Figure 5-12 for grab bar locations. See "Handrails, Grab Bars, and Tub and Shower Seats" later in this chapter for the structural requirements of the seat and grab bars. Faucets shall have controls as described previously for lavatories and the controls shall be located as shown in Figure 5-28. Tub enclosures may not obstruct the controls or restrict transfer onto the seat. Shower spray units are required to be on 60" (1525 mm) minimum hose. They are required to be mounted in such a manner as to be usable as a hand-held unit or a wall-mounted unit. If an unmonitored shower site is subject to vandalism, a fixed shower head mounted 48" (1220 mm) above the floor can be used in lieu of a hand-held unit.

Curbs are not allowed at the entry to a 30 × 60" (760 × 1525 mm) shower. If a curb is installed at a 36 × 36" (915 × 915 mm) shower, it may not be higher than 1/2" (13 mm). The requirement not to have a curb or to have a very small curb means that the structural floor must be recessed at the shower to accommodate the slope of the shower floor.

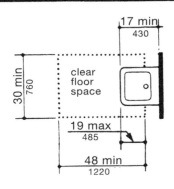

Lavatory Clearances

Clear Floor Space at Lavatories

FIGURE 5-23 ■ Placement of lavatories and mirrors. *Courtesy Uniform Federal Accessibility Standards.*

Toilet Room Accessories

At least one paper towel dispenser, air hand dryers, toilet seat cover dispensers and other similar fixtures in each restroom shall be located from 48" to 54" (1220 to 1370 mm) from the floor, as shown in Figure 5-29. Wall-mounted accessories need to be detailed for height and reach limit. Check the approach to accessories to verify whether front or side reach limits should be used.

Medicine Cabinets

If medicine cabinets are provided, they shall be located with a usable shelf no higher than 44" (1120 mm) above the floor. They shall have an accessible floor space in front of them.

Handrails, Grab Bars, and Tub and Shower Seats

Handrails and grab bars shall be 1 1/4" to 1 1/2" (32 to 38 mm) in diameter or a shape that provides an equivalent gripping surface and shall have a 1 1/2" (38 mm) clear space between them and adjacent surfaces, as shown in Figure 5-12. Handrails, grab bars and the surface behind them shall be free of sharp edges. Edges shall have a minimum radius of 1/8" (3 mm).

Grab bars and shower seats shall be capable of safely supporting a 250 lb (113 kg) load at any point. This will require a secure surface for mounting on and secure mounting brackets. Standard details should be developed to illustrate the required attachment. Grab bars may not rotate within their fittings.

Fixtures

The effect of ADA on drinking fountains, telephones, tables, and counters should be considered in the design process.

Drinking Fountains

At least half the drinking fountains provided on an accessible floor shall meet the requirements illustrated in Figure 5-30. These drinking fountains must also have a proper water flow direction and height. Because of the water flow requirements, it is important to choose a drinking fountain or water cooler that is certified by the manufacturer to meet ADA requirements. There must also be drinking fountains at a height appropriate for those people who cannot stoop over.

Telephones

If public telephones are provided, at least one shall meet the following requirements: The telephone shall be installed as shown in Figure 5-31. It shall be hearing aid compatible and have push button controls. The cord shall be at least 29" (735 mm) long.

Tables and Counters

If fixed tables and counters are provided, 5% of them must be designed for the disabled, as shown in Figure 5-32. Their height shall be between 28" and 34" (710 and 865 mm). The knee space shall be a minimum of 27" (430 mm) high, 30" (760 mm) wide and 19" (485 mm) deep.

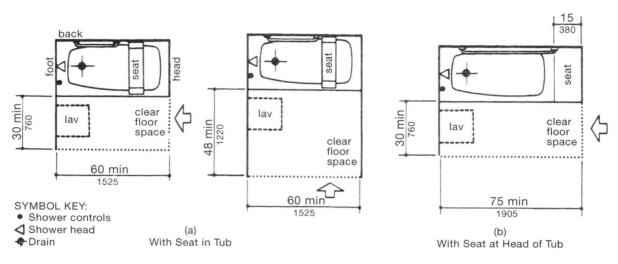

SYMBOL KEY:
● Shower controls
◁ Shower head
✚ Drain

(a)
With Seat in Tub

(b)
With Seat at Head of Tub

Clear Floor Space at Bathtubs

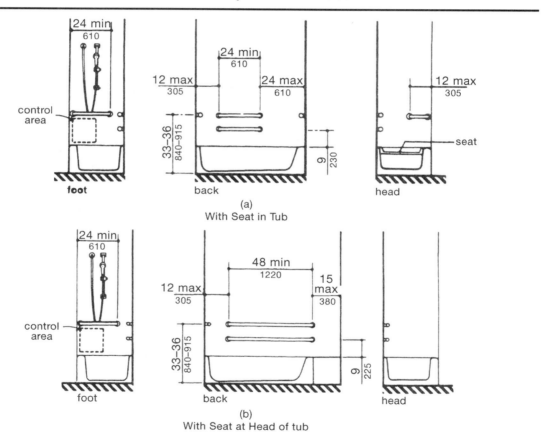

(a)
With Seat in Tub

(b)
With Seat at Head of tub

Grab Bars at Bathtubs

FIGURE 5-24 ■ Required sizes and locations of bathtubs and grab bars. *Courtesy Uniform Federal Accessibility Standards.*

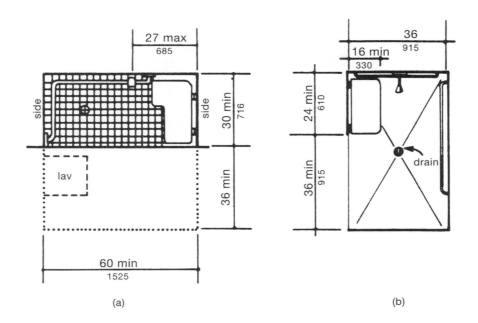

Roll–In Shower with Folding Seat

FIGURE 5-25 ■ A roll-in shower with folding seat. *Courtesy Uniform Federal Accessibility Standards.*

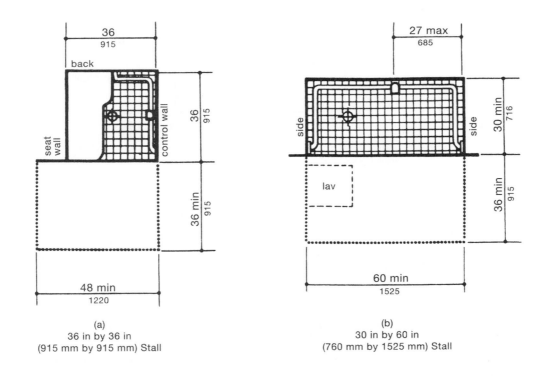

(a)
36 in by 36 in
(915 mm by 915 mm) Stall

(b)
30 in by 60 in
(760 mm by 1525 mm) Stall

Shower Size and Clearances

FIGURE 5-26 ■ Minimum shower size and clearances. *Courtesy Uniform Federal Accessibility Standards.*

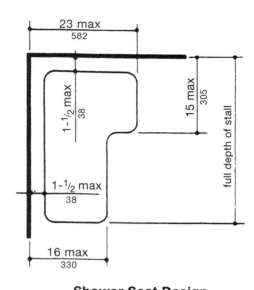

Shower Seat Design

FIGURE 5-27 ■ Shower seat design. *Courtesy Uniform Federal Accessibility Standards.*

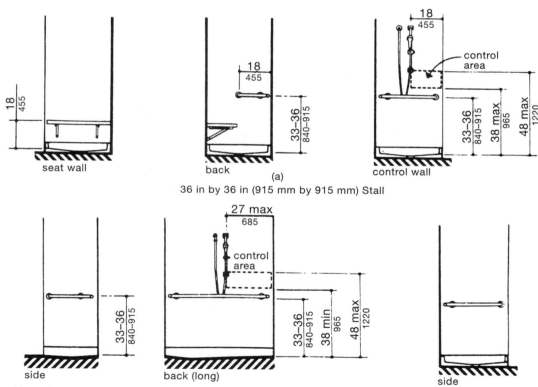

Grab Bars at Shower Stalls

FIGURE 5-28 ■ Minimum requirements for grab bars in showers. *Courtesy Uniform Federal Accessibility Standards.*

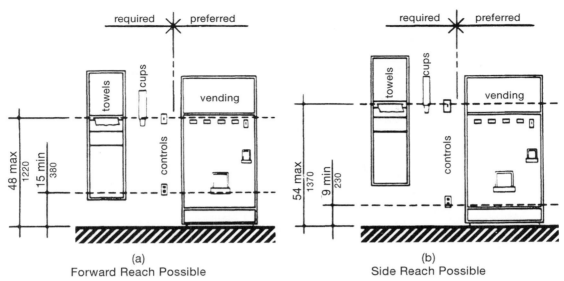

(a)
Forward Reach Possible

(b)
Side Reach Possible

FIGURE 5-29 ■ Mounting guidelines for bathroom accessories. *Courtesy Uniform Federal Accessibility Standards.*

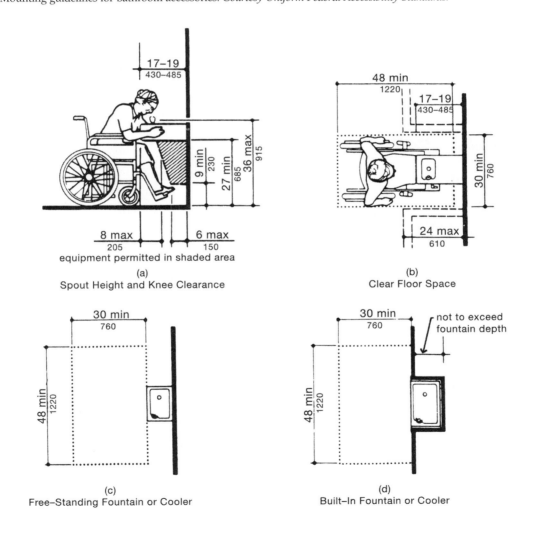

(a)
Spout Height and Knee Clearance

(b)
Clear Floor Space

(c)
Free–Standing Fountain or Cooler

(d)
Built–In Fountain or Cooler

Drinking Fountains and Water Coolers

FIGURE 5-30 ■ Requirements for placement of a drinking fountain. *Courtesy Uniform Federal Accessibility Standards.*

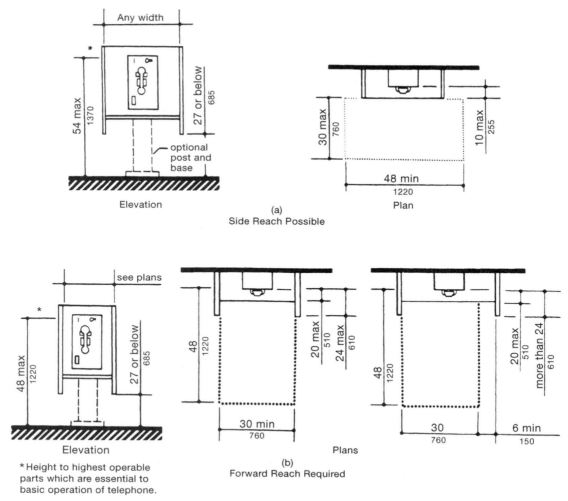

FIGURE 5-31 ■ If public telephones are provided, as least one must meet these minimum standards for placement. *Courtesy Uniform Federal Accessibility Standards.*

REQUIREMENTS BASED ON OCCUPANCY

Design parameters are also established based on the usage of a structure. Occupancies addressed by ADA include assembly areas, restaurants and cafeterias, medical care facilities, business and mercantile, libraries, and transient lodging.

Assembly Areas

Assembly areas such as theaters and auditoriums shall provide seating for people with disabilities as follows:

Capacity of Seating in Assembly Areas	Number of Required Wheelchair Locations
4 to 25	1
26 to 50	2
51 to 300	4
301 to 500	6

If there is seating for over 500, six wheelchair locations plus one additional space for each 100 spaces, or portion there of, over 500 shall be provided. These spaces shall conform to the shape shown in Figure 5-33. These seating locations shall be distributed in such a manner as to provide the disabled with a choice of admission prices and line of sight. Seating must be accessible and adjacent to at least one companion fixed seat. When the seating capacity exceeds 300, wheelchair spaces shall be provided in more than one place. Removable seats can be temporarily placed in wheelchair spaces. In addition to the wheelchair spaces, 1% of all fixed seats must be aisle seats with no armrest on the aisle side.

Accessible viewing positions can be clustered for bleachers, balconies, and other areas having sight lines that require slopes of greater than 5%. Equivalent accessible viewing positions can be located on levels having accessible egress.

Restaurants and Cafeterias

Where tables or counters are provided for dining or drinking, 5% shall be accessible as described in "Tables and Counters." Where

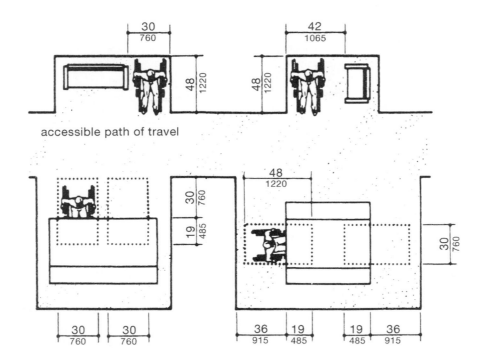

FIGURE 5-32 ■ If fixed tables are provided, a minimum of 5% must meet ADA standards. *Courtesy Uniform Federal Accessibility Standards.*

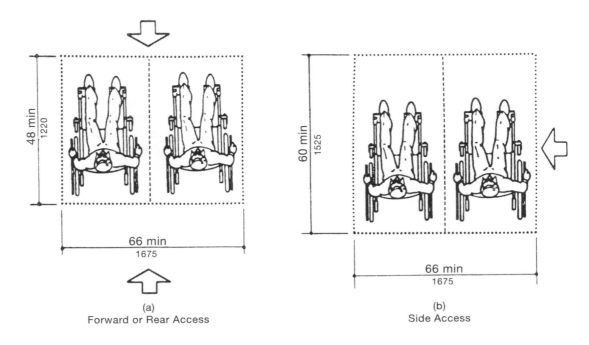

FIGURE 5-33 ■ Seating requirements for wheelchairs. *Courtesy Uniform Federal Accessibility Standards. .*

food or drink is served at counters higher than 34" (865 mm), a 5' (1525 mm) section shall be provided matching the requirements of Figure 5-32 or at a table in the same area. Seating for people with disabilities should be distributed in smoking and nonsmoking areas. All dining and drinking areas, except some mezzanines, shall be accessible. This includes raised, sunken, and outside areas. Raised platforms such as those provided for the head table at a banquet shall be accessible to the disabled. Mezzanines that are not in an elevator-serviced building are not required to be accessible if they meet the following conditions:

■ They are smaller than 33% of the area of the accessible area.

■ The same level of service and decor is available in the accessible area.

■ The accessible area is not limited to the disabled only.

Food-service lines shall be as shown in Figure 5-34. If self-service shelves are provided, they shall be within the reach range of the disabled as shown in Figure 5-35.

Medical Care Facilities

This section covers medical facilities where a patient might stay for a period of 24 hours or more. All public- and common-use areas are required to be accessible to people with disabilities. In general-purpose hospitals, psychiatric facilities, and detoxification facilities, 10% of the patient bedrooms and toilets are required to be accessible. In hospitals and rehabilitation facilities that specialize in treating conditions that affect mobility, all patient bedrooms and toilets are required to be accessible. At least 50% of patient bedrooms and toilets shall be accessible in long-term care facilities and nursing homes. Entrances to these facilities shall be covered and shall incorporate a loading zone per Figure 5-36.

Doors to accessible bedrooms in acute-care hospitals shall be exempted from the requirements for maneuvering space at the latch side of the door if the door is at least 44" (1120 mm) wide. Accessible bedrooms shall have a maneuvering space per Figure 5-20, and a 36" (915 mm) wide access route around three sides of the

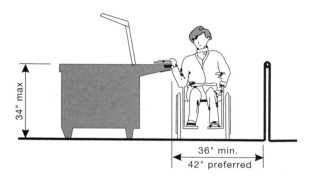

Food Service Lines

FIGURE 5-35 ■ Minimum sizes for food service areas.

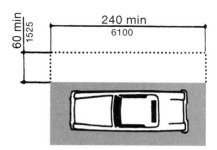

Access Aisle at Passenger Loading Zones

FIGURE 5-36 ■ Loading zones for patients at a medical facility. *Courtesy Uniform Federal Accessibility Standards.*

bed. Where patient toilets or bathrooms are provided, they shall meet the requirements discussed earlier in "Restrooms and Bathing Facilities."

Business and Mercantile

A 36" (915 mm) portion of the counter shall be provided at a maximum height of 36" (915 mm) in retail stores where counters have cash registers, at ticket counters, at teller stations in a bank, and at registration counters in hotels and motels. In lieu of this requirement, an auxiliary counter with a maximum height of 36" (915 mm) can be provided in close proximity to the main counter. Accessible checkout aisles shall be provided per Figure 5-37 unless the selling space is less than 5000 sq ft. When there is less than 5000 sq ft of selling space, one accessible checkout aisle shall be provided.

Security bollards or similar devices used to prevent the removal of shopping carts from the store shall not prevent access or egress to people in wheelchairs. An alternate entry that is equally convenient to that provided for the ambulatory population is acceptable.

Libraries

In reading and study areas, at least 5% of fixed seating, tables, and study carrels shall meet the requirements for tables and counters.

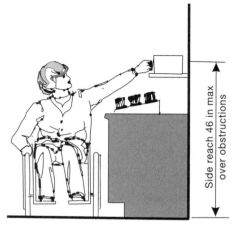

Tableware Area

FIGURE 5-34 ■ Guidelines for tableware areas.

Check-Out Aisles

(1) In new construction, accessible check-out aisles shall be provided in confomance with the table below.

Total Check-out Aisles of Each Design	Minimum Number of Accessible Checkout Aisles (of each design)
1–4	1
5–8	2
8–15	3
over 15	3, plus 20% of additional aisles

FIGURE 5-37 ■ Accessible checkout aisles. *Courtesy Uniform Federal Accessibility Standards.*

At least one lane at each checkout area shall meet the requirements for counters with cash registers, as described previously in "Business and Mercantile" above. Card catalogs and magazines shall comply with Figure 5-38 and stacks shall comply with Figure 5-39.

Transient Lodging

All public- and common-use areas in hotels, motels, inns, boarding houses, dormitories, resorts, and other similar lodging places are required to be accessible. A portion of the sleeping rooms or suites are required to be accessible per Figure 5-40. The roll-in shower required by this table shall conform to Figure 5-26. Accessible rooms or suites shall be dispersed among the various classes of accommodations. Selection can be limited to rooms for multiple occupancy if they are offered at single-occupancy prices for people with disabilities needing single occupancy.

Accessible rooms shall be on an accessible route. They shall have maneuvering space as shown in Figure 5-19. There shall be a 36" (915 mm) space on both sides of a single bed or a 36" (915 mm) space between two beds. In accessible suites, living rooms, dining areas, covered parking areas, one bathroom, and one sleep-

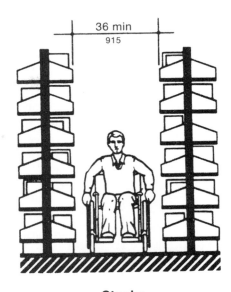

Stacks

FIGURE 5-39 ■ Placement requirements for book stacks in a library. *Courtesy Uniform Federal Accessibility Standards.*

ing area shall be accessible. Balconies, terraces or balconies shall be accessible unless higher thresholds or a change in level is necessary to protect the unit from wind or water damage (equivalent facilities must be provided). If storage cabinets are provided, at least one type of each shall be constructed per Figure 5-41.

Kitchens and wet bars, when provided, shall have counters no higher than 34" (865 mm) and at least 50% of the cabinets and refrigerator/freezer space shall be within the reach range shown in Figures 5-42 and 5-43. The space shall be designed to allow for the operation of cabinet and/or appliance doors so that all cabinets and appliances are accessible and usable.

Rooms shall be provided per Figure 5-44 for the hearing impaired, in addition to the rooms required for the disabled by Figure 5-40. Rooms or suites required to accommodate persons with hearing impairments shall have auxiliary visual devices to alert occupants of danger and phone calls. Phones shall have volume controls and an electrical outlet within 4' (1220 mm) of the telephone connection.

ADDITIONAL READING

The following Web sites can be used as a resource to help get additional information about the ADA and to keep you current with changes.

ADDRESS	COMPANY/ORGANIZATION
www.access-board.gov/adaag/html/adaag.htm	The Federal Government ADA Accessibility Guidelines for Buildings and Facilities
www.usdoj.gov/crt/ada/adahom1.htm	The Federal Government ADA Home Page

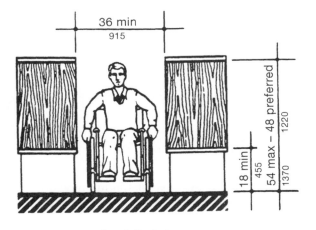

Card Catalog

FIGURE 5-38 ■ Standards for placement of card catalogues or other freestanding displays. *Courtesy Uniform Federal Accessibility Standards.*

Number of Room	Accessible Rooms	Roll in Showers
1 to 25	1	
26 to 50	2	
51 to 75	3	1
76 to 100	4	1
101 to 150	5	2
151 to 200	6	2
201 to 300	7	3
301 to 400	8	4
401 to 500	9	4 plus one for each additonal 100 over 400
501 to 1000	2% of total	
1001 and over	20 plus 1 for each 100 over 1000	

FIGURE 5-40 ■ Required accessible rooms are based on the total number of rooms. *Courtesy Uniform Federal Accessibility Standards.*

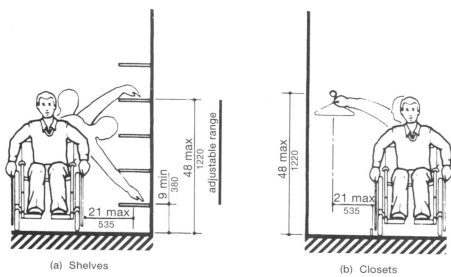

(a) Shelves   (b) Closets

FIGURE 5-41 ■ Guidelines for construction of storage cabinets. *Courtesy Uniform Federal Accessibility Standards.*

CHAPTER 5
Access Requirements for People with Disabilities

CHAPTER QUIZ

Answer the questions on a separate sheet of paper. Print the chapter title, question number, and a short complete statement for each question. If math is required to answer a question, show the work.

Question 5-1 Does a two-story office building require an elevator for people with disabilities?

Question 5-2 What is the minimum width required for a single van-accessible parking space?

Question 5-3 What is the minimum depth required for a disabled-accessible toilet space?

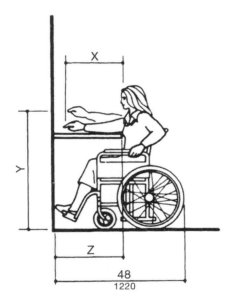

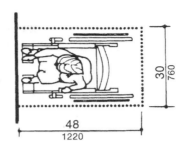

(a)
High Forward Reach Limit

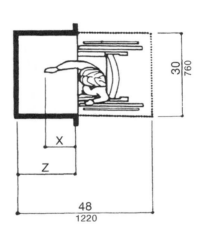

NOTE: X shall be ≤ 25 in (635 mm); Z shall be ≥ X. When X < 20 in (510 mm), then Y shall be 48 in (1220 mm) maximum. When X is 20 to 25 in (510 to 635 mm), then Y shall be 44 in (1120 mm) maximum.

(b)
Maximum Forward Reach over an Obstruction

Forward Reach

FIGURE 5-42 ■ Design standards for forward reach. *Courtesy Uniform Federal Accessibility Standards.*

Question 5-4 What is the maximum allowable cross fall, in decimals, of a foot for a 4'–0" wide accessible route?

Question 5-5 What is the maximum height allowed for a single run of a ramp?

Question 5-6 Why are laws providing access for the disabled more important today than they might have been 60 years ago?

Question 5-7 What is the minimum length of a 26" high ramp?

Question 5-8 What is the required height of a toilet seat?

Question 5-9 Explain the difference between the minimum turning room and the desired turning room for a 180° turn.

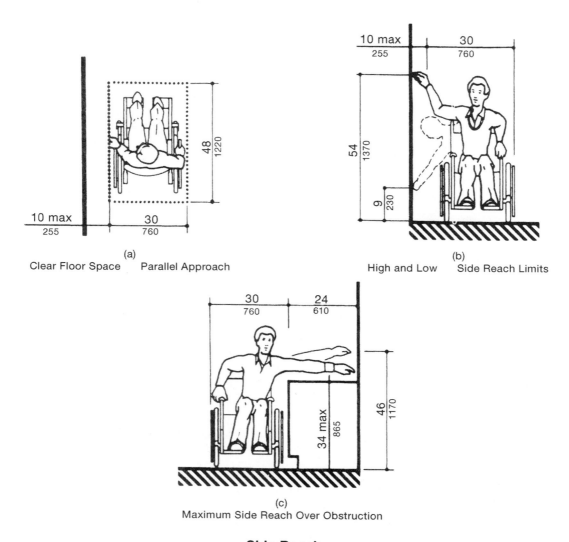

(a)
Clear Floor Space   Parallel Approach

(b)
High and Low   Side Reach Limits

(c)
Maximum Side Reach Over Obstruction

Side Reach

FIGURE 5-43 ■ Design standards for side reach. *Courtesy Uniform Federal Accessibility Standards.*

Number of Elements	Accessible Elements
1 to 25	1
26 to 50	2
51 to 75	3
76 to 100	4
101 to 150	5
151 to 200	6
201 to 300	7
301 to 400	8
401 to 500	9
501 to 1000	2% of total
1001 and over	20 plus 1 for each 100 over 1000

FIGURE 5-44 ■ Required number of rooms to accommodate the hearing impaired. *Courtesy Uniform Federal Accessibility Standards.*

Question 5-10 Give two reasons why an access route should be wider than the required 36".

Question 5-11 Define a public building within the scope of ADA.

Question 5-12 What aids does the ADA require for the hearing impaired?

SECTION 2

Building Methods & Materials of Commercial Drafting

CHAPTER 6

Wood Framing Methods and Materials

Wood and engineered wood products are a major material of construction for light commercial projects similar to the structure in Figure 6-1. To effectively work with the drawings that reflect wood construction, a thorough understanding of the materials and terminology of wood construction must be understood. This chapter will examine the terms that describe wood, uses of wood in light construction, specific code requirements for wood construction, and common terms relating to wood construction.

USES OF WOOD IN LIGHT CONSTRUCTION

Commercial construction is much less dependent on wood and engineered wood products because of the greater loads that must be supported and the greater need to protect the occupants of the structure. The use of wood is also restricted by the IBC based on the size of the structure and the occupancy of the structure. Multifamily housing projects, individual retail sales establishments, and office structures are common uses of wood construction.

FIGURE 6-1 ■ Wood is used to frame the skeleton and as a finishing material for many structures. Design by MLA Architects using redwood siding, photo by Wes Thompson. *Courtesy California Redwood Association.*

Building Code Restriction of Wood

Chapter 4 introduced the effect of codes on the selection of building materials. Wood construction with no special covering is considered *type V-B*. Notice in Figure 4-11 that a wood office structure (B occupancy) has an allowable basic area of 9,000 sq ft per floor. Chapter 11 will introduce materials that can be used to upgrade wood construction to a more fire-resistant type of construction. As materials are substituted to increase the fire resistance to *type III-B*, the size is increased to 19,000 sq ft per floor. Because more products must be used to enclose the wood frame and protect it from fire, the use of wood is typically not an economical choice for framing materials in structures that demand greater fire safety. Not only is the size of each floor restricted by the type of construction, but the number of floors allowed is limited.

Loading Limitations

Notice from Figure 4-11, a *type V-B* office structure has a basic allowable height of 2 levels high. The use of *type III-B* construction increases the basic allowable height to four floors or a maximum of 65' high (19800 mm). The size and height of a structure will also dictate the selection of building products because of the loads that will be generated. The IBC restricts unengineered wood bearing walls to a maximum of 10' (3050 mm) in height. Codes also require vertical wood wall members to be a minimum of 2 × 6 (50 × 150 mm) if more than one floor level and roof level is being supported. In addition to the loads generated from gravity, the structure must resist lateral loads. Loads imposed on a mid-rise or high-rise structure from wind, seismic, and overturning are typically too severe to be resisted by wood structures at an economical price. Because of its flexibility, availability, and economic features, wood is a popular building material. The natural beauty and warmth make wood is the material of choice for many low-level structures similar to Figure 6-2.

LUMBER TERMINOLOGY

An understanding of common wood terminology is essential for a thorough understanding of wood construction. The terms wood, lumber, and timber are often used interchangeably, but each has it own distinct meaning. *Wood* is the material that forms the trunk of trees. *Lumber* is produced as a result of milling wood by sawing, resawing, and planing. This would include materials of 2" and 4" (50 and 100 mm) thickness and 4", 6", 8", 10", 12" and 14" (100,

FIGURE 6-2 ■ Both hardwoods and softwoods can be found in this design by Sandy and Babcock, AIA. Photo by Jeremiah O. Bragstad. *Courtesy California Redwood Association.*

150, 200, 250, 300, 350 mm) depths. Several different terms based on the nominal size of the lumber are used to describe lumber. These terms include the following:

■ *Rough lumber* is lumber that has only been sawn, edged, and trimmed on each of its four longitudinal surfaces.

■ *Dressed lumber* is lumber that has been planed on one side (S1S), two sides (S2S) or a combination of sides and edges (S1S1E), (S1S2E), or (S4S).

■ *Boards* are wood products that are less than 2" (50 mm) thick but are wider than 2" (50 mm).

■ *Structural lumber* is lumber that is 2" (50 mm) thick or wider nominal but less than 5" (130 mm) wide for use where working stresses are required.

■ *Dimension lumber* is lumber that is 2" (50 mm) thick but less than 5" (130 mm) thick and is at least 2" (50 mm) wide.

■ *Timber* is lumber that is 5" (130 mm) or thicker and 5" (130 mm) or wider.

■ *Millwork* is a term used to described products made of wood such as door and window frame components, trim, mantels, and moldings.

Lumber and timber are typically milled from softwoods. Softwoods are needle-bearing trees such as firs, pines, and spruces. Millwork is typically made from hardwoods, which come from leaf-bearing trees such as oak and maple. It is important to remember that the terms hardwood and softwood do not refer to the hardness or softness of the wood but to the leaf or needle. Some softwoods are harder than some hardwoods and vice versa.

Wood Structure

Each year of growth is marked by a new layer of wood being added to the outer layer of the tree. This new growth is reflected by the addition of a growth ring each year. The size of the ring is affected by the growth conditions for that year. Short, dry growing seasons result in less growth and smaller rings, while larger rings result from warmer, moist weather. Each growth ring is also divided into two portions. The inner portion of the growth ring represents the growth in springtime. This wood is usually lighter in color and the wood is composed of large thin-walled cavities. The outer portion of each ring represents summer growth. This portion of the ring is a darker color and is denser wood. Growth rings can be seen in the ends of lumber and timber.

The rate of growth affects the strength of wood. Wood with narrow growth rings typically has higher strength ratings than wood with wider rings. Wood with a high proportion of summer growth is stronger than spring-growth wood. The grading of wood for structural purposes takes into account the rate of growth and the proportion of spring to summer wood.

Wood Grain

Grain is used to describe the pattern of wood fibers in wood products. Wood that has slow growth will have narrow rings and is said to be close grain. Course grain is used to describe wood with wide rings. Lumber grain is also referred to as straight grain or cross grain. These terms relate to the direction of the wood fibers in relation to the sides of a piece of lumber and are not related to the growth rings. *Straight grain* lumber has a grain pattern that runs parallel to the piece of lumber. The grain of *cross grain* lumber runs at an angle to the edge of the lumber.

Moisture Content

Building codes and the engineering specifications set limits for the moisture content of lumber to be used throughout the project. Before wood can be used as a lumber product, it must be dried. Freshly cut wood is described as *green wood*. Green wood contains water within each cell, known as free water. Green wood also contains moisture within the walls of each cell, which is referred to as absorbed water. As wood dries, the moisture within the cell evaporates first. As the evaporation of free water is completed, the absorbed water begins to evaporate. The completion of freewater evaporation is called the *fiber saturation point*, and results when approximately 25 to 30% of total moisture content has been removed. Once the fiber saturation point is reached, wood will begin to shrink and the strength will be increased. Wood dried to a

5% moisture content can range from two to three times its original strength in bending and crushing strength. A trade-off to the increase in strength is an increase of splitting along the grain.

Removal of moisture from wood will also increase its resistance to fungus. Moisture content of wood is expressed as a percentage and is determined by dividing the original weight by the oven-dried weight. Dry lumber is defined as lumber that has a moisture content of 19% or less and is represented by the symbol S-Dry. Wood with a higher moisture content is considered green lumber by the International Building Code.

Units of Measurement

Lumber is measured in nominal units expressed as board feet. One board foot is equal to 1" thick, 12" wide and 12" (25 × 300 × 300 mm) long. Lumber less than 1" (25 mm) thick is considered to be 1" (25 mm) thick. The number of board feet in a piece of lumber is determined by multiplying the nominal thickness in inches by the nominal width in feet by the length in feet. The nominal size of a piece of wood is larger than the actual size. Comparable sizes of nominal, seasoned and dry wood are shown in Table 6-1:

TABLE 6-1 ■ Board Thickness (Metric Sizes Given as Hard Conversions)

Nominal	Seasoned	Dry
1" (25 mm)	25/32" (20 mm)	3/4" (19 mm)
2" (50 mm)	19/16" (40 mm)	1 1/2" (38 mm)
3" (75 mm)	29/16" (65 mm)	2 1/2" (64 mm)
4" (100 mm)	39/16" (90 mm)	3 1/2" (65 mm)

Face Width		
Nominal	Seasoned	Dry
2" (50 mm)	19/16" (40 mm)	1 1/2" (38 mm)
4" (100 mm)	39/16" (90 mm)	3 1/2" (89 mm)
6" (150 mm)	5 5/8" (143 mm)	5 1/2" (140 mm)
8" (200 mm)	7 1/2" (191 mm)	7 1/4" (184 mm)
10" (250 mm)	9 1/2" (241 mm)	9 1/4" (235 mm)
12" (300 mm)	11 1/2" (292 mm)	11 1/4" (286 mm)
14" (350 mm)	13 1/2" (343 mm)	13 1/4" (337 mm)
16" (400 mm)	15 1/2" (394 mm)	15 1/4" (387 mm)

Standard lengths of lumber come in 2' (600 mm) increments starting in 4' (1200 mm) long material. Most lumber yards typically stock lumber in sizes from 8' through 20' (2400 through 6100 mm). Lumber in lengths of 22' (6700 mm) is available in some sizes.

Grading

Because different species of wood have such a wide variety of qualities, lumber must be graded to be certain that the structural requirements of the building code are met. *Grading* is the process of specifying the specific strength of a certain species of wood after it has been sawn, planed, and seasoned. Wood can be inspected by either visual or machined methods. Visual grading occurs when lumber is sawn at a mill. With visual inspection methods, wood is evaluated by ASTM standards in cooperation with the U.S. Forest Products laboratory. Structural lumber that is tested nondestructively by machine and graded is referred to as mechanically evaluated lumber (MEL). Graded lumber will be stamped with a grading symbol that represents the quality, species, use, strength and grading authority. Grading usually takes into account the size of knots or holes and their location, the size and location of splits, and the amount of warp in a piece of lumber. Most lumber is visually graded and given a designation of select structural, no. 1, no. 2, no. 3, construction, standard, utility, and stud. Wood is also divided into three classifications depending on how the lumber is to be used and its size These classifications are as follows:

Dimension—Dimensional lumber is either 2" or 4" (50 or 100 mm) thick and 2" (50 mm) wide used for joist or planks.

Beams and stringers—Lumber that has a nominal size of 5 × 8" (130 × 200 mm) or larger and is graded for strength in bending when loaded on the narrow face.

Post and timbers—Lumber that is square or nearly square in cross section with a nominal size of 5 × 5" (130 × 130 mm) or larger and graded for use as a post or column or other uses in which bending strength is not important.

WOOD FRAMING

Wood frame construction is often referred to as *stick construction* because the frame is made by adding one stick at a time. Most accurately known as western platform framing, the framing system is widely used throughout the United States. Framing members are typically 2" (50 mm) thick with a spacing of 12", 16", 19.2", and 24" (300, 400, 490, and 600 mm) (o.c.). Use of 24" (600 mm) spacings is limited to one-level bearing walls with roof framing members directly aligned with the studs. The length of the vertical members is determined by the distance from floor to floor. Figure 6-3 shows an example of western platform construction for a two-story structure.

Alternative lumber products known as *engineered lumber* are used throughout western platform construction to replace naturally grown lumber products. In order to understand the drawings involved with commercial drafting, an understanding of common framing terms is essential. These terms will involve those used for floor, wall, and roof construction.

Floor Construction

It is important to remember that most lower floor systems used with commercial construction are concrete slabs. Wood floors are used in cooler, damper areas of the country to frame the lower floor system for multifamily dwellings and retail sales outlets. Concrete floors and foundations will be covered in Chapter 10. Basic terms used to describe ground-level wood floor systems include mudsill, floor joist, girder, and rim joist. Each can be seen in Figure 6-4. Joist, rim joist, and blocking are also used to describe upper floors framed with wood members.

FIGURE 6-3 ■ Western platform construction for a multilevel office structure. *Courtesy Southern Forest Products Association.*

Mudsills

The first piece of wood that comes in contact with the concrete foundation is the *mudsill.* It is also referred to as a *base plate* or *sill* in different areas of the country. In order to be protected from moisture in the concrete, the mudsill is required by code to be pressure treated or made from foundation grade cedar or redwood. A 2 × 6 (50 × 150) is typically used for a mudsill. This is specified on the foundation plan with a note and shown on foundation details.

Anchor Bolts

The mudsill is bolted to the foundation with anchor bolts to resist uplifting and sliding. *Anchor bolts* are L-shaped bolts ranging in size from 1/2" to 3/4" (13 to 19 mm) diameter for most applications. Maximum spacing of anchor bolts allowed by code is 6'–0" (1800 mm) o.c. with a minimum length of 10" (250 mm). The loads to be resisted will affect the size and spacing of the bolts. A 2" (50 mm) diameter washer is placed beneath the nut to keep the mudsill from lifting off the bolts when forces of uplifting are applied. In seismic zones, metal anchors or straps similar to those shown in Figure 6-5 will also be specified on the foundation and framing details.

Girders

Once the mudsill is in place, the girders can be set. *Girders* are beams used to support floor joists as they span across the foundation. The floor joists typically rest on top of the girder. A girder can also be set so that the top of the girder aligns with the top of the floor joists. When a girder is set level with the joists it is called a *lush* girder. The weight of the floor joist must be supported by metal hangers. Each type of girder can be seen in Figure 6-6.

Girders are typically made from 4× or 6× (100× or 150×) members but they can also be constructed by 2× (50×) members joined together. If only two members are required to form the girder, they can be nailed together. If three or more 2× (50×) members are to be used, they must be bolted together with bolts that pass through the girder. Where supports need to be kept to a minimum, laminated wood beams called *glu-lams* can be used for girders. Glu-lams are made from sawn lumber that is glued together under pressure to form a beam that is stronger than its lumber counterpart. Beam widths of 3 1/8", 5 1/8", 6 3/4", and 8 3/4" (80, 130, 170 and 220 mm) are typically used. Chapter 7 will provide

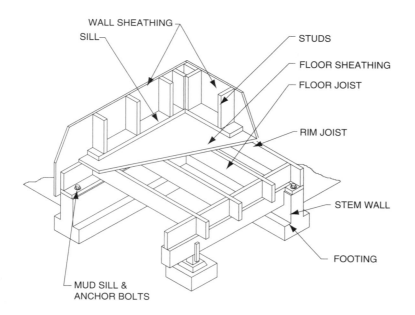

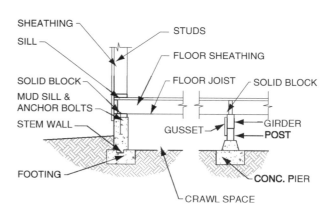

FIGURE 6-4 ■ Conventional floor framing members.

FIGURE 6-5 ■ Metal anchors are often used to bond wood framing members to the foundation. *Courtesy Western Wood Products Association.*

(PSL) is laminated from veneer strips of fir and southern pine, coated with resin and then compressed and heated. Typical widths include 3 1/2", 5 1/4", and 7" (90, 130, 180 mm), with depths ranging from 7" to 18" (180 to 460 mm). Figure 6-7 shows examples of PSL beams.

Steel girders are often used where foundation supports must be kept to a minimum. When steel girders are used, steel floor joists are also typically used. Steel framing methods will be introduced in Chapter 8.

FIGURE 6-7 ■ Beams made of laminated plywood veneers offer superior performance and durability over sawn lumber for supporting heavy loads. *Courtesy Louisiana-Pacific Corporation.*

an in-depth explanation of the use of laminated beams in construction. Regardless of the type of girder used, the width and depth are determined by the load to be supported and the span to be covered.

Girders can also be formed from a wide variety of engineered wood products. Unlike laminated beams that are laminated from sawn lumber, such as a 2 × 4 (50 × 100), *parallel strand lumber*

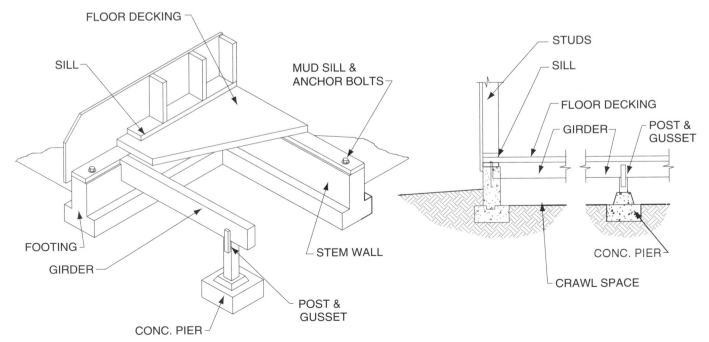

FIGURE 6-6 ■ A girder is used to provide support for floor framing members.

Posts

Wood posts are typically used to support girders made from wood products. As a general rule of thumb, a 4 × 4 (100 × 100) post is typically used below a 4" (100 mm) wide girder, and a 6 × 6 (150 × 150) wide post is used below a 6" (150 mm) wide girder. Post sizes vary based on the load and the height of the post. Solid wood posts are limited by code to a maximum height to depth ratio (*l/d*). That is the quotient resulting from the length or height of the column expressed in inches, divided by the width of the column. It cannot exceed 50. This is referred to as the slenderness ratio. Wood posts and columns fall into three different categories based on their slenderness ratio. Short columns fail in crushing and long columns fail in bending. For intermediate columns, the failure mode is indeterminate. The engineer will determine the required size and connectors to be used.

Posts made of laminated veneers of plywood range in sizes from 3 1/2" to 7" (90 to 180 mm), and are being used increasingly for their ability to support loads. A minimum bearing surface of 1 1/2" (38 mm) is required by code to support a girder resting on a wood support, and a 3" (75 mm) bearing surface is required if the girder is resting on concrete. Because a wooden post will draw moisture out of the foundation, it must rest on 55# felt although sometimes an asphalt shingle is used. If the post is subject to uplift or lateral forces, a metal post base or strap may be specified by the engineer

to firmly attach the post to the concrete. Steel columns can also be used to support girders. Columns are typically rectangular, round, or in the shape of an I. Chapter 8 will discuss the use of steel in construction.

Floor Joists

With the mudsills and girders in place, the framing crew can set the floor joists in place. Floor joists are the structural members used to support the subfloor or rough floor with a maximum span of approximately 20' (6100 mm). The exact span will depend on the size, spacing, and load to be supported. Floor joists range in size from 2 × 6 to 2 × 14 (50 × 150 to 50 × 360) and are typically placed at either 12", 16", 19.2" or 24" (300, 400, 490, or 600 mm) o.c. The spacing depends on the load to be supported and distance the joist will span. Figure 6-8 shows an example of a typical detail that a drafter would be expected to complete to represent western platform floor construction.

When a bearing wall is to be supported by floor joists that are parallel to the wall, building codes require the floor joists directly below the wall to be doubled as a minimum standard, and they often require the loads to be analyzed. Because of the decreasing supply and escalating price of sawn lumber, each of the national wood manufacturers have developed several alternatives to sawn lumber for floor joists. Three of the most common substitutes to

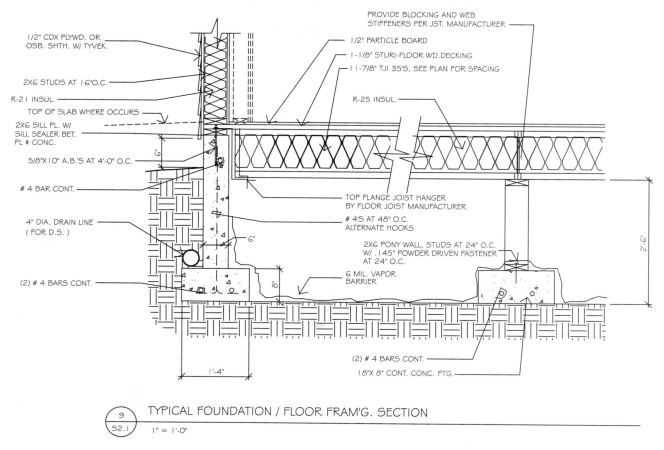

FIGURE 6-8 ■ A detail of a floor framed using truss-joists. *Courtesy Scott R. Beck, Architect.*

sawn floor joists include open-web trusses, I-joists, and laminated veneer lumber.

Open-web trusses are a common alternative to using 2 × (50×) sawn lumber for floor and roof joists. In addition to being able to support greater loads over longer spans, open-web trusses allow for easy routing of HVAC ducts and electrical conduit. Open-web trusses are typically spaced at 24", 32" or 48" (600, 800, or 1220 mm) for light commercial floors. Depths range from 14" to 60" (360 to 1520 mm) depending on the manufacturer and material used to fabricate the truss. Open-web trusses are typically available for floor spans of up to approximately 60' (18300 mm) depending on the load to be supported. Figure 6-9 shows an open-web truss supported by laminated beams. The horizontal members of the truss are called top and bottom *chords* respectively, and they are typically made from lumber or laminated veneer lumber (LVL). The diagonal members are called webs and are made of approximately 1" (25 mm) diameter tubular steel or 2× (50×) sawn lumber. When drawn in section or details, the webs are typically represented with a bold centerline drawn at a 45° angle. The exact angle is determined by the truss manufacturer and is usually unimportant to the detail. As the span or load on the joist increases, the size and type of material used is increased. When drawing sections showing open-web trusses, it is important to work with the manufacturer's details to find exact sizes and truss depth. Most truss manufacturers supply electronic copies of common truss connections. Figure 6-10 shows an example of a truss detail that can be used for a base to complete the truss-to-wall detail.

Trusses referred to as *I*-joists are a high-strength, lightweight, cost-efficient alternative to sawn lumber. I-joists (see Figure 6-11) form a very quiet floor system because they are exceptionally stiff and uniform in size, have no crown, and do not shrink. I- joists come in depths ranging from 9 1/2" to 37" (240 to 940 mm). I-joists are also able to span greater distances than comparably sized sawn joist and are suitable for spans up to approximately 40' (12 000 mm). Webs can be made from plywood or oriented strand

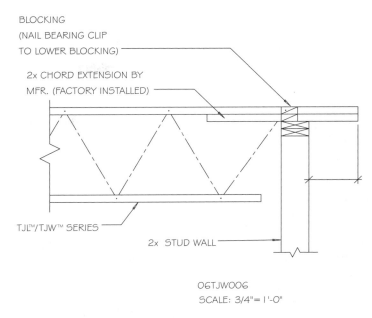

FIGURE 6-10 ■ Many lumber and truss manufacturers supply electronic copies of common connections. *Courtesy Trus-Joist MacMillan.*

board (OSB). It is made from wood fibers that are arranged in a precisely controlled pattern, coated with resin, and compressed and cured under heat. Figure 6-12a shows an example of a solid-web truss with a plywood web. Figure 6-12b shows a truss made with OSB. Holes can be placed in the web to allow for HVAC ducts and electrical requirements based on the manufacturer's specifications.

Laminated veneer lumber (LVL) is made from ultrasonically graded Douglas fir veneers that are laminated with exterior grade adhesives under heat and pressure, with all grains parallel to each other. LVL joist are 1 3/4" (45 mm) wide and range in depths of 5 1/2" to 18" (140 to 450 mm) deep. LVL joists offer the superior performance and durability of other engineered products and are

FIGURE 6-9 ■ Open-web trusses supported on laminated beams. *Courtesy Trus-Joist MacMillan.*

FIGURE 6-11 ■ I-Joists are a high strength, lightweight, cost-efficient alternative to sawn lumber.

FIGURE 6-12A ■ Solid plywood web trusses. *Courtesy Georgia-Pacific Corporation.*

FIGURE 6-12B ■ Trusses made with OSB (oriented strand board). *Courtesy Louisiana-Pacific Corporation.*

designed for single-span or multispan uses that must support heavy loads. Regardless of the material, floor joists are typically shown and specified on the appropriate floor plan and in framing details.

Floor Bracing

Because of the height-to-depth proportions of a joist, it will tend to roll over onto its side as loads are applied. To resist this tendency, a rim joist or blocking is used to support the edge of each joist at its exterior edge. A *rim joist* (sometimes referred to as a band or header) is usually aligned with the outer face of the foundation and mudsill. Sometimes a rim joist will be set around the entire perime-

ter and then end nailed to the perpendicular floor joist. *Solid blocking*, an alternative to a rim joist, is a block of wood used to span between two floor joists to transfer lateral loads. The International Building Code requires blocking between floor joists at 10' (3400 mm) maximum spacings and beneath any bearing walls resting on the floor. The exact spacing depends on the depth of the joist and if there is a finish on one or both sides. Solid blocking is also placed to help provide support for plumbing lines and HVAC ducts, as seen in Figure 6-13. Lightweight metal cross bracing can be substituted for blocking to resist lateral loads in floor joist if the blocking is not required for a fire stop. Floor blocking is typically specified in the written specifications and will be shown in the framing details.

Floor sheathing is installed over the floor joist to form the subfloor. The subfloor provides a surface for the walls to set on. Plywood laid perpendicular to the floor joist is usually used for the subfloor. Plywood ranging from 1/2 to 1 1/8" (13 to 30 mm) thick with an American Plywood Association (APA) grade of EXP 1 or 2, EXT, STRUCT 1 EXP 1 or STRUCT 1 EXT are most typically used for floor sheathing. EXT represents exterior grade, STRUCT represents structural and EXP represents exposure. Plywood is also printed with a number to represent the *span rating*, which represents the maximum spacing from center to center of supports. The

FIGURE 6-13 ■ Blocking is used to provide stiffness to a floor, resist lateral loads, and provide support to plumbing lines. *Courtesy John Black.*

span rating is listed as two numbers separated by a slash (for example, 32/16). The first number represents the maximum recommended spacing of supports if the panel is used for roof sheathing and the long dimension if the sheathing is placed across three or more supports. The second number represents the maximum recommended spacing of supports if the panel is used for floor sheathing and the long dimension of the sheathing is placed across three or more supports. Sheathing will be specified by the engineer in the calculations and must be specified by the drafter in the written specifications, the wall sections, and specific details.

Blocking is also used to provide added support to the floor sheathing. Often, walls will be reinforced to resist lateral loads. These loads are, in turn, transferred to the foundation through the floor system and are resisted by the diaphragm. A diaphragm is a rigid plate that acts similarly to a beam and can be found in the roof, wall, or floor system. In a floor diaphragm, $2\times$ ($50\times$) blocking is typically laid on edge between the floor joists to allow the edges of plywood panels to be supported. Nailing all edges of a plywood panel allows for approximately one-half to two times the design load to be resisted over unblocked diaphragms. The engineer will determine the size and spacing of nails and blocking required, and the drafter will make the notations on details similar to Figure 6-14.

Waferboard or *oriented strand board (OSB)* is a common alternative to traditional plywood subfloors. OSB is made from three or more layers of small strips of wood that have been saturated with a binder and compressed under heat and pressure. The exterior layers of strips are parallel to the length of the panel, and the core layer is laid in a random pattern.

Once the subfloor has been installed, an underlayment for the finish flooring is laid. The underlayment to the finish flooring is not installed until the roof, doors and windows have been installed, making the structure weather tight. The underlayment provides a smooth surface on which to install the finished floor and is usually 3/8" or 1/2" (10 or 13 mm) APA underlayment GROUP 1, EXPOSURE 1 plywood, hardboard or waferboard. Hardboard is referred to as medium- or high-density fiberboard (MDF or HDF) and is made from wood particles of various sizes that are bonded together with a synthetic resin under heat and pressure. APA STURD-I-FLOOR rated plywood of 19/32" to 13/32" (15 to 30 mm) thick can be used to eliminate the underlayment.

Framed Wall Construction

Walls are framed using a sole plate, studs, and a top plate. Walls are assembled in a horizontal position on the floor and then tilted into place. Exterior and load-bearing walls are usually assembled and located first. Interior nonbearing walls are usually assembled after the shell of the structure is completed.

Studs

Studs are the vertical members of the wall and are typically spaced at 16" or 24" (400 or 600 mm) o.c. Occasionally a spacing of 12" (300 mm) or smaller may be used, depending on the loads to be resisted. Typically 2×4s and 2×6s (50×100s or 50×150s) are used to frame walls, but 2×8s (50×200s) or larger can be used to support larger loads or hide other structural members such as a steel column within the wall. Stud-length lumber can be cut to either 88 5/8 " or 92 5/8 " or 96" (2250, 2350 or 2440 mm) long material. If a suspended ceiling is to be used, walls are often framed at 10' (3400 mm) or taller to allow for HVAC ducts and lighting fixtures. Walls over 10' (3400 mm) high must be blocked at mid-height. *Engineered studs* are made from short sections of stud-grade lumber that have had the knots and splits removed. Sections of wood are joined together with 5/8" (16 mm) finger joints. Engineered studs in 2×4 or 2×6 (50×100 or 50×150) are available in 8', 9', and 10' (2440, 2740 or 3400 mm) lengths. Laminated veneer studs and PSL studs are also available. Individual studs are not shown in plan view, but their location and spacing is typically indicated on the framing plan by the use of dimensions and notations.

Sheathing

In addition to transferring loads downward, studs are also used to support the interior and exterior wall materials. Interior material is typically 1/2" or 5/8" (13 or 16 mm) gypsum board for most light construction projects. Exterior coverings are typically supported by 3/8" or 1/2" (9 or 13 mm) thick plywood in colder regions. Plywood sheathing should be APA rated EXP 1 or 2, EXT, STRUCT 1, EXP 1, or STRUCT 1 EXT. Fiberboard or 1/2" (50 mm) OSB can also be used for sheathing in place of plywood. Composite products such as fiber-reinforced gypsum panels are also used for sheathing interior and exterior walls. In temperate regions, unless sheathing is required for structural reasons, the exterior siding can be applied to the studs. In either case, a vapor barrier such as Tyvek is applied to the exterior side of the studs prior to installing the exterior siding. Gypsum board can be added to the exterior side of an exterior wall to increase the fire rating of the wall, as described in Chapter 11. Figure 6-15 shows the components of an exterior

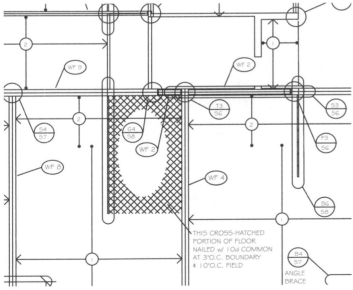

FIGURE 6-14 ■ Blocking and special nailing patterns are often shown throughout drawings by a drafter. *Courtesy Dean Smith, StructureForm Masters, Inc.*

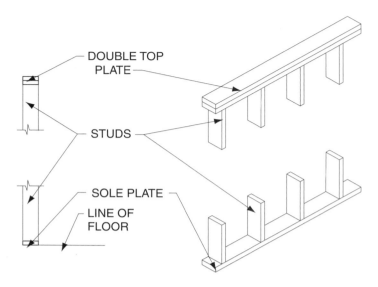

FIGURE 6-15 ■ Standard wall construction uses a double top plate on the top of a wall and a single sole plate at the bottom of the wall, and the vertical members are called studs.

wall. Stud size and spacing is usually specified on the framing plans, sections, and the framing details.

Bracing

In temperate regions, and where sheathing is not required to resist lateral loads, the studs can be kept in a vertical position by the use of a let-in brace. A *let-in brace* is a diagonal board that is placed in a diagonal position across the studs. Typically a 1 × 4 or 1 × 6 (25 × 100 or 25 × 150) is placed in a notch that is placed in the studs so that the brace is flush with the exterior face of the studs. A metal strap, which lies diagonally across the studs, can usually be substituted for the let-in brace. The IBC typically requires the brace or strap to cross a minimum of three studs and tie into the top and bottom plates of the wall. If plywood siding rated APA STURD-I-WALL is used, no underlayment or let-in braces are required. Bracing is shown on a framing plan. Let-in braces are shown on the framing plan and occasionally shown on the exterior elevations.

Sole Plate

The *sole plate* or *bottom plate* supports the studs and is used to help disperse the weight of the wall across the floor. The sole plate is also used to help hold the wall together as it is moved into its vertical position. A 2× or 3× (50× or 75×) is usually used for a sole plate. The size is dictated by the loads being supported and the material used for the finished floor. If a lightweight concrete is to be used over the plywood subfloor, the base plate may be thicker and made from pressure-treated material. Figure 6-16 shows a detail of a wall constructed with a double sole plate. The upper plate holds the studs together, and the lower pressure-treated plate protects the wall from the moisture in the concrete floor. The sole plate is end nailed into the studs while the wall is horizontal, and then nailed to the floor system. The sole plate is nailed into the plywood and

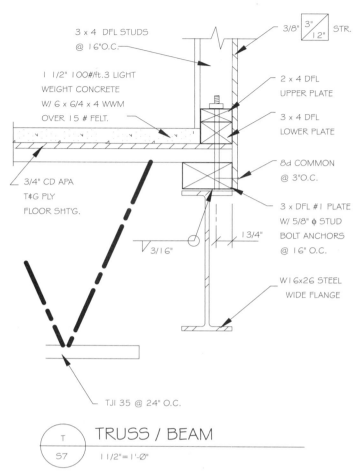

FIGURE 6-16 ■ When a concrete slab rests beside the sill, a pressure-treated sill is used below a standard sill. *Courtesy Ginger Smith, StructureForm Masters, Inc.*

rim joist below the exterior walls and into solid blocking or double joist below interior partitions. Nailing is specified in details or by a nailing schedule. Special nailing that is required to transfer lateral loads through the sole plate to the floor system is specified by the engineer in the calculations and indicated on the framing plan in a method similar to Figure 6-17, or shown in framing details.

Top Plate

The *top plates* are the horizontal members used at the top of a wall to tie the studs together and to provide a bearing surface for the roof or upper floors. Two top plates are required for most bearing walls. The lower plate is used to tie walls to each other. The upper top plate must lap the lower plate at splices by 48" (1200 mm). The upper plate can be omitted if the single plate is tied together by a steel strap at each splice. As seen in Figure 6-18, both plates can be eliminated and substituted with a flush header, if the header is connected to the top plate at each end by a steel strap. Although this situation is not typically used, it allows a window to be set at its highest position in a wall. Top plates are usually shown on wall sections. A flush header is usually specified on the framing plan and in the framing details.

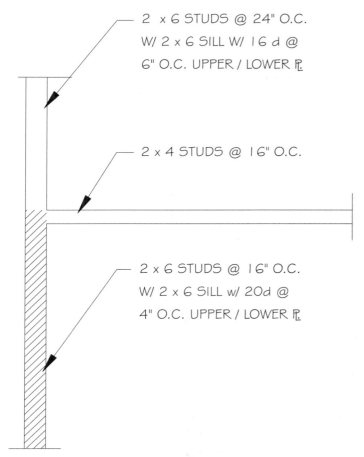

2 x 6 STUDS @ 24" O.C.
W/ 2 x 6 SILL W/ 16 d @
6" O.C. UPPER / LOWER PL.

2 x 4 STUDS @ 16" O.C.

2 x 6 STUDS @ 16" O.C.
W/ 2 x 6 SILL w/ 20d @
4" O.C. UPPER / LOWER PL.

FIGURE 6-17 ■ When special wall construction is required, the drafter will need to specify construction methods throughout the drawings.

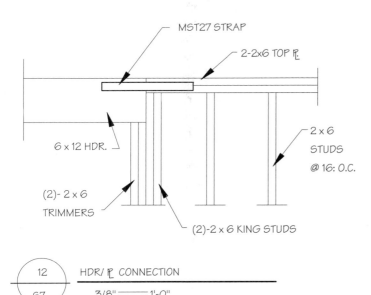

MST27 STRAP

2-2x6 TOP PL.

6 x 12 HDR.

2 x 6 STUDS @ 16: O.C.

(2)- 2 x 6 TRIMMERS

(2)-2 x 6 KING STUDS

12	HDR/ PL. CONNECTION
S7	3/8" = 1'-0"

FIGURE 6-18 ■ A header can be set level with the top of the top plates if a metal strap is used to tie the header to the top plate. The strap can be placed beside or on top of the header depending on the design of the engineer.

Framing Components for Wall Openings

Several additional terms need to be understood to frame an opening in a wall. These terms include header, trimmer, king stud, subsill, and jack stud. Each component can be seen in Figures 6-3 and 6-19. The drafter typically shows each component on framing details and sections.

Header

A *header* is the horizontal piece of lumber used to support the loads over an opening in a wall. When a hole is framed for a door or window, one or more studs must be omitted. A header carries the weight that the missing studs would have carried. Headers are usually the same width as the studs they replace. Headers can be made from sawn lumber, laminated lumber, parallel-strand lumber, or steel. The load to be supported and the span will usually dictate which material is used. The size of the header is specified on the framing plan as well as in the sections and specific details.

Opening Supports

The header is supported at each end by a *trimmer*. It is like a regular stud except that it has been cut shorter to fit beneath the header. A trimmer can be made from one or more 2× or one 4× (50× or 100×), depending on the loads to be supported. In addition to supporting the weight of the header, a trimmer provides a nailing edge for the door or window frame, as well as the interior and exterior finish materials. A *king stud* is normally the same length and size as other full studs and is placed beside the trimmer to help resist the tendency of the trimmer to bend under its load. The king stud also provides stability from lateral pressure applied to the side of the header, which causes a hinge point at the header/trimmer connection. (See Figure 6-20.) Single trimmers and king studs are not specified on the framing plans. Larger or double trimmers and king studs that are required to support the header are specified on the framing plan. King studs and trimmers may also be shown on the framing details if connections are required from one level to another level, as shown in Figure 6-21.

The subsill is a horizontal support placed between the trimmers to support the lower edge of a window. The subsill will be the same width as the studs it is replacing. The subsill is supported on *jack studs*, which are studs that are less than full height. Neither component is typically shown on the framing plan, but they are often shown in the architectural and structural details.

Roof Construction

Wood roof framing includes both conventional and truss framing methods. Each has its own special terminology, but many terms also apply to each type of construction. Roof terms common to conventional and trussed roofs include eave, cornice, eave blocking, baffle, fascia, ridge, sheathing, finished roofing, flashing, and roof pitch. Each is illustrated in Figure 6-22. Each component is usually shown in sections and framing details.

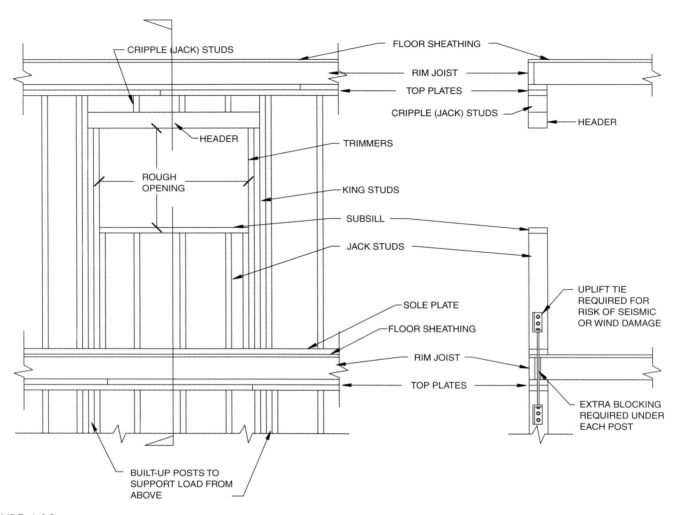

FIGURE 6-19 ■ Construction components of a wall opening.

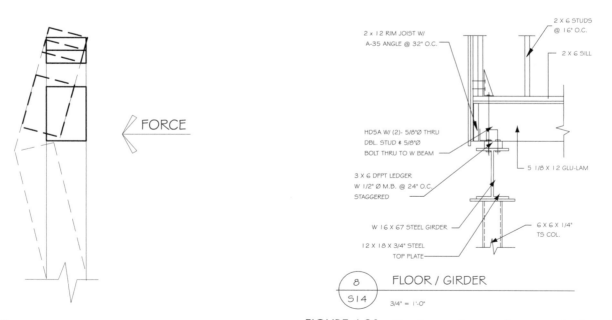

FIGURE 6-20 ■ King studs are placed beside the trimmers to support the header to resist lateral pressure.

FIGURE 6-21 ■ Framing members are often required to be anchored to members of lower levels to resist seismic activity.

Common Roof Terms

The *eave* is the portion of the roof that extends beyond the walls. The *cornice* is the covering that is applied to the eaves. A common method of enclosing the eave is shown in Figure 6-22. When a cornice is provided, a vent must be provided to allow cool air into the attic or rafter space. When the eave is not enclosed, an *eave block* or *bird block* must be provided to keep birds and small animals from entering the space between the framing members. The block also keeps the spacing of framing members uniform and provides an even termination to the top of the siding. To allow cool air into the attic or rafter space, a vent must be provide in the bird blocking. To keep roof insulation from plugging the eave vent, a *baffle* is placed between the insulation and the vent. Baffles are typically a piece of scrap wood or plywood.

The *fascia* is a trim board made from 1× or 2× (25× or 50×) material that is used to hide the rafter or truss tails. The fascia is often 2" (50 mm) deeper than the members it hides and runs parallel to the wall and perpendicular to the roof framing members.

The fascia also serves as a support to a gutter in wetter climates, although gutters are available that replace the fascia. The fascia is shown on the elevations, sections, and framing details. At the opposite end of the roof framing member is the ridge. The ridge is the highest part of a roof and is formed by the intersection of the rafters or the top chords of a truss.

Roof sheathing is used to cover the structural members and provide a base for the exterior finishing material. Either solid or skip roof sheathing is used, depending on the type of roofing material to be used and the area of the country where the structure will be built. Solid sheathing is typically 1/2" (13 mm) thick CDX plywood with span ratings of 24/16, 32/16, 40/20 and 48/24. OSB is also used in many areas of the country for roof sheathing that is not over an exposed eave. Because the material at the eave will be affected by moisture, CCX plywood or 1" (25 mm) T & G lumber must be used.

Skip sheathing is used on sloped roofs to support tile or cedar shake roofing. In warm, temperate climates 1 × 4s (25 × 100s) at 7" (180 mm) o.c. are placed directly over the roof framing members, as seen in Figure 6-23. In cool, damp climates the skip sheath-

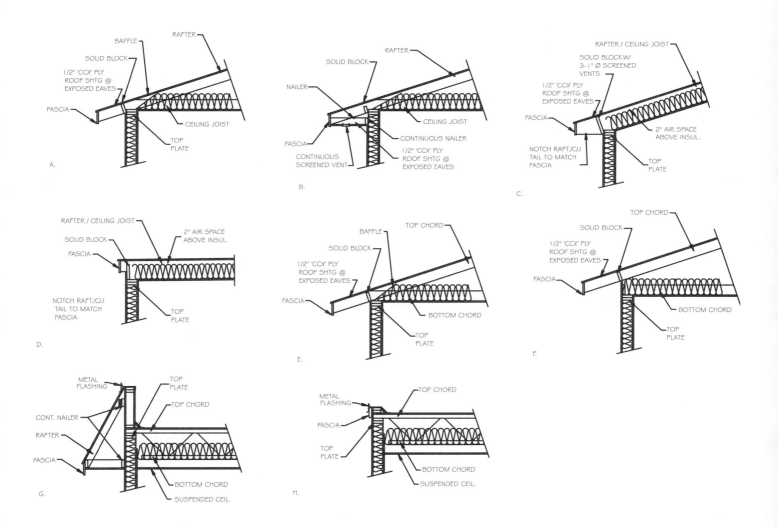

FIGURE 6-22 ■ Common methods of eave construction.

FIGURE 6-23 ■ Skip or spaced sheathing is used under shakes and tile roofs.

ing is applied over continuous plywood sheathing. Once the skip sheathing is in place, 15# building paper is applied as a base for the finished roofing.

The *finished roofing* is the weather protection system and includes materials such as built-up hot asphalt, asphalt and fiberglass shingles, cedar shakes, concrete and clay tile, sprayed and poured concrete, foam, and metal panels. Chapter 17 will review each roofing material and the information that the drafter will need to specify on the architectural and structural drawings.

Two other considerations of roof design are pitch and span. Each is illustrated in Figure 6-24. *Pitch* is used to describe the slope of the roof and is listed in terms of *rise over run*. A 3/12 roof will rise 3" for each 12" horizontal run. The type of roofing material and the local building code will affect the minimum and maximum pitch to be used. A minimum pitch of 1/4" per 12" is required for flat roofs with a built-up hot asphalt roof. A 3/12 pitch is the minimum slope recommended by most manufacturers for other roofing materials without requiring special waterproofing materials under the finished roofing. Pitch is specified on the elevations, sections, details, and roof plan.

Span is used to determine the horizontal run of a piece of lumber or truss. Span is the distance from the center of required bearing points.

Conventionally Framed Roof Terms

Conventional or stick framing involves the use of wood members placed in repetitive fashion one board at a time. Key terms to be understood and represented on the structural drawings include rafters, ridge board, and ceiling joist. Each component is shown in section and framing details. Ceiling joists, rafters and rafter/ceiling joists are also shown on the appropriate framing plan.

Rafters

Rafters (raft.) are the inclined members used to support the weight of the finished roofing. Rafters are typically spaced at 24" (600 mm) o.c.; however, 12" 16" and 19.2" (300, 400, and 490 mm) spacings are also used. The span, rafter size, and material being supported will determine the spacing. Rafters are typically 2× (50×) material with depths ranging from 6" to 12" (150 to 300 mm) deep. Just as with floor joists, engineered products such as I-joists, open-web trusses, OSB, and plywood trusses can also be used for rafters. The engineer will determine the type and size of the rafter, based on the span and the material to be supported. The location of the rafter in the roof will affect the name of the rafter. Figure 6-25 shows several different terms given to rafters based on their use in the roof frame.

A notch called a *bird's mouth* is cut into a sawn rafter where the rafter is to be supported on a wall or beam. The bird's mouth increases the bearing surface of the rafter and spreads the weight of the roof evenly over the wall. A bird's mouth is not specified within the drawings but is placed by normal construction practice. A bird's mouth is not placed in engineered materials. The materials are often required to sit on a top plate that has been cut to provide an inclined surface.

Roof Support

The *ridge board* is a horizontal board placed at the ridge to support the rafters. The ridge board is typically 2" (50 mm) deeper than the rafters. The ridge board does not support the weight carried by the

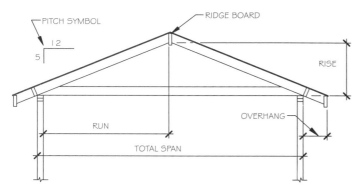

FIGURE 6-24 ■ Roof dimensions needed for construction.

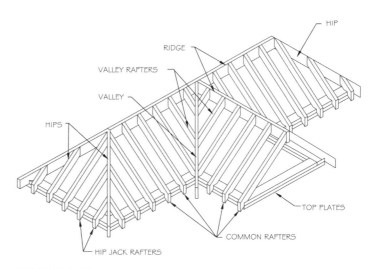

FIGURE 6-25 ■ Roof members in conventional construction.

rafters but provides a bearing surface to transfer force from a rafter to an opposing rafter. The ridge board is supported by a 2× (50×) brace that rests on a bearing wall. The brace resists any downward force and is usually a 2× (50×) member placed at 48" (1200 mm) o.c. The ridge brace must be within 45° of vertical to effectively transfer weight downward. A *purlin* can be used to reduce the span of a rafter. A purlin is placed below the rafter and supported by a purlin brace. The purlin is usually the same size as the rafters being supported. A *purlin brace* transfers the weight from the rafter and purlin into a bearing wall. A purlin brace must be set within 45° of vertical. If no bearing wall is near, a strongback can be used to support the weight from the rafters. A *strongback* is a beam placed in or above the ceiling to support roof weight. Each of these roof components is illustrated in Figure 6-26. Each term is specified in the sections.

The lower end of a rafter is nailed to the top plate of a wall while the upper end rests against the ridge. The natural tendency of the rafter is to rotate downward around the fixed lower end of the rafter, as seen in Figure 6-27. The downward rotation is resisted by the rafter on the opposite side of the ridge. If the rafter cannot rotate downward, the force of the rafter will attempt to force the bearing walls outward. This outward thrust of the rafters is resisted by a *ceiling joist*. Ceiling joists (C.J.) rest on the top plate of the wall and span between the bearing walls. The connection of the ceiling joist to the top plate and the rafters is necessary to keep the roof from sagging and the walls from separating. In addition to resisting the outward force of the rafters, ceiling joists also provide support for the finished ceiling.

If the forces of outward thrust are great enough, a collar tie may be added to the upper third of the roof. A *collar tie* attaches to two opposing rafters and is used to resist the outward force of the rafters. The collar tie is usually the same size as the rafters, but the engineer will determine exact sizes based on the spans and loads supported by the rafters. Ceiling joists and collar ties are illustrated in Figure 6-26.

Vaulted Ceilings

The size of a rafter must be increased to support both the weight of the roof and the finished ceiling if a roof is to be vaulted. The rafters

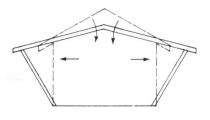

(a)
GRAVITY AND THE WEIGHT OF ROOFING MATERIALS TEND TO TRY TO FORCE THE RAFTERS DOWNWARD AND OUTWARD.

(b)
A CEILING JOIST IS USED TO RESIST THE OUTWARD FORCES FROM THE RAFTERS.

FIGURE 6-27 ■ Roof actions and reactions.

used with a vaulted ceiling are called *rafter/ceiling joists*. Rafter/ceiling joists (raft./C.J.) must be of sufficient size to contain the required insulation and still provide a minimum of 2" (50 mm) of air space above the insulation for ventilation. Rigid insulation or foam sandwich panels can be installed over the rafter/ceiling joists as an alternative method of placing the insulation. Because of the weight to be supported and the lack of interior supports, the rafter/ceiling joists will be supported by a *ridge beam* rather than just a ridge board. Because there are no ceiling joists to resist the outward force from the rafter/ceiling joists on the bearing walls, the rafters must be securely anchored to the ridge beam. Figure 6-28 shows two common methods of attaching the rafter/ceiling joists to the ridge beam. In areas subject to high winds or seismic activity, the lower end of the rafter/ceiling joist will be connected to the top plate with a metal angle. This angle helps to form a secure bond between the roof and wall. Figure 6-29 is an example of a detail showing the rafter/plate connection.

Roof Openings

If an opening is to be framed in the roof, a header and trimmers will need to be specified. Each can be seen in Figure 6-30. A *header* at the roof level consists of two or more members nailed together to support the rafters at the upper and lower edges of the opening. The headers are usually the same size of the rafters. The *trimmers*

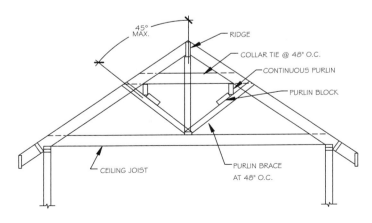

45° MAX.
RIDGE
COLLAR TIE @ 48" O.C.
CONTINUOUS PURLIN
PURLIN BLOCK
CEILING JOIST
PURLIN BRACE AT 48" O.C.

FIGURE 6-26 ■ Common roof supports.

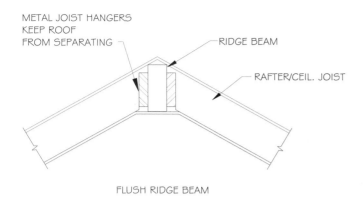

FLUSH RIDGE BEAM

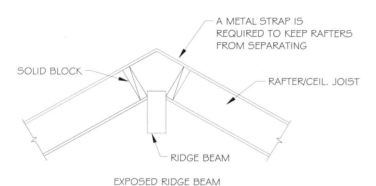

EXPOSED RIDGE BEAM

FIGURE 6-28 ■ Connections between the ridge beam and rafters.

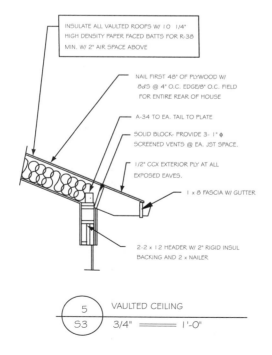

VAULTED CEILING

5
S3   3/4" ════ l'-0"

FIGURE 6-29 ■ Drafters often draw details showing roof-to-wall connections.

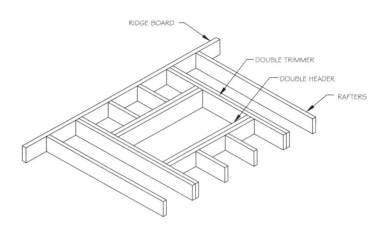

FIGURE 6-30 ■ Framing around an opening in the roof requires the use of headers and trimmers.

consist of two or more rafters nailed together to support the sides of the opening.

Truss Construction Terms

Although truss construction is considered non-conventional framing, most roofs are framed with trusses. A *truss* is a premanufactured component composed of triangular shapes arranged in a single plane used to span large distances. Because of the triangular shapes, loads will result in stress at the intersection of members in either compression or tension. Loads are typically spread throughout the truss so that the ends of the chords are the only portion of the truss that require support. When a truss is used, rafters, ceiling joists, purlins, and braces are eliminated for that portion of the structure. Trusses and conventional framing systems can both be combined in the same structure to form irregular shapes.

Truss Components

Major components of a truss include the top and bottom chords, the webs, and gussets. Figure 6-31 shows common truss components. The *top chords* of a truss are the members that support the loads from the finished roofing material. The *bottom chord* is the lower member of a truss and is typically a horizontal member. The *webs* are the interior members that transfer the loads between the top and bottom chords. Because a truss is formed as a single component, no ridge board or beam is required for support at the ridge. A *ridge block* is place between individual trusses to maintain a uniform spacing, resist the tendency of the trusses to fall sideways, and provide a nailing surface for the roof sheathing. Components of trusses framed with $2\times(50\times)$ material are typically not specified on the structural drawings but are shown on the drawings supplied by the manufacturer.

The spacing of the trusses will determine the size of the chords and webs. Common spacing for trusses include 24" and 32" (600 and 800 mm) o.c. and allow the chords and webs to be made of $2\times$ ($50\times$) material. As the spacing of the trusses increases beyond 32" (800 mm) o.c., timber or steel is used to frame the trusses. A *gusset* is a side plate that is often used at the intersection of a chord to

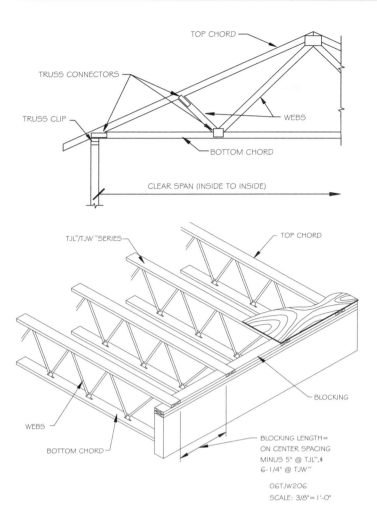

FIGURE 6-31■ ■ Standard truss connection members.

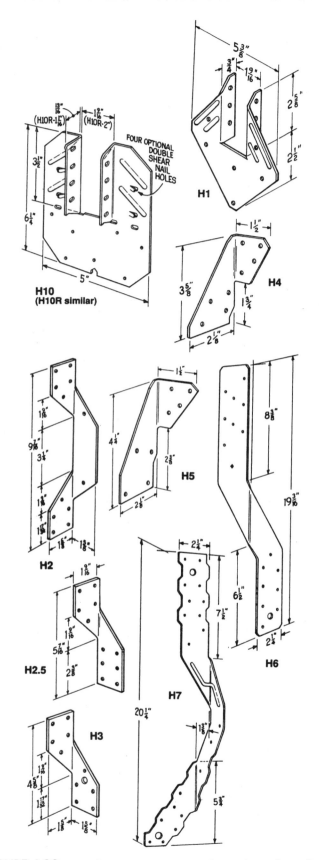

FIGURE 6-32 ■ Metal connectors are required to tie the roof to walls to resist wind and seismic loads. *Courtesy Simpson Strong-Tie Co., Inc.*

a web to form a secure connection. On trusses framed from 2× (50×) material, the gusset is typically a lightweight metal plate. The gusset and other truss components are typically sized by the truss manufacturer for trusses framed from 2" (50 mm) lumber. Timber-framed trusses are typically designed by the structural engineer. Chapter 7 will provide examples of timber trusses.

Although not actually a component of the truss, a *truss clip* or *hurricane tie* is a critical element in attaching the truss to its bearing surface. Figure 6-32 shows examples of common truss clips that are used to resist uplift and seismic motion. The type and size of the clip will be determined by the engineer, based on the amount of force of uplift to be resisted. The clip can be specified by a note on the framing plan and is also represented in details showing the truss bearing points.

Common Types of Trusses

Figure 6-33 shows common truss shapes used in wood frame commercial construction. Each type of truss shown with parallel chords can be built with an inclined top chord to facilitate roof design and water drainage without altering the design of the truss. It is important to remember that the exact number of webs will vary for each application and be determined by the span of the truss. Whatever the configuration of the truss, it is shown on the framing plan as shown in Figure 6-34.

A *fink* or standard truss (at A in Figure 6-33) is used to frame a gable roof for projects such as multifamily dwellings and retail sales outlets. The top chord is always inclined, but the interior webs can be arranged in different configurations to allow for more storage

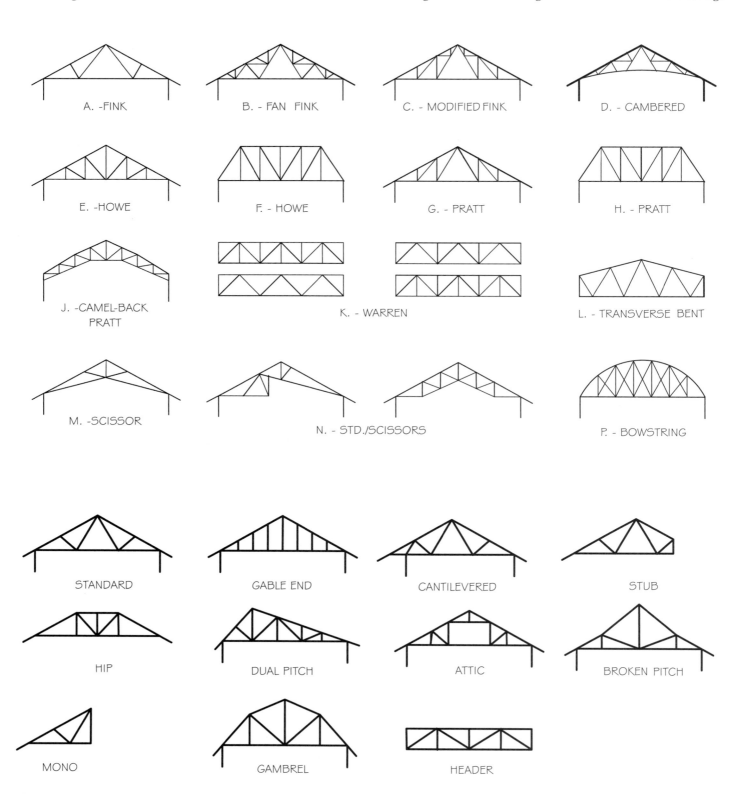

FIGURE 6-33 ■ Common shapes and types of trusses.

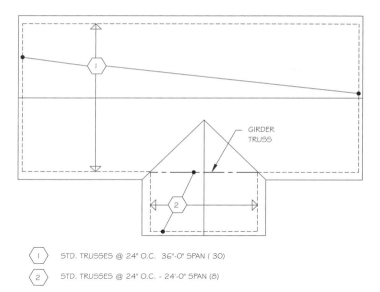

FIGURE 6-34 ■ Trusses are represented on framing plans by showing the span and limits of each type of truss.

space in the center of the truss (see B and C in Figure 6-33). The openness of the modified fink can be useful for attic storage as well as a location for HVAC equipment. A *camber*, which is a curve, can also be built into the bottom chord (see D in Figure 6-33) to assist in supporting loads.

Similar to a fink truss are *howe* and *pratt trusses* (E and G in Figure 6-33). Each of the trusses can be framed with an inclined top chord or a partially inclined top chord with a horizontal chord (F and H in Figure 6-33). The inclined chords are used for framing roofs, and the horizontal top chords are typically used for floor framing. The difference between pratt and howe trusses lies in the arrangement of the webs and the method of transferring loads throughout the truss. The partial flat top chord can be used to form hip roofs. (See Chapter 17.) Pratt trusses can also combine inclined and flat bottom chords as seen at J in Figure 6-33 to form a *camel-back pratt truss* Other flat top chord trusses include *warren trusses* (at K in Figure 6-33). Warren trusses are typically used for floor trusses. The top chord can be slightly inclined for use as roof trusses that are especially useful in dry climates where roof drainage is not heavy. A *transverse bent* style truss (L in Figure 6-33) has top chords with two slightly inclined top chords to facilitate drainage.

A *scissor truss* has top and bottom chords that are inclined, allowing for a vaulted ceiling to be framed. The bottom chord usually must be at a minimum of two pitches less than the top chord. If the top chord is to be set at a 6/12 pitch, the bottom chord cannot be steeper than a 4/12 pitch. This guideline can vary as the chords and webs are strengthened. Figure 6-33 at M shows a scissor truss with equal interior pitches. A scissor can be combined with a standard truss to create a flat ceiling with a vaulted area as shown at N in Figure 6-33. As the configuration of the scissor is altered, the intersections between flat and inclined bottom chord members may require support from interior walls or beams. Additional bearing points of the truss are also based on the span.

A *mono truss* (Figure 6-33) resembles half of a standard truss. The *mono truss* has an inclined and horizontal chord as well as one vertical exterior web. A mono truss can be used to blend a one-level structure into a two-level structure. A mono type truss can be altered to have a dual pitch unequal as illustrated in Figure 6-33. The short, steep top chord works well to support solar collectors in temperate climates. In large structures such as factories or warehouses, mono trusses are often combined to form a *sawtooth truss*.

Because of the triangular patterns that are used to form a truss, the overall configuration of trusses is almost unlimited. One additional truss that may be specified on a framing plan is a *girder truss*. A girder truss is a stronger version of any of the shapes previously mentioned, and it is used to support the weight of other trusses. A girder truss allows interior support walls to be removed by providing the support for other trusses. Figure 6-34 shows a framing plan using trusses set perpendicular to each other. The girder truss lies between the two areas of trusses. The trusses that are perpendicular to the girder truss are hung from the girder truss by metal hangers. Figure 6-35 shows an example of a detail of the intersection of a girder truss with other trusses. Notice that the void area that results in the roof between the trusses is filled in with conventional framing material.

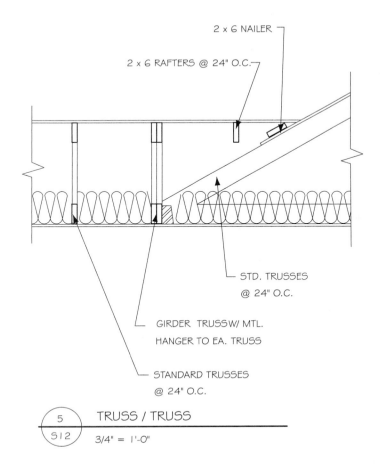

FIGURE 6-35 ■ Truss connection at a girder truss.

STRUCTURAL CONNECTORS

Regardless of the materials being connected, the type of connection is typically of the utmost importance. Connections must be fast, economical, and safe. The connection of walls to floors and ceilings is critical if each element is being used to transfer lateral loads from one to another. It is specified on the framing plan, wall sections, and specific details. Wood members are held together by the use of adhesives, nails, staples, and screws.

Adhesives

Adhesives are a key ingredient in the building industry. Materials such as plywood, OSB, I-joists, and laminated beams all depend on adhesives. Adhesives are also used within the structure to strengthen wood framing connections. Plywood subfloors are often glued to floor joists to stiffen the floor system and help eliminate squeaking. Gusset plates for light trusses are often glued at chord and web connections to strengthen joints. Adhesives are also used to attach plastic piping and finishing materials such as paneling, moldings, and counter tops. Structural adhesives should be specified in the written specifications, the framing plan, and specific details.

Nails

Nails are often used for the fabrication of wood to wood members when their thickness is less than 1 1/2" (40 mm). Thicker wood assemblies are normally bolted. The type and size of a nail to be used, as well as the placement and nailing pattern, will all affect the strength of the joint. Consideration should also be given to how and where nailing specifications will be placed.

Types of Nails

Nails specified by engineers in the calculations for framing include common, deformed, box, and spikes. Most nails are made from stainless steel, but copper and aluminum are also used. Figure 6-36 shows typical types of nails used for construction. Only nails used for structural purposes will be considered at this time. A *common* nail is used for most rough framing applications. Common nails have a single head, smooth shank and diamond tip. *Box* nails are slightly thinner and have less holding power than common nails. Box nails are used because they generate less resistance in penetra-tion and are less likely to split the lumber. Box nails come in sizes up to 16d. *Spikes* are nails larger than 20d. Nailing specifications are found throughout the written specifications, on framing plans, and in details.

Nail Components

Nails are comprised of a head, shank and tip. The head and the shank are often specified by the engineer in the calculations to meet a specific need.

Head

The *head* of most nails used for framing is usually a flat round head. Siding and wood trim are typically installed with nails that have a small cylindrically shaped head. Nails intended for easy removal typically have two heads—a lower head approximately 3/8" (10 mm) below the upper head. The nail is driven in until the first head makes contact, and the second head protrudes for future removal. Boards used for concrete forms are often nailed with double-headed nails.

Shank

Nails are specified by the length of the shank. The *shank* of a nail is the body of a nail. It is usually smooth, but ribs or deformations can be added to the shank to increase holding power. Deformations are typically placed perpendicular to the length of the shaft. If material is removed from the shank, a valley is created. Fluted nails have valleys that spiral around the length of the shank.

Tip

The *tip* of most nails has a diamond-shaped point, but a chiseled point or a flat tip can be used. The sharper the nail tip, the less damage done to the surrounding wood fiber, resulting in greater resistance to withdrawal. A pilot hole can be predrilled to further reduce wood damage and further increase holding strength. Notes for predrilling are often found on architectural drawings where appearance is critical.

Finish

The type of finish of the nail depends on how the nail is manufactured and how it will be used. Most nails have a smooth finish and are referred to as *bright finished nails*. Nails that are tempered by a heating process have a slightly blue finish. A resin coating called a cement finish can be applied to a nail to increase the holding power of the nail. A zinc coating is added to nails to increase the holding power and to increase resistance to corrosion. Zinc-coated or galvanized nails reduce the stains that result on siding when uncoated nails are used. If a special finish is required, it is typically added as a footnote to the nailing schedule or placed in the written specifications.

Nail Sizes

Nail sizes are described by the term *penny*, which is represented by the symbol *d* and used to describe standard nails from 2d through

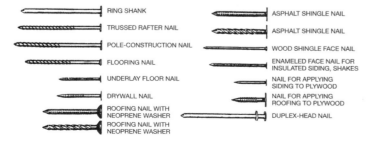

FIGURE 6-36 ■ Common types of nails used in construction.

60d. Penny is a weight classification of nails, which compares the number of pounds of nails per 1000 nails. One thousand 8d nails weigh 8 pounds. Figure 6-37 shows common sizes of nails. The size, spacing, and quantity of nails to be used are typically specified in structural details.

Nailing Specifications

Nails smaller than 20d are typically specified by the penny size and spacing. They are used for continuous joints such as a plate attached to a floor system. An example of a nailing note found on a framing plan or detail would be as follows:

2 X 6 DFL SILL W/ 20d's @ 4" o.c.

Nails are specified by a quantity and penny size for repetitive joints, such as a joist to a sill. The specification on a section or detail would simply read (3)-8d's and point to the area in the detail where the nails will be placed. Depending on the scale of the drawing, the nails may or may not be shown. Specifications for spikes are given in a similar manner to nails, but the penny size is replaced by the spike diameter.

Specifications are often given for the location of nails to the end or edge of a piece of lumber. If nails are too close to the edge or end of a piece of lumber, splitting will occur. Minimum edge distances are often included in the nailing specifications to reduce wood failure.

Nailing Placement

Nailing placements are described by the manner in which they are driven into the members being connected. Common methods include face, end, toe, and blind nailing. Figure 6-38 shows each method of driving nails. The type of nailing to be used is determined by how accessible the head of the nail is during construction and the type of stress to be resisted. The engineer will specify if nailing other than that recommended by the nailing schedule of the prevailing code is to be used. Nails that are required to resist shear are strongest when perpendicular to the grain. Nails that are placed parallel to the end grain such as end nailing, are weakest, and tend to pull out as stress is applied. Common nailing options include the following:

■ *Face nailing* is the process of driving a nail through the face or surface of one board into the face of another. Face nailing is used in connecting sheathing to rafters or studs, nailing a plate to the floor sheathing or in nailing a let-in brace to a stud.

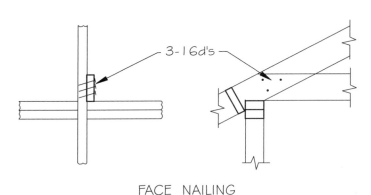

FACE NAILING

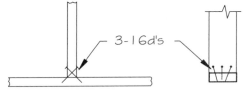

TOE NAILING

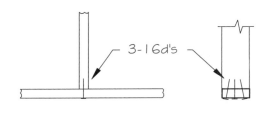

END NAILING

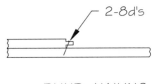

BLIND NAILING

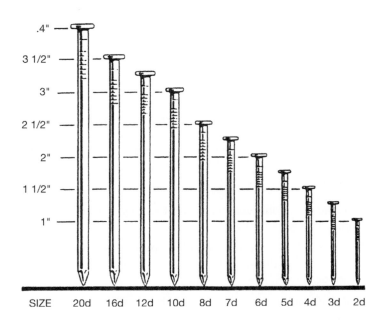

FIGURE 6-37 ■ Common sizes of nails

FIGURE 6-38 ■ Four methods of placing nails.

■ *End nailing* is the process of driving a nail through the face of one member into the end of another member. A plate is typically end nailed into the studs as a wall is assembled.

■ *Toe nailing* drives a nail through the face of one board into the face of another. With face and end nailing, nails are typically driven in approximately 90° to the face. Toe nailing drives the nail at approximately a 30° angle. Rafter-to-top plate or header-to-trimmer connections are typically toe-nailed joints.

■ *Blind nailing* is used where it is not desirable to see the nail head. Attaching wood flooring to the subfloor is typically done with blind nailing. Nails are driven at approximately a 45° angle through the tongue of tongue-and-groove flooring and hidden by the next piece of flooring.

Nailing Patterns

Nails for sheathing and other large areas of nailing often refer to nail placement along an edge, boundary or field nailing. Figure 6-39 shows an example of each location. *Edge nailing* refers to nails placed at the edge of a sheet of plywood. *Field nailing* refers to nails placed into the supports for a sheet of plywood excluding the edges. *Boundary nailing* is the nailing at the edge of a specific area of plywood.

Nailing Schedules

Regardless of the method used for nailing, a nail is required to penetrate into the supporting member by half the depth of the supporting member. If two 2× (5×) members are being attached, a nail would be required to penetrate 3/4" (20 mm) into the lower member. Figure 6-40 shows the IBC schedule for specifying the method, quantity, and size of nailing to be used.

Staples

Hammer-driven nails have been used for centuries. The development of power-driven nails has greatly increased the speed and ease in which nails can be inserted. As technology advanced, staples started to replace nailing for some applications. Staples are most often used for connecting asphalt roofing and attaching sheathing to roof, wall, and floor supports.

Power Driven Anchors

Metal anchors can be used to attach wood or metal to masonry. They range from 1/4" to 1/2" (6 to 13 mm) diameter, in lengths from 3/4" to 6" (20 to 150 mm). *Power driven anchors* are made from heat-treated steel and inserted by a powder charge from a gun-like device. They are typically used where it would be difficult to insert the anchor bolts at the time of the concrete pour. Anchors can also be used to join wood-to-steel construction.

Screws

Screws are often specified for use in wood connections that must be resistant to withdrawal. Three common screws are used

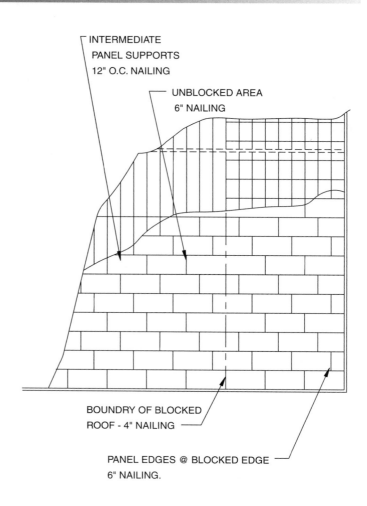

INTERMEDIATE
PANEL SUPPORTS
12" O.C. NAILING

UNBLOCKED AREA
6" NAILING

BOUNDRY OF BLOCKED
ROOF - 4" NAILING

PANEL EDGES @ BLOCKED EDGE
6" NAILING.

5/8" STD. GRADE C-D, 42/20
INT. APA PLY W/ EXT. GLUE.
LAY PERP. TO TRUSSES, STAGGER
SEAMS @ EA. TRUSS. NAIL W/ 10d
COMMON NAILS @ 4" O.C. @ ALL
PANEL EDGES & BLOCKED AREAS,
6" O.C. ALL SUPPORTED PANELS
EDGES @ UNBLOCKED AREAS &
12" O.C @ ALL INTERMEDIATE SUPPORTS.

PLYWOOD SPECIFICATION

FIGURE 6-39 ■ Nail specifications often refer to placement as edge, boundary, and field or intermediate nailing.

throughout the construction industry. Each is identified by its head shape. (See Figure 6-41.)

■ A *flathead or countersunk* screw is typically specified in the architectural drawings for finish work where a nail head is not desirable.

■ *Roundhead* screws are used at lumber connections where a head is tolerable. Roundhead screws are also used to connect lightweight metal to wood.

NAILING SCHEDULE

CONNECTION	NAILING
1. JOIST TO SILL OR GIRDER, TOE NAIL	3- 8d 3- 3" 14 ga. STAPLE
2. BRIDGING TO JOIST, TOENAIL EA. END	2- 8d 2- 3" 14 ga. STAPLE
3. 1 x 6 (25 x 150) SUBFLOOR OR LESS TO EACH JOIST, FACE NAIL	2-8d
4. WIDER THAN 1 x 6 (25 x 150) SUBFLOOR TO EACH JOIST, FACE NAIL	3-8d
5. 2" (50) SUBFLOOR TO JOIST OR GIRDER BLIND AND FACE NAIL	2-16d
6. SOLE PLATE TO JOIST OR BLOCKING FACE NAIL SOLE PLATE TO JOIST OR BLOCKING AT BRACED WALL PANELS	16d @ 16" (400 mm) O.C. 3" 14 ga. STAPLE @ 12" 3- 16d PER 16" (400 mm) 3" 14 ga. STAPLE PER 16"
7. TOP OR SOLE PLATE TO STUD, END NAIL	2-16d 3- 3" 14 ga. STAPLE
8. STUD TO SOLE PLATE, TOE NAIL	4-8d, TOENAIL 3- 3" 14 ga. STAPLE or 2-16d END NAIL 3- 3" 14 ga. STAPLE
9. DOUBLE STUDS, FACE NAIL	16d @ 24" (600 mm) O.C. 3" 14 ga. STAPLE @ 8' O.C.
10. DOUBLE TOP PLATE, FACE NAIL DOUBLE TOP PLATE, LAP SPLICE	16d @ 16" (400mm) O.C. 3" 14 ga. STAPLE @ 12" O.C. or 8-16d 12- 3" 14 ga. STAPLE
11. BLOCKING BTWN. JOIST OR RAFTERS TO TOP PLATE, TOENAIL	3- 8d 3- 3" 14 ga. STAPLE
12. RIM JOIST TO TOP PLATE, TOENAIL	8d @ 6" (150 mm) O.C. 3" 14 ga. STAPLE @ 6" O.C.
13. TOP PLATES, TAPS & INTERSECTIONS, FACE NAIL	2-16d 3- 3" 14 ga. STAPLE
14. CONTINUED HEADER, TWO PIECES	16d @ 16" (400 mm) O.C. ALONG EDGE
15. CEILING JOIST TO PLATE, TOE NAIL	3-8d 5- 3" 14 ga. STAPLE
16. CONTINUOUS HEADER TO STUD, TOE NAIL	4-8d
17. CEILING JOIST, LAPS OVER PARTITIONS, FACE NAIL	3-16d (SEE TABLE 2308.10.4.1) 4- 3" 14 ga. STAPLE
18. CEILING JOIST TO PARALLEL RAFTERS, FACE NAIL	3-16d 4- 3" 14 ga. STAPLE
19. RAFTERS TO PLATE, TOE NAIL	3-8d 3- 3" 14 ga. STAPLE
20. 1" (25mm) BRACE TO EA. STUD & PLATE FACE NAIL	2-8d 2- 3" 14 ga. STAPLE
21. 1 x 8 (25 x 203mm) SHEATHING OR LESS TO EACH BEARING, FACE NAIL	2-8d
22. WIDER THAN 1 x 8 (25 X 203mm) SHEATHING TO EACH BEARING, FACE NAIL	3-8d
23. BUILT-UP CORNER STUDS	16d @ 24" (600 mm) O.C. 3" 14 ga. STAPLE @ 16" O.C.
24. BUILT-UP GIRDER AND BEAMS	20d @ 32" (800 mm) O.C. FACE NAIL @ TOP/BOTTOM 3" 14 ga. STAPLE @ 24" O.C. & STAGGER 2-20d @ ENDS & @ EACH SPLICE. 3- 3" 14 ga. STAPLE
25. 2" PLANKS	2-16d AT EACH BEARING
26. COLLAR TIE TO RAFTER	3-10d / 4-3" 14 ga. STAPLE
27. JACK RAFTER TO HIP	3-10d / 4-3" 14 ga. STAPLE 2-16d / 3-3" 14 ga. STAPLE
28. RAFT.TO 2x RIDGE BEAM, TOE OR FACE NAIL.	2-16d / 3-3" 14 ga. STAPLE
29. JOIST TO RIM JOIST	3-16d / 5-3" 14 ga. STAPLE
30. LEDGER STRIP	3-16d / 4-3" 14 ga. STAPLE

CONNECTION	NAILING
31. WOOD STRUCTURAL PANELS AND PARTICLEBOARD:[2] SUBFLOOR, ROOF AND WALL SHEATHING (TO FRAMING):	
1/2" OR LESS	6d[3] / 1 3/4" 16 ga.[15]
19/32 - 3/4"	8d[4] OR 6d[5] / 2" 16 ga.[16]
7/8 - 1"	8d[3]
1 1/8 - 1 1/4"	10d[4] OR 8d[5]
COMBINATION SUBFLOOR-UNDERLAYMENT (TO FRAMING) 1" = 25.4 mm)	
3/4 AND LESS	6d[5]
7/8" - 1"	8d[5]
1 1/8 - 1 1/4"	10d[4] OR 8d[5]
32. PANEL SIDING (TO FRAMING): 1/2" (13 mm)	6d[6]
5/8" (16 mm)	8d[6]
33. FIBERBOARD SHEATHING: [7] 1/2" (13 mm)	No. 11 ga.[8] 6d[4]
25/32" (20 mm)	No. 16 ga.[9] No. 11 ga.[8] 8d[4] No. 16 ga.[9]
34. INTERIOR PANELING 1/4" (6.4 mm) 3/8" (9.5 mm)	4d[10] 6d[11]

1. COMMON OR BOX NAILS MAY BE USED EXCEPT WHERE OTHERWISE STATED.

2. NAILS SPACED @ 6" (150 mm) ON CENTER @ EDGES, 12" (300 mm) AT INTERMEDIATE SUPPORTS EXCEPT 6" (150 mm) AT ALL SUPPORTS WHERE SPANS ARE 48" 1200 mm OR MORE. FOR NAILING OF WOOD STRUCTURAL PANEL AND PARTICLEBOARD DIAPHRAGMS AND SHEAR WALLS, REFER TO SECTION 2305. NAIL FOR WALL SHEATHING MAY BE COMMON, BOX OR CASING.

3. COMMON OR DEFORMED SHANK.

4. COMMON.

5. DEFORMED SHANK..

6. CORROSION-RESISTANT SIDING OR CASING NAILS

7. FASTENERS SPACED 3" (75 mm) O.C. AT EXTERIOR EDGES AND 6" (150 mm) O.C. AT INTERMEDIATE SUPPORTS.

8. CORROSION-RESISTANT ROOFING NAILS W/ 7/16" & (11 mm) HEAD & 1 1/2" (38 mm) LENGTH FOR 1/2" (13 mm) SHEATHING AND 1 3/4" (44 mm) LENGTH (FOR 25/32" (20 mm) SHEATHING).

9. CORROSION-RESISTANT STAPLES WITH NOMINAL 7/16" (11 mm) CROWN AND 1 1/8" (28 mm) LENGTH FOR 1/2" (13 mm) SHEATHING AND 1 1/2" (38 mm) LENGTH (FOR 25/32" (20 mm) SHEATHING. PANEL SUPPORTS @ 16" O.C.(400 MM) (20" (500 mm) IF STRENGTH AXIS IN THE LONG DIRECTION OF THE PANEL, UNLESS OTHERWISE MARKED.)

10. CASING OR FINISHED NAILS SPACED 6" (150 mm) ON PANEL EDGES, 12" (300 mm) @ INTER. SUPPORTS.

11. PANEL SUPPORTS @ 24" (600 mm). CASING OR FINISH NAILS SPACED 6" (150 mm) ON PANEL EDGES, 12" (300 mm) AT INTERMEDIATE SUPPORTS.

12. FOR ROOF SHEATHING APPLICATIONS, 8d NAILS ARE THE MINIMUM REQUIRED FOR WOOD STRUCTURAL PANELS.

13. STAPLES SHALL HAVE A MINIMUM CROWN OF 7/16 (11 mm).

14. FOR ROOF SHEATHING APPLICATIONS, FASTENERS SPACED 4" (100 mm) O.C. AT EDGES, 8" (200 mm) O.C. AT INTERMEDIATE SUPPORTS

15. FASTENERS SPACED 4" (100 mm) O.C. AT EDGES, 8" (200 mm) AT INTERMEDIATE SUPPORTS FOR SUBFLOOR AND WALL SHEATHING AND 3" (75 mm) O.C. AT EDGES, 6" (150 mm) AT INTERMEDIATE SUPPORTS FOR ROOF SHEATHING.

16. FASTENERS SPACED 4" (100 mm) O.C. AT EDGE, 8" (200 mm) AT INTERMEDIATE.

FIGURE 6-40 ■ Drafters are often required to consult a nailing schedule to determine the size and method of nailing for wood connections. Based on nailing requirements of the IBC.

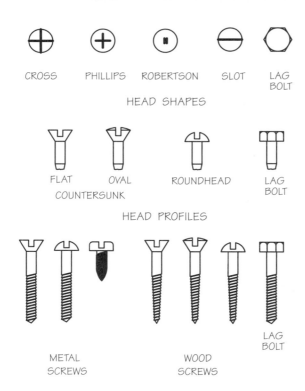

FIGURE 6-41 ■ Common types of screws used in construction.

■ Both flathead and roundhead screws are designated by a gauge that specifies a diameter in inches, the length in inches, and a head shape. A typical specification for a flathead wood screw might be as follows:

#10 × 3" F.H.W.S.

Lengths of screws range from 1/4" in. to 6" (6 to 150 mm).

■ *Lag* screws have a hexagon- or square-shaped head that is designed to be tightened by a wrench rather than a screwdriver. Lag screws are used for lumber connections 1 1/2" (40 mm) and thicker. Lag bolts are available in lengths from 1" to 12" (25 to 300 mm) with diameters ranging from 1/4" to 1 1/4" (6 to 30 mm).

A washer is used with a lag bolt to guard against crushing wood fibers near the bolt. A pilot hole that is approximately three-fourths of the shaft diameter is often specified to reduce wood damage and increase the resistance to withdrawal.

Metal Framing Connectors

Pre-manufactured metal connectors by companies such as Simpson or Tyco are used at many wood connections to strengthen nailed connections. Joist hangers, post caps, post bases, and straps are some of the most common lightweight metal hangers used with wood construction. Each is shown in Figure 6-42. Each connector comes in a variety of gauges of metal, ranging from 12 to 16 gauge, with sizes to fit a wide variety of lumber. Metal connectors are typically specified on the framing plans, sections, and details by listing the model number and type of connector. A metal

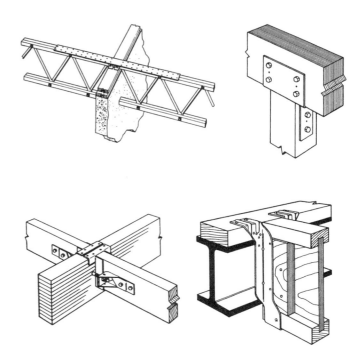

FIGURE 6-42 ■ A wide variety of metal connectors are used to strengthen wood connections. *Courtesy Simpson Strong-Tie Co., Inc.*

connector specification for connecting 2- 2x12 joists to a beam would be

HHUS212-2TF JST. HGR.

The supplier is typically specified in the written specifications or in a general note such as the following:

ALL METAL HANGERS AND CONNECTORS TO BE SIMPSON STRONG-TIE COMPANY, INC. OR EQUAL.

Depending on the connection, the nails or bolts used with the metal fastener may or may not be specified. If no specification is given, it is assumed all nail holes in the connector will be filled. If bolts are to be used, they will normally be specified with the connector, based on the manufacturer's recommendations. Nails associated with metal connectors are typically labeled with the letter *n* instead of *d*. These nails are equal to their *d* counterpart, but the length has been modified by the manufacturer to fit the metal hanger.

ADDITIONAL READING

The following Web sites can be used as a resource to help you keep current with changes in building materials.

ADDRESS	COMPANY/ORGANIZATION
www.ahardbd.org	American Hardboard Association
www.nlga.org	National Lumber Grades Authority
www.bc.com	Boise Cascade Corporation
www.csa.ca	Canadian Standards Association
www.cwc.ca	Canadian Wood Council

www.celotex   Celotex Corporation (Insulating sheathing)

www.gp.com   Georgia-Pacific Corporation

www.hpva.org   Hardwood Plywood & Veneer Association

www.internationalpaper.com

                           International Paper (high performance building products)

www.owenscorning.com   Owens Corning (house wrap and insulation)

www.reemay.com   Reemay (Typar house wrap)

www.strongtie.com   Simpson Strong-Tie Co., Inc.

www.sfpa.org   Southern Forest Products Association

www.trimjoist.com   TrimJoist Engineered Wood Products

www.weyerhaeuser.com   Weyerhaeuser

www.wii.com   Willamette Industries, Inc.

www.woodtruss.com   Wood Truss Council of America

CHAPTER 6

Wood Framing Methods and Materials

CHAPTER QUIZ

Answer the following questions on a separate sheet of paper. Some answers may require the use of vendor catalogs or seeking out local suppliers.

Question 6-1 How wide are girders with a conventional floor system?

Question 6-2 Define the following abbreviations: PSL, OSB, LVL, MDF, EXP, HDF, APA, EXT, STRUCT.

Question 6-3 List four common materials suitable for beams and girders.

Question 6-4 List three types of truss web materials.

Question 6-5 What two elements are typically applied to wood to make engineered wood products?

Question 6-6 What is the advantage of providing blocking at the edge of a floor diaphragm?

Question 6-7 List the common sizes of engineered wood studs.

Question 6-8 What are the common span ratings for plywood, suitable for roof sheathing?

Question 6-9 What grades of plywood are typically used for floor sheathing?

Question 6-10 List eight different qualities typically given in an APA wood rating.

Question 6-11 What resists uplift when a truss roof system is used?

Question 6-12 List six different methods of planing lumber.

Question 6-13 What is the difference between timber and lumber?

Question 6-14 Describe the two portions of a growth ring and explain their significance to strength.

Question 6-15 How would lumber with a moisture content of 20 be rated?

Question 6-16 What is the dry face width of a 2 × 12?

Question 6-17 List three shapes for steel columns.

Question 6-18 What is the difference between a king stud and a trimmer?

Question 6-19 List four different types of rafters.

Question 6-20 What is the common spacing of skip sheathing?

Question 6-21 Explain the meaning of the numbers 6/12 in relation to a roof.

Question 6-22 Explain the difference between a ridge, ridge board, and a ridge beam.

Question 6-23 List two common methods of resisting the outward thrust of a rafter.

Question 6-24 If a 8/12 pitch is used for a scissor truss, what is the maximum pitch for the bottom chord?

Question 6-25 List seven materials used for finished roofing.

Question 6-26 List two common uses for adhesives in the construction industry.

Question 6-27 What type of nail is most typically used for wood framing?

Question 6-28 What would cause a nail to have a bluish finish?

Question 6-29 How long is an 8 penny nail?

Question 6-30 Based on the IBC, how will a rafter be connected to a plate?

Question 6-31 Describe the difference between the terms boundary, edge, and field.

Question 6-32 List the nailing the IBC recommends for securing 3/4" plywood to the floor joists.

Question 6-33 List two common uses for roundhead screws.

Question 6-34 What purpose does a washer serve?

Question 6-35 Use the Internet to research and list the names of five national lumber suppliers.

DRAWING PROBLEMS

Unless other instructions are given by your instructor, complete the following details and save them as wblocks. Skeletons of most details can be accessed from http://www.delmar.com/resources/ocl.html. Use these drawings as a base to complete the assignement.

Show all required views to describe each connection. Draw each detail at a scale of 3/4" = 1'–0" unless noted. Show and specify all connecting materials based on pages 465, 714, and 721 of *Architectural Drafting and Design* Fourth edition. Base nailing on IBC standards unless your instructor tells you otherwise. Specify all material based on common local practice.

■ Represent structural wood members that the cutting plan has passed through with bold lines.

■ Use separate layers for wood, concrete, text, and dimensions.

■ Use dimensions for locations where possible, instead of notes.

■ Provide notation to specify that all metal hangers are to be provided by Simpson Strong-Tie Company or equal. Obtain the needed catalogue from a local supplier or from their Web site.

■ Hatch each material with the appropriate hatch pattern.

■ Use an appropriate text font to label and dimension each drawing as needed. Refer to *Sweet's Catalogs*, vendor catalogs, or the Internet to research needed sizes and specifications. Keep all text 3/4" minimum from the drawing, and use an appropriate architectural style leader line to reference the text to the drawing.

■ Provide a detail marker, with a drawing title, scale, and problem number below each detail.

■ Assemble the details required by your instructor into a appropriate template sheet for plotting. Arrange the drawings using the guidelines presented in Chapters 2 and 3. Assign a drawing sheet number of sheet S1 of 6.

Problem 6-1 Draw a detail showing a truss with a 6/12 roof pitch (assume 2 × 6 chord material) resting on 2 × 4 studs at 16" o.c. Cover the interior with 1/2" gypsum board and the exterior side with 3/4" exterior stucco or CD-EIFS. Specify a concrete tile roof.

Problem 6-2 Draw a detail showing a scissor truss with a 6/12 roof pitch (assume 2 × 4 chord material and a 3/12 pitch for the bottom chord) resting on 2 × 6 studs at 16" o.c. Cover the interior with 3/4" gypsum board and the exterior side with horizontal beveled siding over 1/2" gypsum board. Specify a 26 ga. metal roof over 3/4" plywood sheathing.

Problem 6-3 Draw a detail showing a 16" deep TJI-35 floor truss by Trus-Joist MacMillan resting on a 2 × 6 stud wall. Set the top plate at 10' high. Show a 6 × 10 DFL # 1 header set for a window with the bottom of the header set at 7'–0" high. Show 7/8" stucco over 1/2" plywood on the exterior side. Show 1 3/4" concrete over 55# felt over 1 3/4" plywood subfloor. Frame an upper wall with 2 × 4 studs at 16" o.c. with the lower pressure treated plate nailed w/ 20d's at 4" o.c. Use break lines so that the full height of each wall is not drawn.

Problem 6-4 Draw a detail with a front view showing double 2 × 6 trimmers and king studs resting on a 2 3/4" plate, 3/4" CD APA 42/20 T&G floor sheathing, 2 × 12 F.J. at 12" o.c. with solid blocking, resting on another 2 × 6 stud wall. Use double solid blocking under each post. Support the trimmers and king stud with a 4 × 6 post below. Use a Simpson Company HD5A hold down and specify all bolt sizes.

Problem 6-5 Draw a detail showing a 6 × 14 DFL #1 beam resting on a 4 × 6 post. Select a metal cap than can resist 3500 pounds in uplift.

Problem 6-6 Show a 2 × 8 DFL rafter at 24" o.c. with a 12" overhang supporting 1/2" plywood sheathing, and composition shingles at a 6/12 pitch resting on (2)- 2 × 6 top plates resting on 2 × 6 studs at 16" o.c. Use 2 × 6 ceiling joists at 16"o.c. and specify a 1 × 6 fascia.

Problem 6-7 A 3 × 12 DFPT ledger will be connected to a concrete wall with a 5/8" dia. × 8" M.B. bolts at 24" o.c. staggered 3" u/d. Simpson Company U210 hangers will be used to support 2 × 12 DFL #1 joist at 16" o.c. The joist will support 3/4" standard grade plywood sheathing.

CHAPTER

Timber Framing Methods and Materials

In addition to standard uses of wood in western platform construction, timber is often used for the structural framework of a structure. Timber construction is used for both appearance and structural reasons. This chapter will examine common uses of timber, heavy timber, laminated beams, and the structural connectors used attach these members.

TIMBER

The term *timber* is used to describe wood members that are 5" (130 mm) and thicker, although 6" (150 mm) is typically the smallest size used. Timber construction is used to form many types of commercial structures because it affords large expanses of glass in the exterior shell. Timber has excellent structural and fire-retardant qualities. In a fire, timber will char on the exterior surfaces but will maintain its structural integrity long after an equally sized steel beam would have failed. Components of timber framing include posts, beams, and planks. In addition to framing plans showing the size of the components, details of connections are of utmost importance with timber construction. Figure 7-1 shows an example of a structure framed using timber construction.

Common Components

Instead of using 2" (50 mm) studs at a 16" or 24" (400 or 600 mm) spacing, vertical supports are provided by posts for which the location is based on the design needs of the structure. Post size typically ranges from 5 × 5 to 12 × 12 (130 × 130 to 300 × 300 mm). Wood posts larger than 12 × 12 (300 × 300 mm) are available, but they are not often used, for economic reasons. Posts larger than 8 × 8 (200 × 200 mm) are typically laminated from solid material or from a combination of lumber and plywood.

Horizontal members called beams are used to span between the post and other beams. Beam size is determined by the engineer based on the span and loads to be supported. Usually 1 1/8" (30 mm) APA STURD-I- FLOOR 48 T & G plywood is used to span between beams. Two-inch (50 mm) T & G planks can also be used as a floor material but the exact size must be based on the live loads to be supported.

Representing Members on Drawings

The size and location of posts and beams are typically specified on the framing plan. Posts are represented using linetypes similar to those used to represent walls. Post size is often specified using diagonal text, so that the specification will be clear. Beams are represented with dashed or center lines. Text to describe the beam is normally placed parallel to the beam or in a schedule. Figure 7-2 shows methods of specifying beams and posts on a framing plan. Decking is typically specified in details and sections.

Details showing the connection of post to foundation, post to beams at floor and roof levels, beam to beam, and decking to

FIGURE 7-1 ■ Timber is often used to frame a structure for its appearance and resistance to fire. *Courtesy Timberpeg.*

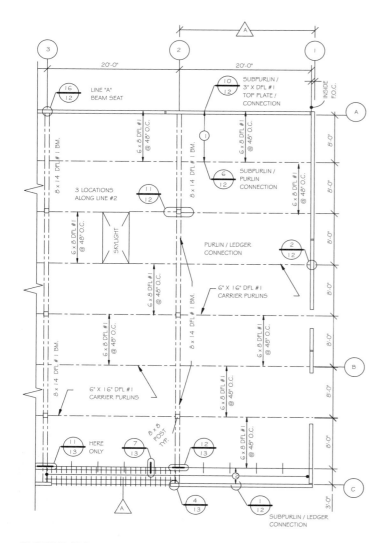

FIGURE 7-2 ■ Wood beams can be shown in plan view by pairs of dashed lines that represent the beam width, or by a polyline. Polylines of continuous, center, or dashed linetype will be used according to office practice.

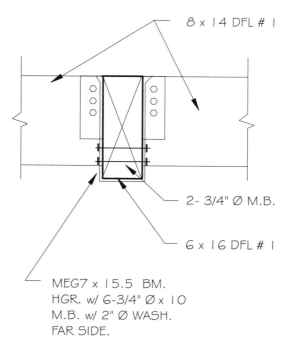

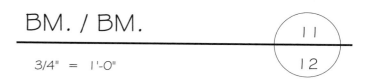

FIGURE 7-3 ■ A beam-to-beam connection detail.

beams, are typically required. Figure 7-3 shows a beam-to-beam connection detail. Beams that are drawn in end view can be drawn using different methods, as shown in Figure 7-4. At a scale of 3/4" = 1'-0" or larger, the x is usually placed to help define the beam. The x is also used to represent the end view of a continuous member such as a beam. One diagonal line is used to represent the end view of non-continuous members such as blocking placed between beams. Many offices use a thick line around the beam perimeter to help distinguish beams.

HEAVY TIMBER

Heavy timber construction is similar to timber construction, except for the size of the members used. The IBC restricts the size of the members that can be used to achieve a heavy timber fire rating. The rating is referred to as a type IV-HT in the occupancy tables of the IBC. In addition to limiting the size of members used,

heavy timber construction eliminates the concealed spaces associated with western platform construction. Space between rafters, joist, or studs that is concealed by the interior finish is eliminated, leaving the framing members and the exterior shell exposed to the interior.

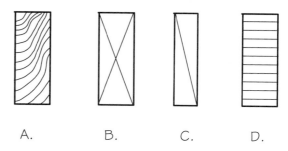

FIGURE 7-4 ■ Beam representation in end view, for details drawn at a scale of 3/4"= 1'-0" or larger: A, End grain; B continuous beam; C, non-continuous beam, blocking; D, laminated beam.

Common Components

Post size for a HT rating requires the use of 8 × 8 (200 × 200 mm) minimum to meet the fire resistance requirements of the building code. If no floor loads are supported, and the beam does not extend below the floor line, a post with a minimum of 6" wide × 8" (150 × 200 mm) deep may be used. If a laminated arch is used to support floor loads, it may not be smaller than 8" (200 mm) in any direction. Laminated arches that do not support floor loads may be as small as 4" in width and 6" (100 × 150 mm) in depth.

The IBC requires beams to be a minimum of 6" (150 mm) wide and 10" (250 mm) deep to provide sufficient fire protection for a floor with a type IV-HT rating. Beams 6" (150 mm) wide and 8" (200 mm) deep are the minimum allowed to support roof loads. Floor decking must be a minimum of 3" (75 mm) deep planks with an overlay of 1" (25 mm) T & G plywood laid crosswise. Floor decking must maintain a 1/2" (13 mm) space between wall members to allow for swelling and shrinkage. Roof decking must be a minimum of 2" (50 mm) thick T & G planks or 11 × 8" (40 mm) T & G plywood. A double layer of 1" (25 mm) boards may also be used if joints are staggered.

Heavy Timber Representation

The size and location of posts and beams are typically specified on the framing plan using methods similar to other timber callouts. Details showing the connection of post to foundation, post to beams at floor and roof levels, beam to beam, and decking to beams, are typically required just as with other types of timber construction. Figure 7-5 shows a beam-to-beam connection detail.

Timber Trusses

Similar in shape to conventional trusses discussed in the previous chapter, trusses can be formed of timber or laminated material. Heavy timber trusses used to support floor loads must have members that are a minimum of 8" (200 mm) nominal in any direction. Connections of truss components are typically achieved by using bolted or ringed connectors. Depending on the type and amount of load to be resisted, plates similar to a gusset can be used at each web-to-chord connection.

LAMINATED BEAMS

Because of the strength limitations of solid wood and limited availability, large beams are typically constructed from smaller members that are laminated together to form a larger beam. Increased length is gained by laminating lumber that is spliced together with scarf or finger joints. Laminated timber is often used for structural framing materials because of its increased strength over sawn lumber, and because of its beauty. (See Figure 7-6.) Laminated timbers are made from dry lumber, which offers a much higher level of dimensional stability over sawn lumber. The use of dry lumber virtually eliminates cracking, twisting, warping, and shrinking. Because of their increased size, laminated beams are installed with mobile construction equipment but can be set using conventional hand tools.

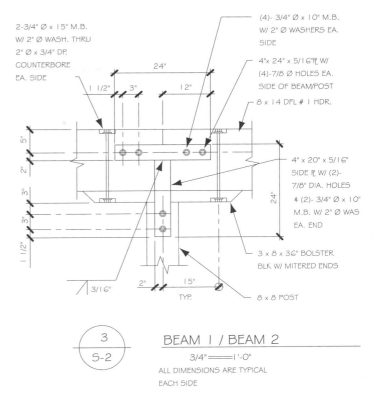

FIGURE 7-5 ■ Heavy timber is referred to as type IV-HT by the IBC and consists of using 8 × 8 (200 × 200) or larger structural members.

FIGURE 7-6 ■ Laminated timbers come in a variety of sizes, strengths and shapes. *Courtesy American Institute of Timber Construction.*

Common Sizes

Laminated beams are made from 2× (50×) material in 4", 6", 8", 10", 12", 14", and 16" (100, 150, 200, 250, 300, 350, and 400 mm) widths. Wider beams are available by special order. Once the members are laminated, the beams are planed to their finish size. Laminated timber comes in finished widths as shown in Table 7-1 below:

TABLE 7-1 ■ Finished Widths for Laminated Timber

Nominal Width (in.)	Western Species	Southern Pine
3	2 1/8	2 1/8
4	3 1/8	3 or 3 1/8
6	5 1/8	5 or 5 1/8
8	6 3/4	6 3/4
10	8 3/4	8 1/2
12	10 3/4	10 1/2
14	12 1/4	12
16	14 1/4	14

Depths for simple beams range from 3" to 84" (75 to 2300 mm) in 1 1/2" (38 mm) increments. One of the features of laminated timbers is that beams do not have to conform to these standard sizes, but can be laminated to specific sizes to fit the design criteria. Figure 7-7a shows the properties of standard laminated beam sizes. These beams all have their wide face parallel to the X-X axis. Laminated beams are also available with their wide face parallel to the Y-Y axis but are not common. Straight beam sizes are available through most lumber yards, and custom shapes are shipped directly from the manufacturer.

Grading

As well as being graded by each of the major building codes, laminated timbers are graded by the American Institute of Timber Construction (AITC). Standards are set to provide minimum standards for production, quality control, inspection, and certification of performance.

Fiber Bending

Figure 7-7b shows a partial listing of design values for laminated timber. Although the engineer will determine all required beam sizes, it is important to understand some of the specifications, so that laminated timbers can be properly specified on the structural drawings. Notice the first beam listed in column 1 of Figure 7-7b is a 24F-V4. The 24F relates to the design value in bending, represented in column 3. In design formulas, this value is usually represented by the symbol fb. Common values listed by most lumber manufactures include 1600, 2000, 2200, and 2400. The higher the value, the more units of stress that can be resisted, and the stronger the beam. The value to be used will depend on the type of lumber to be used and will be selected by the engineer.

Grading Method

The second portion of the specification represents the method used to grade the beam. The letters V or E may be used to describe the grading procedure. V represents wood that has been visually inspected and E specifies that the electronic methods of nondestructive testing have been used. The number that follows the ratings specification represents a specific combination of grades and species of material to be used to form the beam.

Material

Column 2 of table 7-7b represents the type or types of material to be laminated together using abbreviations such as DF/HF. Common abbreviations include the following:

DF   Douglas Fir

DFS   Douglas Fir South

HF   Hem Fir

WW   Soft woods

ES   Eastern Spruce

AC   Alaska Cedar

CSP   Canadian Spruce-Pine

SP   Southern Pine

The first group of letters represents the species of wood used for the outer laminations, and the second letters represent the species used for the beam core or inner layers.

Appearance

In addition to the number/letter grading, beams are also described by their appearance. Common grades include industrial, architectural, and premium. These terms apply to the exposed surfaces of the laminated member and regulate items such as growth characteristics, inserts, wood fillers, and surfacing operations. These ratings in no way affect the structural quality of the beam. For beams that will not be exposed, industrial grades are typically specified. Architectural grade beams can be used in exposed situations, but minor flaws in the beams will be seen. Premium-grade beams are intended for high visibility uses where no flaws are desired.

Finish

Laminated beams can be finished like their wood counterparts. Sealers, stains, and paint products are common finishing products. Surface sealers are used to resist soiling, to help control grain separation, and to reduce moisture absorption. In addition to the finish that can be applied at the job site, the laminator can also apply preservative treatments. Water-borne salt chemicals or oil-borne chemicals are used to treat individual pieces of lumber prior to lamination. Oil-based products such as creosote can be used to treat beams after gluing.

Table 5: Section Properties
Based on 1-1/2 in. thick laminations

No. of lams	d Depth (in.)	A Area (in.²)	I Moment of Inertia (in.⁴)	S Section Modulus (in.³)
2 1/2 in. Widths				
6	9	22.50	151.9	33.75
7	10-1/2	26.25	241.2	45.94
8	12	30.00	360.0	60.00
9	13-1/2	33.75	512.6	75.64
10	15	37.50	703.1	93.75
11	16-1/2	41.25	935.9	113.4
12	18	45.00	1215	135.0
13	19-1/2	48.75	1545	158.4
14	21	52.50	1929	183.8
15	22-1/2	56.25	2373	210.9
3 1/8 in. Widths				
4	6-	18.75	56.25	18.75
5	7-1/2	23.44	109.9	29.30
6	9	28.13	189.9	42.19
7	10-1/2	32.81	301.5	57.42
8	12	37.50	450.0	75.00
9	13-1/2	42.19	640.7	94.92
10	15	46.88	878.9	117.2
11	16-1/2	51.56	1170	141.8
12	18	56.25	1519	168.8
13	19-1/2	60.94	1931	198.0
14	21	65.63	2412	229.7
15	22-1/2	70.31	2966	263.7
16	24	75.00	3600	300.0
17	25-1/2	79.70	4318	338.7
18	27	84.40	5126	379.7
19	28-1/2	89.10	6028	423.0
5 1/8 in. Widths				
4	6	30.75	92.25	30.75
5	7-1/2	38.44	180.2	48.05
6	9	46.13	311.3	69.19
7	10-1/2	53.81	494.4	94.17
8	12	61.50	738.0	123.0
9	13-1/2	69.19	1051	155.7
10	15	76.88	1441	192.2
11	16-1/2	84.56	1919	232.5
12	18	92.95	2491	276.8
13	19-1/2	99.94	3167	324.8
14	21	107.6	3955	376.7
15	22-1/2	115.3	4865	432.4
16	24	123.0	5904	492.0
17	25-1/2	130.7	7082	555.4
18	27	138.4	8406	622.7
19	28-1/2	146.1	9887	693.8
20	30	153.8	11530	768.8
21	31-1/2	161.4	13350	847.5
22	33	169.1	15350	930.2
23	34-1/2	176.8	17540	1017
24	36	184.5	19930	1107
6 3/4 in. Widths				
5	7-1/2	50.63	273.3	63.28
6	9	60.75	410.1	91.13
7	10-1/2	70.88	651.2	124.0
8	12	81.00	972.0	162.0
9	13-1/2	91.33	1384	205.0

No. of lams	d Depth (in.)	A Area (in.²)	I Moment of Inertia (in.⁴)	S Section Modulus (in.³)
6 3/4 in. Widths (cont.)				
10	15	101.3	1898	253.1
11	16-1/2	111.4	2527	306.3
12	18	121.5	3281	364.5
13	19-1/2	131.6	4171	427.8
14	21	141.8	5209	496.1
15	22-1/2	151.9	6407	569.5
16	24	162.0	7776	648.0
17	25-1/2	172.1	9327	731.5
18	27	182.3	11070	820.1
19	28-1/2	192.4	13020	913.8
20	30	202.5	15190	1013
21	31-1/2	212.6	17581	1116
22	33	222.8	20210	1225
23	34-1/2	232.9	23100	1339
24	36	243.0	26240	1458
25	37-1/2	253.1	29660	1582
26	39	263.3	33370	1711
27	40-1/2	273.4	37370	1845
28	42	283.5	41670	1985
29	43-1/2	293.6	46300	2129
30	45	303.8	51260	2278
31	46-1/2	313.9	56560	2433
32	48	324.0	62210	2592
8 3/4 in. Widths				
6	9	78.75	531.6	118.1
7	10-1/2	91.88	844.1	160.8
8	12	105.0	1260	210.0
9	13-1/2	118.1	1794	265.8
10	15	131.3	2461	328.1
11	16-1/2	144.4	3276	397.0
12	18	157.5	4253	472.5
13	19-1/2	170.6	5407	554.5
14	21	183.8	6753	643.1
15	22-1/2	196.9	8306	738.3
16	24	210.0	10080	840.0
17	25-1/2	223.1	12090	948.3
18	27	236.3	14350	1063
19	28-1/2	249.4	16880	1185
20	30	262.5	19690	1313
21	31-1/2	275.6	22790	1447
22	33	288.8	26200	1588
23	34-1/2	301.9	29940	1736
24	36	315.0	34020	1890
25	37-1/2	328.1	38450	2051
26	39	341.3	43250	2218
27	40-1/2	354.4	48440	2392
28	42	367.5	54020	2573
29	43-1/2	380.6	60020	2760
30	45	393.8	66440	2953
31	46-1/2	406.9	73310	3153
32	48	420.0	80640	3360
33	49-1/2	433.1	88440	3573
34	51	446.3	96720	3793
35	52-1/2	459.4	105500	4020
36	54	472.5	114800	4253

No. of lams	d Depth (in.)	A Area (in.²)	I Moment of Inertia (in.⁴)	S Section Modulus (in.³)
8 3/4 in. Widths (cont.)				
37	55-1/2	485.6	124654	4492
38	57	498.8	135037	4738
39	58-1/2	511.9	145980	4991
40	60	525.0	157500	5250
41	61-1/2	538.1	169610	5516
42	63	551.3	182326	5788
10 3/4 in. Widths				
7	10-1/2	112.9	1037	197.5
8	12	129.0	1548	258.0
9	13-1/2	145.1	2204	326.5
10	15	161.3	3023	403.1
11	16-1/2	177.4	4024	487.8
12	18	193.5	5225	580.5
13	19-1/2	209.6	6642	681.3
14	21	225.8	8296	790.1
15	22-1/2	241.9	10200	907.0
16	24	258.0	12380	1032
17	25-1/2	274.1	14850	1165
18	27	290.3	17630	1306
19	28-1/2	306.4	20740	1455
20	30	322.5	24190	1613
21	31-1/2	338.6	28000	1778
22	33	354.8	32190	1951
23	34-1/2	370.9	36790	2133
24	36	387.0	41800	2322
25	37-1/2	403.1	47240	2520
26	39	419.3	53140	2725
27	40-1/2	435.4	59510	2939
28	42	451.5	66370	3161
29	43-1/2	467.6	73740	3390
30	45	483.8	81630	3628
31	46-1/2	499.9	90070	3874
32	48	516.0	99070	4128
33	49-1/2	532.1	108600	4390
34	51	548.3	108800	4660
35	52-1/2	564.4	129600	4938
36	54	580.5	141000	5225
37	55-1/2	596.6	153100	5519
38	57	612.8	165900	5821
39	58-1/2	628.9	179300	6132
40	60	645.0	193500	6450
41	61-1/2	661.1	208400	6777
42	63	677.3	224000	7111
43	64-1/2	693.4	240400	7454
44	66	709.5	257500	7805
45	67-1/2	725.6	275500	8163
46	69	741.8	294300	8530
47	70-1/2	757.9	313900	8905
48	72	774.0	334400	9288
49	73-1/2	790.1	355700	9679
50	75	806.3	377900	10000
51	76-1/2	822.4	401000	10500
52	78	838.5	425100	10900
53	79-1/2	854.6	450100	11300
54	81	870.8	476000	11800

FIGURE 7-7A ■ Common sizes and properties of laminated beams. *Courtesy American Institute of Timber Construction.*

Specifications

The specifications required for a laminated beam on a drawing will differ from a sawn beam. Laminated beams are referenced on the framing plans, sections, and details, with complete information given in the project manual that accompanies the drawings. Common information included for a laminated beam specification on a framing plan includes the beam size, beam type, the grade, and the type of material used to form each layer. A typical specification would resemble

Design Values for Structural Glued Laminated Timber from *AITC 117-93--Design* Table 1
For normal duration of load and dry conditions of use [a]

Combination Symbol[c]	Species-Outer Laminations/Core Laminations[d]	Bending About X-X Axis						Bending About Y-Y Axis					Axially Loaded		
		Loaded Perpendicular to Wide Faces of Laminations 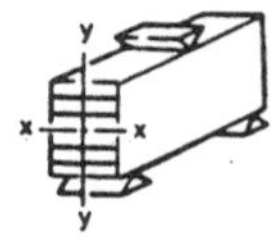						Loaded Parallel to Wide Faces of Laminations 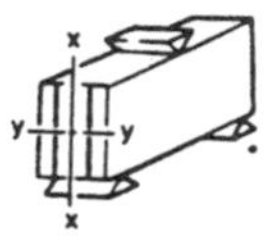							
		Extreme Fiber in Bending, F_{bx}		Compression Perpendicular to Grain, $F_{c\perp x}$		Shear Parallel to Grain (Horizontal) F_{vx}	Modulus of Elasticity, E_x	Extreme Fiber in Bending, F_{by}	Compression Perpendicular to Grain, $F_{c\perp y}$	Shear Parallel to Grain (Horizontal) F_{vy}	Shear Parallel to Grain (Horizontal) (For members with multiple piece laminations which are not edge glued), F_{vy}	Modulus of Elasticity, E_y	Tension Parallel to Grain, F_t	Compression Parallel to Grain, F_c	Modulus of Elasticity, E
		Tension Zone Stressed in Tension psi	Compression Zone Stressed in Tension psi	Tension Face psi	Compression Face psi	psi	Million psi	psi	psi	psi	psi	Million psi	psi	psi	Million psi
1	2	3	4	5	6	7	8	9	10	11	12	13	14	15	16
Visually Graded Western Species															
The following two combinations are not balanced and are for either dry or wet use.															
24F-V4	DF/DF	2400	1200	650	650	165	1.8	1500	560	145	75	1.6	1100	1600	1.6
24F-V5	DF/HF			650	650	155	1.7	1350	375	140	70	1.5	1100	1450	1.5
The following combination is balanced and is intended for members continuous or cantilevered over supports and provide equal capacity in both positive and negative bending.															
24F-V8	DF/DF	2400	2400	650	650	165	1.8	1450	560	145	75	1.6	1100	1650	1.6
Visually Graded Southern Pine															
The following two combinations are not balanced and are for either dry or wet use.															
24F-V1	SP/SP	2400	1200	650	560	200	1.7	1500	560	175	90	1.5	1100	1350	1.5
24F-V3	SP/SP			650	650	200	1.8	1600	560	175	90	1.6	1150	1700	1.6
The following combination is balanced and is intended for members continuous or cantilevered over supports and provides equal capacity in both positive and negative bending.															
24F-V5	SP/SP	2400	2400	650	650	200	1.7	1600	560	175	90	1.5	1150	1700	1.5
Wet-use factors[b]		0.8	0.8	0.53	0.53	0.875	0.833	0.8	0.53	0.875	0.875	0.833	0.8	0.73	0.833

Footnotes for Example (See *AITC 117-93--Design*)

[a] The combinations in this table are applicable to members consisting of 4 or more laminations and are intended primarily for members stressed in bending due to loads applied perpendicular to the wide faces of the laminations. Design values are tabulated, however, for loading both perpendicular and parallel to the wide faces of the laminations. For combinations and design values applicable to members loaded primarily axially or parallel to the wide faces of the laminations, and for members of 2 or 3 laminations, see Table 2, *AITC 117-93--Design*.

[b] Wet-use factors apply to all species.

[c] The combinations symbols relate to a specific combination of grades and species in *AITC 117—Manufacturing* that will provide the design values shown for the combination. The first two numbers in the combination symbol correspond to the design value in bending shown in Column 3. The letter in the combination symbol (either a "V" or an "E")indicates whether the combination is made from visually graded (V) or E-rated (E) lumber in the outer zones.

[d] The symbols used for species are DF = Douglas Fir-Larch; HF = Hem-Fir; and SP = Southern Pine.

FIGURE 7-7B ■ A partial listing of design values for laminated timbers. *Courtesy American Institute of Timber Construction.*

5 1/8 x 13 1/2 PURLIN DF/HF 2400 fb –V5

In addition to the information specified on the drawing by the beam, the appearance grade and information about the finish are specified in the written specifications. A typical notation for laminated beams would resemble

ALL LAMINATED BEAMS TO BE INDUSTRIAL APPEARANCE UNLESS NOTED.

Drawing Representation

Laminated beams are represented in plan view just as sawn beams are with the required text parallel to the beam. Figure 7-8 shows an example of referencing laminated beams on a framing plan. Depending on the complexity of the plan, a schedule can be used to list the specifications for laminated timbers. If a schedule is used,

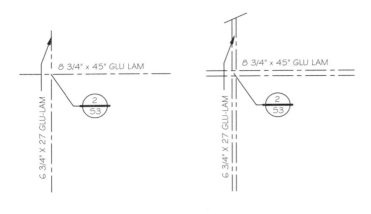

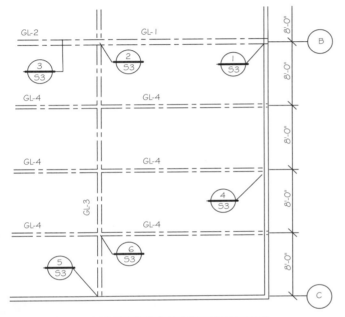

GLU-LAM BEAM SCHEDULE		
GL-1	8 3/4" x 45" GLU LAM	DF/HF 24fb-V5
GL-2	8 3/4" x 37" GLU LAM	DF/HF 24fb-V5
GL-3	6 3/4" X 27 GLU-LAM	DF/DF 20fb-V7
GL-4	5 1/8 x 15 GLU-LAM	DF/DF 20fb-V7

MANUFACTURER TO CERTIFY IN WRITING PRIOR TO
INSTALLATION OF GLU-LAMINATED BEAMS THAT
MINIMUM DESIGN VALUES HAVE BEEN PROVIDED.
ALL BEAMS UNLESS NOTED SHALL BE Fv=165,
Fc=450, fb AS NOTED, INDUSTRIAL APPEARANCE
WITH EXTERIOR GLUE. PROVIDE "A.I.T.C. CERTIFICATE
OF CONFORMANCE TO BUILDING DEPT. PRIOR
TO ERECTION.

FIGURE 7-8 ■ Laminated beams are represented similar to the method used to represent wood beams. As plans become more complicated, schedules are often used to maintain drawing clarity.

a beam symbol is placed near the beam so that appropriate material can be easily referenced. Notice that many of the specifications are given in a general note. Sections and details may only refer to the beam size, grade, and material.

Similar to the use of sawn lumber, laminated beams will require the use of many details to explain connection. Typically details showing beam-to-beam, beam-to-post, truss-to-beam, or rafter-to-beam connections, and column-to-support connections are required. When the end of a beam is represented in a detail, the lamination lines should be shown as at D in Figure 7-4. When the beam is seen in a side view, the lam lines should be represented with thin light lines. This can be done using an AutoCAD plot style such as Grayscale instead of Monochrome. Figure 7-9 shows an example of a beam-to-beam connection.

Beam and Framing Types

Laminated beams can be configured into almost any shape to meet the criteria of the design team. Common shapes include straight, cambered, and arches. Figure 7-10 shows the common shapes into which laminated beams are fabricated.

Single Span

Single-span laminated beams similar to those in Figure 7-11 are used to span between two or more supports for large openings. Because of their structural capabilities, a laminated beam can span much larger distances than a similarly sized sawn beam. Figure 7-12 shows a comparison of sawn, laminated, steel, and LVL beam sizes. Just as with sawn beams, the spacing of laminated beams is based on the strength of the beam being considered, as well as the strength of the material being supported. The architectural team will dictate the use of timber or steel, based on the design criteria.

Cantilevered

A cantilever is a beam that extends past its supports. See Figure 7-13. Although sawn lumber can be cantilevered, laminated members are much more likely to be cantilevered because of the greater distances they typically span. The cantilever is used to decrease the loads on the center of the beam, by increasing the downward action on the ends of the beam. As the ends of the beam are forced downward, the center is forced up, which reduces its natural tendency to sag. Figure 7-14 compares the reactions of a single-span beam and a cantilevered beam. This loading pattern is often used on timber roofs forming a framing system called a panelized roof system.

Panelized Framing

A panelized floor or roof system typically is comprised of beams placed in parallel patterns with a spacing of approximately 20' to 30' (6100 to 9100 mm). (See Figure 7-15.) Smaller beams called purlins are used to span between the beams. A spacing of 8' (2400 mm) is typically used for purlins. Joists of 2" or 3" (50 or 75 mm) width lumber is used to span between the purlins and support the floor or roof sheathing. Figure 7-2 shows an example of how a panelized roof is referenced on a framing plan. Figure 7-16 shows common details for beam connections for a panelized roof.

Curved Beams

As loads and spans are increased, a camber or bend is built into the beam. The camber, which is also known as a crown, is built into

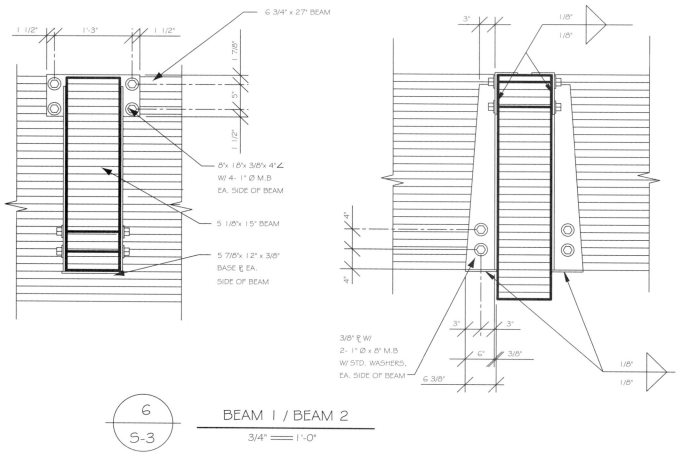

6 3/4" x 27" BEAM

1 1/2" 1'-3" 1 1/2"

1 7/8"

5"

1 1/2"

8"x 18"x 3/8"x 4" ∠
W/ 4- 1" Ø M.B
EA. SIDE OF BEAM

5 1/8"x 15" BEAM

5 7/8"x 12" x 3/8"
BASE ℞ EA.
SIDE OF BEAM

3"

1/8"
1/8"

4"

4"

3/8" ℞ W/
2- 1" Ø x 8" M.B
W/ STD. WASHERS,
EA. SIDE OF BEAM

3" 3"

6" 3/8"

6 3/8"

1/8"
1/8"

6
S-3

BEAM 1 / BEAM 2

3/4" = 1'-0"

FIGURE 7-9 ■ Laminated beam representation in detail.

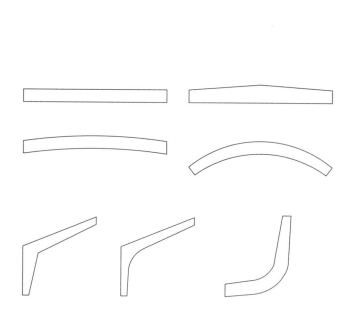

FIGURE 7-10 ■ Common shapes of laminated beams.

FIGURE 7-11 ■ A sawn lumber post supports laminated beams, which
support OSB solid-web trusses. *Courtesy Georgia-Pacific Corporation. All
rights reserved.*

DESIGN CONVERSION TABLES[1] F_b = 2400 psi   E = 1,800,000 psi

EQUIVALENT GLULAM SECTIONS FOR SOLID SAWN BEAM

Sawn[4] Section Nominal Size	Roof Beams[1,2]				Floor Beams[1,3]			
	Select Structural		No. 1		Select Structural		No. 1	
	Douglas Fir	Southern Pine[7]	Douglas Fir	Southern Pine[7]	Douglas Fir	Southern Pine[7]	Douglas Fir	Southern Pine[7]
3 x 8	3-1/8 x 6	3 x 6-7/8	3-1/8 x 6	3 x 5-1/2	3-1/8 x 7-1/2	3 x 6-7/8	3-1/8 x 7-1/2	3 x 6-7/8
3 x 10	3-1/8 x 7-1/2	3 x 8-1/4	3-1/8 x 6	3 x 6-7/8	3-1/8 x 9	3 x 9-5/8	3-1/8 x 9	3 x 9-5/8
3 x 12	3-1/8 x 9	3 x 9-5/8	3-1/8 x 7-1/2	3 x 8-1/4	3-1/8 x 10-1/2	3 x 11	3-1/8 x 10-1/2	3 x 11
3 x 14	3-1/8 x 9	3 x 11	3-1/8 x 7-1/2	3 x 9-5/8 *	3-1/8 x 13-1/2	3 x 13-3/4 *	3-1/8 x 13-1/2	3 x 12-3/8
4 x 6	3-1/8 x 6	3 x 6-7/8	3-1/8 x 6	3 x 5-1/2	3-1/8 x 6	3 x 6-7/8	3-1/8 x 6	3 x 6-7/8
4 x 8	3-1/8 x 7-1/2	3 x 8-1/4	3-1/8 x 6	3 x 6-7/8	3-1/8 x 9	3 x 8-1/4	3-1/8 x 7-1/2	3 x 8-1/4
4 x 10	3-1/8 x 9	3 x 11 *	3-1/8 x 7-1/2	3 x 8-1/4	3-1/8 x 10-1/2	3 x 11 *	3-1/8 x 10-1/2	3 x 9-5/8
4 x 12	3-1/8 x 10-1/2	3 x 12-3/8	3-1/8 x 9	3 x 9-5/8	3-1/8 x 12	3 x 12-3/8	3-1/8 x 12	3 x 12-3/8
4 x 14	3-1/8 x 12	3 x 13-3/4	3-1/8 x 10-1/2	3 x 11	3-1/8 x 15	3 x 15-1/8	3-1/8 x 15	3 x 13-3/4
4 x 16	3-1/8 x 13-1/2	3 x 15-1/8	3-1/8 x 10-1/2	3 x 12-3/8	3-1/8 x 16-1/2	3 x 16-1/2	3-1/8 x 16-1/2	3 x 16-1/2
6 x 8	5-1/8 x 7-1/2	5 x 6-7/8	5-1/8 x 7-1/2	5 x 6-7/8	5-1/8 x 7-1/2	5 x 8-1/4	5-1/8 x 7-1/2	5 x 8-1/4
6 x 10	5-1/8 x 9	5 x 8-1/4	5-1/8 x 7-1/2	5 x 8-1/4	5-1/8 x 10-1/2	5 x 9-5/8	5-1/8 x 10-1/2	5 x 9-5/8
6 x 12	5-1/8 x 10-1/2	5 x 9-5/8	5-1/8 x 9	5 x 9-5/8	5-1/8 x 12	5 x 12-3/8	5-1/8 x 12	5 x 12-3/8
6 x 14	5-1/8 x 12	5 x 12-3/8 *	5-1/8 x 10-1/2	5 x 11	5-1/8 x 13-1/2	5 x 13-3/4	5-1/8 x 13-1/2	5 x 13-3/4
6 x 16	5-1/8 x 13-1/2	5 x 13-3/4	5-1/8 x 12	5 x 12-3/8	5-1/8 x 15-1/8	5 x 15-1/8	5-1/8 x 15-1/8	5 x 15-1/8
6 x 18	5-1/8 x 15	5 x 15-1/8	5-1/8 x 13-1/2	5 x 13-3/4	5-1/8 x 18	5 x 17-7/8	5-1/8 x 18	5 x 17-7/8
6 x 20	5-1/8 x 16-1/2	5 x 16-1/2	5-1/8 x 16-1/2	5 x 15-1/8	5-1/8 x 19-1/2	5 x 19-1/4	5-1/8 x 19-1/2	5 x 19-1/4
8 x 10	6-3/4 x 9	6-3/4 x 8-1/4	6-3/4 x 9	6-3/4 x 8-1/4	6-3/4 x 10-1/2	6-3/4 x 9-5/8	6-3/4 x 10-1/2	6-3/4 x 9-5/8
8 x 12	6-3/4 x 10-1/2	6-3/4 x 9-5/8	6-3/4 x 10-1/2	6-3/4 x 9-5/8	6-3/4 x 12	6-3/4 x 12-3/8	6-3/4 x 12	6-3/4 x 12-3/8
8 x 14	6-3/4 x 12	6-3/4 x 12-3/8	6-3/4 x 12	6-3/4 x 11	6-3/4 x 13-1/2	6-3/4 x 13-3/4	6-3/4 x 13-1/2	6-3/4 x 13-3/4
8 x 16	6-3/4 x 13-1/2	6-3/4 x 13-3/4	6-3/4 x 13-1/2	6-3/4 x 12-3/8	6-3/4 x 16-1/2	6-3/4 x 15-1/8	6-3/4 x 16-1/2	6-3/4 x 15-1/8
8 x 18	6-3/4 x 15	6-3/4 x 15-1/8	6-3/4 x 15	6-3/4 x 13-3/4	6-3/4 x 18	6-3/4 x 17-7/8	6-3/4 x 18	6-3/4 x 17-7/8
8 x 20	6-3/4 x 18	6-3/4 x 16-1/2	6-3/4 x 16-1/2	6-3/4 x 16-1/2	6-3/4 x 19-1/2	6-3/4 x 19-1/2	6-3/4 x 19-1/2	6-3/4 x 19-1/4
8 x 22	6-3/4 x 19-1/2	6-3/4 x 17-7/8	6-3/4 x 18	6-3/4 x 17-7/8	6-3/4 x 21	6-3/4 x 22	6-3/4 x 21	6-3/4 x 22

EQUIVALENT GLULAM SECTIONS FOR STEEL BEAMS

Steel[5] Section	Roof Beams[1,2]				Floor Beams[1,3]			
	Douglas Fir		Southern Pine[7]		Douglas Fir		Southern Pine[7]	
W 6 x 9	3-1/8 x 10-1/2 or	5-1/8 x 9	3 x 11 or	5 x 8-1/4	3-1/8 x 10-1/2 or	5-1/8 x 9	3 x 11 or	5 x 9-5/8
W 8 x 10	3-1/8 x 12	5-1/8 x 9	3 x 12-3/8	5 x 9-5/8	3-1/8 x 13-1/2	5-1/8 x 12	3 x 13-3/4	5 x 11
W 12 x 14	3-1/8 x 16-1/2	5-1/8 x 13-1/2	3 x 16-1/2	5 x 13-3/4 *	3-1/8 x 18	5-1/8 x 15	3 x 17-7/8	5 x 15-1/8
W 12 x 16	3-1/8 x 18	5-1/8 x 13-1/2	3 x 17-7/8	5 x 13-3/4	3-1/8 x 19-1/2	5-1/8 x 16-1/2	3 x 19-1/4	5 x 16-1/2
W 12 x 19	3-1/8 x 19-1/2	5-1/8 x 16-1/2	3 x 20-5/8 *	5 x 15-1/8	3-1/8 x 21	5-1/8 x 18	3 x 20-5/8	5 x 17-7/8
W 10 x 22	3-1/8 x 21	5-1/8 x 16-1/2	3 x 20-5/8	5 x 16-1/2	5-1/8 x 19-1/2	5-1/8 x 16-1/2	3 x 20-5/8	5 x 17-7/8 *
W 12 x 22	5-1/8 x 18	6-3/4 x 15	3 x 22	5 x 17-7/8 *	5-1/8 x 19-1/2	6-3/4 x 16-1/2	5 x 19-1/4	5 x 16-1/2
W 14 x 22	5-1/8 x 18	6-3/4 x 16-1/2	3 x 23-3/8	5 x 17-7/8	5-1/8 x 21	6-3/4 x 18	5 x 20-5/8	6-3/4 x 17-7/8
W 12 x 26	5-1/8 x 19-1/2	6-3/4 x 18	5 x 19-1/4	6-3/4 x 16-1/2	5-1/8 x 21	6-3/4 x 19-1/2	5 x 20-5/8	6-3/4 x 19-1/4
W 14 x 26	5-1/8 x 21	6-3/4 x 18	5 x 20-5/8	6-3/4 x 17-7/8	5-1/8 x 21	6-3/4 x 19-1/2	5 x 22	6-3/4 x 19-1/4
W 16 x 26	5-1/8 x 21	6-3/4 x 19-1/2	5 x 20-5/8	6-3/4 x 17-7/8	5-1/8 x 22-1/2	6-3/4 x 21	5 x 23-3/8	6-3/4 x 20-5/8
W 12 x 30	5-1/8 x 21	6-3/4 x 19-1/2	5 x 20-5/8	6-3/4 x 17-7/8	5-1/8 x 21	6-3/4 x 19-1/2	5 x 22	6-3/4 x 19-1/4
W 14 x 30	5-1/8 x 22-1/2	6-3/4 x 19-1/2	5 x 22	6-3/4 x 19-1/4	5-1/8 x 22-1/2	6-3/4 x 21	5 x 23-3/8	6-3/4 x 20-5/8
W 16 x 31	5-1/8 x 24	6-3/4 x 21	5 x 23-3/8	6-3/4 x 20-5/8	5-1/8 x 25-1/2	6-3/4 x 22-1/2	5 x 24-3/4	6-3/4 x 23-3/8
W 14 x 34	5-1/8 x 24	6-3/4 x 21	5 x 23-3/8	6-3/4 x 20-5/8	5-1/8 x 24	6-3/4 x 22-1/2	5 x 24-3/4	6-3/4 x 22
W 18 x 35	5-1/8 x 27	6-3/4 x 24	5 x 26-1/8	6-3/4 x 22	5-1/8 x 27	6-3/4 x 25-1/2	5 x 27-1/2	6-3/4 x 24-3/4
W 16 x 40	5-1/8 x 28-1/2	6-3/4 x 25-1/2	5 x 27-1/2	6-3/4 x 23-3/8	5-1/8 x 27	6-3/4 x 25-1/2	5 x 27-1/2	6-3/4 x 24-3/4
W 21 x 44	5-1/8 x 33	6-3/4 x 28-1/2	5 x 31-5/8 *	6-3/4 x 27-1/2	5-1/8 x 33	6-3/4 x 30	5 x 33	6-3/4 x 30-1/4
W 18 x 50	5-1/8 x 34-1/2	6-3/4 x 30	5 x 33 *	6-3/4 x 28-7/8	5-1/8 x 31-1/2	6-3/4 x 28-1/2	5 x 31-5/8	6-3/4 x 28-7/8
W 21 x 50	5-1/8 x 34-1/2	6-3/4 x 31-1/2		6-3/4 x 28-7/8	5-1/8 x 34-1/2	6-3/4 x 31-1/2	5 x 34-3/8	6-3/4 x 31-5/8
W 18 x 55	5-1/8 x 36	6-3/4 x 31-1/2	5 x 34-3/8 *	6-3/4 x 30-1/4	5-1/8 x 33	6-3/4 x 30	5 x 33	6-3/4 x 30-1/4
W 21 x 62		6-3/4 x 36	5 x 34-3/8	6-3/4 x 34-3/8		6-3/4 x 34-1/2		6-3/4 x 34-3/8

EQUIVALENT GLULAM SECTIONS FOR LVL BEAMS

LVL[6] Section	Roof Beams[1,2]				Floor Beams[1,3]			
	Douglas Fir		Southern Pine[7]		Douglas Fir		Southern Pine[7]	
2 pcs. 1-3/4 x 9-1/2	3-1/8 x 10-1/2 or	5-1/8 x 9	3 x 11 or	5 x 8-1/4	3-1/8 x 10-1/2 or	5-1/8 x 9	3 x 11 or	5 x 9-5/8
2 pcs. 1-3/4 x 11-7/8	3-1/8 x 13-1/2	5-1/8 x 10-1/2	3 x 13-3/4	5 x 11	3-1/8 x 13-1/2	5-1/8 x 12	3 x 13-3/4	5 x 11
2 pcs. 1-3/4 x 14	3-1/8 x 15	5-1/8 x 12	3 x 15-1/8	5 x 12-3/8	3-1/8 x 16-1/2	5-1/8 x 13-1/2	3 x 16-1/2 *	5 x 13-3/4
2 pcs. 1-3/4 x 16	3-1/8 x 18	5-1/8 x 13-1/2	3 x 17-7/8	5 x 13-3/4	3-1/8 x 18	5-1/8 x 15	3 x 17-7/8	5 x 15-1/8
2 pcs. 1-3/4 x 18	3-1/8 x 19-1/2	5-1/8 x 15	3 x 19-1/4	5 x 15-1/8	3-1/8 x 19-1/2	5-1/8 x 16-1/2	3 x 20-5/8	5 x 16-1/2 *
3 pcs. 1-3/4 x 9-1/2	3-1/8 x 13-1/2	5-1/8 x 10-1/2	3 x 13-3/4	5 x 11	3-1/8 x 12	5-1/8 x 10-1/2	3 x 12-3/8	5 x 11
3 pcs. 1-3/4 x 11-7/8	3-1/8 x 16-1/2	5-1/8 x 13-1/2	3 x 16-1/2	5 x 12-3/8	3-1/8 x 15	5-1/8 x 13-1/2	3 x 15-1/8	5 x 13-3/4
3 pcs. 1-3/4 x 14	3-1/8 x 19-1/2	5-1/8 x 15	3 x 19-1/4	5 x 15-1/8	3-1/8 x 18	5-1/8 x 15	3 x 17-7/8	5 x 15-1/8
3 pcs. 1-3/4 x 16	5-1/8 x 19-1/2	6-3/4 x 16-1/2	3 x 22	5 x 16-1/2	3-1/8 x 21	5-1/8 x 18	3 x 20-5/8	5 x 17-7/8
3 pcs. 1-3/4 x 18	5-1/8 x 24	6-3/4 x 19-1/2	3 x 24-3/8 *	5 x 19-1/4	3-1/8 x 22-1/2	5-1/8 x 19-1/2	3 x 23-3/8	5 x 19-1/4

FIGURE 7-12 ■ A comparison of simple-span laminated beams to wood steel beams and LVL beams.

FIGURE 7-13 ■ A cantilevered beam is a beam that extends past its supports. The cantilever is used to decrease the loads on the center of the beam, by increasing the downward action on the ends of the beam. As the ends of the beam are forced downward, the center is forced up, which reduces its natural tendency to sag. *Courtesy LeRoy Cook.*

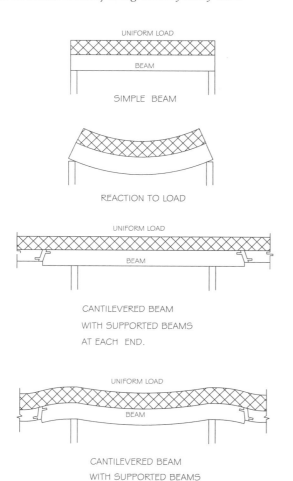

FIGURE 7-14 ■ Beam reactions of simple and cantilevered beams. A simple beam will sag in the center. If each end of the beam is extended past its supports, the loads on the end of the beam force the center of the beam upward, helping to resist the tendency to sag in the center.

FIGURE 7-15 ■ The framing of a panelized roof. *Courtesy American Institute of Timber Construction.*

beams to allow for the natural deflection that occurs in a beam as loads are applied. Laminated beams can be curved so that the beam can be used in domes or they can be made so that the top of the beam is inclined to provide the pitch for a roof. The bottom of the beam can either be flat or curved. The size of the curve will depend on the length of the beam and the size of the material used to fabricate the beam. Beams can be curved and formed into almost any shape to meet the design criteria of the architectural team.

Arches, Vaults, and Domes

Arches, vaults, and domes have been used for centuries to span large areas. Arches come in many shapes and can be thought of as structural ribs with a skin placed between them. Vaults and domes can be either a ribbed structure with a skin between ribs, or they can be created as a shell. When they are created in a shell form, concrete is most often the material of choice.

Arches

Timber, steel, or concrete can be used to form an arch. Common arch shapes include circular or elliptical. The most common arches in construction include the fixed arch, a two-hinged arch, and a three-hinged arch. Each can be seen in Figure 7-17. Fixed arches are typically associated with smaller spans, and double and triple-hinged arches are used for long spans. There are no real hinges used with arches. The number of hinges refers to the point where forces causing rotation are resisted by the connection method. A two-hinge arch uses one beam to form the required arch. A three-hinged arch uses two beams to complete the arch to be formed. Triple-

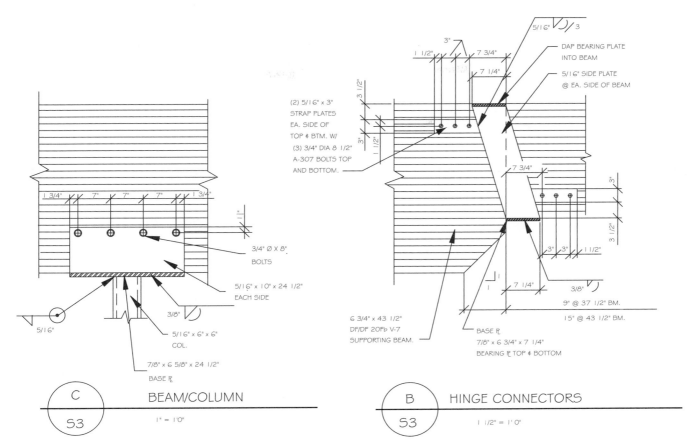

FIGURE 7-16 ■ A beam-to-column and a beam-to-beam connection. The cantilevered beam supports the smaller beam by use of a metal support called a saddle.

hinged arches are often used to provide a roof structure for arenas and convention centers. As with other types of wood construction, intersections of structural members must be specified on the framing plans and details. Figure 7-18 shows an arch foundation connection.

Vaults

Vaults are produced by placing arches next to each other to form the elongated shapes shown in Figure 7-19. When adjacent arches are placed perpendicular to their planes, a ribbed ceiling as found in most European cathedrals is constructed. Vaults can be constructed between rectangular shapes in plan view.

Domes

When an arch is rotated around itself at its crown, a dome is formed. A dome structure will form a circular shape in plan view. Figure 7-20 shows an example of a dome framed with timber.

STRUCTURAL CONNECTORS

Just as with lumber connections, timber connections must also be able to resist the stress placed on them. Bolts, metal connectors, and timber connectors are the most common methods of connecting timbers.

Bolts and Washers

Bolts used in the construction industry include anchor bolts, carriage bolts, and machine bolts. Each can be seen in Figure 7-21. Washers are used under the head and nut for most bolting applications. Washers keep the head and nut from pulling through the lumber and reduce lumber damage by spreading the stress from the

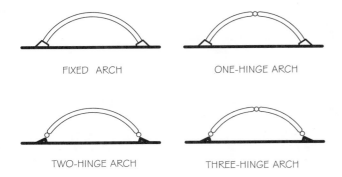

FIGURE 7-17 ■ Common arches.

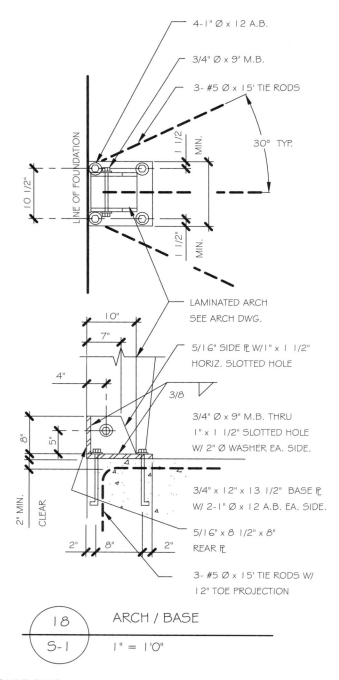

FIGURE 7-18 ■ An arch-to-foundation detail.

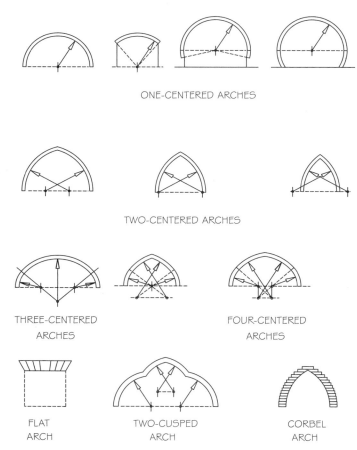

FIGURE 7-19 ■ Common methods of forming arches.

to receive a nut. A 2" (50 mm) washer is typically used with anchor bolts. Anchor bolts can be abbreviated on the structural drawings with the letters A.B., along with a specification including the size of the member to be connected, the bolt diameter, length, embedment into concrete, spacing, and washer size. A typical note might resemble the following:

FIGURE 7-20 ■ A lamella frame is a dome framed from a series of perpendicular ribs. *Courtesy American Institute of Timber Construction.*

bolt across more wood fibers. Typically a circular washer, specified by its diameter, is used.

Anchor Bolt

An anchor bolt is an L-shaped bolt that is inserted into concrete and used to bolt lumber to the concrete (at A in Figure 7-21). The short leg of the L is inserted into the concrete and provides resistance from withdrawal. The upper end of the long leg is threaded

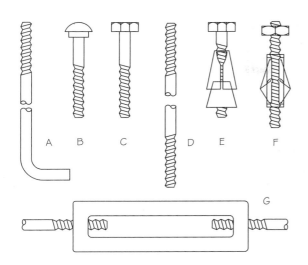

FIGURE 7-21 ■ Common bolts used in construction: A, Anchor bolt; B, Carriage bolt; C, Machine bolt; D, Drift bolt (threaded rod); E, Expansion bolt; F, Toggle bolt; G, Turnbuckle and threaded rods.

2×6 DFPT SILL W/5/8" Ø. × 12" A.B. @ 48" O.C. W/ 2"

DIA. WASHERS. PROVIDE 9" MIN. EMBEDMENT.

When anchor bolts are used to attach lumber to the side of a concrete wall, a specification is usually given to stagger the bolts in relation to the edge of the lumber to be attached as well as the bolt spacing. A typical specification might resemble

3 × 12 DFPT LEDGER W/3/4" Ø × 8" A.B. @ 32" O.C.

STAGGERED 3" UP/DN WITH 2" DIA. WASHERS.

Carriage Bolts

A carriage bolt is used for connecting steel and other metal members as well as timber connections. Carriage bolts (at B in Figure 7-21) have a rounded head with the lower portion of the shaft threaded. Directly below the head, at the upper end of the shaft is a square shank. As the shank is pulled into the lumber, it will keep the bolt from spinning as the nut is tightened. Diameters range from 1/4" to 1" (6 to 25 mm) with lengths typically available to 12" (300 mm). The specification for a carriage bolt will be similar to an anchor bolt except for the designation of the bolt type.

Machine Bolts

Bolts with a hexagonal head and a threaded shaft are described as machine bolts (M.B.). Machine bolts (at C in Figure 7-21) are divided into the classifications of unfinished or high strength bolts. Common bolts are made from A-307 low-carbon steel and used for attaching steel-to-steel, steel-to-wood or wood-to-wood connections. Common bolts are often used in steel joints to provide a temporary connection while field welds are completed. Bolts are assumed to be common unless high-strength bolts are specified and are referred to with a note such as:

Use (3)-3/4" Ø × 8" M.B. @ 3" o.c. w/ 1 1/2" Ø washers.

The strength of bolts will be called out in a general note on the framing plans or on pages of details containing bolted connections. Although the engineer will determine the bolt locations based on the stress to be resisted, bolts for wood members are usually placed:

- 1 1/2" (40 mm) from an edge parallel to the grain
- 3" (75 mm) minimum from the edge when perpendicular to the grain
- 1 1/2" (40 mm) from the edge of steel members
- 2" (50 mm) from the edge of concrete

The pilot hole for the bolts in wood is usually not specified but is assumed to be 1/16" (2 mm) bigger than the bolt shaft. If an elongated hole is used to allow slight movement for field adjustment, the hole and the bolt will be specified with a note such as

3/4" Ø × 8" M.B. THRU 7/8" × 1 1/2" SLOTTED HOLE W/

2" Ø WASHER EA. SIDE.

High Strength Bolts

Bolts used for connecting timber and structural steel are referred to as high-strength bolts and come in several common compositions. These bolts are manufactured with an ASTM (American Society for Testing Materials) or Institute of Steel Construction Specifications grading number on the head. Common specifications used in commercial construction include the following:

- A-325 high-strength bolts
- A-490 high-strength bolts, medium carbon steel
- A-441 high-strength, low-alloy steel bolts
- A-242 corrosion-resistant high-strength low-alloy steel bolts

Bolts made of A-325 steel come in three classifications:

- Type 1 bolts are made from medium-carbon steel in sizes from 1/2" to 1 1/2" (13 to 40 mm) diameter.
- Type 2 bolts are made from low-carbon steel in sizes of 1/2" to 1" (13 to 25 mm).
- Type 3 bolts are made of weathering steel. Bolts labeled A-490 are made from medium-carbon steel.

High-strength bolts are normally specified to be tightened with a pneumatic impact wrench. Chapter 8 will further discuss steel requirements and specifications.

Miscellaneous Bolt Types

Several other types of bolts are used for special construction circumstances. These include studs, drift bolts, expansion bolts, and toggle bolts.

- A stud is a bolt that has no head. A stud is welded to a steel beam so that a wood plate can be bolted to the beam. A drift bolt (at D in Figure 7-21) is a steel rod that has been threaded.

■ Threaded rods can be driven into one wood member with another member bolted to the threaded protrusion. Threaded rods can also be used to span between metal connectors on two separate beams (as seen in Figure 7-22) or to connect concrete panels. Threaded rods are also used with steel construction to provide lateral support. (See Figure 7-23.)

■ Expansion bolts are used for connecting lumber to masonry. Expansion bolts have a special expanding sleeve (at E in Figure 7-21) that will expand once inserted into a hole to increase holding power.

■ A toggle bolt has a nut that is designed to expand once inserted through a hole, so that it cannot be removed. Toggle bolts (at F in Figure 7-21) are used where one end of the bolt may not be accessible due to construction parameters.

Timber Connectors

Timber connections are often reinforced by metal rings, plates, and disks that are embedded into each piece of lumber at the joint to resist sliding. The engineer will determine the size and number of connectors per joint based on the stress that must be resisted. A groove is typically precut into adjoining pieces of timber to hide the metal connector. The connector is placed between the members to be joined and forced into the groove as bolts are tightened. A toothed ring is forced into each member by pressure, usually from

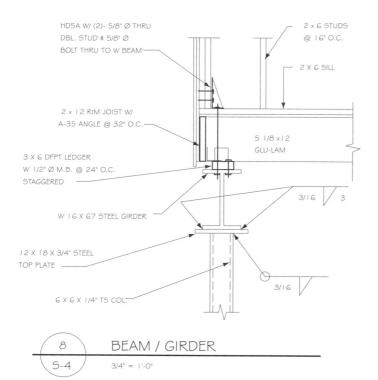

FIGURE 7-22 ■ Framing anchors connected by drift bolts or metal straps are often used to tie structural members of one level to those of another level.

FIGURE 7-23 ■ A drift bolt and turnbuckle used to provide lateral support between steel columns. *Courtesy Megan Jefferis.*

hammering. Shear plates are set in pairs to join lumber. The face of a shear plate is set flush with each timber to be joined and then bolted together.

Metal Framing Connectors

Pre-manufactured metal connectors made by companies such as Simpson or Tyco are used at many timber-to-timber connections to strengthen nailed connections. Beam hangers, post caps and bases, and straps are some of the most common metal connectors used with timber construction. Examples of pre-manufactured connectors can be seen in Figure 7-24. These connectors are usually made from metal, ranging from 3- to 7-gauge. Timber connectors are often held in place by bolting rather than nailing because of the heavy loads being carried. Figure 7-25 shows a detail specifying a pre-manufactured metal connector and the required nailing. Because the hanger is pre-manufactured, the drafter must only specify the size of the hanger and the required nailing or bolting size and quantity. The location of bolts is not required because the holes for them are predrilled. Metal connectors are typically specified on the framing plans, sections, and details by listing the model number and type of connector, just as with lightweight metal connectors. When seen in end view, the metal connector is represented by pairs of parallel diagonal lines.

If pre-manufactured connectors that can support the required loads are not available, the engineer must design a steel connector to be fabricated for a specific joint. These connectors are assembled in a shop and shipped to the job site. The drafter will need to specify the material, welds, and exact bolt locations. Figure 7-26 shows an example of a fabricated metal connector.

Additional Reading

The following Web sites can be used as a resource to help you keep current with changes in timber framing materials and methods.

ADDRESS	COMPANY/ORGANIZATION
www.afandpa.org	American Forest & Paper Association

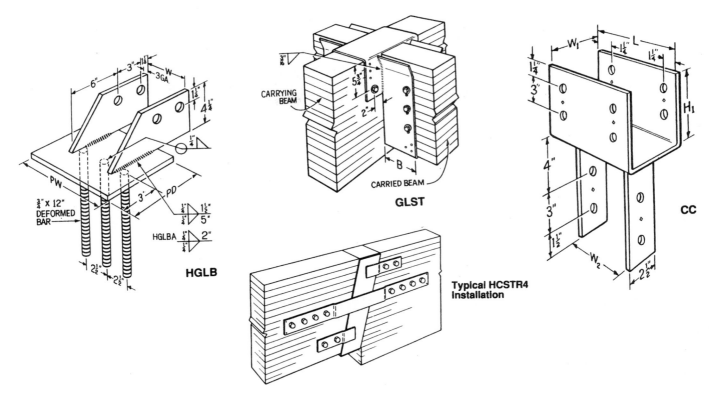

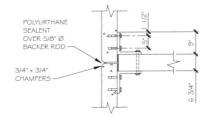

FIGURE 7-24 ■ Common pre-manufactured metal connectors. *Courtesy Simpson Strong-Tie Co., Inc.*

www.aitc-glulam.org   American Institute of Timber Construction

www.apawood.org   APA - The Engineered Wood Association

www.awc.org   American Wood Council

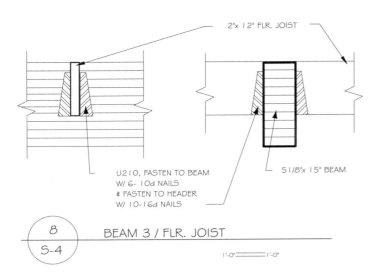

FIGURE 7-25 ■ A beam-to-beam detail with premanufactured metal hangers does not require dimensions to locate hangers' nails or bolts.

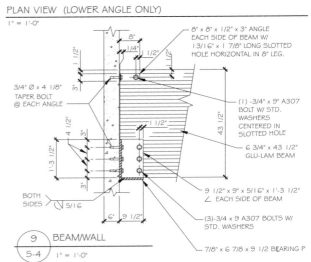

FIGURE 7-26 ■ Metal connectors that are to be fabricated require all steel sizes and hole locations to be specified and dimensioned.

www.bc.com	Boise Cascade Corporation	www.sfpa.org	Southern Forest Products Association
www.bcewp.com	Boise Cascade engineered wood products	www.southernpine.com	Southern Pine Council
www.calredwood.org	California Redwood Association	www.tfguild.org	Timber Framers Guild
www.cwc.ca	Canadian Wood Council	www.timberpeg.com	Timberpeg
www.forestdirectory.com	Directory of Products, Wood Science and Marketing	www.todayscedar.org	Western Red Cedar Lumber Association
www.iwpawood.org	International Wood Products Association	www.wwpa.org	Western Wood Products Association
www.lpcorp.com	Louisiana-Pacific Corporation	www.weyerhaeuser.com	Weyerhaeuser
www.strongtie.com	Simpson Strong-Tie Co., Inc.	www.wii.com	Willamette Industries, Inc.

CHAPTER 7

Timber Framing Methods and Materials

CHAPTER QUIZ

Answer the questions on a separate sheet of paper. Print the chapter title, question number, and a short complete statement.

Question 7-1 List three advantages of using timber construction.

Question 7-2 List five types of laminated beams and show a sketch of how they could be used in a structure.

Question 7-3 List six common widths of laminated beams.

Question 7-4 At what size do sawn posts become uneconomical and difficult to guarantee quality?

Use the appropriate charts from this chapter or reference material to answer the following questions.

Question 7-5 A laminated beam is to be 24-V5. What are the options for materials?

Question 7-6 If the fir option is used, what will be the value for E?

Question 7-7 If the other option is used, what is the category listed in column 6 of Figure 7-7b, and what is the value?

Question 7-8 A plan calls for a 6 × 14 DFL #1 floor beam or equal. Give the alternative laminated and steel beam sizes.

Question 7-9 List two alternatives for describing beams on a framing plan.

Question 7-10 List seven types of information typically specified for a laminated beam and then provide an example of a beam specification.

Question 7-11 What properties make heavy timber construction more fire-resistant than western platform construction?

Question 7-12 What is the minimum size of a post used to support floor loads using heavy timber framing methods?

Question 7-13 What would be the depth of beams with 9, 15, and 24 laminations?

Question 7-14 List and explain the eight symbols typically used to describe the materials used to fabricate laminated beams.

Question 7-15 What grade of beam should be used for a beam that will be in a highly visible location?

Question 7-16 Why is it possible, given an equal length and load, to use a smaller cantilevered beam than a single-span beam?

Question 7-17 What is a hinge as the term relates to arches?

Question 7-18 What are the minimum sizes allowed for timber trusses?

Question 7-19 Describe how a panelized roof is made.

Question 7-20 What is the maximum allowable square footage and height for an office building constructed of heavy timber construction?

Question 7-21 A 3/4" dia. lag bolt is going to be used to join a 2 × 8 ledger to a concrete wall. What size pilot hole would you recommend?

Question 7-22 List the names of the following bolts. A.B., H.S., C.B., and M.B.

Question 7-23 What allows a carriage bolt to be tightened?

Question 7-24 List five pieces of information that might be included in a bolt specification.

Question 7-25 List the types of steel from which high-strength bolts are typically made.

Question 7-26 How is a split-ring connector used to hold timber together?

Question 7-27 List two major suppliers of timber connectors.

Question 7-28 Give the gauge of a Simpson HD5A hold down anchor.

Question 7-29 Visit the Web site for the American Institute of Timber Construction and list at least ten other associations that are related to timber.

Question 7-30 Use the Internet and visit five sites that deal with timber. Order information related to timber framing.

DRAWING PROBLEMS

Unless other instructions are given by your instructor, complete the following details and save them as wblocks. Skeletons of most details can be accessed from http://www.delmar.com/resources/ocl.html. Use these drawings as a base to complete the assignment.

■ Show all required views to describe each connection.

■ Draw each detail at a scale of 1" = 1'–0" unless noted.

■ Represent structural wood members that the cutting plan has passed through with bold lines. Specify each laminated beam as Fb2400 DF/HF.

■ Use separate layers for wood, concrete, text, laminations, and dimensions.

■ Use dimensions for locations where possible instead of notes.

■ Show and specify all connecting materials. Base nailing on IBC standards unless your instructor tells you otherwise. Specify all material based on common local practice.

■ Provide notation to specify that all metal hangers are to be provided by Simpson Strong-Tie Company or equal. Hatch each material with the appropriate hatch pattern.

■ Use an appropriate text font to label and dimension each drawing as needed. Refer to *Sweet's Catalogs*, vendor cata-

logs, or the Internet to research needed sizes and specifications. Keep all text 3/4" minimum from the drawing, and use an appropriate architectural style leader line to reference the text to the drawing.

■ Provide a detail marker, with a drawing title, scale, and problem number below each detail.

■ Assemble the details required by your instructor into a appropriate template sheet for plotting. Arrange the drawings using the guidelines presented in Chapters 2 and 3. Assign a drawing sheet number of sheet S2 of 6.

Problem 7-1 Draw a detail showing a beam-to-beam connection. Use a 6 3/4 × 15 glu-lam beam to support 6 × 12 purlins. Select and specify an appropriate metal hanger to support a 4000 lb roof load.

Problem 7-2 Draw a detail showing an 8 × 8 post connecting to concrete with an appropriate column base to resist 6,000# in uplift.

Problem 7-3 Draw a detail showing a beam-to-beam connection. Use a 6 3/4 × 27 glu-lam beam to support a 5 1/8 × 15 purlins on each side of the carrier beam. Select and specify an appropriate metal hanger to support a 9,500 lb load with provisions for seismic ties.

Problem 7-4 Draw a detail showing an 8 × 8 post supporting an 8 × 16 header. Use a suitable CC cap to connect the beams to the post and suitable hangers to provide support for 6 × 12 purlins 48" o.c. supporting 5,000# max. (1 on each side and perpendicular to the 8 × 16).

Problem 7-5 Draw a detail showing the end-to-end connection of a 5 1/8 × 36 glu-lam to a 5 1/8 × 27. Use an HCA connector capable of supporting 20,000 lb.

Problem 7-6 Two 6 3/4 × 19 1/2 fb 2400 DF/HF beams will be resting on a 6 × 6 DFL #1 post. Draw a detail using an appropriate metal column cap for the connection. Use a 12-ga. strap (1 each side) 3" down from the top of the beams with a minimum of 12" overlap for each beam.

Problem 7-7 An 8 3/4 × 22 1/2 glu-lam will cantilever 6 ft. past a 6 × 8 post and support a 8 3/4 × 16 1/2 beam. Draw the beam-to-beam connection based on the following assumptions:

a) The center of all bolts are to be 1 1/2" from the edge of steel, with 3" minimum spacing.

b) All welded connections to be 5/16" fillet connections.

c) Side plates to be made from 5/16" × 8" wide steel.

d) Base and top plates to be 3/4" × 9 5/8" × 8 1/4" steel.

e) 3 1/2" wide × 5/16" side strap with (2)-3/4" diameter × 11" M.B. Bolts to be a minimum of 10" from beam splice. Length based on bolt placement. Place the centerline of straps 4" up from bottom, 3" down from top.

Problem 7-8 Draw a side and front view showing a 6 × 6 post hidden in a 2 × 6 stud wall on the upper floor and resting on a 2 × 6/3 × 6 base plate. The lower plate is to be DFPT. The plates will rest on 1 1/8" plywood, and 2 × 12 DFL #2 floor joists at 12" o.c. The lower wall will be framed out of 2 × 6 studs with a double top plate. A 6 × 6 post will be placed directly below the upper post. A hold down anchor that can resist at least 7000 pounds in uplift will be required.

Problem 7-9 Draw a plan view for a 48' × 60' structure framed with 2 × 6 studs at 16" o.c. at a scale of 1/8" = 1'-0". Place a 6 3/4" × 16 1/2" ridge beam in the center of the structure 24' from the side walls. Specify a 6 × 6 wood post at 20' intervals along this beam with 4 × 6 posts at the ends. Specify 4 × 6 or 4 × 4 post at each beam/wall intersection. Show splices in the ridge beam 48" from the post so that the center beam is 12' long. Use 13 1/2" deep beam for the center span.

Hang 5 1/8 × 15 glu-lam beams at 15'-0" o.c. on each side of the ridge beam. Hang 3 1/8 × 13 1/2 glu-lams @ 8'-0" o.c. between each 15" glu-lam. Support these beams with (2)-2 × 6 at each wall. Hang 2 × 6 rafters at 24" o.c. between each 13 1/2" glu-lam with appropriate hangers. Show enough of the roof framing to represent the framing patterns. Provide grids to represent each wall and post so that four vertical and three horizontal grids are provided. Specify the location of all members. Use a schedule wherever possible. Assume the sheet to be S-2.

Problem 7-10 Draw details to represent the post-to-beam connections and all beam-to-beam connections represented in problem 7-9. Assume the sheet to be S-3. Place all details in a logical order based on Chapter 2 and place detail reference numbers by each detail and on the framing plan.

CHAPTER 8

Steel Framing Methods and Materials

Steel is used throughout the construction industry in many sizes, shapes, and types of construction. In addition to steel, there are many metals and metal alloys used throughout the construction of a structure. The type of metal or steel that is used will affect the type of framing used, the design of the structure, and what the drafter is required to place on the architectural and structural drawings. Four of the most common uses of metal and steel in construction include the use of steel framing members, metal buildings, specialty structures such as space frames, and steel rigid frames. Figure 8-1 shows examples of four different uses of steel in a multilevel structure.

STEEL IN CONSTRUCTION

Although a drafter will not specify the contents of the steel on the structural drawings, a knowledge of the contents will aid in the understanding the steel used throughout a structure. Steel is a metal made from various elements, but it consists of approximately 98% iron. The other major element in steel is carbon. Although steel is composed of a maximum of only 1.5% carbon, it is carbon that controls the stiffness and hardness of steel. Steel is classified by the carbon content of the steel. The carbon contents of various steels referenced on structural drawings based on the American Society of Testing Materials (ASTM) are as follows:

Carbon Content

Low-carbon steel	0.06 – 0.30
Medium-carbon steel	0.30 – 0.50
High-carbon steel	0.50 – 0.80

In addition to carbon and iron, steel contains phosphorus and sulfur. Both are limited in steel because they affect the brittleness of the steel. Brittle steel tends to crack as loads are applied. Common elements such as silicon, manganese, copper, nickel, chromium, tungsten, molybdenum, and vanadium are also added to steel. These elements comprise a very small percentage of the chemical composition of steel but affect qualities such as strength, hardness, and corrosion resistance. The elements used for making steel are strictly controlled by the ASTM.

Structural Steel

Varied amounts of elements will produce different types of steel. Strength, hardness, corrosion resistance, brittleness, and ductility are specified on the drawings. They are all controlled by varying the elements introduced into steel. Before examining various types of steel, it is important to realize how steel reacts to stress induced from a load. Although the engineer will determine the strength of steel to be used, the drafter must understand the values that are contained in the calculations. Figure 8-2 shows the stages that a low-carbon steel member will go through as it resists a load.

- In the *elastic range*, a member resisting a load will deflect, but maintain its structural integrity. This area corresponds to the modulus of elasticity of a wood member.

- The *yield point* is the measurement of stress in a material, which corresponds to the limiting value of the usefulness of the material. Once steel is stressed past its yield point, it will not return to its original shape or size. The yield point is represented by the letters Fy in beam formulas and is placed in the written specifications. The higher the yield point, the less ductile the steel.

If a material is ductile, it will deflect large amounts before breaking. Materials that are supported by a beam that has gone past its yield point may crack, but the beam itself will not actually fail.

Once the stress has entered the *work hardening range*, steel actually increases its stress carrying capability. As the steel deforms, the internal structure of the steel is dislocated, which makes it harder for future dislocations to occur.

The final stage in resisting stress is when the member reaches its *ultimate tensile strength* or breaking point. Once the ultimate tensile strength has been exceeded, the beam will cause structural failure.

Structural steels are identified on structural drawings by their ASTM designation number. Common numbers and their metric

FIGURE 8-1 ■ Many multilevel structures include the steel for the frame, trusses, decking, and wall studs. *Courtesy Jordan Jefferis.*

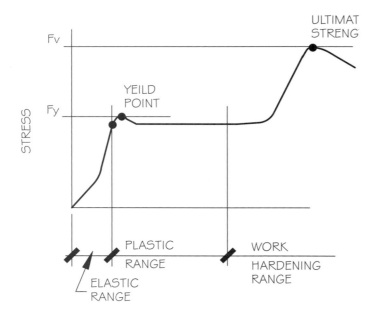

FIGURE 8-2 ■ Typical reactions to stress in mild-carbon steel.

equivalent used for construction include A-36 (A36M), A-242 (A242M, A-441 (A441M), A-529 (A529M), A-572 (A572M), and A-588 (A588M.

ASTM A-36—Steel with the designation ASTM A-36 is the most widely used structural steel. It is a mild-carbon steel with a yield point of 36 ksi (36,000 pounds per square inch) (250MPa).

ASTM A-242—Steel with a designation of ASTM A-242 represents a high strength, low alloy steel with a yield point of 50 ksi (345 MPa). It has limited amounts of carbon and manganese, which increase it welding capabilities. It is also used where self-weathering is required because of its high resistance to corrosion. A-242 steel has four-times the atmospheric corrosion resistance of carbon steel without copper.

ASTM A-441—ASTM A-441 is a high-strength, weldable, low-alloy structural manganese vanadium steel. Its yield point is 42 ksi (290 Mpa).

ASTM A-529—A-529 steel is a high-strength carbon steel with a minimum yield point of 42 ksi (290 Mpa). With the addition of 0.02% copper to the content, A-529 steel achieves twice the atmospheric corrosion resistance of A-36 steel.

ASTM A-588—A-588 is a high-strength, low-alloy steel with a minimum yield point of 50 ksi (345 Mpa). It is often referred to as weathering steel because it is left unpainted. It has greater durability, lighter weight and is four times greater in weather resistance than A-36 steel. Self-weathering steel is used for exposed steel structures such as access bridges or other exposed areas where maintenance would be costly.

Common Steel Shapes

Steel on the structural drawings can be divided into structural shapes, plates, bars, and cables. Each has its own distinctive use, shape, and specifications that must be understood by the drafter to complete the drawings. The standard designations of structural shapes of steel adopted by the American Institute of Steel Construction include *W, S, M, C, MC, L, WT,* and *MT.* These common shapes are illustrated in Figure 8-3a. Figure 8-3b shows sev-

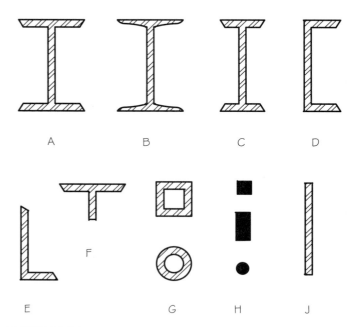

FIGURE 8-3A ■ Common steel shapes in construction include A, wide flange; B, I-beam or American standard beam; C, M shape; D, channel; E, angle; F, tee; G, tubes; H, bars; J, plates.

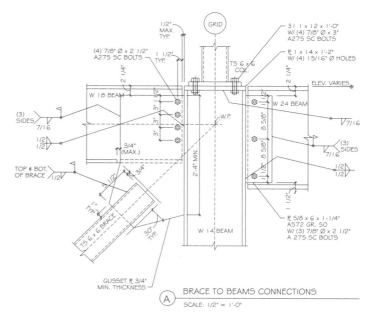

FIGURE 8-3B ■ Common shapes are often represented in detail as well as bolt and weld size, location, and spacing. *Courtesy Jim Ellsworth.*

eral of these common shapes as well as common methods of representing bolting and welding information.

Wide Flange

The *W* or *wide flange* seen in Figure 8-4 is used for columns and beams. It has the cross-sectional shape of the letter **I** and is comprised of two horizontal surfaces called flanges, and a vertical surface called the web. **W**-shaped members are relatively narrow when compared to their depth and are used for beams and girders. Shapes that are closer to square are generally used for columns. *W-shaped* steel comes in a minimum depth of 4" with a weight of 7.5 pounds per linear foot (plf) (100 mm–0.01 kg/mm), and a maximum depth of 48" and a weight of 848 plf (1200 mm–1.26 kg/mm). The *S-shaped* beams, which are sometimes called *I-beams* or American standard beams, have a narrower flange than a W-shape and have a slope of 1/6 for the inner face of the flange. Sizes range from a minimum of 3" deep with a weight of 5.7 plf (75 mm–0.008 kg/mm) to 24" with a weight of 120 plf (610 mm–0.18 kg/mm). The **M** designation refers to miscellaneous shapes that can't be classified as a **W** or **S** shape.

A **W** or **S** beam can be represented on a framing plan as illustrated in Figure 8-5. The specification for the beam can be placed

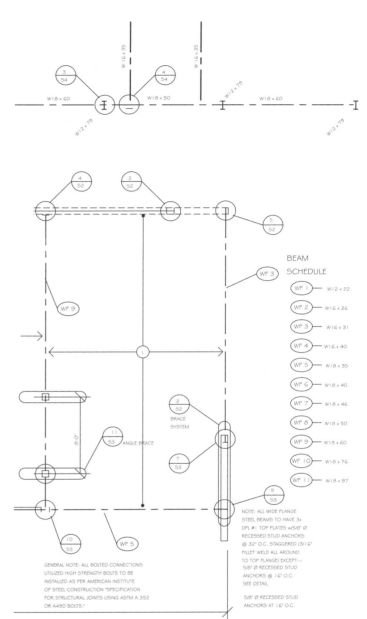

FIGURE 8-5 ■ Representing steel beam and columns in plan view. Steel members can be represented in a schedule to keep the framing plan uncluttered.

parallel to the beam or represented with a reference symbol and referred to a beam schedule. The same procedure is used for sawn and laminated beams. If a schedule is used, separate schedules should be used for timber and steel beams because they will come from different suppliers. When steel products thicker than 1/8" (3 mm) are represented in end view in a section or detail, they are cross-hatched with pairs of parallel lines at a 45° angle, as shown in Figure 8-6.

Wide-flange members are specified by numbers representing the approximate depth and weight of the beam. For example, the W18 × 46 beam in Figure 8-6 has an actual depth of 18.06" (460 mm). Other characteristics can be seen in Figure 8-7, a sample of a steel

FIGURE 8-4 ■ Wide flange steel used for beams and columns. *Courtesy Bethlehem Steel Corporation.*

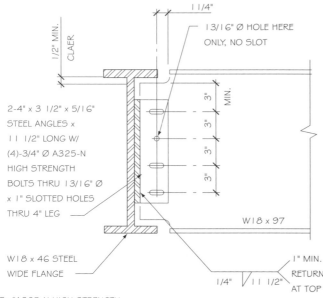

2-4" x 3 1/2" x 5/16"
STEEL ANGLES x
11 1/2" LONG W/
(4)-3/4" Ø A325-N
HIGH STRENGTH
BOLTS THRU 13/16" Ø
x 1" SLOTTED HOLES
THRU 4" LEG

1/2" MIN.
CLAER.

1 1/4"

13/16" Ø HOLE HERE
ONLY, NO SLOT

3"
3"
3"
3"
3"
MIN.

W18 x 97

W18 x 46 STEEL
WIDE FLANGE

1" MIN.
RETURN
AT TOP

1/4"   1 1/2"

NOTE: #A325-N HIGH STRENGTH
BOLTS REQUIRE SPECIAL INSPECTION
SECTION 1704 2000 IBC

5
S5

BEAM /BEAM

1 1/2"=1'-0"

FIGURE 8-6 ■ When drawn in end view in details, steel members are hatched with pairs of diagonal lines.

table published by the American Institute of Steel Construction. Drafters will have to refer to tables like these to gain the information needed to draw accurate details. The actual flange and web thickness are usually slightly exaggerated so that the beam can be easily seen.

Channels

A channel is shaped like half of an American standard beam (at D in Figure 8-3a). Channels are represented by the letters **C** for a standard channel and **MC** for miscellaneous shapes. Standard shapes range in size from 3" and 4.1 plf (76 mm–0.006 kg/mm) to 18" deep and 58 plf (460 mm–0.09 kg/mm). The inner face of each flange has a slope matching that of an S-shaped beam. Channels are represented on the plan views similar to a W-beam. The text is placed parallel to the channel and numbers are used to represent the approximate depth and pounds per linear foot of beam. A typical specification would resemble

C 12 × 20.7

Angle

An angle is a piece of steel that has a 90° bend in it (at E in Figure 8-3a). Angles may have legs of equal or unequal length. Equal-length angles range from 1 × 1 × 1/8"/0.80 plf to a maximum of 8 × 8 × 1 1/8"/56.9 plf (25 × 25 × 3/0.009 kg/mm to 200 × 200 × 29/0.08 kg/ mm). Angles are represented by the letter **L** followed

by numbers representing the height of each leg followed by the thickness. The weight is usually not specified, but a length may be specified after the thickness is listed. A typical specification would resemble

L 6 × 6 × 3/8 × 10'–0"

Angles with unequal leg lengths range in size from 1 3/4 × 1 1/4/1.23 to 8 × 6 × 1/44.2 plf (45 × 45 × 32/0.002 kg/ mm to 200 × 160 × 25/0.07 kg/mm).

When a wide-flange beam is used to support another wide flange beam, an **L** is typically welded to the supporting beam and bolted to the beam being supported. Figure 8-6 shows an example of an angle used in a beam-to-beam detail that a drafter might be expected to draw. Bolting and welding procedures used for steel connections will be introduced later in this chapter. If a wood member is to be joined to a steel beam, it is typically done by welding bolts with no heads—called *studs*—to the flange or web of the steel member. The studs are then used to bolt a wood plate or ledger to the beam. *Plates* are wood members bolted to the top of a web to support wood joists and facilitate nailing. *Ledgers* are wood members that are bolted to the flange of a steel beam. Wood joists can then be hung from the ledger with a metal hanger. The engineer will determine the size and spacing of the studs based on the type and amount of loads to be resisted, as well as the size of the weld used to connect the stud to the steel beam. Figure 8-8 shows an example of a top plate bolted to a beam with a stud.

Tees

A tee (at F in Figure 8-3a) has a cross-sectional area shaped like the letter T. Often referred to as structural tees, they are made from cutting the web of wide flange or American standard beam in half. The size of the tee is based on the beam that it is cut from and is represented by the letters **WT**, **ST**, or **MT** followed by the height and weight of the tee. The specification for a tee would resemble

WT 12 × 38

Steel Tubes and Columns

Steel columns or pipes are hollow circular shapes of steel used to support loads (at G in Figure 8-3a). Steel tubes are hollow square or rectangular shapes of steel. Figure 8-9 shows both tubes and steel plates, which provide support to steel beams. Square tubes are available in sizes ranging from 2 × 2 to 10 × 10 (50 × 50 to 250 × 250 mm) with the wall thickness ranging from 3/16" through 5/8" (5 to 16 mm) thick. Rectangular tubes range from 2 × 3 to 8 × 12 (50 × 75 to 200 × 300 mm) with a wall thickness of 3/16" to 1/2" (5 to 13 mm). Circular steel columns range from 1/2" to 12" (13 to 300 mm) with wall thickness ranging from 0.216" to 0.50" (5 to 13 mm) thick.

Tubes and columns are shown in plan view just as a wood post would be represented. When seen in a section or detail, the interior surface of the tube is represented by a hidden line as shown in Figure 8-10. The actual wall thickness is usually slightly exaggerated so it can be easily seen. Square and rectangular tubes are specified on drawings by the letters **TS** followed by the overall size and thickness. Examples of tube specifications include

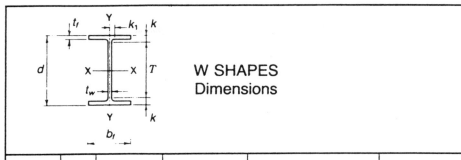

**W SHAPES
Dimensions**

Desig-nation	Area A (In.²)	Depth d (In.)		Web Thickness t_w (In.)		t_w/2 (In.)	Flange Width b_f (In.)		Flange Thickness t_f (In.)		Distance T (In.)	k (In.)	k₁ (In.)
W 18×311ᵃ	91.5	22.32	22⅜	1.520	1½	¾	12.005	12	2.740	2¾	15½	3⁷/₁₆	1³/₁₆
×283ᵃ	83.2	21.85	21⅞	1.400	1⅜	¹¹/₁₆	11.890	11⅞	2.500	2½	15½	3³/₁₆	1³/₁₆
×258ᵃ	75.9	21.46	21½	1.280	1¼	⅝	11.770	11¾	2.300	2⁵/₁₆	15½	3	1⅛
×234ᵃ	68.8	21.06	21	1.160	1³/₁₆	⅝	11.650	11⅝	2.110	2⅛	15½	2¾	1
×211ᵃ	62.1	20.67	20⅝	1.060	1¹/₁₆	⁹/₁₆	11.555	11½	1.910	1¹⁵/₁₆	15½	2⁹/₁₆	1
×192	56.4	20.35	20⅜	0.960	1	½	11.455	11½	1.750	1¾	15½	2⁷/₁₆	¹⁵/₁₆
×175	51.3	20.04	20	0.890	⅞	⁷/₁₆	11.375	11⅜	1.590	1⁹/₁₆	15½	2¼	⅞
×158	46.3	19.72	19¾	0.810	¹³/₁₆	⁷/₁₆	11.300	11¼	1.440	1⁷/₁₆	15½	2⅛	⅞
×143	42.1	19.49	19½	0.730	¾	⅜	11.220	11¼	1.320	1⁵/₁₆	15½	2	¹³/₁₆
×130	38.2	19.25	19¼	0.670	¹¹/₁₆	⅜	11.160	11⅛	1.200	1³/₁₆	15½	1⅞	¹³/₁₆
W 18×119	35.1	18.97	19	0.655	⅝	⁵/₁₆	11.265	11¼	1.060	1¹/₁₆	15½	1¾	¹⁵/₁₆
×106	31.1	18.73	18¾	0.590	⁹/₁₆	⁵/₁₆	11.200	11¼	0.940	¹⁵/₁₆	15½	1⅝	¹⁵/₁₆
× 97	28.5	18.59	18⅝	0.535	⁹/₁₆	⁵/₁₆	11.145	11⅛	0.870	⅞	15½	1⁹/₁₆	⅞
× 86	25.3	18.39	18⅜	0.480	½	¼	11.090	11⅛	0.770	¾	15½	1⁷/₁₆	⅞
× 76	22.3	18.21	18¼	0.425	⁷/₁₆	¼	11.035	11	0.680	¹¹/₁₆	15½	1⅜	¹³/₁₆
W 18× 71	20.8	18.47	18½	0.495	½	¼	7.635	7⅝	0.810	¹³/₁₆	15½	1½	⅞
× 65	19.1	18.35	18⅜	0.450	⁷/₁₆	¼	7.590	7⅝	0.750	¾	15½	1⁷/₁₆	⅞
× 60	17.6	18.24	18¼	0.415	⁷/₁₆	¼	7.555	7½	0.695	¹¹/₁₆	15½	1⅜	¹³/₁₆
× 55	16.2	18.11	18⅛	0.390	⅜	³/₁₆	7.530	7½	0.630	⅝	15½	1⁵/₁₆	¹³/₁₆
× 50	14.7	17.99	18	0.355	⅜	³/₁₆	7.495	7½	0.570	⁹/₁₆	15½	1¼	¹³/₁₆
W 18× 46	13.5	18.06	18	0.360	⅜	³/₁₆	6.060	6	0.605	⅝	15½	1¼	¹³/₁₆
× 40	11.8	17.90	17⅞	0.315	⁵/₁₆	³/₁₆	6.015	6	0.525	½	15½	1³/₁₆	¹³/₁₆
× 35	10.3	17.70	17¾	0.300	⁵/₁₆	³/₁₆	6.000	6	0.425	⁷/₁₆	15½	1⅛	¾
W 16×100	29.4	16.97	17	0.585	⁹/₁₆	⁵/₁₆	10.425	10⅜	0.985	1	13⅝	1¹¹/₁₆	¹⁵/₁₆
× 89	26.2	16.75	16¾	0.525	½	¼	10.365	10⅜	0.875	⅞	13⅝	1⁹/₁₆	⅞
× 77	22.6	16.52	16½	0.455	⁷/₁₆	¼	10.295	10¼	0.760	¾	13⅝	1⁷/₁₆	⅞
× 67	19.7	16.33	16⅜	0.395	⅜	³/₁₆	10.235	10¼	0.665	¹¹/₁₆	13⅝	1⅜	¹³/₁₆
W 16× 57	16.8	16.43	16⅜	0.430	⁷/₁₆	¼	7.120	7⅛	0.715	¹¹/₁₆	13⅝	1⅜	⅞
× 50	14.7	16.26	16¼	0.380	⅜	³/₁₆	7.070	7⅛	0.630	⅝	13⅝	1⁵/₁₆	¹³/₁₆
× 45	13.3	16.13	16⅛	0.345	⅜	³/₁₆	7.035	7	0.565	⁹/₁₆	13⅝	1¼	¹³/₁₆
× 40	11.8	16.01	16	0.305	⁵/₁₆	³/₁₆	6.995	7	0.505	½	13⅝	1³/₁₆	¹³/₁₆
× 36	10.6	15.86	15⅞	0.295	⁵/₁₆	³/₁₆	6.985	7	0.430	⁷/₁₆	13⅝	1⅛	¾

ᵃFor application refer to Notes in Table 2.

AMERICAN INSTITUTE OF STEEL CONSTRUCTION

FIGURE 8-7 ■ Drafters often need to research the sizes of steel shapes to complete details. *Courtesy American Institute of Steel Construction.*

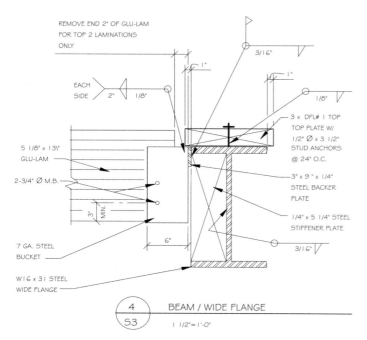

REMOVE END 2" OF GLU-LAM
FOR TOP 2 LAMINATIONS
ONLY

3/16"

EACH
SIDE 2" 1/8"

1"

1"

1/8"

3 x DFL# 1 TOP
TOP PLATE W/
1/2" Ø x 3 1/2"
STUD ANCHORS
@ 24" O.C.

5 1/8" x 13¹/₂"
GLU-LAM

3" x 9 " x 1/4"
STEEL BACKER
PLATE

2-3/4" Ø M.B.

1/4" x 5 1/4" STEEL
STIFFENER PLATE

3" MIN.

6"

3/16"

7 GA. STEEL
BUCKET

W16 x 31 STEEL
WIDE FLANGE

(4 / S3)   BEAM / WIDE FLANGE
1 1/2" = 1'-0"

FIGURE 8-8 ■ Studs (bolts with no head) are welded to a steel member to allow a wood plate to be bolted to the top of a steel beam. The plate provides a nailing surface so that other wood members can be attached. If the wood member is bolted to the side of the beam (to the flange) the wood member is called a ledger.

$$TS\ 4'' \times 4'' \times 3/8''\ or\ TS\ 4'' \times 1/4''$$

Round columns are specified by the nominal diameter, and a specification of standard, extra strong, or double extra strong. A pipe would be represented with as

$$TS\ STD.\ 4''\ DIA.\ PIPE$$

Chapter 21 will include a further discussion of steel columns and tubes.

Bars

Steel bars (at H in Figure 8-3a) are solid members available in circular, square, and flat shapes that are used to provide reinforcing to other members. Bars are often used for lateral bracing, or as hangers to support horizontal members. Figure 8-9 shows a round bar used to stabilize a column. Steel is classed as a bar if it is less than 8" (200 mm) wide. Bars are specified in 1/8" (3 mm) increments for thickness and 1/4" (6 mm) increments for widths (see H in Figure 8-3a). Depending on the shape, bars will be specified on drawings in one of three possible methods:

Square bar	1" sq. bar
Flat	2" × 3/8" bar
Round	1" dia. bar

See Chapter 9 for information about bars used for masonry and concrete reinforcing. Threaded bars used for fastening devices will be discussed later in this chapter.

Plates

A plate (at J in Figure 8-3a) is a flat piece of steel 8" (200 mm) or wider ranging in thickness from 1/2" to 2" (13 to 50 mm) thick. Common uses of plates include top, bottom, gusset, end, stiffener plates, or as part of a fabricated hanger.

■ A *top plate* provides a bearing surface for a steel beam resting on a column (see Figure 8-9). The plate is welded to the top of the column, and the beam to be supported is bolted and welded to the plate.

■ A *base plate* provides support for a column resting on another steel member or for a column resting on concrete. Figure 8-10 shows a column welded to a base plate, with the plate attached to the concrete with anchor bolts.

■ A *gusset plate* is a plate added to an intersection of structural members to provide support and stiffness. Figures 8-9 and 8-10 each show a *gusset plate*, which is used to provide a welding surface for the diagonal steel tube.

■ An *end plate* is used to provide an attachment surface at the end of a W or similar shaped beam.

FIGURE 8-9 ■ Steel tubes, columns, rods, and plates are used to transfer loads into the foundation. *Courtesy Zachary Jefferis.*

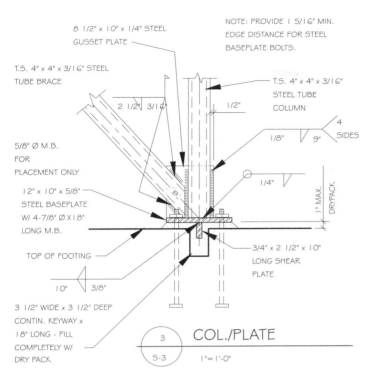

FIGURE 8-10 ■ In the side view of a detail, steel tubes are drawn similar to wood and timber post. The interior surface of a steel column is represented by a hidden line in order to distinguish a steel tube from a wood column.

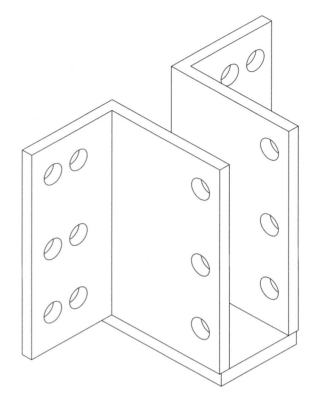

FIGURE 8-11 ■ Steel plates used to fabricate a beam hanger.

■ The flanges of a W-beam are often reinforced by a *stiffener plate* to provide support for members that are to be hung from the beam.

■ Fabricated angles for beam hangers similar to Figure 8-11 can be constructed from flat steel plates.

A plate is referenced by its size followed by its thickness. A typical callout would be

12 X 12 X 3/8 PL

Cable

Cables can be used in place of solid bars for lateral bracing because of their high resistance to forces in tension. A cable can be used to form an *X* between two parallel members. The cable is tightened by a turnbuckle placed at a convenient place in the cable span. Cable diameters range from 1/2" to 3 5/8" (13 to 90 mm). When specified on plans, cable smaller than 1" (25 mm) diameter is specified by giving the diameter and safe working stress limits of the cable. Larger cable sizes are specified by providing the diameter, the number of strands, and the number of wires per strand. Figure 8-12 shows an example of specifying cables on a framing plan.

TYPES OF METAL AND ALLOYS

Some of the most common alloys used in the construction include aluminum, copper, lead, stainless steel, and tin.

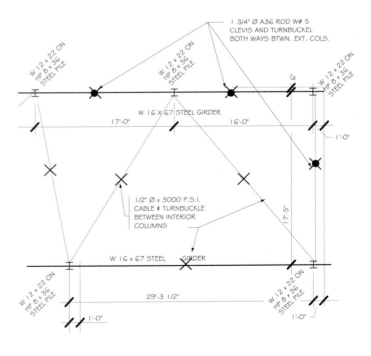

FIGURE 8-12 ■ Representing steel columns and support cables in plan view.

Aluminum

Aluminum is lightweight, noncorrosive, a good heat conductor, and very strong for its weight. Common uses of aluminum are roofing material, moldings, window mullions, and window frames. The architectural drawings often include details similar to Figure 8-13 to show how the aluminum window frame will intersect with the structural shell. Adding one or more elements to aluminum improves both the hardness and strength of aluminum. The Aluminum Association classifies aluminum alloys using a series code number. If a specific class of aluminum is to be used it would be referenced in the project manual. Classes include the following:

2000 series	adds copper
3000 series	adds manganese
4000 series	adds silicone
5000 series	adds magnesium

Aluminum alloys are used for such products as roof and siding panels, trim, nails, structural members such as truss webs or space frames, railings, electrical wiring, and door and window frames.

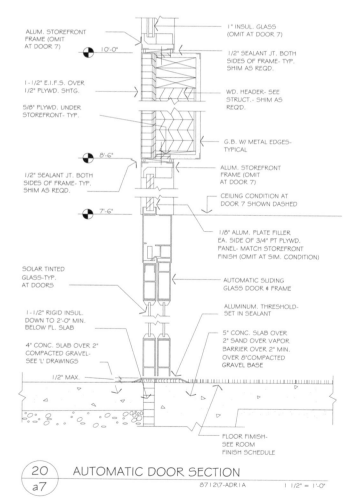

ALUM. STOREFRONT FRAME (OMIT AT DOOR 7)

10'-0"

1-1/2" E.I.F.S. OVER 1/2" PLYWD. SHTG.

5/8" PLYWD. UNDER STOREFRONT- TYP.

1/2" SEALANT JT. BOTH SIDES OF FRAME- TYP. SHIM AS REQD.

8'-6"

7'-6"

SOLAR TINTED GLASS-TYP. AT DOORS

1-1/2" RIGID INSUL. DOWN TO 2'-0" MIN. BELOW FL. SLAB

4" CONC. SLAB OVER 2" COMPACTED GRAVEL- SEE 'L' DRAWINGS

1/2" MAX.

1" INSUL. GLASS (OMIT AT DOOR 7)

1/2" SEALANT JT. BOTH SIDES OF FRAME- TYP. SHIM AS REQD.

WD. HEADER- SEE STRUCT.- SHIM AS REQ'D.

G.B. W/ METAL EDGES- TYPICAL

ALUM. STOREFRONT FRAME (OMIT AT DOOR 7)

CEILING CONDITION AT DOOR 7 SHOWN DASHED

1/8" ALUM. PLATE FILLER EA. SIDE OF 3/4" PT PLYWD. PANEL- MATCH STOREFRONT FINISH (OMIT AT SIM. CONDITION)

AUTOMATIC SLIDING GLASS DOOR & FRAME

ALUMINUM. THRESHOLD- SET IN SEALANT

5" CONC. SLAB OVER 2" SAND OVER VAPOR BARRIER OVER 2" MIN. OVER 8"COMPACTED GRAVEL BASE

FLOOR FINISH- SEE ROOM FINISH SCHEDULE

20 / a7   AUTOMATIC DOOR SECTION

87127-ADR1A   1 1/2" = 1'-0"

FIGURE 8-13 ■ An aluminum window frame in detail. *Courtesy Architects Barrentine, Bates & Lee, A.I.A.*

Stainless Steel

Stainless steel is used where appearance and maintenance are a priority for both interior and exterior uses. It has a nonstaining finish and is used for kitchen equipment for food establishments, elevator and doorway trim, or other metal products subject to abuse or in need of a high polish. Stainless steel is usually specified on drawings by a gauge number, grade, type, and finish. Material to be made from stainless steel that is 3/16" (5 mm) or thicker or widths of greater than 10" (250 mm), is specified as a plate. Material that is thinner than 3/16" (5 mm) and wider than 24" (600 mm) is referred to as a sheet. Material that is less than 3/16" (5 mm) thick and less than 24" (600 mm) wide is a strip.

Grades

Two grades of stainless steel can be included in the written specifications of a drawing. These include austenitic and ferritic grades. Austenitic grades of stainless steel have the highest corrosion resistance and contain both chromium and nickel. This grade is nonmagnetic. Ferritic grades are always magnetic. Both ferritic and austenitic grades can be hardened by heat treatment.

Types

Although there are more than forty types of stainless steel, only five types are used in the construction industry. Each type is established by the American Iron and Steel Institute and given a designation number. The five common types of stainless steel used in construction and their uses include the following:

Type 301	Used for structural applications such as door frames.
Type 302	Used for architectural applications of stainless steel such as exterior paneling, false columns, soffits, fascias, and gutters.
Type 304	Has a lower carbon content than type 302 stainless steel. Used for many of the same applications.
Type 316	Molybdenum is added to provide a high resistance to atmospheric corrosion.
Type 430	Used for interior applications. Resistance to corrosion is less than type 302.

Finish

In addition to the grade and type of stainless steel, a finish is often specified in the project manual. Common finishes include the following:

No. 2D Special	A frosty matte finish
No. 4	A dense, bright, highly reflective finish
No. 6	A soft satin finish, moderately reflective
No. 7	Highly reflective finish

Copper

Copper is used throughout the construction industry for its ability to conduct electricity and its resistance to corrosion. Copper is used for electrical wiring, pipes for fresh water supply, sheet metal panels for roofing, shingles, gutters, and flashing.

Lead

Although not widely used throughout the construction of a structure, lead is used for flashing as well as specialty items such as shower pans or other areas that are required to be watertight. Lead is also used in hospitals and labs for protection from x-rays and for plating to provide a resistance to acid.

Tin

Tin is very resistant to corrosion. The major use of tin is as a coating for sheet metal used for roofing or siding panels.

LIGHTWEIGHT STEEL FRAMING

Lightweight steel products are used throughout the framing of many structures that require increased fire protection over type V construction. Common uses of steel include studs, joists, trusses, decking, and lath.

Steel Studs

Steel studs are often used to help meet the requirements of type I, II, and III construction methods. Depending on the covering material, steel studs can achieve fire ratings of between one and three hours. (See Chapter 11.) Steel studs offer lightweight, noncombustible, corrosion-resistant framing for interior partitions and interior and exterior load-bearing walls up to five stories in height. Steel products also offer greater dimensional stability and a level surface without problems such as attack by termites, rotting, shrinkage, splitting, or wrapping associated with wood construction.

Studs are designed for rapid assembly and are predrilled for electrical and plumbing conduits. (See Figure 8-14.) The standard 24" (600 mm) spacing reduces the number of studs required by about one-third when compared with traditional wood framing. Stud width ranges from 3 5/8" to 10" (90 to 250 mm) but can be manufactured in any width. Stock lengths include 8', 9', 10', 12', and 16' (2400, 2700, 3100, 3700, and 4900 mm) lengths, with custom lengths typically available up to 40' (12 200 mm). Studs are made of steel, ranging from 12 to 25 gauge, with a yield point of 40 ksi. The engineer selects the gauge to be used based on the loads to be supported and the usage. The shape of the studs varies with the load to be supported. Figure 8-15 shows common shape and size variations of studs used for bearing and nonbearing walls.

Steel studs are mounted in a channel at the top and bottom of the wall. This channel serves a similar function to the plates and sill of wood frame construction. The material used for channels typi-

FIGURE 8-14 ■ Steel studs are designed for rapid assembly and are predrilled for electrical and plumbing materials.

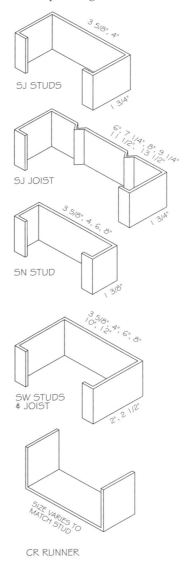

FIGURE 8-15 ■ Common sizes and shapes of steel studs and channels.

cally has a yield point of 33 ksi. Horizontal bridging may be placed through the predrilled holes in the studs and then welded to the stud to serve the same function as blocking in a stud wall.

Stud Specifications

Section properties and steel specifications vary among manufacturers; therefore, the material to be specified on the plans will vary. Specific catalogs developed by the Metal Stud Manufacturers Association should be consulted to determine the structural properties of the desired stud. The architect or engineer determines the size of stud to be used, but the drafter often needs to consult vendor materials to completely specify the studs. As a minimum, the gauge and usage are specified on the framing plan, details, and sections. If a manufacturer is to be specified, the callout will include the size, style, gauge, and manufacturer of the stud. An example of a steel stud specification would read

362SJ20 STEEL STUDS BY UNIMAST

Steel Joists

Steel joists offer the same qualities as steel studs over their wood counterparts. Joists are available in 3 5/8", 4", 6", 7 1/4", 8", 9 1/4", 10", 11 1/2", 12" and 13 1/2" (90, 100, 150, 185, 200, 235, 250, 290, 305 and 340 mm) depths. The gauge and yield point for joists are similar to these values for studs. Many joists are manufactured so that a joist may be placed around another joist, or *nested*. Nested joists allow the strength of a joist to be greatly increased without increasing the size of the framing members. A nested joist is often used to support a bearing wall above the floor. If a post is to be supported by a metal joist, a steel stiffener plate can be placed between the flanges of the joist at the bearing point to prevent bending of the web (see Figure 8-16). Joists are available in lengths up to approximately 40' (12 200 mm) depending on the manufacturer.

Open-Web Steel Joist

Open-web steel joists are manufactured in the same configurations as open-web wood trusses. In comparison to their wood counterparts, open-web steel joists offer the advantage of being able to support greater loads over longer spans with less dead load than solid members. Figure 8-17 shows an example of light steel trusses used to support floor loads. These joist are usually made of small angles, round bars, or tees, in depths ranging from 8" to 72" (200 to 1800 mm).

Open-web joists are referred to by their series number, which defines its use. The Steel Joist Institute (SJI) designations include K, LH, and DLH. All sizes listed below are based on SJI specifications for steel joists. The K series joists are parallel-web members that range in depth from 8" to 30" (200 to 760 mm) in 2" (50 mm) increments, for spans up to approximately 60' (18 300 mm). LH (longspan) and DLH (deep longspan) series of joists are made from steel with a yield point of 50 ksi. Depths range from 18" to 48" (460 to 1200 mm) for LH series and from 52" to 72" (1320 to 1830 mm) for DLH series joists. Spans range up to 96' (29 300 mm) for LH series joists to up to 144' (43 900 mm) for DLH joists.

FIGURE 8-16 ■ A stiffener plate is inserted into channels when a post must be supported. Notice the channel nested below the floor joist to help spread loads from the stiffener plate the floor joist. *Courtesy Unimast Incorporated.*

FIGURE 8-17 ■ Steel open-web trusses and steel decking are often used to support floor and ceiling loads. *Courtesy Megan Jefferis.*

Steel joists are specified on the framing plans, structural details, and sections by providing the approximate depth, the series designation, the chord size, and the spacing. An open-web steel joist specification would resemble

14K4 OPEN-WEB STEEL JOIST AT 32" O.C.

Just as with a wood truss, span tables are available to determine safe working loads of steel joists. The engineer determines the truss to be used, but the drafter is often required to find information to complete the truss specification.

Steel Decking

Corrugated steel decking is supported by steel channels or open-web joists used for most lightweight floor systems. (See Figure 8-17.) Decking is produced in a variety of shapes, depths, and gauges. Common shapes are shown in Figure 8-18. Common depths include 1 1/2", 3", and 4 1/2" (38, 75, and 115 mm) deep with lengths available up to 30' (9140 mm) long. The engineer determines the required rib depth based on the load to be supported and the lateral load to be resisted. Steel decking is attached to the floor or roof framing system with screws or welds. Decking is covered with a minimum of 1 1/2" (38 mm) of concrete above the ribs to provide the finished floor. Decking is specified in the project manual and in sections and details by giving the manufacturer, type of decking, depth, and gauge. A typical specification would resemble METAL DECKING TO BE VERCO, TYPE 'N', 3"–20GA. GALV. DECKING

Steel Lath

Galvanized metal lath in 27" × 96" (690 × 2440 mm) sheets are typically used for backing for plaster, gypsum lath, fiberboard, or similar plaster bases used to cover steel products. Other metal products include drywall clips, drywall corner bead, wire, and screws.

METAL BUILDINGS

Prefabricated or rigid frame buildings are common in many areas of the country because they provide fast erection time compared with other types of construction. The frame, roof and wall components are standardized and sold as modular units with given spans, wall heights, and lengths, to reduce cost and eliminate wasted materials. Figure 8-19 shows a steel framed structure being erected.

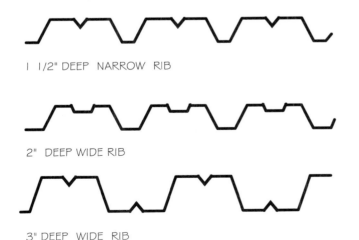

1 1/2" DEEP NARROW RIB

2" DEEP WIDE RIB

3" DEEP WIDE RIB

FIGURE 8-18 ■ Common corrugated decking shapes.

FIGURE 8-19 ■ A rigid frame structure consists of the frame, girts, and metal siding. *Courtesy David Jefferis.*

Typical Components

The structural system is made up of a frame that supports the walls, roof, and all externally applied loads. There are several different types of structural systems used. The two most common shapes are shown in Figure 8-20. Frame members are normally made from plate and bar stock with a yield point of 50 ksi (345 Mpa). The size of members is determined by engineers working for the fabricator. Tapering of members allows the minimum amount of material to be used, while still maintaining the required area to resist the loads to be supported. End plates are welded or bolted to each end of the vertical members. The plate at the base is predrilled to allow for bolting the steel to the concrete support. (See Figure 8-21.) The plate at the top of the frame is used to bolt the vertical to the inclined member. The vertical wall member is bolted to what will become the inclined roof member to form one rigid member. This frame member is similar to the three-hinged laminated arch discussed in Chapter 7.

The wall system is made of horizontal girts attached to the vertical frame. Usually the girts are a channel or a Z that is welded to the frame. Girts are typically welded to the outer flange of the frame, but they can also be welded between the frames if the wall thickness must be kept to a minimum. Girts are shown in Figure 8-22. Metal siding is screwed to the girts to complete the wall. The roof is made with steel purlins spaced at 24", 32", 48" or 60" (600, 800, 1200, 1500 mm) o.c. Purlins provide support for the roofing material. The spacing of the purlins depends on the sheet metal used to complete the roof. Both the roof and wall frames are reinforced with steel cable **x**-bracing to resist lateral loads.

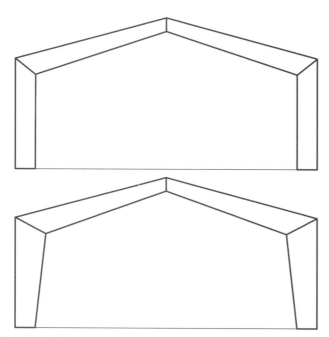

FIGURE 8-20 ■ Common shapes of fabricated metal frames.

Representation on Drawings

The engineering team working with the fabricator will complete the structural drawing required to build the structure. This includes drawings showing exact locations of all framing members and connections similar to those shown in Figure 8-23. Figure 8-24 shows an example of a section for a steel structure.

FIGURE 8-21 ■ Predrilled end plates are welded to the top and bottom of each portion of the frame. *Courtesy Michael Jefferis.*

FIGURE 8-22 ■ Steel girts are installed between the frame to support metal siding. *Courtesy Aaron Jefferis.*

MULTILEVEL STEEL FRAMES

Because of their ability to span large distances, steel is used to frame large sports and assembly arenas, exhibition centers, and multilevel structures. Multilevel structures are built in tiers consisting of vertical columns, horizontal girders, and intermediate support beams.

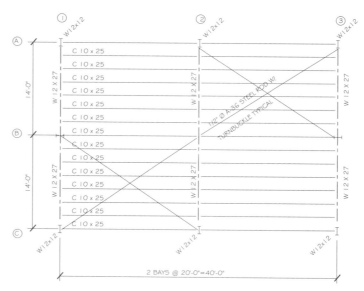

FIGURE 8-23 ■ The framing plan for a prefabricated metal building.

Multilevel structures framed with steel are similar to the elements of post-and-beam timber construction.

Framing Plans

As with other methods of construction, plans are required to show the location and size of each column, girder, beam, and joist. Columns are usually W-shaped but M and S shapes are also used. Steel columns are shown on the framing plan by representing the proper shape, the location from center to center, and a specification for the size and strength to be used. Steel columns are typically placed in two-story sections. Although smaller columns can be used on upper levels because of decreased loads to be supported,

column size is typically kept uniform to make splicing easier. Splices are usually kept about 24" (600 mm) above a floor level so that the connection can be made without interfering with the girder-to-column connection. The load to be supported and the method used for lateral bracing will affect how the columns are attached. With W shapes, a steel plate is often added to the outer edge of each column flange to provide stiffness at the connection. On rectangular shapes, the columns can be directly attached with no side plate (Figure 8-25). Tube steel (TS) can be used for columns for heights below three stories depending on the loads to be supported. Steel studs are used to frame between vertical supports but are not load bearing when used in conjunction with steel columns.

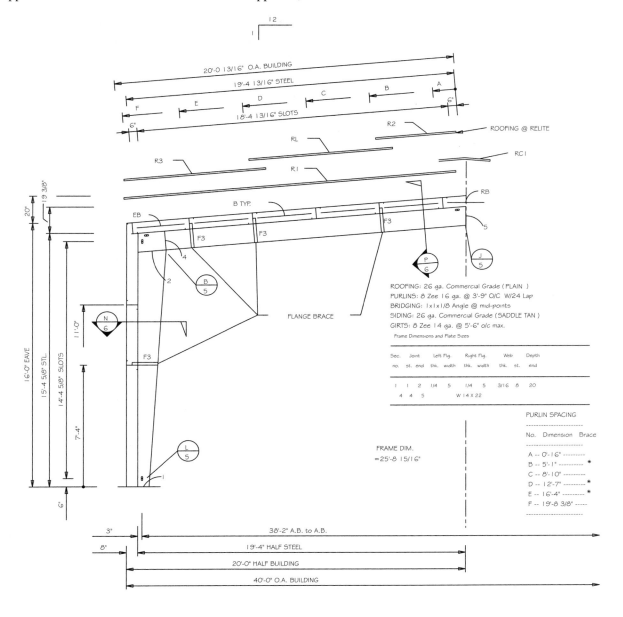

FIGURE 8-24 ■ A rigid frame seen in section.

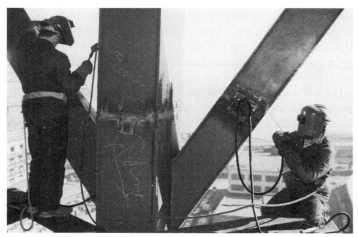

FIGURE 8-25 ■ Welded joints in a steel framed structure. As loads on columns increase, the joint can be reinforced with a side plate. *Courtesy Bethlehem Steel Corporation.*

Girders are also normally formed from W shapes because of the ease of making connections. Intermediate beams may be either W, S, M, or C. Depending on the loads to be supported, girders can be built up to increase the bearing capacity. A built-up girder can be composed of standard shapes or fabricated from steel plates. Figure 8-26 shows common methods of built-up girders based on AISC standards. On the framing and floor plans, girders are usually represented by a single line placed between, but not touching, the columns that will be used for support. The size and spacing of the girders are typically specified. Steel joists and decking covered with lightweight concrete can be used to span between intermediate beams to form floor and roof levels. Reinforced concrete slabs are also used to span between intermediate support beams and will be discussed in Chapter 10. An example of a steel framing plan can be seen in Figure 8-27.

Detail Drawings

Details are required to show each intersection of structural members and the connection method of various materials to the steel frame. Details showing the connections of structural materials will be referenced on the framing plan and sections. Welded joints and

FIGURE 8-26 ■ Built-up steel girders based on the American Institute of Steel Construction recommendations.

bolts are used almost exclusively for steel frames. Bolting is discussed in Chapter 7, and welding methods and symbols are discussed later in this chapter. Common types of connections can be seen in Figure 8-28.

Exterior Coverings

Materials such as stone, marble, granite, brick, and precast concrete panels are used to face steel structures. Units can be assembled at the job site or premanufactured and shipped to the site. Panels typically have steel connection plates that can be welded to the steel frame. Lightweight insulated sheet metal panels are also used for an exterior covering. In addition, tempered glass panels set in metal frames are a major exterior covering. Details for connecting the exterior coverings to the frame must also be provided. Details similar to Figure 8-29 are typically referenced on the floor plan and elevations.

Trusses

In addition to the steel trusses already mentioned to support floor and roof loads, steel can be used to form trusses that act as a girder, such as the girder trusses in Figure 8-1. When steel shapes are combined to form a truss, drafters working with the engineering team will draw the details to specify the shapes to be used and the types of connections. Figure 8-30 shows an example of the detail drawing required for the trusses in Figure 8-1.

Space Frames

Three-dimensional trusses are used to cover large areas of open space. Figure 8-31 shows an example of the use of a space frame for supporting the exterior shell of a structure. Drafters working for the architect, the structural engineer, and the fabricator will be involved in the detailing of space frames.

WELDING

Welding is the method of providing a rigid connection between two or more pieces of steel. Through the process of welding, metal is heated to a temperature high enough to cause melting of metal. The parts that are welded become one, with the welded joint actually stronger than the original material. Welding offers better strength, better weight distribution of supported loads, and a greater resistance to shear or rotational forces than a bolted connection. There are many other welding processes available to the industry, but the most common welds in the construction field are shielded metal arc welding, gas tungsten arc welding, and gas metal arc welding. In each case, the components to be welded are placed in contact with each other, and the edges are melted to form a bond. Additional metal is also added to form a sufficient bond.

Welds are specified in details similar to Figure 8-32. A horizontal reference line is connected to the parts to be welded by an inclined line with an arrow. The arrow touches the area to be welded. It is not uncommon to see the welding line bend to point to places that are difficult to reach. The welding symbol may also have

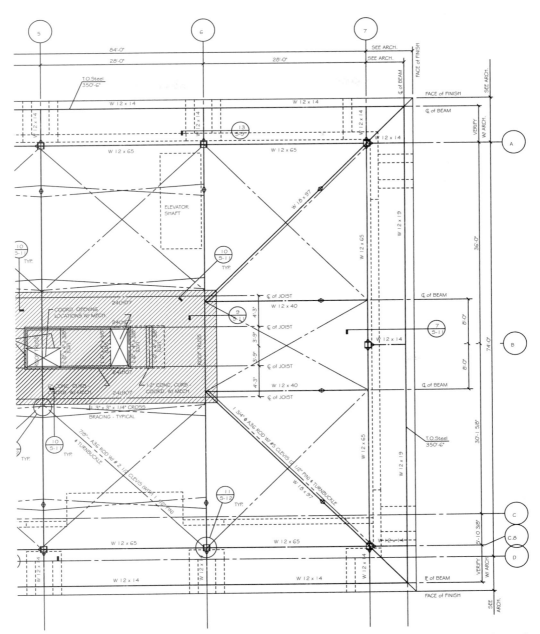

FIGURE 8-27 ■ Steel joist, columns, and rods shown on a framing plan. *Courtesy Van Domelen/Looijenga/McGarrigle/Knauf Consulting Engineers.*

more than one leader line extending from the reference line. Information about the type of weld, the location of the weld, the welding process, and the size and length of the weld is all specified on or near the reference line. Figure 8-33 shows a welding symbol and the proper location of information.

Welded Joints

The way that steel components intersect greatly influences the method used to weld the materials together. The welding method and the joint to be used are often included in the specification. Common joints include butt, lap, tee, outside corner, and edge joints. Examples of each joint are illstrated in Figure 8-34.

Types of Welds

The type of weld is distinguished by the weld shape and/or the type of groove in the metal components that receive the weld. Welds specified on structural and architectural drawings include fillet, groove, and plug welds. The methods of joining steel and the symbol for each method can be seen in Figure 8-35.

The most common weld used in construction is a fillet weld. A *fillet weld* is formed at the internal corner of two intersecting pieces of steel. The fillet can be applied to one or both sides and can be continuous or a specified length and spacing. Five other welds common to construction are specified by the shape of the metal to be joined.

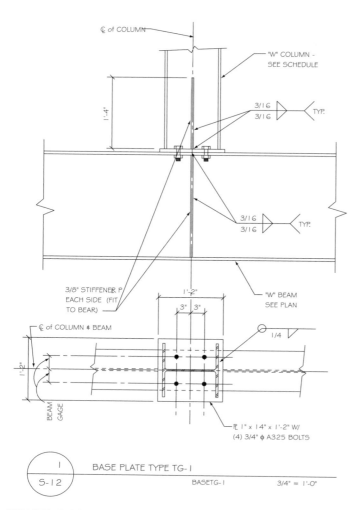

FIGURE 8-28 ■ A detail of a steel column and beam showing both welded and bolted connections. *Courtesy Van Domelen/Looijenga/ McGarrigle/Knauf Consulting Engineers.*

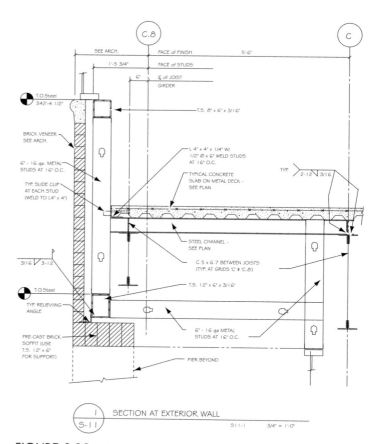

FIGURE 8-29 ■ Details showing the connections of the exterior skin to the structural skeleton. *Courtesy Van Domelen/Looijenga/McGarrigle/Knauf Consulting Engineers.*

■ A *square groove weld* is applied when two pieces with perpendicular edges are joined end to end. The spacing between the two pieces of metal is called the *root opening*. The root opening is shown to the left of the symbol.

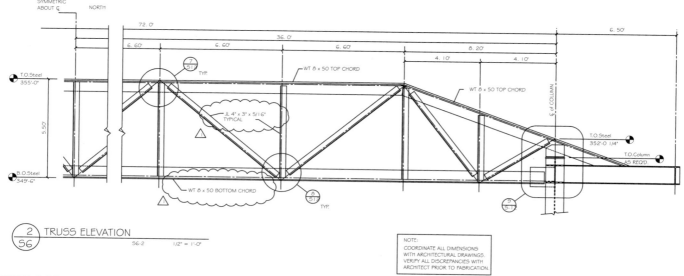

FIGURE 8-30 ■ Steel truss shapes are often used to form a beam for long spans with smaller trusses set perpendicular to provide intermediate support similar to a panelized framing system used with wood members. *Courtesy Van Domelen/Looijenga/McGarrigle/Knauf Consulting Engineers.*

FIGURE 8-31 ■ Three-dimensional trusses formed from lightweight metal are called a space frame. This space frame is used to support 8' 3 8' diamond-shaped glass modules used to form the roof structure for the Interfirst II structure in Dallas, Texas. Courtesy MERO Structures, Inc.

■ A *V-groove weld* is applied when each piece of steel to be joined has an inclined edge that forms the shape of a **V**. The included angle, as well as the root opening, is often specified.

■ A *beveled weld* is created when only one piece of steel has a beveled edge. An angle for the bevel and the root opening is typically given.

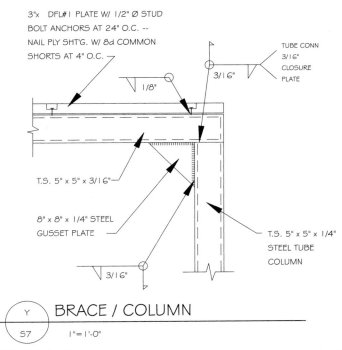

3"x DFL#1 PLATE W/ 1/2" Ø STUD BOLT ANCHORS AT 24" O.C. -- NAIL PLY SHT'G. W/ 8d COMMON SHORTS AT 4" O.C.

1/8"

3/16"

TUBE CONN 3/16" CLOSURE PLATE

T.S. 5" x 5" x 3/16"

8" x 8" x 1/4" STEEL GUSSET PLATE

3/16"

T.S. 5" x 5" x 1/4" STEEL TUBE COLUMN

Y 57 BRACE / COLUMN 1" = 1'-0"

FIGURE 8-32 ■ Welds are specified in details by the use of specialized symbols pointing to the area to be welded.

■ A *U-groove weld* is created when the groove between the two mating parts forms a **U**.

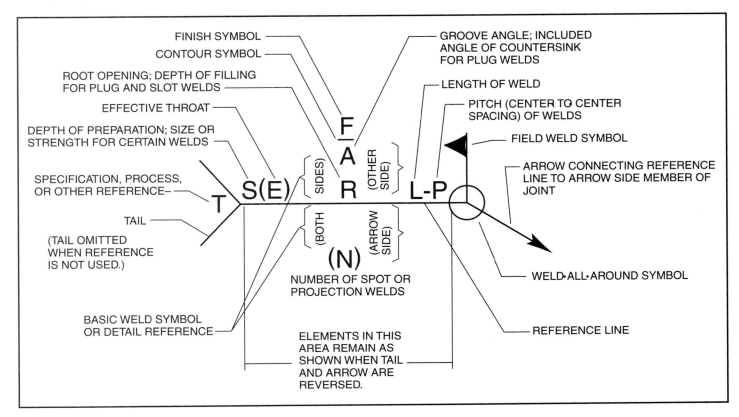

FIGURE 8-33 ■ Common locations of each element of a welding symbol. *Courtesy American Welding Society.*

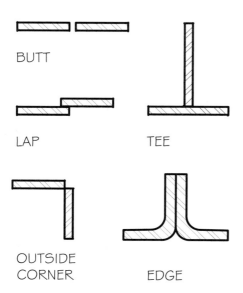

FIGURE 8-34 ■ Types of welded joints.

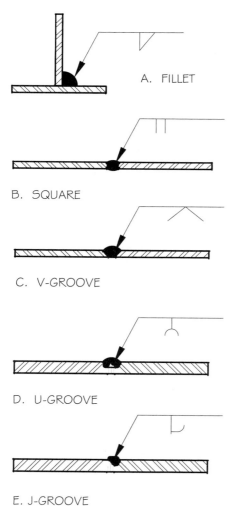

FIGURE 8-35 ■ The type of weld specified by the engineer is based on the shape of the material being welded. Common shapes of materials to be welded and the symbol used to represent the weld.

■ A *J-groove weld* results when one piece has a perpendicular edge and the other has a curved grooved edge. The included angle, the root opening, and the weld size are given for both types of welds.

Welding Location

Welds specified on the structural drawings may be done in a shop away from the job site, and the components are then shipped, ready to be installed. For an item that can be easily shipped such as metal beam connector, shop assembly is less expensive and more efficient. For large components that must be assembled at the job site, the welding is said to be *field welded*. Two symbols used to refer to a field weld are shown in Figure 8-36. When field welds are specified, the engineer may require that these welds are to be inspected by a representative of the engineer in addition to the inspection provided by the building department. The drafter will need to provide a specification by each welding detail indicating the need for special inspection.

Weld Placements

The placement of the welding symbol in relation to the reference line is critical in indicating where the actual weld will take place. Options include *this side, other side,* and *both sides.* Figure 8-37 shows the effects of placing a fillet weld symbol on the reference line. The distinction of symbol placement can be quite helpful to the drafter if adequate space is not available on the proper side of the detail to place a symbol. The welding symbol can be placed on either side of the drawing and the relationship to the reference line can be used to clarify the exact location.

All Around

The all around symbol is used to indicate that a weld is to be placed around the entire intersection. A circle placed at the intersection of the leader and reference line indicate that a feature is to be welded *all around.* Figure 8-32 shows a column–beam intersection joined with welds requiring an all around weld.

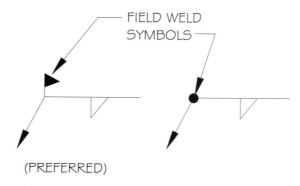

FIGURE 8-36 ■ The welding symbol can be used to distinguish if the weld will be performed on site or off site by the use of a small flag or a solid circle placed at the intersection of the leader line.

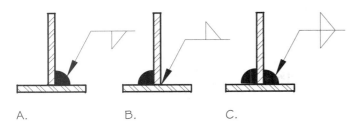

A.   B.   C.

FIGURE 8-37 ■ The welding symbol can be used to describe the relationship of the weld to the materials being welded by the placement of the symbol to the line: A, symbol below the line means the weld occurs on this side of the material; B, symbol placed above the line means the weld is to be placed on the other side; C, symbol placed on each side of the reference line means the weld is to occur on both sides of the material.

Weld Length and Increment

If a weld does not surround the entire part, the length and spacing of the weld should be placed beside the weld symbol. The number preceding the weld symbols indicates the size of the weld. When weld sizes are expressed in millimeters, they should be designated in 1 millimeter increments up to 8 mm, 2 millimeter increments from 8 to 20 mm, 5 millimeter increments from 20 to 40 mm and 10 millimeter increments beyond 40 mm. The number following the weld symbol indicates the length each weld is to be, and the final size indicates the spacing of the weld along a continuous intersection of two mating parts.

ADDITIONAL READING

The following Web sites can be used as a resource to help you keep current with changes in steel framing materials and methods.

ADDRESS	COMPANY/ORGANIZATION
www.aluminum.org	The Aluminum Association, Inc.
www.aisc.org	American Institute of Steel Construction
www.steel.org	AISI—American Iron & Steel Institute
www.astm.org	ASTM—American Society for Testing and Materials
www.americanwelding.com	American Welding Services, Inc.
www.aws.org	American Welding Society
www.bethsteel.com	Bethlehem Steel Corporation
www.copper.org	Copper Development Association
www.lgsea.com	Light Gauge Steel Engineers Association
www.mbma.com	Metal Building Manufacturers Association
www.steelframingalliance.com	North American Steel Framing Alliance
www.steelfoundation.org	AISE Steel Foundation
www.ssina.com	Specialty Steel Industry of North America
www.steeljoist.org	Steel Joist Institute
www.steellinks.com	Steel links search engine
www.spfa.org	Steel Plate Fabricators Association

CHAPTER 8

Steel Framing Methods and Materials

CHAPTER QUIZ

Answer the following questions on a separate sheet of paper. Print the chapter title, the question number, and a short complete statement for each question. Some answers may require the use of vendor catalogs or seeking out local suppliers.

Question 8-1 What are the two major elements in steel?

Question 8-2 What element must be controlled to keep steel from being brittle?

Question 8-3 If steel is in its elastic stage, is it safe to use and how will it react to a load?

Question 8-4 What happens to steel as it is stretched past its yield point?

Question 8-5 What organization governs the designations of structural steel?

Question 8-6 What are the five most common steel designations used in the construction industry?

Question 8-7 Describe the most common type of steel used in the construction industry in terms of its ASTM designation, yield point, and its maximum carbon content for all shapes.

Question 8-8 List nine common shapes of steel used in construction.

Question 8-9 What does the ultimate tensile strength of steel represent?

Question 8-10 What are the components, cross-sectional shape, and the minimum size and weight of a W shape?

Question 8-11 List the area and size descriptions of a W16 × 50 in decimal inches.

Question 8-12 What information should be given when a channel is specified on a drawing.

Question 8-13 List the three common shapes of steel tubes and their range in sizes.

Question 8-14 What is the difference in size between a bar and a plate?

Question 8-15 List common alloys used throughout the construction of a structure.

Question 8-16 What material is required to specify steel studs on a drawing?

Question 8-17 What is the common depth of steel joist?

Question 8-18 What information about steel joists must be specified on a drawing?

Question 8-19 What are the major components of the walls and roof of a metal framed structure?

Question 8-20 What are the most common methods of connecting large steel shapes to each other?

Question 8-21 List the three possible locations for placing a weld and give the symbol that represents each location.

Question 8-22 List five types of joints where welds can be applied.

Question 8-23 Sketch the symbol for a 3/16" fillet weld, applied to the other side, all around a column, and applied at the job site.

Question 8-24 Sketch the symbol used for a 1/4" × 3" long V groove weld applied to this side.

Question 8-25 Sketch the symbol for a 1/8" field, fillet weld on both sides.

DRAWING PROBLEMS

Use the attached isometric drawings and create the required views. Use scale factors suitable for plotting each detail at a scale of 3/4" = 1'–0" minimum. Label all material with generic notes that can be altered with each use. Select all sizes based on material found in *Sweet's Catalogs*, *Sweet Spot*, vendor catalogs or Web sites. Save each detail as a wblock.

Problem 8-1 Draw a side and end view showing the necessary materials.

Problem 8-2 Draw an end view showing the necessary materials.

Problem 8-3 Show a generic end view of the joist/beam intersection.

Problem 8-4 Show a view of the wall reinforcement.

Problem 8-5 Draw a side and end view to show typical opening framing.

Problem 8-6 Draw the end view of each typical steel shape with appropriate hatching and save each shape as a wblock, named for the specific shape.

Unless other instructions are given by your instructor, complete the following details and save them as wblocks. Skeletons of most details can be accessed from http://www.delmar.com/resources/ocl.html. Use these drawings as a base to complete the assignment

■ Show all required views to describe each connection.

■ Draw each detail at a scale of 1" = 1'-0" unless noted.

■ Use separate layers for wood, concrete, steel, hatch, text, laminations, and dimensions. Hatch each material with the appropriate hatch pattern.

■ Use dimensions for locations where possible instead of notes.

■ Provide notation to specify that manufactured metal hangers are to be provided by Simpson Strong-Tie Company or equal.

■ Use an appropriate text font to label and dimension each drawing as needed. Refer to *Sweet's Catalogs*, vendor catalogs, or the Internet to research needed sizes and specifications. Keep all text 3/4" minimum from the drawing, and use an appropriate architectural style leader line to reference the text to the drawing.

■ Provide a detail marker, with a drawing title, scale, and problem number below each detail.

■ Assemble the details required by your instructor into a appropriate template sheet for plotting. Arrange the drawings using the guidelines presented in Chapters 2 and 3. Assign a drawing sheet number of sheet S4 of 6.

Problem 8-7 Show a plan view showing two steel columns in a 2 × 6 stud wall. Show the columns as 36" from center to center. Specify TS 5 × 5 × 3/8" columns of ASTM A500, grade B steel. Cover the exterior side of the wall with 7/8" exterior plaster over 5/8" metal lath with 5/8" type X gyp board on exterior side of each stud cavity containing the steel columns.

Problem 8-8 Show a W12 × 22 with a 3 × 6 DFL #1 top plate with 1" minimum overhang on each side with a 5/8" diameter 4" stud anchor at 32" o.c. welded to top flange with a 3/16" field fillet weld. Hang 2 × 12 DFL #1 floor joists at 12" o.c. from the top plate with an appropriate hanger for supporting 1300 lbs. Cover the bottom of the floor and beam with 7/8" exterior stucco over 3/8" metal lath. Provide 1 1/2" lightweight concrete flooring over 3/4" T & G plywood subfloor.

Problem 8-9 Show a W12 × 30 supporting steel joist. Weld a 925SJ16 floor joist at 16" o.c. to the top flange with a 1/8" × 3" field fillet weld at each side. Provide a web stiffener plate where the joists rest on the beam and weld with fillet weld. Cover the bottom of the floor and beam with 7/8" exterior stucco over 3.4# metal lath. Provide 1 1/2" lightweight concrete flooring over 3/4" T & G plywood subfloor.

Problem 8-10 Show a W14 × 120 supporting a TJ60/24 joist at 32" o.c. support on a 3 × 6 DFPT sill w/5/8"∅ studs @ 32" o.c. welded to the top flange with a 1/8" field fillet weld.

Problem 8-11 Use the attached sketch to complete the required drawing. Use a TS 4 × 4 3/16" column to support a W16 × 31 beam with a 1/4" end stiffener plate welded with a 3/16" fillet weld. Rest beam on a 5 1/2" × 11" × 1/2" steel cap plate with (4)-7/8" diameter bolts to beam and weld plate to column with 3/16" fillet weld, all around. Provide TS 4 × 4 × 3/16" diagonal brace welded to 14 × 11 × 3/8" gusset plate welded to column and cap plate with 3/16" fillet weld at all contact points. Provide a plan view showing the relationship of the bolts, the column, and the gusset plate.

Bolt 3 × 4 DFL #1 top plate with a 5/8"∅ stud anchor at 32" o.c. welded to top flange with a 3/16" field fillet weld. Set a 6 × 12 DFL #1 beam level with the top plate and support beam with a 5 × 7 × 7 ga. steel bucket connection with a 5/8" dia. M.B. 1 1/2" down and in from edge. Weld bucket to end plate w/1/8" fillet all around. Tie beam to plate with MST27 strap by Simpson Company.

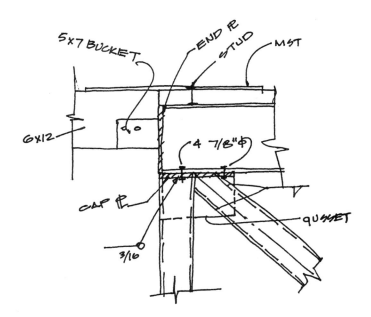

Problem 8-12 Use the attached drawing and a scale of 1 1/2" = 1'–0" to complete the required drawing. Use a TS 5" × 5" × 3/8" column to support a W18 × 76 beam bolted to 11" × 10" × 1 1/4" steel cap plate w/ (4)-5/8"∅ machine bolts to beam and weld plate to column with 3/16" fillet weld all around. Butt W18 × 97 beam into left side and W18 × 60 on right. Support each with (2)-4 × 3 1/2 × 12 × 1/4" angles welded to W18 × 76 with 1/4 × 11 1/2" fillet weld and (4)-3/4"∅ machine bolts through 4" leg through WF. Set bolts 1 1/2" from angle edge @ 3" o.c. Set top bolt 3" down from top flange.

Use a scale of 3/4" = 1'–0" to draw the following details. Base nailing on IBC standards unless your instructor tells you otherwise. Specify all material based on common local practice. Show and specify all connecting materials. Show a front and side view or a side and top view to describe each connection.

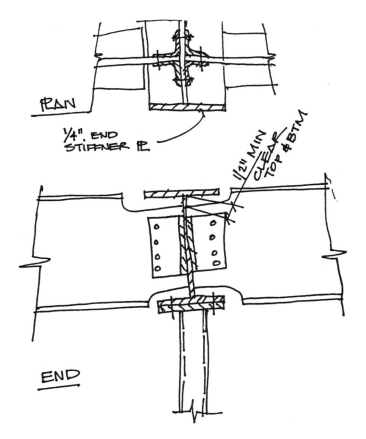

Problem 8-13 A 6" × 6" × 3/8" steel column will be supported by a 3/4" × 10" × 10" base plate. Set on a 1" mortar base. Use a 3/16" fillet weld all around to form the connection. The plate will be attached to a concrete footing using (4)-3/4" dia. × 12" A.B. laid out on a 7" grid that is laid out diagonally to the column.

Problem 8-14 Show a 3 × 6 DFL #1 sill supported on a W12 × 30. Use 3/4"Ø stud anchors at 36" o.c. staggered 1 1/2" from plate edge. Use a 1/8" fillet, field weld all around. Use 2"Ø washers. Provide 2 × 12 DFL #1 floor joist @ 16" o.c. to the plate. Provide solid blocking between the floor joist, and provide 3/4" std. grade plywood nailed with 10d @ 4" / 8" o.c. to support 1 1/2" lightweight concrete over 55 # felt.

Problem 8-15 Show a 6" × 6" × 3/16" steel column with an Simpson Co. HL46 angle welded on two opposing sides with a 3/16" fillet weld all around in the field. Support a 5 1/8 × 15 glu-lam beam on ea. angle w/ 1/4" clear to beam. Bolt the beam to the angle following the manufacturer's recommendations. Use a Simpson Co. MST 27 strap on each side of the beams, bolted to each beam and welded to the steel column with a 1/8" × 3" long fillet field weld. Place the strap so that minimum recommended sizes for wood and steel will be met.

CHAPTER 9

Unit Masonry Methods and Materials

Masonry construction consists of setting building materials in a bed of mortar. Brick, stone, glass, and concrete masonry units (CMUs) are the materials commonly used in masonry construction. This chapter examines methods of construction typically used with brick and concrete masonry units—including mortar and reinforcement.

BRICK CONSTRUCTION

Bricks are made in varied colors and textures by pressing various types of clay, shale, and a combination of oxides in a mold of a desired shape and size. Brick is usually produced in red, brown, and gray tones. Common surfaces include smooth, water or sandstruck, scored, wire-cut, combed, and roughened. Bricks can also be finished with a ceramic glaze—providing a polished finished in any color.

Brick Types

Bricks are produced in solid, cored, and hollow units. Cored units are considered to be solid if a minimum of 75% of the cross-sectional area is solid. A brick is hollow if at least 60% of the cross-sectional area is solid. Vertical cores are placed in bricks to reduce the weight of the brick.

Common or building brick is most widely used in the construction industry. Face brick is produced to standards that produce units in specific sizes, textures, and colors. Other types of brick used in construction include the following:

Adobe brick is made from a mixture of natural clay and straw that is placed in molds and dried in the sun. These units require protection from rain and subsurface moisture unless a moisture-proofing agent is added.

Back-up brick is inferior units used in applications where they can't be seen, for example, behind face brick.

Fire brick is brick with a high resistance to high temperature, used for the facing material of the firebox of a fireplace.

Hollow brick tile is masonry units that are cored in excess of 25% of the gross cross-sectional area.

Nail-on brick is flat units used for interior applications where solid masonry can't be structurally supported.

Paving brick is masonry units with a hard, dense surface used for floor applications.

Sewer brick is units with a low absorption rate used for sewer or storm drain applications.

The type of brick to be used is normally indicated by the drafter on the floor and elevations as well as in written specifications. The surface of the brick to be exposed is also often specified. Each surface of a unit has a specific name, as shown in Figure 9-1. The surface name is often referred to in written specifications to determine which surface will be displayed or cut. For some applications, the mason must cut units to meet the design criteria. Common methods of cutting brick are illustrated in Figure 9-2.

Brick Quality

Strength, appearance, and durability determine the quality of brick. The quality is specified in the written specifications. Common brick is classified into three grades, determined by the weather conditions it will be required to withstand. Grade SW is for severe weather conditions, which include heavy rains or sustained below-freezing conditions. Grade MW is for moderate weather conditions for use in areas with moderate rainfall and some freezing conditions. Grade NW is for use in areas with negligible weathering from minimal rainfall and above freezing temperatures.

Face brick is rated for durability of exposure and is available in grades SW and MW. It is also identified by the ASTM specifications, which dictate the range allowed in size, texture, color, and structural quality. FBA grade brick is non-uniform in size, color and texture of each unit. Grade FBS brick allows for variations in mechanical perfection with a wide range of color variation. Grade FBX has

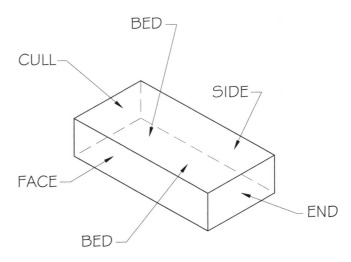

FIGURE 9-1 ■ Each surface of a brick is named to indicate how it is to be positioned.

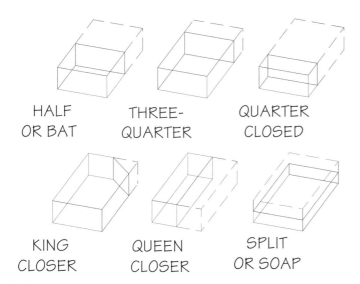

HALF OR BAT   THREE-QUARTER   QUARTER CLOSED

KING CLOSER   QUEEN CLOSER   SPLIT OR SOAP

FIGURE 9-2 ■ Brick can be cut into varied shapes based on the pattern used to position each unit.

the highest degree of mechanical perfection and is the most controlled for color variation.

Hollow brick is classified by SW or MW as well as by factors that affect it appearance. Graded *HBA* is non-uniform in size, color, and texture. Grade *HBS* is more controlled than HBA brick but allows for some size variation and a wide color range. Grade HBX has the highest degree of mechanical perfection with the least variation in size and the smallest range in color.

Brick Sizes

To effectively control the cost in using brick and to minimize labor, structures using masonry units should conform to the size of the unit being used. Bricks vary in size not only because of the variety of bricks made, but also because of shrinkage in the drying process. Modular bricks were developed to regulate the coursing dimensions. An economy-8 modular brick, is 3 1/2" high and 7 1/2" long laid with a 1/2" joints. It will produce courses exactly 4" high and 8" long. Brick sizes are controlled by the *Standard Guide for Modular Construction of Clay and Concrete Masonry Units* (ASTM document E835/E835M). This document establishes metric dimensions for clay and concrete products based on the basic building module of 100 mm. Many common brick sizes are within a millimeter or two of metric module sizes and nearly all can fit within a 100 mm module vertically by using 10 mm joint widths. Drafters should verify exact sizes of masonry units and their availability with local suppliers. Common nominal sizes of standard bricks are listed in Table 9-1.

Brick Shapes and Placement

Common shapes of brick are shown in Figure 9-3. One of the most popular features of brick is the wide variety of positions and patterns that can be created. These patterns are achieved by placing the brick in various positions relative to each other. The position in

TABLE 9-1 ■ Common Nominal Masonry Unit Sizes (*Courtesy Construction Metrication*)

Unit Name		Size (in.)	Actual Size (mm)	Actual Metric Size (mm)	Modular Metric Vertical Coursing
Modular	width	3 1/2"	89	90	3:8"
Metric Modular	height	2 1/4"	57	57	3:200 mm
	length	7 1/2"	190	190	
Engineer	width	3 1/2"	89	90	5:16"
Modular	height	2 3/4"	70	70	5:400 mm
	length	7 1/2"	190	190	
Roman	width	3 1/2"	89	90	4:8"
	height	1 5/8"	41	40	4:200 mm
	length	11 1/2"	292	290	
Norman	width	3 1/2"	89	90	3:8"
Metric Norman	height	2 1/4"	57	57	3:200 mm
	length	11 1/2"	292	290	
Engineer	width	3 1/2"	89	90	5:16"
Norman	height	2 3/4"	70	70	5:400 mm
	length	11 1/2"	292	290	
Utility	width	3 1/2"	89	90	2:8
Metric Jumbo	height	3 1/2"	89	90	5:400 mm
	length	11 1/2"	292	290	
Standard	width	3 5/8"	92	90	3:16"
	height	2 1/4"	57	57	3:200 mm
	length	8"	203	200	
*Engineer	width	3 5/8"	92		5:16"
Standard	height	21 3/16"	71	70	5:400 mm
	length	8"	203		
* King	width	3"	76		5:16"
	height	2 3/4"	70	70	5:400 mm
	length	9 5/8"	245		

*Note: bricks identified within an * are not a modular unit in either inch-pound or the metric system. Vertical coursings are based on 9 to 11 mm mortar joint, depending on the brick height.*

which the brick is placed will alter what the brick course is called. Figure 9-4a illustrates common brick positions and their names. Figure 9-4b shows examples of how these patterns would be represented. Common patterns for laying masonry units are illustrated in Figure 9-5. The brick pattern is represented and specified on the elevation.

Wall Construction

Bricks can be placed in any of the positions shown in Figures 9-4 and 9-5 to form a wall, as well as a variety of bonds and patterns.

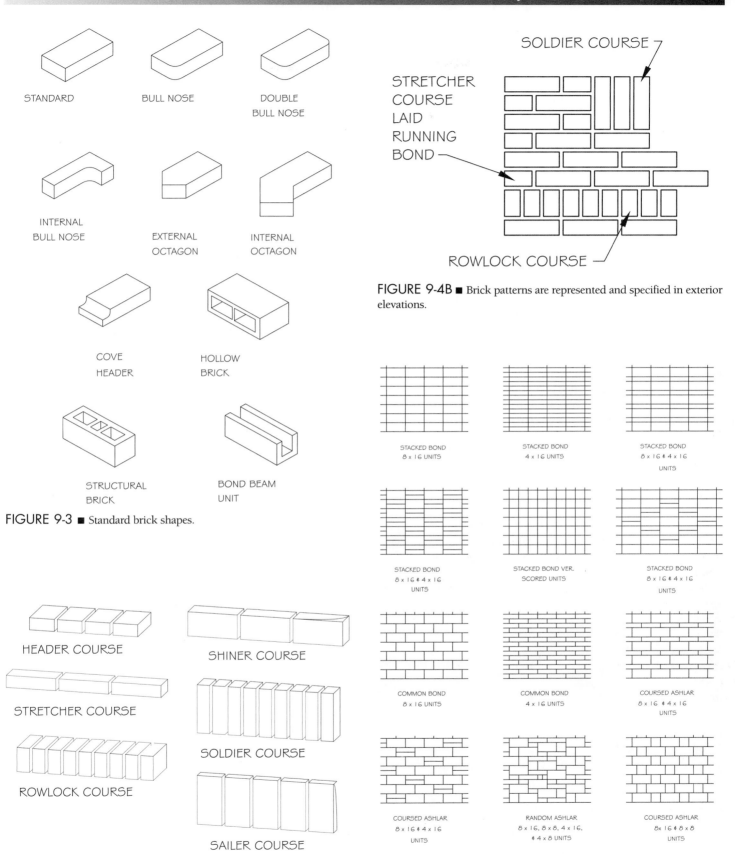

STANDARD

BULL NOSE

DOUBLE
BULL NOSE

INTERNAL
BULL NOSE

EXTERNAL
OCTAGON

INTERNAL
OCTAGON

COVE
HEADER

HOLLOW
BRICK

STRUCTURAL
BRICK

BOND BEAM
UNIT

FIGURE 9-3 ■ Standard brick shapes.

SOLDIER COURSE

STRETCHER
COURSE
LAID
RUNNING
BOND

ROWLOCK COURSE

FIGURE 9-4B ■ Brick patterns are represented and specified in exterior elevations.

HEADER COURSE

SHINER COURSE

STRETCHER COURSE

SOLDIER COURSE

ROWLOCK COURSE

SAILER COURSE

FIGURE 9-4A ■ Names of common brick placement patterns.

STACKED BOND
8 x 16 UNITS

STACKED BOND
4 x 16 UNITS

STACKED BOND
8 x 16 ‡ 4 x 16
UNITS

STACKED BOND
8 x 16 ‡ 4 x 16
UNITS

STACKED BOND VER.
SCORED UNITS

STACKED BOND
8 x 16 ‡ 4 x 16
UNITS

COMMON BOND
8 x 16 UNITS

COMMON BOND
4 x 16 UNITS

COURSED ASHLAR
8 x 16 ‡ 4 x 16
UNITS

COURSED ASHLAR
8 x 16 ‡ 4 x 16
UNITS

RANDOM ASHLAR
8 x 16, 8 x 8, 4 x 16,
‡ 4 x 8 UNITS

COURSED ASHLAR
8 x 16 ‡ 8 x 8
UNITS

FIGURE 9-5 ■ Common patterns for placing masonry units.

Masonry units are laid in rows called *courses* and in vertical planes called *wythes*. Figure 9-6 shows an example of single wythe construction with the temporary supports still in place. Single wythe walls are made with structural bricks (see Figure 9-3) with steel reinforcing placed as per local codes to resist lateral loads. Brick walls normally consist of two wythes, with an air space of two to three inches between each wythe. This type of construction is referred to as a *cavity wall*. Rigid foam insulation can be placed against the inner wythe to increase the R-value of the wall. If the space is filled with grout, it is called a *grouted cavity* wall. Reinforcing steel can be placed in the air space between wythes and then surrounded by solid grout.

If the wythes are connected, the wall is referred to as a *solid wall*. A bond is the connection between the wythes to add stability to the wall so that the entire assembly acts as a single structural unit. It can be made by either wire ties or by masonry members. The Flemish and English bonds of Figure 9-7 are the most common methods of bonding two wythes together. The Flemish bond consists of alternating headers and stretchers in each course. (A course of masonry is one unit in height.) An English bond consists of alternating courses of headers and stretchers. In each case the header spans between wythes to keep the wall from separating. The bonding course is usually placed at every sixth course.

The cavity between wythes is typically 2" (50 mm) wide and creates approximately a 10" (250 mm) wide wall with masonry exposed on both the exterior and interior surfaces. The air space between wythes provides an effective barrier to moisture penetration to the interior wall. Weep holes must be provided in the lower course of the exterior wythe to allow moisture to escape. Rigid insulation can be applied to the interior wythe to increase the insulation value of the air space in cold climates. Care must be taken to

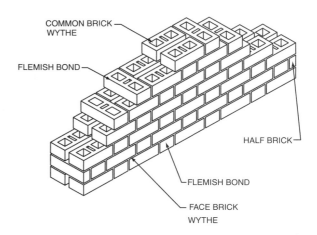

FLEMISH BOND

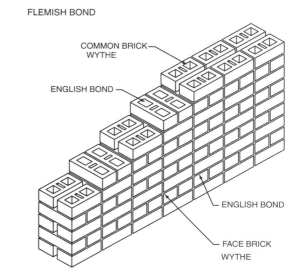

ENGLISH BOND

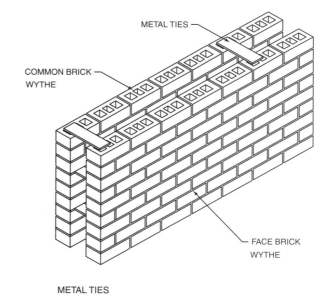

METAL TIES

FIGURE 9-6 ■ A single wythe masonry wall can be constructed of structural brick with steel reinforcing. Masonry over opening is temporarily supported by wood members until the grout can cure and achieve its full strength.

FIGURE 9-7 ■ Brick walls can be reinforced by using metal ties or by placing brick bonds to connect each wythe. Common bonds include the Flemish and English.

keep the insulation from touching the exterior wythe so that moisture is not transferred to the interior. Metal ties are embedded in mortar joints at approximately 16" (400 mm) to tie each pair of wythes together.

If a masonry wall is to be used to support a floor, a space one-wythe wide will be left to support the joist similar to Figure 9-8. Joists are usually required to be strapped to the wall so that the wall and floor will move together under lateral stress. The strap will be specified on the framing plan and the details. The end of the joist must be cut on an angle. This cut is called a *fire cut* and will be specified in the details. If the floor joists were to be damaged by fire, the fire cut will allow the floor joist to fall out of the wall, without destroying the wall. A fire cut is shown in Figure 9-9.

Brick is very porous and absorbs moisture easily. Some method must be provided to protect the end of the joist from absorbing moisture from the masonry. The end of the joist is wrapped with 55# felt and set in a 1/2"(13 mm) air space. The interior of a masonry wall also must be protected from moisture. A layer of hot asphaltic emulsion can be applied to the inner side of the interior wythe and a furring strip can be attached to the wall. In addition to supporting sheet rock or plaster, the space between the furr strips can be used to hold batt or rigid insulation.

An alternative method of attaching a floor to a brick wall is with the use of a pressure-treated ledger bolted to the wall with the bolts tied to the wall reinforcing. The joists are connected to the ledger with metal hangers. Figure 9-10 shows an example of the use of a ledger. When a roof framing system is to be supported on masonry, a pressure-treated plate is usually bolted to the brick in a manner similar to the way a plate is attached to a concrete foundation. Figure 9-11 shows how roof members are typically attached to masonry.

Brick Joints

The joints between each course of brick must be specified on the construction drawings. Common methods of finishing the mortar in joints are illustrated in Figure 9-12. In addition to the type of joint being specified for decorative purposes, joints must also be used to relieve stress in the wall. Walls longer than 200' (60 960 mm) or in a building having two or more wings must have expan-

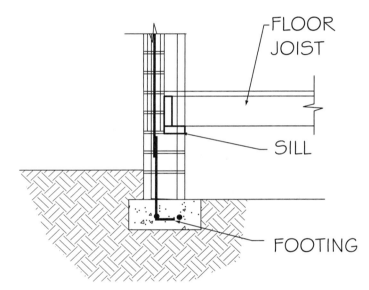

FIGURE 9-8 ■ If a masonry wall is to be used to support a floor, a space one wythe wide will be left to support the joist.

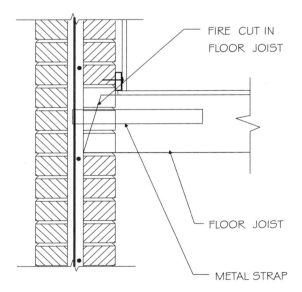

FIGURE 9-9 ■ A fire cut is placed in floor and roof members to prevent wall damage in case of a fire.

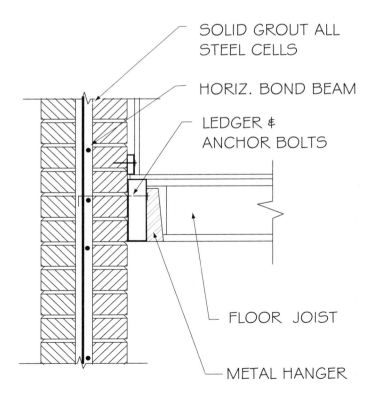

FIGURE 9-10 ■ A double-wythe brick wall with a floor system support by a wood ledger.

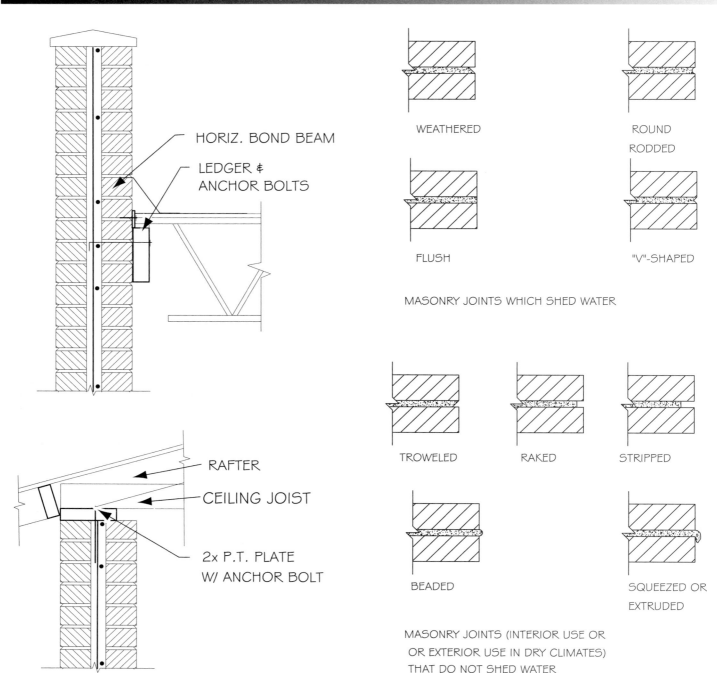

FIGURE 9-11 ■ Roof trusses are attached to a masonry wall using a ledger.

FIGURE 9-12 ■ Common methods of finishing mortar joints.

sion joints. An *expansion joint* is a seam placed in a wall to relieve cracking caused by expansion or contraction. Rather than causing random cracks throughout the brick, the stress is relieved in the expansion joint. Two common methods of constructing an expansion joint in brick walls are shown in Figure 9-13.

Brick Reinforcement

Load-bearing masonry walls must be reinforced. The architect or engineer will determine the size and spacing of the rebar to be used

based on the loads to be supported and the stress from wind or seismic loads. Rebar specifications will be discussed later in this chapter. Most brick walls are reinforced with a wire reinforcement pattern similar to those shown in Figure 9-14. The IBC requires wire that meets ASTM-A82 specifications. Depending on the manufacturer, wire is usually 8 or 9 gauge or 3/16" diameter. Welded wire reinforcement and wire ties do not change size for metric use. The engineer will specify the required type and size based on the vertical loads to be supported, and the seismic and wind loads will affect the placement. Wire reinforcement similar to that shown in Figure

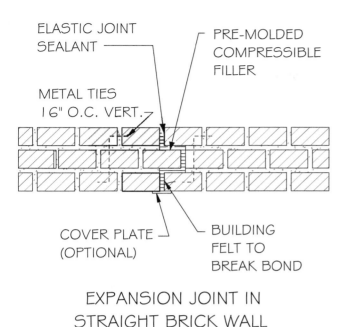

ELASTIC JOINT SEALANT

PRE-MOLDED COMPRESSIBLE FILLER

METAL TIES 16" O.C. VERT.

COVER PLATE (OPTIONAL)

BUILDING FELT TO BREAK BOND

EXPANSION JOINT IN STRAIGHT BRICK WALL

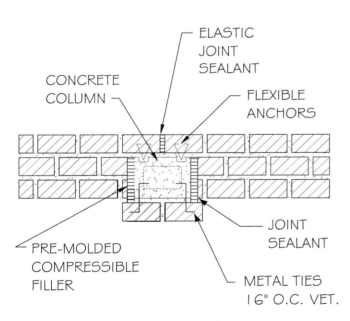

ELASTIC JOINT SEALANT

CONCRETE COLUMN

FLEXIBLE ANCHORS

JOINT SEALANT

METAL TIES 16" O.C. VET.

PRE-MOLDED COMPRESSIBLE FILLER

EXPANSION JOINT IN CONCEALED COLUMN

FIGURE 9-13 ■ Two methods of constructing expansion joints to relieve cracking in masonry walls.

9-14 is usually placed continuously in horizontal mortar joints at six-course intervals. The top course and the first two courses above and below any wall opening should be reinforced in all walls. The reinforcement should extend 24" (600 mm) minimum beyond each side of the opening. The cavity space between wythes is filled with grout and normally reinforced with vertical and horizontal steel. Steel reinforcement will be introduced later in this chapter.

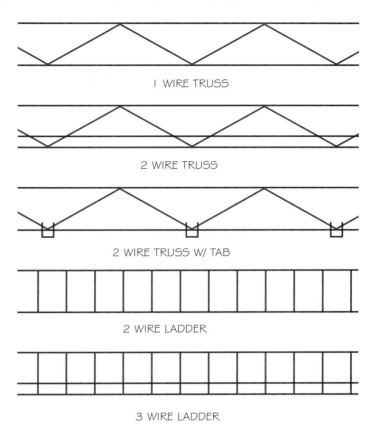

I WIRE TRUSS

2 WIRE TRUSS

2 WIRE TRUSS W/ TAB

2 WIRE LADDER

3 WIRE LADDER

FIGURE 9-14 ■ Standard wire reinforcement patterns for masonry walls.

Brick Veneer Construction

Using brick as a veneer over a wood, steel, or concrete masonry bearing wall allows the amount of brick required to be cut in half, greatly reducing the cost of construction. Care must be taken to protect the wood or steel frame from moisture in the masonry. Brick is installed over a 1" (25 mm) air space and a 15# layer of felt is applied to the framing. The veneer is attached to the framing with 26-gauge metal ties at 24" (600 mm) o.c. along each stud. Figure 9-15 shows an example of how masonry veneer is attached to a wood or steel frame wall.

Masonry can also be attached to a steel frame as shown in Figure 9-16. When the flange is parallel to the masonry, straps measuring approximately 2 × 7 × 1/8" (50 × 180 × 3 mm) are used to bond the masonry to the steel. When the web is parallel to the masonry, 16-gauge straps are typically used to connect the masonry to the steel frame.

CONCRETE MASONRY CONSTRUCTION

Concrete masonry units (CMUs) similar to Figure 9-17 provide a durable, economical building material with excellent structural and insulation values. CMU walls are usually single wythe, but they may be combined with a wythe of decorative stone or brick. Figure

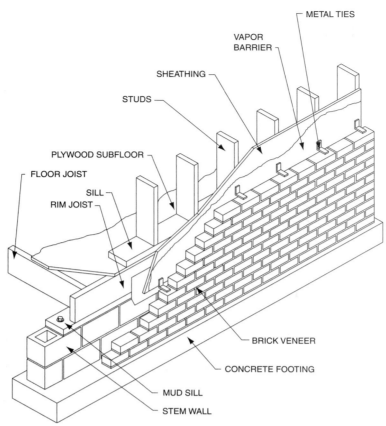

FIGURE 9-15 ■ Brick is attached to wood or steel stud walls by the use of metal ties placed on each stud.

9-18 shows standard wall construction. The IBC does not allow concrete-block walls to exceed 35' (10 700 mm) in height between diaphragms. CMUs can be waterproofed with clear waterproof sealers, cement-based paints can be used as the exterior finish, or the walls can be covered with stucco. Pressure-treated wood furring strips can be attached to the interior side of the block to support sheet rock, or the interior surface can be left exposed. Concrete blocks are also used for below-grade construction, which is discussed in Chapter 23.

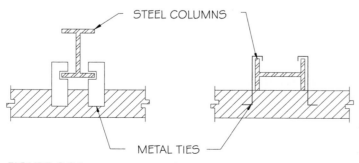

FIGURE 9-16 ■ Brick can be attached to a steel skeleton by the use of metal ties that attach to the steel column.

FIGURE 9-17 ■ Concrete masonry units provide a durable, economical building material that provides excellent structural and insulation values.

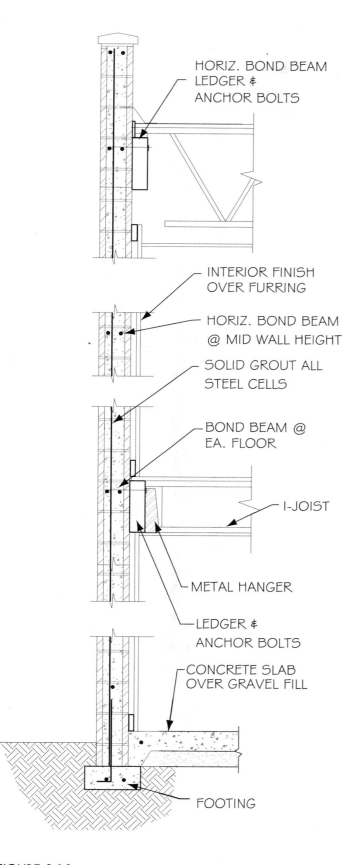

HORIZ. BOND BEAM
LEDGER & ANCHOR BOLTS

INTERIOR FINISH
OVER FURRING

HORIZ. BOND BEAM
@ MID WALL HEIGHT

SOLID GROUT ALL
STEEL CELLS

BOND BEAM @
EA. FLOOR

I-JOIST

METAL HANGER

LEDGER &
ANCHOR BOLTS

CONCRETE SLAB
OVER GRAVEL FILL

FOOTING

FIGURE 9-18 ■ Standard concrete block wall construction.

Grades

Four classifications are used to define concrete blocks for construction. These include hollow, load bearing (ASTM C 90), solid load bearing (ASTM C 145), and non-load-bearing blocks (ASTM C 129) which can be either solid or hollow blocks. Solid blocks must be 75% solid material in any cross-sectional plane. Blocks are also classified by their weight as normal, medium, and lightweight blocks. The weight of the block is affected by the type of aggregate used to form the unit. Normal aggregates such as crushed rock and gravel produce a block weight of between 40 and 50 lbs (18 and 23 kg) for an 8 × 8 × 16" block. Lightweight aggregates include: coal cinders, shale, clay, volcanic cinders, pumice, and vermiculite. Use of a lightweight aggregate will produce approximately a 50% savings in weight.

Walls made of CMUs are able to qualify for up to a 4-hour fire rating. The type of aggregate used in the block and the thickness of the block dictate the fire rating. Blocks made with pumice in a minimum of 6" (150 mm) wide units produce a four-hour rating.

Common Sizes

CMUs come in a wide variety of patterns and shapes. In the United States, the standard module for concrete masonry units is 4 inches. Standard inch-pound blocks are manufactured to 4", 6", 8", 10", and 12" nominal widths, 4" and 8" heights and 8" and 16" lengths. The most commonly used sizes are 8" and 16" blocks with 8 × 8 × 16 the industry standard. Each dimension of a concrete block is actually 3/8" smaller to allow for a 3/8" mortar joint in each direction

Metric Block Sizes

The size of metric block will depend on whether the block size is based on soft or hard conversion. The initial switch to the metric blocks for many U.S. manufacturers was to relabel their inch-pound products with the equivalent metric dimensions. Soft conversion requires no physical change in product size. An 8 × 8 × 16 would be relabeled as a 203 × 203 × 406. This relabeling of inch-pound block is referred to as a *soft conversion*. A designer working with a 4" design module would now be working with a design module of 101.6.

With *hard metric conversion*, masonry blocks are manufactured to metric specifications based on a 100 mm module. Hard metric conversion requires the use of new molds to produce blocks that are 200 × 200 × 400 mm. This requires block manufacturers to purchase new block molds, maintain dual inventories, and develop ways of distinguishing between similar looking metric and inch-pound blocks during storage and shipping. Metric (hard) blocks are manufactured to nominal dimensions of 100, 200, 250, and 300 mm widths, 100 and 200 mm heights, and 200 and 400 mm lengths. Actual dimensions are 10 mm smaller than the nominal size to allow for the vertical and horizontal mortar joints. A designer would now be working with a design module of 100.

Metric Layout

The designer and the drafter must take care in the planning and drafting a CMU structure. Severe problems will result if the hard conversion metric blocks are used in a building designed with soft metric dimensions and vise versa. With soft conversion, a 4" module will be 101.6 mm or 1.6% larger than a 100 mm block. This may seem unimportant, but the difference of 1.6 mm (approximately 1/16") will accumulate to become 3.2 mm (1/8") in 8 inches, 6.4 mm (1/4") in 16" and 19.2 mm (3/4") in 48 inches. This size difference will be huge in a wall of several hundred feet. Because concrete block is too big to fit within the standard 100 mm module without changing all block sizes, federally funded block structures can be drawn in either inch/pound or metric sizes. The U.S. Congress allows the use of either inch-pound blocks or metric blocks for federal construction projects. The National Concrete Masonry Association makes the following recommendations for those involved in the design of a metric structure:

1. The architect should verify the availability of the size and type of building components before establishing if the design is to be based on soft or hard conversions.

2. Hard conversions should only be used when all components are readily available in metric modular sizes from multiple suppliers.

3. Soft metric design should be used when hard metric components are not readily available.

4. Soft metric structures should be designed in inch-pound units and then all dimensions should be converted to their metric equivalents.

5. If hard metric design methods are to be used, special consideration should be given to the use of custom 7.5" (190.5 mm) high units and the placement of openings.

Figure 9-19 shows the difference in layouts between hard and soft conversions for horizontal and vertical coursings.

Modular Block Layout

To minimize cutting and labor cost, concrete block structures should be kept to their modular size. When using the standard 8 × 8 × 16" (200 × 200 × 400 mm) CMU, structures that are an even number of feet long should have dimensions that end in 0" or 8". For instance, structures that are 24'–0" or 32'–8" are each modular. Walls that are an odd number of feet should end with 4". Walls that are 3'–4", 9'–4" and 15'–4" are all modular, but a wall that is 15'–8" requires blocks to be cut.

Common Shapes

In addition to the exterior patterns, concrete blocks also come in a variety of shapes, as shown in Figure 9-20. Individual blocks are manufactured with two or three cores. Solid blocks are also manufactured. Two-core concrete blocks reduce heat conduction by approximately 4%, have more space for mechanical pipes and conduits, and weigh about 4 lbs (1.8 kg) less than a three-core block. These cores also allow for the placement of steel reinforcement bars, although steel reinforcing mesh—similar to the mesh used for

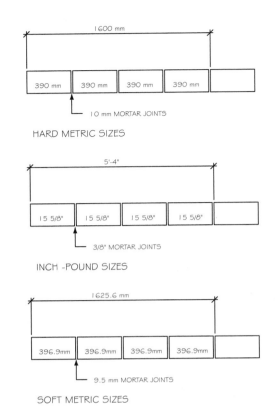

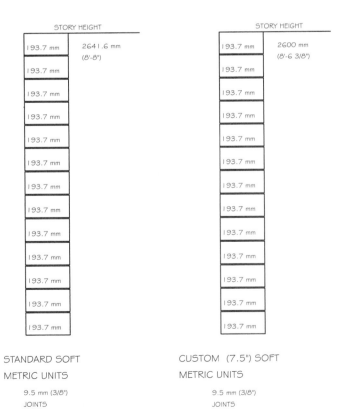

FIGURE 9-19 ■ A comparison of hard and soft metric conversions for horizontal and vertical coursings.

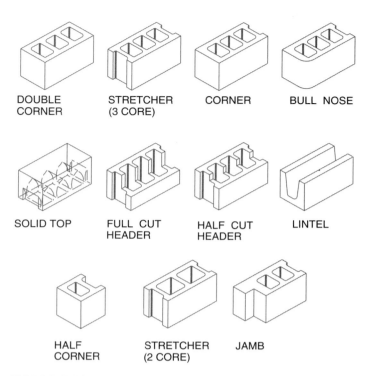

FIGURE 9-20 ■ Standard shapes of concrete masonry units.

brick—can also be used. Reinforcement is discussed later in this chapter.

Concrete blocks come in a variety of surface finishes. A standard block has a smooth but porous finish. The texture can be modified by changing the coarseness of the aggregate used to make the block. *Grid face* or *scored face* blocks are made with seams cut into the block, so that they resemble 8 × 8 × 8" (200 × 200 × 200 mm) blocks rather than 16" (400 mm) blocks. Seams are also available at 4" (100 mm) o.c. Blocks are also available with a raised or recessed geometric shape pressed into the face.

Masonry Representation

Brick and concrete block are represented on plan views with bold lines used to represent the edges of the masonry. Thin lines for hatching are placed at a 45° angle to the edge of the wall using approximately 0.125 spacing. The AutoCAD ANSI31 hatch pattern is the common method of hatching masonry. The type, size, and reinforcement are specified in note form but are not represented on the drawings. See Figure 9-21. If brick and concrete block walls are to be represented, the hatch patterns are placed at opposing angles.

The size and location of concrete-block walls are dimensioned on the floor plan. Concrete block is dimensioned from edge to edge of walls, as shown in Figure 9-22. Openings in the wall are also located by dimensioning to the edge.

The scale of the drawing affects the drawing method used to represent masonry shown in sections and details. At scales under 1/2" = 1'–0", the wall is usually drawn just as in plan view. At larger scales, details typically reflect the cells of the block. Figure 9-23

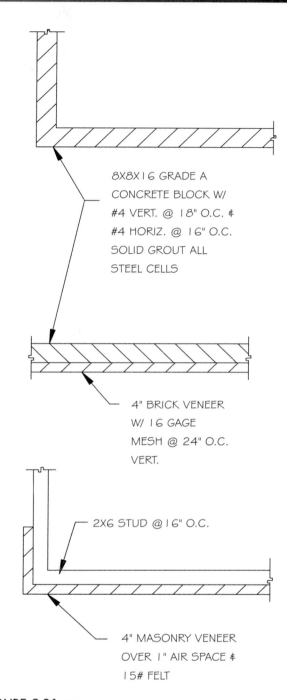

8X8X16 GRADE A CONCRETE BLOCK W/ #4 VERT. @ 18" O.C. & #4 HORIZ. @ 16" O.C. SOLID GROUT ALL STEEL CELLS

4" BRICK VENEER W/ 16 GAGE MESH @ 24" O.C. VERT.

2X6 STUD @ 16" O.C.

4" MASONRY VENEER OVER 1" AIR SPACE & 15# FELT

FIGURE 9-21 ■ Representing masonry units in plan view.

shows methods of representing concrete block in cross section. Figure 9-24 shows an example of a section and the notes that usually accompany concrete-block construction.

When CMUs have to support loads from a beam, a pilaster is placed in the wall to carry the loads. A *pilaster* is a thickened area of wall used to support gravity loads or to provide lateral support to the wall when the wall is to span long distances. Figure 9-25 shows a detail showing the size and thickness of a pilaster. Figure 9-22 shows how a pilaster would be represented in the plan view.

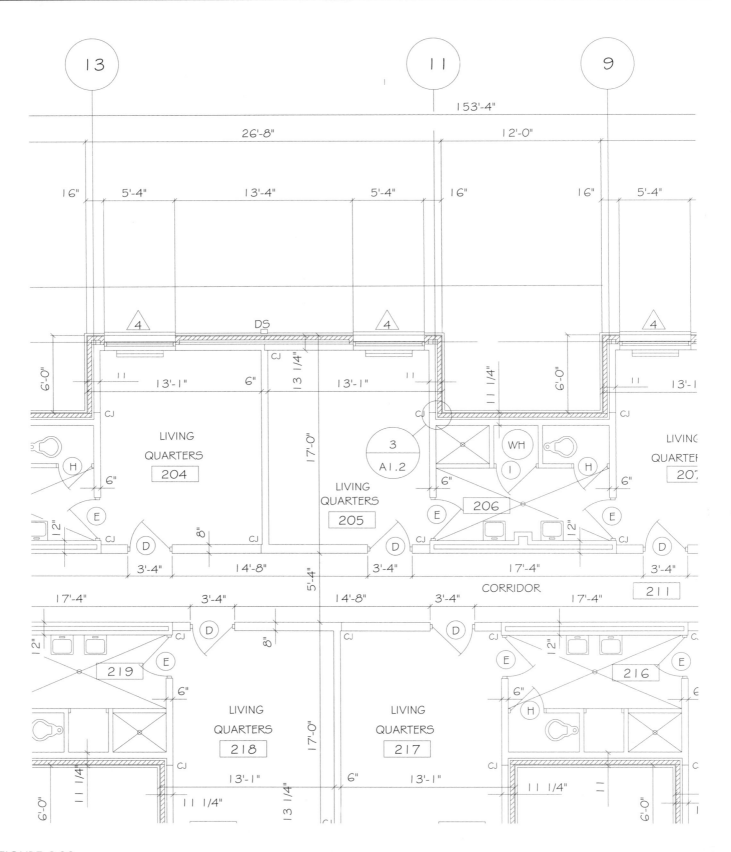

FIGURE 9-22 ■ Structural brick and concrete masonry units are dimensioned from exterior face to exterior face. Brick veneer is not dimensioned in plan view. The drafter should take great care to maintain modular sizes as masonry walls are dimensioned. *Courtesy G. Williamson Archer A.I.A., Archer & Archer P.A.*

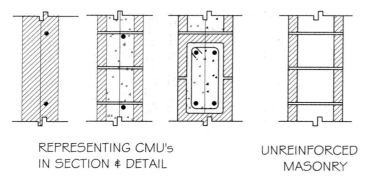

REPRESENTING CMU's
IN SECTION & DETAIL

UNREINFORCED
MASONRY

FIGURE 9-23 ■ ■ Methods of representing CMUs in sections and details.

Details similar to Figure 9-26 are also required to show reinforcement connections at each intersection and where different materials are joined to the blocks. This will require the drafter to draw wall-to-footing, wall-to-slab, wall-to-wall, wall-opening, and wall-to-roof details. Figure 9-23 shows common block wall details.

When a wood floor or roof is to be supported by the block, a ledger is bolted to the block. Holes are punched in the block to allow an anchor bolt to penetrate into the cell, where it is wired to the wall reinforcing. The block is reinforced and the bolt and rebar is held in place by filling the entire cell with grout. Figure 9-27 shows details for floor and roof connections to a block wall. Care

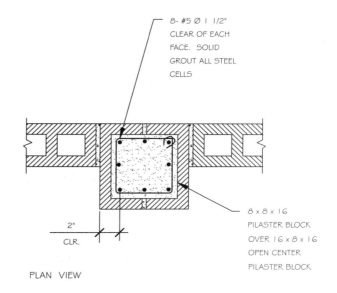

8- #5 Ø 1 1/2"
CLEAR OF EACH
FACE. SOLID
GROUT ALL STEEL
CELLS

8 x 8 x 16
PILASTER BLOCK
OVER 16 x 8 x 16
OPEN CENTER
PILASTER BLOCK

2"
CLR.

PLAN VIEW

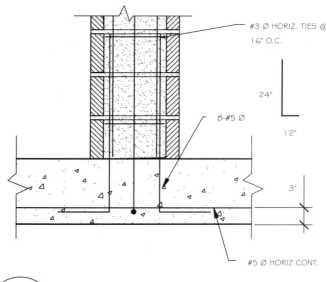

#3 Ø HORIZ. TIES @
16" O.C.

24"

12"

8-#5 Ø

3"

#5 Ø HORIZ CONT.

┌─────┐
│ 6 │ WALL SECTION DETAIL
│ A6.3│ SCALE: 1" = 1'-0"
└─────┘

SEE 3/A6.3 FOR
NOTES & DIM.

DOWEL @ 24" o.c.

FILL CELL W/ CONC.

6" CMU

FLOOR/BASE
AS SCHED.

FLASHING

1" RIGID INSULATION

4" FACE BRICK

FLASHING

┌─────┐
│ 5 │ PILASTER / FND.
│ A-8 │
└─────┘

FIGURE 9-24 ■ Representing brick and CMUs in detail. *Courtesy G. Williamson Archer A.I.A., Archer & Archer P.A.*

FIGURE 9-25 ■ Construction detail of a concrete block pilaster.

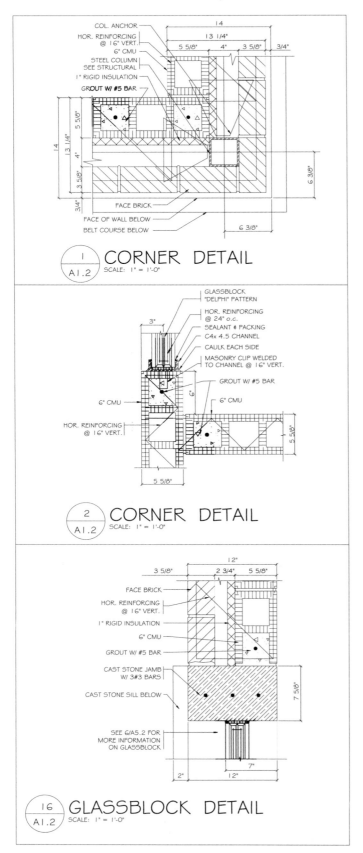

FIGURE 9-26 ■ Details are required to show intersections with each material to be used as well as intersections of floor and roof systems.

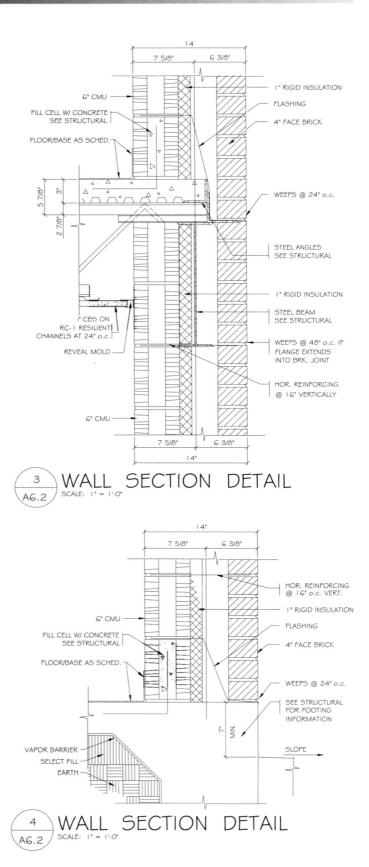

FIGURE 9-27 ■ Connecting a floor system to a masonry wall. *Courtesy G. Williamson Archer A.I.A., Archer & Archer P.A.*

must be taken to tie the ledger bolts into the wall reinforcement, as well as securely tying the floor joist to the ledger. The same procedure must be followed when a roof is tied into a wall. In addition, methods of keeping the intersection waterproof must be implemented. The roofing is usually extended up the wall over a cant strip that is covered with metal flashing, which sheds water from the joint. Figure 9-28 shows examples of how the roof intersections are represented in detail. Figure 9- 29 shows examples of common mortar wall caps that are used to protect the top of the masonry from moisture

STEEL REINFORCEMENT

Masonry units are excellent for resisting forces from compression, but they are poor for resisting forces in tension. Steel is excellent for resisting forces of tension, but it tends to buckle under forces of compression. The combination of these two materials forms an excellent unit for resisting great loads from lateral and vertical forces. Reinforced masonry structures are stable because the masonry, steel, grout, and mortar bond together so effectively. Loads that create compression on the masonry are transferred to the steel. By careful placement of the steel, the forces will result in tension on the steel and be safely resisted

Reinforcing Bars

Placing steel in the masonry creates a wall known as *reinforced masonry construction.* An intersecting horizontal and vertical grid of concrete reinforced with steel runs throughout a concrete-block wall, forming the frame for a structure. Although wire mesh simi-

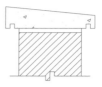

FIGURE 9-29 ■ Common masonry wall caps.

lar to that used with brick can be used in concrete block, steel reinforcing bars are the common method of reinforcing concrete block. The IBC sets specific guidelines for the size and spacing of rebar based on the seismic zone of the construction site. Reinforcement is specified by the engineer throughout the calculations and sketches based on the building code and the Portland Cement Association (PCA) and the Concrete Reinforcing Steel Institute (CRSI) guidelines, and it must be accurately represented by the drafter.

Rebar Sizes and Shapes

Reinforcing bars can be either smooth or deformed. Smooth bars are primarily used for joints in floor slabs. Because the bars are smooth, the concrete of the slab does not bond easily to the bar. This allows the concrete to expand and contract. Bars used for reinforcing masonry walls are usually deformed similar to those shown in Figure 9-30 so that the concrete will bond more effectively with the bar. This prevents slippage as the wall flexes under pressure. The deformations are small ribs that are placed around the surface of the bar. Deformed bars range in size from 3/8" to 2 1/4" (10 to 57 mm) diameter. Bars are referenced on plans by a number rather than a size. A number represent the size of the steel in approximately 1/8" increments. Inch sizes don't readily convert to metric sizes. Steel reinforcing bars conform to ASTM A615-78A or to ASTM A615M standard for metric reinforcing bar. Each standard governs the size and the physical characteristics of rebar. Steel may be either grade 40, with a minimum yield strength of 40 ksi (275 MPa), or grade 60, with a minimum yield strength of 60 ksi (415 MPa). Common sizes of steel are shown in Table 9-2:

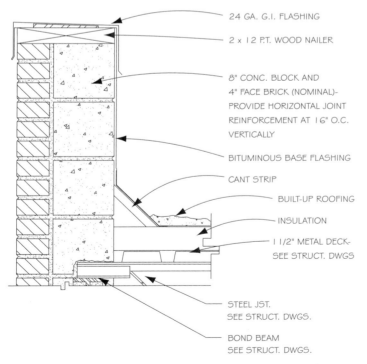

24 GA. G.I. FLASHING

2 x 12 P.T. WOOD NAILER

8" CONC. BLOCK AND 4" FACE BRICK (NOMINAL)- PROVIDE HORIZONTAL JOINT REINFORCEMENT AT 16" O.C. VERTICALLY

BITUMINOUS BASE FLASHING

CANT STRIP

BUILT-UP ROOFING

INSULATION

1 1/2" METAL DECK- SEE STRUCT. DWGS

STEEL JST. SEE STRUCT. DWGS.

BOND BEAM SEE STRUCT. DWGS.

FIGURE 9-28 ■ The method used to waterproof the wall/roof intersection must be represented in detail as well as the written specifications.

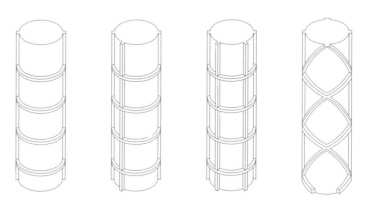

FIGURE 9-30 ■ Common deformed steel reinforcing patterns.

TABLE 9-2 ■ Common Bar Sizes of Steel (Inch-Pound and Metric)

Bar Size	Inch-Pound	Nominal Diameter (in / mm)	Bar Size Metric
# 3	0.375	9.5	# 10
# 4	0.50	12.7	# 13
# 5	0.625	15.9	# 16
# 6	0.75	19.1	# 19
# 7	0.875	22.2	# 22
# 8	1.00	25.4	# 25
# 9	1.13	28.7	# 29
# 10	1.27	32.3	# 32
# 11	1.41	35.8	# 36
# 14	1.69	43.0	# 43
# 18	2.26	57.3	# 57

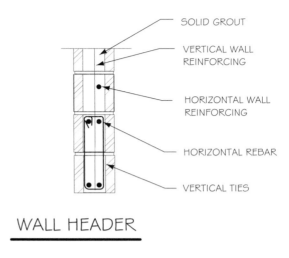

WALL HEADER

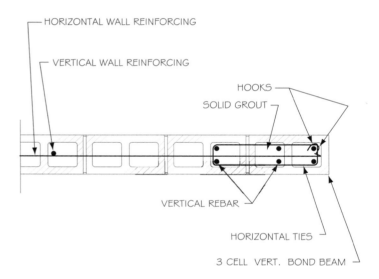

WALL OPENING

FIGURE 9-31 ■ Concrete block over and beside openings requires special reinforcement. Above an opening, horizontal steel is held together with vertical ties. Vertical reinforcing beside the opening is held in place by horizontal ties.

Steel Placement

Reinforcing steel is required in all masonry walls in seismic zones C, D, E, and F. Wall reinforcing is held in place with wire ties and by filling each masonry cell containing steel with grout. When exact locations within the cell are required by design, the steel is wired into position, so that it can't move as grout is added to the cell. The placement of steel varies with each application, but steel reinforcement is placed on the side of the wall that is in tension. Because an above ground wall will receive pressure from each side, masonry walls have the steel centered in the wall cavity. For cantilevered retaining walls, steel is placed near the soil side of the wall. For full-height retaining walls anchored at the top and bottom, the steel is placed near the side that is opposite the soil. The location of steel in relation to the edge of the wall must be specified in the wall details.

Vertical Reinforcing

Vertical reinforcing is required at 48" (1200 mm) o.c. with a maximum spacing of 24" (600 mm) if stacked bond masonry is used. IBC requires a vertical piece of steel to be placed within 16" from each end of a wall. Vertical steel can be represented in details similar to Figure 9-18. Additional vertical reinforcing is usually required at the edges of openings in walls to serve the same function as a post in wood construction. For small loads two vertical bars in the same cell may be sufficient. As loads increase, the size of the vertical bond and the number of bars used will increase. Horizontal ties are normally added when four or more vertical bars are required to keep the bars from separating. Figure 9-31 shows an example of a reinforced vertical column located between two wall openings.

Horizontal Reinforcing

Horizontal reinforcement is approximately 16" (400 mm) o.c., but the exact spacing depends on the seismic or wind loads to be resisted and is designed by the engineer for each specific use. In addition

to the horizontal steel required by the building code, extra steel is usually placed at the mid-point of a wall between each floor level. See Figure 9-18. Generally two bars are placed at the mid-point of each wall level, forming a reinforced area referred to as a bond beam. Extra reinforcing is also added where each floor or roof level intersects the wall. A 16" deep bond beam with two pieces of steel at the top and bottom is a standard method of constructing the bond beam at floor and roof levels. An 8" deep bond beam with two pieces of steel is also placed at the top of concrete block walls. Another common location for a bond beam is over the openings for doors and windows. A 16" minimum depth with four pieces of steel is normally provided similar to Figure 9-31. The exact depth, quantity of rebar, and ties will be specified by the engineer based on the load to be supported and the span of the bond beam.

Steel Overlap

Because of the limits of construction, steel must often be lapped to achieve the desired height or length. In-line steel bars are lapped and wired together so that individual bars will act as one. Figure 9-32 shows an example of a detail with lapped steel. Depending on the loads to be resisted, the engineer may require laps to be welded. The amount of lap required at steel intersections can be found in the details or in the written specifications of the drawings. Where the information is placed depends on the complexity of the project. The intersection of a wall at the foundation, for example, is a typical place where steel is lapped. The engineer usually places a note in the calculations indicating the desired overlap of steel to be specified by the drafter. If the engineer specifies the use of #5 bar with a 36 diameter lap, the drafter would need to show the steel overlapping 23" on the drawing. Laps of steel based on diameter are listed in Table 9-3:

TABLE 9-3 ■ Inches of Lap Corresponding to the Number of Bar Diameters

Number of Diameters	Size of Bar								
	#3	#4	#5	#6	#7	#8	#9	#10	#11
20	6	10	13	15	18	20	23	26	29
22	8	11	14	17	20	22	25	28	32
24	9	12	15	18	21	24	28	31	34
30	12	15	19	23	27	30	34	39	43
32	12	16	20	24	28	32	37	41	46
36	14	18	23	27	32	36	41	46	51
40	15	20	25	30	35	40	46	51	57
48	18	24	30	36	42	48	55	61	68

Steel Bends and Hooks

A *bend* allows steel reinforcement to be continuous at a corner. A *hook* ties intersecting members to each other. For example, the horizontal steel in a footing may be hooked around the vertical steel in an intersecting footing to ensure that the horizontal can't move in a lateral direction. Unless the engineer has provided a detail, hooks and bends should be detailed to match the minimum standards set by the Concrete Reinforcing Steel Institute (CRSI). Figure 9-33 shows the minimum standards for hooks and bends. The location and length of the hook is normally placed on the details and in the specifications.

Rebar Representation

The drafter needs to specify the quantity of bars, the bar size by number, the direction the steel runs, and the grade on the drawings where steel is referenced. On drawings such as the framing plan, walls are drawn, but the steel is not drawn. Steel specifications are

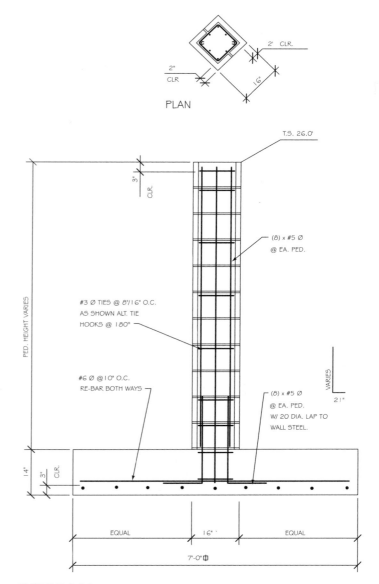

FIGURE 9-32 ■ Representing steel placement and overlap in detail.

generally included in the wall reference as shown in Figure 9-21. When shown in section or detail, steel is represented by a bold line, which can be solid or dashed depending on office practice. When steel is shown in end view, it is represented by a solid circle. The AutoCAD **DONUT** command is effective for drawing rebar in end view. Guidelines for representing reinforcing steel in detail include the following:

■ On drawings plotted at a scale smaller than 3/4" = 1'–0", a constant symbol size is used to represent various steel bar sizes.

■ With larger scaled details, variations in bar sizes can be represented.

■ Steel is often drawn at an exaggerated size so that it can be seen more clearly. Rebar should not be so small that it

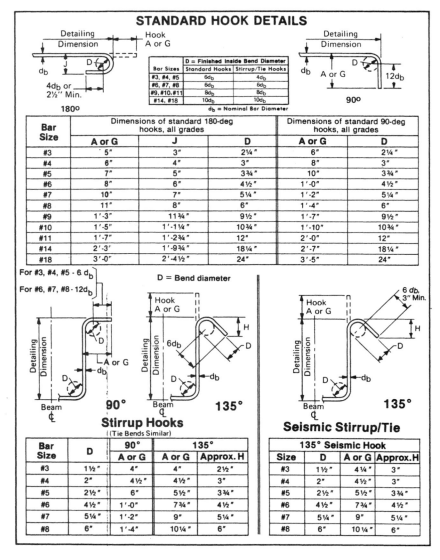

FIGURE 9-33 ■ Minimum standards for reinforcement hooks. *Courtesy Concrete Reinforcing Steel Institute, Schaumberg, IL.*

blends with the hatch pattern used for the grout, or so large that it is the first thing seen in the detail.

Figures 9-23 through 9-27 show common examples of representing steel in detail. The direction the steel is to be laid is specified to add clarity to the drawing. Steel direction is specified using the terms vertical, horizontal, or diagonal lines. Occasionally the letters *E.W.* or *E/W* are used to indicate that the steel is running each way (horizontal and vertical).

When hooks or ties are represented in a detail, they should be shown to extend past the steel they are surrounding, as shown in Figure 9-32. Spacing of ties often varies within a detail such as a pilaster or column. A specification such as

#3 TIES W/3 @ 8/16" O.C.

indicates that the ties are to be a #3 diameter with spacings of 8" and 16". The pattern is then shown in a detail similar to Figure 9-32. Notice that the three bars at the top and bottom are placed at

8" spacings and the bars in the middle of the column are spaced at 16" o.c.

Locating Steel

Dimensions are required to show the location of the steel from the edge of the masonry. The engineer may provide a note in the calculations such as

7– #5 HORIZ. @ 3" UP/DN.

Although this note could be placed on the drawings exactly as is, a better method is shown in Figure 9-32. The quantity and size of the steel are specified in the detail, but the locations are placed using separate dimensions. This requires slightly more work on the part of the drafter, but the visual specifications will be less likely to be overlooked or misunderstood. Depending on the engineer and the type of stress to be placed on the masonry, the location may be given from edge of masonry to edge of steel or from edge of masonry to center of steel. To distinguish the edge of steel location, the term *clear* or *CLR.* is added to the dimension. Figure 9-32 shows examples of each method of locating concrete. In addition to the information in the details, steel is also referenced in the written specifications. The written specifications detail the grade and strength requirements for general areas of the structure such as walls, floors, columns, or retaining walls. Figure 9-34 shows an example of steel specifications that accompany a small office structure. See Chapter 10 for additional types of steel reinforcing used with other types of concrete construction.

MORTAR AND GROUT

Mortar used to bond masonry products and steel is composed of portland cement, sand, lime, and clean water. Other materials can be added to the mix to increase its ability to bond to masonry units and steel. Mortar mix is normally governed by ASTM C144. The strength of the mortar is of critical importance to the strength of the masonry wall and is specified on the architectural drawings and the written specifications. Type M and S are most often used for exterior walls and are suitable for walls above or below grade. Type N or S mortar is normally used for load-bearing walls, while type O mortar is used for nonbearing walls. The mortar is specified by the engineer in the calculations and specified by the drafter in the masonry wall details.

In addition to the strength and makeup of the mortar, the type of joint is usually specified. Joints used with masonry units are similar to the joints used with brick construction and are shown in Figure 9-12.

Grout is composed of the same components as mortar, but the grading is slightly different. Grout is specified as fine or coarse and governed by ASTM C33. Specifications are placed on the drawings or in the project manual to control the proportions of elements in grout. Figure 9-34 shows an example of common mortar notes placed on the working drawings of a concrete block structure.

ADDITIONAL READING

The following Web sites can be used as a resource to help you keep current with changes in unit masonry.

ADDRESS	COMPANY/ORGANIZATION
www.ambrico.com	American Brick Company
www.bia.org	Brick Industry Association
www.confast.com	Concrete Fastening Systems, Inc.
www.crsi.org	Concrete Reinforcing Steel Institute
www.generalshale.com	General Shale Brick
www.imiweb.org	The International Masonry Institute
www.maconline.org	Masonry Advisory Council
www.masonrysociety.org	The Masonry Society
www.masonryinstitute.org	Masonry Institute of America

REINFORCING

1. ALL REINFORCING STEEL TO BE ASTM A615 GRADE 60, EXCEPT TIES, STIRRUPS & DOWELS TO MASONRY TO BE GRADE 40. W. W. F. SHALL CONFIRM TO ASTM A185 AND SHALL BE 6X6-W1.4XW1.4 WWF MATS.
2. FABRICATE AND INSTALL REINF. STEEL ACCORDING TO THE "MANUAL OF STANDARD PRACTICE FOR DETAILING REINFORCED CONCRETE STRUCTURES" ACI STANDARD 315.
3. PROVIDE 2'-0" X 2'-0" CORNER BARS TO MATCH HORIZ. REINFORCING IN POURED IN-PLACE WALLS AND FTGS. @ ALL CORNERS AND INTERSECTIONS.
4. SPLICES IN WALL REINFORCING SHALL BE LAPPED 30 DIAMETERS (2'-0" MINIMUM AND SHALL BE STAGGERED AT LEAST 4' AT ALTERNATE BARS.
5. ALL OPENINGS SMALLER THAN 30" X 30" THAT DISRUPT REINFORCING SHALL HAVE AN AMOUNT OF REINFORCING EQUAL TO THE AMOUNT DISRUPTED PLACED BOTH SIDES OF OPENING & EXTENDING 2'-0" EA. SIDE OF OPENING.
6. PROVIDE THE FOLLOWING REINFORCING AROUND WALL OPENINGS LARGER THAN 30" x 30":
   A. (2) #5 OVER OPENING X OPENING WIDTH PLUS 2'-0" EACH SIDE.
   B. (2) #5 UNDER OPENING X OPENING WIDTH PLUS 2'-0" EACH SIDE.
   C. (2) #5 EACH SIDE OF OPENING X FULL STORY HEIGHT.
   D. PROVIDE 90 DEGREE HOOK FOR BARS AT OPENINGS IF REQUIRED EXTENSION PAST OPENING CANNOT BE OBTAINED.
7. PROVIDE (2) #4 CONTINUOUS BARS AT TOP AND BOTTOM AND AT DISCONTINUOUS ENDS OF ALL WALLS.
8. PROVIDE (2) #5 X OPENING DIMENSION PLUS 2'-0" EACH SIDE AROUND ALL EDGES OF OPENINGS LARGER THAN 15" X 15" IN STRUCTURAL SLABS AND PLACE (1) #4 X 4'-0" AT 45 DEGREES TO EACH CORNER.
9. PROVIDE DOWELS FROM FOOTINGS TO MATCH ALL VERTICAL WALL, PILASTER, AND COLUMN REINFORCING (POURED-IN-PLACE COLUMNS AND WALLS). LAP 30 DIAMETERS OR 2'-0" MINIMUM.
10. ALTERNATE ENDS OF BARS 12" IN STRUCTURAL SLABS WHENEVER POSSIBLE.
11. ALL WALL REINFORCING TO BE PLACED IN CENTER OF WALL UNLESS SHOWN OTHERWISE ON THE DRAWINGS.

CONCRETE BLOCK

1. DESIGN FM = 1500 PSI FOR SOLID GROUTED WALLS, 1350 PSI FOR WALLS WITH REINFORCED CELLS ONLY GROUTED.
2. ALL CMU UNITS TO BE GRADE N TYPE 1 LIGHTWEIGHT UNITS PER ASTM C90 DRY TO INTERMEDIATE HUMIDITY CONDITION PER TABLE NO. 1.
3. MORTAR TO BE PER IBC TABLE 2103.7(1) TYPE S OR PER MANUFACTURER'S RECOMMENDATION TO REACH CORRECT STRENGTH. GROUT TO HAVE MIN. STRENGTH TO MEET DESIGN STRENGTH AND MADE WITH 3/8" MINUS AGG.
4. CONTRACTOR TO PREPARE AND TEST 3 GROUTED PRISMS & 3 UNGROUTED PRISMS PER EVERY 5000 SQUARE FEET OF WALL, DURING CONSTRUCTION.
5. ALL WORK SHALL CONFORM TO SECTION 2103 THROUGH SECTION 2109 OF THE LATEST INTERNATIONAL BUILDING CODE.
6. REINFORCING FOR MASONRY TO BE ASTM A615 GR. 60 PLACED IN CENTERS OF CELLS AS FOLLOWS (UNLESS OTHERWISE INDICATED):
   A. VERTICAL: 1- #5 @ 48" O.C. PLUS 2- #5'S FULL HEIGHT EACH SIDE OF OPENINGS, UNLESS SHOWN OTHERWISE ON DRAWINGS.
   B. HORIZ.: 2- #4'S @ 48" O.C. (FIRST BOND BEAM 48" FROM GROUND FLOOR) PLUS 2- #4'S AT TOP OF WALL AND AT EACH INTERMEDIATE FLOOR LEVEL.
   C. LINTELS: LESS THAN 4'-0" WIDE---2- #4'S IN BOTTOM OF 8"DEEP LINTEL. 4'-0" TO 6'-0" WIDE----2- #4'S IN BOTTOM OF 16" DEEP LINTEL.
   D. CORNERS AND INTERSECTIONS: 1- 24" X 24" CORNER BAR AT EACH BOND BEAM SAME SIZE AS HORIZONTAL REINFORCING.
7. GROUT ALL CELLS FULL @ EXTERIOR OR LOAD-BEARING WALLS. GROUT ALL CELLS THAT CONTAIN REINFORCING OR EMBEDDED ITEMS.
8. ELECTRICAL BOXES, CONDUIT AND PLUMBING SHALL NOT BE PLACED IN ANY CELL THAT CONTAINS REINFORCING.
9. SEE DRAWINGS FOR ADDITIONAL REINFORCING.

BRICK

1. ASSUMED FM 1800 PSI MINIMUM.
2. EXTERIOR BRICK TO BE GRADE SW AND INTERIOR BRICK TO BE GRADE MW.
3. MORTAR TO BE TYPE S AT BRICK VENEER AND TYPE M AT REINFORCED BRICKWORK, GROUT TO HAVE MINIMUM COMPRESSIVE STRENGTH AT 28 DAYS OF 2000 PSI AND BE MADE WITH 3/8" MINUS AGGREGATE.
4. ALL WORK SHALL CONFORM TO CHAPTER 21 OF THE IBC.

STRUCTURAL & MISCELLANEOUS STEEL

1. DETAILING, FABRICATION & ERECTION SHALL CONFORM TO THE STEEL CONSTRUCTION MANUAL OF AISC.
2. ALL STEEL TO BE A36 EXCEPT AS NOTED.
3. ALL WELDS TO BE MADE BY CERTIFIED WELDERS TO AWS STANDARDS WITH E70XX ELECTRODES.
4. UNLESS NOTED OTHERWISE, BOLTS TO BE A325N FOR STEEL TO STEEL CONNECTIONS & A449 FOR A. B. . DRILLED A. B.TO BE "PARABOLT", "KWIKBOLT" OR APPROVED EQUAL.
5. ALL STRUCTURAL TUBING TO BE ASTM A500 GRADE B. FY=46 KSI. ALL STEEL PIPE TO BE ASTM A501 (FY = 36 KSI) OR ASTM A53, TYPE E OR S, GRADE B (FY = 35 KSI).

FIGURE 9-34 ■ Written specifications for steel reinforcing, concrete block and brick construction for a multilevel office structure are often placed with the structural drawings. *Courtesy Van Domelen/Looijenga/McGarrigle/Knauf Consulting Engineers.*

www.ncma.org   National Concrete Masonry   www.portcement.org   Portland Cement Association
                   Association

Unit Masonry Methods and Material

CHAPTER QUIZ

Answer the following questions on a separate sheet of paper. Use local vendor catalogs as needed.

Question 9-1 A structure is to be 33'–8" wide, and covered with baby Roman brick. Is the structure modular? Explain your answer.

Question 9-2 Explain the difference between a course and a wythe.

Problem 9-3 A wall is to be built of Roman bricks, approximately 12'–0" high. How high must the wall be to be modular?

Question 9-4 List two common methods of bonding wythes together.

Question 9-5 List the two most common methods of masonry bonds.

Question 9-6 What minimum code is wire reinforcement required to meet?

Question 9-7 Use local vendor catalogs to find a suitable connector for attaching masonry veneer-to-steel stud framing.

Question 9-8 List four classifications of CMUs.

Question 9-9 What factors dictate the fire rating of a block wall?

Question 9-10 List the approximate spacing of concrete block reinforcement.

Question 9-11 Explain the guidelines for the length of modular walls.

Question 9-12 What is the name of the piece of wood that attaches to the side of concrete block walls to support wood framing? To the top of concrete block?

Question 9-13 Explain how beams are supported with masonry walls.

Question 9-14 What do the numbers of rebar correspond to?

Question 9-15 What would be the benefit of having smooth rebar?

DRAWING PROBLEMS

Unless other instructions are given by your instructor, complete the following details and save them as wblocks using a name that represents the contents of the drawing. Skeletons of most details can be accessed from http://www.delmar.com/resources/ocl.html. Use these drawings as a base to complete the assignment.

■ Show all required views to describe each connection.

■ Draw each detail using a minimum scale of 3/4" = 1'–0" unless noted.

■ Unless noted, all wall construction is to be CMU. Specify all steel to be centered and all steel cells to be solid grouted for all details unless noted.

■ Use separate layers for wood, concrete, concrete block, text, laminations, and dimensions.

■ Represent structural members that the cutting plan has passed through with bold lines.

■ Use dimensions for locations where possible instead of notes.

■ Show and specify all connecting materials. Base nailing on IBC standards unless your instructor tells you otherwise. Specify all material based on common local practice.

■ Provide notation to specify that all metal hangers are to be provided by Simpson Strong-Tie Company or equal. Hatch each material with the appropriate hatch pattern.

■ Use an appropriate text font to label and dimension each drawing as needed. Refer to *Sweet's Catalogs*, vendor catalogs, or the Internet to research needed sizes and specifications. Keep all text 3/4" minimum from the drawing, and use an appropriate architectural style leader line to reference the text to the drawing.

■ Provide a detail marker, with a drawing title, scale, and problem number below each detail.

■ Assemble the details required by your instructor into an appropriate template sheet for plotting.

■ Arrange the drawings using the guidelines presented in Chapters 2 and 3. Assign a drawing sheet number of sheet S5 of 6.

Problem 9-1 Draw a detail to represent a brick veneer wall on a 18 × 8" concrete footing for a two-level apartment project. Reinforce the footing with (1)-#5, 3" up from the bottom. Form the wall with 8 × 8 × 16 concrete blocks. Reinforce the wall with #5 rebar at 16" o.c. horizontal and 48" o.c. vertical. Use a 2 × 6 DFPT sill supporting 2 × 8 floor joist, supporting 3/4" ply floor sheathing, and 2 × 6 studs at 16" o.c. Cover the wall with brick veneer over 1/2" wafer board and Tyvek. Specify the appropriate connection of the brick to the wall.

Problem 9-2 Show standard trusses with a 24" overhang, and a 4/12 pitch supported by a 2 × 6 DFPT plate w/ 5/8" A.B. @ 36" o.c. Support on a 8 × 8 × 16 CMU wall. Reinforce with #5 @ 48" o.c. each way.

Problem 9-3 Show an 8 × 16" wide footing 18" into the grade to support 8 × 8 × 16" grade A concrete blocks. Show a 5" concrete slab at grade level on gravel fill. Thicken the slab to 10" × 10" at the edge and show a #5 × 24" × 12" L @ 48" o.c. from wall to slab. Reinforce the footing with (2)-#5 continuous, 3" up from the bottom of the footing. Show a #5 × 12" long L projecting from the footing. Extend the footing steel far enough into the wall to get a 24 diameter lap. Reinforce the wall with #5 vert. at 48" o.c. and #5 horizontal at 16" o.c. Specify to solid grout all steel cells. See Table 10-1 for minimum concrete coverage and placement of foundation steel.

Problem 9-4 Show an 8 × 16" bond beam for a door header in an 8" block wall. Use (4)-#6 horizontal bars continuous with #3 ties at 24" o.c. Provide #5 vertical at 48" o.c. and #5 @ 24" o.c. horizontal.

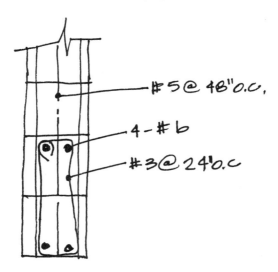

Problem 9-5 Show a plan view of a wall opening. Reinforce the last three cells with (2)-#5 vertical rebar. Tie with #3 ties at 16" o.c. Specify horizontal steel to be #5 at 48" o.c. with standard hook around the #5 at 24 o.c. vertical steel.

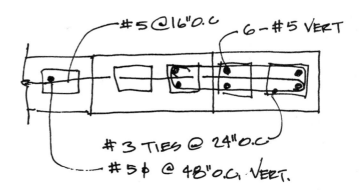

Problem 9-6 Show the intersection of a wood floor and a block wall formed from 8 × 8 × 16" CMUs. Use a 3 × 12 DFPT ledger with 3/4" dia. ×12" A.B. @ 32" o.c. staggered 2" up/dn, supporting 2 × 12 DFL #1 joist @ 16" o.c. with appropriate metal hangers. Reinforce wall w/ #5 @ 32" o.c. vertical and #5 @ 16" o.c. horizontal. Provide a 8 × 16" deep bond beam w/ (4)-#5 @ ledger. Bolt a 3" high × 4" wide × 3/8" plate above floor joist with 3/4" diameter M.B. @ 48" o.c. and connect an 18 ga. Simpson strap capable of resisting 1000 # floor loads below 3/4" plywood sheathing. Weld the strap to the wall plate with 1/8" fillet field weld.

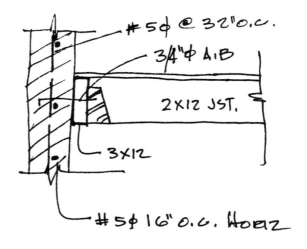

Problem 9-7 Draw a similar detail to Problem 9-6 with the joist parallel to the wall at 16" o.c. Use 5/8" dia. A.B. at 48" o.c. staggered 2" up/dn. In addition to the minimum wall steel allowed by the IBC, provide an 8 × 8" deep bond beam w/ (2)-#5 @ ledger. Provide solid blocking at 48" o.c. for 48" out from wall. Bolt a 3" high × 4" wide

× 3/8" plate above floor joist with 1/2" diameter A.B. @ 48" o.c. and connect 1 1/2" × 48" × 3/16" strap below 3/4" plywood sheathing. Weld the strap to the wall plate with 1/8" fillet field weld and nail strap to blocks with (48)-10d nails.

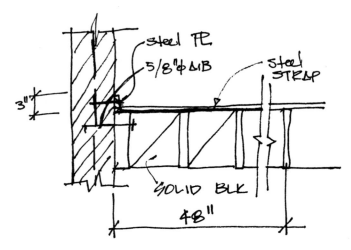

Problem 9-8 Show a plan view of a 16 × 16" pilaster. Reinforce with (8)-#5 vertical 2 1/2" clear of outside face. Tie with #4 horizontal at 16" o.c. Show #5 verticals @ 48" o.c. and #5 horizontal @ 16" o.c. for the wall.

Problem 9-9 Show a side view of the pilaster drawn in Problem 9-8. Use an appropriate GLB beam seat to hold a 6 3/4 × 15 fb 2400 glu-lam beam and support 9000# @ 375 psi minimum. Weld the seat to wall reinforcement with 1/8" fillet. Specify beam bolting based on the manufacturer of the seat.

Problem 9-10 Show a 3 × 12 DFPT ledger with 3/4" dia. A.B. @ 32" o.c. staggered 2" up/dn supporting TJW 24 open-web truss joist @ 32" o.c. Reinforce wall w/ #5 @ 32" o.c. vertical and #5 @ 16" o.c. horizontal. Reinforce the ledger with (4)-#5 in an 8 × 16 bond beam. Bolt a 4" high × 6" wide × 12" plate above trusses with 1/2" diameter A.B. @ 64" o.c. and connect Simpson Co. MST27 @ 64" o.c. Connect to plate with 1/8" fillet field weld. Show a 30" high parapet above the roof trusses and cap with a double-sloping masonry cap. Show a 6" wide cant strip covered with 24-ga. flashing and built-up roofing.

Problem 9-11 Draw a CMU wall with brick veneer resting on a 20" wide × 12" footing that extends 18" minimum into natural grade. Provide (3)-#5 continuous 3" up from the bottom of the footing and a #5 × 24 × 6" bend @ 48" o.c. Extend to the steel with a minimum of 20 dia. lap. Reinforce the wall with #5 at 48" o.c. vertical and #5 @ 24" o.c. horizontal steel. Provide a 4" concrete slab over 4" gravel fill. Reinforce the slab with w.w.m. Attach the brick veneer to the CMU with metal ties at 24" o.c. each direction.

Problem 9-12 Combine details 9-3 and 9-10 to create a partial wall section. Set the top of ledger at 12"–0" above the finish floor. Provide (2)-#5 at 48" o.c. horizontal and (2)-#5 horizontal at the wall top.

CHAPTER 10

Concrete Methods and Materials

Concrete is the most versatile material used in the construction industry. In its liquid or plastic state, its shape is limited only by the limitations of the materials used to create the forms that mold it. It can be finished in a variety of colors, textures, and finishes to withstand the most severe weather conditions. Because concrete comes to the construction site in a plastic state, it is subject to change during the construction process. This requires on-site control and inspections to ensure that all of the engineer's specifications have been met. Engineering specifications often include information on how to mix, transport, prepare forms, place, reinforce, finish, and cure concrete so that all design values can be achieved. More so than any other material, the drafter must pay strict attention to detailing to ensure the building materials will provide a strong, durable construction method. This chapter will introduce the basic components of concrete, common construction methods, and reinforcing materials. Figure 10-1 shows concrete being poured, as well as many of the elements of concrete construction that will be discussed in this chapter.

PHYSICAL PROPERTIES OF CONCRETE

Concrete can be greatly altered depending on the materials and proportions of materials used to make the concrete. The methods of pouring, setting, and curing will each be affected by the weather at the site where concrete is poured, the temperature, the rate of placement, and the size of the area to be poured. All of these factors can affect the appearance, strength, and weight of the finished concrete. Most of the qualities of concrete are governed by ACI 318 (American Concrete Institute) standard, which the engineer will specify to govern the quality best suited to the project. Although the contractor is responsible for verifying that the standards of the mix and pouring procedures are met, an understanding of concrete and the pouring process will help the drafter better understand the contents of the specifications and details being provided.

Concrete Materials

Concrete is a mixture of portland cement, aggregate, miscellaneous additives called admixtures, and water, and it is cast while in a plastic or liquid state. The written specifications in the drawings and in the project manual specify the proportions of the mixture and the conditions under which they can be used.

Cement

Portland cement is a mixture of lime, silica sand, iron oxide, and alumina. When mixed with clean water, this mixture will harden and set into the form in which it is molded. Portland cement is usually required to conform to ASTM C150, which is the standard specification for portland cement. Five different types of portland cement are used in the construction industry.

- *Type 1* is referred to as standard or normal cement. Common uses include beams, columns, and floor slabs

- *Type II* cement is used in localities where a strong resistance to sulfate is not required. It is used primarily for

FIGURE 10-1 ■ Construction methods, reinforcement, and placement must all be considered as drawings detailing concrete structures are completed. *Courtesy American Concrete Institute.*

members that require a lower heat of hydration such as large footings or foundations. The hydration process will be discussed as the water mixture is considered. Type II cement has a lower early strength than type I cement. A cement designated type I/II meets the specifications for both type I and type II and may be used where either is specified.

■ *Type III* cement achieves a high early strength, which continues to accelerate as it cures. Type III cement has the same strength after one day that type I cement has after three days. After 28 days, type I and III cements will have achieved about the same strength. Type III cement is often used for precast concrete and for cold-weather pours because of its high heat hydration. This quality also makes type III cement unsuitable for use with large pour members.

■ *Type IV* cement is used for massive pour construction and is generally considered low-heat cement. Type IV cement is excellent where cracking must be kept to a minimum.

■ *Type V* cement is used in areas that will be exposed to water and high amounts of sulfate content.

Water

The amount of water is an important element in making concrete. The quantity to be used must be specified on the working drawings. *Potable water* that is drinking-water quality and free of oils or excessive amounts of acid or alkali must be used. Although water from a municipal water supply is usually adequate, water must conform to ASTM C94 *Standard Specification for Ready Mixed Concrete*. In addition to controlling the water in the mix, the engineer may provide specifications for the soaking of the forms prior to use, as well as methods of keeping the concrete damp during the early stages of the curing process.

When water is added to cement, a chemical reaction known as *hydration* takes place, causing a bonding of the molecules. As the mixture begins to harden, it releases heat in a process known as *heat of hydration*. Methods of controlling the rate of evaporation of water from the exposed surfaces, and hydration within the concrete, are often specified by the engineer and must be included in the written specification for the job. Control of the amount of water affects not only the hydration but also the workability of the concrete. Excess water is held to a minimum to keep the concrete from becoming porous. This keeps cracks from developing, in addition to limiting voids in the concrete.

Aggregates

Although the aggregates do not affect the chemical process that causes the cement to harden, they are controlled in the specifications because they play an important role in the quality of concrete. Sand, gravel, and crushed rock are used for the aggregate in the concrete mix. Aggregates should be rounded, insoluble, clean, and free of dust or other substances so that the cement paste can bond to the aggregate. It is also important that the aggregate not contain

elements that will chemically react when added to water or cement. Aggregate materials are divided into fine and course aggregates based on the size of the material used. Fine material is usually sand, but rocks small enough to pass through a 1/4" (6 mm) sieve are also used. Anything that will not pass through a no. 4 sieve is considered course aggregate.

The aggregate for most construction is between 3/4" and 1 1/2" (20 and 40 mm) washed gravel or crushed stone conforming to ACI 318–3-3. The ACI does not allow the size of the aggregate to exceed 1/5 of the pour width between forms; 1/3 of the slab depth; or 3/4 of the clear rebar spacing. The aggregate can be altered, affecting the weight of the mixture. Aggregate will be discussed further in the Lightweight Concrete section of this chapter. The drafter needs to note the ingredients, proportions, or the size of the aggregate specified by the engineer on the drawings or in the written specifications.

Admixtures

Any substance other than cement, water, or aggregates that is added to concrete to affect its workability, accelerate its setting time, or alter its hardness is known as an *admixture*. Admixtures are listed in the written specifications for concrete drawings and are used to increase early strength, increase ultimate strength, accelerate or retard setting time, or increase workability. ACI318-3.6 governs chemical admixtures. Common admixtures include hydrated lime, calcium chloride, and kaolin. When calcium chloride is added to the water, concrete with a fast setting time is produced. This concrete is known as *high-early-strength concrete*. A high degree of strength is obtained quickly, allowing for protection from rain or freezing conditions. High-early-strength concrete allows for rapid removal of forms that can be reused again, further reducing the cost of construction. The engineer or contractor determines the admixtures required for the mix, based on expected weather conditions at the time of the pour, and the set speed.

Air-Entrained Concrete

Air can be injected into the water before it is added to the mixture, forming microscopic air bubbles in the mix. Air-entrained cement meeting ASTM 360 and C33 is effective for resisting deterioration that results from contraction and expansion caused by freezing and thawing. Adding air to the cement mixture also improves its workability and lowers the tendency for mixture separation. Air-entrained cement is identified as types IA, IIA, and IIIA and is used in ways similar to its non-air-containing counterpart. The air bubbles that form in the hardened concrete provide space for expansion of moisture during freezing conditions.

Mixture Ratios

The ratio of the concrete mix is very important to the setting, curing, and ultimate strength of the concrete. A common ratio for concrete is 1:2:4, which represents 1 part of cement, 2 parts of sand, and 4 parts of gravel. Other common mixes are 1:3.75:5, and 1:2.5:3. Concrete is usually specified by its batch weight; giving a ratio of water to cement (w/c) stated in pounds. If a batch of concrete contains 200 pounds of water and 500 pounds of cement, the w/c would be

$$200/500=0.40 \text{ (lb of water per lb cement)}$$

The ACI limits the water/cement ratio based on the desired strength and the air content. The engineer will specify the exact proportions of water to cement to be used to achieve the highest strength and maintain a workable mixture. If too much water is added, the strength and durability of the concrete will be compromised. The strength of concrete is measured in terms of its ultimate compressive strength in pounds per square inch (psi) or megapascals (MPa) and represented by the symbol F'c. A concrete with a compressive strength of 2500 psi is referred to as 2500 pound concrete. Because of the variations in producing concrete, the working stress of concrete is used for design purposes. The design strength is established by the engineer and is based on a portion of the ultimate strength and expressed as a fraction of F'c. Concrete is considered to reach its ultimate compression strength at 28 days after pouring based on ASTM C192, but concrete actually continues to get harder at a very slow rate. Forms are typically removed after approximately seven days.

Concrete Testing

To determine if the concrete meets the requirements of the engineer, slump, compressive strength, and air content can be tested. A *slump test* determines the consistency and workability of concrete in its plastic state. This test can be preformed at the mixing site or at the job site, with results specified by ASTM C143.

The ultimate compressive unit strength of the concrete (F'c) is tested by taking samples of the concrete in its plastic state. Concrete samples are placed in 6" diameter × 12" long molds. The concrete is allowed to harden and is tested at various stages of the curing process. Cores are generally tested at days 3, 7, and 28, to see the amount of compression that can be resisted. If the test results fall below the design values intended for the concrete, the engineer will need to determine if the pour is suitable for design conditions. Careful planning by the engineer and quality control at the mixing site typically provide concrete well above the minimum design value.

TYPES OF CONCRETE CONSTRUCTION

Common methods of concrete construction include plain, reinforced, prestressed, precast, prefabricated, lightweight, thin shell, and pneumatically applied concrete. With each method, the engineer will provide calculations and sketches for the drafter to complete the finished drawing.

Plain Concrete

Concrete that contains no reinforcement, or contains less reinforcement than the minimum requirements of the ACI 318, is considered plain concrete. Plain concrete must have a minimum compressive strength of 2500 psi (17.24 MPa) and is used primarily for foundations and slabs of office structures. Unreinforced walls are

required by code to be a minimum of 5 1/2" (140 mm) thick, and slabs must be a minimum of 3 1/2" (90 mm) thick. Walls must be connected at each floor and roof level. The engineer will determine the exact thickness of the concrete member based on the amount of loads to be resisted.

Reinforced Concrete

Concrete is extremely strong in resisting loads that cause compression, but very weak at resisting forces that cause tension. Reinforced concrete contains steel reinforcement to increase the tensile strength of the concrete. Typically the concrete reinforcement is rebar (introduced in Chapter 9). The bond of the concrete to the surface of the reinforcing bars and the resistance provided by the deformations make the concrete and steel act as one material. Figure 10-2 shows an example of reinforced concrete construction. Welded wire fabric is also used to reinforce concrete and carry stress from tension. Methods of reinforcement specific to concrete construction will be introduced later in this chapter.

Prestressed Concrete

Most concrete members are stressed by loads that will cause bending. To resist the forces of bending, members are frequently prestressed. Thin horizontal members such as beams and slabs are most likely to be prestressed to resist tension in the member. In

FIGURE 10-2 ■ The type of cement, amount of water, aggregates, admixtures, air, and reinforcing all contribute to the quality of concrete. Each will be addressed throughout the project manual, but the drafter will be required to draw many details showing reinforcement patterns. *Courtesy Dick Schmitke.*

reinforced construction, the reinforcing steel carries all of the forces of tension. In prestressed construction, the entire concrete member is effective at resisting the forces of tension. Prestressed members are used for pilings, columns, wall panels, floor slabs, and roof panels. The stressing is applied by placing high-strength steel reinforcing bars or cables that have been stretched. The bars can be stretched prior to the concrete being poured in a process called pretensioning. Steel can also be stretched after the concrete has been poured and hardened around the steel for a process called post-tensioning.

In pretensioning, high-strength stranded steel wires are stretched into position before the concrete is poured. After the concrete is poured and allowed to harden, the tension on the wires is released. As the steel contracts and attempts to regain its original shape, the wires conduct compressive forces throughout the concrete. The compression in the concrete helps prevent cracking from deflection and allows the size of the member to be reduced when compared to non-stressed members.

In the post-tensioning process, large-diameter steel rods or cables are placed in conduit that is embedded within the concrete, with one end of the steel anchored into the concrete. After the concrete has hardened, the steel is stretched by hydraulic jacks (Figure 10-3) and then attached to the concrete so that the steel remains under tension.

Precast Concrete

Precast concrete members are cast and cured in an off-site location and then transported to the job site. Figure 10-4 shows an example of a precast floor panel being positioned. When a large number of identical units are to be built, time, money, and space at the job site are saved when members are cast off site. Members can be built ahead of time and shipped as needed to the construction site. Beams, columns, and small wall panels are usually precast. Panels are normally welded to the building frame at the top and bottom of the panel, allowing multilevel structures to be built one floor at a time. Figure 10-5 shows the welded connection of a precast concrete panel to a steel frame.

Prefabricated Concrete Units

Prefabricated units are formed at the job site and then lifted into place. This form of construction is also know as tilt-up construction in many parts of the country. Tilt-up construction is usually used for large single-level structures. Large industrial complexes or retail sales outlets represent ideal uses of concrete tilt-up structures. Wall panels are cast in a horizontal position over the floor slab or in a bed of sand. Once they have cured, they are lifted into place by a crane. Panels can be attached to precast columns or welded to

FIGURE 10-3 ■ Concrete can be reinforced by placing tension on steel cables placed in a concrete member. When the cables are tensioned after the concrete has cured, the process is referred to as post-tensioning. *Courtesy Portland Cement Association.*

FIGURE 10-4 ■ Precast concrete members can be cast and tensioned off site and then positioned as needed. *Courtesy Ken Stead.*

FIGURE 10-5 ■ Precast concrete members can be welded to a steel skeleton. *Courtesy Cindy Stead.*

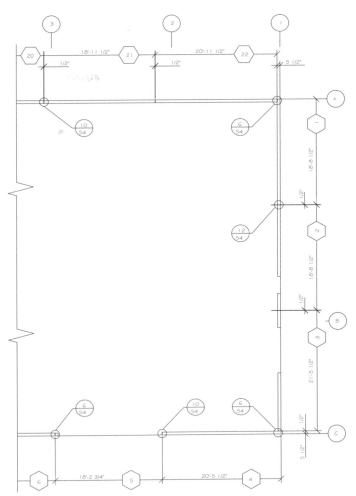

FIGURE 10-6 ■ Precast members are often identified on the floor plan. The number in the hexagons refers to a specific concrete panel shown in a drawing similar to Figure 10-6. *Courtesy Ginger Smith, StructureForm Masters, Inc.*

adjacent panels. Panel construction requires a plan view similar to Figure 10-6 showing the location of each panel; elevations similar to Figure 10-7, showing the location of each opening and placement of the reinforcing steel; and details to show typical connections such as panel to panel, panel to foundation, slab to panel, and roof to panel. An example of the intersection of two beams to a wall can be seen in Figure 10-8. Precast panels will be discussed further in the section on methods of wall construction later in this chapter.

Lightweight Concrete

Concrete weighs approximately 150 pounds per cubic foot (2400 kg/m³). The weight can be lessened depending on the aggregate used or chemicals that are injected into the mix. Lightweight concrete is classified in three groups according to the unit weight per cubic foot:

■ The lightest concrete has a unit weight of between 20 and 70 lb/ft³ (320 and 1120 kg/m³) and is classed as *insulating lightweight concrete*. Because its compressive strength is usually below 1000 psi (6.9 MPa), it is not used as a structural material, but as a protective covering for fireproofing. It is frequently poured over a wood or steel decking to form a floor.

■ *Structural lightweight concrete* has a weight of up to 115 lb/ft³ (1840 kg/m³) with a compressive strength of 2000 psi (13.8 MPa).

■ *Semi-lightweight concrete* has a unit weight of from 115 to 130 lb/ft³ (1842 to 2080 kg/m³). Its ultimate strength is comparable to that of normal concrete.

Thin Shell

Thin-shell or eggshell construction can be used to form three-dimensional structures made of one or more curved slabs or folded plates. Concrete as thin as 2 1/2" (65 mm) thick with a compressive strength 3000 psi (20.7 MPa) can be used to form the shell. The strength of the shell is achieved by the shape of the concrete into which it is molded. In addition to the material saved in

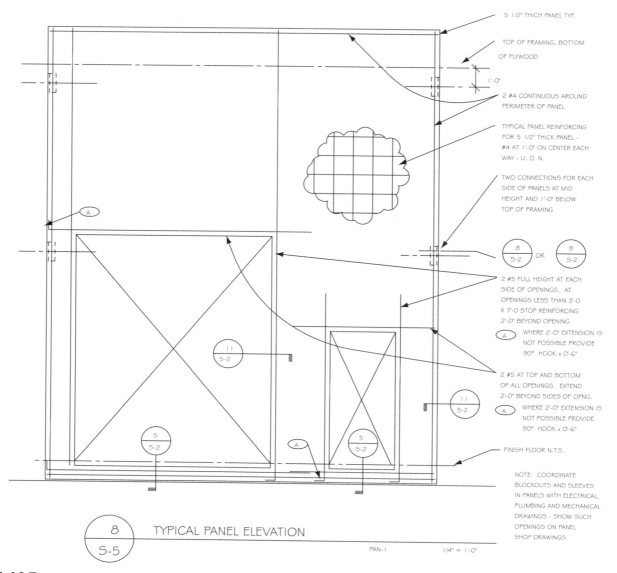

5 1/2" THICK PANEL TYP.

TOP OF FRAMING, BOTTOM
OF PLYWOOD

1'-0"

2 #4 CONTINUOUS AROUND
PERIMETER OF PANEL

TYPICAL PANEL REINFORCING
FOR 5 1/2" THICK PANEL -
#4 AT 1'-0" ON CENTER EACH
WAY - U. O. N.

TWO CONNECTIONS FOR EACH
SIDE OF PANELS AT MID
HEIGHT AND 1'-0" BELOW
TOP OF FRAMING

8 / S-2   OR   9 / S-2

2 #5 FULL HEIGHT AT EACH
SIDE OF OPENINGS. AT
OPENINGS LESS THAN 3'-0
X 7'-0 STOP REINFORCING
2'-0" BEYOND OPENING

A   WHERE 2'-0" EXTENSION IS
NOT POSSIBLE PROVIDE
90° HOOK x 0'-6"

2 #5 AT TOP AND BOTTOM
OF ALL OPENINGS. EXTEND
2'-0" BEYOND SIDES OF OPNG.

A   WHERE 2'-0" EXTENSION IS
NOT POSSIBLE PROVIDE
90° HOOK x 0'-6"

FINISH FLOOR N.T.S.

NOTE: COORDINATE
BLOCKOUTS AND SLEEVES
IN PANELS WITH ELECTRICAL,
PLUMBING AND MECHANICAL
DRAWINGS - SHOW SUCH
OPENINGS ON PANEL
SHOP DRAWINGS.

8 / S-5   TYPICAL PANEL ELEVATION

PAN-1   1/4" = 1'-0"

FIGURE 10-7 ■ An elevation is drawn for each member to show size, shape, locations of openings, and required reinforcing steel. *Courtesy Van Domelen/Looijenga/McGarrigle/Knauf Consulting Engineers.*

the roof, supporting walls, columns, and foundations can also be reduced because of the reduced weight of the roof. Arcs or barrels, hyperbolic paraboloid, domes, and folded plates are common shapes used for thin-plate construction because of their inherent strengths; but there are many variations that can be achieved with these basic shapes.

Shotcrete

Portland cement or plaster that is applied with a compressed air gun is considered shotcrete or pneumatic concrete. Shotcrete is extremely dense and strong. It tends to have a high resistance to weathering and is an excellent waterproofing material because of its low absorption rate.

REINFORCEMENT PLACEMENT AND SPECIFICATIONS

As with concrete block, deformed steel bars and welded wire mesh are the major reinforcing materials of concrete construction. As seen in Figure 10-9, concrete reinforcement patterns and specifications are usually more complicated than masonry construction.

Reinforcing Bars

Steel bars having the same characteristics as those used with masonry are used to reinforce concrete construction, although the

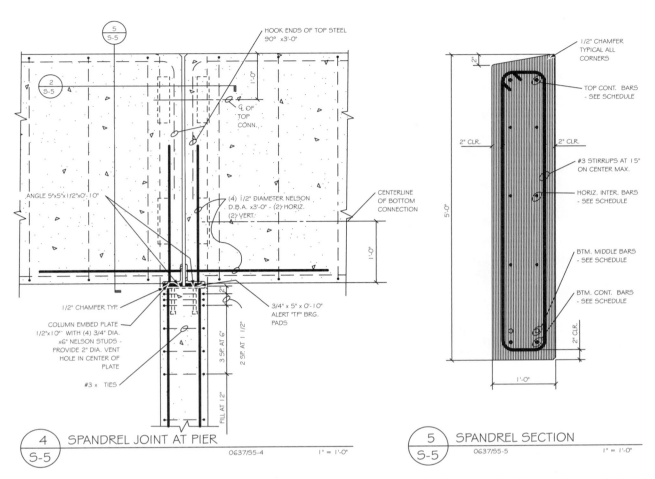

FIGURE 10-8 ■ Details of connections for each concrete member will need to be drawn to detail connection methods. *Courtesy Van Domelen/Looijenga/McGarrigle/Knauf Consulting Engineers.*

drafter will need to be mindful of different configurations, spacings, and clearances. Steel reinforcing is required to meet the ASTM standards listed in the IBC. Non-deformed bars can be used for spiral reinforcing of columns and must conform to ASTM standards. Structural steel shapes, steel pipes, and steel tubing are also used for concrete reinforcement and must meet the specifications of ASTM required by the code. The engineer will select the standard to be met, and the drafter will place the required standards on the concrete drawings and written specifications. Figure 10-10 shows the reinforcing notes that supplement the drawings for the tilt-up structure used throughout this text.

Unlike masonry construction, reinforcement in poured concrete also requires the use of *bolsters* and *chairs* to maintain a uniform positioning. Each is made from steel or plastic wire, but protective coatings or a galvanized finish can be applied to steel wires. The finish is often specified in the written specifications that accompany the drawings.

Figure 10-11 shows several common types of supports that are designed to place the steel in the desired finished position. *Bolsters* come in lengths up to 10' (3000 mm) long and are designed to hold the lower steel-reinforcing mat in position. *Chairs* are designed to hold upper-level pieces of steel in position within a slab. These

members are not specified on the structural drawings but are referenced on fabrication drawings. A minimum of 1" (25 mm) is required by the IBC/ACI 318-7.6 between parallel layers of reinforcing, and each layer must be staggered so parallel members do not align. The spacing of steel mats must be specified on the structural drawings using methods similar to Figure 10–12.

Welded Wire Fabric

Wire fabric, or mesh as it is sometimes called, is laid in perpendicular patterns and is used for reinforcement of floor, wall, and roof slabs. Wire meshes are required to meet ASTM standards of the IBC or of those specified by the engineer. Patterns are made in one-way rectangular and two-way square patterns.

One-way mesh is made of heavy longitudinal wires that are spaced from 2" to 16" (50 to 400 mm) apart. A lighter tie wire is spaced from 1" to 18" (25 to 460 mm) apart and welded at each intersection. Common wire fabric dimensions can be seen in Figure 10-13. The specification of 6 × 18-W8 × D4 represents *welded wire fabric* with a spacing of 6" × 18" using no. 8-gauge smooth wire for the longitudinal wires with no. 4-gauge tie wires. The letter *D* can be substituted for the *W* to represent deformed

FIGURE 10-9 ■ The type, size, strength, and location of all reinforcement must be specified in the project manual and throughout the structural drawings. *Courtesy American Concrete Institute.*

REINFORCING:

1. ALL REINFORCING STEEL TO BE ASTM A615 GRADE 60 EXCEPT TIES, STIRRUPS AND DOWELS TO MASONRY TO BE GRADE 40. WELDED WIRE FABRIC SHALL CONFORM TO ASTM A-185 AND SHALL BE 6x6-W1.4 x W1.4 WWF MATS.

2. FABRICATE AND INSTALL REINFORCING STEEL ACCORDING TO THE "MANUAL OF STANDARD PRACTICE FOR DETAILING REINFORCED CONCRETE STRUCTURES" ACI STANDARD 315.

3. PROVIDE 2'-0" x 2'-0" CORNER BARS TO MATCH HORIZONTAL REINFORCING IN POURED-IN-PLACE WALLS AND FOOTINGS AT ALL CORNERS AND INTERSECTIONS.

4. SPLICES IN WALL REINFORCING SHALL BE LAPPED 30 DIAMETERS (2'-0" MINIMUM) AND SHALL BE STAGGERED AT LEAST 4'-0" AT ALTERNATIVE BARS.

5. ALL OPENINGS SMALLER THAN 30" x 30" THAT DISRUPT REINFORCING SHALL HAVE AN AMOUNT OF REINFORCING EQUAL TO THE AMOUNT DISRUPTED PLACED BOTH SIDES OF OPENING AND EXTENDING 2'-0" EACH SIDE OF OPENING.

6. PROVIDE THE FOLLOWING REINFORCING AROUND WALL OPENINGS LARGER THAN 30" x 30":
   A. (2) #5 OVER OPENING x OPENING WIDTH PLUS 2'-0" EACH SIDE.
   B. (2) #5 UNDER OPENING x OPENING WIDTH PLUS 2'-0" EACH SIDE.
   C. (2) #5 EACH SIDE OF OPENING x FULL STORY HEIGHT.
   D. PROVIDE 90 DEGREE HOOK FOR BARS AT OPENINGS IF REQUIRED EXTENSION PAST OPENING CANNOT BE OBTAINED..

7. PROVIDE (2) #4 CONTINUOUS BARS AT TOP AND BOTTOM AND AT DISCONTINUOUS END OF ALL WALLS.

8. PROVIDE (2) #5 x OPENING DIMENSION PLUS 2'-0" EACH SIDE AROUND ALL EDGES OF OPENINGS LARGER THAN 15" x 15" IN STRUCTURAL SLABS AND PLACE (1) #4 x 4'-0" AT 45 DEGREES TO EACH CORNER.

9. PROVIDE DOWELLS FROM FOOTING TO MATCH ALL VERTICAL WALL, PILASTER, AND COLUMN REINFORCING (POURED-IN-PLACE COLUMNS AND WALLS). LAP 30 DIAMETERS OR 2'-0" MINIMUM.

10. ALTERNATE ENDS OF BARS 12" IN STRUCTURAL SLABS WHENEVER POSSIBLE.

11. ALL WALL REINFORCING TO BE PLACED IN CENTER OF WALL UNLESS SHOWN OTHERWISE ON THE DRAWINGS.

FIGURE 10-10 ■ Many building departments require reinforcement specifications to be placed with the structural drawings rather than in the project manual. *Courtesy H.D.N. Architects A.I.A.*

wire. The wire mesh is represented and specified in detail as seen in Figure 10-12.

Two-way wire fabric is arranged in a square pattern with equal-sized wires used in each direction. The designation and representation is similar to one-way wire. A 6 × 6-W8 × W8 represents welded-wire mesh made with a 6" spacing with no. 8-gauge wires.

Reinforcement Coverage

Steel reinforcing bars are typically placed as close as possible to the outer edge of the member being reinforced to resist the forces of tension and resist surface cracking. The distance between the outer edge of the steel and the outside edge of the concrete member is referred to as the *cover*. The engineer will specify exact requirements for coverage, based on the requirements for weather, fire protection and the space required to develop a bond between the concrete and the steel to develop adequate bending resistance within the component being detailed. The drafter should specify the distance, as shown in Figure 10-14. Depending on the experience level of the drafter, the engineer may not always provide a specifi-

cation for coverage, intending the minimum recommended coverage to be followed. These minimum standards can be seen in Table 10-1.

FLOOR SYSTEMS

Concrete floor systems are typically constructed as cast-in-place, one- and two-way reinforced slab systems, or as lift systems.

Cast-in-Place Floors

Cast-in-place floor systems can be formed over a wood or steel decking. Wood is a common forming material for cast-in-place reinforced slabs. Wood decking is used to support floors that will be finished with concrete. Concrete is placed over wood decking for the floor system for many retail sales or apartment structures. The concrete is typically 1 1/2" (40 mm) thick. Steel decking (introduced in Chapter 8) is a common method of supporting concrete floor slabs. Ribbed decking is welded to steel trusses that provide the support for the floor system. Ribs with a depth of 1.5", 3"

TABLE 10-1 ■ Minimum Concrete Cover

Concrete Exposure	Minimum Cover in inches /mm
Concrete cast against and permanently exposed to earth	3 / 75
Concrete exposed to earth or weather	
No. 5 bars, W31 or D31 wire and smaller	1 1/2 / 38
No. 6 through No. 18 bars	2 / 50
Concrete not exposed to weather or in contact with ground	
Slabs, walls and joists:	
No. 11 bars and smaller	3/4 / 20
No 14 and No 18 bars	1 1/2 / 38
Beams and columns:	
Primary reinforcement, ties, stirrups and spirals	1 1/2 / 38
Shells, folded plate members:	
No. 5 bar, W31 or D31 wire and smaller	3/4 / 20
No. 6 bars and larger	1/2 / 13

or 4.5" (40, 80, or 115 mm) are most often used for floor slabs. Figure 10-15 shows an example of a poured-in-place floor system.

One-Way Reinforced Floor Systems

Self-supporting concrete slabs are often used with larger commercial construction projects. A *one-way system* consists of a slab supported by parallel reinforced concrete beams supported on reinforced-concrete columns. This system of construction is normally considered feasible for slabs spanning from 10' to 35' (3000 to 10 500 mm). Figure 10-16 shows an example of a one-way floor system. Intermediate beams placed at right angles to the main support beams can also be used for longer spans or heavier loads. A versatile system called a *one-way joist* or *one-way ribbed floor* uses narrow beams at close repetitive spacing. Joists are typically limited to a minimum width of 4" (100 mm) with a maximum spacing of 30" (750 mm) clear. The depth of the slab can range from 1 1/2" to 3" (38 to 75 mm) depending on the span with the minimum depth of the joist required to be three times the depth of the slab. The floor slab and support beams are poured to make a monolithic structure. *Monolithic* is used to describe the process of pouring several components in one pour so that all of the units act as one. The system is usually constructed using metal forms to construct voids between the joist. The system also can be constructed using precast, pretensioned floor units. Figure 10-17 shows examples of common precast floor shapes. Figure 10-18 shows an example of a detail drawn to show support for floor panels.

Two-Way Reinforced Floor Systems

As the name implies, the *two-way system* has major slab reinforcement running in two directions. Figure 10-19 shows three common methods of forming a two-way system. With each system, columns are usually placed in square patterns. In its simplest form

(at top in Figure 10-19), a solid slab is placed over the support columns. This method of construction would be suitable for floors with light loads such as apartment or small office structures. A flat plate with internal two-way reinforcing (at middle in Figure 10-19) can be placed over the columns to support the slab loads and better distribute lateral loads from the floor into the column. The plate allows the overall depth of the floor to remain thin, while still providing ample area to resist shear forces. Shear is created in the floor slab as the gravity loads try to push the slab downward, forcing the post to punch a hole in the slab. The plate can be further reinforced by placing a crown at the top of the column. The crown allows necessary depth to help transfer lateral loads from the slab into the column while still keeping the slab thickness to a minimal depth.

As loads increase, beams are placed below the slab and supported on reinforced columns at the intersection of the beams (at bottom in Figure 10-19). Beams can either be precast or formed in monolithic construction. A fourth method of two-way reinforcement is called a *waffle-flat-plate* system. A two-way grid of narrow beams is placed below the slab, with a void formed by metal pans placed below the slab. Figures 10-20a and 10-20b show example of a waffle-flat-plate slab system and typical section. Common spacing of the cores range from 20" to 30" (500 to 750 mm) with a joist width of 4", 5" and 6" (100, 125 or 150 mm) typical. The joist depth ranges in 2" (50 mm) increments from 6" to 14" (150 to 350 mm) depending on the load, span, and manufacturer of the metal pan. Most heavy metal pans can be removed after the concrete has cured—allowing reuse. Lightweight metal pans can be left in place and then covered with a layer of metal lath to support a finish coat of plaster.

Lift Slabs

With the lift-slab method of construction, floor slabs for a multi-level structure are poured on the ground, one above another.

SYMBOL	BAR SUPPORT ILLUSTRATION	TYPE OF SUPPORT	STANDARD SIZES
SB	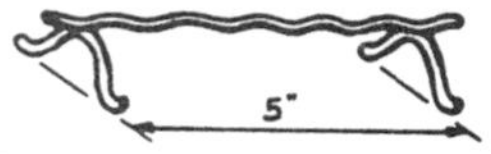	Slab Bolster	¾, 1, 1½, and 2 inch heights in 5 ft. and 10 ft. lengths
SBU*		Slab Bolster Upper	Same as SB
BB	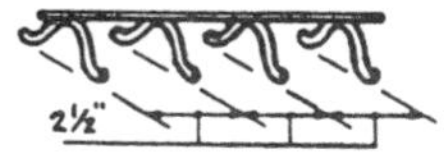	Beam Bolster	1, 1½, 2; over 2" to 5" heights in increments of ¼" in lengths of 5 ft.
BBU*	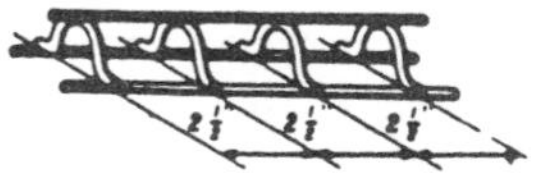	Beam Bolster Upper	Same as BB
BC		Individual Bar Chair	¾, 1, 1½, and 1¾" heights
JC	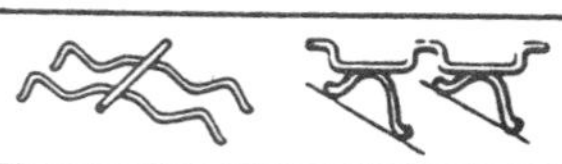	Joist Chair	4, 5, and 6 inch widths and ¾, 1, and 1½ inch heights
HC		Individual High Chair	2 to 15 inch heights in increments of ¼ in.
HCM*		High Chair for Metal Deck	2 to 15 inch heights in increments of ¼ in.
CHC	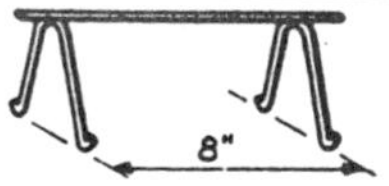	Continuous High Chair	Same as HC in 5 foot and 10 foot lengths
CHCU*	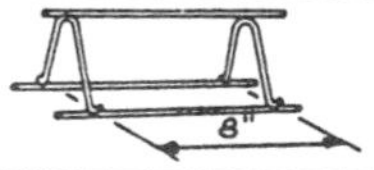	Continuous High Chair Upper	Same as CHC
CHCM*		Continuous High Chair for Metal Deck	Up to 5 inch heights in increments of ¼ in.
JCU**	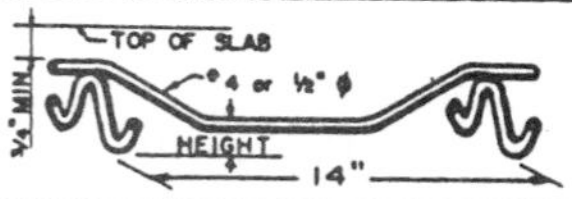	Joist Chair Upper	14" Span. Heights —1" through +3½" vary in ¼" increments

*Available in Class 3 only, except on special order.
**Available in Class 3 only, with upturned or end bearing legs.

FIGURE 10-11 ■ Typical supports used to hold reinforcing in place. *Courtesy Concrete Reinforcing Steel Institute, Schaumberg IL.*

Support columns are fabricated prior to pouring the slabs, and the slabs, ranging in thickness from 6" to 10" (150 to 250 mm) deep, are formed around the columns. The depth is determined by the spacing of supports and the loads to be supported. Once the slabs have cured, the slabs are lifted into place by hydraulic jacks and fastened to supporting columns. This system is used primarily for what are classed as lightweight loads, encountered in office structures, apartments, or dormitories.

WALLS

Concrete walls can either be poured in the vertical or horizontal position. Walls taller than 10' (3000 mm), which can't easily be formed with plywood, are typically poured in a horizontal position and lifted into place.

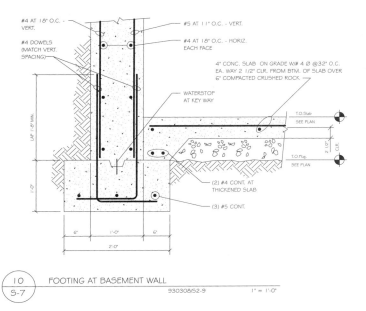

FIGURE 10-12 ■ The placement of reinforcing steel is indicated in details. The method used to hold the reinforcement in place is referenced in the project manual. *Courtesy Van Domelen/Looijenga/McGarrigle/Knauf Consulting Engineers.*

Poured-in-Place Concrete Walls

Poured-in-place walls are formed by placing plywood forms in the desired shape, as shown in Figures 10-21a and 10-21b. Figure 10-21c shows an example of a detail used to explain cast-in-place concrete. Reinforcing is wired into position when one side of the form has been installed. Once the reinforcing has been placed in its proper location, the forms can be completed. Metal ties are used to keep the forms in their proper alignment. Some types of ties remain in the wall; others are removed and leave a small V-shaped indentation in the wall. The indentation can either be left exposed or hidden with mortar. Notes in the written specifications will indicate how tie holes will be treated after tie removal. Methods for placement of concrete are also typically provided. Figure 10-22 shows how poured walls are represented in plan views.

Tilt-Up Walls

Casting concrete wall units in a horizontal position and lifting them into position is an economical method of concrete construction. Figure 10-23 shows a precast wall being lifted into position. The casting surface is usually the concrete floor slab of the structure. A liquid bond breaker is sprayed over the floor slab to prevent the wall slab from bonding to the floor. Door and window openings are framed in the wall prior to pouring.

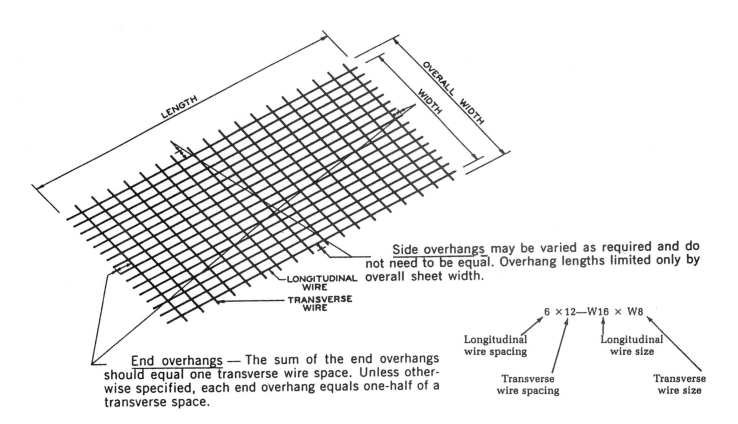

FIGURE 10-13 ■ Material referenced for welded wire fabric. *Courtesy American Concrete Institute.*

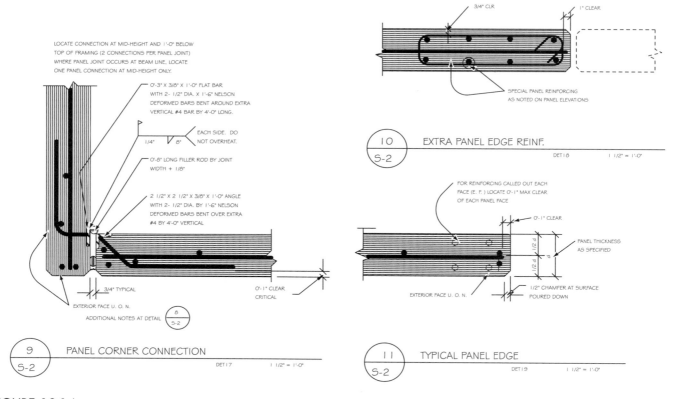

LOCATE CONNECTION AT MID-HEIGHT AND 1'-0" BELOW
TOP OF FRAMING (2 CONNECTIONS PER PANEL JOINT)
WHERE PANEL JOINT OCCURS AT BEAM LINE, LOCATE
ONE PANEL CONNECTION AT MID-HEIGHT ONLY.

0'-3" X 3/8" X 1'-0" FLAT BAR
WITH 2- 1/2" DIA. X 1'-6" NELSON
DEFORMED BARS BENT AROUND EXTRA
VERTICAL #4 BAR. BY 4'-0" LONG.

EACH SIDE. DO
NOT OVERHEAT.

0'-8" LONG FILLER ROD BY JOINT
WIDTH + 1/8"

2 1/2" X 2 1/2" X 3/8" X 1'-0" ANGLE
WITH 2- 1/2" DIA. BY 1'-6" NELSON
DEFORMED BARS BENT OVER EXTRA
#4 BY 4'-0" VERTICAL

3/4" TYPICAL

EXTERIOR FACE U. O. N.

0'-1" CLEAR
CRITICAL

ADDITIONAL NOTES AT DETAIL ⑧ S-2

⑨ S-2   PANEL CORNER CONNECTION   DET 17   1 1/2" = 1'-0"

3/4" CLR   1" CLEAR

SPECIAL PANEL REINFORCING
AS NOTED ON PANEL ELEVATIONS

⑩ S-2   EXTRA PANEL EDGE REINF.   DET 18   1 1/2" = 1'-0"

FOR REINFORCING CALLED OUT EACH
FACE (E. F.) LOCATE 0'-1" MAX CLEAR
OF EACH PANEL FACE

0'-1" CLEAR

PANEL THICKNESS
AS SPECIFIED

1/2 d   d   1/2 d

1/2" CHAMFER AT SURFACE
POURED DOWN

EXTERIOR FACE U. O. N.

⑪ S-2   TYPICAL PANEL EDGE   DET 19   1 1/2" = 1'-0"

FIGURE 10-14 ■ Specifying reinforcement coverage in detail. *Courtesy KPFF Consulting Engineers.*

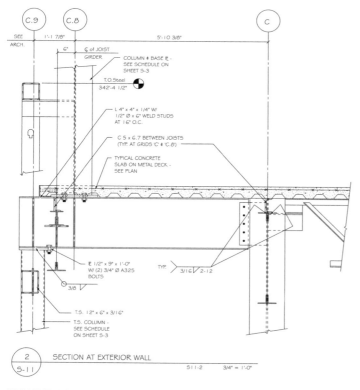

SEE ARCH.   1'-1 7/8"   5'-10 3/8"   C.9 C.8 C

6"   ₵ of JOIST
GIRDER

COLUMN & BASE ₽ -
SEE SCHEDULE ON
SHEET S-3

T.O.Steel
342'-4 1/2"

L 4" x 4" x 1/4" W/
1/2" Ø x 6" WELD STUDS
AT 16" O.C.

C 5 x 6.7 BETWEEN JOISTS
(TYP. AT GRIDS 'C' & 'C.8')

TYPICAL CONCRETE
SLAB ON METAL DECK -
SEE PLAN

₽ 1/2" x 9" x 1'-0"
W/ (2) 3/4" Ø A325
BOLTS

3/8   TYP.   3/16   2-12

T.S. 12" x 6" x 3/16"

T.S. COLUMN -
SEE SCHEDULE
ON SHEET S-3

② S-11   SECTION AT EXTERIOR WALL   S11-2   3/4" = 1'-0"

FIGURE 10-15 ■ A cast-in-place floor system placed over metal decking. *Courtesy Van Domelen/Looijenga/McGarrigle/Knauf Consulting Engineers.*

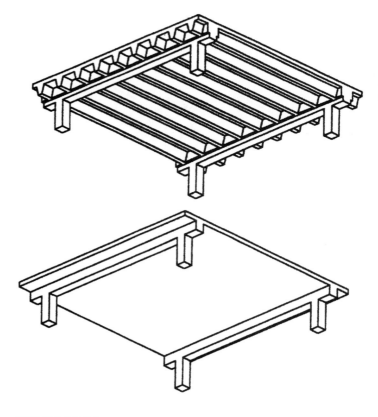

FIGURE 10-16 ■ One-way slab and a one-way joist floor systems.

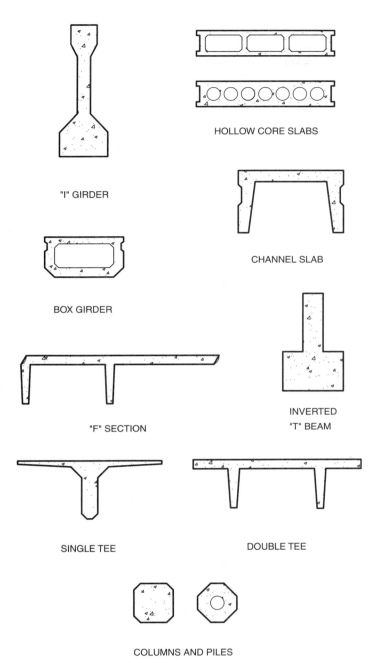

FIGURE 10-17 ■ Common precast shapes.

Concrete walls must be a minimum width of 4" (100 mm). The minimum required steel reinforcing is based on a percentage of the wall thickness. Thinner walls normally have a single layer of horizontal and vertical steel at the center of the wall. Thicker walls normally have a double layer of horizontal and vertical steel with each mat placed as close to the surface as coverage standards will allow. Frequently there is a horizontal bar referred to as a *chord bar*, placed near the roof diaphragm. The chord bars in each panel are joined together by welding a transfer angle from panel to panel, as shown in Figure 10-24. These angles tie all panels together to resist horizontal forces from the roof diaphragm.

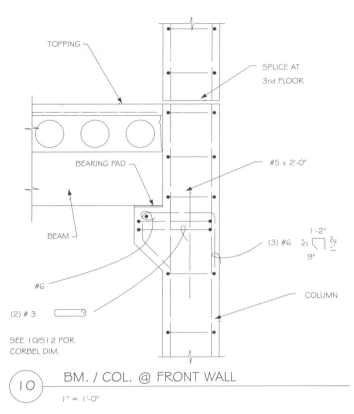

FIGURE 10-18 ■ The connection of hollow-core floor slabs with supporting columns. *Courtesy H.D.N. Architects A.I.A.*

The engineer determines the required wall thickness and steel reinforcement, which the drafter must reflect on the floor, framing plans similar to Figure 10-6, and wall details. Precast concrete construction requires details showing the reinforcement of panels similar to Figure 10-7, connection of panels to each other, placement and reinforcement of openings similar to Figure 10-8, anchors for connecting members, as well as fabrication and erection methods. Figure 10-25 shows an example of a fabrication drawing for placing panel attachments for lifting.

BEAMS

All structural members are subject to bending stresses, which result in forces of compression, tension, and shear within the member. The surface nearest the load is in compression and the surface away from the load is in tension. Concrete beams require details to locate the placement, size and quantity of steel needed to resist the forces causing bending. Because the side of the beam closest to the load is not considered to be resisting any tension, steel can often be omitted. The beam shape can be altered from rectangular to T shape to increase the compression surface and decrease the tension surface. If a beam is continuous over a support, the surface in tension will change to the upper surface over the support. Bars from the lower surface are often bent on an angle to reinforce the top surface of the beam as it passes over a support, as seen in Figure 10-26.

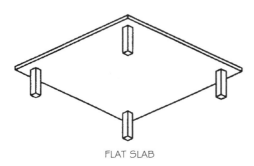

FLAT SLAB

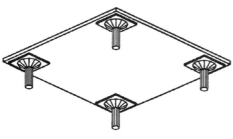

FLAT SLAB
WITH DROP PANELS.

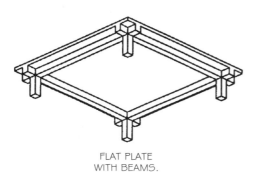

FLAT PLATE
WITH BEAMS.

FIGURE 10-19 ■ Three standard methods of constructing two-way floor systems include flat slab, flat slab with drop panels, and flat plate with beams.

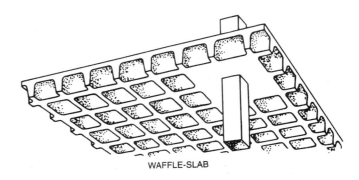

WAFFLE-SLAB

FIGURE 10-20B ■ A waffle plate system is a common method of forming a two-way floor system. *Courtesy Portland Cement Association.*

FIGURE 10-20A ■ The forming of a waffle plate system.

FIGURE 10-21A ■ Poured in place walls are formed by placing plywood forms in the desired shape. *Courtesy Megan Jefferis.*

FIGURE 10-21B ■ A poured in place wall after the forms are removed. *Courtesy Aaron Jefferis.*

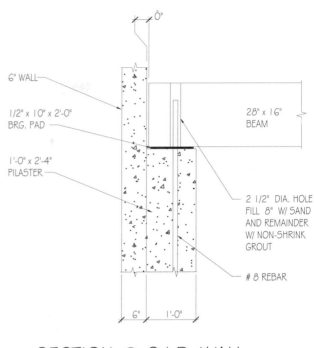

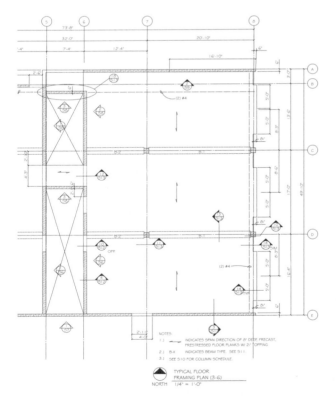

FIGURE 10-22 ■ A floor plan for a multilevel structure framed with stacked concrete columns and permanent concrete walls. *Courtesy KPFF Consulting Engineers.*

SECTION @ C.I.P. WALL
SCALE: "" = 1'-0"

FIGURE 10-21C ■ Cast in place wall to beam detail.

Horizontal and vertical shear stresses also tend to make a beam fail. These forces can be resisted by bending some of the top bars downward, as seen in Figure 10-26, to form an angle. In addition to the shear forces being resisted at the end of the beam, more steel is added to the bottom of the center of the beam to resist forces causing bending. Another method of resisting the shear forces is with the use of *stirrups*, which are U-shaped rebar, hung from the compression bars around the tension bars.

COLUMNS

Columns must withstand heavy compression loads. As the height of the column increases, stress from rotation and bending must also be resisted by the column. To reduce the size of the concrete, and reduce cracking, steel reinforcement is added to a column near the surface. In its simplest form, a poured concrete column can be square, rectangular, or round, with steel reinforcement placed in an arrangement similar to the reinforcement patterns in Figure 10-27. Vertical steel is placed near the surface of the column and then tied with horizontal ties at approximately 8" to 12" (200 to 300 mm) intervals to restrict spreading of the steel. Figure 10-28 shows common methods used for column reinforcement.

Circular Columns

Circular columns are formed in pre-manufactured fiber tubes. Because of the difficulty of making circular ties to reinforce the ver-

FIGURE 10-23 ■ Precast concrete walls are lifted into place and welded to adjoining panels. *Courtesy Jordan Jefferis.*

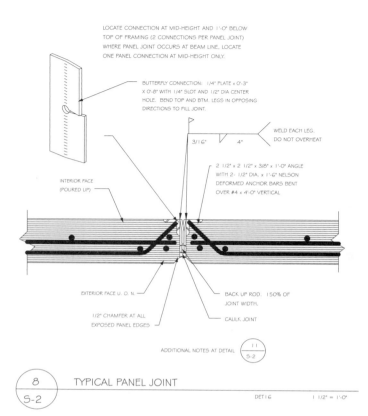

LOCATE CONNECTION AT MID-HEIGHT AND 1'-0" BELOW
TOP OF FRAMING (2 CONNECTIONS PER PANEL JOINT)
WHERE PANEL JOINT OCCURS AT BEAM LINE, LOCATE
ONE PANEL CONNECTION AT MID-HEIGHT ONLY.

BUTTERFLY CONNECTION: 1/4" PLATE x 0'-3"
X 0'-8" WITH 1/4" SLOT AND 1/2" DIA CENTER
HOLE. BEND TOP AND BTM. LEGS IN OPPOSING
DIRECTIONS TO FILL JOINT.

WELD EACH LEG,
DO NOT OVERHEAT

3/16"   4"

2 1/2" x 2 1/2" x 3/8" x 1'-0" ANGLE
WITH 2- 1/2" DIA. x 1'-6" NELSON
DEFORMED ANCHOR BARS BENT
OVER #4 x 4'-0" VERTICAL

INTERIOR FACE
(POURED UP)

EXTERIOR FACE U. O. N.

BACK UP ROD. 150% OF
JOINT WIDTH.

1/2" CHAMFER AT ALL
EXPOSED PANEL EDGES

CAULK JOINT

ADDITIONAL NOTES AT DETAIL   11 / S-2

8 / S-2   TYPICAL PANEL JOINT

DET16   1 1/2" = 1'-0"

FIGURE 10-24 ■ A connection detail for two wall panels. *Courtesy Van Domelen/Looijenga/McGarrigle/Knauf Consulting Engineers.*

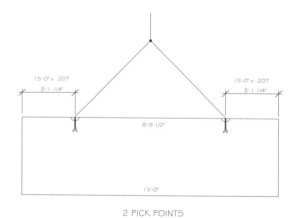

15'-0" x .207   15'-0" x .207
3'-1 1/4"   3'-1 1/4"

8'-9 1/2"

15'-0"

2 PICK POINTS

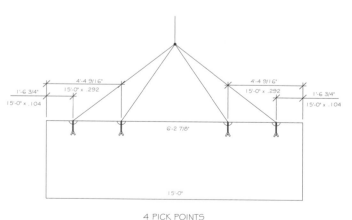

4'-4 9/16"   4'-4 9/16"
1'-6 3/4"   15'-0" x .292   15'-0" x .292   1'-6 3/4"
15'-0" x .104   15'-0" x .104

6'-2 7/8"

15'-0"

4 PICK POINTS

FIGURE 10-25 ■ Details by the fabricator are required to explain each step of the forming and moving process. *Courtesy Jim Ellsworth.*

tical steel in the field, circular columns are reinforced with continuous spiral hoops that wrap around the vertical steel. These hoops can be shipped to a job site or bent at the site with special equipment. Spiral reinforcing steel is often used for square columns (see Figure 10-27) because it is better than individual ties at resisting loads. Spiral reinforcing for cast-in-place concrete is required by the IBC/ACI 318-7.6 to be a minimum of No. 3 bar with a minimum spacing of 1" (25 mm) and a maximum spacing of 3" (75 mm). Columns with an irregular shape are used throughout the construction industry. Details will need to be drawn of each column to clarify the placement of reinforcement and ties.

Bundled Bars

Small bars of steel are often bundled in groups to form larger members that can be used to reinforce concrete as seen in Figure 10-29. When steel is wired or welded together, the bundle of rebar acts as one unit. The IBC / ACI 318-7.7.4 limits bundle bars to the following:

■ Four pieces of equal-sized rebar per bundle.

■ Laps in individual bundles in flexural members must terminate at different points with a minimum of 40 db (bar diameters) stagger.

■ Concrete-beam bundles are limited to the use of No. 11 or smaller bars.

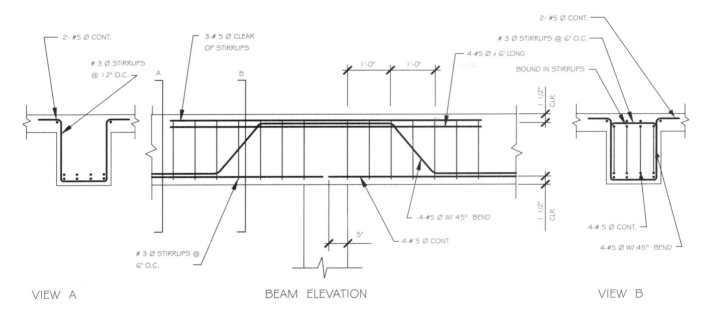

VIEW A   BEAM ELEVATION   VIEW B

FIGURE 10-26 ■ Details are required to show steel placement in continuous concrete beams.

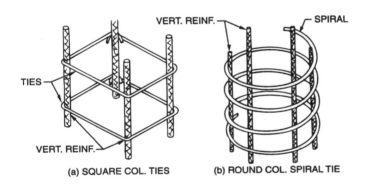

(a) SQUARE COL. TIES   **(b) ROUND COL. SPIRAL TIE**

FIGURE 10-27 ■ Steel reinforcement is often made from ties placed perpendicular to or in a spiral pattern to the main reinforcing steel.

Composite Columns

As the need to resist greater loads is increased, columns are reinforced with solid steel shapes. A *composite column* uses a steel tube filled with concrete in the center of the column, surrounded with reinforced concrete around the outer surface (at left in Figure 10-30). The metal core is not allowed to account for more than 20 percent of the total area of the column. A minimum of 3" (75 mm) of clearance between the core and the spiral steel must be provided.

Combination Columns

A *combination column* is made by using a steel shape such as a W-column (at middle in Figure 10-30) encased in wire mesh and surrounded in concrete. The mesh must be a minimum of 4 × 8/W1.4 × W1.4 with the 8" wires parallel to the axis of the column. Typically a minimum of 2 1/2" (65 mm) of concrete is required to surround the column. Both composite and combination columns are used for the lower floors of multilevel structures. The steel is used as part of the concrete support system rather than just reinforcing the concrete.

JOINTS IN CONCRETE

Three types of joints used in concrete are control joints, isolation joints, and construction joints. The engineer will specify the type of joint and require the drafting team to provide the details needed. In addition to the care that must be given in placing reinforcing at joints, a drafter must also specify the type of joint to be used, where it will be located, and how it will be constructed. Joint locations in slabs are specified on the slab on grade and foundation plans. These will be explained further in Chapter 23. Placement of these joints is often specified in the soils report. Joints in walls are normally specified on the elevations or on panel elevations for tilt-up structures. Details similar to Figure 10-31 show the type of joint to be used and are typically placed with or near the slab plan.

Control Joints

Control joints, or *contraction joints*, are joints that are placed in concrete members to allow for movement resulting from temperature change, shrinkage, or deformation. Control joints are sawn, formed, or tooled part-way through the concrete. By placing the joint partially through a slab or wall, structural stability and water tightness are maintained. The joint is one-third the depth of the member. The joint weakens the surface of the concrete slightly, causing any cracks that develop from movement to occur along the control joint. The roughness of the cracks that develop and the reinforcing steel keep the wall from moving independently once cracks occur. The engineer determines the size of joints and where they are to be placed, based on the size of the member and the loads

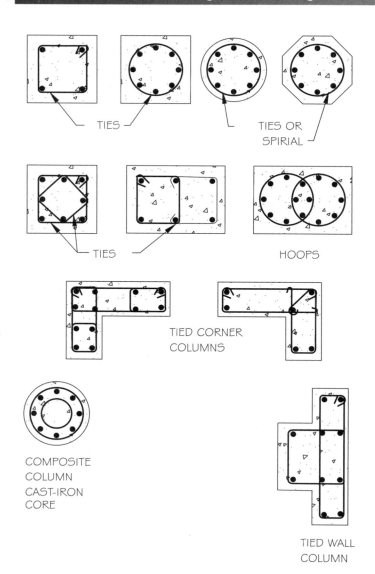

FIGURE 10-28 ■ Standard placements of steel reinforcement for irregular shaped columns.

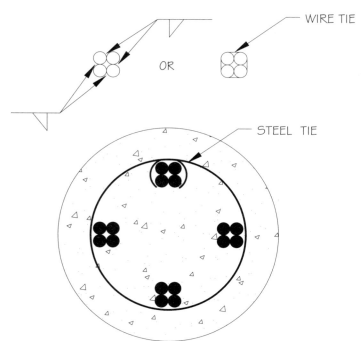

FIGURE 10-29 ■ Steel bars can be bundled in groups to form larger members by the use of wire ties or welding.

FIGURE 10-30 ■ Placement of steel reinforcement for composite columns.

that it will be subjected to. Joints are placed at a maximum of 20' (6000 mm) intervals in walls unless the wall contains multiple openings. Other guidelines based on the American Concrete Institute are shown in Table 10-2:

TABLE 10-2 ■ Recommended Joint Spacing

Wall Height Recommended	Vertical Control Joint Spacing
2–8 ft (610–2440 mm)	three times the wall height
8–12 ft (2440–3660 mm)	two times the wall height
>12 ft (3660 mm)	equal to wall height

Joints are typically provided within 10 to 15' (3000 to 4500 mm) of building corners. For multistory structures, joints are often centered above an opening. To induce cracking in the control joint, one-half of the horizontal reinforcement is often stopped at the

joint. Documentation by the engineer will specify any special treatment of steel reinforcing as well as the location of each joint. Although the engineer may specify the location of joints in a note within the structural calculations, the drafter should physically represent the joint on the drawings rather than merely placing a note.

Isolation Joints

Joints that permit differential movement of various parts of a floor slab or wall panel are referred to as isolation or expansion joints. These joints will fully penetrate through the member. Joints are placed at intersections of members such as walls to slabs, or at stress points within a slab. Figure 10-32 shows a partial slab on grade plan with isolation joints placed between slabs. Isolation joints are also provided around each column base so that forces causing the column to rotate will not affect the slab. This joint also allows for settling of the column footing due to normal shrinkage.

Care must be taken when representing mesh in details that show the intersections of slabs. As a general rule, smooth steel dowels are often placed across the joint to allow lateral movement of the joint and to prevent out of plane movement. No mesh should be allowed to cross the joint because the mesh would restrict expansion and contraction. Figure 10-33 shows an example of a slab floor joint placed around a concrete pedestal. Material used for a filler is also specified in the joint detail. The exterior side is usually caulked with a weather-resistant caulking that will remain flexible. The caulking is placed over a flexible gasket material to further seal the seam.

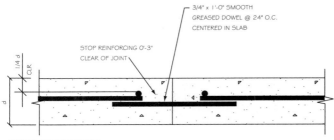

DOWELED JOINT INSTRUCTIONS:
1. GRIND ENDS OF SHEARED DOWEL TO REMOVE DEFORMITIES.
2. CAREFULLY REALIGN BARS AFTER TROWELING SLAB TO INSURE BAR IS PERPENDICULAR TO JOINT HORIZONTALLY AND VERTICALLY.
3. GREASE ONE END OF BAR.
4. ALL CONSTRUCTION JOINTS TO BE COLD CONTROL JOINTS.
5. SEE PLANS AND SPECIFICATIONS FOR TYPE AND LOCATION OF JOINTS.

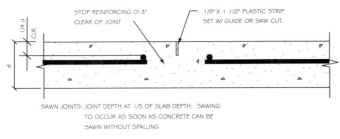

SAWN JOINTS: JOINT DEPTH AT 1/5 OF SLAB DEPTH. SAWING TO OCCUR AS SOON AS CONCRETE CAN BE SAWN WITHOUT SPALLING

WET CONTROL JOINT (C.J.)

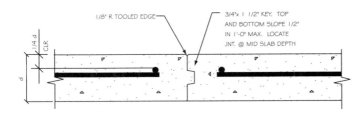

COLD KEYED CONTROL JOINT (C.J.)

TYPICAL CONTROL JOINTS

1 / S-2   DLFDN-8   1 1/2" = 1'-0"

FIGURE 10-31 ■ A control joint is placed partially through a slab or wall to allow stress to be relieved in the joint. An isolation joint is placed completely through a concrete member to allow for expansion and contraction. *Courtesy Michael & Kuhns Architects, P.C.*

Construction Joints

Joints provided to facilitate the construction process are referred to as *construction* or *pour joints.* They are located where one day's placement will end or where work has been interrupted long enough that new concrete will not bond with old concrete. Unlike expansion joints, the drafter needs to show the reinforcement that will extend into the next phase of work. Construction joints in floors, beams, and girders are restricted to the middle third of the span. Horizontal construction joints are usually placed in vertical members for each floor level. Vertical joints are usually located at corners, pilaster or column edges, or on other elements where they will be hidden.

CONDUIT

In addition to the structural considerations of planning and drawing the floor slabs and walls, provision must also be made for providing conduits for the passage of electrical, plumbing and HVAC ducts. Plastic or fiber conduit is fastened to the reinforcing material to allow for passage of electrical wiring and plumbing lines. Copper and aluminum should not be connected to reinforcement because of electrolysis. The engineer is responsible for determining the size and location of all penetrations into concrete members, and the drafter details and specifies the required material. Care must be taken to avoid using aluminum conduit or piping in concrete unless the member has been coated to prevent a reaction with either the concrete or steel. The IBC / ACI 318-6.3 limits the material, size, and placement of the conduit. Restrictions include the following:

■ Conduits or pipes in a column can't displace more than 4 percent of the surface area of the column.

■ The outer diameter of the conduit can't be greater than one-third the thickness of the member it passes through without the approval of the building department.

■ Conduit can't be spaced closer than three conduit diameters or widths on center.

■ The conduit must be of a size that will not require the cutting of the reinforcement. Conduit can be considered as replacing structurally in compression the displaced concrete under the following conditions.

   ■ The conduit must be not be exposed to rusting or other deterioration.

   ■ The conduit is un-coated or galvanized iron or steel with a minimum thickness of schedule 40 steel.

■ The conduit has a nominal interior diameter of 2" (50 mm) maximum with a minimum spacing of 3 diameters on center.

When placed in below-grade concrete or concrete that will be exposed to weather conditions, the conduit must have a minimum of 1 1/2" (40 mm) of concrete cover. Conduit not exposed to either condition only requires a 3/4" (19 mm) cover. Whenever possible,

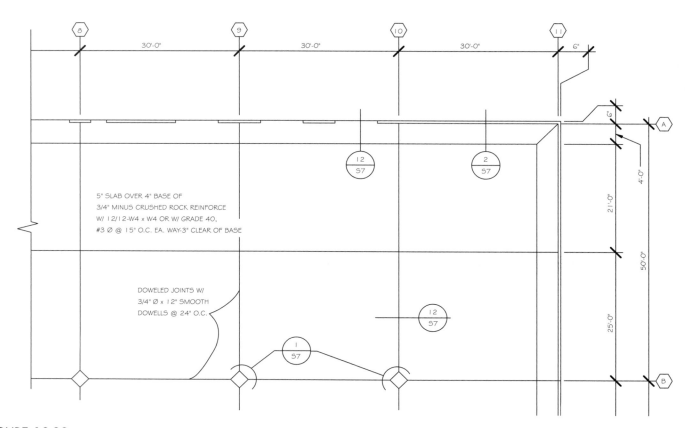

FIGURE 10-32 ■ Representing joints on a concrete slab on grade plan. *Courtesy David Ambler, StructureForm Masters, Inc.*

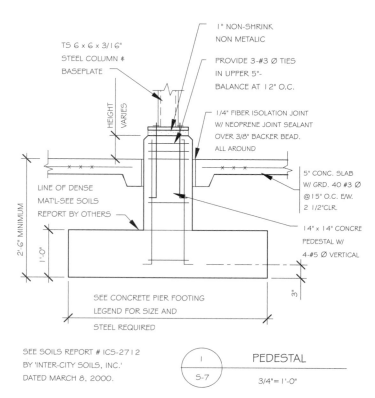

FIGURE 10-33 ■ Representing joints in detail. *Courtesy Gisela Smith, StructureForm Masters, Inc.*

electrical, plumbing, and HVAC lines and equipment are placed in the space below the floor slabs for a suspended ceiling.

ADDITIONAL READING

The following Web sites can be used as a resource to help you keep current with changes in concrete materials.

ADDRESS	COMPANY/ORGANIZATION
www.aci-int.org	American Concrete Institute
www.bluemaxxaab.com	AAB Building System Inc. (ICFs)
www.concretenetwork.net	Concrete Information and Resources
www.cmpc.org	Concrete Masonry Promotions Council
www.crsi.org	Concrete Reinforcing Steel Institute
www.portcement.org	Portland Cement Association
www.rrc-info.org.uk	Reinforced Concrete Council
www.soils.org	Soil Science Society of America
www.precast.org	National Precast Concrete Association
www.post-tensioning.org	Post-Tensioning Institute

CHAPTER 10

Concrete Methods and Materials

CHAPTER QUIZ

Answer the following questions on a separate sheet of paper. Some answers may require the use of vendor catalogs or seeking out local suppliers.

Question 10-1 List six areas that are often addressed by concrete specifications to ensure that all design values are met.

Question 10-2 What standard controls portland cement?

Question 10-3 What type of cement should be specified if an early high strength is needed?

Question 10-4 What type of cement should be used for large pour members requiring low heat of hydration?

Question 10-5 What is type IV cement used for?

Question 10-6 Define potable water.

Question 10-7 List elements that should be kept from water to be used for concrete.

Question 10-8 What two qualities of concrete will the water affect?

Question 10-9 What size aggregate is used for most construction projects?

Question 10-10 List three guidelines required by the IBC for sizing aggregate.

Question 10-11 List two methods of describing the contents of a concrete mixture.

Question 10-12 What does the water/cement ratio affect?

Question 10-13 When will concrete mixed to ASTM C192 standards reach its ultimate strength?

Question 10-14 List and briefly describe the eight major areas of concrete construction.

Question 10-15 What does ASTM A-36 cover?

Question 10-16 What is the minimum spacing between parallel strands of rebar?

Question 10-17 How are perpendicular strands of rebar held in place?

Question 10-18 What holds the upper layer of a double-layered rebar mesh?

Question 10-19 Describe the note: W6 × 12 10/10.

Question 10-20 Define monolithic.

Question 10-21 What keeps tilt-up wall panels from sticking to the floor slab?

Question 10-22 Describe general steel requirements for tilt-up walls thicker than 10 inches.

Question 10-23 Describe two methods of reinforcing round columns.

Question 10-24 Describe the difference between a composite and a combination column.

Question 10-25 List and describe three different types of concrete joints.

DRAWING PROBLEMS

Unless other instructions are given by your instructor, complete the following details and save them as wblocks using a name that represents the contents of the drawing. Skeletons of most details can be accessed from http://www.delmar.com/resources/ocl.html. Use these drawings as a base to complete the assignment.

■ Show all required views to describe each connection.

■ Draw each detail using a minimum scale of 3/4" = 1'–0" unless noted.

■ Unless noted, all wall construction is to be poured concrete.

■ Unless noted, specify all walls to be 6" thick 4000 psi concrete with grade 60, #5 bars at 10" o.c. E.W. All steel is to extend to be within 2 inches of wall edges. Steel is to be 2" clear of exterior wall surface. Place unlocated steel as per minimum specifications suggested in this chapter.

■ Use separate layers for wood, concrete, steel, laminations, text, and dimensions. Represent structural members that the cutting plan has passed through with bold lines.

■ Use dimensions for locations where possible instead of notes.

■ Show and specify all connecting materials. Base nailing on IBC standards unless your instructor tells you otherwise. Specify all material based on common local practice.

■ Provide notation to specify that all metal hangers are to be provided by Simpson Strong-Tie Company or equal. Hatch each material with the appropriate hatch pattern.

■ Use an appropriate text font to label and dimension each drawing as needed. Refer to *Sweet's Catalogs*, vendor catalogs, or the Internet to research needed sizes and specifications. Keep all text 3/4" minimum from the drawing, and use an appropriate architectural style leader line to reference the text to the drawing.

■ Provide a detail marker, with a drawing title, scale, and problem number below each detail.

■ Assemble the details required by your instructor into an appropriate template sheet for plotting. Arrange the drawings using the guidelines presented in Chapters 2 and 3. Assign a drawing sheet number of sheet S6 of 6.

Problem 10-1 Use the attached sketch to draw a section view showing TJ/60/36 open-web trusses at 32" o.c. intersecting a 6" wide concrete wall. Wall steel will be shown in other details, so no steel will be shown in this detail. Support the trusses on a 3 × 12 DFPT ledger bolted with 3/4" × 8" A. B. at 48" o.c. through 1/8" × 2"Ø washers staggered 3" up/dn. Provide 2 × 4 solid blocking between the trusses at the ledger and anchor each block with a Simpson Co. A-35 at each end of each block. Provide a 3/4" × 4" × 3" steel plate bolted to the wall with 5/8" × 4 1/8" M.B. at each third truss. Weld a MST27 strap to plate with 1/8" fillet field weld. Nail the strap to the truss as per the manufactures specifications. Cover the roof with 5/8" plywood sheathing.

Problem 10-2 Use the attached sketch to draw a section view showing a 6" wall panel intersecting a 30" × 12" deep concrete footing that extends 18" into the natural grade. No steel needs to be drawn in the wall and footing. Reference the architectural drawings for steel location. Show the base of the footing extending 18" into the natural grade. Show a 3 × 3 × 9" sleeve at 48" o.c. Support the wall on 1" non-shrink grout with 7/8" diameter × 14" structural connector by Richmond or equal centered in wall and footing.

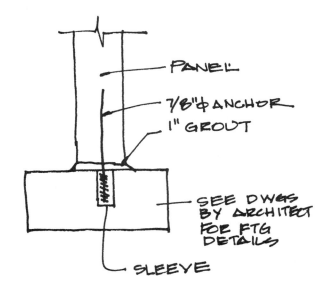

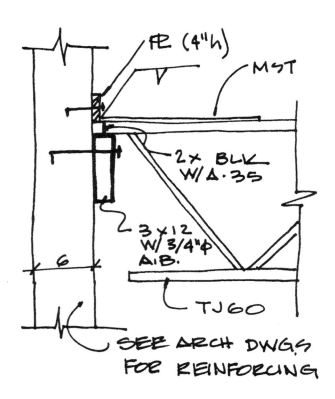

Problem 10-3 Use the attached sketch and a scale of 1/4" = 1'–0" to draw an elevation of a concrete wall panel. Show a portion of the wall detailing steel placement. Use grade 60, #5 @ 10" o.c. vertical & grade 40 #4 bars @ 12" o.c. horizontal, centered in wall. Extend all wall steel to within 2" clear of panel edge. Indicate that the left side of the panel lies on grid A and the right side lies on grid B.

Draw a detail of each door using a scale of 1/2"=1'–0" to locate and represent all door steel. Reinforce beside each door with (3)-grade 60, #6 @ 8" o.c. vertical bars 1" clear of the interior face and 1 1/2" clear of doorway opening each side. Use (3)-grade 60, #4 @ 10" o.c. horizontal above and below each door and grade 40, #4 @12" o.c. × 36" vertical above and below door. Reinforce each corner with (2)-grade 60, #5 × 36" diagonal bars at 8" o.c.

Problem 10-4 Use the attached sketch to draw a section view showing a slab to wall intersection at a doorway. Show a 5" thick slab. Use W12 × 12 4/4 mesh 3" from top surface. Provide a 3" wide × 3/4" × 3/4" chamfer in the slab at door. Provide a 1/4 × 1 1/4 fiber isolation joint between slab and the wall. Thicken the slab to 10" × 10" wide at the wall and reinforce with (2)-#7 @ 6" o.c. continuous.

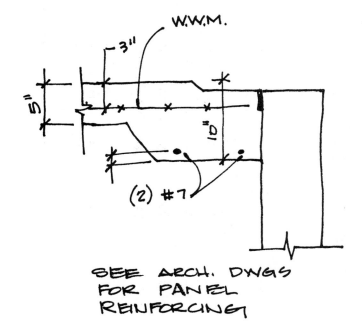

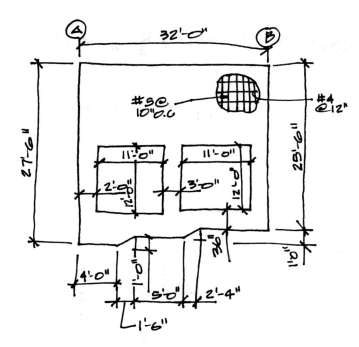

Problem 10-5 Use the attached sketch to draw a section view showing an 8" wide × 48" maximum high retaining wall. Support on a 12" deep footing that tapers from 30" to 54". Show the footing extending 8" to 12" at soil side, with a 4" diameter French drain in 8" × 24" –3/4" gravel bed. Extend the wall 6" min above the finish grade. Show a 5" thick slab reinforced with W12/12 × 4/4 mesh 3" from top surface over 4" gravel fill. Reinforce the footing with grade 40, #4 @ 8" o.c. each way 3" up/dn. Show a grade 60, #5 @ 10" o.c. × 24" L into footing, with a 40Ø lap to the vertical wall steel. Reinforce the wall with grade 60, #5 @10" o.c. each way, 2" clear of wall.

Problem 10-6 Use the attached sketch and a scale of 1" = 1'–0" to draw a section view showing a 14" × 14" × 42" pedestal. Show a 5" thick slab with W12 × 12 / 4 × 4 mesh 3" from top surface with a 1/4" × 1 1/4" fiber filled isolation joint filled with a neoprene backer bead level with the top of the pedestal. Support the pedestal on a 14" × 6'–6" × 6'–6" concrete footing. Reinforce the footing with a Grade 40, #5 @12" o.c. each way 3" up from bottom. Use Grade 40, #4 @ 8" o.c. each way 2" down from top. Extend the (8)-grade 60, #5 × 6" leg in the footing into the pedestal and lap with the pedestal steel with a 20 diameter lap. Reinforce the pedestal with (8)-grade 60, #5 vertical 1 1/2" clear of face. Tie with #3 @ 1 1/2" up/dn. @ 5/10" o.c. Provide (4)- 3/4" dia. × 16" A.B. on a 8" pattern with a 2 1/2" minimum projection. Show a plan view of the pedestal and footing and indicate that the pedestal is rotated 45° to the footing.

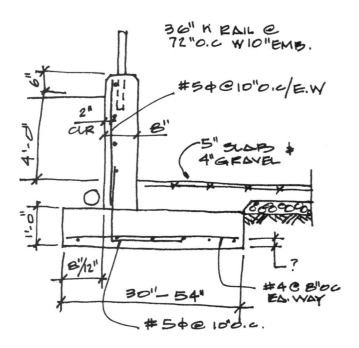

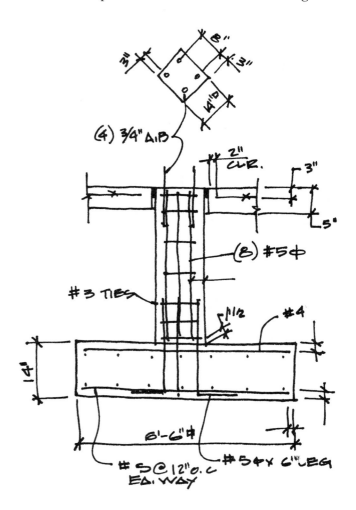

Problem 10-7 Use the attached sketch and a scale of 1" = 1'–0" to draw a section and plan view showing the intersection of wall panels. Use (2)-3/4"ø × 2 1/2" coil inserts through 1" diameter holes in an 11"h × 10"w × 1/2" plate. Inserts to be 1 1/2" minimum from plate edge and 3" minimum from concrete edge. Attach the edge of the plate to 3" × 13" × 1/2" plate embedded into concrete panel with 3" minimum edge clearance. Provide (2)-3/4"ø × 3" studs at 8" o.c. centered and welded to back of the 13" plate with 1/8" fillet all around weld. Provide 1 1/2" plate overlap and weld with 1/4" fillet weld. Place connectors at 60" o.c. along wall seam. Seal the exterior side of the wall with neoprene sealant and 3/4" backer bead.

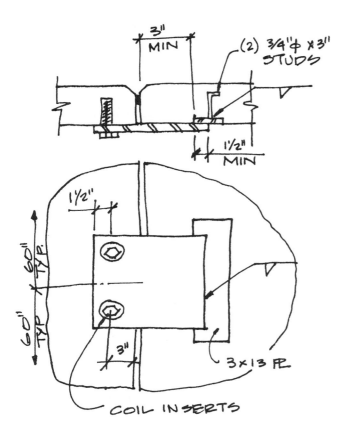

Problem 10-8 Draw a section view of an 18" wide × 36" deep concrete beam. Specify (6)-grade 60, #8 horizontal 3" up from bottom and (4)- #6 horizontal bars 2" down from top. Wrap the horizontal steel with #4 ties at 12" o.c. and stagger each lap 180°. Ties to be 1 1/2" minimum clear.

Problem 10-9 Draw a section view to show 2 × 8 studs at 16" o.c. resting on 7 1/2" × 12" deep stem wall. Support stem wall on a 24" × 36" deep concrete footing. Provide a 5" concrete slab at each side and 6" below the top of the stem wall, and thicken as required. Reinforce the slab with W12 × 12/ 4 × 4 mesh 3" from top surface with a 1/4" × 1 1/4" fiber filled isolation joint with a neoprene backer bead. Use 3/4" × 12" anchor bolts through a 2 × 8 DFPT sill w/ 2"ø washers at 48" o.c. Cover the wall with 2 layers of 5/8" type X gypsum board on each side, with joints laid perpendicular. Place (2)-grade 60, #5 rebar 3" up/down in stem wall. Use (5)-grade 60, #5 rebar × 20'–6" long (approximately) with standard hook, 3" up and 1 1/2" down. Use grade 40, #3 ties at 16" o.c. with 2" clear of sides, stagger all ties 180°. Use grade 40, #3 tie at 12" o.c. around upper stem wall steel to lower footing steel.

Problem 10-10 Draw a section, a plan view of a 16" diameter concrete column, and a plan view of the base plate. Show (8)-grade 60, #6 vertical bars evenly spaced wrapped with grade 40, #4 spiral tie at 6" o.c. 2" clear of exterior face. Reinforce the column with T.S. 6 × 6 × 3/8" column. Weld column to 11 × 11 × 1/2" base plate with 3/16" fillet weld. Attach the base plate to the foundation with (4)-3/4" × 12" anchor bolts 1 1/2" clear of plate edge. Foundation to be 16" deep × 36" square with grade 60, #6 horizontal bars each way at 8" o.c. 1 1/2" down from top. Provide grade 60, #7 bars each way at 6" o.c. 3" up from the bottom. Provide (8)-#6 × 8" leg in the foundation extending to the column with 20 diameter lap to vertical steel.

Problem 10-11 Use a suitable scale to draw an enlarged elevation of the reinforcing steel for an 8' × 8' door in a 6" thick concrete tilt-up wall panel. Place the door 2'–0" from the panel edge and 4'–10" from the bottom of the panel. Unless noted, all steel is to be 1" clear of the exterior face. Steel to be 1 1/2" clear of the edges of the opening.

■ Use (3)-grade 60 #5 Ø @ 10" o.c. vertical bars on each side of the door. Extend each piece to within 2" clear of the top and bottom of the panel.

■ Use (3)-grade 60 #5 Ø @ 8" o.c. horizontal bars above and below the opening.

■ Use grade 40 #5 Ø @ 10" o.c. × 36" long vertical bars above and below the opening.

■ Provide (2)-grade 40, #5 × 48" diagonal rebar 2"clear of each corner, and 10" o.c.

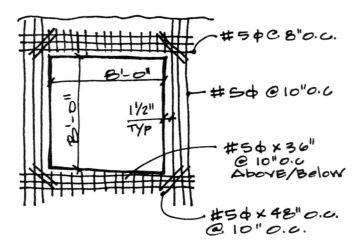

Problem 10-12 Draw a section that can be used to show six different widths × 14" deep footing with rebar that is 3" up from the bottom of the footing and 2" clear of each edge. Provide a 3" × 4" × 9" sleeve at 6'–0" o.c. All steel is to be ASTM A615 grade 40. Dimension the width of the footing with a letter to represent each of the following sizes. Create a table to represent the following sizes and reinforcing:

Width	Reinforcing
20" × 20"	(3)-#5 Ø rebar each way
22" × 22"	(3)-#5 Ø rebar each way
24" × 24"	(4)-#5 Ø rebar each way
32" × 32"	(5)-#5 Ø rebar each way
34" × 34"	(5)-#5 Ø rebar each way
36" × 36	(6)-#5 Ø rebar each way

Problem 10-13 Draw a section view of a joint in a 5" thick slab over 4" gravel fill. Provide 12 × 12/6 × 6 w.w.m. centered in the slab. Provide a 1/4" fiber isolation slab joint with neoprene joint sealant over a 3/8" backer bead. Keep the mesh 2" minimum clear of the slab joint. Reinforce the joint with #6 Ø × 18" smooth rebar with a paper shield placed at 18" o.c. placed 2" clear of the bottom of the slab.

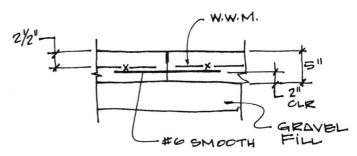

Problem 10-14 Draw a section view of a perimeter joint in a 5" thick slab that is 48" from a 6" exterior concrete tilt up wall. Place the slab over 4" gravel fill. Provide 12 × 12/6 × 6 w.w.m. centered in the slab. Provide a 1/4" fiber isolation slab joint with neoprene joint sealant over a 3/8" backer bead. Keep the mesh 2" minimum clear of the slab joint. Reinforce the joint with #6 Ø × 36" smooth rebar with a paper shield placed at 15" o.c. placed 2" clear of the bottom of the slab. Tie the slab to the wall with a 3/4" Ø × 24" long coil rods @ 45" with 3" wall penetration and 2" clear of the bottom of the slab.

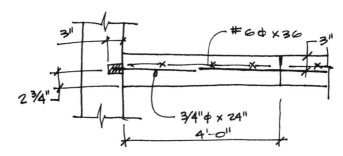

Problem 10-15 Draw plan and elevation views of a concrete tilt-up panel corner joint. Use a 4" × 15" × 3/8" plate inserted flush into a panel 3 1/2" from the panel joint. Attach (2)-3/4" Ø stud anchors at 9" o.c centered on the plate and attached with 1/8" fillet weld, all around into the panel.

Provide (2)-3/4"Ø × 2 3/4" long coil inserts at 6" o.c. and 6" from panel joints in the opposite panel. Use a 4 × 8 × 10h × 3/8" angle with (2)-1" Ø holes, 1 1/2" from each plate edge to allow connection to the coil inserts. Lap this plate over the recessed 4" × 15" plate with a 1 1/2" minimum lap and weld at the job site with 3/16" fillet weld. Provide connection plates 1'-0" from top and bottom of each panel and at 6'-0" o.c. maximum along panel joint.

Problem 10-16 Draw a plan and section view of a 6 × 6 × 3/16 T.S. column resting on a 16" square concrete pedestal. Support the column on a 12 × 12 × 1/2 base plate on 1" dry pack with (4)-3/4" Ø × 15" A.B. in a 9" grid. Bolts to have a 3" projection through the slab. Attach the column to the base plate with 3/16 fillet weld, all around.

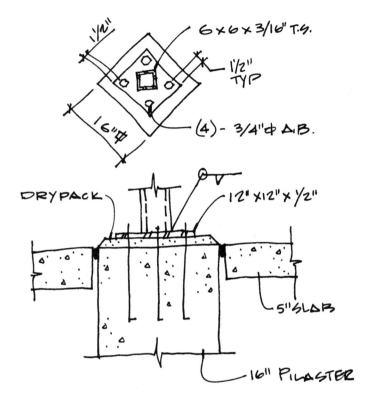

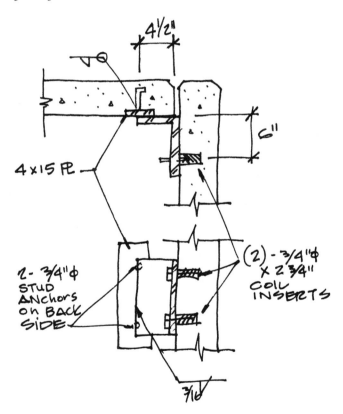

Problem 10-17 Use the attached sketch to draw a plan and section view of a 6 3/4" × 43 1/2" laminated beam intersecting a concrete tilt-up wall. Support the beam with (2)-9" × 10" × 15 1/2" × 5/16" angles. Bolt angles to wall with (6)-3/4" Ø × 4 1/8" taper bolts at each angle and use (3)-3/4" Ø × 9" A307 bolts with standard washers. Provide a 6 7/8" × 10" × 7/8" bearing plate welded to angles with 5/16" fillet welds.

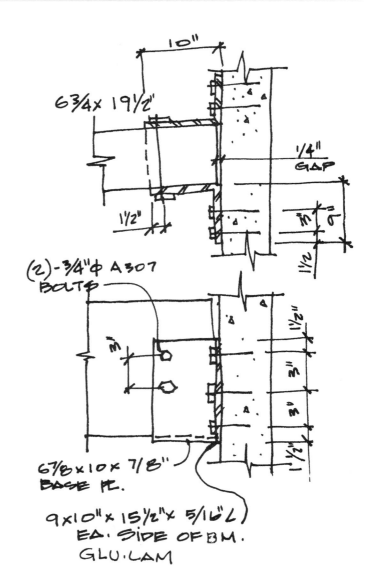

CHAPTER 11

Fire-Resistive Construction

The International Building Code (IBC) has many requirements for incorporating fire-resistive construction in a portion of a building. Shafts, hallways or corridors, parapets, exterior walls, and structural components have to be of fire-resistive construction under some conditions. Sometimes entire buildings have to be built of fire-resistive construction. Some buildings must not only be fire-resistive, but they must also be built entirely of noncombustible materials such as steel or concrete.

The most common methods of fire protection are covering wood or metal members with drywall or plaster, encasing metal members in concrete, and spraying metal beams and post with fire-resistant materials. This chapter will explore when and how to add fire protection to a building.

OBJECTIVES OF FIRE PROTECTION

The primary reason that the code requires fire protection is not to protect property (the building) but to protect human lives. The code attempts to do this by

- Protecting structural members from damage by fire and therefore preventing a structural collapse

- Containing the fire and preventing it from spreading to other portions of the structure or to other structures

- Limiting the amount of fuel available to the fire

Fire Resistance

Fire resistance is measured in time. If an assembly (see assembly below) is able to resist fire for an hour, it is a 1-hour assembly. The code generally discusses 1-, 2-, 3- and 4-hour assemblies.

Fire-Resistive Assembly

An example of a fire-resistive assembly is shown in Figure 11-1. This particular assembly is for an interior metal stud wall and is rated for 2-hour fire protection. The wall has two layers of drywall on each side, which must be of a certain type and thickness and must be installed in a specific way. Detailed instructions for installation, such as direction of layers and method of attachment, etc., are usually called out in the specifications (see Chapter 13). Metal studs used in an assembly must also meet certain requirements of width, gauge, and spacing. All of the parts must be in place (for

example, drywall must be on both sides of the studs) to be a fire-resistive assembly. Figure 11-2 shows two examples of a 1-hour floor/ceiling assembly.

Product manufacturers hire testing labs to perform tests on assemblies using their products. The results of these tests are published by Standard Fire Test, Fire Prevention Research Institute, Underwriters' Laboratories, etc. A list of these tests is available in Chapter 7 of the IBC, from The Gypsum Association, and from other building-related organizations. If manufactured joists are to be used on a project, the joist manufacturer can often supply the architectural team with the necessary report to draft details and specifications for the assembly. The building codes usually contain the necessary information for detailing a lumber assembly.

BASIC METHODS OF PROTECTING MEMBERS

The method of protecting an assembly varies depending on its location in the building, its location relative to the property lines and the construction material that is being protected. The code lists assembly requirements for walls, floor and ceiling assemblies, roof and ceiling assemblies, columns, shafts, corridors, and wall openings. The code also lists requirements for penetrations through each of these assembly types.

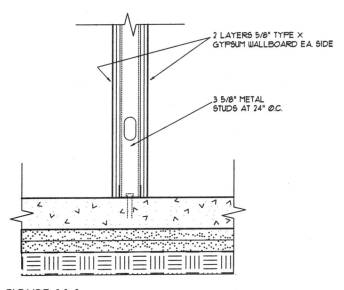

2 LAYERS 5/8" TYPE X
GYPSUM WALLBOARD EA. SIDE

3 5/8" METAL
STUDS AT 24" O.C.

FIGURE 11-1 ■ 2-hour fire-rated wall assembly.

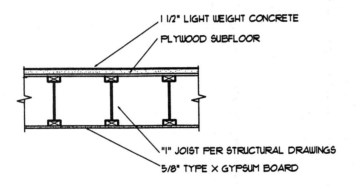

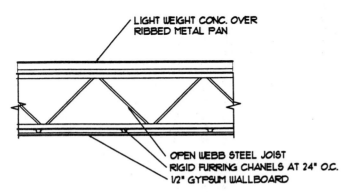

FIGURE 11-2 ■ 1-hour fire-rated floor assemblies.

Walls

The material used to form the wall will affect the fire resistance of the assembly. The code addresses concrete, masonry, wood, and steel assemblies.

Concrete

Concrete is a fire-resistive material; therefore, concrete walls are automatically fire resistive. The amount of fire resistivity provided by a concrete wall is a function of the type of concrete used and the thickness of the wall, as shown in Figure 11-3. In this chart the five columns on the right show the required thickness of concrete necessary for 1-, 1 1/2-, 2-, 3- or 4-hour fire walls. Four different types of concrete are listed. It is important to know what kind of concrete is used in the area where the project is to be constructed so that the proper thickness of concrete can be specified.

Masonry

Masonry is also a naturally fire-resistive material and its resistance depends on its thickness and the type of masonry used, as shown in Figures 11-4 and 11-5. Figure 11-4 deals with hollow masonry units. The table refers to minimum equivalent thickness of these units. As stated in the notes below the chart, the equivalent thickness is the average thickness of the solid material in the unit. This number can be obtained from the concrete block manufacturer.

Fire ratings are listed from 0.50 or one-half hour to four hours. The chart lists the minimum equivalent thicknesses required to achieve these fire ratings for five different types of material that hollow concrete masonry units can be made of. Figure 11-5 shows the fire resistivity of other types of masonry units.

Wood and Metal

With the exception of heavy timber (discussed later in this chapter) wood is not considered a fire-resistive material. Steel is never considered to be fire resistive because it weakens and melts under relatively low temperatures. When these materials are used for studs in a wall, any necessary fire resistance is achieved by covering the studs with a fire-resistive material such as type X gypsum board. There are many variations to fire assemblies for steel and wood stud walls. Good sources for these assemblies are the code, the Gypsum Association's *Fire Resistance Design Manual*, and *The Plaster Design Manual*.

Floor/Ceiling Assemblies

These assemblies are typically considered as concrete, wood, or metal.

Concrete

A structural concrete floor can be a floor/ceiling fire assembly if it is the proper thickness. Figure 11-6 is a table from the IBC that shows required thicknesses for these assemblies. If the concrete is something other than a slab with a uniform thickness, its equivalent thickness can be calculated. A method of calculating equivalent thickness is shown in Figure 11-7. The concrete provides the necessary protection for both the floor side and ceiling side of the assembly. A suspended ceiling is frequently placed below the concrete to conceal utilities such as air conditioning ducts, plumbing, and electrical lines. In most cases the suspended ceiling is not considered part of the fire assembly.

Wood and Metal

If wood or metal is used for the structural system, it must be protected by other materials to achieve a fire rating. Figure 11-2 at A is an example of this. A thin layer of lightweight concrete serves as protection on the floor side of the assembly. Both examples provide fire protection on the ceiling side with a layer of type X drywall. Figure 11-2 at B shows an example using factory-built wood I-joists for structural support. The concrete floor is poured over a plywood subfloor and the gypsum board is attached directly to the bottom chord of the joist. When open-web steel joists are used, the concrete is poured over fluted metal sheathing. The gypsum board is fastened to metal furring channels that are wire tied to the bottom chord of the truss. Both these details require coordination with the structural engineer to ensure that the trusses meet depth and spacing requirements for the selected fire assembly. The type of lightweight concrete used will vary depending on whether the concrete will be the wear surface (exposed) or not. Another factor for selecting the type of lightweight concrete is the type of activity taking place on the floor. Office floors do not receive the same wear and

TABLE 719.1(2)
RATED FIRE-RESISTANCE PERIODS FOR VARIOUS WALLS AND PARTITIONS [a,o,p]

MATERIAL	ITEM NUMBER	CONSTRUCTION	MINIMUM FINISHED THICKNESS FACE-TO-FACE[b] (inches)			
			4 hour	3 hour	2 hour	1 hour
1. Brick of clay or shale	1-1.1	Solid brick of clay or shale[c]	6	4.9	3.8	2.7
	1-1.2	Hollow brick, not filled.	5.0	4.3	3.4	2.3
	1-1.3	Hollow brick unit wall, grout or filled with perlite vermiculite or expanded shale aggregate.	6.6	5.5	4.4	3.0
	1-2.1	4" nominal thick units at least 75 percent solid backed with a hat-shaped metal furring channel $3/4$" thick formed from 0.021" sheet metal attached to the brick wall on 24" centers with approved fasteners, and $1/2$" Type X gypsum wallboard[e] attached to the metal furring strips with 1"-long Type S screws spaced 8" on center.	—	—	5[d]	—
2. Combination of clay brick and load-bearing hollow clay tile	2-1.1	4" solid brick and 4" tile (at least 40 percent solid).	—	8	—	—
	2-1.2	4" solid brick and 8" tile (at least 40 percent solid).	12	—	—	—
3. Concrete masonry units	3-1.1[f,g]	Expanded slag or pumice.	4.7	4.0	3.2	2.1
	3-1.2[f,g]	Expanded clay, shale or slate.	5.1	4.4	3.6	2.6
	3-1.3[f]	Limestone, cinders or air-cooled slag.	5.9	5.0	4.0	2.7
	3-1.4[f,g]	Calcareous or siliceous gravel.	6.2	5.3	4.2	2.8
4. Solid concrete[h,i]	4-1.1	Siliceous aggregate concrete.	7.0	6.2	5.0	3.5
		Carbonate aggregate concrete.	6.6	5.7	4.6	3.2
		Sand-lightweight concrete.	5.4	4.6	3.8	2.7
		Lightweight concrete.	5.1	4.4	3.6	2.5
5. Glazed or unglazed facing tile, nonload-bearing	5-1.1	One 2" unit cored 15 percent maximum and one 4" unit cored 25 percent maximum with $3/4$" mortar-filled collar joint. Unit positions reversed in alternate courses.	—	$6^{3/8}$	—	—
	5-1.2	One 2" unit cored 15 percent maximum and one 4" unit cored percent maximum with $3/4$" mortar-filled collar joint. Unit positions side with $3/4$" gypsum plaster. Two wythes tied together every fourth course with No. 22 gage corrugated metal ties.	—	$6^{3/8}$	—	—
	5-1.3	One unit with three cells in wall thickness, cored 29 percent maximum.	—	—	6	—
	5-1.4	One 2" unit cored 22 percent maximum and one 4" unit cored 41 percent maximum with $1/4$" mortar-filled collar joint. Two wythes tied together every third course with 0.030-inch (No. 22 galvanized sheet steel gage) corrugated metal ties.	—	—	6	—
	5-1.5	One 4" unit cored 25 percent maximum with $3/4$" gypsum plaster on one side.	—	—	$4^{3/4}$	—
	5-1.6	One 4" unit with two cells in wall thickness, cored 22 percent maximum.	—	—	—	4

FIGURE 11-3 ■ The amount of fire resistance produced by a concrete wall is a function of the wall thickness and the type of concrete used. This partial table contains footnotes that are important to its use. Reproduced from the 2000 edition of the International Building Code, copyright © 2000, with the permission of the publisher, the *International Conference of Building Officials*, under license from the International Code Council. The 2000 IBC is a copyrighted work of the International Code Council.

TABLE 720.3.2
MINIMUM EQUIVALENT THICKNESS (inches) OF BEARING
OR NONBEARING CONCRETE MASONRY WALLS[a,b,c,d]

TYPE OF AGGREGATE	FIRE-RESISTANCE RATING (hours)														
	1/2	3/4	1	1 1/4	1 1/2	1 3/4	2	2 1/4	2 1/2	2 3/4	3	3 1/4	3 1/2	3 3/4	4
Pumice of expanded slag	1.5	1.9	2.1	2.5	2.7	3.0	3.2	3.4	3.6	3.8	4.0	4.2	4.4	4.5	4.7
Expanded shale, clay or slate	1.8	2.2	2.6	2.9	3.3	3.4	3.6	3.8	4.0	4.2	4.4	4.6	4.8	4.9	5.1
Limestone, cinders, or unexpanded slag	1.9	2.3	2.7	3.1	3.4	3.7	4.0	4.3	4.5	4.8	5.0	5.2	5.5	5.7	5.9
Calcareous of siliceous gravel	2.0	2.4	2.8	3.2	3.6	3.9	4.2	4.5	4.8	5.0	5.3	5.5	5.8	6.0	6.2

For SI: 1 inch = 25.4 mm.

a. Values between those shown in the table can be determined by direct interpolation.
b. Where combustible members are framed into the wall, the thickness of solid material between the end of each member and the opposite face of the wall, or between members set in from opposite sides, shall not be less than 93 percent of the thickness shown in the table.
c. Requirements of ASTM C 55, C 73 or C 90 shall apply.
d. Minimum required equivalent thickness corresponding to the hourly fire-resistance rating for units with a combination of aggregate shall be determined by linear interpolation based on the percent by volume of each aggregate used in manufacture.

FIGURE 11-4 ■ Table 720.3.2 in the *International Building Code* can be used to determine the fire resistance for hollow masonry units. Reproduced from the 2000 edition of the International Building Code, copyright © 2000, with the permission of the publisher, the *International Conference of Building Officials*, under license from the International Code Council. The 2000 IBC is a copyrighted work of the International Code Council.

tear that an industrial area would. An industrial area would require a stronger lightweight concrete with a tougher wear surface than an office area.

Gypsum board is used for the fire-rated ceiling material in Figure 11-2 to provide fire protection for the structural members. Both details depend on additional information in the specifications for proper drywall application. Other materials that could be used effectively are plaster or a suspended ceiling system. When using steel joists, the ceiling is sometimes left off. When this is done, a fire-protective coating is sprayed on the joists and the underside of the deck. There are three basic types of coatings that are used to protect structural steel: Two types, cementitous and sprayed min-

TABLE 720.4.1(1)
FIRE-RESISTANCE PERIODS OF CLAY MASONRY WALLS

MATERIAL TYPE	MINIMUM REQUIRED EQUIVALENT THICKNESS FOR FIRE RESISTANCE[a,b,c] (inches)			
	1 hour	2 hour	3 hour	4 hour
Solid brick of clay or shale[d]	2.7	3.8	4.9	6.0
Hollow brick or tile of clay or shale, unfilled	2.3	3.4	4.3	5.0
Hollow brick or tile of clay or shale, grouted or filled with materials specified in Section 720.4.1.1.3	3.0	4.4	5.5	6.6

For SI: 1 inch = 25.4 mm.

a. Equivalent thickness as determined from Section 720.4.1.
b. Calculated fire resistance between the hourly increments listed shall be determined by linear interpolation.
c. Where combustible members are framed in the wall, the thickness of solid material between the end of each member and the opposite face of the wall, or between members set in from opposite sides, shall not be less than 93 percent of the thickness shown.
d. For units in which the net cross-sectional area of cored brick in any plane parallel to the surface containing the cores is at least 75 percent of the gross cross-sectional area measured in the same plane.

FIGURE 11-5 ■ Table 720.4.1(1) of the *International Building Code* can be used to determine the fire resistance for masonry walls. Reproduced from the 2000 edition of the International Building Code, copyright © 2000, with the permission of the publisher, the *International Conference of Building Officials*, under license from the International Code Council. The 2000 IBC is a copyrighted work of the International Code Council.

TABLE 720.2.1.2(1)
VALUES OF $R_n^{0.59}$ FOR USE IN EQUATION 7-4

TYPE OF MATERIAL	THICKNESS OF MATERIAL (inches)											
	1½	2	2½	3	3½	4	4½	5	5½	6	6½	7
Siliceous aggregate concrete	5.3	6.5	8.1	9.5	11.3	13.0	14.9	16.9	18.8	20.7	22.8	25.1
Carbonate aggregate concrete	5.5	7.1	8.9	10.4	12.0	14.0	16.2	18.1	20.3	21.9	24.7	27.2[c]
Sand-lightweight concrete	6.5	8.2	10.5	12.8	15.5	18.1	20.7	23.3	26.0[c]	Note c	Note c	Note c
Lightweight concrete	6.6	8.8	11.2	13.7	16.5	19.1	21.9	24.7	27.8[c]	Note c	Note c	Note c
Insulating concrete[a]	9.3	13.3	16.6	18.3	23.1	26.5[c]	Note c	Note c	Note c	Note c	Note c	Note c
Air space[b]	—	—	—	—	—	—	—	—	—	—	—	—

For SI: 1 inch = 25.4 mm, 1 pound per cubic foot = 16.02 kg/m³.

  a. Dry unit weight of 35 pcf or less and consisting of cellular, perlite, or vermiculite concrete.
  b. The $R_n^{0.59}$ value for one 1/2" to 3 1/2" air space is 3.3. The $R_n^{0.59}$ value for two 1/2" to 3 1/2" air spaces is 6.7.
  c. The fire-resistance rating for this thickness exceeds 4 hours.

FIGURE 11-6 ■ The type of concrete and the thickness affects the fire resistance of a floor assembly. Reproduced from the 2000 edition of the International Building Code, copyright © 2000, with the permission of the publisher, the *International Conference of Building Officials*, under license from the International Code Council. The 2000 IBC is a copyrighted work of the International Code Council.

eral fibers, are rough and somewhat uneven and meant to be hidden. These are the least expensive types of fireproofing. The third type is intumescent paint. This material looks like ordinary paint but when it is exposed to heat, it swells or expands to form a thick protective layer. This type of fireproofing is usually used when the structural steel is to be left exposed to make an architectural statement.

TABLE 720.2.2.1
MINIMUM SLAB THICKNESS (inches)

CONCRETE TYPE	FIRE-RESISTANCE RATING (hour)				
	1	1 ½	2	3	4
Siliceous	3.5	4.3	5.0	6.2	7.0
Carbonate	3.2	4.0	4.6	5.7	6.6
Sand-lightweight	2.7	3.3	3.8	4.6	5.4
Lightweight	2.5	3.1	3.6	4.4	5.1

For SI: 1 inch = 25.4 mm.

720.2.2.1.2 Slabs with sloping soffits. The thickness of slabs with sloping soffits (see Figure 720.2.2.1.2) shall be determined at a distance $2t$ or 6 inches (152 mm), whichever is less, from the point of minimum thickness, where t is the minimum thickness.

720.2.2.1.3 Slabs with ribbed soffits. The thickness of slabs with ribbed or undulating soffits (see Figure 720.2.2.1.3) shall be determined by one of the following expressions, whichever is applicable:

For $s \geq 4t$, the thickness to be used shall be t
For $s \leq 2t$, the thickness to be used shall be t_e
For $4t > s > 2t$, the thickness to be used shall be

$$t + \left(\frac{4t}{s} - 1\right)(t_e - t) \qquad \text{(Equation 7-5)}$$

where:
s = Spacing of ribs or undulations.
t = Minimum thickness.
t_e = Equivalent thickness of the slab calculated as the net area of the slab divided by the width, in which the maximum thickness used in the calculation shall not exceed $2t$.

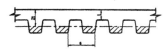

NEGLECT SHADED AREA IN CALCULATION OF EQUIVALENT THICKNESS

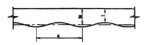

For SI: 1 inch = 25.4 mm.

FIGURE 720.2.2.1.3
SLABS WITH RIBBED OR UNDULATING SOFFITS

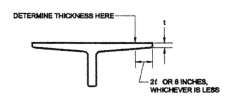

DETERMINE THICKNESS HERE
t
2t OR 6 INCHES, WHICHEVER IS LESS

For SI: 1 inch = 25.4 mm.

FIGURE 720.2.2.1.2
DETERMINATION OF SLAB THICKNESS
FOR SLOPING SOFFITS

FIGURE 11-7 ■ Calculation method for fire rating equivalent thickness of ribbed concrete. Reproduced from the 2000 edition of the International Building Code, copyright © 2000, with the permission of the publisher, the *International Conference of Building Officials*, under license from the International Code Council. The 2000 IBC is a copyrighted work of the International Code Council.

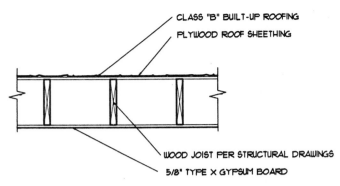

FIGURE 11-8 ■ Fire-rated roof assembly.

Roof/Ceiling Assemblies

When working with the roof, a floor/ceiling assembly is selected and the materials above the rafters are replaced with appropriate roofing materials, as shown in Figure 11-8. The roofing materials are required to be fire retardant.

Beams

Metal beams that project outside the floor/ceiling assembly are required to have their own fire protection. Steel beams are normally wrapped with drywall or plaster, or sprayed with a fire-protective coating. Figure 11-9 shows an example of a beam that extends below the rated ceiling and needs its own fire protection. Wood beams are frequently large enough to be considered heavy timber and would not have to be protected. Concrete beams are considered to be fire protected.

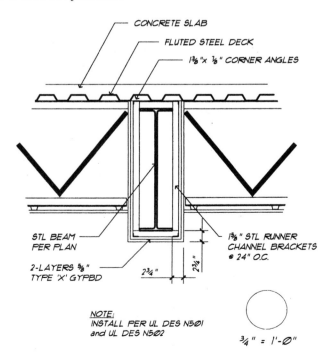

FIGURE 11-9 ■ Fire-protected floor beams.

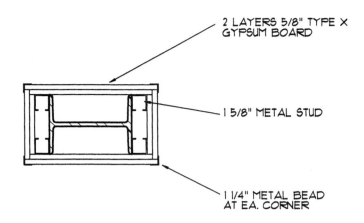

FIGURE 11-10 ■ Fire-protected column.

Columns

Columns are normally fire protected in the same way as beams are fire protected. Figures 11-10 and 11-11 show protected columns. The amount of protection required for a steel column depends on the weight of the steel member. Two layers of drywall provide 2-hour protection for a lightweight column and 3-hour protection for a heavy column (Figure 11-11). Coordination with the structural engineer is necessary to achieve the necessary protection in the most economical manner.

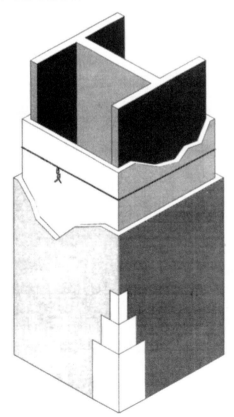

FIGURE 11-11 ■ Fire-protected columns.

Another consideration for fire protection at columns is that they are frequently subject to impact from cars, people, carts, forklifts and other moving objects. The material that provides the fire protection for columns will frequently need to be protected from such impact. This can be accomplished in several ways: Barriers can be placed between the column and the traffic, the fire-resistive material can be covered by another material such as sheet metal, or the column can be set upon a concrete pedestal.

Shafts

Vertical shafts extending through floors can play a key part in aiding the spread of fire. A shaft can act like a chimney, pulling smoke and fire up from floor to floor. For this reason, the construction of any shaft that penetrates through more than one floor is highly regulated by the model codes. Openings into shafts are required to have fire dampers, a device that closes off the opening when fire or smoke is detected.

Parapets

Many fire walls are required to have a parapet, a portion of the wall extending above the roof. The parapet is required to be as fire resistive as the lower portion of the wall. It is required to extend at least 30" above the roof. If the roof slopes at a rate of 2 in 12 or greater, the parapet may be required to extend more than 30' above the roof.

Exterior walls that are required to be fire protected are often required to have a parapet, but there are many exceptions to this rule. Area separation walls are usually required to have a parapet. Occupancy separation walls usually do not require parapets

Figures 11-12 and 11-13 are fire-resistive parapet details. Take note of the continuation of the fire-resistive material. In Figure 11-12, solid wood blocking forms part of the fire-resistive materials. In Figure 11-13, the parapet wall sits on top of the roof sheathing. This is a design that is frequently avoided, because the parapet has to be braced to prevent it from being pushed over by wind or other forces. Figure 11-14 shows a concrete wall with a parapet. This detail could easily be altered to apply to masonry. Note that if the roofing material is combustible, it must stop 18" minimum from the top of the wall.

Corridors

Corridors are frequently required to be fire rated in order to protect the occupants of a building while they exit. When the corridor is required to be fire rated, walls are required to be of 1-hour construction. Corridor ceilings are required to be constructed as they would be for a 1-hour floor/ceiling assembly. The doors opening onto these fire-rated corridors are required to be 20-minute assemblies.

The basic concept here is to give the occupants of a building an exit path that is safe from fire and smoke. This requires that the doors be tightly sealed and that any penetrations such as air-conditioning ducts be properly treated. Air-conditioning ducts serving these corridors will normally be required to have fire dampers

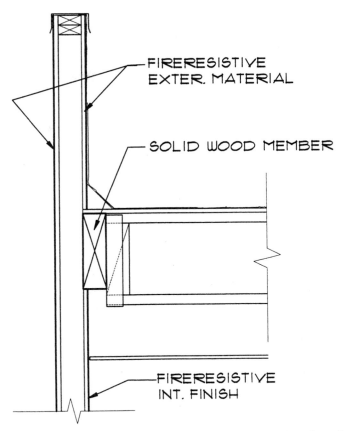

FIGURE 11-12 ■ Fire-rated parapet constructed with a balloon-framed wall.

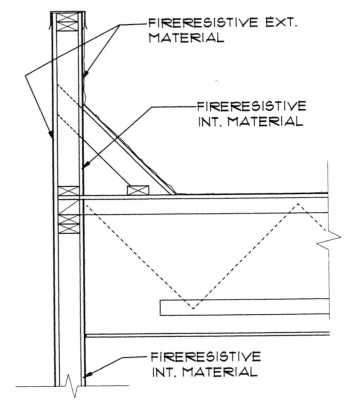

FIGURE 11-13 ■ Fire-rated braced parapet wall.

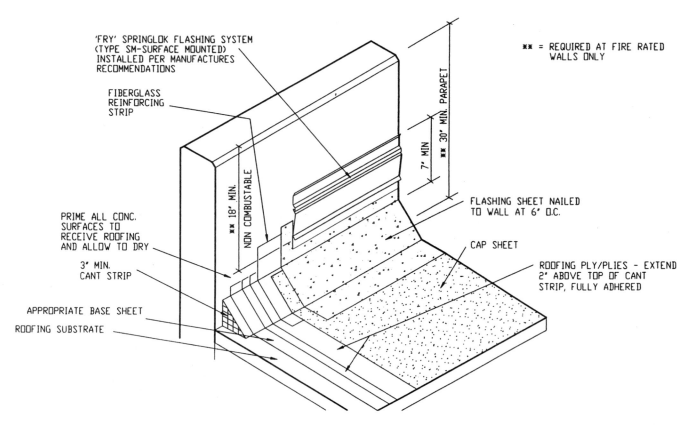

FIGURE 11-14 ■ Concrete fire rated parapet.

installed in them. A fire damper is a device that shuts off the air duct when a fire is detected so that smoke is not pumped into the corridor from other areas of the building. Figures 11-15 through 11-17 illustrate basic principals in designing fire-rated corridors. All three of the examples shown in these figures are for buildings that can be built of combustible construction and are not otherwise fire rated. Buildings that are required to be fire rated or that are required to be constructed of noncombustible construction will require slightly different details.

Figure 11-15 shows a section of a typical fire-rated corridor. The walls are constructed as typical 1-hour walls with 2 × 4 studs and 5/8" type X drywall. These walls extend all the way up to the underside of solid wood floor joists and require solid blocking to extend from the top of the walls to the underside of the floor sheathing. The undersides of the floor joists have drywall installed as if they were a part of a 1-hour floor/ceiling assembly. Fire-proofing methods may be slightly different if the floor joists are trusses instead of solid wood members. A 1-hour surface is not required on the upper side of the floor joists. A non–fire-rated ceiling may be installed below the fire-rated ceiling if it is noncombustible or made of fire-retardant treated wood. Ducts that serve the corridor and penetrate the fire-rated wall or ceiling must have fire dampers. Ducts that penetrate the optional ceiling do not require fire dampers. In this system and the following two, illustrated in Figures 11-16 and 11-17, ducts that do not serve the corridor may pass through the fire-rated assemblies if they are constructed of heavy-gauge metal.

Another way to construct a fire-rated corridor is to build a tunnel as shown in Figure 11-16. In this case the corridor walls do not extend to the underside of the floor joists. A ceiling is built across the top of the walls. This ceiling is constructed exactly as required for a 1-hour wall, with a fire-resistive finish on both sides of the ceiling joists. This system offers advantages if the structural floor system is not suitable to receive a ceiling finish or if mechanical systems need to pass over the corridor.

Figure 11-17 combines some elements of the previous two systems. It has a lower fire-rated ceiling surface for cases where the floor joists are not suitable to receive a ceiling. The walls must extend to the floor joists and solid blocking must extend from the top of the wall to the underside of the floor sheathing. On the outside of the corridor, the fire-resistive wall finish extends to the underside of the floor joists. In the corridor, the fire-resistive wall finish need only extend to the fire-resistive ceiling.

Wall Openings

Firewalls are built for two primary reasons. The first is to prevent structural collapse that would occur if the building were to burn down. The second is to prevent the spread of fire into other areas of the building or to other buildings. When the purpose of the firewall is to prevent the spread of fire and smoke, openings in that wall must receive special treatment.

If openings were to remain open during a fire, the purpose of the wall would be defeated by smoke and fire pouring through the

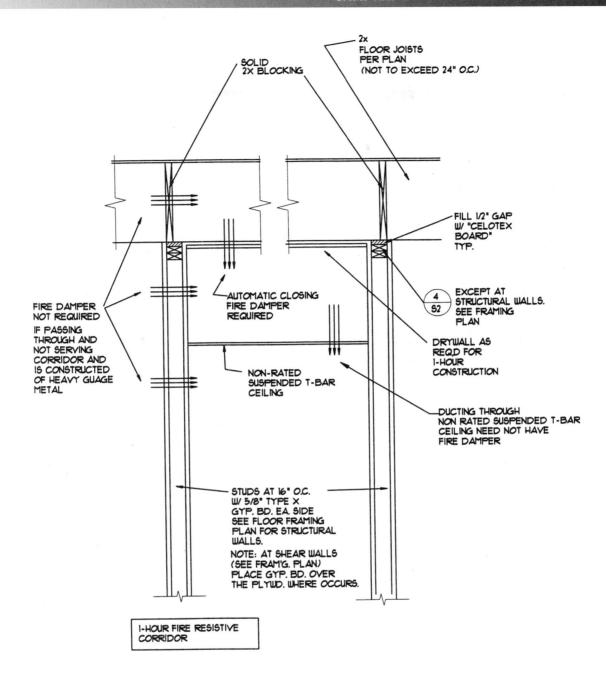

SOLID
2X BLOCKING

2x
FLOOR JOISTS
PER PLAN
(NOT TO EXCEED 24" O.C.)

FILL 1/2" GAP
W/ "CELOTEX
BOARD"
TYP.

FIRE DAMPER
NOT REQUIRED
IF PASSING
THROUGH AND
NOT SERVING
CORRIDOR AND
IS CONSTRUCTED
OF HEAVY GUAGE
METAL

AUTOMATIC CLOSING
FIRE DAMPER
REQUIRED

EXCEPT AT
STRUCTURAL WALLS.
SEE FRAMING
PLAN

DRYWALL AS
REQ.D FOR
1-HOUR
CONSTRUCTION

NON-RATED
SUSPENDED T-BAR
CEILING

DUCTING THROUGH
NON RATED SUSPENDED T-BAR
CEILING NEED NOT HAVE
FIRE DAMPER

STUDS AT 16" O.C.
W/ 5/8" TYPE X
GYP. BD. EA SIDE
SEE FLOOR FRAMING
PLAN FOR STRUCTURAL
WALLS.
NOTE: AT SHEAR WALLS
(SEE FRAM'G. PLAN)
PLACE GYP. BD. OVER
THE PLYWD. WHERE OCCURS.

1-HOUR FIRE RESISTIVE
CORRIDOR

FIGURE 11-15 ■ A 1-hour fire rated corridor constructed with wood walls and 5/8" type X gypsum board. *Courtesy Dean K. Smith.*

opening. If a window or door in the opening were to be readily destroyed by heat or if it were easy to break, the purpose of the wall would also be defeated. For this reason, doors and windows installed in firewalls to contain fire and smoke must meet certain requirements.

Door and window assemblies must carry labels stating that they have been tested and have passed fire and/ or smoke tests. Different firewall ratings and types require different ratings for openings. These assemblies are usually labeled as 20-minute, 3/4-hour, 1-hour, 1 1/2-hour, 2-hour or 3-hour assemblies. A fire-rated door assembly consists of a fire-rated and labeled door and frame, prop-

er hinges and a closing device. The closing device will either be self-closing or automatically closing, as required by the code. In some cases, the amount of opening that is allowed in a fire-rated wall is limited by area. In some firewalls, openings are prohibited altogether.

Frequently a firewall has an opening that needs to remain open all of the time. The opening at a counter between a school cafeteria and a kitchen is a good example of this type of condition. The code defines the cafeteria as an assembly area, and as such there must be a fire-separation wall between the cafeteria and the kitchen, which is a very likely place for a fire to start. Figure 11-18

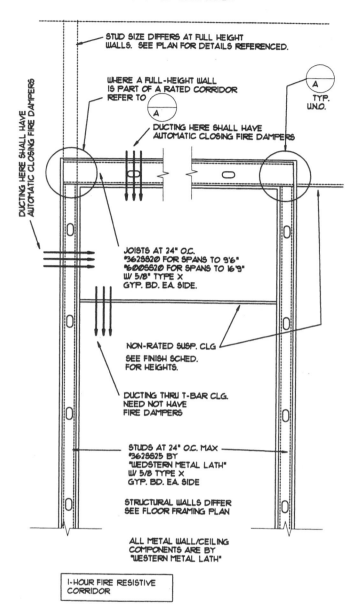

NON RATED FLOOR SYSTEM

STUD SIZE DIFFERS AT FULL HEIGHT WALLS. SEE PLAN FOR DETAILS REFERENCED.

WHERE A FULL-HEIGHT WALL IS PART OF A RATED CORRIDOR REFER TO

A

A TYP. U.N.O.

DUCTING HERE SHALL HAVE AUTOMATIC CLOSING FIRE DAMPERS

DUCTING HERE SHALL HAVE AUTOMATIC CLOSING FIRE DAMPERS

JOISTS AT 24" O.C.
#362S520 FOR SPANS TO 9'6"
#600S520 FOR SPANS TO 16'9"
W/ 5/8" TYPE X
GYP. BD. EA. SIDE.

NON-RATED SUSP. CLG
SEE FINISH SCHED. FOR HEIGHTS.

DUCTING THRU T-BAR CLG. NEED NOT HAVE FIRE DAMPERS

STUDS AT 24" O.C. MAX #362S625 BY "WESTERN METAL LATH" W/ 5/8 TYPE X GYP. BD. EA SIDE

STRUCTURAL WALLS DIFFER SEE FLOOR FRAMING PLAN

ALL METAL WALL/CEILING COMPONENTS ARE BY "WESTERN METAL LATH"

1-HOUR FIRE RESISTIVE CORRIDOR

FIGURE 11-16 ■ A fire-rated corridor constructed with walls that do not extend to the underside of the floor joists. *Courtesy Dean K. Smith.*

shows such an opening. In the case of a fire, a fire door would automatically roll down and meet the counter top to close off the opening and prevent the spread of smoke and fire.

Penetrations

Fire-resistive surfaces will normally have to be penetrated to allow for the installation of building equipment such as electrical, plumbing and HVAC. These penetrations must be properly dealt with in order to preserve the integrity of the fire-resistive system.

Plumbing

Plumbing penetrations are usually easy to avoid. If they are not avoidable, metal piping will probably have to be used in lieu of plastic. The pipe will not normally completely fill the hole it goes through. The area between the pipe and the wall material will have to be filled with special fire-resistant caulking.

HVAC

Ductwork for heating and air conditioning will frequently have to penetrate firewalls and ceilings, as shown in Figures 11-15, 11-16, and 11-17. The duct penetration itself is not usually a problem if it is constructed of the proper gauge of metal. It is the air register (outlet) or return air (inlet) register that presents the problem. These openings into the ductwork normally remain open at all times. Fire dampers are required at these openings. A *fire damper* is an air register that has a fire-sensing device attached to it. An example of a fire damper is shown if Figure 11-19. The requirements for fire dampers are shown in Figure 11-20. When a fire is detected, the openings to the ducts are automatically closed. This prevents the spread of fire and smoke through the HVAC system.

Electrical

Electrical fixtures create more wall and ceiling penetrations than any other building system. These penetrations normally require special consideration and detailing. Figures 11-21 and 11-22 illustrate some of these considerations when nonmetallic wiring is used. These examples apply to wiring for television, phones, computers, and sound systems, as well as electrical wiring. Wiring passing through such walls should be placed in metal conduit as shown in Figure 11-21. Note that the metal conduit extends 5' beyond the firewall in each direction to reduce the likelihood of fire entering and passing through the conduit. Electrical convenience outlets that are recessed into fire walls also require special detailing, as shown in Figures 11-21 and 11-22.

A classic penetration of a fire-resistive assembly is a drop-in light fixture in a fire-rated suspended ceiling system. Figure 11-23 illustrates methods of preserving the integrity of the fire-rated ceiling at lighting fixtures in a suspended ceiling. The methods illustrated in this figure have been carefully tested by an approved testing laboratory.

Heavy Timber

Under some conditions, the code will allow heavy timber construction to be substituted for fire-rated construction. Thick and deep wood beams, thick post, and thick floor or roof sheathing can be left unprotected under this condition. The code describes the minimum dimensions required for a wood member to be considered heavy timber (see Chapter 7). The minimum required size varies for a vertical member (post) and a horizontal member (beam). Thick wood members will char on the outside but not burn through for an extended period of time. In a fire, the outside of the post or beam will be charred and become structurally useless. The inner portion of the wood will remain sound and have adequate strength to keep the structure from collapsing.

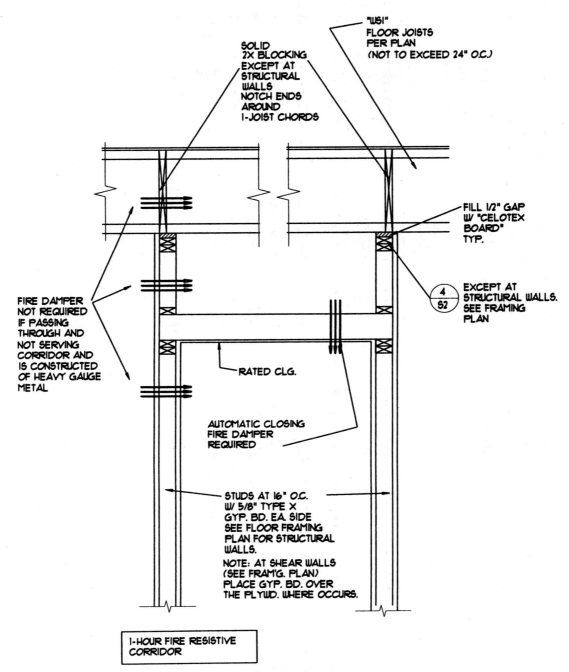

SOLID 2X BLOCKING EXCEPT AT STRUCTURAL WALLS NOTCH ENDS AROUND I-JOIST CHORDS

"WSI" FLOOR JOISTS PER PLAN (NOT TO EXCEED 24" O.C.)

FILL 1/2" GAP W/ "CELOTEX BOARD" TYP.

FIRE DAMPER NOT REQUIRED IF PASSING THROUGH AND NOT SERVING CORRIDOR AND IS CONSTRUCTED OF HEAVY GAUGE METAL

EXCEPT AT STRUCTURAL WALLS. SEE FRAMING PLAN

4 / S2

RATED CLG.

AUTOMATIC CLOSING FIRE DAMPER REQUIRED

STUDS AT 16" O.C. W/ 5/8" TYPE X GYP. BD. EA. SIDE SEE FLOOR FRAMING PLAN FOR STRUCTURAL WALLS. NOTE: AT SHEAR WALLS (SEE FRAM'G. PLAN) PLACE GYP. BD. OVER THE PLYWD. WHERE OCCURS.

I-HOUR FIRE RESISTIVE CORRIDOR

FIGURE 11-17 ■ Construction of a fire-rated corridor with floor joists that are not suitable for receiving a ceiling. *Courtesy Dean K. Smith.*

Fire and Draft Stops

Long, concealed spaces can aid the spread of fire, especially in combustible construction. For this reason the code requires these concealed spaces, both horizontal and vertical, to be blocked off at certain points and intervals. Figure 11-24 shows typical fire stops in a stud wall. The fire stops are located at strategic points to prevent fire from spreading from one part of the structural system to another.

Spaces between the ceiling and the floor above need to be divided into controlled areas by draft stops. The allowable amount of square footage in each controlled area is spelled out in the codes and is different for sprinklered and unsprinklered buildings. An attic must be divided up in a similar manner. Allowable areas for the attic are usually larger than the allowable area between the ceiling and floor. Figure 11-25 shows an example of a draft stop above a suspended ceiling system.

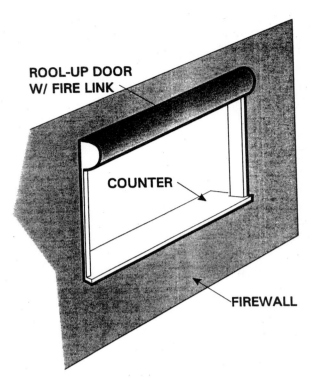

ROOL-UP DOOR
W/ FIRE LINK

COUNTER

FIREWALL

FIGURE 11-18 ■ Fire-rated doors with a closing device that is activated by heat or smoke are used at openings that need to remain open for normal business operations.

ADDITIONAL READING

The following Web sites can be used as a resource to help you find additional information.

ADDRESS	COMPANY/ORGANIZATION
www.gypsum.org	Gypsum Association
www.nfpa.org	*NFPA*—National Fire Protection Association

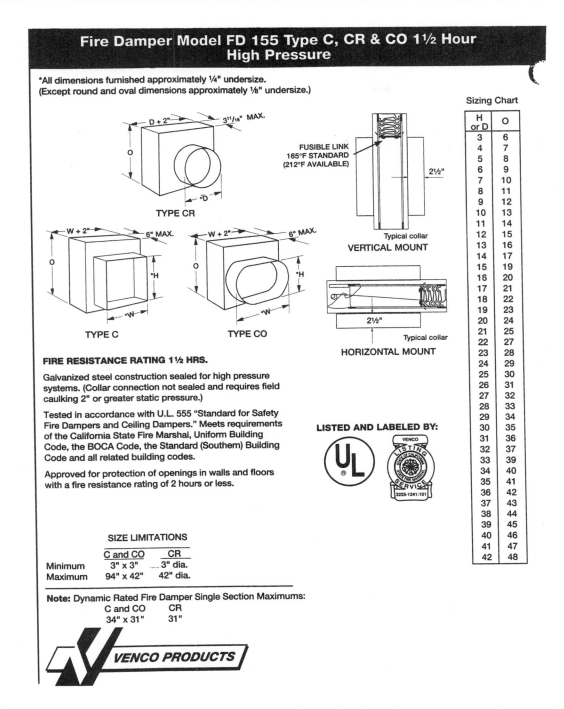

Fire Damper Model FD 155 Type C, CR & CO 1½ Hour High Pressure

*All dimensions furnished approximately ¼" undersize.
(Except round and oval dimensions approximately ⅛" undersize.)

D + 2"   3¹¹/₁₆" MAX.
O
*D

TYPE CR

FUSIBLE LINK
165°F STANDARD
(212°F AVAILABLE)

2½"

Typical collar
VERTICAL MOUNT

W + 2"   6" MAX.
O
*H
*W

TYPE C

W + 2"   6" MAX.
O
*H
*W

TYPE CO

2½"

Typical collar
HORIZONTAL MOUNT

FIRE RESISTANCE RATING 1½ HRS.

Galvanized steel construction sealed for high pressure systems. (Collar connection not sealed and requires field caulking 2" or greater static pressure.)

Tested in accordance with U.L. 555 "Standard for Safety Fire Dampers and Ceiling Dampers." Meets requirements of the California State Fire Marshal, Uniform Building Code, the BOCA Code, the Standard (Southern) Building Code and all related building codes.

Approved for protection of openings in walls and floors with a fire resistance rating of 2 hours or less.

LISTED AND LABELED BY:

UL   VENCO LISTING SERVICE
3225-1241:101

SIZE LIMITATIONS

	C and CO	CR
Minimum	3" x 3"	3" dia.
Maximum	94" x 42"	42" dia.

Note: Dynamic Rated Fire Damper Single Section Maximums:

	C and CO	CR
	34" x 31"	31"

VENCO PRODUCTS

Sizing Chart

H or D	O
3	6
4	7
5	8
6	9
7	10
8	11
9	12
10	13
11	14
12	15
13	16
14	17
15	19
16	20
17	21
18	22
19	23
20	24
21	25
22	27
23	28
24	29
25	30
26	31
27	32
28	33
29	34
30	35
31	36
32	37
33	39
34	40
35	41
36	42
37	43
38	44
39	45
40	46
41	47
42	48

FIGURE 11-19 ■ Fire dampers come in several shapes and sizes and must be carefully chosen from vendor catalogs. Reproduced from *Dampers and Louvers* catalog. *Courtesy Venco Products.*

TABLE 715.31
FIRE DAMPER RATING

TYPE OF PENETRATION	MINIMUM DAMPER RATING (hour)
Less than 3-hour fire-resistance-rated assemblies	1.5
3-hour or greater fire-resistance-rated assemblies	3

FIGURE 11-20 ■ Fire dampers have different ratings based on the ratings of the assembly they are to penetrate. Reproduced from the 2000 edition of the International Building Code, copyright © 2000, with the permission of the publisher, the *International Conference of Building Officials*, under license from the International Code Council. The 2000 IBC is a copyrighted work of the International Code Council.

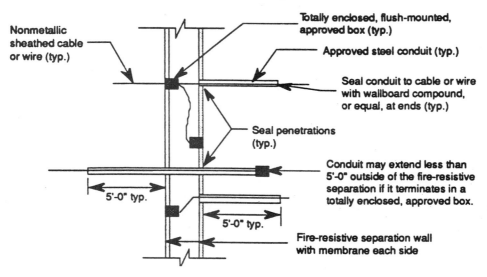

Wood or steel frame one- or two-hour fire-resistive separation wall (cross-section)
Note: Requirements for fire-resistive separation floor/ceiling assemblies are similar.

Nonmetallic sheathed cable or wire (typ.)

Totally enclosed, flush-mounted, approved box (typ.)

Approved steel conduit (typ.)

Seal conduit to cable or wire with wallboard compound, or equal, at ends (typ.)

Seal penetrations (typ.)

Conduit may extend less than 5'-0" outside of the fire-resistive separation if it terminates in a totally enclosed, approved box.

5'-0" typ.

5'-0" typ.

Fire-resistive separation wall with membrane each side

FIGURE 11-21 ■ Electrical penetrations in fire-resistive walls must be carefully considered. *Fire-resistive Separation Wall reproduced courtesy the City Of San Diego Building News Letter.*

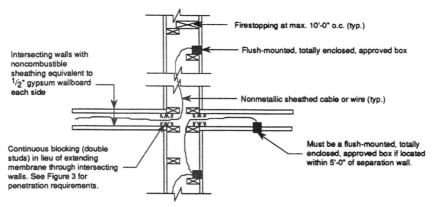

Wood or steel frame one- or two-hour fire-resistive separation wall at intersection with fire-rated or nonfire-rated walls (plan view)

Firestopping at max. 10'-0" o.c. (typ.)

Flush-mounted, totally enclosed, approved box

Intersecting walls with noncombustible sheathing equivalent to 1/2" gypsum wallboard each side

Nonmetallic sheathed cable or wire (typ.)

Continuous blocking (double studs) in lieu of extending membrane through intersecting walls. See Figure 3 for penetration requirements.

Must be a flush-mounted, totally enclosed, approved box if located within 5'-0" of separation wall.

Note: This type of installation may be used in steel frame separation walls, however, for steel frame construction, membranes must be continuous through intersecting walls.

FIGURE 11-22 ■ Electrical penetrations of fire-resistive walls must be carefully detailed in the architectural drawings. *Courtesy City of San Diego Building News Letter.*

Fire and Sound Accessories

Light Fixture Protection for Fire-Rated Ceilings

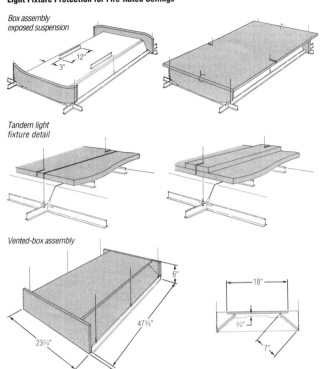

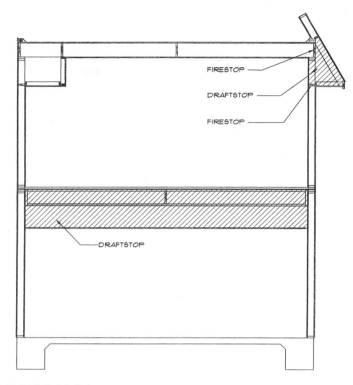

FIGURE 11-24 ■ Fire and draft stops.

FIGURE 11-23 ■ Fire-rated assemblies need special treatment at penetrations. This figure shows some ways to maintain a fire rating when a light fixture penetrates a fire-rated suspended-ceiling system. *Courtesy United States Gypsum Company.*

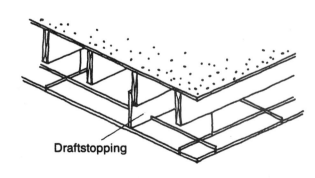

FIGURE 9. Drop T-bar Acoustical Floor-Ceiling Assembly

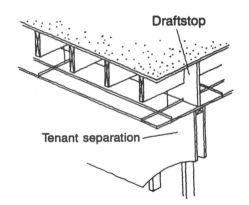

FIGURE 11. Floor-Ceiling Assembly in Multifamily Dwellings, Motels, Hotels

FIGURE 11-25 ■ Draft stops at suspended ceilings. Reproduced from *Wood Frame Design. Courtesy Western Wood Products Association.*

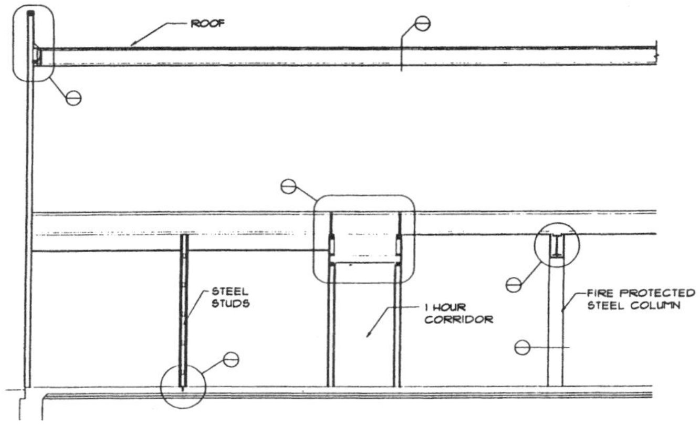

FIGURE 11-26 ■ Use this drawing to answer Question 10.

CHAPTER

11 *Fire-Resistive Construction*

CHAPTER QUIZ

Answer the questions on a separate sheet of paper. Print the chapter title, question number, and a short complete statement for each question. If math is required to answer a question, show the work.

Question 11-1 What is fire-resistive construction trying to protect?

Question 11-2 Describe three ways that fire protection of a structural member is achieved.

Question 11-3 How thick does a concrete wall need to be to be rated as a 4-hour firewall?

Question 11-4 What special consideration would be given to the fire protection of a steel column in a parking garage?

Question 11-5 What is a fire damper?

Question 11-6 What are the two primary purposes of a firewall?

Question 11-7 What is the primary purpose of a 20-minute fire door?

Question 11-8 What function does the parapet serve on a fire wall?

Question 11-9 Wood burns and metal does not; but, in a fire-resistive building, a 8 × 8" wood post might not have to be protected from fire while a 8" steel column would. Explain why.

Question 11-10 Identify the six figures from this chapter that are referenced in the building section shown in Figure 11-26.

DRAWING PROBLEMS

Problem 11-1 Draw and letter a parapet detail for a concrete block wall located 1' from a property line. Use orthographic projection and include appropriate callouts on the detail, assuming that there will be specifications accompanying the working drawings. Make sure that the surface materials within the top 18" of the parapet are noncombustible.

SECTION 3

Engineering Design of Structural Members

CHAPTER 12

Structural Considerations Affecting Design

Buildings should stand up. They should remain standing regardless of the forces applied to them. Not many would argue with these statements. There have been a few exceptions to this principle over the years, such as the houses of Japan that were made of bamboo and paper. They were designed to be lightweight so that when they fell over, due to the shaking of an earthquake, no one was hurt; the occupants just put the pieces back together. This is not a practical approach for today's factories or office buildings.

LOADS AFFECTING STRUCTURES

Today's buildings need to resist changing vertical loads from people, furniture, machinery, snow, and so on, which are called *live loads*. They also need to support vertical loads from their own weight. The weight of the rafters, the roofing material, the walls, and so on are referred to as *dead loads*. Some buildings need to resist *dynamic loads*, which are loads imposed by a moving object such as a car. Buildings also need to resist *horizontal loads* from wind and earthquakes.

The information in this chapter is intended to broaden your knowledge of the structures you will be drawing. It is not intended to make you an engineer. It is not essential that a drafter have an understanding of loads, but it is very helpful. If the drafter can visualize how a load affects the building and can follow that load's path from its origin to the foundation, it will help the drafter visualize conditions that need to be detailed on the plans. The way that building components are put together may have a major impact on how the structure behaves. As an example, hanging a load off the side of a column instead of directly on top of the column would introduce a bending moment into the column that it might not have been designed for. A drafter can better serve as a safeguard for the engineer and the designer.

Dead Loads

Dead loads are determined by the actual weight of the materials from which the building is constructed. Asphalt roof shingles and the felts under them have a given weight. The weight of air-conditioning units can be readily found in the manufacturer's literature. Plywood, framing lumber, glass, and gypsum board all have a given weight. The weights of these materials are readily available from manufacturer's literature or reference books, similar to the table seen in Figure 12-1. These materials are frequently combined into commonly used assemblies as shown in Figure 12-2. Loads

from the materials are considered to be in place for the life of the building.

Live Loads

Live loads are much less exacting than dead loads. They are not static, but rather, constantly changing. Live loads are spelled out by the building code. At the floor, live loads are based on occupancy type (see Chapter 4) as shown in Figure 4-4. Live loads are approximations. The live load on a roof may assume that a worker is carrying a load of roofing material across it. The live load on a ceiling may assume that someone is crawling across it to fix an electrical connection. People may be packed tightly on a balcony, watching a performance below.

Building codes do not try to anticipate every conceivable live load condition. A laborer working on the roof, loading a wheelbarrow full to the brim with heavy material, may cause a failure. The engineer and architect need to anticipate unusual loads that may occur in the building they are designing, such as a room full of heavy file cabinets. They may need to design for loads beyond those called for in the code. Live loads are a judgment call by the code and by the professional.

Snow Loads

Snow creates a live load. Snow loads vary greatly across the country. Some areas experience heavy snowfall, while others don't have any snow. A typical snow load map is shown in Figure 12-3. The snowfall in one area may be substantially different from the snowfall a few miles down the road. The shape of a structure may cause increased localized loading from snowdrifts. Specific design loads for snow should be obtained from the local building department.

Lateral Loads

Lateral loads are usually caused by wind, earthquakes, or earth and ground water pressure on retaining walls.

Earth Pressure

Retaining and restraining walls are designed to resist the lateral force from the earth piled up against it. As seen in Figure 12-4, restraining walls act like a beam spanning between the ground and the floor or roof of the building. The floor or roof diaphragm will need to be designed to resist the loads that the restraining wall will transfer to it. As seen in Figure 12-5, retaining walls resist loads by

Weights of Building Materials

Material	PSF	Material	PSF
Ceilings		**Roofing**	
Acoustical tile	1.0	Asbestos, corrugated 1/4"	3.0
Channel suspended	1.0	15 lb felt	0.85
1/2" gypsum board	2.2	3 ply and gravel	5.5
5/8" gypsum board	2.8	4ply and gravel	6.0
		5ply and gravel	6.5
Decking			
2"	4.4	**Shingles**	
		Asphalt	2.0
Floor Finishes		Asbestos cement	4.0
Asphalt tile	2.0	Book tile 2"	12.0
Brick pavers	10.0	Book tile 3"	20.0
Ceramic tile 3/4"	10.0	Cement tile	16.0
Concrete (lightweight per in.)	10.0	Clay tile	14.0
Concrete (reinforced per in.)	12.0	Fiberglass	0.5
Hardwood 1"	4.0	Ludowici	10.0
Linoleum	2.0	Roman	12.0
Marble	30.0	Slate 1/4"	10.0
Subfloor per in. of depth	3.0	Spanish	19.0
Quarry tile 3/8"	5.0	Wood 3.0	
Terrazzo 1"	13.0		
Vinly asbestos tile	1.3	**Walls**	
Wood floor joists12" O.C.16" O.C.		4" glass block	18.0
2 × 66.0	5 .0	Glass 1/4" plate	3.3
2 × 86.0	6.0	Window (glass, frame, sash)	8.0
2 × 107.0	6.0	Glazed tile	18.0
2 × 128.0	7.0	Gypsum board 1/2"	2.2
		Gypsum board 5/8"	2.8
Insulation		Marble	15.0
Rigid 1"	1.5		
Fiberboard 1"	2.0	**Masonry 4" thick unless noted**	
Foam board per in.	0.2	Brick	38.0
Poured in place	2.0		
4" batts	1.7	**Concrete block**	
6" batts	2.5	4" hollow	30.0
10" batts	4.5	6" hollow	42.0
		6" hollow	55.0
Plywood Sheathing		Hollow clay tile	18.0
3/8"	1.1	Hollow gypsum tile	13.0
1/2"	1.5	Limestone	55.0
5/8"	1.8	Terra-cotta tile	25.0
3/4"	2.2	Stone	55.0
1"	3.0	Plaster 1"	8.0
1 1/8"	3.3	Plaster 1" w/wood lath	10.0
		Porcelain enamel, steel	3.0
		Stucco 7/8"	10.0

For more exact weights, see manufacturers' specifications.

FIGURE 12-1 ■ Standard weights of common building materials.

cantilevering off their footings. The footing transfers the horizontal load from the earth they are retaining into the soil below. The further the footing extends from the wall the less the pressure is on the soil below.

The earth produces a relatively static load against the wall. It is assumed to act like a fluid with a weight of about one-third its actual weight, or about 30 lbs per cu ft (4.71 Kn/m^3) but it is subject to some change. Soil is heavier when it is wet. Different soil types

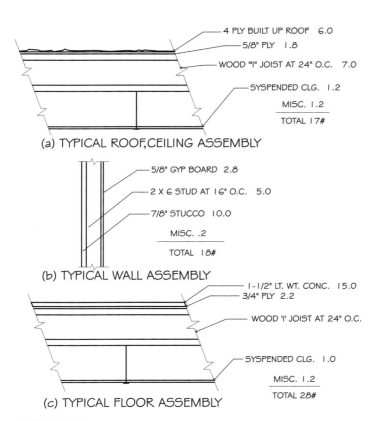

(a) TYPICAL ROOF, CEILING ASSEMBLY

4 PLY BUILT UP ROOF 6.0
5/8" PLY 1.8
WOOD "I" JOIST AT 24" O.C. 7.0
SYSPENDED CLG. 1.2
MISC. 1.2
TOTAL 17#

(b) TYPICAL WALL ASSEMBLY

5/8" GYP BOARD 2.8
2 X 6 STUD AT 16" O.C. 5.0
7/8" STUCCO 10.0
MISC. .2
TOTAL 18#

(c) TYPICAL FLOOR ASSEMBLY

1-1/2" LT. WT. CONC. 15.0
3/4" PLY 2.2
WOOD 'I' JOIST AT 24" O.C.
SYSPENDED CLG. 1.0
MISC. 1.2
TOTAL 28#

FIGURE 12-2 ■ Weight of typical assemblies.

absorb water differently. In order to minimize the effects of moisture, efforts are made to remove the water. This is done by installing a drain line behind the wall, as shown in Figure 12-4, or by letting the water flow through weep holes at the base of the wall, shown in Figure 12-6. The pressure also changes when another load, such as a car in Figure 12-7, or the footing for a building is placed on top of the earth, as in Figure 12-8. This type of loading is called a *surcharge*.

Load Design Considerations

Buildings are not designed to withstand every conceivable force. If they were, they would not be affordable. Buildings are designed to resist all dead loads and all anticipated live loads and dynamic loads. They are designed to resist most anticipated lateral forces with no damage. They should perform well, but some nonstructural damage may occur under severe conditions such as very high winds or moderate earthquakes. Under the most severe storm or earthquake conditions, the building might have structural problems beyond repair, but it should not fail in such a manner as to cause loss of life. Details should be designed to be *ductile*, that is, to give rather than fail suddenly and explosively. Examples of ductile details are given in other chapters of this book.

LOAD DISTRIBUTION

The contractor builds the building from the ground up. The engineer and drafter need to work in the opposite direction. Starting from the top of the building, the live loads, dead loads, and lateral loads are collected. Members are designed to support those loads. These members then transfer their loads to other members until the load is transferred into the ultimate supporting member—the earth. The primary reason for the structural portion of plans is to illustrate this transfer. Load transfers are depicted on the plans in plan views, sections, and details. Figure 12-9 depicts a simple two-story structure that will be examined throughout this chapter. You will follow the loads depicted in this figure, from the roof to the soil.

Roof

The roof shown in Figure 12-9 has two uniformly loaded joist spans, one of 40' and one of 20'. Because these are simple-span members, they will send half their load to each end. This means that the center wall carries half the total load (30'), and this load is transferred down to the floor joists below. The left exterior wall carries 20' of roof load and the right wall carries 10' of roof load. Note that adding overhangs and balconies changes the reaction at each end of a joist or beam.

Floor

At the middle floor, the left wall supports the 20' of roof load discussed above, plus the load of the wall above and 10' of floor load. The right wall is a bit more complicated. It supports 10' of roof load as discussed above, 20' of floor load and the wall above. The right wall also receives a point load from the center wall. The point load consists of the weight of the wall and the roof it supports. Because this wall falls at the center of the 40' floor joist span, the floor joists transfer half the weight of the point load to the right wall and half to the center support on the lower floor.

The center support at the lower floor for the floor above is a beam and three columns. Just like the right wall, the beam receives loads from the center wall above, half the wall load and half the roof load, supported by the wall. It also receives half the left floor joist load (10') and half the right joist load (20').

Foundation

The continuous left footing receives the weight of the left wall plus all of the load supported by the left wall. The continuous right footing does the same. It receives all the load the right wall supports plus the load of the right wall. The center support is not continuous and does not have a continuous footing under it. The beam transfers the loads from above and its own weight to the post. These posts then transfer their loads to pier footings. The weight of the footing will need to be considered along with the other loads to arrive at a total load to be supported by the soil.

Soil

All loads are eventually transferred into the earth or the soil. To determine the footing size, the engineer will need to know the bearing value of the soil. The building code provides a table giving

FIGURE 1608.2—continued
GROUND SNOW LOADS, p_g, FOR THE UNITED STATES (psf)

FIGURE 12-3 ■ A ground snow load chart and the local building department must be consulted to determine design snow loads. Reproduced from the 2000 edition of the International Building Code, copyright © 2000, with the permission of the publisher, the *International Conference of Building Officials*, under license from the International Code Council. The 2000 IBC is a copyrighted work of the International Code Council.

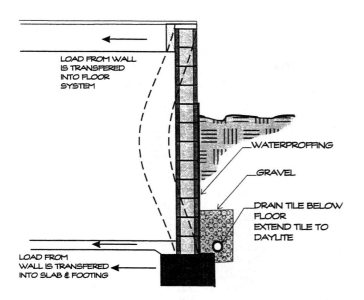

FIGURE 12-4 ■ Restraining walls act like a beam spanning between each floor level. The lateral load of the soil causes the "beam" to bend in the middle.

FIGURE 12-5 ■ Retaining walls resist loads by cantilevering from the footing. Forces from the exposed side of the wall are not considered because they are much smaller than the forces caused by the pressure of the soil on the wall.

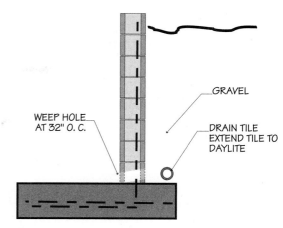

FIGURE 12-6 ■ To minimize the effect of water on the soil load, a weep hole is placed in the bottom of the retaining wall.

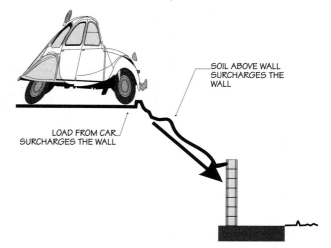

FIGURE 12-7 ■ The pressure resisted by a wall is altered when a load is supported by the soil. The load is transferred through the soil to the wall, causing a condition referred to as a surcharge.

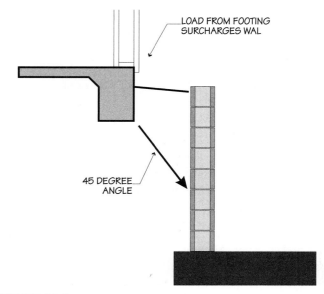

FIGURE 12-8 ■ Loads from the footing surcharge the wall below a point extending at a 45° angle from the bottom of building footing.

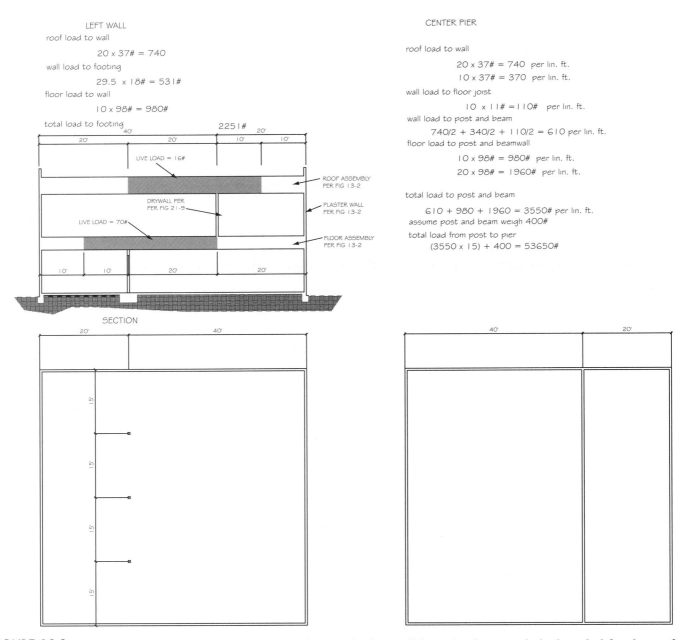

LEFT WALL

roof load to wall

20 x 37# = 740

wall load to footing

29.5 x 18# = 531#

floor load to wall

10 x 98# = 980#

total load to footing   2251#

LIVE LOAD = 16#

DRYWALL PER
PER FIG 21-9

LIVE LOAD = 70#

ROOF ASSEMBLY
PER FIG 13-2

PLASTER WALL
PER FIG 13-2

FLOOR ASSEMBLY
PER FIG 13-2

SECTION

CENTER PIER

roof load to wall

20 x 37# = 740 per lin. ft.

10 x 37# = 370 per lin. ft.

wall load to floor joist

10 x 11# = 110# per lin. ft.

wall load to post and beam

740/2 + 340/2 + 110/2 = 610 per lin. ft.

floor load to post and beamwall

10 x 98# = 980# per lin. ft.

20 x 98# = 1960# per lin. ft.

total load to post and beam

610 + 980 + 1960 = 3550# per lin. ft.

assume post and beam weigh 400#

total load from post to pier

(3550 x 15) + 400 = 53650#

FIGURE 12-9 ■ Transferring loads to the soil. The calculations shown in this figure will be used to determine the loads on the left and center footing.

allowable allowable bearing values in pounds per square foot for various soil types. These values can be used without a soils report, but, this is seldom the wise thing to do unless you are supporting very light loads. A soils investigation and perhaps a geological investigation should be done. To do a soils report, an engineer visits the site and obtains soil samples at various depths and locations on the site. These soil samples will then be taken back to a soils lab, tested, and classified. After this is done, the soils engineer will prepare a soils report. A good soils report will contain, among other things, soil bearing values and construction design recommendations. There will be recommendations for minimum footing width, depth and reinforcing, slab thickness, underlayment and reinforc-

ing and paving materials. It will also give fluid pressure and friction coefficient values for designing retaining walls, expansive soils information, and grading recommendations. Usually values for the soil at most building sites will be higher than the 1000 lbs the code allows, but not always. The soil might be uncompacted fill or alluvial soil (soil deposited by a stream). In these cases, corrective measures would have to be taken. The site might have to be regraded and compacted before it is suitable to support a building or the building might have to be supported by piles or caissons. For the following examples, it will be assumed that the soils report gives a value of 1500 psf for the allowable soil bearing value and a minimum footing depth of 18".

DETERMINING LOADS

The balance of this chapter will provide an introduction to compiling loads. Before the actual loads are examined, the drafter must again be reminded that determining the loads is the responsibility of the engineer. The drafter's role is to carefully specify all of the engineer's calculations.

Footing Calculations—Left Wall

The loads for the left wall transfer straight down, so they are relatively simple to calculate. The roof weighs 37 lbs per square foot (per (a) in Figure 12-2, 17 lbs D.L.+20 lbs L.L). Multiplying 37 lbs by 20', the portion of the roof loading the left wall, equals 740 lbs per lineal foot. The second floor weighs 98 lbs (per (c) in Figure 12-2, 28 lbs D.L.+70 lbs L.L.). Multiplying 98 lbs by 10', the portion of the floor loading the left wall, equals 980 lbs per lineal foot. The wall is 29.5' tall (6+10+3.5+10) and weighs 18 lbs per square foot (see (b) in Figure 12–2). Multiplying the height by the weight per square foot results in a weight of 531 lbs per lineal foot. The total load to the top of the footing is 740 lbs plus 980 lbs plus 531 lbs, or 2251 lbs per lineal foot.

The soil also has to support the load of the footing. The exact size of the footing is not yet known, but you can calculate how much it will weigh per square foot. If this weight is then subtracted from the soil bearing value, the weight of the footing can then be ignored for the rest of the calculation. The footing will be 18" thick and concrete weighs 150 lbs per cubic foot. Using Figure 12-10, it can be seen that the footing will weigh 225 lbs (1.5 × 150) per square foot. This weight can be subtracted from the soil bearing value, leaving a value of 1275 PSF (1500–225). Dividing the 2251 lb load to the top of the footing by the remaining soil bearing value of 1275 lbs results in a required footing width of 1.77' or 1'–9 1/4". Because this would not be a convenient size to trench, a 2' wide footing would probably be specified for this condition.

Footing Calculation–Center Support

The loads for this support are not as simple to calculate. The loads are not stacked directly above. The roof will transfer 1110 lbs (37lbs × ((40/2) + (20/2))) to the interior stud wall. The center wall weighs 110 lbs (11lbs × 10'). The wall transfers its weight and the weight of the roof load, a total of 1220 lbs, to the floor joists. Floor joists on the right side of the center support beam transfer half the load from the wall above to the right exterior wall and half, 610 lbs, to the center beam. The floor itself weighs 2940 lbs (98 lbs × ((20/2)+(40/ 2))). This means that the floor joists transfer 3550 lbs per lineal foot to the center beam. Assuming that the post and beam weigh 400 lbs, 53650 lbs (400+(3550 × 15)) will be transfered to the pier footing. Next, subtract the weight per square foot of the pier from the soil-bearing value. As shown in the previous example, the weight of an 18" deep footing is 225 lbs per square foot. Subtracting this value from the allowable soil bearing value of 1500 lbs leaves an adjusted value of 1275 lbs per square foot. Dividing the load of 53,650 lbs by the adjusted soil- bearing value gives a value of 42.08 sq ft. Now take the square root of this number to find the size of a square pier. This results in a pier size of 6.48 ft square. A pier 6.5 ft square would likely be specified for this condition.

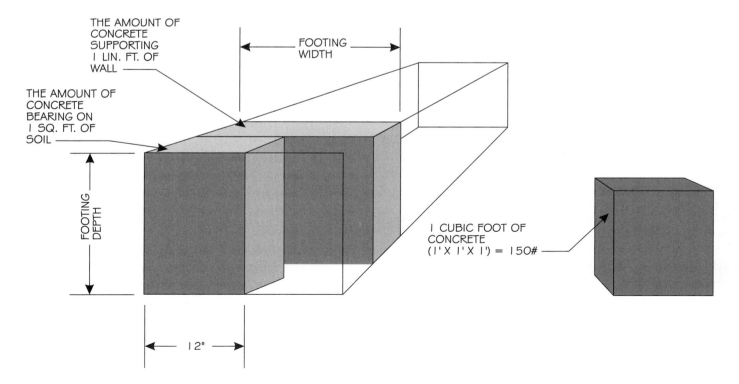

FIGURE 12-10 ■ Weight of concrete footings. *Courtesy Ginger M. Smith.*

THE EFFECTS OF SEISMIC AND WIND FORCES ON A STRUCTURE

So far only vertical loads have been explored. Forces that push sideways are somewhat less obvious. These sideway forces are referred to as *lateral forces*. Every building experiences lateral forces. The two most common sources of lateral forces are seismic forces that result from earthquakes and wind. Buildings must be constructed to resist these forces as well as the vertical forces discussed in earlier chapters. The building code contains maps to classify the intensity of wind and earthquake forces that are likely to occur in various parts of the country. It also contains minimum design requirements to resist these forces.

Earthquakes

Earthquakes happen without warning and last for a very short period of time, perhaps 10 to 20 seconds. A shift in the earth's crust sets off shock waves that shake and rattle buildings. Most earthquakes in North America occur in California and Alaska, but they have occurred in other parts of the country. The code contains maps, similar to the snow load and wind speed maps, showing the expected ground motion caused by an earthquake. The shock wave from an earthquake causes a building to move laterally and vertically. It is assumed that the weight of the building and normal design for vertical forces will take care of the vertical motion, so vertical forces are generally ignored in seismic design. The lateral movement is very dramatic. It is not a constant or uniform force. It shakes the building back and forth with different intensities. The type of structure (solid masonry, moment-resisting frame, and so on) and the nature of the soil under the building are major factors in the effect that an earthquake will have on a building.

Buildings are not designed to be earthquake proof. They are designed to be earthquake resistant. Protecting the people inside the building is more important than protecting the building itself. It is better for a building to twist and bend in a severe earthquake than it is for it to suddenly break apart. With this in mind, engineers try to design connections that will bend or stretch rather than snap. This type of connection is called a ductile connection. Ductile connections give the building a better chance of remaining standing during and after an earthquake and give the occupants of a building a better chance of evacuating. The building may be as useless as if it had collapsed, but there will be fewer lives lost.

Wind Loads

Every portion of the world has wind, and wind patterns are far more predictable than earthquakes. There are warning systems in place for high-wind conditions. Even so, hurricanes and tornadoes are very destructive forces. Because destructive wind events happen for a longer period of time and over a larger portion of North America, they cause a much greater loss of life and property than earthquakes. Figure 12-11 is a wind speed map of the United States. It has wind contour lines ranging from 85 to 150 miles per hour.

Wind causes both horizontal and vertical movement and both components are considered by the engineer. The vertical forces are called uplift. As seen in Figure 12-12, uplift is generally caused in one of two ways. Wind blowing into a structure through an opening increases the pressure and tries to lift the roof. The wind blowing across the top of a structure creates a negative pressure that tries to lift or suck the roof off its supports. Both these vertical forces require that the roof be firmly anchored to the supporting elements. The horizontal forces are treated in very much the same manner as seismic forces. Wind speeds can reach as high as 200 miles per hour in a hurricane and 600 miles per hour in a tornado. As with earthquakes, buildings are not designed to withstand the worst wind event without damage but are designed to protect human life.

GENERAL DESIGN CONSIDERATIONS FOR HORIZONTAL FORCES

Figure 12-13 at A shows an open frame stud wall with a lateral force pushing on it as indicated by the arrow. If the joints between the top plate and the studs are not stiff enough, the wall will try to fold into a parallelogram as shown at B in Figure 12-13. If the wall is stiff enough to resist these forces but is not anchored well to the foundation, it will tend to slide along the foundation or will try to lift up and turn over, as shown at C in Figure 12-13.

The wall can be made stiff by applying a rigid sheathing to it, such as plywood (at A in Figure 12-14), or by adding diagonal straps or wood let-in-braces (at B in Figure 12-14). A wall with rigid sheathing is called a *shear wall*. Figure 12-15 shows a typical frame and plywood shear wall. An alternate material could be used for the wall, such as concrete block that makes up a rigid solid wall as shown at C in Figure 12-14, or by using a steel rigid frame as shown at D in Figure 12-14. Figure 12-16 shows a shear wall made of steel studs and straps. Connections for this wall are shown in Figures 12-17 and 12-18.

Stiffening the wall is only a part of the solution. The wall must also be tied to the foundation with anchor bolts or similar devices to prevent sliding, as shown in Figures 12-15 and 12-16. Shear walls also need adequate weight at the end to prevent overturning. The footing itself may be heavy enough to prevent overturning, or a pier may need to be added to the end, as shown at A in Figure 12-14. Special bolting, called a hold-down, shown in Figures 12-15, 12-16, and 12-17, may be needed on the end of the wall to firmly tie the wall to the footing. The end of a wall may also be held down by a load being applied to the top of the wall such as a header, shown at B in Figure 12-14.

Figures 12-13 and 12-14 show the lateral force acting in one direction. This is never the actual case. Wind loads or seismic loads can come from any direction. For engineering design purposes, the lateral forces are calculated as if they act perpendicular to the building's surfaces. The building must be designed to resist lateral forces from all sides. Figure 12-19 shows a tall thin and very rigid wall that resists a large lateral force. The wall has hold-down bolts on

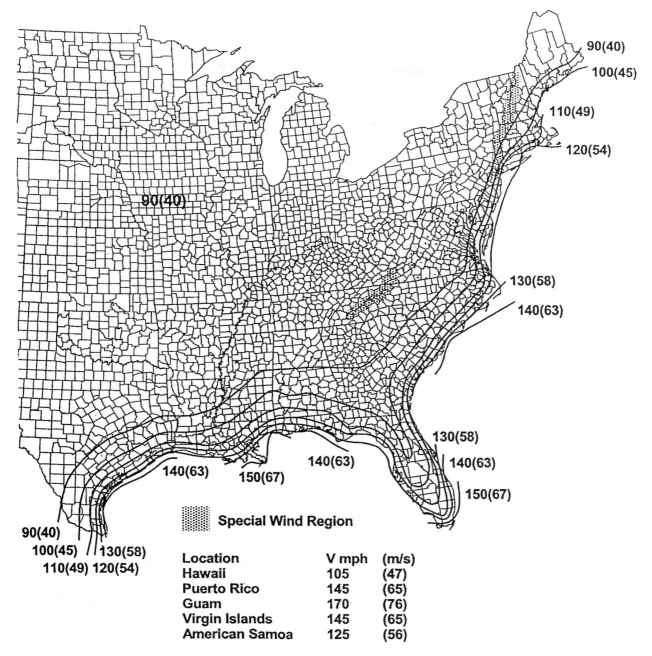

90(40)
100(45)
110(49)
120(54)

90(40)

130(58)
140(63)

130(58)
140(63)
150(67)

140(63)
150(67)

140(63)

90(40)
100(45) 130(58)
110(49) 120(54)

▒ **Special Wind Region**

Location	V mph	(m/s)
Hawaii	105	(47)
Puerto Rico	145	(65)
Guam	170	(76)
Virgin Islands	145	(65)
American Samoa	125	(56)

Notes:
1. Values are nominal design 3-second gust wind speeds in miles per hour (m/s) at 33 ft (10 m) above ground for Exposure C category.
2. Linear interpolation between wind contours is permitted.
3. Islands and coastal areas outside the last contour shall use the last wind speed contour of the coastal area.
4. Mountainous terrain, gorges, ocean promontories, and special wind regions shall be examined for unusual wind conditions.

FIGURE 1609—continued
BASIC WIND SPEED (3-SECOND GUST)

FIGURE 12-11 ■ Basic Wind Speed Map, Reproduced from the 2000 edition of the International Building Code™, copyright © 2000, with permission of the publisher, the International Code Council, Inc.

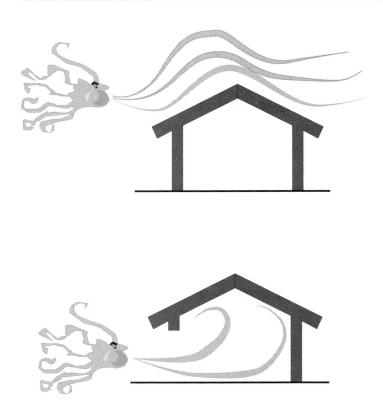

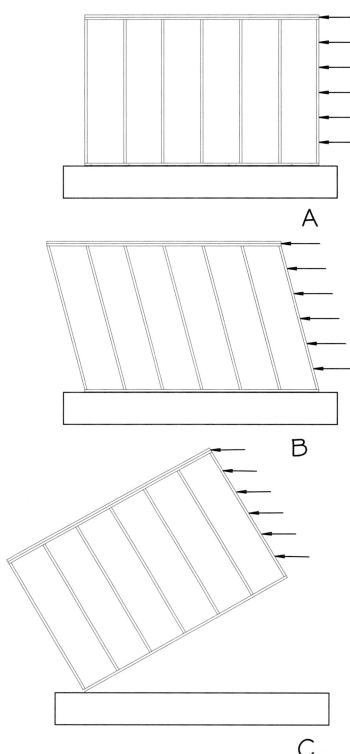

FIGURE 12-12 ■ Uplift is caused by wind blowing into the structure and attempting to lift the structure, or by creating a negative pressure as it blows over the structure attempting to suck the roof off the structure.

each side of it and is anchored to a footing that projects beyond the wall on each side. This extended footing would have extra reinforcing steel encased in the concrete and is called a *grade beam*. The grade beam will resist overturning by changing the leverage point and by adding weight.

Collecting and Resisting Loads

So far, we have talked about the wall resisting loads, but where do the loads come from? It isn't from the wind blowing on the narrow end of the wall itself. Figure 12-20 at A shows a simple building with the wind blowing on one of the long sides of the building. The wall acts as a beam with a uniform load and transfers half the wind load to the ground. The other half of the wind load is transferred to the roof (at B in Figure 12-20). Figure 12-21 is the same building with the wind-loaded wall removed for clarity. The arrows indicate the portion of the wind load that was transferred to the roof. This load pushes sideways on the roof.

The roof is made rigid in one of several ways: a layer of plywood nailed to the joists; metal decking welded to open web steel joists; cable X bracing tied to metal rigid frames; or by another method similar to these. A rigid roof surface is called a *roof diaphragm* and is discussed in more detail later in this chapter. It is very similar to a shear wall, except that it is a horizontal surface instead of a vertical surface. A roof diaphragm acts like a beam spanning sideways,

FIGURE 12-13 ■ (A) Lateral forces acting on a wall. (B) Lateral forces will try to flatten or turn the wall into a parallelogram or (C) lift and rotate the wall.

with the shear walls below acting as the reaction points. The shear walls hold the diaphragm in place on each end and the lateral force tries to bend it. The lateral force is acting as a uniform load applied

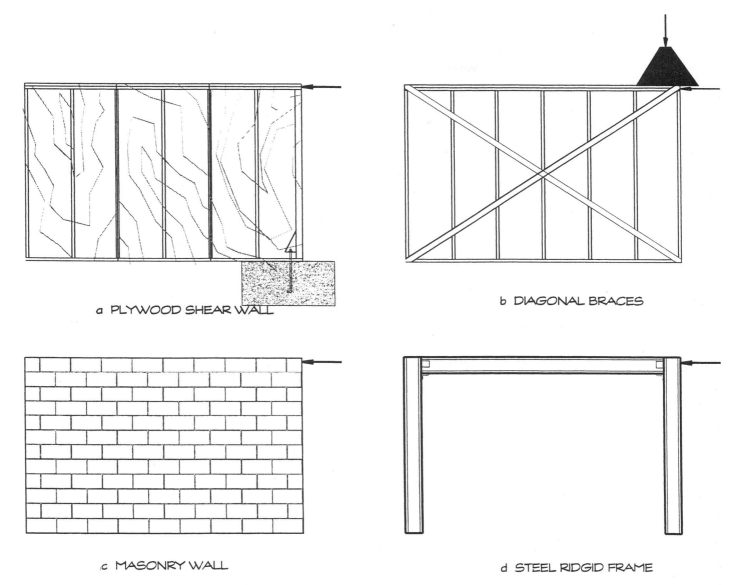

a PLYWOOD SHEAR WALL

b DIAGONAL BRACES

c MASONRY WALL

d STEEL RIDGID FRAME

FIGURE 12-14 ■ Shear wall examples. Shear walls can be made of many materials including (A) studs and plywood, (B) studs with diagonal strapping, or (C) masonry. A steel rigid frame (D) serves the same function as a shear wall. Shear walls depended on weight from foundation below or roof and floor loads from above to keep them from lifting and rotating.

sideways to the roof. The roof acts like a uniform loaded beam. It has a horizontal reaction at each end that is transferred to the walls.

If the lateral force were to push on the narrow wall, as shown in Figure 12-22, it would try to bend the roof diaphragm in the other direction, as shown in Figure 12-23. The loads are easier to resist in this direction because the beam formed by the roof diaphragm is deeper and the wall that the load is transferred to is longer.

Diaphragms

Floor and roof diaphragms act like a beam to resist the forces transferred to them by the walls. The engineer will calculate the magnitude of this force and specify a system to resist that force.

If plywood is used for the sheathing, its ability to resist the forces is dependent on the grade and thickness of the plywood, the size

and spacing of the supporting members, whether or not the edges of the plywood are blocked, the pattern the plywood is laid in, and the type, size, and spacing of the nails. Figure 12–24 is a table from the IBC that gives these values. According to this table, 15/32" structural 1 plywood laid per case 3 and nailed to 3 × 6 joists with a single row of 10d nails at

■ 4" O.C. at the diaphragm boundary

■ 4" O.C. at continuous panel edges parallel to the load

■ 6" O.C. at all other panel edges

would be capable of resisting 480 pounds per square foot. Figure 12–25 shows common placement patterns for placing plywood over framing members.

Diaphragm values for metal decks are available from their man-

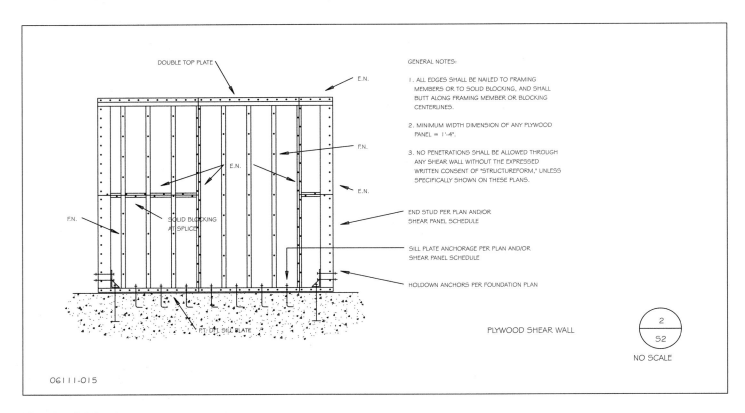

FIGURE 12-15 ■ Typical plywood shear wall detail.

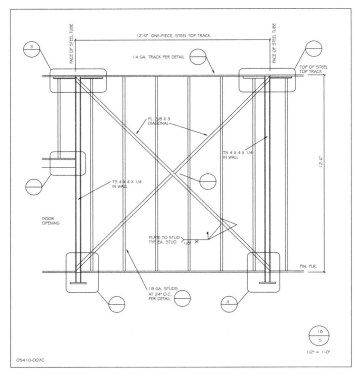

FIGURE 12-16 ■ Steel stud shear wall.

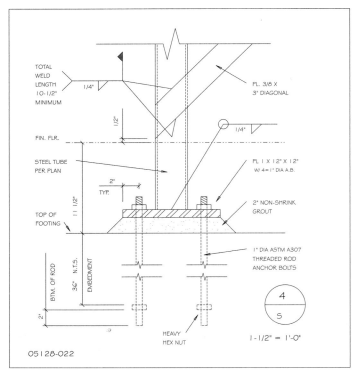

FIGURE 12-17 ■ Steel stud shear-wall connection.

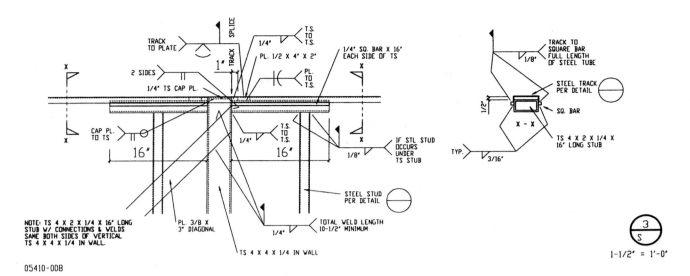

FIGURE 12-18 ■ Steel stud shear-wall connection.

ufacturer. Metal sheathing is either welded or screwed to its supporting member. Frequently, metal decks have concrete poured in them to add rigidity. Concrete floors and roofs make excellent diaphragms.

Chord Forces

As discussed earlier, a floor or roof diaphragm acts like a horizontal beam. When the diaphragm bends as shown by the dotted lines in Figure 12-21, one side is in compression and the other side is in tension. The plywood and framing members on the compression side normally need no reinforcing, but tension on the far side is trying to tear the diaphragm apart, and the individual pieces of sheathing on the diaphragm need to be held together. The diaphragm is held together by using a chord. A *chord* is a continuous piece of material, or several pieces properly joined together, that resist the forces of tension. Chords are normally made of steel. Chords may be in the roof or in the wall that the roof attaches to. Figure 12-26 shows a chord that is made of reinforcing steel cast into a tilt-up wall panel. The walls are stood up and a piece of angle iron is weld-

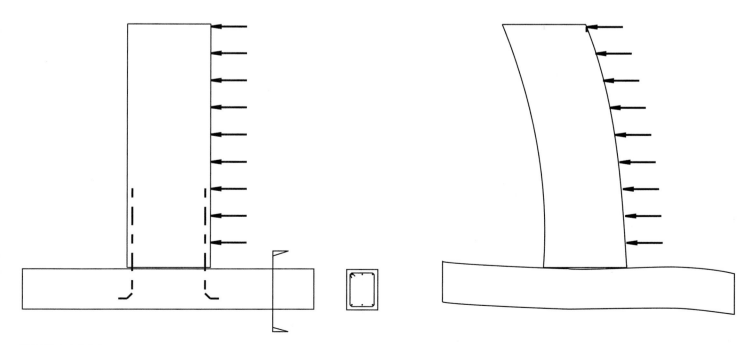

FIGURE 12-19 ■ Lateral forces transfer from the wall to the footing and apply pressure to the soil.

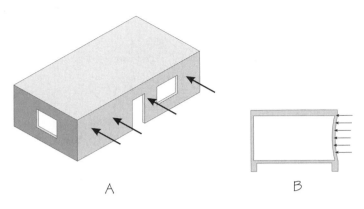

FIGURE 12-20 ■ Lateral forces put pressure on the walls and try to bend them.

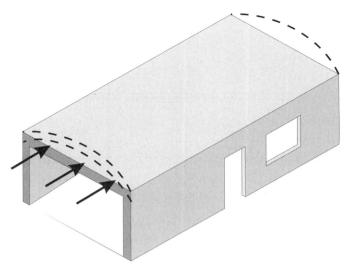

FIGURE 12-23 ■ The loads are easier to resist in this direction because the beam formed by the roof diaphragm is deeper than that shown in Figure 12-22 and the wall that the load is transferred to is longer.

ed to the chord bars to join them together so that they act as one continuous piece. In a concrete block structure, the bond beam at the top of the wall (shown in Figure 12-27) acts as the chord. The steel is lapped in the wall so that it acts as one continuous piece. Figure 12-28 shows a block wall with two #5 bars acting as a chord. It also shows the edge nailing of the plywood to transfer the diaphragm loads to the ledger and the bolting required to transfer the diaphragm loads from the ledger to the wall.

Shear Walls

Shear walls may be made of many different materials. Common materials are concrete, concrete block, brick, and plywood over wood frame. Shear walls are tied to the floor or roof diaphragm above. They transfer the loads from this diaphragm to the foundation. In the case of a multistory building, shear walls transfer loads to shear walls below and then to the foundation. It is important that the wall be adequately tied to the diaphragm above and the footing below. Lateral loads on the shear wall try to lift one end of the wall and footing, while pushing down on the other end of the wall and footing, as shown in Figure 12-19. The footing must be sized to support the lateral loads from the shear wall as well as the normal vertical loads. Frequently these walls will be extended to the floor sheathing or roof sheathing, as shown in Figure 12-29. This allows both the plywood floor diaphragm and the plywood shear panel to be nailed to the top plate to transfer lateral loads from the diaphragm to the shear wall.

Concrete

Tilt-up concrete walls often have additional steel added at each end. This steel is welded to matching steel projecting from the footing, as shown in Figure 12-30. These matching bars are each welded to a piece of angle iron to transfer the forces to the footing.

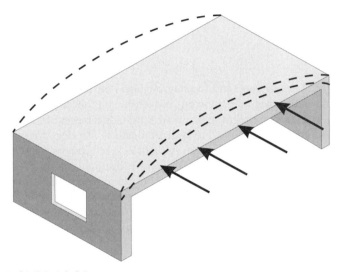

FIGURE 12-21 ■ Lateral forces on the short wall of building are generally easier to resist because they are transferred into the long walls.

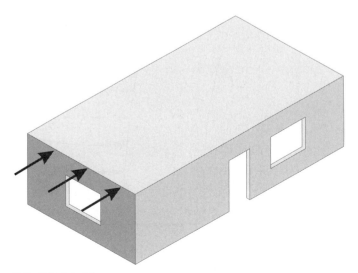

FIGURE 12-22 ■ Lateral forces are transferred from walls to the roof causing the roof diaphragm to bend. The roof diaphragm transfers the forces back to the perpendicular walls and then to the footings.

TABLE 2306.3.1
RECOMMENDED SHEAR (POUNDS PER FOOT) FOR WOOD STRUCTURAL PANEL DIAPHRAGMS
WITH FRAMING OF DOUGLAS-FIR-LARCH, OR SOUTHERN PINE[a] FOR WIND OR SEISMIC LOADING

PANEL GRADE	COMMON NAIL SIZE OR STAPLE[f] LENGTH AND GAGE	MINIMUM FASTENER PENETRATION IN FRAMING (inches)	MINIMUM NOMINAL PANEL THICKNESS (inch)	MINIMUM NOMINAL WIDTH OF FRAMING MEMBER (inches)	6	4	2-1/2[c]	2[c]	Case 1 (No unblocked edges or continuous joints parallel to load)	All other configurations (Cases 2, 3, 4, 5 and 6)
					6	6	4	3		
Structural I Grades	6d[e]	1¼	5/16	2	185	250	375	420	165	125
				3	210	280	420	475	185	140
	1½ 16 Gage	1		2	155	205	310	350	135	105
				3	175	230	345	390	155	115
	8d	1³⁄₈	3/8	2	270	360	530	600	240	180
				3	300	400	600	675	265	200
	1½ 16 Gage	1		2	175	235	350	400	155	115
				3	200	265	395	450	175	130
	10d[d]	1½	15/32	2	320	425	640	730	285	215
				3	360	480	720	820	320	240
	1½ 16 Gage	1		2	175	235	350	400	155	120
				3	200	265	395	450	175	130
Sheathing, Single Floor and Other Grades Covered in DOC PS 1 and PS 2	6d[e]	1¼	5/16	2	170	225	335	380	150	110
				3	190	250	380	430	170	125
	1½ 16 Gage	1		2	140	185	275	315	125	90
				3	155	205	310	350	140	105
	6d[e]	1¼	3/8	2	185	250	375	420	165	125
				3	210	280	420	475	185	140
	8d	1³⁄₈		2	240	320	480	545	215	160
				3	270	360	540	610	240	180

(continued)

FIGURE 12-24 ■ Allowable stress for wood floor and roof diaphragms. Reproduced from the 2000 edition of the International Building Code, copyright © 2000, with the permission of the publisher, the *International Conference of Building Officials*, under license from the International Code Council. The 2000 IBC is a copyrighted work of the International Code Council.

Concrete Block

Concrete block walls have vertical steel placed closer together and all of the cells solid grouted. All of the vertical steel extends into the footing. The steel from the footing laps the steel in the wall and the concrete grout transfers the loads from one bar to the other.

Wood Shear Walls

Plywood is nailed to one or both sides of the studs. If the plywood is nailed to both sides of a wall, the studs need to be thicker than the normal 2" (50 mm) nominal size so that the nails penetrating from both sides do not split the stud. Figure 12-31 shows values for particleboard shear walls for different thickness and nailing conditions. The code has similar tables for plywood and plaster.

A typical hold-down detail that might be used at each end of Figure 12-15 is shown in Figure 12-32. A manufactured bracket with tested values is attached to the stud. A bolt is cast into the footing and attached to the bracket.

Braced Frames

Braced frames made of steel are frequently used instead of shear walls as forces increase. They consist of vertical, horizontal, and diagonal members forming triangles. Figure 12-33 shows two examples of braced frames. Figures 12-34 and 12-35 shows typical braced frame details.

Moment Frames

Moment frames are made of steel members and depend on a very rigid connection of the horizontal members to the vertical members. Figure 12-36 illustrates a connection as it would be done with a bolted connection. A similar connection can be accomplished with rivets or by welding. When connected, the members all act as one. Lateral forces induce bending in the horizontal and vertical members, as illustrated in Figure 12-37.

TABLE 2306.3.1—continued
RECOMMENDED SHEAR (POUNDS PER FOOT) FOR WOOD STRUCTURAL PANEL DIAPHRAGMS
WITH FRAMING OF DOUGLAS-FIR-LARCH, OR SOUTHERN PINE[a] FOR WIND OR SEISMIC LOADING

PANEL GRADE	COMMON NAIL SIZE OR STAPLE[f] LENGTH AND GAGE	MINIMUM FASTENER PENETRATION IN FRAMING (inches)	MINIMUM NOMINAL PANEL THICKNESS (inch)	MINIMUM NOMINAL WIDTH OF FRAMING MEMBER (inches)	BLOCKED DIAPHRAGMS — Fastener spacing (inches) at diaphragm boundaries (all cases) at continuous panel edges parallel to load (Cases 3, 4), and at all panel edges (Cases 5 and 6)[b]				UNBLOCKED DIAPHRAGMS — Fasteners spaced 6" max. at supported edges[b]	
					6	4	2-1/2[c]	2[c]	Case 1 (No unblocked edges or continuous joints parallel to load)	All other configurations (Cases 2, 3, 4, 5 and 6)
					Fastener spacing (inches) at other panel edges (Cases 1, 2, 3 and 4)[b]					
					6	6	4	3		
Sheathing, Single Floor and other grades covered in DOC PS 1 and PS 2 (continued)	1½ 16 Gage	1	3/8	2	160	210	315	360	140	105
				3	180	235	355	400	160	120
	8d	1 3/8	7/16	2	255	340	505	575	230	170
				3	285	380	570	645	255	190
	1½ 16 Gage	1	7/16	2	165	225	335	380	150	110
				3	190	250	375	425	165	125
	8d	1 3/8	15/32	2	270	360	530	600	240	180
				3	300	400	600	675	265	200
	10d[d]	1 1/2	15/32	2	290	385	575	655	255	190
				3	325	430	650	735	290	215
	1½ 16 Gage	1	15/32	2	160	210	315	360	140	105
				3	180	235	355	405	160	120
	10d[d]	1 1/2	19/32	2	320	425	640	730	285	215
				3	360	480	720	820	320	240
	1¾ 16 Gage	1	19/32	2	175	235	350	400	155	115
				3	200	265	395	450	175	130

FIGURE 12-24 ■ Allowable stress for wood floor and roof diaphragms. Reproduced from the 2000 edition of the International Building Code, copyright © 2000, with the permission of the publisher, the *International Conference of Building Officials*, under license from the International Code Council. The 2000 IBC is a copyrighted work of the International Code Council.

Drags

When buildings have irregular shapes, as shown in Figure 12-28, there is frequently a need to transfer or drag the forces from a floor or roof diaphragm into the shear wall. This can be done through a beam or joist as shown in Figure 12-28. A *drag* is a beam that is attached to the edge of a diaphragm to transfer lateral loads from the diaphragm into a shear wall or braced frame. The beam must then be fastened to the shear wall in a secure manner. Figure 12-29 is a typical detail showing the drag and the attachment to the shear wall. The bracket has steel rebar welded to it and extending back into the concrete wall to transfer the loads. Bolts through the bracket and the beam transfer the load to the beam. The roof diaphragm is securely fastened to the beam with nails.

ADDITIONAL READING

The following Web sites can be used as a resource to help obtain additional information of structural issues:

ADDRESS	COMPANY/ORGANIZATION
www.seaint.prg	The Structural Engineers Association – International
www.acec.org/programs/case.htm	Council of American Structural Engineers

TABLE 2306.3.1—continued
RECOMMENDED SHEAR (POUNDS PER FOOT) FOR WOOD STRUCTURAL PANEL DIAPHRAGMS
WITH FRAMING OF DOUGLAS-FIR-LARCH, OR SOUTHERN PINE[a] FOR WIND OR SEISMIC LOADING

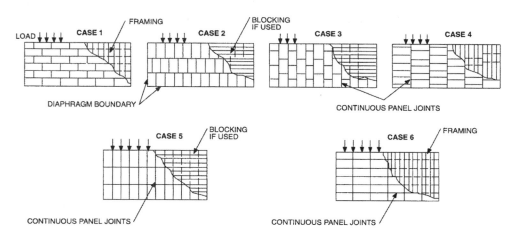

For SI: 1 inch = 25.4 mm, 1 pound per foot = 14.5939 N/m.

a. For framing of other species: (1) Find specific gravity for species of lumber in AFPA National Design Specification. (2) For staples find shear value from table above for Structural I panels (regardless of actual grade) and multiply value by 0.82 for species with specific gravity of 0.42 or greater, or 0.65 for all other species. (3) For nails find shear value from table above for nail size for actual grade and multiply value by the following adjustment factor: Specific Gravity Adjustment Factor = [1-(0.5 - SG)], Where SG = Specific Gravity of the framing lumber. This adjustment factor shall not be greater than 1.

b. Space nails maximum 12 inches o.c. along intermediate framing members (6 inches o.c. where supports are spaced 48 inches o.c.).

c. Framing at adjoining panel edges shall be 3-inches nominal or wider, and nails shall be staggered where nails are spaced 2 inches o.c. or $2^1/_2$ inches o.c.

d. Framing at adjoining panel edges shall be 3-inch nominal or wider, and nails shall be staggered where both of the following conditions are met: (1) 10d nails having penetration into framing of more than $1^1/_2$ inches and (2) nails are spaced 3 inches o.c. or less.

e. 8d is recommended minimum for roofs due to negative pressures of high winds.

f. Staples shall have a minimum crown width of $^7/_{16}$ inch.

FIGURE 12-25 ■ Sheathing and framing configurations. Reproduced from the 2000 edition of the International Building Code, copyright © 2000, with the permission of the publisher, the *International Conference of Building Officials*, under license from the International Code Council. The 2000 IBC is a copyrighted work of the International Code Council.

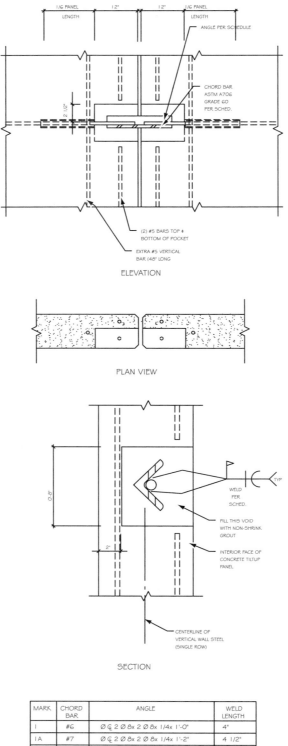

MARK	CHORD BAR	ANGLE	WELD LENGTH
I	#6	Ø ℄ 2 Ø 8x 2 Ø 8x 1/4x 1'-0"	4"
I A	#7	Ø ℄ 2 Ø 8x 2 Ø 8x 1/4x 1'-2"	4 1/2"
I B	#8	Ø ℄ 2 Ø 8x 2 Ø 8x 3/8x 1'-3"	5"
I C	#9	Ø ℄ 2 Ø 8x 2 Ø 8x 3/8x 1'-3"	5 1/2"
I D	#10	Ø ℄ 3x 3x 1/2x 1'-6"	6 1/2"

03472009

N.T.S.

FIGURE 12-26 ■ A chord in a tilt-up wall. A chord keeps the edges of
diaphragms from tearing apart when the diaphragm tries to bend.

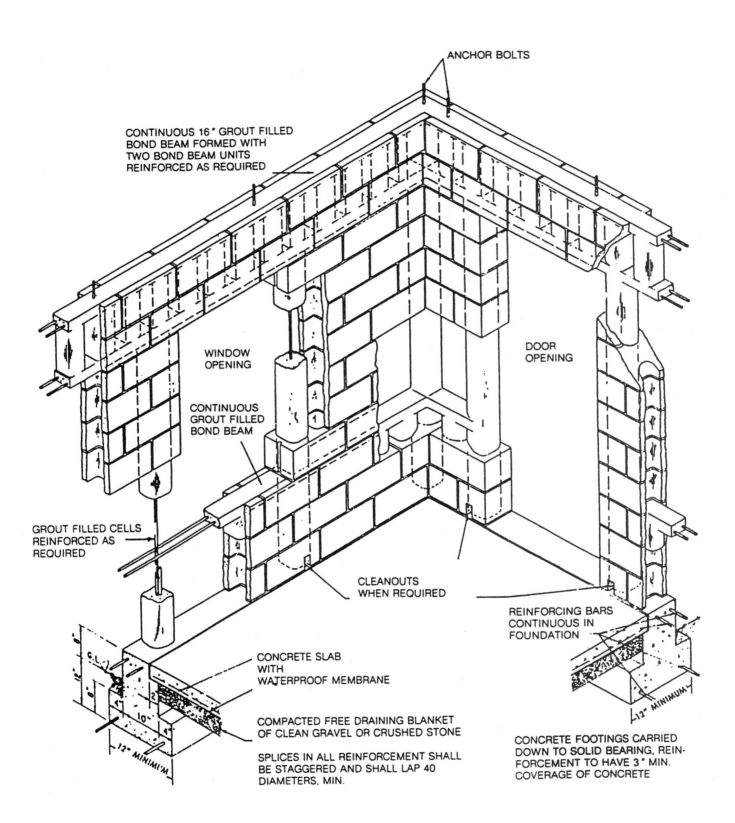

ANCHOR BOLTS

CONTINUOUS 16″ GROUT FILLED
BOND BEAM FORMED WITH
TWO BOND BEAM UNITS
REINFORCED AS REQUIRED

WINDOW
OPENING

DOOR
OPENING

CONTINUOUS
GROUT FILLED
BOND BEAM

GROUT FILLED CELLS
REINFORCED AS
REQUIRED

CLEANOUTS
WHEN REQUIRED

REINFORCING BARS
CONTINUOUS IN
FOUNDATION

CONCRETE SLAB
WITH
WATERPROOF MEMBRANE

COMPACTED FREE DRAINING BLANKET
OF CLEAN GRAVEL OR CRUSHED STONE

SPLICES IN ALL REINFORCEMENT SHALL
BE STAGGERED AND SHALL LAP 40
DIAMETERS, MIN.

12″ MINIMUM

CONCRETE FOOTINGS CARRIED
DOWN TO SOLID BEARING, REIN-
FORCEMENT TO HAVE 3″ MIN.
COVERAGE OF CONCRETE

12″ MINIMUM

FIGURE 12-27 ■ A chord in block wall. *Courtesy the Concrete Masonry Association of California and Nevada.*

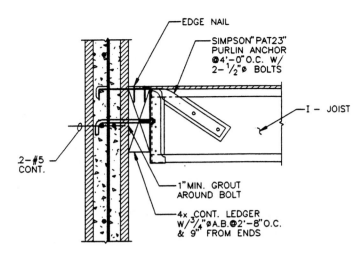

FIGURE 12-28 ■ Diaphragm transfer to block wall. *Courtesy the Concrete Masonry Association of California and Nevada.*

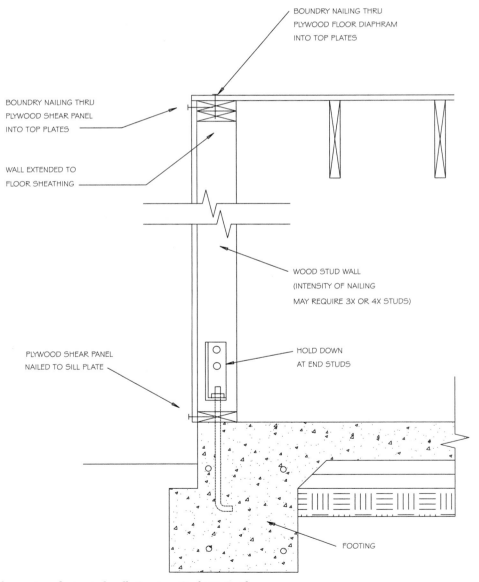

FIGURE 12-29 ■ Diaphragm transfer to stud wall. *Courtesy Gisela M. Smith.*

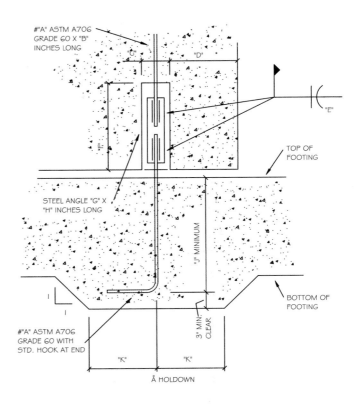

#"A" ASTM A706
GRADE 60 X "B"
INCHES LONG

"C"   "D"

TOP OF
FOOTING

STEEL ANGLE "G" X
"H" INCHES LONG

"J" MINIMUM

BOTTOM OF
FOOTING

#"A" ASTM A706
GRADE 60 WITH
STD. HOOK AT END

3" MIN
CLEAR

"K"   "K"

Ⓐ HOLDOWN

HOLDOWN SCHEDULE											
MARK	"A"	"B"**	"C"	"D"*	"E"	"F"	"G"	"H"	"J"	"K"	REMARKS
8A	6	60	5	6	4	16	2-1/2" X 2-1/2" X 1/4"	12	15	18	
8B	7	72	5	6	4	18	2-1/2" X 2-1/2" X 1/4"	14	20	20	
8C	8	84	5	6	5-1/2	19	2-1/2" X 2-1/2" X 3/8"	15	26	22	
8D	9	96	5	6	6	19	2-1/2" X 2-1/2" X 3/8"	15	32	25	
8E	10	108	6	6	6-1/2	22	3" X 3" X 1/2"	18	38	28	
8F											
8G											

* NOTE: IN THE CASE THAT AN ABUTING CONC. PANEL THICKER THAN 7" WOULD HINDER ACCESS TO THE POCKET,
"D" SHALL BE ADJUSTED TO EQUAL THE WIDTH OF THE ABUTTING PANEL SO AS TO AVOID CONFLICT.

** NOTE: SEE PANEL ELEVATION DRAWING WHERE A LONBER "B" MAY BE REQUIRED.

1" = 1'-0"

03473-009

FIGURE 12-30 ■ Hold-down to foundation detail, tilt-up. *Courtesy Dean Smith.*

TABLE 2306.4.3
ALLOWABLE SHEAR FOR PARTICLEBOARD SHEAR WALL SHEATHING

PANEL GRADE	MINIMUM NOMINAL PANEL THICKNESS (inch)	MINIMUM NAIL PENETRATION IN FRAMING (inches)	Nail size (common or galvanized box)	PANELS APPLIED DIRECT TO FRAMING Allowable shear (pounds per foot) nail spacing at panel edges (inches)[a]			
				6	4	3	2
M-S "Exterior Glue" and M-2 "Exterior Glue"	$3/8$	$1\,1/2$	6d	120	180	230	300
	$3/8$	$1\,1/2$	8d	130	190	240	315
	$1/2$			140	210	270	350
	$1/2$	$1\,5/8$	10d	185	275	360	460
	$5/8$			200	305	395	520

For SI: 1 inch = 25.4 mm, 1 pound per foot = 14.5939 N/m.

a.   Values are not permitted in Seismic Design Category D, E or F.

FIGURE 12-31 ■ Shear wall table. Reproduced from the 2000 edition of the International Building Code, copyright © 2000, with the permission of the publisher, the *International Conference of Building Officials*, under license from the International Code Council. The 2000 IBC is a copyrighted work of the International Code Council.

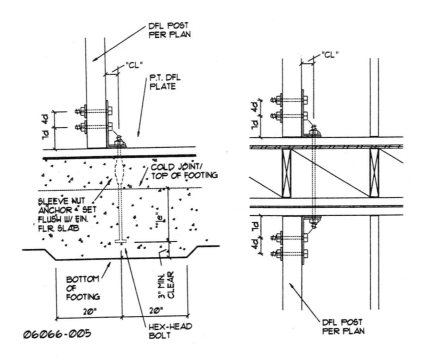

06066-005

"SIMPSON" HOLDOWN	MINIMUM STUD SIZE	STUD BOLTS	ANCHOR BOLTS	DIMENSIONS	
				"CL"	"le"
HD2A	4 x	(2) 5/8"	5/8"	1-1/2"	15"
HD5A	4 x	(2) 3/4"	3/4"	2-1/16"	15"
HD6A	4 x	(2) 7/8"	7/8"	2-1/16"	15"
HD8A	4 x	(3) 7/8"	7/8"	2-1/16"	15"
HD10A	4 x	(4) 7/8"	7/8"	2-1/16"	15"
HD20A	4 x	(4) 1"	1-1/4"	2-3/8"	20"

GENERAL NOTES:

1. MINIMUM CONCRETE END DISTANCE TO BE 12".
2. MINIMUM CONCRETE STRENGTH TO BE 2000 PSI, U.N.O.
3. 'd' = DIAMETER OF STUD BOLT.

HOLDOWN SCHEDULE

1" = 1'-0"

FIGURE 12-32 ■ Hold-down to foundation detail. *Courtesy Dean Smith.*

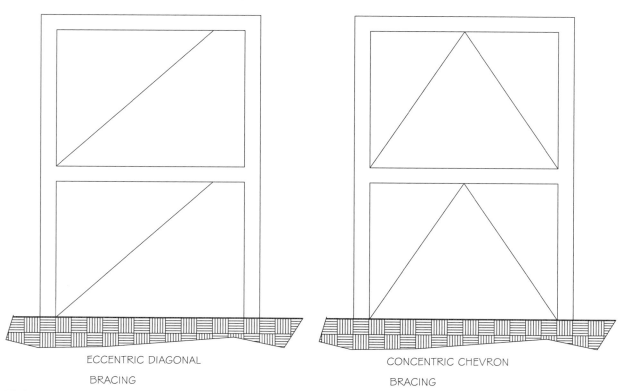

ECCENTRIC DIAGONAL
BRACING

CONCENTRIC CHEVRON
BRACING

FIGURE 12-33 ■ Examples of braced frames.

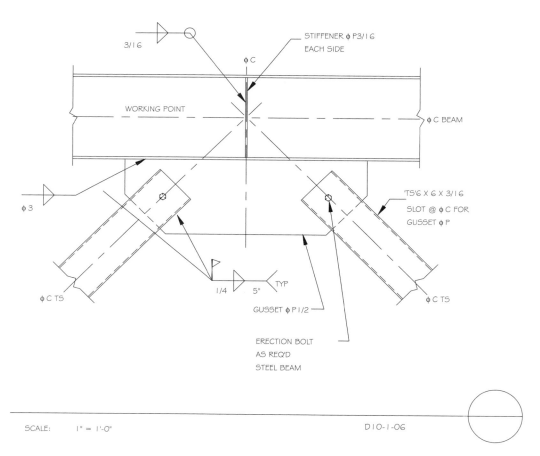

FIGURE 12-34 ■ Braced frame detail.

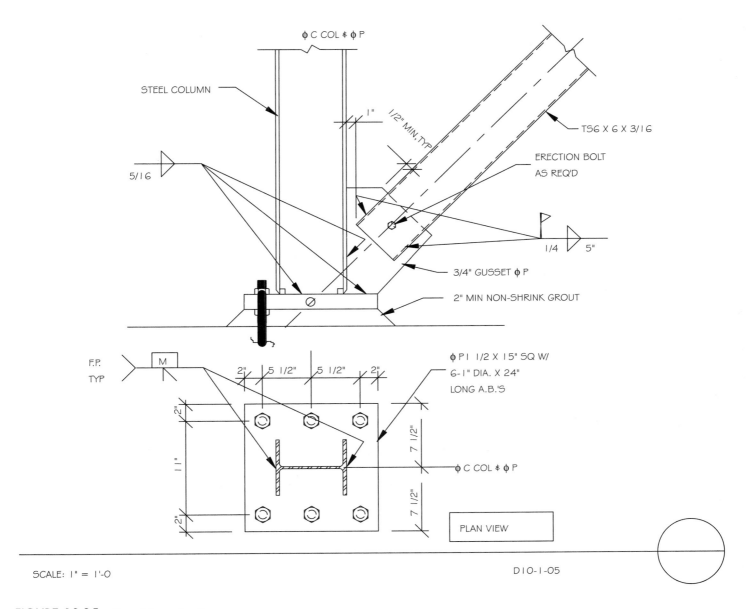

φ C COL ⊈ φ P

STEEL COLUMN

1"

1/2" MIN. TYP

TS6 X 6 X 3/16

ERECTION BOLT
AS REQ'D

5/16

1/4   5"

3/4" GUSSET φ P

2" MIN NON-SHRINK GROUT

F.P.
TYP

M

2"   5 1/2"   5 1/2"   2"

φ PI 1/2 X 15" SQ W/
6-1" DIA. X 24"
LONG A.B.'S

2"

11"

2"

7 1/2"

7 1/2"

φ C COL ⊈ φ P

PLAN VIEW

SCALE: 1" = 1'-0

D10-1-05

FIGURE 12-35 ■ Braced frame detail.

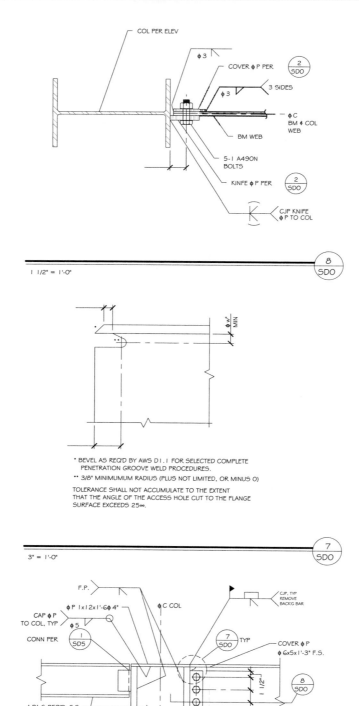

FIGURE 12-36 ■ Moment connection detail. The connection is carefully designed to be ductile. If it fails under extreme conditions, it is designed to give way gradually, not suddenly.

FIGURE 12-37 ■ Vertical and lateral forces acting on a moment frame. *Courtesy Ginger Smith.*

CHAPTER

12 *Structural Considerations Affecting Design*

CHAPTER QUIZ

Answer the questions on a separate sheet of paper. Print the chapter title question number and a short complete statement for each question. If math is required to answer a question, show the work.

Question 12-1 What type of load does a desk produce on a floor?

Question 12-2 What is the wind speed in Detroit?

Question 12-3 What live load is produced by the occupants of a restaurant?

Question 12-4 How much does a 4 × 12 DFL beam weigh per linear foot?

Question 12-5 How much weight is transferred to the top of the footing at the right wall in Figure 12-12?

Question 12-6 What is the actual psf transferred to the soil for the left wall of the structure examined in this chapter?

Question 12-7 What is the significance of triangles in braced frames?

Question 12-8 What causes seismic forces? In what direction do they try to move a structure?

Question 12-9 What does a hold-down do?

Question 12-10 How can 640 lbs per foot be resisted if 15/32" Structural 1 plywood is used for a roof diaphragm?

Question 12-11 What is the difference between a case 1 and a case 5 roof diaphragm?

Question 12-12 What is an importance factor?

Question 12-13 What is a drag?

Question 12-14 What is the importance of a chord?

CHAPTER 13

Project Manuals
and Written Specifications

Construction documents consist of the data needed to build a structure and are used to clearly convey the intentions of the owner, the architectural team, and each consultant to be involved with the project. Written specifications provide a method of supplementing the working drawings regarding the quality of materials to be used and labor to be supplied. The drafter needs to remember that both the drawings and the written specifications are legal documents that must be very carefully researched and prepared. Both may someday be used in a court of law as a standard for quality for a specific structure.

The construction drawings and the written specifications must be compatible so that the project can be accurately bid and built. They must be clear in order to avoid misinterpretation. The drawings visually define the relationships between materials, products, and systems within the structure by showing the location and size of each element. Drawings should show the following:

- Location of materials, fixtures, and equipment

- Sizes, quantities, and the physical relationship

- Dimensions to locate the size and location of each material and component

- The size, type, finish, and hardware associated with rooms, windows, and doors

Written specifications express the requirements in words. Specifications describe the following:

- The quality and type of material

- Required gauge, size, or capacity of material and equipment

- The quality of workmanship

- Methods of fabrication

- Methods of installation

- Test and code requirements

Specifications provide information regarding the quality of materials and workmanship, methods of installation, the desired performance to be achieved at completion, and how performance is to be measured. The drafter must understand the role of each type of specification to fully understand notations that are placed on the drawings. This would include knowledge of placement, formats, and types of specifications.

SPECIFICATION PLACEMENT

The specifications for a project are either found in a project manual or placed on one or more sheets of the construction drawing. The AIA recommends that the documentation for every project contain a project manual. In order to reduce cost, many owners request the architect to substitute less detailed specifications into the drawings than are found in a project manual. Projects that are funded by federal, state, or municipal funds are required to have a project manual.

Specifications within the Construction Drawings

Many owners of small, privately funded projects opt for specifications that are less specific than those found in a project manual. These specs are placed on one or more sheets of the construction drawing. Many engineering firms place the written specifications for structural material within the drawings to meet building department requirements and to establish minimum standards for construction materials. An example of the drawing specifications that are part of the steel structure used throughout this text are shown in Figure 13-1.

Project Manual

The term *specs* has often been used by contractors and their subcontractors to refer to the written specifications that are used to supplement the *plans*. The terms *project manual* and *drawings* have become more representative of the contents of each aspect of the construction documents. The *project manual* contains many documents related to a specific construction project in addition to the written specifications. The manual is produced by the combined efforts of the owner, the architect, and a representative for each of the consulting engineers involved with the project. Several portions are developed under the supervision of attorneys. A project manual often contains the following separate documents that cover a specific portion of the construction process:

- Invitation to bid

- Instruction to bidders

- Bid form

- Bond form

GENERAL STRUCTURAL NOTES

CODE REQUIREMENTS:

Conform to the IBC 2000 edition, as amended by the state of Oregon.

TEMPORARY CONDITIONS:

The contractor shall be responsible for structural stability during construction. The structure shown on the drawings has been designed for stability under the final configuration only.

EXISTING CONDITIONS:

All existing conditions, dimensions and elevations are to be field verified. The contractor shall notify the architect of any significant discrepancies from that shown on the drawing.

DESIGN CRITERIA:

Design was based on the strength and deflection criteria of the 2000 IBC. In addition to the dead loads, the following loads and allowables were used for design, with live loads reduced per IBC.

_____ Roof: - 25 PSF L.L.(snow drift as shown on plans.

_____ Floors: Retail - 75 PSF L.L.

_____ Allowable soil bearing pressure: 2000, 2500 PSF (per soils report)

_____ Retaining walls: 40 PCF (equivalent fluid pressure)

_____ Wind: 80 MPH - Exposure B

_____ Earthquake design is based on the following:

_____ Z= 0.30, I= 1.0, C= 2.75, Rw= 6

Design and detailing based on criteria for SEISMIC ZONE C.

SUBMITTALS:

Shop drawings shall be submitted to the architect prior to fabrication and construction for all structural items including the following:

Concrete mix designs, concrete and masonry reinforcement, embedded steel items, structural steel, glued laminated members, and prefabricated wood joist.

If the shop drawings differ from or add to the design of the structural drawings, they shall bear the seal and signature of a structural engineer, registered in the state of Oregon. Any changes to the structural drawings are the subject of review and acceptance of the architect.

Design drawings, shop drawings, and calculations for the design and fabrication of items that are designed by others including: Prefabricated wood joist, Skylights, Window wall and all other glazing systems, and seismic restraints for mechanical, plumbing, electrical, equipment, machinery, and associated piping shall bear the seal and signature of a structural engineer, registered in the state of Oregon and shall be submitted to the architect prior to fabrication.

Calculations are to be included for all connections to the structure, considering localized effects. Design is to be based on the requirements of the 2000 IBC for the following:

Earthquake zone C   Wind zone 80 MPH, Exposure B

Field engineered details, developed by the contractor, that differ from or add to the structural drawings shall bear the seal and signature of a structural engineer registered in the state of Oregon and shall be submitted to the architect prior to construction.

INSPECTION:

Special inspection, by an approved special inspector, shall be performed for soils compaction, concrete and reinforcement placement, embedded bolt placement, expansion anchor placement, and structural masonry placement. All soil-bearing surfaces shall be inspected by the soils engineer prior to placement of reinforcing steel.

CONCRETE:

Concrete work shall conform to chapter 26 of the IBC. Concrete strengths shall be verified by standard 28-day cylinder test per ASTM C39 and shall be as follows:

FIGURE 13-1 ■ Structural specifications governing major components of the skeleton of a structure may be placed in the project manual or on the structural drawings. *Courtesy KPFF Consulting Engineers.*

Absolute Water-Cement Ratio

f'c PSI	By Weight	By Use	
	Non Air-Entrained	Air-Entrained	
3000	.58	.46	Slab-on-grade
4000	.44	.35	All uses unless noted

Higher water/cement ratios than shown above may be used if substantiated in accordance with ACI 318-89, Chapter 5. Minimum cement content per cubic yard shall be as follows:

f'c PSI	Minimum Cement
	Per Cubic yard
3000	470 lbs.
4000	550 lbs

Fly ash conforming to the IBC standard No. 26-9, Type F or Type C, may be used to replace up to 20% of the cement content provided that the mix strength is substantiated by test data. The contractor shall submit concrete mix designs, along with test data as required, a minimum of two weeks prior to placing concrete.

Water-reducing admixture conforming to ASTM C494, used in strict accordance with the manufacturer's recommendations, shall be incorporated in concrete design mixes. A high-range water-reducing (HRWR) admixture conforming too ASTM C494, Type F or G may be used in concrete mixes providing that the slump does not exceed 10". An air-entraining agent, conforming to ASTM C260, shall be used in concrete mixes for exterior horizontal surfaces exposed to weather. The amount of entrained air shall be 5% +/- 1% by volume.

Sleeves, openings, conduit and other embedded items not shown on the structural drawings shall be approved by the structural engineer before pouring. Conduits embedded in slabs shall not be larger in outside dimension than 1/3 of the thickness of the slab and shall be spaced not closer than three diameters on center. Provide 3/4" chamfers on all exposed concrete edges unless noted otherwise.

REINFORCING STEEL:

Reinforcing steel shall conform to ASTM A615, including S1, grade 60, for deformed bars and ASTM A185 for smooth welded wire fabric (WWF), unless otherwise noted. Reinforcing steel to be welded shall conform to ASTM A706. Reinforcing steel shall be securely tied in place with #16 annealed iron wire. Bars in slabs shall be supported on well-cured concrete blocks or approved metal chairs as specified by the CRSI manual of standard practice, MSP-1. Reinforcing steel shall be detailed in accordance with the "ACI Manual of Standard Practice For Detailing Reinforced Concrete Structures," ACI 315. Lap all reinforcing bars at splices 36 diameters, with a minimum lap of 18", except as noted.

Reinforcing steel shall have protection as follows:

Use	Cover
Column bars:	1-1/2"(to stirrups or ties)
Wall bars: interior faces	3/4"
Exposed to earth	-1/2" (#5 and smaller)
or weather	2" (#6 and larger)
Footing bars	3"

REINFORCED BRICK MASONRY:_

Hollow clay masonry units shall comply with ASTM C 652, grade SW. Assemblies shall have a minimum compressive strength of f'm = 2500 PSI as verified by Prism Tests. Masonry walls shall be reinforced as shown on the plans and details, and if not shown shall be as noted under "Masonry Reinforcing Steel."

CONCRETE MASONRY:

Concrete masonry units shall comply with ASTM C90, grade

N-1. Assemblies shall have a minimum compressive strength of f'm = 1500 PSI as verified by Prism Tests before and during construction. Concrete masonry walls shall be reinforced as shown on the plans and details and if not shown shall be as noted under "Masonry Reinforcing Steel."

FIGURE 13-1 ■ *continued*

MORTAR:

Mortar shall be type S with a minimum compressive strength at 28 days of 1800 PSI and shall conform to IBC section 2402 and 2403.

MASONRY GROUT:

Grout shall have a minimum compressive strength at 28 days of 2000 PSI and shall conform to IBC sections 2402 and 2403. Grout shall consist of a mixture of cemtentitious materials and aggregate to which sufficient water has been added to cause the mixture to flow without segregation of constituents. All cells containing vertical bars and all bond beams shall be filled with grout.

MASONRY REINFORCING STEEL:

Reinforcing shall conform to IBC section 2402(b)10, deformed bars shall be grade 60, and shall be securely placed in accordance with IBC section 2404(c). Unless otherwise noted on the plans the minimum wall reinforcement shall be as follows

Wall Thickness	Vertical Bars	Horizontal bars (in bond beams)	
		Running Bond	Stack Bond
8"	#6 @ 48" o.c.	(2)#4 @ 48" o.c.	(2)#5 @ 48" o.c.
12"	#7 @ 48" o.c.	(2)#5 @ 48" o.c.	(2)#6 @ 48" o.c.

Bond beams with (2) #5 bars horizontally shall be provided at all floor and roof lines and at the tops of walls. Provide a bond beam with (2) #5 bars horizontally above and below all openings, and extend these bars 2'-0" past the opening at each side. Provide one bar, matching the vertical bar size, for the full height of the wall at each side of openings, wall ends and intersections. Dowels to masonry walls shall be embedded a minimum of 1'-6"or hooked into supporting structure and be of the same size and spacing as the wall reinforcing. Provide corner bars to match the horizontal wall reinforcing at wall intersections. Lap all bars at splices 60 diameters, with a minimum lap of 18", except as noted.

STRUCTURAL STEEL:

Structural steel shall be ASTM A36. Tubes shall be A500 grade B (Fy =46 KSI). Pipes shall be ASTM A501 or ASTM A53, grade B. Design, fabrication and erection shall be in accordance with the "AISC Specifications For the Design, Fabrication and Erection of Structural Steel for Buildings". Welding shall conform to the AWS codes for arc and gas welding in building construction. Welds shall be made using E70XX electrodes and shall be 3/16" minimum unless otherwise noted. Welding shall be by AWS certified welders. Pre-qualified welding procedures are to be used unless AWS qualification is submitted to the architect prior to fabrication.

SWAN LUMBER:

Sawn lumber shall conform to West Coast Lumber Inspection Bureau or Western Wood Products Association grading rules. Lumber shall be the species and grade noted below:

Use	Grade	Fb (PSI) (Single Use)
Joist, Planking, Blocking, & Wall Studs. 2"-4" Thick, 5" and wider	DFL No. 2	1250

All lumber in contact with concrete or CMU is to be pressure treated unless an approved barrier is provided. Framing accessories and structural fasteners shall be manufactured by Simpson Company (or approved equal) and of the size and type shown on the drawings. Hangers not shown shall be Simpson HU of size recommended for member. All other framing nails shall be common nails and shall be of the size and number indicated on the drawing. Nailing not shown shall be as indicated on IBC table 25-Q.

FIGURE 13-1 ∎ *continued*

PLYWOOD:

Plywood panels shall conform to the requirements of "U.S. Product Standard PS 1 for Construction and Industrial Plywood" or APA PRP-108 Performance Standards. Unless noted, panels shall be APA rated sheathing, Exposure 1, of the thickness and span rating shown on the drawings.

Plywood installation is to be in conformance with APA recommendations. Allow 1/8" spacing at panel ends and edges unless otherwise recommended by the panel manufacturer.

All roof sheathing and sub-flooring shall be installed face grain perpendicular to supports except as indicated on the drawings. Roof sheathing shall be either blocked, tongue and groove, or have edges supported by plyclips. Sub-floor sheathing shall be unblocked except as indicated on drawings. Shear wall sheathing shall be blocked with 2x framing at all panel edges. Nailing not shown shall be as indicated on IBC Table 25-Q. All nails shall be common nails except use ring shank for roof sheathing.

GLUED LAMINATED MEMBERS:

Glued laminated members shall be fabricated in conformance with U.S. Product Standards PS 56, "Structural Glued Laminated Timber" and American Institute of Timber Construction AITC 117. Each member shall bear an AITC identification mark and be accompanied by an AITC Certificate of Conformance. One coat of end sealer shall be applied immediately after trimming in either shop or field. Beams shall be visually graded Western Species Industrial Grade, and of the strength indicated below:

	Combination		
Depth	**Symbol**	**Species**	**Use**
Less than 16"	20F-V3	DF/DF	(simple span)
Less than 16"	20F-V7	DF/DF	(cont. or cantilever)
Greater than 16"	24F-V4	DF/DF	(simple span)
Greater than 16"	24F-V8	DF/DF	(cont. or cantilever)

Glulam hangers not shown shall be Simpson EG.

Adhesive shall be wet-use exterior waterproof glue.

PREMANUFACTURED WOOD JOIST:

Premanufactured wood joist shall be the size and type shown on the drawings, manufactured by the Trus-Joist Corporation, or an approved equal. Provide bridging in conformance with the manufacturer's recommendations. Joist and bridging shall be capable of resisting the wind uplift noted on the drawings. The joist manufacturer shall visit the job site as required and verify in writing to the architect the proper installation of joist. Premanufactured wood joist alternates will be considered, provided the alternate is compatible with the load capacity, dimensional and fire rating requirements of the project, and is ICBO approved.

FIGURE 13-1 ■ *continued*

■ Form of agreement

■ General conditions of the contract for construction

■ Supplementary conditions

■ Specifications

■ Addenda

Each document is a legal contract in itself, which as a whole constitutes the project manual. Whenever possible, the use of AIA forms for each document should be used because they are well known to the construction industry. These documents have been tested and interpreted in the courts and are widely understood by the construction industry.

Invitation to Bid

An *invitation to bid* is a summary of the bidding and construction procedures for a project. It is usually about one page in length and is used to advise potential bidders about the existence of a project. If a contractor has already been secured for the project prior to bidding, an invitation is not a part of the project manual. On privately funded projects, based on the request of the owner, the architect may mail the invitation to a few selected contractors known to have a proven track record of success with this type of project. On publicly funded projects, laws require the invitation to be published in a prescribed form in a newspaper.

Instruction to Bidders

The purpose of the instruction to bidders is to tell each bidder the format to be used for the reviewing of bids. The AIA standard form provides information about the bidding process and procedure including the following:

■ Definitions

■ Bidder's representations

- Bidding documents
- Bidding procedures
- Consideration of bids
- Post-bid information
- Performance bond and payment bond
- Form of agreement between owner and contractor

Bid Forms

The purpose of a *bid form* is to provide a uniform presentation of the cost associated with the construction project. The bid form provides blank lines where the bidder is to insert the price for specific portions of the project. Using a uniform bid form limits bidders in the use of *exclusions* and *substitutions*, which often make it unclear what cost the bid actually covers.

Bond Forms

A *bond* is a legal document, which assures that the contractor will provide the goods or services represented in the contract agreement. Three types of bonds used in the construction industry include a bid, performance, and a labor and materials payment bond. A *bid bond* provides assurance that a contractor will sign the contract if the firm is awarded the bid. A *performance bond* is posted by the winning bidder as a guarantee that the firm will complete the job and not become defunct during the process. The *labor and material payment bond* is posted by the winning bidder as a guarantee that all bills for material and labor used for this project will be paid by the contractor.

Form of Agreement

The *Form of Agreement* is a legal contract that includes the who, what, how much, and when of the project. These aspects of the project are usually represented in the following five elements of the agreement:

- Identification of the parties involved
- Statement of the work to be performed
- Statement of the consideration
- Time of performance
- Signatures of the parties involved

General Conditions of the Contract for Construction

This document specifies the relationship of each party that signed the contracts and how they will be administered. The AIA format of general conditions includes fourteen categories overseeing the contract. These include the following:

- General provisions
- Owner
- Contractor

- Administration of the contract
- Subcontractors
- Construction by owner or by separate contractors
- Changes in work
- Time
- Payments and completion
- Protection of persons and property
- Insurance and bonds
- Uncovering and corrections of work
- Miscellaneous provisions
- Termination or suspension of the contract

Supplementary Conditions

The general conditions are used to address elements that are common to all construction. The supplementary or special conditions are used to address the elements that are only found within a specific project. The *supplementary conditions* are written to address portions of the project that are special to that specific project which do not fit into the preceding section of the contract.

Specifications

Although only one of many documents, the *written specifications* comprise the bulk of a project manual; it is this portion of the project manual that affects the drafter. Depending on the size of the project and the firm, an experienced drafter may be required to alter or update specifications. Every drafter involved with the project needs to be aware of the contents of the written specifications so that the drawings and the *specs* can be coordinated. The following sections of this chapter address this aspect of the project manual.

Addenda

Although not always used, the *addenda* is used to amend the contract during the bidding process but prior to the awarding of the contract. Changes issued after the contract has been awarded are referred to as a *change notice* or *change order* and are not considered in this chapter.

CSI MASTERFORMAT

Several formats have been used to write specifications during the last fifty years. An alphabetical listing of materials, listings based on the order of use, and hundreds of personal styles were common prior to the development of the standard for today's construction industry. First published in 1963, most construction specifications are now written according to the *MasterFormat* system published by the Construction Specification Institute (CSI) in the United States, and by Construction Specifications Canada (CSC) in Canada. This system consists of seventeen major divisions, which can be used for production, distribution, filing, and retrieval of construction documents. Each major division of the *MasterFormat* is related to a major grouping of the construction process. The major listings for *MasterFormat* include the following:

Division 00   Bidding and contract requirements
Division 01   General requirements
Division 02   Sitework
Division 03   Concrete
Division 04   Masonry
Division 05   Metals
Division 06   Wood and Plastics
Division 07   Thermal and moisture protection
Division 08   Doors and windows
Division 09   Finishes
Division 10   Specialties
Division 11   Equipment
Division 12   Furnishings
Division 13   Special construction
Division 14   Conveying system
Division 15   Mechanical
Division 16   Electrical

These same divisions are used in the *Sweet's Catalogs* as well as most other vendor material. Each division is in turn divided into numbered sections represented by five-digit numbers and is referred to as the *Uniform Construction Index*. Figure 13-2 shows a listing of the sections of each of the seventeen divisions. Notice that numbers are placed in order, but many unassigned numbers are available for individual projects. Each section is further divided into three subsections that deal with general, materials, and execution.

Part One

Part One, *General* deals with the scope of a section by describing work, related definitions, quality control, submittals, shop drawings, guarantees, and warranties. Common components of the General Section include the following:

1.01   Summary
1.02   References
1.03   Definitions
1.04   System Descriptions
1.05   Submittals
1.06   Quality Assurance
1.07   Delivery, Storage, and Handling
1.08   Project and Conditions
1.09   Sequencing and Scheduling
1.10   Warranty
1.11   Maintenance

Part Two

The Materials section describes materials, products, and equipment that is to be used. Common components of the Materials Section include the following:

2.01   Manufacturers
2.02   Materials
2.03   Manufactured Units
2.04   Equipment
2.05   Components
2.06   Accessories
2.07   Mixes
2.08   Fabrication
2.09   Source Quality Control

Part Three

Execution describes the method that is to be used to install materials or products specified in Part Two, and how work is to be performed. Common components of the Execution Section include the following:

3.01   Examination
3.02   Preparation
3.03   Erection/Installation/Application
3.04   Field Quality Control
3.05   Adjusting
3.06   Cleaning
3.07   Demonstration
3.08   Protection
3.09   Schedules

Figure 13-3 shows the three-part specification for Exterior Insulation and Finish System (EIFS) for the structure shown in Chapters 15 and 16. Keep in mind that the following are the specifications for only one component of the structure. Each component of the structure is listed in the project manual using the three-part system. Components are placed within the manual according to their CSI number. Notice that at the start of each page of the specifications, the job number, project name, CSI section number, the component name and the page number of the section are listed.

TYPES OF SPECIFICATIONS

Regardless of where the specifications are contained, six common methods are used to present technical specifications. These methods are cash allowance, descriptive, open, performance, proprietary, and reference specifications. A construction document can contain each type of specification and should not be limited to one specific type.

DIVISION 0 - BIDDING AND CONTRACT REQUIREMENTS

00010	PRE-BID INFORMATION
00100	INSTRUCTIONS TO BIDDERS
00200	INFORMATION AVAILABLE TO BIDDERS
00300	BID/TENDER FORMS
00400	SUPPLEMENTS TO BID/TENDER FORMS
00500	AGREEMENT FORMS
00600	BONDS AND CERTIFICATES
00700	GENERAL CONDITIONS OF THE CONTRACT
00800	SUPPLEMENTARY CONDITIONS
00950	DRAWINGS INDEX
00900	ADDENDA AND MODIFICATIONS

SPECIFICATIONS—DIVISIONS 1-16

DIVISION 1 - GENERAL REQUIREMENTS

01010	SUMMARY OF WORK
01020	ALLOWANCES
01030	SPECIAL PROJECT PROCEDURES
01040	COORDINATION
01050	FIELD ENGINEERING
01060	REGULATORY REQUIREMENTS
01070	ABBREVIATIONS AND SYMBOLS
01080	IDENTIFICATION SYSTEMS
01100	ALTERNATES/ALTERNATIVES
01150	MEASUREMENT AND PAYMENT
01200	PROJECT MEETINGS
01300	SUBMITTALS
01400	QUALITY CONTROL
01500	CONSTRUCTION FACILITIES AND TEMPORARY CONTROLS
01600	MATERIAL AND EQUIPMENT
01650	STARING OF SYSTEMS
01660	TESTING, ADJUSTING, AND BALANCING OF SYSTEMS
01700	CONTRACT CLOSEOUT
01800	MAINTENANCE MATERIALS

DIVISION 2 - SITE WORK

02010	SUBSURFACE INVESTIGATION
02050	DEMOLITION
02100	SITE PREPARATION
02150	UNDERPINNING
02200	EARTHWORK
02300	TUNNELLING
02350	PILES, CAISSONS AND COFFERDAMS
02400	DRAINAGE
02440	SITE IMPROVEMENTS
02480	LANDSCAPING
02500	PAVING AND SURFACING
02580	BRIDGES
02590	PONDS AND RESERVOIRS
02600	PIPED UTILITY MATERIALS AND METHODS
02700	PIPED UTILITIES
02800	POWER AND COMMUNICATION UTILITIES
02850	RAILROAD WORK
02880	MARINE WORK

DIVISION 3 - CONCRETE

03010	CONCRETE MATERIALS
02050	CONCRETING PROCEDURES
03100	CONCRETE FORMWORK
03150	FORMS
03180	FORM TIES AND ACCESSORIES
03200	CONCRETE REINFORCEMENT
03250	CONCRETE ACCESSORIES
03300	CAST-IN-PLACE CONCRETE
03350	SPECIAL CONCRETE FINISHES
03360	SPECIALLY PLACED CONCRETE
03370	CONCRETE CURING
03400	PRECAST CONCRETE
03500	CEMENTITIOUS DECKS
03600	GROUT
03700	CONCRETE RESTORATION AND CLEANING

DIVISION 4 - MASONRY

04050	MASONRY PROCEDURES
04100	MORTAR
04150	MASONRY ACCESSORIES
04200	UNIT MASONRY
04400	STONE
04500	MASONRY RESTORATION AND CLEANING
04550	REFRACTORIES
04600	CORROSION RESISTANCE MASONRY

DIVISION 5 - METALS

05010	METAL MATERIALS AND METHODS
05050	METAL FASTENING
05100	STRUCTURAL METAL FRAMING
05200	METAL JOISTS
05300	METAL DECKING
05400	COLD-FORMED METAL FRAMING
05500	METAL FABRICATIONS
05700	ORNAMENTAL METAL
05800	EXPANSION CONTROL
05900	METAL FINISHES

DIVISION 6 - WOOD AND PLASTICS

06050	FASTENERS AND SUPPORTS
06100	ROUGH CARPENTRY
06130	HEAVY TIMBER CONSTRUCTION
06150	WOOD-METAL SYSTEMS
06170	PREFABRICATED STRUCTURAL WOOD
06200	FINISH CARPENTRY
06300	WOOD TREATMENT
06400	ARCHITECTURAL WOODWORK
06500	PREFABRICATED STRUCTURAL PLASTICS
06600	PLASTIC FABRICATIONS

DIVISION 7 - THERMAL AND MOISTURE PROTECTION

07100	WATERPROOFING
07150	DAMPPROOFING
07200	INSULATION
07250	FIREPROOFING
07300	SHINGLES AND ROOFING TILES
07400	PREFORMED ROOFING AND SIDING
07500	MEMBRANE ROOFING
07570	TRAFFIC TOPPING
07600	FLASHING AND SHEET METAL
07800	ROOF ACCESSORIES
07900	SEALANTS

DIVISION 8 - DOORS AND WINDOWS

08100	METAL DOORS AND FRAMES
08200	WOOD AND PLASTIC DOORS
08250	DOOR OPENING ASSEMBLES
08300	SPECIAL DOORS
08400	ENTRANCES AND STOREFRONTS
08500	METAL WINDOWS
08600	WOOD AND PLASTIC WINDOWS
08650	SPECIAL WINDOWS
08700	HARDWARE
08800	GLAZING
08900	GLAZED CURTAIN WALLS

DIVISION 9 - FINISHES

09100	METAL SUPPORT SYSTEMS
09200	LATH AND PLASTER
09230	AGGREGATE COATINGS
09250	GYPSUM WALLBOARD
09300	TILE
09400	TERRAZZO
09500	ACOUSTICAL TREATMENT
09550	WOOD FLOORING
09600	STONE AND BRICK FLOORING
09650	RESILIENT FLOORING
09680	CARPETING
09700	SPECIAL FLOORING

FIGURE 13-2 ■ Major subsections within the seventeen sections of the CSI *MasterFormat* system. Reproduced from *MasterFormat*. *Courtesy Construction Specifications Institute.*

09760	FLOOR TREATMENT	
09800	SPECIAL COATINGS	
09900	PAINTING	
09950	WALL COVERING	

DIVISION 10 - SPECIALTIES

10100	CHALKBOARDS AND TACKBOARDS
10150	COMPARTMENT AND CUBICLES
10200	LOUVERS AND VENTS
10240	GRILLES AND SCREENS
10250	SERVICE WALL SYSTEMS
10260	WALL AND CORNER GUARDS
10270	ACCESS FLOORING
10280	SPECIALTY MODULES
10290	PEST CONTROL
10300	FIREPLACES AND STOVES
10340	PREFABRICATED STEEPLES, SPIRES, AND CUPOLAS
10350	FLAGPOLES
10400	IDENTIFYING DEVICES
10450	PEDESTRIAN CONTROL DEVICES
10500	LOCKERS
10520	FIRE EXTINGUISHERS, CABINETS, AND ACCESSORIES
10530	PROTECTIVE COVERS
10550	POSTAL SPECIALTIES
10600	PARTITIONS
10650	SCALES
10670	STORAGE SHELVING
10700	EXTERIOR SUN CONTROL DEVICES
10750	TELEPHONE ENCLOSURES
10800	TOILET AND BATH ACCESSORIES
10900	WARDROBE SPECIALTIES

DIVISION 11 - EQUIPMENT

11010	MAINTENANCE EQUIPMENT
11020	SECURITY AND VAULT EQUIPMENT
11030	CHECKROOM EQUIPMENT
11040	ECCLESIASTICAL EQUIPMENT
11050	LIBRARY EQUIPMENT
11060	THEATER AND STAGE EQUIPMENT
11070	MUSICAL EQUIPMENT
11080	REGISTRATION EQUIPMENT
11100	MERCANTILE EQUIPMENT
11110	COMMERCIAL LAUNDRY AND DRY CLEANING EQUIPMENT
11120	VENDING EQUIPMENT
11130	AUDIO-VISUAL EQUIPMENT
11140	SERVICE STATION EQUIPMENT
11150	PARKING EQUIPMENT
11160	LOADING DOCK EQUIPMENT
11170	WASTE HANDLING EQUIPMENT
11190	DETENTION EQUIPMENT
11200	WATER SUPPLY AND TREATMENT EQUIPMENT
11300	FLUID WASTE DISPOSAL AND TREATMENT EQUIPMENT
11400	FLOOD SERVICE EQUIPMENT
11450	RESIDENTIAL EQUIPMENT
11460	UNIT KITCHENS
11470	DARKROOM EQUIPMENT
11480	ATHLETIC, RECREATIONAL, AND THERAPEUTIC EQUIPMENT
11500	INDUSTRIAL AND PROCESS EQUIPMENT
11600	LABORATORY EQUIPMENT
11650	PLANETARIUM AND OBSERVATORY EQUIPMENT
11700	MEDICAL EQUIPMENT
11780	MORTUARY EQUIPMENT
11800	TELECOMMUNICATION EQUIPMENT
11850	NAVIGATION EQUIPMENT

DIVISION 12 - FURNISHINGS

12100	ARTWORK
12300	MANUFACTURED CABINETS AND CASEWORK
12500	WINDOW TREATMENT
12550	FABRICS
12600	FURNITURE AND ACCESSORIES
12670	RUGS AND MATS
12700	MULTIPLE SEATING
12800	INTERIOR PLANTS AND PLANTINGS

DIVISION 13 - SPECIAL CONSTRUCTION

13010	AIR SUPPORTED STRUCTURES
13020	INTEGRATED ASSEMBLIES
13030	AUDIOMETRIC ROMS
13040	CLEAN ROOMS
13050	HYPERBARIC ROOMS
13060	INSULATED ROOMS
13070	INTEGRATED CEILINGS
13080	SOUND, VIBRATION, AND SEISMIC CONTROL
13090	RADIATION PROTECTION
13100	NUCLEAR REACTORS
13120	PRE-ENGINEERED STRUCTURES
13130	SPECIAL PURPOSE ROOMS AND BUILDINGS
13140	VAULTS
13150	POOLS
13160	ICE RINKS
13170	KENNELS AND ANIMAL SHELTERS
13200	SEISMOGRAPHIC INSTRUMENTATION
13210	STRESS RECORDING INSTRUMENTATION
13220	SOLAR AND WIND INSTRUMENTATION
13410	LIQUID AND GAS STORAGE TANKS
13510	RESTORATION OF UNDERGROUND PIPELINES
13520	FILTER UNDERDRAINS AND MEDIA
13530	DIGESTION TANK COVERS AND APPURTENANCES
13540	OXYGENATION SYSTEMS
13550	THERMAL SLUDGE CONDITIONING SYSTEMS
13560	SITE CONSTRUCTED INCINERATORS
13600	UTILITY CONTROL SYSTEMS
13700	INDUSTRIAL AND PROCESS CONTROL SYSTEMS
13800	OIL AND GAS REFINING INSTALLATIONS AND CONTROL SYSTEMS
13900	TRANSPORTATION INSTRUMENTATION
13940	BUILDING AUTOMATION SYSTEMS
13970	FIRE SUPPRESSION AND SUPERVISORY SYSTEMS
13980	SOLAR ENERGY SYSTEMS
13990	WIND ENERGY SYSTEMS

DIVISION 14 - CONVEYING SYSTEMS

14100	DUMBWAITERS
14200	ELEVATORS
14300	HOISTS AND CRANES
14400	LIFTS
14500	MATERIAL HANDLING SYSTEMS
14600	TURNTABLES
14700	MOVING STAIRS AND WALKS
14800	POWERED SCAFFOLDING
14900	TRANSPORTATION SYSTEMS

DIVISION 15 - MECHANICAL

15050	BASIC MATERIALS AND METHODS
15200	NOISE, VIBRATION, AND SEISMIC CONTROL
15250	INSULATION
15300	SPECIAL PIPING SYSTEMS
15400	PLUMBING SYSTEMS
15450	PLUMBING FIXTURES AND TRIM
15500	FIRE PROTECTION
15600	POWER OR HEAT GENERATION
15650	REFRIGERATION
15700	LIQUID HEAT TRANSFER
15800	AIR DISTRIBUTION
15900	CONTROLS AND INSTRUMENTATION

DIVISION 16 - ELECTRICAL

16050	BASIC MATERIALS AND METHODS
16200	POWER GENERATION
16300	POWER TRANSMISSION
16400	SERVICE AND DISTRIBUTION
16500	LIGHTING
16600	SPECIAL SYSTEMS
16700	COMMUNICATIONS
16850	HEATING AND COOLING
16900	CONTROLS AND INSTRUMENTATION

FIGURE 13-2 ■ *continued*

EXTERIOR INSULATION AND FINISH SYSTEM

Part 1 - General

1.1 WORK INCLUDED

A. Provide field applied exterior insulation and finish system as indicated on Drawings and specified herein.

1.2 SYSTEM DESCRIPTION

A. Structural Requirements: Deflection of exterior wall substrate shall not exceed L/240 or L/360.

B. Substrate Requirements: Substrates shall be as indicated on the Drawings.

1.3 QUALITY ASSURANCE

A. Manufacturer's Qualifications:

1. Manufacturer shall have successful performance history over at least five (5) years in the local geographic area.

2. Manufacturer shall have established on-going design and specification assistance to Architects and Engineers.

3. Manufacturer shall have established on-going contractor / applicator training programs for at least five (5) years in local geographic area.

4. Manufacturer shall have established warranty program and a five (5) warranty on this Project.

5. Manufacturer shall have a full-scale fire test reports and documentation of IBC compliance.

6. Manufacturer shall have resident Field Technical Service personnel.

7. Manufacturer shall attend job site pre-application conference and provide inspection of work in progress and completed work.

8. Manufacturer shall have local inventory of products.

B. Applicator's Qualifications:

1. Applicator shall be licensed, bonded and insured.

2. Applicator shall have successful performance history with Exterior Insulation and Finish Systems over at least five (5) years in the local geographic area.

3. Applicator shall be trained and approved by manufacturer for at least five (5) years.

4. Applicator shall have an established on-going training program for workmen, including manufacturer's training.

5. Applicators shall attend Pre-application conference and provide periodic inspection of work in progress by principal/officer of firm. Job field foreman shall attend pre-application conference.

6. Applicator shall provide three (3) year minimum warranty workmanship.

7. Applicator shall provide acceptable 320 square feet, or approved mock up of finished system for Architect's review and acceptance. Accepted mocks-ups may be incorporated into the Project.

FIGURE 13-3 ■ Written specifications that are several hundred pages long are a major component of a project manual. The specification contained on the following pages is the specification for *one* component. Each fixture, unit, and system in the structure must be specified. *Courtesy Ron Lee, Architects Barrentine, Bates & Lee, A.I.A.*

8712   CCC Wilsonville   Section 07240
Page 2

EXTERIOR INSULATION AND FINISH SYSTEM

Pre-Application Conference:

Approximately five (5) days prior to scheduled start of installation, schedule and participate in an on-site review of Drawings, Specifications and job-site conditions with the following participants:

- Owner's representative.
- Architect.
- General Contractor.
- Subcontractor Responsible for:
    Substrates.
    Exterior Insulation and Finishing System.
    Sealants.
    Flashings.
- Manufacturer's Representative.

D. Supervision/Inspection:

1. Provide continually daily supervision of working crew.
2. Secure daily inspection by General Contractor.
3. Provide periodic inspection by a principal/officer of the installing firm.
4. Secure periodic inspection by the manufacturer's representative, including inspection of completed work.

E. Substrate Protection:

1. Protect exterior plywood and other "moisture sensitive" substrate materials from exposure to adverse weather; replace all substrate materials evidencing adverse effects of weathering.
2. Protect unfinished areas of installed Exterior Insulation and Finish System from exposure to adverse weather; ensure that no water is trapped behind or within the Exterior Insulation and Finish System.

1.4 SUBMITTALS

A Office Sample: Submit one sample of the exterior synthetic plaster finish. Architect will review sample for color and texture only.

1.5 DELIVERY, STORAGE, AND HANDLING

Deliver materials in original unopened packages. Deliver and maintain all materials free of damage and contamination.

B. Store materials in dry areas, protected from moisture and temperatures less than 40 degrees F. and above 110 degrees F. Protect materials from direct sunlight.

C. Handle, mix and apply materials as recommended by manufacturer.

1.6 SITE CONDITIONS

Existing Substrate Conditions: Planar Irregularities not greater then 1/4". Structurally stable substrate free of high temperatures, releasing agents and residue.

FIGURE 13-3 ■ *continued*

8712 CCC Wilsonville Section 07240

Page 3

EXTERIOR INSULATION AND FINISH SYSTEM

B. Application Conditions.

1. Apply all materials when ambient temperatures are 40 degrees Fahrenheit, or above, and rising, but under 110 degrees Fahrenheit.

2. Apply all material when rain forecast is ten percent or less for the next 48 hours.

3. Apply all materials when relative humidity is below 75 percent and expected to remain so, or drop, over the next 48 hours. Application up to 90% RH may take place if wind velocity exceeds 7-1/2 mph.

4. Apply all materials to substrate that is clean, dry and otherwise suitable for covering.

5. Moisture content of sheathing at time of insulation board installation must be 20% surface dryness or less.

1.7 WARRANTIES

A. Provide manufacturer's five (5) year warranty.

B. Provide installing firm's three (3) year warranty on workmanship.

C. Provide General Contractor's three (3) wear warranty covering substrate, sealants, and EIFS installation.

D. Receipt of Warranties is a prerequisite to final payment.

PART 2 - PRODUCTS

2.1 ACCEPTABLE MANUFACTURERS

A. Synthetic Plaster:

1. Dryvit Systems, Inc.

2. Senergy.

3. ThoroWall A by Thoro System Products.

4. STO.

5. Or accepted substitute.

B. Insulation Board:

1. Western Insul-foam Corp.

2. Northwest ESP, Inc.

3. Or accepted suitable.

2.2 MATERIALS

A. **Adhesive Primer:** Manufacturer's standard synthetic adhesive appropriate to substrate encountered.

B. **Portland Cement:** ASTM C-150, Type 1.

C. **Insulation Board:** Expanded Polystyrene, FS HH-I-524C, Type I, class A. Flame spread less than 25. Average density 1.0 pounds per cubic foot. K-value of 0.26 per inch; R value of 3.85 per inch. Provide thickness indicated on Drawings.

FIGURE 13-3 ■ *continued*

8712   CCC Wilsonville   Section 07240
Page 4

EXTERIOR INSULATION AND FINISH SYSTEM

D. **Fasteners:** As required and recommended by the manufacturer.

E. **Mesh Reinforcement:** As recommended by the EIFS manufacturer. ("Panzer" mesh is to be the "heavy" mesh).

    1. At Non-Impact areas: "Standard/Detail."

    2. At Light to moderate Impact Areas: "Intermediate."

    3. At Moderate to high Impact Areas: "Panzer"; provide Hi-Standard II mesh over "Panzer" at impact areas.

    4. At Exposed outside corners in Moderate to High Impact Areas:"Panzer Corner Mesh."

F. **Synthetic Plaster:** Manufacturer's standard Quartztone finish mix.Two (2) custom colors as selected by Architect.

    1. Color #1: <u>Match Miler Paint Company #5363M CountryTwill</u>.

    2. Color #2: <u>Match Miler Paint Company # 5371W Ancestral</u>.

2.3 MIXES

A. **Base Coat:** Mix one part by weight fresh Portland Cement to one part Adhesive Primer.

B. **Synthetic Plaster:** Mix factory prepared materials with high speed mixer until uniform consistency. Clean water may be added to improve workability in accordance with the manufacturer's recommendations.

PART 3 - EXECUTION

SITE EXAMINATION

A. Following the Pre-Application Conference, but prior to actual start of work, examine site and surfaces to verify suitability for commencing application. Verify sheathing and substrate surface dryness.

B. Notify General Contractor and Architect of any unsuitable conditions.

3.2 PREPARATION

A. Insulation Board: Air dry insulation boards for six weeks prior to installation.

B. Precut insulation Board: As required to fit substrate.

C. Verify that the surfaces to be covered are clean, dry, and otherwise acceptable.

D. Place protective coverings over adjoining surfaces and elements.

3.3 GENERAL APPLICATION REQUIREMENTS

A. Apply all materials in strict accordance with manufacturer's recommendations; maintain printed recommendations on site throughout duration of work.

B. Apply all materials in compliance with manufacturer's conditions precedent to warranty.

C. Provide uniform 11/2" wide joints for backer rods and sealant where indicated on Drawings or as determined at the Pre-Application Conference.

D. Consider quantities of wet goods recommended by manufacturer as minimum quantities; application of lesser amounts or amounts thicker than 1/4" is not acceptable.

FIGURE 13-3 ■ *continued*

8712   CCC Wilsonville   Section 07240

Page 5

EXTERIOR INSULATION AND FINISH SYSTEM

E. Use tools and application techniques and methods approved by the manufacturer.

F. Perform all application under favorable conditions conducive to optimum long-term performance of EIFS.

G. Perform all work in a good workmanlike manner, detailing as required.

H. Wrap, coat, and finish all edges and interruptions as recommended or detailed by manufacturer.

3.4 INSTALLATION OF INSULATION BOARD

A. Begin installation at base of vertical walls and provide temporary or permanent support and bottom edge wrap.

B. Stagger vertical joints and butt together without gaps.

C. By trowel or extrusion, apply 2" wide by 1/4" to 3/8" thick ribbon of adhesive primer cement mix to perimeter and at 8" on center to field of each insulation board.

D. Install board with pressure over entire surface.

E. Make joints between insulation boards tight and flush.

3.5 INSTALLATION OF REINFORCING FABRIC

A. With stainless steel trowel apply base coat mix to installed insulation board at 1/16" thickness or sufficient to completely embed mesh. Water may be added to adhesive primer cement mix to improve workability.

B. Install corner bead and edge stops as required.

C. Immediately place reinforcing fabric against applied base coating and trowel from center to edges until reinforcing fabric is embedded in coating.

D. Install reinforcing fabric continuous at corners and lap edges 2 1/2" at fabric edges.

E. Apply additional base coat mix where required to fully embed the reinforcing fabric.

3.6 APPLICATION OF SYNTHETIC PLASTER

A. With stainless steel trowel, apply coat of synthetic plaster on reinforced base coating.

B. Texture surface by rolling trowel on plaster aggregate to match reviewed office and field samples.

C. Final thickness not to exceed 1 1/2" times the diameter of largest aggregate in synthetic plaster.

D. Do not use base coat mix which has exceeded the manufacturer's recommended pot life.

3.7 ADJUSTING AND CLEANING

A. Replace finish not matching review sample.

B. Remove material drippings from the Site and clean adjacent surfaces where required.

C. Cleanup Site, leaving free of excess materials, packaging and other debris.

END OF SECTION

FIGURE 13-3 ■ *continued*

Cash Allowance Specifications

Cash allowance specifications are used when the information regarding quality or quantity has not been determined. These specifications can be found in a project manual or within the working drawings. The cash allowance temporarily replaces drawings for a specified portion of work and instructs all bidders to plan for a specified amount of money to cover that work. This money is then dispersed under the architect's direction when the work is completed. Finish items such as hardware, carpet, and lighting fixtures are often handled by this type of specification. Although exact quantities can be determined from the drawings, the owner may wish to determine quality at a later date in the building process.

Cash allowances are also used because the owner may not know who will occupy the structure. For example, a structure can be built for the intended use of medical professionals. Because there is such diversity in the profession, individual occupants will have specific needs. These needs can best be met by cash allowances. When cash allowances are given, information should include the dollar amount of the allowance, installation methods, the method of measuring cost to be applied against the allowance. When work covered by the allowance is complete, savings can be credited to the contractor or owner. If allowance costs are exceeded, the contractor is usually entitled to charge additional fees. An example of a cash allowance specification is shown in Figure 13-4.

Descriptive Specifications

Descriptive specifications are the most detailed type of specification and are used when the architect assumes total responsibility for the performance of a system. Descriptive specifications are exclusively used within a project manual and describe the method of assembly, physical properties, installation, arrangement, or any other category of information need to fulfill the manufacturer's requirements, or owner's demands for a specific system. The specification for EFIS in Figure 13-3 is an example of a descriptive specification.

Performance Specifications

Specifications that define the desired results of a product or system are referred to as *performance specifications*. Using this type of specification, the exact makeup of individual components is not described, allowing suppliers maximum input. Performance specifications describe the conditions or parameters a product or system must function in, as well as the method of measurement to be used

to judge compliance. An example of a performance specification can be seen in Figure 13-5.

Proprietary Specifications

Specifications that call for materials by their trade name and model name are referred to as *proprietary specifications*. Individual products, specific pieces of equipment, or entire systems can be referenced using this method. Material provided by the manufacturer's literature forms the basis for the specification that can be found in both project manuals and drawing specifications.

Proprietary specifications may be written in either a closed or open form. *Closed* or *sole source* specifications require that only one specific manufacturer be allowed to supply the desired product. Closed specifications are often used on renovations where a specific product is required to match existing conditions. An example of a proprietary specification can be seen in the specification 3.4 B for finish hardware in Figure 13-6.

SECTION 07900- CAULKING
PART 1- GENERAL

1.01 SUMMARY
A. The purpose of the caulking work is to provide a positive barrier against penetration of air and moisture at the joints between items where caulking is essential to continued integrity of the barrier.
B. Comply with governing codes and regulations.

1.02 MATERIALS
A. Caulking material shall be properly selected to resist the type of abuse it will receive. It shall have expansion and contraction characteristics suitable to the installed conditions. Use only material that is best suited to the installation and is so recommended by the caulking material manufacturer.
B. Provide sealants in colors suitable for the location. Colors are to be selected from the manufacturer's standards.

1.03 INSTALLATION
A. Beginning work means acceptance of substrates.
B. Install caulking in strict accordance with the manufacture's recommendations. Caulking bead shall be true to line and surface, of uniform width and smooth, solid and slightly below the surface or adjacent materials.
C. Cure and protect sealants as directed by the manufacturer. Replace or restore damaged sealants. Clean adjacent surfaces.
D. Caulking is to be protected from the work of other trades during and after installation.

Allow $6500.00 for wall brackets and hanging incandescent fixtures. All others fixtures indicated on drawings to be included in bid.

FIGURE 13-4 ■ A cash allowance specification is used if information regarding quality or quantity can't be determined until a building tenant is selected. *Courtesy StructureForm Masters, Inc.*

FIGURE 13-5 ■ A performance specification describes the conditions or parameters a product or system must function in.

3.4 MANUFACTURERS

A. Hinges: Stanley (Specification base), McKinney, Hagger, or accepted substitute.

B. Locks and Latchsets: Russwin, no other substitutions.

C. Exit Devices: Von Duprin or accepted substitute.

FIGURE 13-6 ■ Proprietary specifications call for materials by their trade name. Portion 3.4-A and C are examples of open specifications. Portion 3.4-B is a closed or sole source specifications. *Courtesy Architects Barrentine, Bates & Lee, A.I.A.*

Open or *equal* bid specifications allow for substitutions of products that are deemed to be equal by the architect. With an open specification, the architect may name several acceptable products, allowing the contractor or owner to choose based on availability or price. Typically the specification names a minimum of three products that could be used when in a project manual. Although the names of three products are listed, the qualifications for only one of the three are listed. When placed within the drawings, the phrase "or equal" or "accepted substitute" is sometimes used to allow the contractor to submit names of products judged to meet the project requirements. An example of an open specification can be seen in Figure 13-6.

Reference Specifications

A minimum standard for quality or performance can be established by using a *reference specification*. Typically used in conjunction with other types of specifications, reference specifications refer to minimum levels of quality established by recognized testing authorities such as UL (Underwriters Laboratories), ASTM (American Society for Testing and Materials), ANSI (American National Standards Institute), ACI (American Concrete Institute), and ICBO (International Conference of Building Officials). Specifications should include a date, to ensure the latest standard is used, and should also reference which portions of the standard are to be applied. An example of a reference specification can be seen in the specification for custom casework in Figure 13-7.

WRITING SPECIFICATIONS

Specifications are written by an architect, engineer, or professional writer with expertise in legal and architectural matters. Rarely is a drafter involved with the actual writing of specifications. Many smaller offices have master sets of specifications, which can be edited for similar projects. These *masters* can serve as a core set of specifications that are edited for each project, saving hours of labor. A senior drafter is often required to update the master specs to meet the requirements of the current project. A wide variety of software

programs are also available to aid in writing specifications. Programs typically offer several options and allow the user to pick the options that meet the needs of the project.

The Relationship of Drawings to Specifications

A key to good construction documentation is to remember the goal of each portion of the documentation. Drawings present pictures of the structure and give the size, form, location, and arrangement of various components by use of lines, symbols, and text. Specifications provide written descriptions of type and quality of materials; labor and components, and methods of testing; fabricating, and installing components within the structure. Each form should complement the other without repeating information.

■ Drawings should be used to depict the physical relationship

■ Specifications should be used to show the quality and types of material in the structure

Duplication between the drawings and specs—unless it is exact word for word— can lead to contradiction, misunderstanding, and a court date. Word-for-word duplication is redundant and harmful to productivity. Because many privately funded projects do not have a project manual, duplication throughout the drawings is often used to avoid misunderstandings.

The drawings and the specs should be developed together to achieve a balanced relationship. At the onset of a project, someone on the architectural team is usually assigned the task of keeping a checklist for the project. The checklist generally includes a schedule of what drawings are required, and what will appear on each drawing, related schedules, or what needs to appear in the specifications. Material that is added to the preliminary spec list can then

PART 2 PRODUCTS

2.1 MATERIALS

A. **Lumber:** Transparent finished wood AWI Premium grade as indicated in Section 100 in AWI Quality Standards.

B. **Plywood:** APA, PS 1 for softwood. PS 1-71and Industry Standard, I.S1 for hardwood 5 ply minimum.

C. **Particleboard:** Mat-formed wood particle board, CS 236-66, Type 1-B-2 except 2-B-2 for sink countertops and similar moisture exposed work, 45 pound density. 3/8" to 1-1/4" thick, Duraflake by Willamette Industries.

D. **Fasteners**: Nails, staples and screws to comply with Section 400 in AWI Quality Standards.

FIGURE 13-7 ■ Reference specifications establish a minimum standard for quality or performance. *Courtesy Architects Barrentine, Bates & Lee, A.I.A.*

be noted on the plans by the drafter in a shortened note form. For instance, a wall can be described in a detail or section as an 8" concrete wall. The project manual would contain the specifications for the type of concrete, how it is to be mixed, when it can be poured, and numerous other features about quality. It is imperative that the drafter be well informed about material placed in the project manual so that materials can be adequately described. Figure 13-8 shows a detail of the EIFS that depends on the specifications seen in Figure 13-3. Structural drawings done by an engineering team tend to be more redundant in callouts to meet the demands of the building department. Figure 13-9 shows an example of a steel detail that is placed with the specifications. The written specifications can be seen in Figure 13-1.

Writing Metric Symbols and Names

Dimensions and sizes found in the specifications should never be expressed in dual units. The same units found throughout the drawings should be expressed throughout the specifications to reduce dimensioning time, reduce chance of conversion errors, and reduce confusion. If units are to be expressed in metric, all sizes should be listed in millimeters except for large distances associated with the site drawings. Notations describing area should be expressed in square meters, and expressions of volume should be expressed in cubic meters. Fluid measurements should be expressed as liters. Loads that have traditionally been expressed as pounds per square foot (psf) should be expressed as kilograms per square meter (kg/m^2) or kilonewtons per square meter (Kn/m^2).

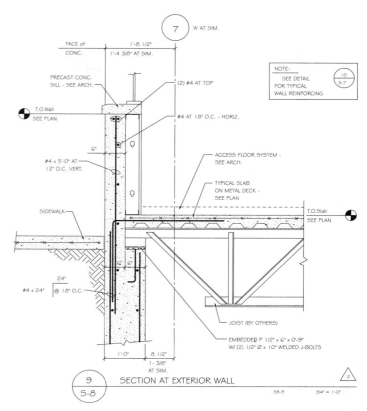

FIGURE 13-9 ■ Because of building department requirements for drawings, structural drawings often duplicate information contained in written specifications. The drafter must carefully coordinate both parts of the project. *Courtesy KPFF Consulting Engineers.*

Kilograms per square meter should be used for expressing floor loads because most live and dead loads are measured in kilograms. Many engineers use kilonewtons per square meter (kN/m^2) or their equivalent, megapascals (MPa) for structural calculations. Additional guidelines for expressing metric units within specifications include the following:

■ Unit names, including those derived from a proper name such as newton or pascal should be given in lowercase letters. Write millimeter, meter, or pascal.

■ Use lowercase letters to represent abbreviations of units such as millimeter, meter, or kilogram. Use capitals letters to represent the abbreviations for proper names such as kelvin, newton or pascal. Write m, kg, K, or Pa.

■ Do not mix names and symbols. Write Nm or newton meter, but not N meter.

■ Leave a space between a unit symbol and its numeric value. Write 10 mm not 10mm.

■ Always use decimals and not fractions. Write .75 g not 3/4 g.

■ Place a zero before the decimal marker for values less than one. Write 0.45 g, not .45 g.

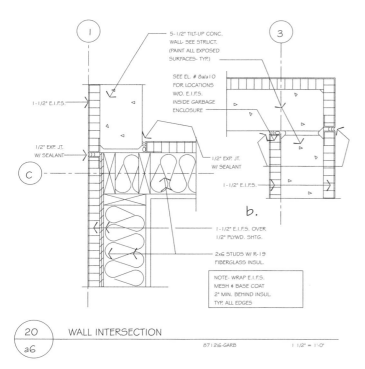

FIGURE 13-8 ■ Drawings and specifications should be developed together so that information is not duplicated. This detail relies on the information contained in Figure 13-3. *Courtesy Architects Barrentine, Bates & Lee, A.I.A.*

■ Use a space instead of a comma to separate blocks of three digits for any number (except for references to money) larger than four digits. Write 0.000 001 or 1 000 000.

■ Do not use a period after a symbol except when the symbol occurs at the end of a sentence.
Write: Mid sentence 10 g
End of sentence 10 g.

■ Write decimal prefixes in lowercase for magnitudes of 10^3 or less. Write prefixes in uppercase for magnitudes of 10^3 and higher.

■ Do not leave a space between a unit symbol and its decimal prefix. Write kg not k g.

■ Use symbols in conjunction with numerals. Write 15 m^2.

■ Write out unit names if numbers are not used.

Additional information on metric conversions can be obtained from the following:
Construction Metrication Council
National Institute of Building Sciences
1201 L Street, N.W., Suite 400
Washington, DC 20005

American Institute of Architects
AIA Bookstore
1735 New York Avenue N.W.
Washington, DC 20006

American National Standards Institute, Inc.
11 W. 42nd Street
New York, NY 10036

Construction Specification Institute
601 Madison Street
Alexandria, VA 22314-1791

ADDITIONAL READING

The following Web sites can be used as a resource to help you keep current with changes in writing specifications.

ADDRESS	COMPANY/ORGANIZATION
www.aiaonline.com	American Institute of Architects
www.ansi.org	*ANSI*—American National Standards Institute
www.nibs.org/cmchome.html	Construction Metrication Council
www.csc-dcc.ca	Construction Specifications Canada
www.scip.com	Specifications Consultants in Independent Practice
www.csinet.org	Construction Specification Institute
www.pcea.org	Professional Construction Estimators Association of America, Inc.

CHAPTER

13 *Project Manuals and Written Specifications*

CHAPTER QUIZ

Answer the following questions on a separate sheet of paper. Print the chapter title, the question number, and a short complete statement for each question.

Question 13-1 What are the two common locations for specifications and what are the major differences between the two systems?

Question 13-2 What areas of information should drawings describe?

Question 13-3 Why is duplication in the drawings and specifications a possible problem?

Question 13-4 What types of information should the specifications include?

Question 13-5 List common documents that are typically part of the manual.

Question 13-6 Name and describe the common format for writing specifications within a manual.

Question 13-7 How are the seventeen major divisions of the CSI system broken down?

Question 13-8 Describe the three major portions of a specification written to meet the CSI format.

Question 13-9 Describe the drafter's role in the writing of specifications.

Question 13-10 List and describe the five major methods of writing a specification.

CHAPTER 14

Land Descriptions and Drawings

Every project to be built requires one or more drawings and a legal description to describe the construction site. This chapter describes land descriptions and the types of drawings used to describe site-related construction.

LEGAL DESCRIPTIONS

For legal purposes, each piece of land is described by a description of the property known as a *legal description*. The legal description may be given in several forms such as a *metes and bounds system*, a *rectangular system*, or a *lot and block system*. The type of description to be used depends on the contour of the area to be described or the requirements of the municipality reviewing the plans.

Metes and Bounds

A metes and bounds description is also referred to as a long description. A copy of the description can be obtained from a title company, or the accessor's office. A complete description can be added to the site plan or attached to the drawings on a separate sheet of paper. This system provides a written description of the property in terms of measurements of distance and angles of direction from a known starting point. The known starting point is referred to as the *true point of beginning* in a legal description. The true point of beginning is usually marked by a steel rod or a benchmark established by the US Geological Society, (USGS). This is referred to as a *monument*.

Metes

The metes are measured in feet, yards, rods, or surveyor's chains. A rod is equal to 16.5 ft (5000 mm) or 5.5 yards. A chain is equal to 66 ft (20 100 mm) or 22 yards. Directions are given from a monument such as a benchmark established by the USGS or an iron rod set from a previous survey. This known location is known as the point of beginning, and may be several hundred feet away from the property to be described. Directions are given from the point of beginning to a specified point on the perimeter of the property to be described. Directions are then given to outline the property with all distances set in units of feet expressed in one hundredth of a foot rather than the traditional feet and inches normally associated with construction. Metric units should be expressed in either millimeters or meters. Centimeters are not used on construction drawings

Bounds

The boundaries of property are described by *bearings*, which are angles referenced to a quadrant on a compass. Figure 14-1 shows the four possible quadrants and descriptions of lines within quadrant. Bearings are always described by starting at north or south and turning to the east or west. Bearings are expressed in angles, minutes, and seconds. Each quadrant of the compass contains 90°, each degree contains 60 minutes and each minute can be divided into 60 seconds. A degree is represented by the ° symbol, a minute is represented by the ' symbol, and seconds are represented by the " symbol. Bearings are used to describe the angular location of a property line. Some properties are also defined by their location to the centerline of major streets. When property abuts a body of water such as a creek or river, a boundary in angles may not be given and the property boundary is defined by the center point of the body of water. A metes and bounds legal description would resemble the description shown in Figure 14-2a. This description would be listed on legal documents describing the property as well as the site plan. Figure 14-2b shows the land described by the legal description. A short description of the property should be copied exactly onto the site.

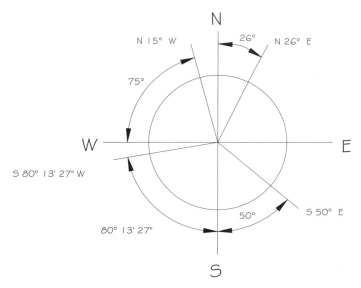

FIGURE 14-1 ■ Bearings are referenced by quadrants on a compass beginning at either north or south.

A tract of land situated in the Southeast quarter of the Northeast quarter of Section 17, T3S, R1W of the Willimette Meridian, Clackamas County, Oregon, being more particularly described as follows, to wit: Beginning at the 5/8 inch iron rod at the Southwest corner of the Northeast quarter of said section 17; thence north 0°06'10" East along the west side of the Southeast quarter of the Northeast quarter, 322.50'; thence leaving said West line North 89°38'15" East 242.00 feet; thence South 0°06'10" West parallel with said West line of the Southeast quarter of the Northeast quarter, 50.00 feet, thence North 89°38'15" East 310.47 feet to the Westerly right of way line of Bell Road No. 113; thence South 3°31' East along said westerly right of way line, 272.91 feet to a 5/8 inch iron rod; thence leaving said westerly right of way line, South 89°38'15" West 569.97 feet to the point of beginning.

FIGURE 14-2A ■ ■ A metes and bounds legal description.

Rectangular Systems

Many areas of the country refer to land based on its location by latitude and longitude. Parallels of latitude and meridians of longitude were used by the United States Bureau of Land Management in states that were originally defined as public land states. As the land was divided, large parcels of land were defined by what are known as *basic reference lines*. There are 31 pairs of standard lines in the continental United States and three in Alaska. Principal meridians and base lines can be seen in Figure 14-3. These divisions of land are described as the *great land surveys*. As the initial division of land was started, the first six principal meridians were numbered. The last numbered meridian passes through Nebraska, Kansas, and Oklahoma. All subsequent meridians are defined by local names. The great land surveys were further divided by surveys that define land by townships and sections.

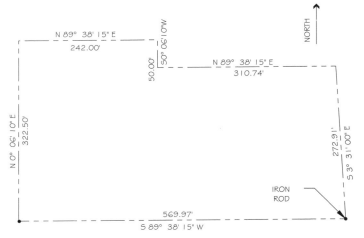

FIGURE 14-2B ■ The plot of land described in the metes and bounds legal description.

Townships

Baselines and meridians are divided into six-mile square parcels of land called *townships*. Each township is numbered based on its location to the principal meridian and base line. Horizontal tiers are numbered based on their position above or below the base line. Vertical tiers are defined by their position east or west of the principal meridian. Township positions can be seen in Figure 14-4. The township highlighted in Figure 14-4 would be described as *Township No. 2 North, Range 3 West* because it is in the second tier north of the baseline and in the third row west of the principal meridian. The name of the principal meridian would then be listed. This information is abbreviated as T2R3W on the site plan.

Sections

Land within the townships can be further broken down into one-mile square parcels known as *sections*. Sections are assigned numbers from 1 to 36 (as shown in Figure 14-5) beginning in the northeast corner of the township. Each section is further broken down into quarter sections. Each section contains 640 acres or 43,560 square feet. Quarter sections can be further broken down as seen in Figure 14-6. The areas are defined by quarters and quarters or halves of quarters. These small segments can be further broken down by quarters or halves again.

Specifying the Legal Description

The legal description, based on the rectangular system, lists the portion of the land to be developed described by its position within the section, the section, the township, and the principal meridian. A typical legal description resembles the following description:

The southeast one quarter, of the southwest one quarter of the southwest one quarter of Section 31, Township No. 2 North, Range 3 West of the San Bernardino meridian, City of El Cajon, in the state of California.

On the site plan the drafter often abbreviates this into a legal description as follows:

THE SE 1/4, OF THE SW 1/4 OF THE SW 1/4, S31, T2N, R3W, SAN BERNARDINO MERIDIAN, EL CAJON, CALIFORNIA.

Because the method of describing quarters of quarters of quarters can become confusing, many municipalities have gone to a labeling system using letters. Quarter sections are labeled as A, B, C, or D. Quarter sections are further divided into quarters by the letters A, B, C, or D. The northeast quarter of the northeast quarter would then be listed as Parcel AA. This method of land description works especially well in areas where the land contour is fairly flat. In areas with irregular land shapes, the rectangular land description can be linked with a partial metes and bounds description to accurately locate the property.

Lot and Block

This system of describing land is usually found within incorporated cities. Most states require the filing of a subdivision map that defines individual lot size and shape as land is being divided. Each lot is defined on a subdivision map by a length and angle of each

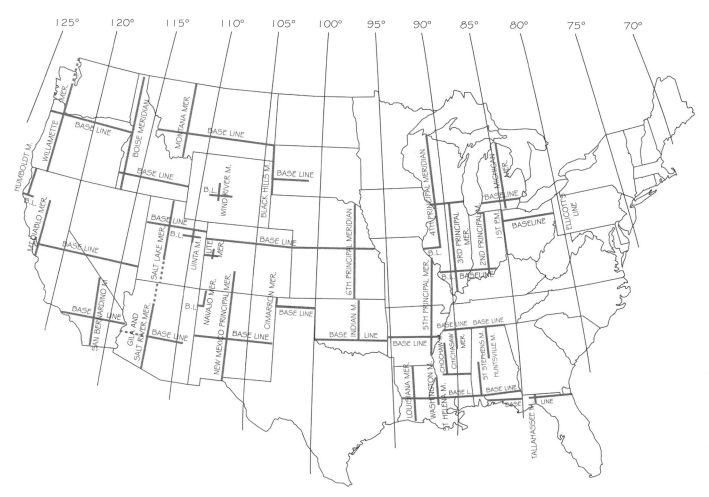

FIGURE 14-3 ■ Principal meridians and basic reference lines are used to divide land in the continental United States.

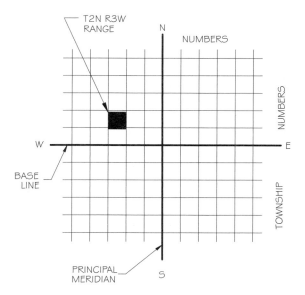

FIGURE 14-4 ■ Townships are a 6 sq. mile portion of land that is defined by its position to a principal meridian and a baseline. The indicated township is referred to as T2NR3W because it is in the second tier north of the baseline and three rows west of the principal meridian.

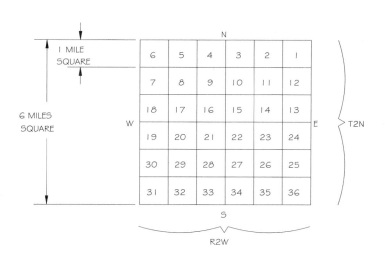

FIGURE 14-5 ■ Land within a township is divided into 36 one-mile square portions called a section.

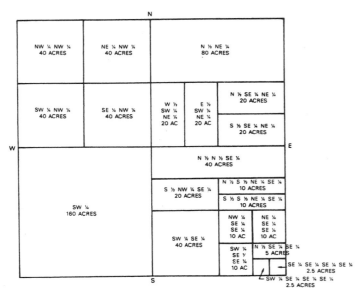

FIGURE 14-6 ■ A section can be further broken into quarter sections and quarter/quarter sections continuously.

property line as well as a legal description. On older subdivision maps, land is first divided into areas based on neighborhood, called *subdivisions*. The subdivision is, in turn, broken into blocks based on street layout. The block is further divided into lots. On newer maps, most municipalities assign a number to a parcel of land that corresponds to either a tax account or a map number. Lots can be irregularly shaped or rectangular. The shape of the lot is often based on the contour of the surrounding land and the layout of streets. A typical legal description for a lot and block description would resemble the following description:

LOTS NO. 1, 2 AND 3 OF MAP #17643 OF THE CITY OF BONITA, COUNTY OF SAN DIEGO, CALIFORNIA.

LAND DRAWINGS

The most common plan for describing property is the site plan. The site plan is started in conjunction with the floor plan, although it probably will not be finished until other architectural drawings are completed. With the property boundaries drawn, the preliminary design for the floor plan can be inserted into the site plan. Preliminary designs for access and parking can then be determined, and adjustments to both the site and floor plans can then be made.

The size of the project and the complexity of the site will dictate the drawings required to describe site-related construction, who will do them, and when they will be done. In addition to the site plan, a topography, grading plan, landscape, sprinkler, freshwater, and sewer plan may be required to describe the alterations to be made to the site. On simple construction projects, all of these plans can be combined into one site plan. This plan would typically be the first sheet of the architectural drawings and labeled **A-1**. On most plans, the site-related drawings are placed at the start of the architectural drawings and listed as civil drawings. The topography, grading plan, demolition, landscape, sprinkler, freshwater, and

sewer plans may comprise the civil drawings (**C-1** or **L-1**) and precede the site plan.

Drawing Origin

The size and complexity of the project will determine who will draw the project. On most projects, each of the site-related drawings will be prepared by a civil engineer, surveyor, or landscape architect and then given to the architect to be incorporated into the working drawings. On simple projects, the drawings may be drawn by the architectural or engineering office under the supervision of a landscape architect and a civil engineer.

VICINITY MAP

A *vicinity map* is used to show the area surrounding the construction site. It is placed near the site plan and is used to show major access routes to the site. This could include major streets, freeway on and off ramps, suggested routes for large delivery trucks, and rail routes. The drafter should prepare a vicinity map showing an area that reflects the size of the project. If building components primarily come from the surrounding area, the map needs to reflect only the immediate construction area. If materials come from several different cities or states, the area of the vicinity map should be expanded to aid drivers who may not be familiar with the area. Figure 14-7 shows an example of a vicinity map.

SITE PLANS

The *site plan* for a commercial project is the basis for all other site-related drawings. It shows the layout and size of the property, the outline of the structure to be built, north arrow, ground and finish floor elevations, setbacks, parking and access information, and information about utilities. The results of engineering studies and soils reports related to the site may be summarized on the site plan.

Common Linetypes

The shape of the construction site can be drawn based on information provided by the legal description or subdivision map. Property lines are represented by a line with a long-short-short-long pattern. PHANTOM2 or PHANTOMX2 from the AutoCAD line file can be used for representing property lines. Centerlines of access roads or easements are represented by a long-short-long line pattern using CENTER, CENTER2, or CENTERX2. Edges of easements and building setbacks are often represented by dashed or hidden lines using DASHED, DASHED2, DASHEDX2, HIDDEN, HIDDEN2, or HIDDENX2. Utility lines are usually represented by thicker lines using either a dashed or hidden pattern. Each line is labeled with a G (gas), S (sewer), W (water), P (power), or T (communications) to denote its usage. Specifications must also be provided to distinguish between existing utilities and those that the contractor is expected to provide. The location of each utility referenced to the property should also be provided. Figure 14-8 shows a partial site plan for a small commercial structure and the appropriate linetypes. Sidewalks, patios, stairs, and parking outlines are

VICINITY MAP

1" = 20'-0"

FIGURE 14-7 ■ A vicinity map is used to show the surrounding area and major access routes to the proposed construction site.

typically represented by a continuous linetype, with each item properly labeled.

Representing the Structure

The structure to be built must be accurately represented and easily distinguishable from the property and utilities. Common methods of representing a structure include the use of a polyline to define the perimeter and a hatch pattern such as ANSI31 or ANSI37 to further highlight the structure. This method can be seen in Figure 14-8. Many firms **XREF** the floor into the site plan. This offers the advantage of an up-to-date site plan each time the drawing is opened. This can be especially important as the footprint of the structure is altered or door locations are moved. If the project consists of more than one structure, each building should be identified by either a building number or name that relates to the corre-

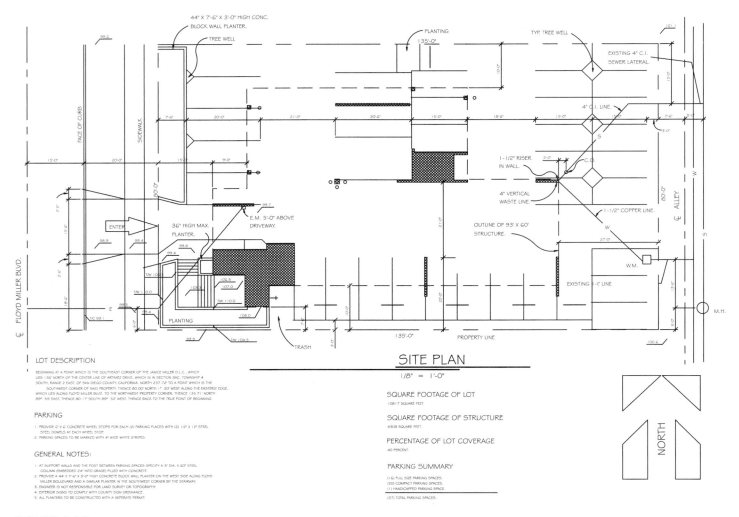

FIGURE 14-8 ■ Varied linetypes and widths are used to represent materials on a site plan.

sponding title placed with the floor and foundation plans. The use of the structure should also be specified within the outline of the structure with titles such as PROPOSED 2 LEVEL OFFICE STRUCTURE. Consideration must also be given by the drafter to accurately distinguish between portions of the structure that are in contact with the ground and those that are supported on columns. Commercial projects often include structures that are to be demolished to make way for the new project. These structures must be accurately located and distinguished from new construction. A separate demolition plan similar to Figure 14-9 may also be used to supplement the site plan.

Parking Information

Information about access, areas to be paved, parking spaces, ramps, and walkways must be shown on the site plan. Because requirements vary from city to city, the drafter must verify requirements for each project. Continuous lines are used to represent each, with the exception of the centerline of access roads. Many offices hatch concrete walkways with a random pattern of dots so they can be easily distinguished from asphalt paving areas. In addition to each parking space being represented, parking areas should be represented using a method similar to Figure 14-10. Many municipalities require a parking schedule to be part of the site plan. A parking schedule can be used to specify the number, type, and size of each type of parking space. Parking spaces are specified as full, compact, handicapped, or van handicapped. Common sizes are 9' × 20' (2740 × 6100 mm) for a full, 8' × 16' (2440 × 4875 mm) for a compact, 14' × 19' (4270 × 5790 mm) for an ADA-approved handicapped space and 16' × 19' (4875 × 5790 mm) minimum for an ADA-approved handicapped van space. It is important to remember that there is a wide variance in sizes depending on the municipality and the direction of entry into the space. Figure 14-11 shows an example of common alternatives for parking based on the angle of entry. Perpendicular spaces can be shorter than spaces that require parallel parking. Spaces that are placed on an angle require different lengths, widths, and driveway size as the entry angle is varied. Parking spaces next to obstructions such as raised planters or building supports should have added width to ease access and allow the minimum required width to be maintained.

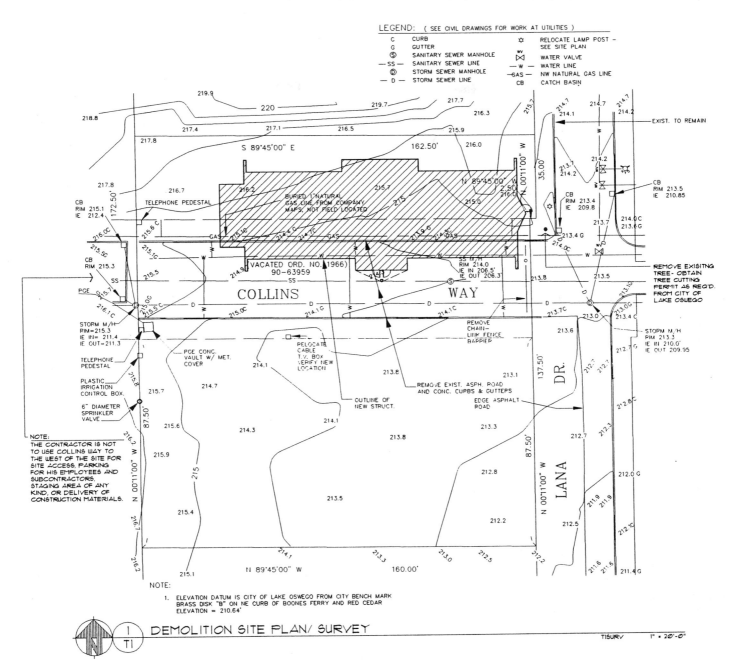

LEGEND: (SEE CIVIL DRAWINGS FOR WORK AT UTILITIES)

C	CURB	☼	RELOCATE LAMP POST – SEE SITE PLAN
G	GUTTER		
Ⓢ	SANITARY SEWER MANHOLE	⋈ WV	WATER VALVE
—SS—	SANITARY SEWER LINE	—W—	WATER LINE
Ⓓ	STORM SEWER MANHOLE	—GAS—	NW NATURAL GAS LINE
—D—	STORM SEWER LINE	CB	CATCH BASIN

DEMOLITION SITE PLAN/ SURVEY

TISURV   1" = 20'-0"

NOTE:
1. ELEVATION DATUM IS CITY OF LAKE OSWEGO FROM CITY BENCH MARK BRASS DISK "B" ON NE CURB OF BOONES FERRY AND RED CEDAR ELEVATION = 210.64'

FIGURE 14-9 ■ When structures must be demolished to make way for a proposed project, they can be indicated on a site plan or a separate demolition plan will be provided. *Courtesy Architects* Barrentine, Bates & Lee. A.I.A.

Building supports often require a permanent protective device to be installed for protection from vehicle damage. The addition of steel columns filled with concrete near each wood column is a common method of protecting wood columns from damage caused by careless drivers. These columns must be represented and specified on the site plan. Many offices have a standard detail in a symbols library that can be inserted into the site plan. Wheel stops are also required to be drawn and specified for individual spaces. Six-foot (1830 mm) long wheel stops are typically used for spaces which are 90° to the access drive. Stops are normally placed 24" (600 mm) from the front end of the stall and straddle the dividing line between spaces so that one stop is shared by two spaces. Building supports and wheel stops can be represented as shown in Figures 14-8 and 14-10.

Site Dimensions

Each item represented on the site plan needs to be located by dimensions. Three types of dimensions are often found on a site plan. Land sizes are represented in feet and hundredths of a foot using notations such as 100.50' or in meters (30.6 m). Property line dimensions are placed parallel to the property line with no use of

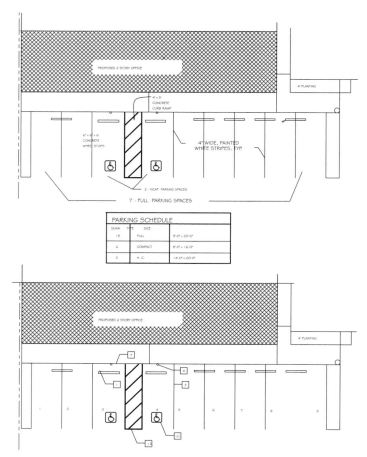

FIGURE 14-10 ■ Parking spaces must be clearly defined on the site plan by either referencing each space or by referencing each group of spaces.

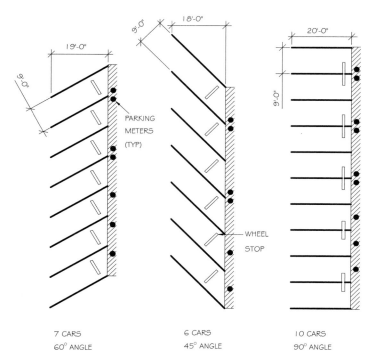

FIGURE 14-11 ■ The size of each parking space will vary depending on the angle of approach. Each city has it own parking requirements, which the drafter should verify prior to starting the site plan.

dimension or extension lines. Overall sizes of structures are often placed parallel to the side of the structure and specified in feet and inches using notations such as 75'–4" or 22.8 m for meters. These overall dimensions are also usually placed without the use of dimension lines. Objects such as light poles, signs, catch basins, parking boundaries, and the structure are located using dimension and extension lines. Objects such as sidewalks or planters can often be described in a note rather than by dimensions. Figure 14-12 shows examples of each type of dimension. Notice the symbol used to designate each property corner. It is often omitted from rectangular lots but is very helpful in locating small changes of angle on irregularly shaped lots.

Specifications

In addition to drawing and locating information, a note must be placed on the site plan to completely specify required construction. General text on a site drawing is typically 1/8" (3 mm) high, with titles approximately 1/4" (6 mm) high. Text describing the names of streets, and describing the proposed structure should be treated as titles. Notations such as A PROPOSED 3000 SQ. FT. TWO-STORY OFFICE COMPLEX are used to define the proposed construction.

Local and general notes are used to define the construction to be completed. General notes should be broken into categories so that specific information is easier to find. Categories such as parking, signs, utilities, flatwork, and miscellaneous are common. In addition to the general note, which will give the full specification, a local note is usually placed on the plan with a partial note to clarify what each symbol represents. Figure 14-13 shows an example of common notes and symbols that are often placed on a site plan. Notice that these notes are referenced to specific portions of the site plan, similar to the site plan shown in Figure 14-10. Abbreviations exclusive to site related drawings are also used. Common abbreviations include the following:

B.F.	Bottom of footing
B.W.	Bottom of wall
C.B.	Catch basin
C.O.	Clean out
D/W	Driveway
G	Gas
G	Gutter
LP	Light pole
EG	Edge of gravel
EP	Edge of pavement
FF	Finish floor
FH	Fire hydrant
FP	Flag pole stanchion

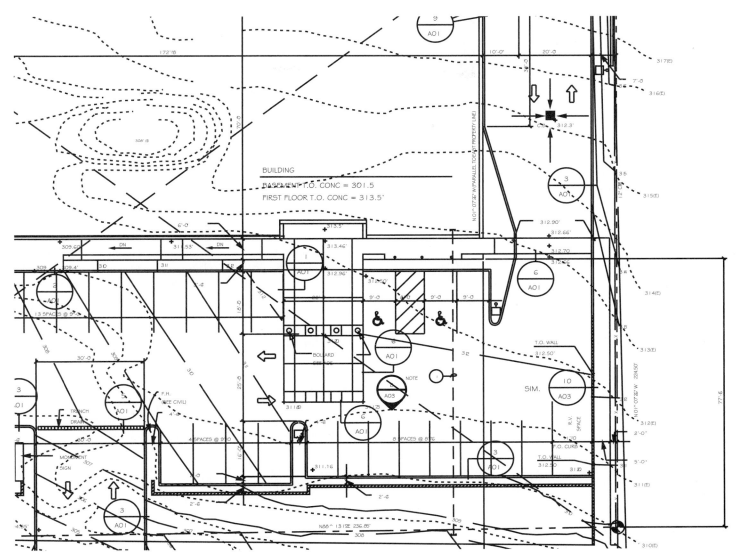

FIGURE 14-12 ■ Site dimensions are expressed in engineering units and placed parallel to the property line they describe. Overall sizes of the structure are placed parallel to the side of the structure they describe and expressed in feet and inches. All signs, parking boundaries, light poles, and drains should be dimensioned using extension and dimension lines with text expressed in architectural units. *Courtesy Peck, Smiley, Ettlin Architects.*

M	Meter
MH	Man hole
O/H	Overhead utility line
PL	Property line
ROW	Right of way
SD	Storm drain
SOV	Shut-off valve
SP	Signal pole
SS	Sanitary sewer line
STP	Steel post
STS	Street sign
TC	Top of curb
TG	Top of grate

TP	Top of paving
TW	Top of wall
UP	Utility pole
W	Water
WCR	Wheel chair ramp
WM	Water meter
WV	Water valve

Many plans include common abbreviations used throughout the drawings on a title page. Common abbreviations of construction drawings can be found at the end of this text.

In addition to information used to describe the construction site, general information is also placed on the site plan. General information might include a table of contents, a list of consultants, and information about the overall construction project. This would

FIGURE 14-13 ■ Notes can either be placed on the site plan as seen in Figure 14-10 or referenced to a list of general notes.

include information about the occupancy, construction type, and building areas per floor. An alternative to placing this general information on the site plan is to provide a title sheet that includes the vicinity map.

Elevation and Swale

Major changes of contour requiring excavation of large quantities of soil will have a grading plan prepared by a civil engineer to reflect site changes. Sites that do not require extensive excavation often reflect finished grade elevations on the site plan. Four common methods of denoting elevation can be seen in Figure 14-14. The elevation of ground level is indicated on the site plan with a note similar to

FINISH FLOOR ELEVATION 100.75'

A symbol similar to a leader line is used to indicate the elevation of each property corner as well as other important features. The elevation of the specified location is indicated above the leader line. Comparisons of elevations can also be indicated by indicators such as TW (top of wall), which are place above the symbol or BF (bottom of footing), which is placed below the symbol.

A third method of showing minor change of elevation is with the use of a swale indicator. A *swale* is a small valley used to divert water away from a structure. The slope of the swale is dependent on the surface material being drained. A minimum slope of 2% should be provided for dirt, and a slope of between 1% and 2% for asphalt or concrete paving areas.

The fourth method of showing a change of contour is with a bank indicator. A *bank indicator* is represented by a V placed between lines that indicate the top and toe of the bank. The bottom of the V is placed at the bottom of the bank. Three short lines are normally placed between the Vs to indicate the top of the bank.

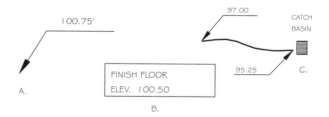

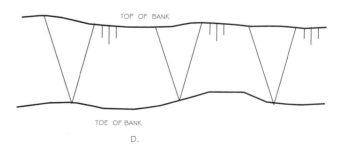

FIGURE 14-14 ■ Four methods of referencing ground elevation include a) an elevation symbol, b) a note to describe the elevation, c) a swale line with elevation symbols, and d) a bank indicator to represent the top and toe of a bank.

Drainage

Once the structure and pavement areas have been determined, a swale for drainage can be indicated on a site plan. On the simplest of projects, swales to divert rainwater from the building perimeter can be shown on the site plan. On complex projects, the drainage system may be shown on a separate drainage plan. An underground concrete box called a *catch basin* is often placed at the low point of swales to divert water from the site. Catch basins act as a funnel to collect water from parking areas and channel it into the storm system. A catch basin is covered by a metal cover that is level with the ground surface, called a *drainage grate*. The grate allows water to flow into the catch basin without allowing anyone to fall in. Water flows through the grate into the catch basin, and it is then funneled into pipes that connect to public storm sewers. The pipe that connects the construction site to the public sewer pipe is referred to as a *lateral*. The location of the lateral should be indicated on the site plan and be noted as *existing* or *new*.

A catch basin is placed at low areas of the structure, such as loading docks where water will naturally collect. The engineer designs the system and determines the required change of elevation to ensure proper runoff. Elevation markers referred to as *spot grades* should be established on the plan to specify the finished grading to ensure runoff. Markers at the corners of paved areas and at each drain should be indicated. The elevation, size, type and location of all grates and drainage lines should also be specified on either the site or drainage plan. The grate elevation is shown using a symbol, as shown at A in Figure 14-14. If a drainage plan is drawn, the elevation of the inside surface of the bottom of the drainage pipe is also specified. This elevation is known as an *invert elevation*. Either the minimum slope required or specific elevations along the drainage pipe are specified on a drainage plan.

Site Plan Development

Figure 14-15 shows an example of a completed site plan for a very simple site. Figure 14-16 shows a site plan that will have supplemental drawings. In its original state, the site plan also included the notes seen in Figure 14-13 and the details seen in Figure 14-17. Figure 14-10 shows the grading information for the site. Other related drawings are introduced throughout this chapter. Because the site plan serves as a basis for so many other drawings, it is important that the drafter have a clear understanding of what additional site drawings are required and who will be completing them. Storing all of the site-related drawings in one file, using external referencing, or using a separate drawing file for each drawings are the most common methods used to develop a site plan. See Chapter 3 for a review of drawing and layer prefixes and titles.

One Drawing File

On small-scale projects, if each of the site drawings is to be prepared by one firm, all site-related drawings can be stored within one drawing file. Storing all site-related drawings in one file can only be done by placing each drawing on separate layers. You should give layers names that describe both the base drawing and

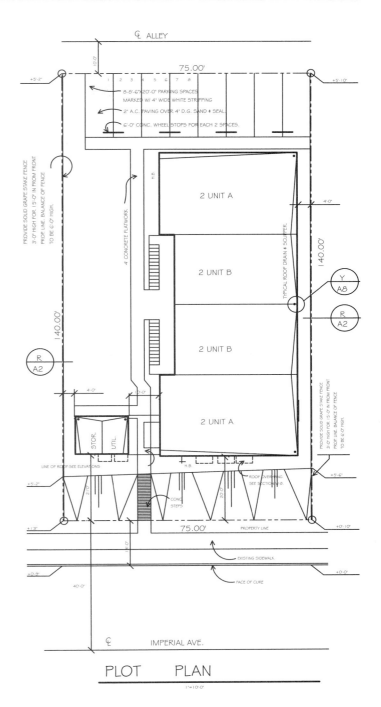

FIGURE 14-15 ■ The site plan for a simple development can often be used to show drainage requirements. *Courtesy StructureForm Masters, Inc.*

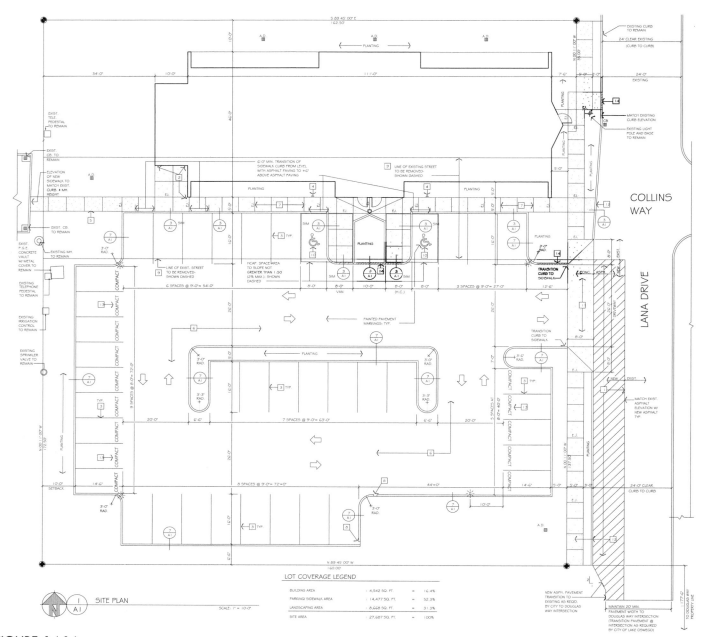

FIGURE 14-16 ■ A site plan for a structure that will have supplemental drawings. See Figure 14-9 and 14-26. *Courtesy Architects Barrentine, Bates & Lee A.I.A.*

the contents of the layer, as described in Chapter 3. Titles such as C-BLDG, C-PROP, or C-ANNO will clearly describe the contents. To plot the topography plan, basic items from the site plan such as the building and the property and all TOPO layers should be set to **THAW**, with all other layers such as site annotation set to **FREEZE**. The use of separate layouts in one drawing file can also be used to store multiple drawings in one file. Separate layouts can be created for the site plan, topography plan, and landscape plan in one drawing, allowing easy selection of the desired drawing for plotting.

Separate Drawing Files

The use of separate drawing files is the least effective use of disk space, but it can speed up the plotting process. To effectively create separate drawing files, place information that is required for each of the related drawing files in a base drawing and store as a wblock. This information can then be reused, saving valuable drafting time on each new drawing. With newer versions of AutoCAD, layouts allow this to be done in one drawing file without wasting disk space.

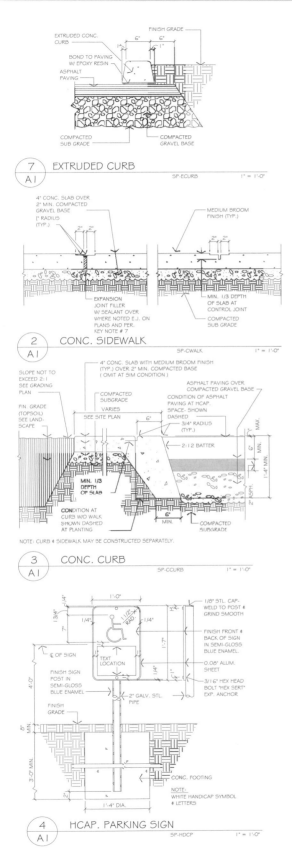

FIGURE 14-17 ■ Typical details can be stored as wblocks associated with the site plan seen in Figure 14-16. *Courtesy Architects Barrentine, Bates & Lee, A.I.A.*

External Reference

Site-related drawings are an excellent example of drawings that can be referenced to other drawings. A drafter working for the architectural team can draw the basic SITE plan information in a base drawing that will be reflected on all other site drawings. Copies of this drawing file can be given to other consulting firms who will develop the landscape, sprinkler, and grading drawings. Using external referencing allows each firm to have a current drawing file as a base, while each firm progresses with its work independently.

Setting Site Plan Parameters

Offices usually have stock template drawings containing common site-related linetypes, dimension and text variables, and symbols. As a student, you may need to develop your own template drawings for site plans. Site plans are often plotted at a scale factor such as 1" = 10' (1:120) or 1" = 20' (1:240). Preferred metric scales for civil drawings are as follows:

■ 1:100 (close to 1/8" = 1'–0")

■ 1:200 (close to 1/16" = 1'–0")

■ 1:500 (close to 1" = 40'–0")

■ 1:1000 (close to 1" = 80'–0")

This requires a template drawing to be developed reflecting engineering values rather than architectural values. This can be done by making a wblock of a title sheet developed in a introductory CAD class. Parameters such as **UNITS**, **LIMITS**, **SNAP** aspect, and **GRID** sizes can then be set to meet the specific needs of the site plan. Common layers can also be set up to divide site information from other information that will be stored with the site file.

Common Site Plan Details

The details required for a site plan vary widely from office to office and will vary based on the complexity of the construction. Common areas requiring details on a site plan include lighting supports, sign supports, parking barriers, parking painting details, sidewalks, curbs, and decorative walls. Figure 14-17 shows an example of common site details. Each is considered a standard detail and could be stored as a wblock in a file labeled as \PROTO\SITE\ (dwg file name) that is edited for specific job requirements. The engineer or project coordinator will generally note for the drafter the details that should be inserted into the site plan, and the drafter is expected to compile and edit the details to match project requirements.

COMPLETING A SITE PLAN

The site plan can be completed by following these steps.

1. Working from a plat map or metes and bounds legal description, lay out the shape of the site.

2. Establish the center of all access roads.

3. Locate all public sidewalks, easements, and curbcuts for driveways.

4. Locate the proposed structure from preliminary floor plans as well as required setback distances.

5. Define all paved areas and locate all required parking spaces, drainage swales, and catch basins.

6. Draw all utilities.

7. Draw all walkways and ramps.

8. Draw required planting areas.

9. Draw signs.

10. Draw other man-made features that are specific to this site, such as trash enclosures, furniture, benches, flagpoles, retaining walls, and fencing.

Steps 1 through 10 can be seen in Figure 14-18. Once all man-made items have been drawn, each should be dimensioned. Place dimensions by locating major items first. Dimensions can be placed by using the following order:

11. Dimension and list bearings of each property surface.

12. Dimension the location of the structure to the property.

13. Dimension the minimum required yard setbacks.

14. Dimension the footprint of the structure.

15. Dimension the locations of streets, public sidewalks, and easements.

16. Dimension all paved areas as well as individual parking spaces and catch basins.

17. Dimension all utility locations where they enter the site.

18. Dimension all other miscellaneous man-made features.

Steps 11 through 18 can be seen in Figure 14-19. With all features drawn and located, add any special symbols necessary to describe the material being used. Typically this would include hatching concrete walkways with a dot pattern. Other symbols to be added might include property corner markers, a north arrow, elevation markers, catch basins, fire hydrants, or clean outs.

The next stage of completing the site plan is to provide notations to define all man-made features that have been added to the site. Care should be taken to distinguish between material that is existing and material that is to be provided. Occasionally, existing material to be removed from the site must also be specified. Notes that should be included on a site plan are as follows:

19. Specify proposed building use and square footage.

20. Include finished floor elevation.

21. Include legal description.

22. Specify all streets, public walkways, curbs, and driveways.

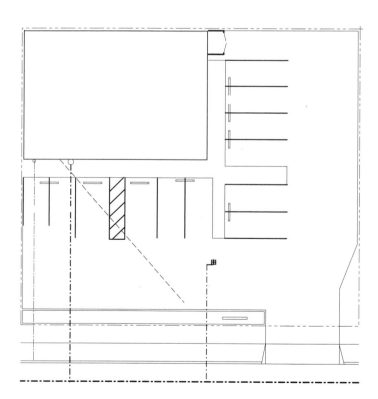

FIGURE 14-18 ■ Steps 1 through 10 can be used to establish the basic parameters of the site plan.

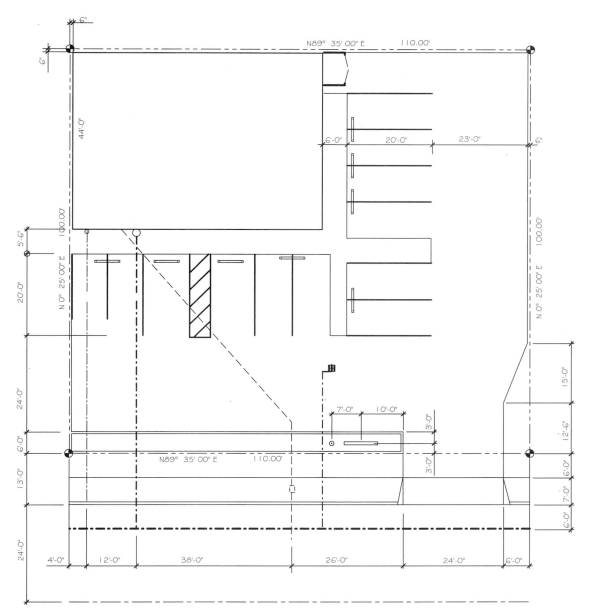

FIGURE 14-19 ∎ Steps 11 through 18 can be used as guidelines to locate parking and other man-made components as well as providing their locations.

23. Describe all utilities.

24. Describe all paved areas and specific parking areas.

25. Describe specific features that have been added to the site, including furniture, flag poles, retaining walls, fencing, and planting areas.

26. Add details to show construction of man-made items.

27. Add general notes to describe typical construction requirements.

28. Specify a drawing title and scale.

Figure 14-20 shows a completed site plan.

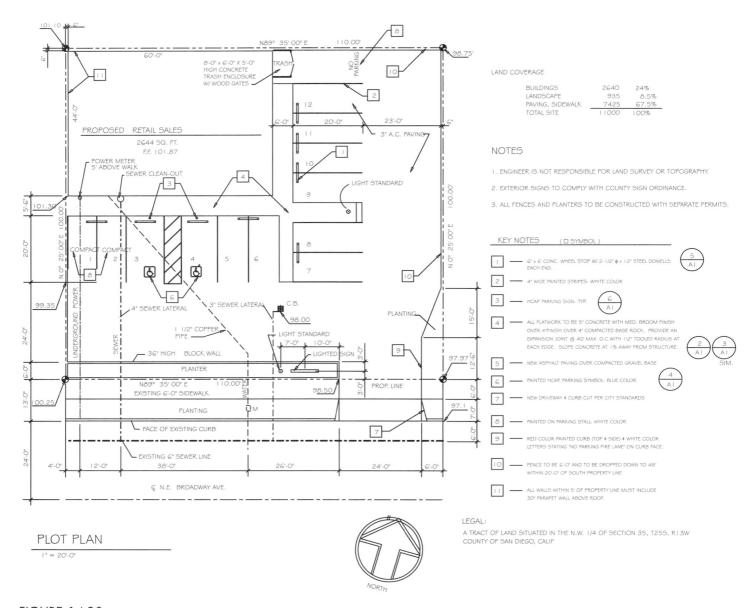

FIGURE 14-20 ■ The completed site plan includes notes and dimensions to describe all man-made features of the project.

RELATED SITE DRAWINGS

In addition to the site and demolition plans, several related drawings can accompany a set of working drawings. Related drawings include a topography plan, grading plans, landscaping plan, and sprinkler plan. A topography plan shows the existing contour of the construction site. This plan is normally prepared by a licensed surveyor based on notes from a field survey or by a civil engineer. It is based on existing municipal drawings describing the site, on measurements taken by the surveyor, or by aerial photographic methods. Once the shape of the site is prepared, the results of the field survey are translated onto the drawing by a drafter working for a civil engineer. A grading plan is used to show the finished configuration of the building site. A grading plan for major projects is designed and completed under the direction of a civil engineer. A landscaping plan similar to Figure 14-21 is completed by drafters working for a landscape architect. On small projects, drafters working for the architectural team may complete the project under the supervision of the project architect. In arid climates, a sprinkler plan similar to Figure 14-22 will be required to maintain landscaping. The same team that provides the landscape plan usually completes the sprinkler drawing. Because the landscape and sprinkler

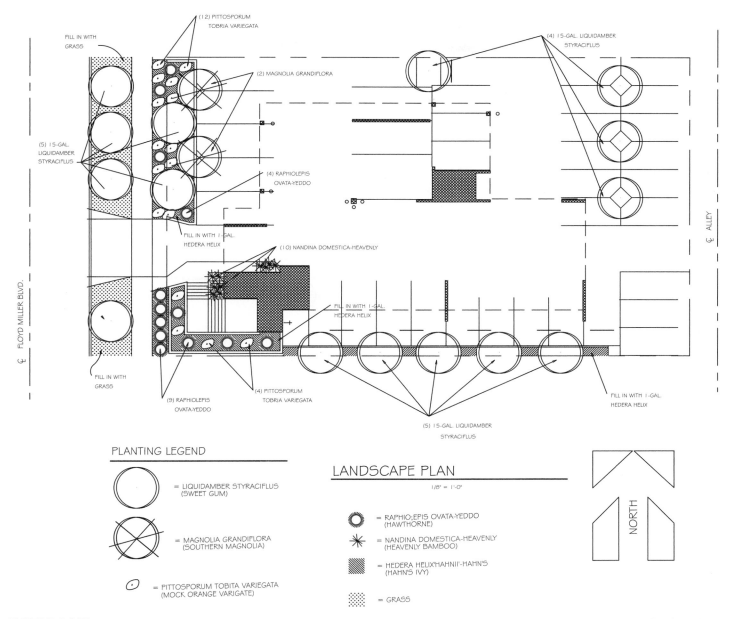

FILL IN WITH
GRASS

(12) PITTOSPORUM
TOBRIA VARIEGATA

(2) MAGNOLIA GRANDIFLORA

(4) 15-GAL. LIQUIDAMBER
STYRACIFLUS

(5) 15-GAL.
LIQUIDAMBER
STYRACIFLUS

(4) RAPHIOLEPIS
OVATA-YEDDO

FILL IN WITH 1-GAL.
HEDERA HELIX

(10) NANDINA DOMESTICA-HEAVENLY

FILL IN WITH 1-GAL.
HEDERA HELIX

FILL IN WITH
GRASS

(9) RAPHIOLEPIS
OVATA-YEDDO

(4) PITTOSPORUM
TOBRIA VARIEGATA

FILL IN WITH 1-GAL.
HEDERA HELIX

(5) 15-GAL. LIQUIDAMBER
STYRACIFLUS

Ȼ FLOYD MILLER BLVD.

Ȼ ALLEY

PLANTING LEGEND

= LIQUIDAMBER STYRACIFLUS
(SWEET GUM)

= MAGNOLIA GRANDIFLORA
(SOUTHERN MAGNOLIA)

= PITTOSPORUM TOBITA VARIEGATA
(MOCK ORANGE VARIGATE)

LANDSCAPE PLAN

1/8" = 1'-0"

= RAPHIO;EPIS OVATA-YEDDO
(HAWTHORNE)

= NANDINA DOMESTICA-HEAVENLY
(HEAVENLY BAMBOO)

= HEDERA HELIX'HAHNII'-HAHN'S
(HAHN'S IVY)

= GRASS

NORTH

FIGURE 14-21 ■ A landscaping plan is completed by drafters working for a landscape architect. On small projects, drafters working for the architectural team may complete the project under the supervision of the project architect.

drawings usually fall under the supervision of the landscape architect, procedures for creating these drawings will not be presented in this chapter.

Topography Plans

Figure 14-23 shows an example of a sketch provided to reflect the existing surface elevations. A grid is drawn to represent each known elevation location. Notice the six elevations in the northwest corner of the survey. Between grid 69.41 and 76.44, seven

contour lines are required to represent the change of elevation. It can be assumed the contours fall at an even spacing because the surveyor did not change the distance between grids to reflect a rise or depression. Six contour lines are required between grids 69.74 and 75.58, and eight contour lines are required to reflect the change in elevation between grids 69.41 and 77.49.

Once the distance between grids has been divided into the required divisions, points of equal elevation can be connected, as seen in Figure 14-24. The use of a polyline aids in finishing the drawing. As the distance between contour lines is decreased, the

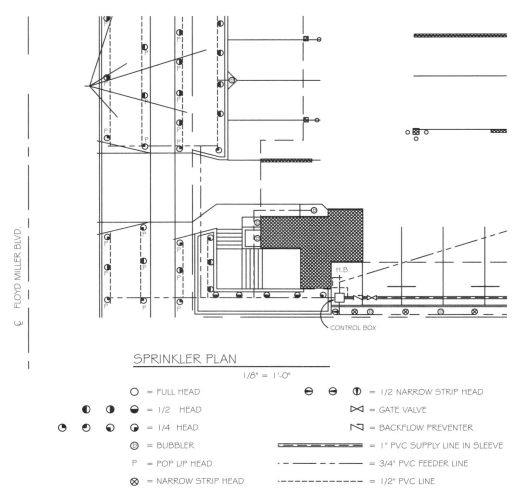

SPRINKLER PLAN

1/8" = 1'-0"

O = FULL HEAD   ◒ ◒ ◐ = 1/2 NARROW STRIP HEAD

◐ ◑ ◒ = 1/2 HEAD   ⋈ = GATE VALVE

◔ ◕ ◓ ◔ = 1/4 HEAD   ⋈ = BACKFLOW PREVENTER

Ⓑ = BUBBLER   ═══ = 1" PVC SUPPLY LINE IN SLEEVE

P = POP UP HEAD   ·—— — · —— = 3/4" PVC FEEDER LINE

⊗ = NARROW STRIP HEAD   ---------- = 1/2" PVC LINE

FIGURE 14-22 ■ In arid climates, a sprinkler plan will be required to maintain landscaping. The same team that provides the landscape plan usually completes the sprinkler drawing. *Courtesy Zachary Jefferis.*

land becomes steeper. As the distances between contour lines increases, the land becomes flatter. The topography plan must be completed before accurate estimates of soil excavation or movement can be planned.

Contour Lines and Symbols

Once the known elevations have been converted to contour lines, the lines can be curved and altered to provide clarity. Unlike the real contour of the site, in the initial layout stage, the lines run from point to point and have distinct directional change at each known point. These angular contour lines can be curved using the FIT CURVE option of the **PEDIT** command. **PEDIT** leaves each vertex in its exact location but changes the straight line to a curved line. The WIDTH option of the **PEDIT** command can also be used to alter the width of the contour lines. Most offices use varied line weights to help distinguish between contours. Depending on the spacing of the contours, lines representing every 5' or 10' (1500/3000 mm) are usually highlighted and labeled as shown in Figure 14-24.

Grading Plans

The grading plan shows the finished configuration of the building site. On simple plans, the finished grades will be placed over the existing grades, or the proposed grades can be shown separately from the existing grades. Figure 14-25 shows an example of a simple grading plan. On simple projects, a drafter working for the architectural team may complete the grading plan. A drafter working for the civil engineer translates preliminary designs by the architectural team and the topography drawings into the finished drawings. The grading plan shows the finished contour lines, areas of cut and fill, building footprint, the outline of parking areas, walkways, patios, and steps, catch basins, and drainage provisions. Figure 14-26 shows a portion of a grading plan.

Common Land Terms

The height of land above a given point is referred to as its *elevation*. The elevation can be referenced to sea level, a USGS bench mark, or to some predominate feature near the job site. Land in its unal-

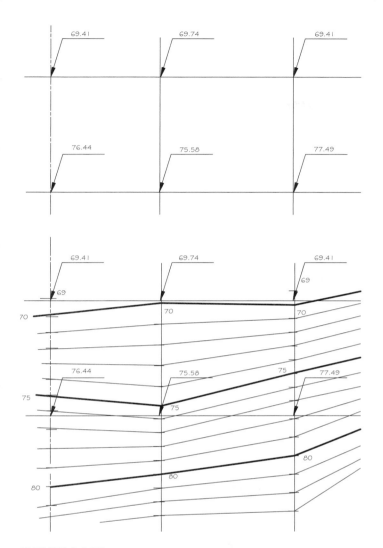

FIGURE 14-23 ■ Existing topography is determined from a sketch or drawing provided by the civil engineer, which lists the known elevation at specific points at the site.

tered state is considered to be the *natural grade*. *Finish grade* refers to the shape of the ground once all excavation and movement of earth has been completed. Earth that is moved is referred to as cut or fill material. *Cut* material is soil that is removed so that the original elevation can be lowered. *Fill* material refers to soil that is added to the existing elevation to raise the height. The point that represents the division between cut and fill is referred to as *day light*. The civil engineer determines the finish elevation of floor levels and major area of concern. It is then typically the drafter's job to indicate the extent of cut and fill material based on the desired angle of repose.

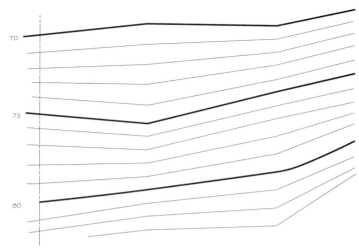

FIGURE 14-24 ■ Once known elevations are located, lines representing a specific elevation can be placed. Grades can be determined by estimating the rise or fall between two known points. Because lines for grades 70 through 76 occur in the upper left grid, the distance between the two points is divided to represent each elevation.

Angle of Repose

The angle of the cut or fill bank that is created is referred to as the *angle of repose*. The maximum angle of repose is usually determined by the municipality that governs the construction project and is based on the soil type. For fill banks, a common angle of repose is often set at a 2:1 angle. For every two units of horizontal measurement, one vertical unit of elevation change can occur. A common angle for cut banks is 1.5:1. The engineer specifies variations in the angle of repose that will be allowed near footings or retaining walls to minimize loads that must be supported.

Representing Contours

If the new and existing contours are to be combined on one plan, care must be taken to distinguish between contours. Thin dashed lines often represent existing contours. New contours can be represented by thickened continuous lines, as seen in Figure 14-25, or by dashed lines, as seen in Figure 14-26. Note that the bank symbol can be added to the contours to further highlight new banks that will be created. The angle of the new bank should also be specified on the grading plan.

Grading Plan Details

Details are often required to show the construction of drainage materials and retaining walls. Figure 14-27 shows an example of each. Both types of details are wblocks, which would be edited by the drafter, based on project requirements specified by the engineer.

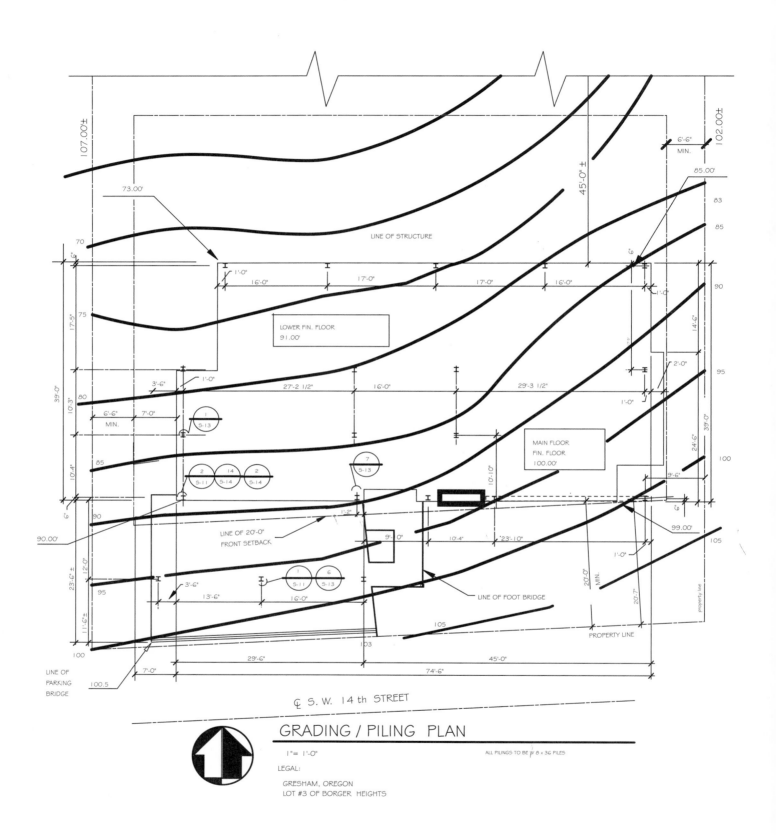

FIGURE 14-25 ■ A simple grading plan showing the existing grades, spot elevations, piling locations, and the finish floor level of each portion of the hillside office.

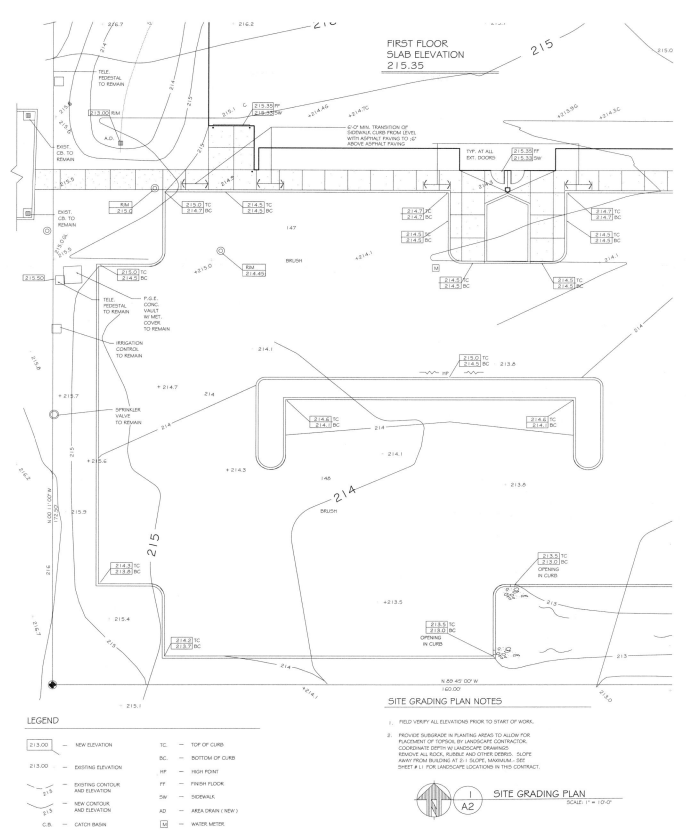

FIGURE 14-26 ■ The grading plan is usually prepared by drafters working for a civil engineer. *Courtesy Van Domelen/Looijenga/McGarrigle/Knauf Consulting Engineers.*

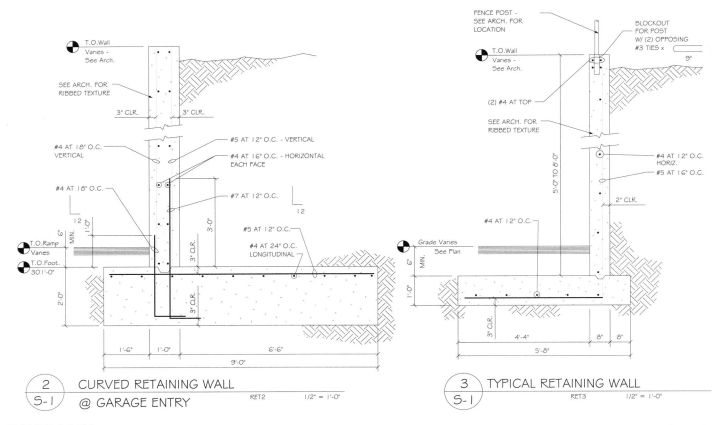

FIGURE 14-27 ■ Retaining walls are often referenced and drawn on the grading plan. *Courtesy Van Domelen/Looijenga/McGarrigle/Knauf Consulting Engineers.*

LAYER GUIDELINES FOR CIVIL DRAWINGS

C-BLDG	Proposed building foot prints
C-COMM	Site communications
C-COMM-OVHD	Overhead communications lines
C-COMM-UNDR	Underground communications
C-FIRE	Fire protection
C-NGAS	Natural gas
C-NGAS-UNDR	Underground natural gas lines
C-PKNG	Parking lots
C-PKNG-CARS	Graphic illustrations of cars
C-PKNG-DRAN	Parking lot drainage slope indications
C-PKNG-ISLD	Parking islands
C-PKNG-STRP	Parking lot striping, handicapped symbols
C-PROP	Property lines
C-PROP-BRNG	Bearings and distance labels
C-PROP-CONS	Construction controls
C-PROP-ESMT	Easements, rights- of-way and setback lines
C-ROAD	Roadways
C-ROAD-CNTR	Centerlines
C-ROAD-CURB	Curbs
C-SSWR	Sanitary sewer
C-SSWR-UNDR	Underground sanitary sewer lines
C-STRM	Storm drainage catch basins and manholes
C-STRM-UNDR	Underground storm drainage pipes
C-TOPO	Proposed contour lines and elevations
C-TOPO-RTWL	Retaining walls
C-TOPO-SPOT	Spot elevations
C-WATR	Domestic water, manholes pumping stations, storage tanks
C-WATR-UNDR	Domestic water—underground lines

Status field modifiers similar to other drawing areas should be used throughout structural drawings. See chapter 3 and the *AIA Layer Guidelines* for additional layer names.

CHAPTER 14

Land Descriptions and Drawings

CHAPTER QUIZ

Use a separate sheet of paper to answer the following questions. Use of local vendor catalogs may be required for some questions.

Question 14-1 What is a monument as it relates to a site drawing?

Question 14-2 List four units of measurement that might be referred to in a legal description.

Question 14-3 What two directions are used to commence a bearing?

Question 14-4 What would the following designation represent on a site plan? 100.67' S37° 30'45"W

Question 14-5 What would the following designation represent on a site plan? S28AA, T 3S,R1E

Question 14-6 List three components of a legal description commonly used for incorporated areas.

Question 14-7 List two common sources of site drawings.

Question 14-8 List and describe three methods of organizing site-related files for a small commercial project.

Question 14-9 Explain the difference between a topography plan and a grading plan.

Question 14-10 What is the purpose of a vicinity map?

Question 14-11 Sketch examples of linetypes to represent the following materials: property line, centerline, easement, sewer, water line, parking stop, and curb edge.

Question 14-12 List and describe two common hatch patterns that can be used to highlight a building.

Question 14-13 List the minimum size requirements for a compact, full, and handicapped parking space that is perpendicular to the access route.

Question 14-14 List two common methods of removing rainwater from a site.

Question 14-15 Give an example of how the top of a wall with an elevation of 100.67' resting on a footing supported by soil with an elevation of 95.75'.

Question 14-16 Sketch a bank indicator representing a slope.

Question 14-17 How are existing and new grades typically represented on a plan?

Question 14-18 Describe the process of changing field sketches to a topography map.

Question 14-19 What is an angle of repose?

Question 14-20 Give the common proportions of cut and fill banks.

Question 14-21 Verify with a local zoning department the minimum size requirement for a compact, full, and handicapped parking space that is perpendicular to the access route.

Question 14-22 Determine with the local zoning agency the required back-out space for perpendicular parking spaces.

Question 14-23 Determine with the local zoning department the legal description of three prominent structures in your area.

Question 14-24 Obtain a land map and legal description from a local title company of the property your school is built on.

Question 14-25 Interview a local landscape architect, civil engineer, and landscape contractor to assess the drafting opportunities in your area and to obtain examples of the types of drawings local professionals expect drafters to draw.

Question 14-26 Take a minimum of 15 photographs representing the installation or completion of work specified on a site, grading, landscape, and sprinkler plan.

DRAWING PROBLEMS

Note: *The sketches in this chapter are to be used as a guide only.* As you progress through the projects contained in Chapters 14 through 23, you will find that some portions of the drawings do not match things that have already been drawn. Each project has errors that will need to be solved. The errors are placed to force you to think in addition to draw. Most of the errors are so obvious that you will have no trouble finding them. If you think you have found an error, do not make changes to the drawings until you have discussed the problem and possible solutions with your engineer (your instructor).

As a drafter, you will need to sort through conflicting information. It is your responsibility to coordinate the material that you draw with other drawings for your selected projected. To sort through conflicts, use the following order of precedence:

1. Written changes given to you by the engineer (your instructor) as change orders

2. Verbal changes given to you by the engineer (your instructor)

3. The engineer's calculations (written instruction in the text)

4. Sketches provided by the engineer (sketches provided in each of the following chapters)

5. Lecture notes and sketches provided by your instructor

6. Sketches that you make to solve problems; verify alternatives with your instructor

Unless other instructions are given by your instructor, draw the site plan that corresponds to the floor plan that will be drawn in Chapter 15, using appropriate line and scale factors for plotting at the designated scale. Skeletons of most details can be accessed from http://www.delmar.com/resources/ocl.html. Use these drawings as a base to complete the assignment. Use appropriate drawing methods to represent and label all material. Provide the following minimum materials for each site plan:

■ Provide and number the specified parking spaces. Areas in front of all overhead doors to be marked "loading zone, no parking."

■ Use the notes in Figures14-13 and 14-20 as a guide unless your instructor gives you other instructions.

■ Assume utilities to be in the street unless noted at the following distances from the property line: storm sewer 3', sewer 6', water 15', telephone 18'. Specify a clean out 24" from structure in the sewer line. All utilities are to be underground.

■ Establish swales to provide drainage to catch basins that drain to the storm sewer.

■ Assume 24' wide curbcuts with streets to be 40' from property line to the centerline with a five-foot wide sidewalk located 3' from the property line.

■ Provide 14'-high light standards to light all public parking areas.

■ Use the lot provided with each project and the legal description provided by a local title company and draw a vicinity map for this project using local references for your community. If a legal description is not provided, call a title company in your area and order a plat map for a local business.

Problem 14-1 Vicinity Map. Draw a vicinity map of your school showing major access routes and important landmarks within a five-mile radius.

Problem 14-2 Legal Descriptions. Use the following description to draw and label a site plan that could be plotted at a scale of 1/8" = 1'–0". Beginning at a point that is the NE corner of the G.M. Smith D.L.C., which lies 250 feet north of the centerline of N.E.122 street which is in Section 1ab, Township 6 north, Range 1 west of Los Angeles county, California thence south 225.00 feet to a point which is the southwest corner of said property, thence North 85.00 feet 1° 25' 30" west along the easterly edge which lies along N.E. Sweeney Drive (25' to street centerline) to the northwest property corner, thence 140.25 feet South 89° 15' 00" East, thence south 85.25 feet 1° 25' 30" east, thence westerly along the north boundary of N.E. 122 street to the true point of **beginning.**

Problem 14-3 Site Plan. (See Problem 15–1.) Use the attached sketch to draw a site plan to be plotted at a scale of 1" = 10'–0". The lot is 120' × 110' and is Lot 4 and 5 of Arrowhead Estates, Rapid City, South Dakota. The project fronts onto Sioux Parkway with 36' from the property line to centerline of the roadway. Tee water to the property 75' east of the westerly property line from a main line 30' south of the property line. Provide a water shutoff valve in the garage of each unit. Access city sewer line, which is 25' south of the property with a new line 25' east of the westerly property line. Provide a clean out in the lower planting strip of each unit. Provide a continuous drainage grate across each driveway. Set each drain 3" below the finish floor level at each door. Provide a 7.5' wide easement along the northerly property line for a telephone easement. Provide each unit with a fenced yard. Dimension each unit as needed and specify all grades. Interpolate as needed to determine the grades for each corner of the structure.

Problem 14-4 Site Plan. Use the attached drawing to complete the site plan. Provide a minimum of 13 parking spaces including (1)-handicapped space, and a maximum of (3)-8 × 16 compact spaces. Provide an 8 × 8 × 6 ft. high enclosed trash area with double wood gates.

PROBLEM 14-3

Problem 14-5 Site Plan. Use the attached drawing to complete the required site plan. Coordinate the parking areas between door openings and ensure that no overhead doors will be blocked by public parking. Provide curved corners on all planting curbs. Locate each catch basin centered in each driveway.

Problem 14-6 Site Plan. Use the attached drawing to complete the required site plan. Coordinate the parking areas between door openings and ensure that no overhead doors will be blocked by public parking. Provide curved corners on all planting curbs. Locate each catch basin centered in each driveway. The elevation marked 14.90' on the east property line is 40' from the north property line. Show grade lines for each 1' grade interval.

Problem 14-7 Topography. Use the attached drawing to lay out the represented grades. Draw the site plan using a scale of 1" = 10'–0". Show grades at 1' intervals and highlight contours every 5'.

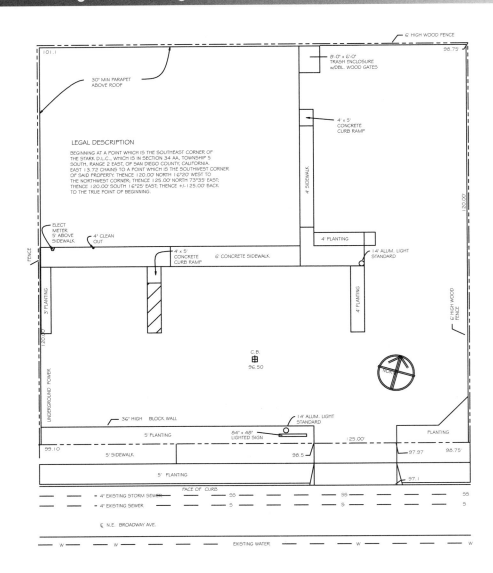

101.1

6' HIGH WOOD FENCE

98.75'

8'-0" x 6'-0"
TRASH ENCLOSURE
w/DBL. WOOD GATES

30" MIN PARAPET
ABOVE ROOF

4' x 5'
CONCRETE
CURB RAMP

4' SIDEWALK

120.00'

LEGAL DESCRIPTION

BEGINNING AT A POINT WHICH IS THE SOUTHEAST CORNER OF
THE STARK D.L.C., WHICH IS IN SECTION 34 AA, TOWNSHIP 5
SOUTH, RANGE 2 EAST, OF SAN DIEGO COUNTY, CALIFORNIA.
EAST 13.72 CHAINS TO A POINT WHICH IS THE SOUTHWEST CORNER
OF SAID PROPERTY; THENCE 120.00' NORTH 16°20' WEST TO
THE NORTHWEST CORNER; THENCE 125.00' NORTH 73°35' EAST;
THENCE 120.00' SOUTH 16°25' EAST; THENCE +/-125.00' BACK
TO THE TRUE POINT OF BEGINNING.

ELECT
METER
5' ABOVE
SIDEWALK

4' CLEAN
OUT

4' x 5'
CONCRETE
CURB RAMP

6' CONCRETE SIDEWALK

4' PLANTING

14' ALUM. LIGHT
STANDARD

FENCE

3' PLANTING

120.00'

4' PLANTING

6' HIGH WOOD
FENCE

UNDERGROUND POWER

C.B.
96.50

NORTH

36" HIGH   BLOCK WALL

14' ALUM. LIGHT
STANDARD

PLANTING

5' PLANTING

84" x 48"
LIGHTED SIGN

125.00'

99.10

98.5

97.97

98.75'

5' SIDEWALK

5' PLANTING

97.1

FACE OF CURB

4" EXISTING STORM SEWER   SS   SS   SS

4" EXISTING SEWER   S   S   S

₵ N.E. BROADWAY AVE.

W   W   EXISTING WATER   W   W

PROBLEM 14-4

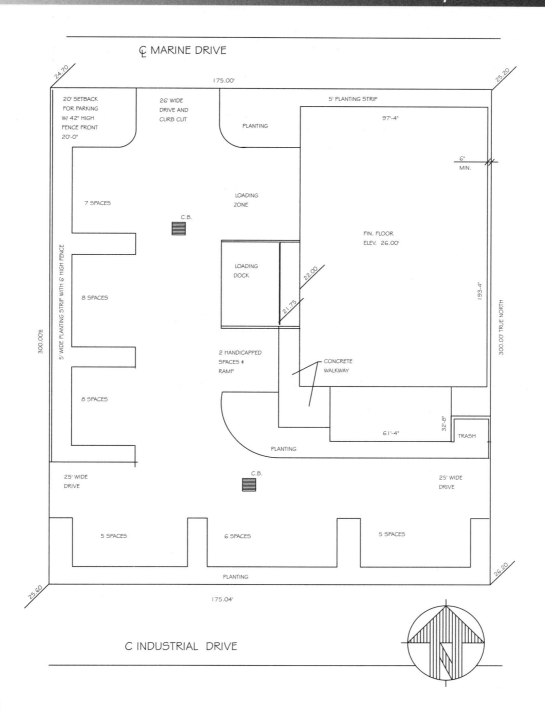

PROBLEM 14-5

PLOT PLAN

PROPOSED WAREHOUSE
FINISH FLOOR 16.00'

48'-0" LOADING DOCK

DRAINAGE GRATE

PLANTING STRIP

CATCH BASIN

9'-4" X 6'-0" X 5'-0"
C.M.U. TRASH
ENCLOSURE

LOADING ZONE

3" ASPHALTIC PAVING OVER
4" COMPACTED GRAVEL

7 PARKING SPACES

8 PARKING SPACES

PLANTING

25'-0"
CURBCUT &
DRIVEWAY

25'-0"
CURBCUT &
DRIVEWAY

SIGN UNDER
SEPARATE CONTRACT

CATCH BASIN

2 PARKING SPACES

3 PARKING SPACES

5' CONC. PLATFORK

PLANTING

10 PARKING SPACES

16'-0" HIGH
LIGHT STANDARD
TYPICAL

8 PARKING SPACES

4 PARKING SPACES

PLANTING

25'-0"
CURBCUT &
DRIVEWAY

₵ ROWELL DRIVE

₵ TOM WHORCHESTER PARKWAY

N.T.S.

NORTH

300.00'

300.00'

185.00'

185.00'

185.00'

97'-0"

71'-0"

69'-0"

48'-6"

32'-0"

87'-6"

50'-6"

32'-0"

55'-0"

5'-0"

14.70'

15.60'

26'-0"

15.35'

15.50'

15.50'

12.75'

12.25'

12.25'

17'-6"

65'-0"

13.00'

5'-0"

32'-0"

6'-0"

42'-6"

9.00'

PROBLEM 14-6

℄   MARGARITA COURT

110.00'

61.75	60.10	58.80	61.00	62.25	
63.40	61.40	59.50	61.50	62.30	68.40
	65.80	64.90	62.90	65.80	73.70
68.25					
	69.74	69.41	70.74	74.40	82.00
73.41					
	74.74	75.49	79.91	88.00	95.08
77.44					
	83.08	84.41	85.58	95.00	99.90
81.25					
	92.5	91.43	95.75	102.75	106.40
94.00					
				P.O.B.	
100.00	101.16	103.33	105.83	108.16	110.24

107.00±

105.00'

PROPERTY LINE

℄   FLORA DRIVE

GRADING   PLAN

1"=10'-0"1'-0"

22'-0" GRIDS EAST / WEST.
17'-0" GRIDS NORTH/SOUTH

PROBLEM 14-7

Floor Plan Components, Symbols, and Development

The floor plan forms the core of the architectural working drawings and is typically the first drawing created as the drawing set is being developed. Information on the floor plan will impact nearly every other drawing in the architectural and structural drawings. A *floor plan* is a view of a structure looking down on the floor and walls. It is created by passing a horizontal cutting plane through the walls at approximately five feet above the floor level and removing the upper portion of the structure. This process is repeated for each level of a structure to reveal the locations of all prominent items for that specific level. The floor plan also serves as a basis for the drawing of the ceiling, framing, mechanical, plumbing, and electrical plans. Key elements that are shown on the floor plan include the walls, doors, windows, and cabinets.

FLOOR PLAN DEVELOPMENT

Initially, the floor plan is developed as a schematic drawing by the architect to meet the design needs of the client. This plan is initially sketched on either graph or sketch paper to define the arrangement of basic shapes. Figure 15-1 shows an example of a preliminary floor plan. Once the best possible solution is determined, the plan is refined and drawn to scale. Depending on the office structure, and the skill of the drafter, this may be the first involvement of a drafter in the development of the floor plan. Generally, a senior drafter or designer who will head the project draws the preliminary floor plans for presentation to the client, planning commissions, and review boards. Figure 15-2 shows an example of the preliminary drawing of the floor plan. During this stage of development, the floor plan and other preliminary drawings are given to other consulting firms for preliminary input on structural, mechanical, and plumbing considerations. Keep in mind that, as a drafter prepares the preliminary drawings, the floor plan is developed in conjunction with the elevations, site and roof plans, and sections. Although the preliminary floor plan may be drawn first, as each of the other preliminary drawings is developed, changes may be required for the floor plan.

Once input from the owner, review boards, and other consultants has been compiled, the floor plan to be used for the construction drawings can be started. This plan is usually completed by a drafter or a team of drafters, depending on the size and complexity of the structure to be drawn. The floor plan will become a reference for all other architectural drawings. Figure 15-3 shows a completed floor plan.

Drawing Scales

The size and complexity of the project and the paper size to be used dictate the scale to be used to draw the floor plan. The paper size usually remains standard within an office, with all projects typically drawn on 24" × 36" or 30" × 42" material. The architect generally sets the scale of the floor plan for an inexperienced drafter. On small structures, a scale of 1/4" = 1'–0" is used. On large simple structures, scales of 1/8" = 1'–0" (1:50) or 1/16" = 1'–0" (1:200) are common. Scales of 3/16" = 1'–0", 3/32" = 1'–0", or 1/16" = 1'–0" are used for large structures with detailed areas of the plan enlarged to scales such as 1/4" = 1'–0", 3/8" = 1'–0" or 1/2" = 1'–0". Because of the few choices of metric scales, large metric structures typically must be divided into zones and spread over two or more drawing sheets. Scales of 1:50 or 1:20 can be used for enlarged metric plans. Examples of enlarged floor plans can be seen later in this chapter. In selecting a scale, the floor plan must be drawn at a scale that is large enough to reflect necessary details of the structure. For large projects, the floor plan is often drawn in segments with the use of a key plan and match lines.

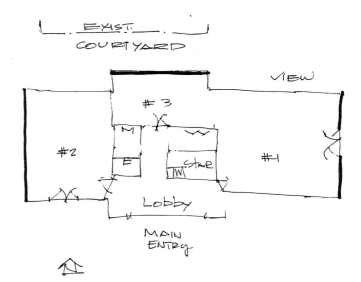

FIGURE 15-1 ■ A sketch is developed by the architect to plan the floor plan.

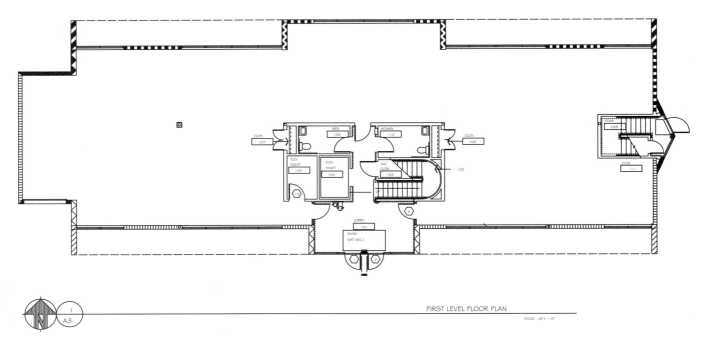

FIRST LEVEL FLOOR PLAN

SCALE: 1/8" = 1'-0"

FIGURE 15-2 ■ The architect's rough sketch is converted to a preliminary floor plan by a project manager or senior drafter. *Courtesy Architects Barrentine, Bates & Lee, A.I.A.*

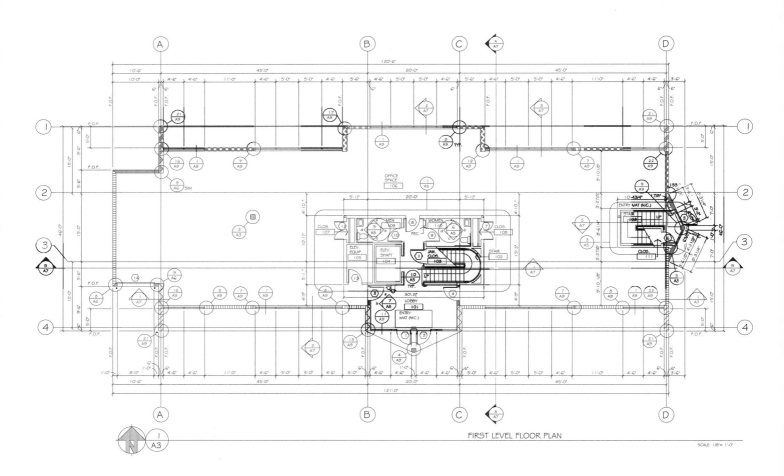

FIRST LEVEL FLOOR PLAN

SCALE: 1/8" = 1'-0"

FIGURE 15-3 ■ A completed floor plan often requires the efforts of several drafters. *Courtesy Architects Barrentine, Bates & Lee, A.I.A.*

Key Plans and Match Lines

A *key plan* is a very small-scale floor plan that shows the overall footprint of the structure, with specific areas of the structure divided into zones. A key plan is used when a structure is too large to fit on one sheet or if a specific portion of the structure is to be enlarged to show details that are too complex for the scale used to show the balance of the structure. The key plan can be placed anywhere on the drawing sheet that space allows, but should always be placed in the same location throughout the drawing set. Figure 15-4 shows a key plan for an office structure. Breaking the floor plan into zones allows a larger scale to be used than would be possible if the structure was drawn as a whole unit.

When the entire floor plan is not shown on one sheet, a *match line* must be provided to show how the portions of the floor plan relate to each other. Figure 15-5 shows Zone B of the structure shown in Figure 15-4. When the drawings are assembled, care should be taken to always use the same order of placing zoned drawings within the drawing set. On multilevel drawings, if the main floor level is divided into east and west zones, all levels should be divided into east and west zones.

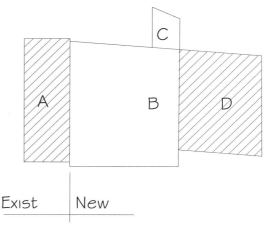

KEY PLAN

FIGURE 15-4 ■ Floor plans that are too large to fit on one sheet can be broken into several components. A key plan is used to indicate where the portion being viewed fits into the overall plan. The floor plan representing portion B can be found in Figure 15-5. *Courtesy Chris DiLoreto, DiLoreto Architects, LLC.*

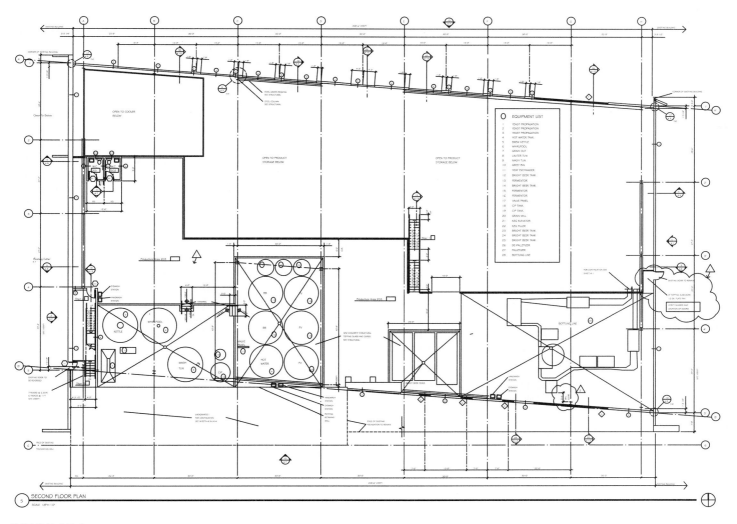

FIGURE 15-5 ■ Portion B of the key plan shown in Figure 15-4. *Courtesy DiLoreto Architects, LLC.*

FLOOR PLAN COMPONENTS

One of the key roles of a floor plan is to show the location and materials used to construct the walls, windows, doors, cabinets, and the dimensions and special symbols. Although the symbols are similar to those used in residential drafting, a wider variety of wall materials must be represented on commercial construction drawings.

Walls

Walls are the vertical members of a structure and can be used to meet each of the three goals of a building component. Although walls are typically vertical to accommodate windows and doors, IBC allows any portion of the structure angled at 60° or greater to be considered a wall. Walls can be constructed of single units forming a framework, as with wood and steel studs, or may be monolithic units, as with concrete walls. Walls are classified as either *interior* or *exterior* and *bearing* or *nonbearing* walls. The IBC also refers to shear, faced, parapet, fire, and retaining walls.

Exterior Walls

Exterior walls help form the basis for the skin of the structure. The architectural team draws the wall locations to meet the needs of the client. The structural team is responsible for detailing the construction of these walls. These walls may be required to

■ Restrict heat loss in winter but allow air flow in summer

■ Contain interior humidity, while repelling exterior humidity

■ Allow for view and access to the outdoors, or totally restrict access for security reasons

■ Welcome guests and inhabitants while keeping intruders out

Exterior walls will be the recipient of external forces from wind and rain and must be capable of transferring this force to the shell of the structure or to the lateral bracing system. Although it is the architect's responsibility to design an exterior covering to protect the structure, the drafter should have an understanding of these forces to understand how the structure resists the forces. An understanding of the forces that affect the building will provide meaning to the drawings that the drafter is required to complete. Exterior walls are typically permanent in nature, but they may be either bearing or nonbearing. The building codes consider walls that separate the structure from interior courtyards as exterior walls. Exterior walls are represented on the floor plan and will be discussed in this chapter. The framing plan shows specific measures that must be taken to build each wall and connect it to the frame of the structure. Framing plans are discussed in Chapter 21.

Interior Walls

Interior walls, or partitions as they are sometimes called, are used to define space. They may be bearing or nonbearing, fixed or removable. Walls that form hallways, enclose elevator shafts, surround bathrooms, or surround a specific tenant space in an office complex are examples of *fixed partitions*. Walls within a specific office or tenant space are examples of partitions that are moveable and nonbearing. Many designs will not locate the interior partitions of an office. This allows the future tenant to design the exact placement based on individual needs. Interior walls are drawn in a similar manner to exterior walls constructed of the same material, and they are generally the responsibility of the architectural team.

Bearing Walls

Bearing walls are defined by the building code as walls that support compression loads transferred from floors, roofs, or other bearing walls. The IBC defines any wall made of wood or steel studs as a bearing wall if it supports more than 100 pounds (1459 N/m) per lineal foot of superimposed load. Any masonry or concrete wall that supports more than 200 pounds (2119 N/m) per lineal foot of superimposed load is considered a bearing wall. Any wall that is more that one story in height is considered a bearing wall, even if it only supports its own weight. Bearing walls are shown by the architectural and structural teams on the floor and framing plans. The structural team is usually responsible for detailing the construction of the bearing walls. Chapter 21 discusses distribution of loads to walls and weights of common building materials. Chapter 18 discusses methods of drawing wall sections to show the construction of bearing walls.

Nonbearing walls

One-story walls that support only their own weight, or that support less than the acceptable lower limits of bearing walls, are considered nonbearing walls. Exterior nonbearing walls are often referred to as *curtain walls*. A load-bearing post or column may be contained in a nonbearing wall without requiring the wall to support any loads other than its own. Nonbearing walls are not shown on the framing plan. A drafter working with the architectural team completes details showing the construction of nonbearing walls.

Shear Walls

In addition to supporting loads that are transferred down through the structure to the foundation, shear walls are used to resist lateral (horizontal) forces transferred through the structure. These forces may come from wind, snow, or seismic activity. Lateral forces attempt to change a rectangular wall into a parallelogram. Although shown on a floor plan like any other wall, the components that comprise a sheer wall are usually specified on the framing plan. The engineer determines which walls will be required to be strengthened to resist lateral forces. Drafters working with the structural team are responsible for specifying the materials on the framing plan and details. Figure 15-6 shows an example of how a shear wall can be represented on a framing plan. Chapter 21 further explains construction and representation of shear walls.

Faced Walls

A *faced wall* is a wall framed with either wood or steel studs and covered with masonry. The masonry facing may be either on the

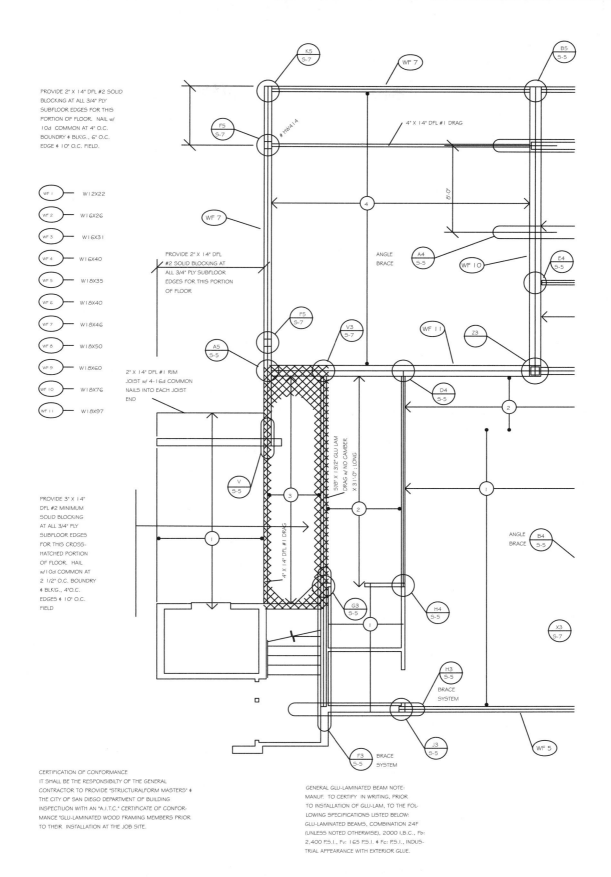

PROVIDE 2" X 14" DFL #2 SOLID BLOCKING AT ALL 3/4" PLY SUBFLOOR EDGES FOR THIS PORTION OF FLOOR. NAIL w/ 10d COMMON AT 4" O.C. BOUNDRY & BLK'G., 6" O.C. EDGE & 10" O.C. FIELD.

WF 1	W12X22
WF 2	W16X26
WF 3	W16X31
WF 4	W16X40
WF 5	W18X35
WF 6	W18X40
WF 7	W18X46
WF 8	W18X50
WF 9	W18X60
WF 10	W18X76
WF 11	W18X97

PROVIDE 2" X 14" DFL #2 SOLID BLOCKING AT ALL 3/4" PLY SUBFLOOR EDGES FOR THIS PORTION OF FLOOR

2" X 14" DFL #1 RIM JOIST w/ 4-16d COMMON NAILS INTO EACH JOIST END

PROVIDE 3" X 14" DFL #2 MINIMUM SOLID BLOCKING AT ALL 3/4" PLY SUBFLOOR EDGES FOR THIS CROSS-HATCHED PORTION OF FLOOR. NAIL w/10d COMMON AT 2 1/2" O.C. BOUNDRY & BLK'G., 4"O.C. EDGES & 10" O.C. FIELD

4" X 14" DFL #1 DRAG

ANGLE BRACE

4" X 14" DFL #1 DRAG

5/8" X 13/2" GLU LAM DRAG w/ NO CAMBER X 31'-0" LONG

ANGLE BRACE

BRACE SYSTEM

BRACE SYSTEM

CERTIFICATION OF CONFORMANCE
IT SHALL BE THE RESPONSIBILTY OF THE GENERAL CONTRACTOR TO PROVIDE "STRUCTURALFORM MASTERS" & THE CITY OF SAN DIEGO DEPARTMENT OF BUILDING INSPECTIIUON WITH AN "A.I.T.C." CERTIFICATE OF CONFOR-MANCE "GLU-LAMINATED WOOD FRAMING MEMBERS PRIOR TO THEIR INSTALLATION AT THE JOB SITE.

GENERAL GLU-LAMINATED BEAM NOTE:
MANUF. TO CERTIFY IN WRITING, PRIOR TO INSTALLATION OF GLU-LAM, TO THE FOL-LOWING SPECIFICATIONS LISTED BELOW: GLU-LAMINATED BEAMS, COMBINATION 24F (UNLESS NOTED OTHERWISE), 2000 I.B.C., Fb: 2,400 P.S.I., Fv: 165 P.S.I. & Fc: P.S.I., INDUS-TRIAL APPEARANCE WITH EXTERIOR GLUE.

FIGURE 15-6 ∎ Walls used to resist lateral forces are usually indicated on the framing walls. *Courtesy Gisela Smith, StructureForm Masters Inc.*

interior or exterior of the wall. The two faces must be joined to move as one unit. Figure 15-7 shows a wall detail used to explain the construction. The architectural team is usually responsible for locating the face material and for detailing the connection to the structural frame.

Parapet Walls

A *parapet* is a wall that extends above the roof line. A parapet wall is typically used to hide mechanical equipment located on low-pitch roofs (as seen in Figure 15-8), or to provide fire protection to the roof from other structures and vice versa. Locations are determined during the design stage by the architectural team. Drafters working with the architectural team are responsible for locating parapet walls on the roof and roof framing plans. Details showing the construction of parapet walls are also usually part of the architectural drawings.

Fire Walls

Chapters 4 and 11 introduce occupancies and fire construction requirements. A *fire wall* is a wall that has a specific fire rating and can resist the spread of fire for a specific amount of time due to the materials that were used to construct the wall. For example, a wood stud wall covered with a layer of 5/8" (16 mm) type X gypsum board on each side has a 1-hour fire resistance rating. A wall covered with 2 layers of 5/8" (16 mm) type X gypsum board on each side has a 2-hour fire rating. A steel beam protected with 2" (50 mm) of carbonate or lightweight concrete has a 4-hour fire rating. Table 719.1.1 of the IBC provides a comprehensive listing of common methods of protecting structural materials with various building materials. The Gypsum Association also has a comprehensive listing of methods of protecting wood and steel members from fire.

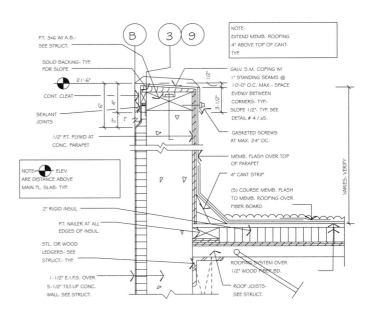

FIGURE 15-8 ■ A parapet wall is used to hide roof-mounted equipment or to provide protection from fire to adjacent properties. *Courtesy Architects Barrentine, Bates & Lee, A.I.A.*

Materials used to achieve a specific fire rating are typically specified on the floor plan by the architectural team. Details similar to Figure 15-9 are used to specify the fireproofing methods.

Retaining Walls

A retaining wall is a masonry wall designed to resist lateral displacement of soil or other materials. Retaining walls are shown on the floor and framing plans if they extend above the floor level, but

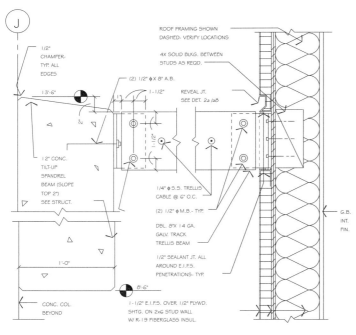

FIGURE 15-7 ■ Details are used to show how exterior finishes are attached to the skeleton. *Courtesy Architects Barrentine, Bates & Lee, A.I.A.*

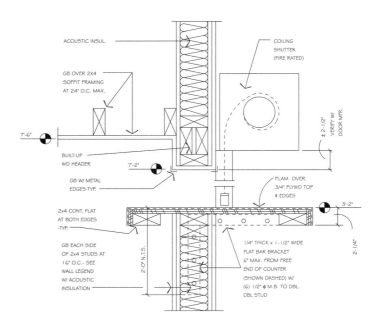

FIGURE 15-9 ■ Details are used to show methods of protecting a structure from the spread of smoke and fire. *Courtesy Architects Barrentine, Bates & Lee, A.I.A.*

they are always represented on the foundation plan. Wall location can be determined by the architectural team or a civil engineer. Details showing the construction of the wall are grouped with other concrete details by the engineering team. Figure 15-10 shows an example of an internal retaining wall completed by an architectural firm.

Representing Walls

The drafter is required to specify and distinguish each type of wall based on office standards for the construction crew. Walls are shown in plan view by pairs of bold parallel lines. The distance between the lines is determined by the type of material being represented. Common materials used for walls were introduced in Chapters 6 through 10. Several different types of walls can be seen on the plan shown in Figure 15-11. A wall legend similar to Figure 15-12 is usually provided to show what each pattern represents. If demolition or remodeling is required to be represented on the floor plan, additional symbols must also be used to distinguish between material to be removed and new construction. A separate floor plan similar to Figure 15-13 is typically used to represent material that must be used.

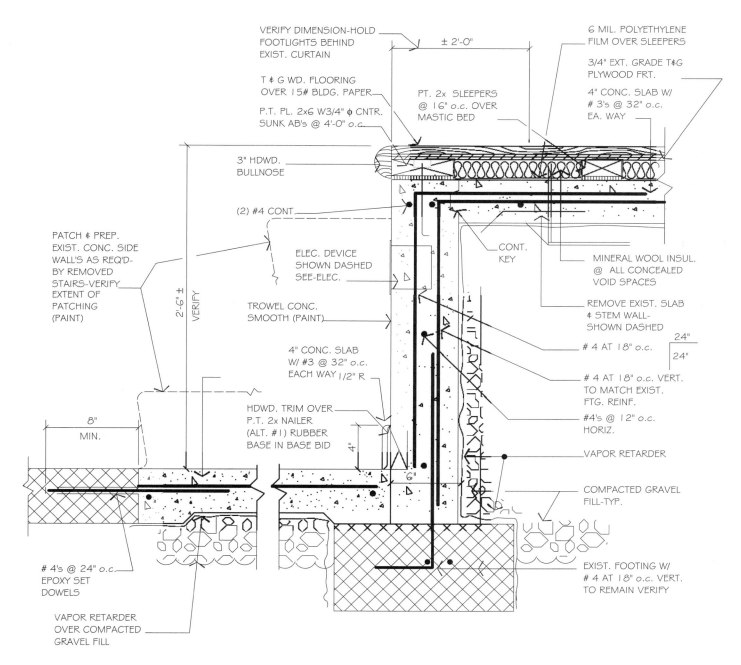

FIGURE 15-10 ■ Retaining wall details are used to show how soil loads are resisted. Drafters working for the architect, structural engineer, or civil engineer can complete the details depending on where the wall is located. *Courtesy Architects Barrentine, Bates & Lee, A.I.A.*

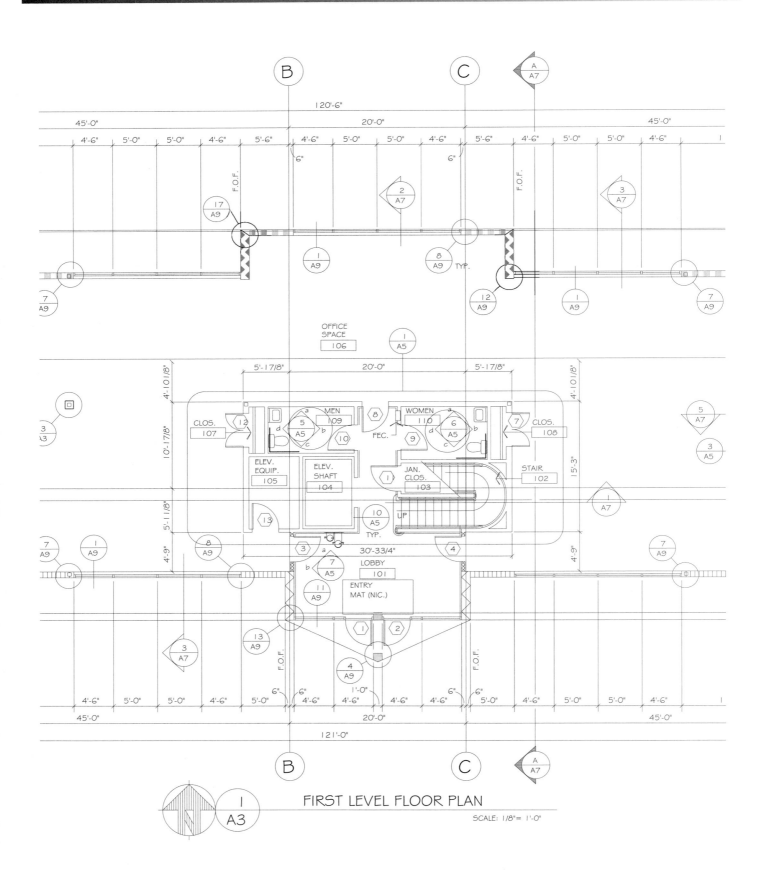

FIRST LEVEL FLOOR PLAN

SCALE: 1/8"= 1'-0"

FIGURE 15-11 ■ The floor plan can be used to represent different wall material. *Courtesy Architects Barrentine, Bates & Lee, A.I.A.*

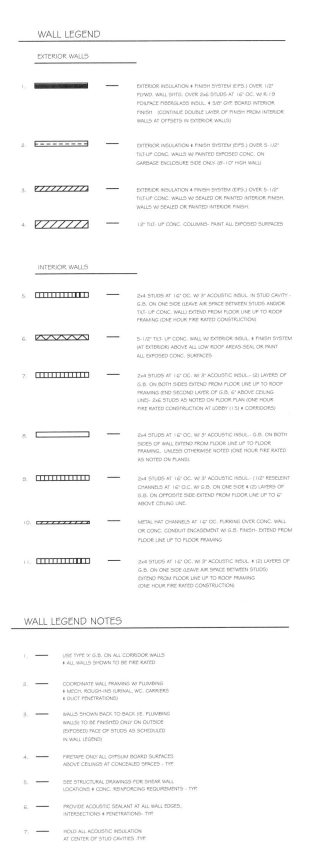

WALL LEGEND

EXTERIOR WALLS

1. — EXTERIOR INSULATION & FINISH SYSTEM (EIFS.) OVER 1/2" PLYWD. WALL SHTG. OVER 2x6 STUDS AT 16" OC. W/ R-19 FOILFACE FIBERGLASS INSUL. & 5/8" GYP. BOARD INTERIOR FINISH (CONTINUE DOUBLE LAYER OF FINISH FROM INTERIOR WALLS AT OFFSETS IN EXTERIOR WALLS)

2. — EXTERIOR INSULATION & FINISH SYSTEM (EIFS.) OVER 5-1/2" TILT-UP CONC. WALLS W/ PAINTED EXPOSED CONC. ON GARBAGE ENCLOSURE SIDE ONLY- (8'-10" HIGH WALL)

3. — EXTERIOR INSULATION & FINISH SYSTEM (EIFS.) OVER 5-1/2" TILT-UP CONC. WALLS W/ SEALED OR PAINTED INTERIOR FINISH. WALLS W/ SEALED OR PAINTED INTERIOR FINISH.

4. — 12" TILT- UP CONC. COLUMNS- PAINT ALL EXPOSED SURFACES

INTERIOR WALLS

5. — 2x4 STUDS AT 16" OC. W/ 3" ACOUSTIC INSUL. IN STUD CAVITY - G.B. ON ONE SIDE (LEAVE AIR SPACE BETWEEN STUDS AND/OR TILT- UP CONC. WALL) EXTEND FROM FLOOR LINE UP TO ROOF FRAMING (ONE HOUR FIRE RATED CONSTRUCTION)

6. — 5-1/2" TILT- UP CONC. WALL W/ EXTERIOR INSUL. & FINISH SYSTEM (AT EXTERIOR) ABOVE ALL LOW ROOF AREAS-SEAL OR PAINT ALL EXPOSED CONC. SURFACES

7. — 2x4 STUDS AT 16" OC. W/ 3" ACOUSTIC INSUL.- (2) LAYERS OF G.B. ON BOTH SIDES EXTEND FROM FLOOR LINE UP TO ROOF FRAMING (END SECOND LAYER OF G.B. 6" ABOVE CEILING LINE)- 2x6 STUDS AS NOTED ON FLOOR PLAN (ONE HOUR FIRE RATED CONSTRUCTION AT LOBBY (15) & CORRIDORS)

8. — 2x4 STUDS AT 16" OC. W/ 3" ACOUSTIC INSUL.- G.B. ON BOTH SIDES OF WALL EXTEND FROM FLOOR LINE UP TO FLOOR FRAMING, UNLESS OTHERWISE NOTED (ONE HOUR FIRE RATED AS NOTED ON PLANS).

9. — 2x4 STUDS AT 16" OC. W/ 3" ACOUSTIC INSUL.- (1/2" RESELEINT CHANNELS AT 16" O.C. W/ G.B. ON ONE SIDE & (2) LAYERS OF G.B. ON OPPOSITE SIDE-EXTEND FROM FLOOR LINE UP TO 6" ABOVE CEILING LINE.

10. — METAL HAT CHANNELS AT 16" OC. FURRING OVER CONC. WALL OR CONC. CONDUIT ENCASEMENT W/ G.B. FINISH- EXTEND FROM FLOOR LINE UP TO FLOOR FRAMING

11. — 2x4 STUDS AT 16" OC. W/ 3" ACOUSTIC INSUL. & (2) LAYERS OF G.B. ON ONE SIDE (LEAVE AIR SPACE BETWEEN STUDS) EXTEND FROM FLOOR LINE UP TO ROOF FRAMING (ONE HOUR FIRE RATED CONSTRUCTION)

WALL LEGEND NOTES

1. — USE TYPE 'X' G.B. ON ALL CORRIDOR WALLS & ALL WALLS SHOWN TO BE FIRE RATED

2. — COORDINATE WALL FRAMING W/ PLUMBING & MECH. ROUGH-INS (URINAL, WC. CARRIERS & DUCT PENETRATIONS)

3. — WALLS SHOWN BACK TO BACK (IE. PLUMBING WALLS) TO BE FINISHED ONLY ON OUTSIDE (EXPOSED) FACE OF STUDS AS SCHEDULED IN WALL LEGEND)

4. — FIRETAPE ONLY ALL GYPSUM BOARD SURFACES ABOVE CEILINGS AT CONCEALED SPACES - TYP.

5. — SEE STRUCTURAL DRAWINGS FOR SHEAR WALL LOCATIONS & CONC. REINFORCING REQUIREMENTS - TYP.

6. — PROVIDE ACOUSTIC SEALANT AT ALL WALL EDGES, INTERSECTIONS & PENETRATIONS- TYP.

7. — HOLD ALL ACOUSTIC INSULATION AT CENTER OF STUD CAVITIES -TYP.

FIGURE 15-12 ■ Wall legends are often used to explain the symbols used on a floor plan. *Courtesy Michael & Kuhns Architects, P.C.*

Windows

The drawing scale to be used and the type of structure affect the method used to represent the windows on a floor plan. Thin lines are used to represent the glass, sill, and jamb. Multifamily residential projects often use window symbols similar to those used with residential construction. Common symbols are shown in Figure 15-14. Many commercial projects use fixed glass set in an aluminum frame, referred to as storefront windows. Storefront windows can be represented in plan view as seen in Figure 15-15. The scale that is used to represent the floor plan alters the exact appearance of storefront windows. Figure 15-16 shows three common methods used to represent storefront windows, based on the scale of the floor plan.

In addition to representing the window in plan view, specifications and details are required to indicate the size, material, and installation of all windows. The size and type of the window are usually represented in a window schedule. A circle is placed by each window on the floor plan, with a letter or number placed in the symbol to reference the window to the schedule. The use of schedules is discussed later in this chapter. Window details are provided to represent how the window jamb, sill, and head attach to the wall. Figure 15-17 shows an example of a typical window detail referenced to the floor plan. Detail symbols are discussed later in this chapter.

Doors

Many of the same types of door symbols used on residential drawings are also used on commercial drawings. Common door symbols can be seen in Figure 15-18. Swinging doors are normally shown open 90° to the wall they are in. A thin line, thick line, or pairs of thin lines can be used to represent the door. The line that represents the door swing is drawn with a thin line. For doors that swing in two directions, the secondary direction is often shown using thin dashed lines. As with windows, doors are also represented in schedules, specifications, and details. Hexagons with either a letter or a numeral are also used to key doors on the floor plan to the door schedule. The same symbol used to key windows to the schedule should never be used for doors. If numbers are used to represent windows, letters should be used to key doors to the door schedule. In structures with a large number of rooms, the door may also be referenced to a specific room.

Openings

In addition to doors and windows, relites, transoms, skylights, archways, cased openings, and pass-throughs must be represented and specified, and detail symbols must be placed on the floor plan. *Relites*, or sidelights as they are often referred to, are windows that are mounted beside a door. They are drawn as a window on the floor plan, but they may be referenced with the door symbol or given a separate symbol and treated as a window. A *transom* is a window placed above a doorway. They are not drawn on the floor plan but are referenced with a separate symbol placed beside the door symbol.

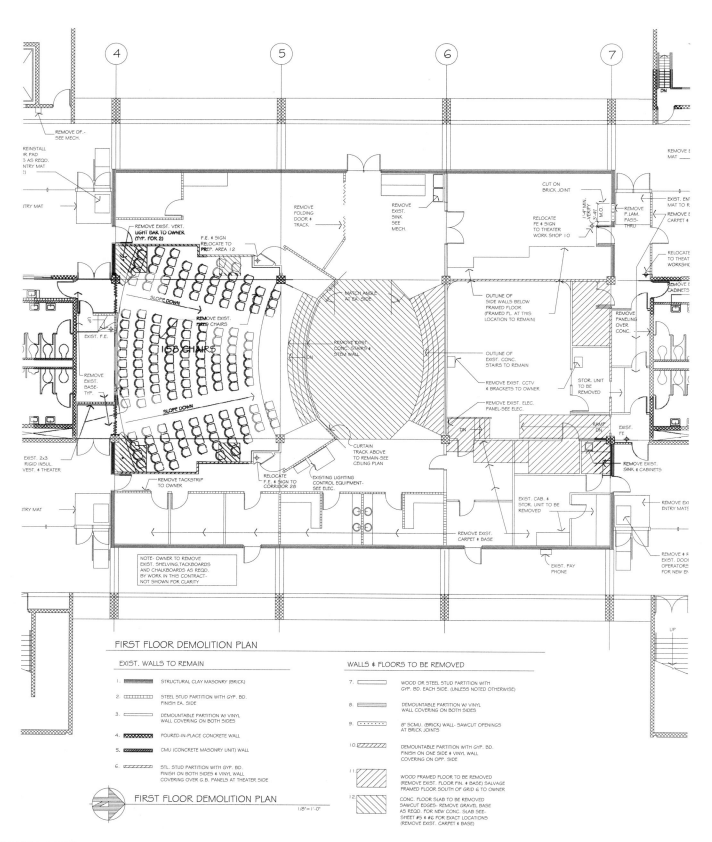

FIRST FLOOR DEMOLITION PLAN

EXIST. WALLS TO REMAIN

1. STRUCTURAL CLAY MASONRY (BRICK)

2. STEEL STUD PARTITION WITH GYP. BD. FINISH EA. SIDE

3. DEMOUNTABLE PARTITION W/ VINYL WALL COVERING ON BOTH SIDES

4. POURED-IN-PLACE CONCRETE WALL

5. CMU (CONCRETE MASONRY UNIT) WALL

6. STL. STUD PARTITION WITH GYP. BD. FINISH ON BOTH SIDES & VINYL WALL COVERING OVER G.B. PANELS AT THEATER SIDE

WALLS & FLOORS TO BE REMOVED

7. WOOD OR STEEL STUD PARTITION WITH GYP. BD. EACH SIDE. (UNLESS NOTED OTHERWISE)

8. DEMOUNTABLE PARTITION W/ VINYL WALL COVERING ON BOTH SIDES

9. 8" SCMU. (BRICK) WALL- SAWCUT OPENINGS AT BRICK JOINTS

10. DEMOUNTABLE PARTITION WITH GYP. BD. FINISH ON ONE SIDE & VINYL WALL COVERING ON OPP. SIDE

11. WOOD FRAMED FLOOR TO BE REMOVED (REMOVE EXIST. FLOOR FIN. & BASE) SALVAGE FRAMED FLOOR SOUTH OF GRID 6 TO OWNER

12. CONC. FLOOR SLAB TO BE REMOVED SAWCUT EDGES- REMOVE GRAVEL BASE AS REQD. FOR NEW CONC. SLAB SEE-SHEET #5 & #6 FOR EXACT LOCATIONS (REMOVE EXIST. CARPET & BASE)

FIRST FLOOR DEMOLITION PLAN

1/8"= 1'-0"

FIGURE 15-13 ■ Dashed lines often represent existing construction that must be removed. When removal is extensive, a separate demolition plan is provided. *Courtesy Architects Barrentine, Bates & Lee, A.I.A.*

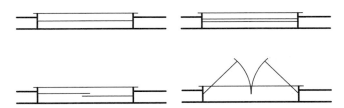

FIGURE 15-14 ■ Windows can be represented on a floor plan using several common symbols. Each can be created as a wblock and inserted as needed.

FIGURE 15-15 ■ Representing storefront glass in plan view.

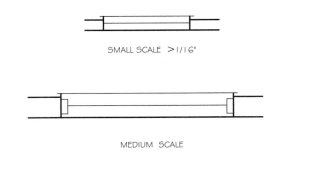

SMALL SCALE >1/16"

MEDIUM SCALE

LARGE SCALE 1/2"<

FIGURE 15-16 ■ The drawing scale will affect the detail used to represent windows.

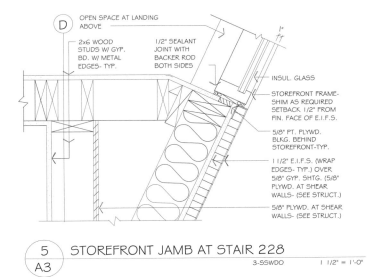

FIGURE 15-17 ■ Window details explain how the window and finish materials are to be placed and are referenced to the floor plan. *Courtesy Architects Barrentine, Bates & Lee, A.I.A.*

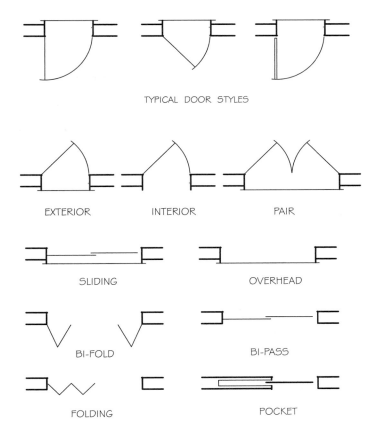

TYPICAL DOOR STYLES

EXTERIOR   INTERIOR   PAIR

SLIDING   OVERHEAD

BI-FOLD   BI-PASS

FOLDING   POCKET

FIGURE 15-18 ■ Representing doors in plan view.

Skylights are often drawn as a square or rectangle on the floor plan with either a thin dashed or continuous line depending on office preference. On a simple plan, the specifications to describe the skylight can be placed on the floor plan. On complicated plans, skylights are referenced to a schedule that explains the size, type, and manufacturer. Details that show how the skylight is installed are referenced on the roof plan. An *archway* is an opening in a wall, which usually has no trim. A *cased opening* is an opening in a wall that has trim similar to a door jamb but has no trim for door stops. A *pass-through* is an opening in a wall that does not go all the way to the floor. It may be trimmed or untrimmed.

Floors

The floor system provides the horizontal surface for each level of the structure. Some occupancies such as theaters or assembly halls have a sloping floor to enhance viewing or sound reproduction. In multilevel structures, the floor of an upper space is also part of the ceiling of the lower level. The architectural team determines the shape and limits of the floor. The structural team designs the necessary supports and the material to construct the floor system. If the floor is carefully connected to supporting walls, the floor system can also be used to resist lateral loads from the walls. The shape and levels of the floor are shown and specified on the floor plan. Construction of the floor is shown on the framing plans and in the sections and details.

Changes in Floor Elevation

Steps, stairs, and ramps are common methods of changing elevation that must be shown on the floor plan. Each is represented on the floor plan by thin lines. *Steps* represent the change in elevation between two or more levels of the same floor. A *ramp* is an inclined floor that connects two different floor elevations of the same floor level. An elevated stage within a room would be an example of where steps or a ramp would be used to travel from one level to another. In addition to representing the steps, specifications are placed on the plan to indicate the number and maximum rise and minimum run of the steps. Ramps or inclined floors are generally represented on the floor plan, as shown in Figure 15-19. Ramps should include elevations of each floor level serviced by the ramp or provide the rise and run in slope of the ramp.

Stairs represent the change in elevation between two or more different floors and require the use of two or more floor plans to show the complete stair. Because stairs connect two different floor levels, the lower portion of the stair is shown on the lower floor and the upper portion of the stair is shown on the upper floor. Figure 15-20 shows how a straight stair run can be represented on a floor plan. Each step is represented by thin lines that are spaced a distance equal to the run. At each edge of the stairs, a thin line should be drawn to represent the handrail. If walls do not surround the stairs, thin lines should be used to represent the guardrail, which would surround the stairs on the upper floor. If the stair has a landing in the run, it should be represented on each floor plan showing the stair, as seen in Figure 15-21. Showing the landing and the next few steps establishes a visual reference point in placing the landing. On structures with more than three levels of stacked stairs, the landing shown on each floor plan represents the landing that is above that floor, except for the uppermost floor plan. A stair for a four-level structure can be seen in Figure 15-22.

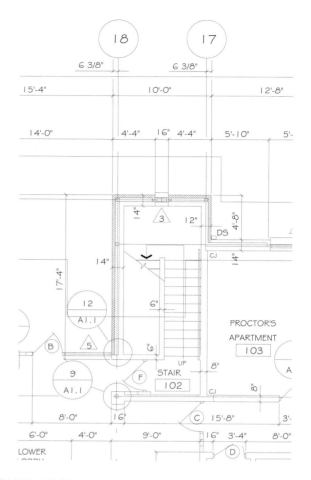

FIGURE 15-20 ■ Representing stairs on a floor plan. *Courtesy G. Williamson Archer A.I.A., Archer & Archer P.A.*

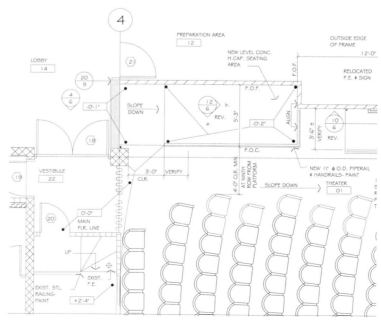

FIGURE 15-19 ■ Ramps and sloping floors and their heights should be represented and specified on the floor plan. *Courtesy Architects Barrentine, Bates & Lee, A.I.A.*

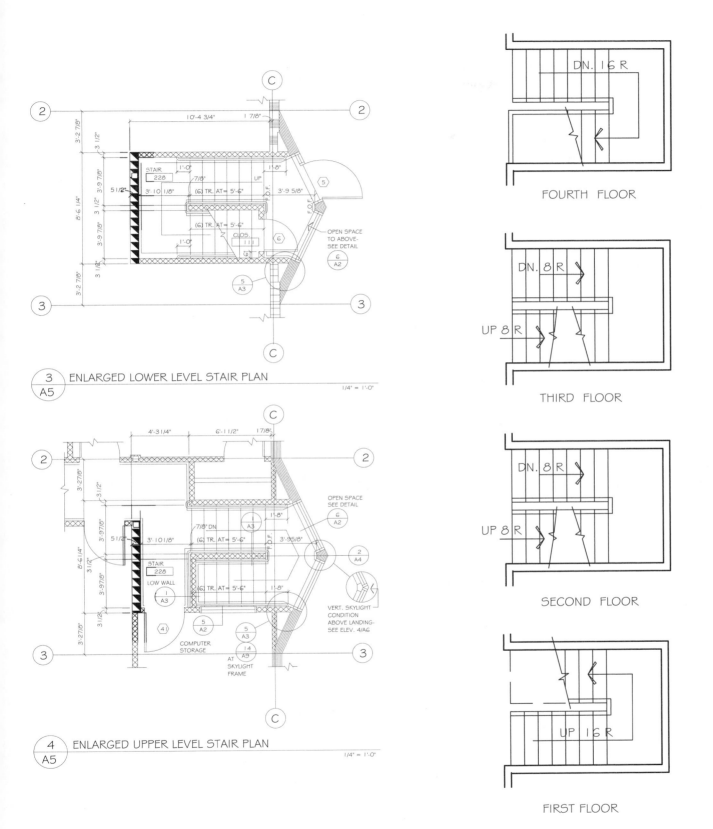

FIGURE 15-21 ■ If a stair has a landing between floors, it must be represented on each level the stair serves. *Courtesy Architects Barrentine, Bates & Lee, A.I.A.*

FIGURE 15-22 ■ Stairs for a multilevel structure must represent each run direction.

Plumbing Symbols

Plumbing symbols are added to the floor plan by the use of blocks. The type of structure greatly influences the plumbing fixtures that must be represented on the floor plan. An apartment may show common items such as tubs, showers, toilets, lavatories, sinks, spas, and pools. A warehouse may require drinking fountains, lavatories, toilets, urinals, and equipment specific to a particular occupancy. Figure 15-23 shows plumbing symbols that may be represented on the floor plan. Although the symbols are self explanatory, many firms label each fixture on the floor plan and show each fixture in the interior elevations. Figure 15-24 shows a bathroom for an office structure. The cabinet elevations that would accompany the plan can be seen in Figure 15-25. The drawing of interior elevations are discussed in Chapter 19.

Interior Equipment and Furnishings

The complexity of the structure, the drawing scale, and the amount and type of cabinets affect where interior equipment and furnishings are shown. Cabinets and other interior furnishing supplied by

FIGURE 15-23 ■ A wide range of plumbing fixtures may need to be represented on the floor plan, depending on the occupancy.

the contractor can be shown on a floor plan, a space plan, or a fixture schedule. Common materials that are represented as the interior furnishing include the following:

- Built-in cabinets
- Base cabinets and countertops, upper cabinets
- Shelves, display racks, and display counters
- Mechanical equipment such as water heaters, washers and dryers, cooking equipment, heating or cooling units, condensers, and compressors.
- Electrical equipment such as cash registers, communications equipment, or computer work stations
- Built-in furnishings such as seating or benches
- Special equipment specific to the occupancy such as churches, schools, restaurants, medical facilities, malls, or theaters

Blocks for some features are available from third-party vendors, but most blocks must be developed by the drafter and stored for future use. Thin lines are used to represent the item being drawn. The method and the detail used to represent each item are affected by the scale of the floor plan. The complexity used to show a feature can be reduced if the item is manufactured and must be installed rather than fabricated according to the working drawings. Features that must be installed are often represented on an enlarged floor plan. Features that must be fabricated at the job site must be explained by a detail, which must be referenced on the floor plan.

DIMENSIONS AND TEXT

Dimensions are used on the floor plan to locate walls, windows, doors, and interior features. Not all windows, doors, and interior furnishing have dimension lines to locate them if they can be located by their position to a known wall. The method of placing dimensions is also affected by the material being represented.

Basic Dimensioning Concepts

Dimensions for commercial drawings build on the skills developed in residential drawing classes. Dimension text is 1/8" (3 mm) high and placed parallel to the surface being defined. Dimensions are centered slightly above a thin, continuous line called a *dimension line*. The dimension line extends between thin, continuous lines called *extension lines*, which extend from the edge of the structure or feature being described. Where the dimension line intersects the extension line, an arrow or tick mark is used as a line terminator. The dimension line often extends past the extension line about 1/8" (3 mm).

Extension lines are normally offset from the feature being described by about 1/8" (3 mm) and extend past the dimension line by about 1/8" (3 mm). If the extension line is describing the edge or face of a surface, a thin, continuous line is used. When a feature is being located by its center point, a thin centerline with a long-short-long line pattern is used. The long line of the pattern is

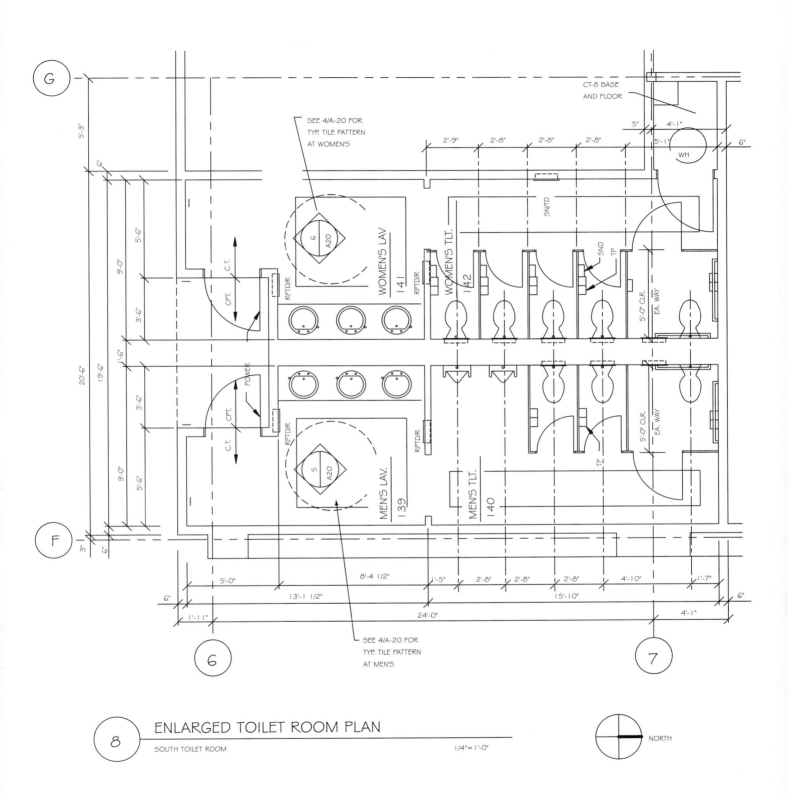

ENLARGED TOILET ROOM PLAN

SOUTH TOILET ROOM   1/4"=1'-0"

NORTH

FIGURE 15-24 ■ Because the floor plan of a bathroom generally must show special finishes and all fixtures, an enlarged plan of the area is often provided. *Courtesy Michael & Kuhns Architects, P.C.*

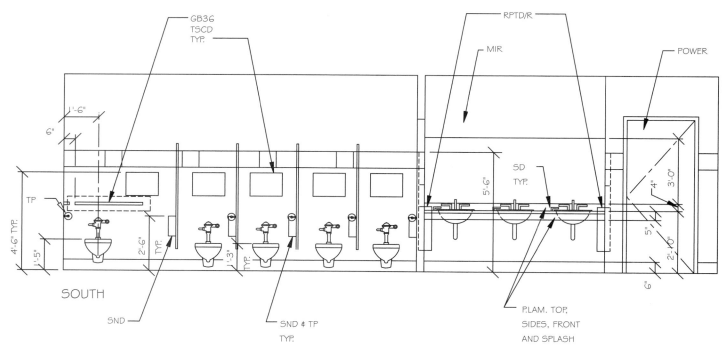

FIGURE 15-25 ■ Cabinet elevations can be used to specify fixtures and finishes. Elevations are often referenced to the floor plan. *Courtesy Michael & Kuhns Architects, P.C.*

3/4 to 1" (20 to 25 mm) long and the short line is about 1/8" (3 mm) long with a 1/16" (1.5 mm) gap between each. Figure 15-26 shows examples of each dimensioning component.

Dimension Placement

Placing dimensions in plan view can be divided into the areas of interior and exterior dimensions. Whenever possible, dimensions should be placed outside the drawing area.

Exterior Dimensions

Most offices start by placing an overall dimension on each side of the structure that is approximately 2" (50 mm) from the exterior wall. Moving inward, approximately 1/2" (13 mm) is placed between lines that are used to describe major jogs in exterior walls, the distance from wall to wall, and the distance from wall to opening. The exact method of placing these dimensions will vary slightly depending on the material being used. Figure 15-27 shows an example of exterior dimensions for a structure.

Two different systems are used to represent dimensions between exterior and interior walls:

■ Architectural firms tend to represent the distance from edge to edge of walls as shown in Figure 15-28

■ Engineering firms tend to dimension walls from edge of exterior walls to the center of interior walls as seen in Figure 15-29

When placing dimensions, it is important to adjust the use of four dimension lines to the shape of the structure. If one side of the

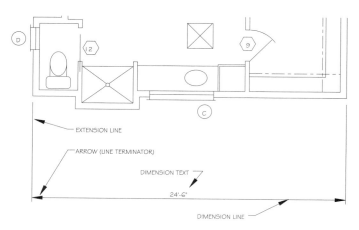

FIGURE 15-26 ■ Dimensions are placed on the floor plan by use of an extension line, dimension line, and a line terminator such as a tick or an arrow.

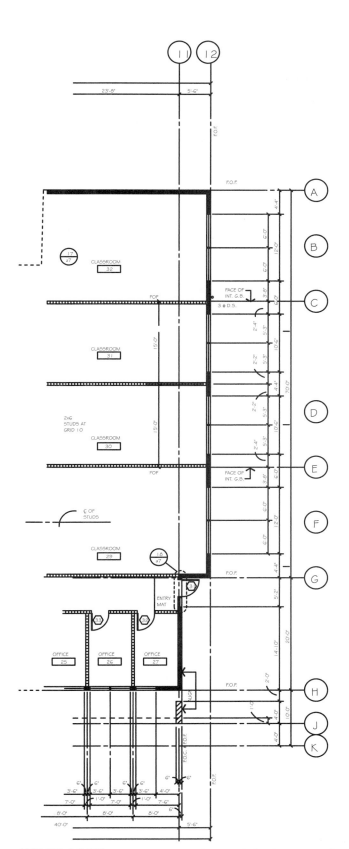

FIGURE 15-27 ■ Dimensions must be provided to locate every feature. Each successive line of dimensions must add up to the dimension on the line above it. *Courtesy Architects Barrentine, Bates & Lee, A.I.A.*

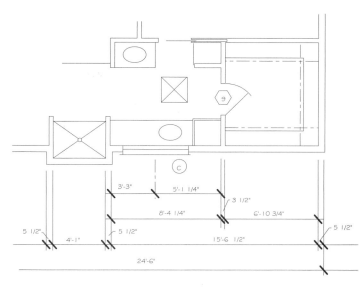

FIGURE 15-28 ■ Many architectural firms dimension walls from edge to edge.

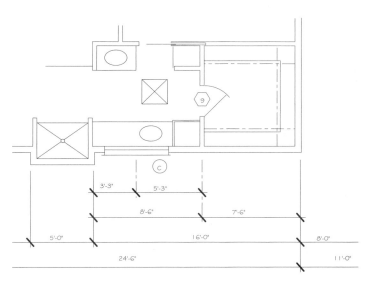

FIGURE 15-29 ■ Many engineering firms dimension walls from the edge of exterior to the center of interior walls. Interior walls made of masonry or concrete are dimensioned to an edge.

structure has no offsets, the wall-to-wall and wall-to-opening lines of dimensions can be moved out from the structure. If one section of a structure has no interior walls and several windows, the dimension for wall-to-windows would be moved out as far as possible from the structure. It is not important that four lines of dimensioning always be maintained. It is critical, however, that each line of dimensions adds up to the corresponding dimension on each succeeding dimension line, as seen in Figure 15-27.

Interior Dimensions

The two main considerations in placing interior dimensions are clarity and groupings. Dimension lines and text must be placed so that they can be read easily and so that neither interferes with other information that must be placed on the drawing. Information should also be grouped together as seen in Figure 15-30 so that construction workers can find dimensions easily. Interior dimensions should be placed so that they extend between the location of a surfaces described by the exterior dimensions. Interior walls are

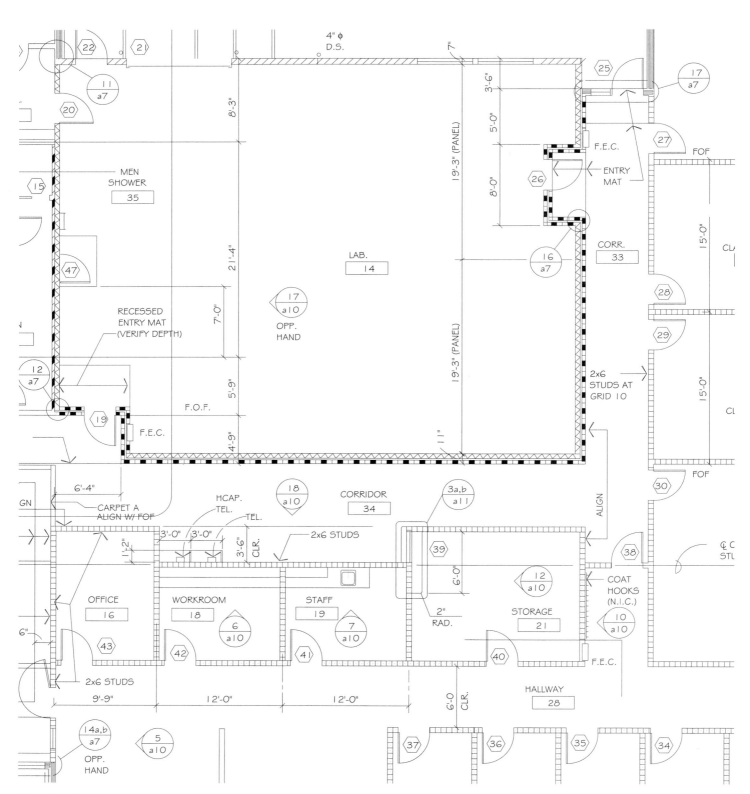

FIGURE 15-30 ■ Interior dimensions should be grouped together whenever possible to ease the job of the print reader. *Courtesy Architects Barrentine, Bates & Lee, A.I.A.*

generally dimensioned to the center of stud. Doors located in the corner or centers of rooms are generally not dimensioned.

Dimensioning Small Spaces

Often small areas in which not enough space is available for the text to be placed between extension lines must be dimensioned. Although options vary with each office, several alternatives for placing dimension in small spaces can be seen in Figure 15-31.

Metric Dimensioning

Structures dimensioned using metric measurement are defined in millimeters. Large distances may be defined in meters. When all units are given in millimeters, the unit of measurement requires no identification. For instance, a distance of 2440 millimeters would be written as 2440. Dimensions other than millimeters should have the unit description by the dimension such as 24 m. See Chapter 3 for metric conversions and equivalents.

Dimensioning Commands

Twelve AutoCAD command options are available for placing dimensions on a drawing. Four of these options are for providing linear dimensioning: Horizontal, Vertical, Aligned, and Rotated. Another four options, used to describe circular features are Angular, Diameter, Radius, and Center. Three other options, Quick, Baseline, and Continue, are used to place dimensions and can be combined with other dimensioning options. The dimension options can be accessed by toolbars, keyboard, or by using the on-screen menus.

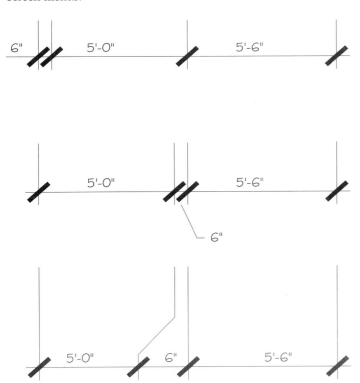

FIGURE 15-31 ■ Dimensions for small spaces should be placed so they can be clearly seen.

Light-Frame Dimensions

Light-frame structures are usually dimensioned by following the four dimension groupings just presented. Dimensions are placed on each side of the structure to define

■ The outer line defines the overall limits of the structure

■ The next line of dimensions is used to describe changes in shape or major jogs. The changes are dimensioned from exterior face of the material used to build one wall to the exterior face of the next wall. The extension line is sometimes labeled with text parallel to the extension line that reads F.O.S. or F.S., representing the face of stud.

■ The third line of dimensions would represent the distance from the edge of exterior walls to the center (edge for some architectural offices) of interior walls

■ The fourth line of dimensions is used to locate openings and extends from a wall to the center of the openings

Figure 15-27 shows dimensioning methods for a light-frame structure.

Masonry Dimensions

Structures made of concrete block, poured concrete, or tilt-up concrete are dimensioned using similar line placement methods as with light-frame construction. Differences include the following:

■ Masonry walls are always dimensioned to edge and never to center

■ Openings for doors and windows are also dimensioned from edge rather than from center

When the structure is made from concrete block, all dimensions should be based on 8" (200 mm) modules. Distances in odd number of feet should always end in a 4" (100 mm) increment such as 15'–4". Even-numbered distances will always end in 0 or 8 inch increments such as 8'–0" or 10'–8" to be modular. Interior light-frame walls are dimensioned to their centers as previously described. Figure 15-32 shows an example of common dimensioning practices for a masonry structure.

Structures made from poured concrete are dimensioned in similar fashion to concrete block but do not need to conform to the 8" (200 mm) module. Tilt-up concrete walls are dimensioned from edge to edge of each panel or from one grid point to the next grid on the architectural drawings. Figure 15-33 shows how concrete tilt-up walls can be dimensioned. The structural drawings include elevations of each panel showing exact panel sizes. Details are typically provided to define the exact space between panels.

Steel and Timber Dimensions

In addition to the four types of dimensions used with other materials, structures framed with steel or timber require dimensions to

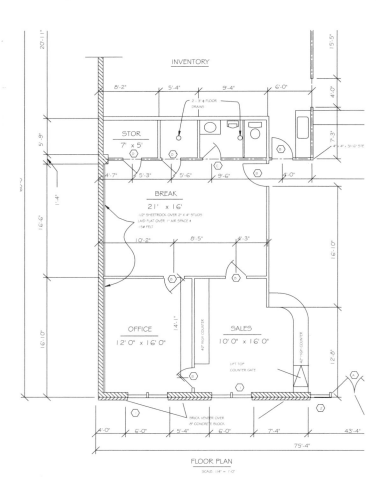

FIGURE 15-32 ■ Dimensions for locating masonry should always be in modular sizes unless the project manager provides other instructions. *Courtesy Wil Warner.*

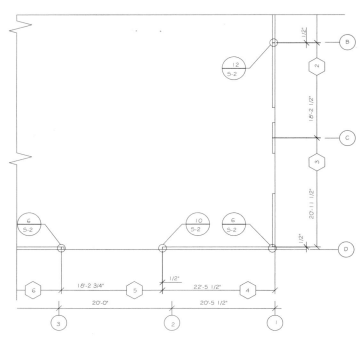

FIGURE 15-33 ■ Concrete tilt-up structures are dimensioned from panel joint to panel joint or from grid to grid. *Courtesy StructureForm Masters, Inc.*

locate the grid placement of the vertical columns. The grid shows the location of each vertical column and is placed after the overall dimensions have been placed. Once the structural shell has been erected, other dimensions can be used to locate walls and openings in the skeleton. Walls are typically referenced back to the grid line of the structural material. Figure 15-34 shows a floor plan for a structure framed of steel.

Drawing Notation

The use of local notes to specify materials on the floor plan has been mentioned throughout this chapter. Items that are inserted into the drawing as a block or wblock often include text as an attribute that can be customized to the usage when inserted. Material that must be identified by notation should be identified after dimensions are placed so that the text does not interfere with dimension placement. Text can be referenced to the item being described by the use of a leader line as seen in Figure 15-35. The leader line should be kept as close to the feature as possible, with-

out crowding the text or other features. If a note is lengthy, a shortened version can be placed on the drawing to identify the feature, with the complete specification given in a general note.

Many professionals place general notes on the floor plan to ensure the compliance of the code that governs construction and to clarify all construction. These general notes can be placed beside the floor plan or on a separate specification page that is included with the working drawings. Figure 15-36 shows common notes that might be included with the floor plan. When a material is specified in a general note, a shortened version of the note should still be given on the floor plan to denote the accurate location of the specified material. A useful order for placing text on a drawing would be

1. Room titles
2. Component identification
3. General notes
4. Drawing titles

DRAWING SYMBOLS

Symbols are added to drawings to add clarity and eliminate notations. Common symbols shown on a floor plan include the north arrow, grid markers, section markers, detail markers, cabinet reference symbols, elevation symbols, and finish symbols. Because there are many variations to symbols used on the floor plan, a legend should be included with the floor plan. Figure 15-37 shows common symbols found on a floor plan.

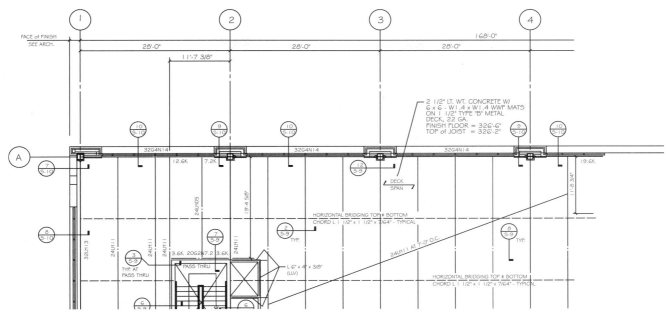

FIGURE 15-34 ■ Structural steel and timber framing require dimensions to represent each member of the skeleton. *Courtesy Peck, Smiley, Ettlin Architects.*

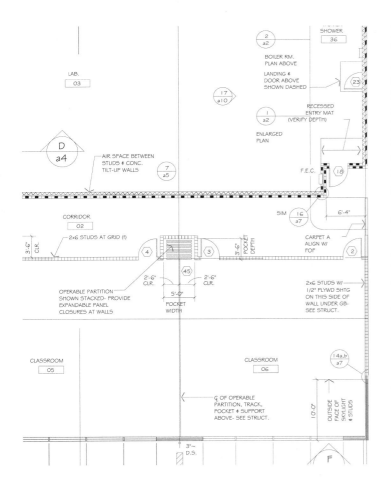

FIGURE 15-35 ■ Referencing local notes to material on the floor plan. *Courtesy Architects Barrentine, Bates & Lee, A.I.A.*

North Arrow

A north arrow should be placed on the floor plan to help orient the structure to the site. The north arrow should be simple and is typically placed in a circle measuring between 3/4" and 1" (20 and 25 mm) diameter. The symbol is often placed near the drawing title. North is placed at the top of the page if possible, although it can be rotated to the right side of the drawing sheet because of project and sheet size limitations. True north can be indicated if the orientation is skewed by use of a north reference arrow. Examples of each are shown at A in Figure 15-37.

Grid Markers

A 3/8" to 1/2" (9 to 13 mm) diameter circle is typically used to represent the grid and column lines of a structure. A medium-width centerline extends from the circle through the drawing. Either 1/4" (6 mm) high text or numbers are used to define each grid. Numbers can be used to define grids from top to bottom, with consecutive letters of the alphabet used to define the grids from left to right. In addition to referencing steel and timber columns, grid markers are used to reference each major change in shape of the structure, as well as interior bearing columns. Figure 15-37 at B shows an example of grid reference symbols. Figure 15-34 and all of the other floor plans in this chapter show the use of grid markers.

Section Markers

Section markers and section lines are used to indicate where each building and wall section is cut. A 3/8" to 1/2" (9 to 13 mm)

WALL LEGEND NOTES

1. COORDINATE WALL FRAMING W/ PLUMBING & MECH. ROUGH-IN (URINAL, WC. CARRIERS & DUCT PENETRATIONS & RETURN AIR DROPS)

2. WALLS SHOWN BACK TO BACK (IE. PLUMBING WALLS) TO BE FINISHED ONLY ON OUTSIDE (EXPOSED) FACE OF STUDS AS SCHEDULED IN WALL LEGEND, EXCEPT @ STRUCT. WALLS)

3. FIRETAPE ALL GYPSUM BOARD SURFACES ABOVE CEIL. AT CONCEALED SPACES - TYP.

4. PROVIDE TYPE 'X' GYP. BD. ON ALL WALLS- TYPICAL- SEE ROOM FINISH SCHEDULE FOR FINISHES NOT INCLUDED IN LEGEND

5. PROVIDE ACOUSTIC SEALANT @WALL EDGES, INTERSECTIONS & PENETRATIONS- TYP.

6. HOLD ALL ACOUSTIC INSULATION AT CENTER OF STUD CAVITIES -TYP.

7. SET ALL 2x SILL PLATES AT EXTERIOR WALLS ON SILL INSULATION- TYPICAL.

PLAN NOTES

1. VERIFY ALL CONDITIONS, DIMENSIONS AND LOCATIONS.

2. PERFORM ALL WORK IN COMPLIANCE WITH APPLICABLE CODES AND STANDARDS.

3. COORDINATE & INSTALL ALL NECESSARY BLOCKING, SHIMS AND BACKING FOR FIXTURES, EQUIPMENT AND ACCESSORIES.

4. COORDINATE ALL WORK WITH STRUCTURAL, CIVIL, LANDSCAPE AND SPECIFICATIONS.

5. PROVIDE 2x FIRE STOPPING PER UBC. SECTION 2516 (f).

6. SEE STRUCTURAL DRAWINGS FOR SHEAR WALL LOCATIONS AND REQUIREMENTS- TYP.

7. EXTERIOR DIM. ARE TO OUTSIDE FACE OF EXTERIOR INSUL. & FINISH SYSTEM (E.I.F.S.) UNLESS NOTED OTHER WISE ON PLANS - INTERIOR DIM. ARE TO FACE OF STUDS, INCLUDING AT OPENINGS (ROUGH OPENING) UNLESS OTHER WISE NOTED ON PLANS.

8. WRAP BACK OF WALL MOUNTED EQUIPMENT W/ GYP. BD. FOR ACOUSTICAL RATING INTEGRITY. VERIFY ALL LOCATIONS.

FIGURE 15-36 ■ Notes that apply to the entire project can be placed in a clear portion of the floor plan. *Courtesy G. Williamson Archer A.I.A., Archer & Archer P.A.*

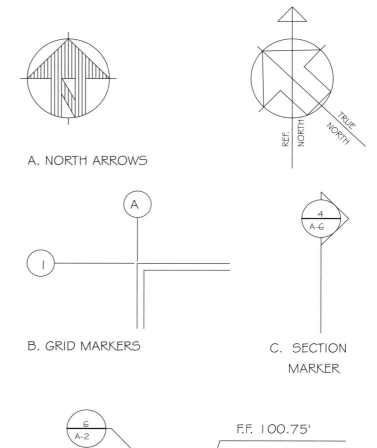

A. NORTH ARROWS

B. GRID MARKERS

C. SECTION MARKER

D. DETAIL MARKER

E. ELEVATION REFERENCE

FIGURE 15-37 ■ Standard symbols used throughout floor plans.

diameter circle is typically used to identify the section. A bold phantom line is used to indicate where the cutting plane has passed through the structure. An arrow is placed around the circle to indicate the viewing direction. As seen at C in Figure 15-37, the circle is divided into two portions, with a letter representing the section placed over the page number. Chapter 18 will discuss methods for drawing sections.

Detail Markers

Detail markers are used to reference construction details to the floor plan. A 3/8" to 1/2" (9 to 13 mm) diameter circle is typically used to identify each detail. A thin line is used to reference where the detail occurs. The circle referencing the detail is divided into two portions, with a number representing the detail placed over the page number. Figure 15-37 at D shows an example of a detail marker and the detail it represents.

Elevation Symbols

An elevation symbol is used to represent changes in height of a specific material. On plan views, the symbol consists of the elevation placed above a horizontal line, which is connected to the material by an inclined line with an arrow terminator. An elevation symbol can be seen at E in Figure 15-37. Elevation symbols are used on floor, framing, site, and foundation plans, and on exterior elevations. Figure 15-19 shows examples of elevations referenced to a floor plan. Chapters 16, 18 and 22 introduce elevation symbols used to specify vertical locations on elevations, sections, and details.

SCHEDULE NOTATIONS

Schedules are used on most architectural and structural drawings to reduce the information that must be placed in and around the drawing. Schedules add clarity to a drawing by removing notations and placing the notations either at the edge of the floor plan or on another sheet. Common schedules associated with floor plans include door, window, finish, hardware, and appliance schedules. Symbols used on the floor plans can be seen in Figure 15-38. Text used in the schedule is approximately 1/8" (3 mm) high, with titles within the schedule slightly larger. For example, a title such as **Door Schedule** is generally about 1/4" (6 mm) high in a bold text font. Lines should be placed between each entry in the schedule to add to clarity, with a minimum of half the height of the lettering provided to avoid crowding.

Door Schedules and Symbols

A door is referenced on the floor plan to a schedule by the use of a hexagon containing a number. (See Figure 15-38 at A.) The text representing the door symbol is typically 1/8" (3 mm) high in a 1/4" (6 mm) hexagon. Depending on office procedure, the number may either be sequential or related to the room containing the door.

■ With sequential numbering, the main entry door is usually assigned the number 1. Other doors may be numbered as they would be encountered if you were to walk through the building.

■ An alternative to sequential numbering would be to assign 1 to the main entry door, 2 to the next exterior door encountered walking around the perimeter of the structure and so on. Once all exterior doors are referenced, interior doors can be referenced based on size and type.

Interior doors are often listed in the schedule from largest to smallest for each specific type. Doors can also be assigned a number based on the location of the door. If you were in a hallway and needed to represent a door going into office 102, the door would receive a symbol of 102. If more than one door enters office 102 from the hall, the doors could be referenced with the symbol 102a and 102b.

Typically, doors are assigned a door symbol on a print of the floor plan. A rough draft of the schedule is established prior to placing the information on the floor plan and schedule. The architect will assign the type of doors to be used and the drafter will develop the schedule based on company policy. Although material will vary depending on the occupancy and complexity of a structure, a schedule will normally contain information related to

■ The reference number

■ The width, height, and thickness

■ Door type and UL rating

■ Door and frame material

■ Door and frame finish

■ References for details showing the jamb, header, and threshold construction

■ Notations to clarify door operation, location, or construction

Figure 15-39 shows a portion of the door schedule, the notes, and door drawings for the doors on the floor plan that will be

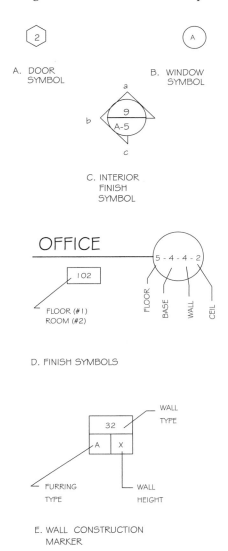

FIGURE 15-38 ■ Symbols are often developed by an office to explain materials referenced to schedules.

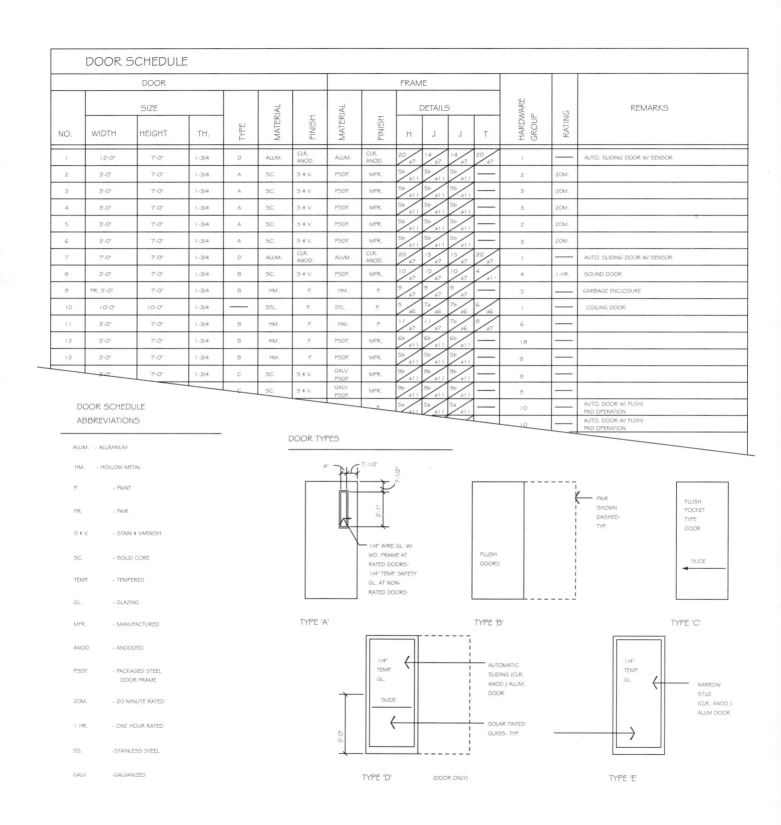

NO.	WIDTH	HEIGHT	TH.	TYPE	MATERIAL	FINISH	MATERIAL	FINISH	H	J	J	T	HARDWARE GROUP	RATING	REMARKS
1	12'-0"	7'-0"	1-3/4	D	ALUM.	CLR. ANOD.	ALUM.	CLR. ANOD.	20 a7	14 a7	14 a7	20 a7	1	—	AUTO. SLIDING DOOR W/ SENSOR
2	3'-0"	7'-0"	1-3/4	A	S.C.	S & V.	PSDF.	MFR.	5b a11	5b a11	5b a11		2	20M.	
3	3'-0"	7'-0"	1-3/4	A	S.C.	S & V.	PSDF.	MFR.	5b a11	5b a11	5b a11		3	20M.	
4	3'-0"	7'-0"	1-3/4	A	S.C.	S & V.	PSDF.	MFR.	5b a11	5b a11	5b a11		3	20M.	
5	3'-0"	7'-0"	1-3/4	A	S.C.	S & V.	PSDF.	MFR.	5b a11	5b a11	5b a11		2	20M.	
6	3'-0"	7'-0"	1-3/4	A	S.C.	S & V.	PSDF.	MFR.	5b a11	5b a11	5b a11		3	20M.	
7	7'-0"	7'-0"	1-3/4	D	ALUM.	CLR. ANOD.	ALUM.	CLR. ANOD.	20 a7	15 a7	15 a7	20 a7	1	—	AUTO. SLIDING DOOR W/ SENSOR
8	3'-0"	7'-0"	1-3/4	B	S.C.	S & V.	PSDF.	MFR.	10 a7	10 a7	10 a7	4 a11	4	1-HR.	SOUND DOOR
9	PR. 5'-0"	7'-0"	1-3/4	B	HM.	P.	HM.	P.	9 a7	9 a7	9 a7		5	—	GARBAGE ENCLOSURE
10	10'-0"	10'-0"	1-3/4	—	STL.	P.	STL.	P.	5 a6	7a a6	7b a6	6 a6	1	—	COILING DOOR
11	3'-0"	7'-0"	1-3/4	B	HM.	P.	HM.	P.	11 a7	11 a7	7b a6	8 a7	6	—	
12	3'-0"	7'-0"	1-3/4	B	HM.	P.	PSDF.	MFR.	6b a11	6b a11	6b a11		18	—	
13	3'-0"	7'-0"	1-3/4	B	HM.	P.	PSDF.	MFR.	5b a11	5b a11	5b a11		8	—	
	3'-0"	7'-0"	1-3/4	C	S.C.	S & V.	GALV. PSDF.	MFR.	9b a11	9b a11	9b a11		9	—	
				C	S.C.	S & V.	GALV. PSDF.	MFR.	9b a11	9b a11	9b a11		9	—	
								P.	5a a11	5a a11	5a a11		10	—	AUTO. DOOR W/ PUSH/ PAD OPERATION
													10	—	AUTO. DOOR W/ PUSH/ PAD OPERATION

DOOR SCHEDULE
ABBREVIATIONS

ALUM. - ALUMINUM

HM. - HOLLOW METAL

P. - PAINT

PR. - PAIR

S & V. - STAIN & VARNISH

SC. - SOLID CORE

TEMP. - TEMPERED

GL. - GLAZING

MFR. - MANUFACTURED

ANOD. - ANODIZED

PSDF. - PACKAGED STEEL DOOR FRAME

20M. - 20 MINUTE RATED

1 HR. - ONE HOUR RATED

SS. - STAINLESS STEEL

GALV. - GALVANIZED

DOOR TYPES

4" 7-1/2"
7-1/2"
2'-1"
1/4" WIRE GL. W/ WD. FRAME AT RATED DOORS
1/4" TEMP. SAFETY GL. AT NON-RATED DOORS

TYPE 'A'

PAIR SHOWN DASHED-TYP.
FLUSH DOORS

TYPE 'B'

FLUSH POCKET TYPE DOOR
SLIDE

TYPE 'C'

1/4" TEMP. GL.
SLIDE
3'-0"
AUTOMATIC SLIDING (CLR. ANOD.) ALUM. DOOR
SOLAR TINTED GLASS- TYP.

TYPE 'D'   (DOOR ONLY)

1/4" TEMP. GL.
NARROW STILE (CLR. ANOD.) ALUM DOOR

TYPE 'E'

FIGURE 15-39 ■ A door schedule supplemented with elevations and notes can be used to keep door information off the floor plan. *Courtesy Architects Barrentine, Bates & Lee, A.I.A.*

developed at the end of the chapter. Doors need to be detailed using details similar to Figure 15-40 to show how the door frame relates to the exterior finish and the skeleton of the structure. Details showing door installation should be placed on or near the sheet containing the floor plan.

Window Schedules and Symbols

The use of a window schedule varies with each office, depending on the type of window to be installed and the amount of detailing provided. On the structure used throughout this chapter, no window schedule is provided. Sizes would be determined based on the dimensions provided on the floor plan (see Figure 15-27), the exterior elevations, and jamb, header, and sill details. Figure 15-41

shows a portion of the exterior elevations that references the height and construction details. If a window schedule is used, it should include

- The reference letter
- The width and height
- Type of window
- Type of glass to be used
- Frame material and finish
- References for details showing the jamb, header, and sill construction
- Notations to clarify operation, location, or construction

Figure 15-42 shows an example of a window schedule with drawings to represent the type of windows.

Interior Reference Symbols

Interior reference markers (at C in Figure 15-38) are used to reference interior elevations to the floor plan. A 3/8" to 1/2" (9 to 13 mm) diameter circle is used to identify each detail. The circle with an arrow is placed in the room to indicate which direction the marker is referencing. Four arrows can be placed around the circle to represent each wall of the room. If an arrow is not placed on a specific side of a circle that indicates that no elevation is to be drawn for that wall. The page number where the detail is drawn is placed in the circle. Each arrow is numbered to represent a specific elevation. Interior elevations will be discussed in Chapter 19.

Finish Symbols and Schedules

To supplement the interior elevations, the use of finish symbols (at D in Figure 15-38) referenced to a finish schedule can be used to

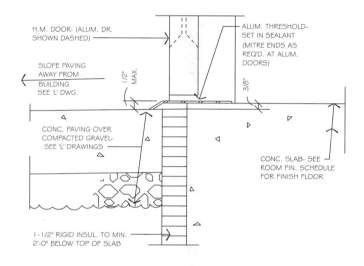

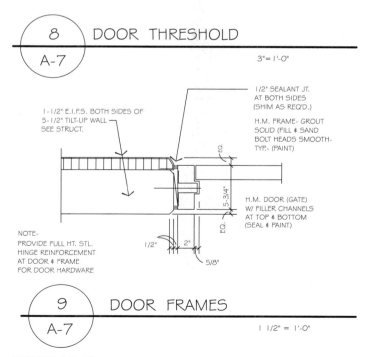

FIGURE 15-40 ■ Details showing door information must be referenced to the floor plan. *Courtesy Architects Barrentine, Bates & Lee, A.I.A.*

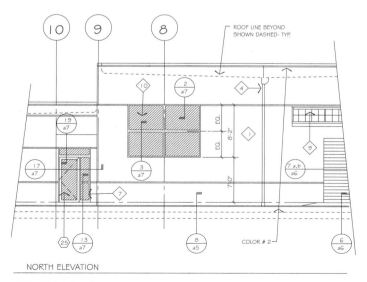

FIGURE 15-41 ■ Information for windows for a simple structure can often be shown on the elevations rather than the floor plan. *Courtesy Architects Barrentine, Bates & Lee, A.I.A.*

WINDOW SCHEDULE

MK	SIZE WIDTH X HEIGHT	MATERIAL FRAME	TYPE	NOTES	HEAD	JAMB	SILL
AA	18'-0" X 6'-11 1/2" M.O. M.O.	ALUM.	FIXED		3/A18	9/A18	9/A18
AB	16'-0" X 6'-11 1/2" M.O. M.O.	ALUM.	FIXED		3/A18	9/A18	9/A18
AC	2'-6" X 13'-9" M.O. M.O.	ALUM.	FIXED	SEE ELEV., SEE JAMB, SILL	3/A18	22/A19 28/A19	16/A18
AD	6'-0" X 6'-11 1/2" M.O. M.O.	ALUM.	FIXED		3/A18	9/A18	9/A18
AE	6'-0" X 3'-10" M.O. M.O.	ALUM.	FIXED		3/A18	9/A18	9/A18
AF	1'-11 1/2" X 8'-0 1/4" R.O. R.O.	TEMPERED GLASS ALUM.	FIXED	SEE ELEVATION, SEE SILL DTL.	10/A18 6/A18	27/A18	16/A18
AG	7'-9 1/2" X 8'-0 1/4" R.O. R.O.	ALUM.	FIXED	SEE ELEVATION, SEE SILL DTL. SEE DTL. 5/A18 FOR MULLION	10/A18 6/A18	26/A18 28/A18	16/A18
AH	8'-9 1/2" X 8'-0 1/4" R.O. R.O.	ALUM.	FIXED	SEE ELEVATION, SEE SILL DTL. SEE DTL. 5/A18 FOR MULLION	10/A18 6/A18	26/A18 28/A18	16/A18
AI	NOT USED	-	-	-	-	-	-
AJ	16'-0" X 6'-11 1/2" M.O. M.O.	ALUM.	FIXED	SEE ELEVATION	3/A18	9/A18	9/A18

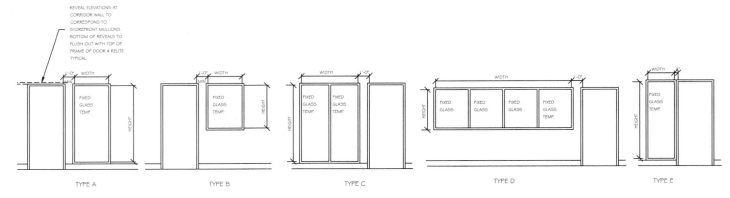

FIGURE 15-42 ■ A window schedule can be used to keep the floor plan from becoming cluttered. *Courtesy Michael & Kuhns Architects, P.C.*

specify interior finishes. When several options for construction material must be specified, finish symbols can be used to add clarity. No matter what the symbol, it is typically placed near the room title or room number. Within the symbol, references for the floor, base, wall, and ceiling material should be given. This information is then specified in a finish schedule. The finish schedule typically includes information concerning

■ The room number and name

■ Columns for the floor, base, each wall, and the ceiling

■ Wall height

■ Notations to clarify materials

Figure 15-43 shows an example of a finish schedule. This schedule represents each finish with a specific number, and then provides a legend of finishes below the schedule. Another common format is to have a column to represent each finish, and provide a check in a box of required finishes. This schedule format would require six columns for the floor to represent each material. Notice in Figure 15-43 that the walls columns are referenced to N, S, E, and W. Several walls also have notations for two types of materials on the same wall.

Wall Construction Symbols

A final symbol that may be shown on the symbol legend is a symbol for wall construction. When required, this symbol will reference several details of wall construction to a specific wall represented on a floor plan. Figure 15-38 at E shows one option for creating a wall construction symbol. This particular symbol references several different wall details, which are shown in Figure 15-44.

ROOM FINISH SCHEDULE

NO.	ROOM NAME	FLOOR	BASE	WALLS N	WALLS E	WALLS S	WALLS W	CEILING	HEIGHT	REMARKS
101	LOBBY	2/5	2	2	2	2/4	2	1	9'-0"	
102	STAIR	1	4	2	2	2	2	1	21'-0"	
103	JANITOR CLOSET	3	1	1	1	1	1	3	9'-0"	
104	ELEVATOR SHAFT	4	—	7	7	7	7	5	29'-0"	CAB ELEVATIONS- SEE INTERIOR ELEVATION 9/A5.
105	ELEV. EQUIPMENT	4	1	1	1	1	1	3	9'-0"	
106	OFFICE SPACE	4	1	1/4	1	1/4	1	1	9'-0"	
107	CLOSET	4	1	1	1	1	1	3	9'-0"	
108	CLOSET	4	1	1	1	1	1	3	9'-0"	
109	MEN	5	2	3	3	3	3	3	9'-0"	
110	WOMEN	5	2	3	3	3	3	3	9'-0"	
111	CLOSET	3	1	1	1	1	1	3	9'-0"	
201	LOBBY	1	4	—	2	—	2	1	9'-0"	
202	RECEPTION	1	4	2	2	2/4	2	1/4	9'-0"	
203	JANITOR CLOSET	3	1	1	1	1	1	3	9'-0"	
204	CONFERENCE	1	4	1	1	1	1	1	9'-0"	
	CLOSET	1	4	1	1	1	1	3	9'-0"	
				1	1	1	1	3	9'-0"	
									9'-0"	

ROOM FINISH KEY

FLOORS

1. CARPET (NIC.)
2. ENTRY MAT (NIC.)
3. SHEET VINYL (NIC.)
4. SEALED CONCRETE
5. CERAMIC TILE (NIC.)
6. HARDWOOD (NIC.)

BASE

1. 4" COVED RUBBER BASE (NIC)
2. CERAMIC TILE (NIC.)
3. HDWD. SKIRT (NIC.)
4. 4" CARPET (NIC.)

NOTES

— NO WORK REQUIRED

WALLS

1. GYPSUM BOARD (PAINT- NIC.)
2. GYPSUM BOARD W/ REVEALS (PAINT- NIC.)
3. CERAMIC TILE WAINSCOT- 2x2 TILE THINSET OVER W.R. G.B. (PAINT- NIC.)
4. STOREFRONT SYSTEM
5. E.I.F.S.
6. METAL RAILING (PAINT)
7. GYPSUM BOARD- ROUGH TAPED

CEILINGS

1. 2X4 SUSP. ACOUSTIC TILE (NON-RATED)
2. SUSPENDED TYPE "X" GYPSUM EXT. SOFFIT BOARD (1 HR. FIRE RATED)
3. SUSPENDED GYP. BD. (NON RATED)
4. SKYLIGHT SYSTEM
5. GYPSUM BOARD- ROUGH TAPED

FIGURE 15-43 ■ A finish schedule is used to group information relating to interior finishes and keep the floor plan uncluttered. *Courtesy Architects Barrentine, Bates & Lee, A.I.A.*

WALL TYPES

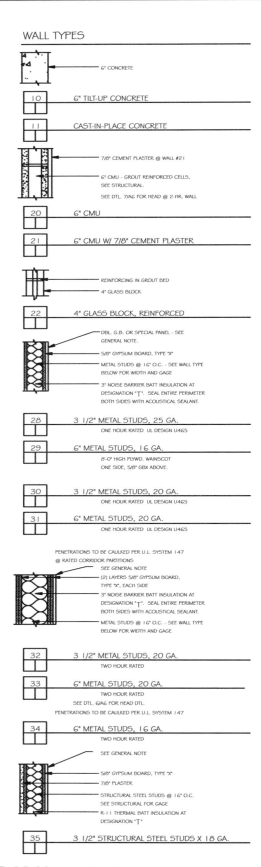

FIGURE 15-44 ■ Details referencing wall construction should be placed on or near the floor plan. *Courtesy Michael & Kuhns Architects, P.C.*

COMPLETING A FLOOR PLAN

The floor plan can be completed using the following steps.

1. Determine the scale to be used and establish the required DIMVARS, LTSCALE, and text scale factors

2. Establish all layers and assign colors and linetypes. Save these settings in a template for future projects

3. Draw all exterior walls

4. Draw all interior walls

5. Draw all openings. Include windows, doors, archways, skylights, and pass-throughs.

6. Draw all required stairs, steps, and ramps

7. Draw all built-ins

8. Draw all required plumbing symbols

9. Draw all required equipment and interior furnishings

10. Draw all required hatch patterns

The floor plan should now resemble the floor plan shown in Figure 15-45. With all required items drawn, the dimensions can be completed using the following steps.

11. Draw all grid lines

12. Dimension all overall dimensions

13. Dimension all major jogs

14. Dimension all walls that intersect an exterior wall

15. Dimension all openings in the exterior walls

16. Dimension all interior walls

The floor plan with completed dimensions will now resemble Figure 15-46. The plan can be completed by adding required symbols and written specifications.

17. Place a title and scale below the drawing

18. Draw a north arrow

19. Provide identification text in each grid symbol

20. Provide room names and numbers

21. Provide all door and window symbols

22. Locate anticipated section and detail markers. Additional symbols can be completed as the sections and details are drawn.

23. Locate interior elevation symbols

24. Specify all elevation changes

25. Draw all required finish and wall construction symbols

26. Draw a schedule to reference all symbols used on the floor plan

27. Provide text to identify all features that are not represented in a schedule

28. List all general notes required to explain the floor plan

29. Provide all material necessary to complete the title block

The complete floor plan will now resemble the plan shown in Figure 15-47.

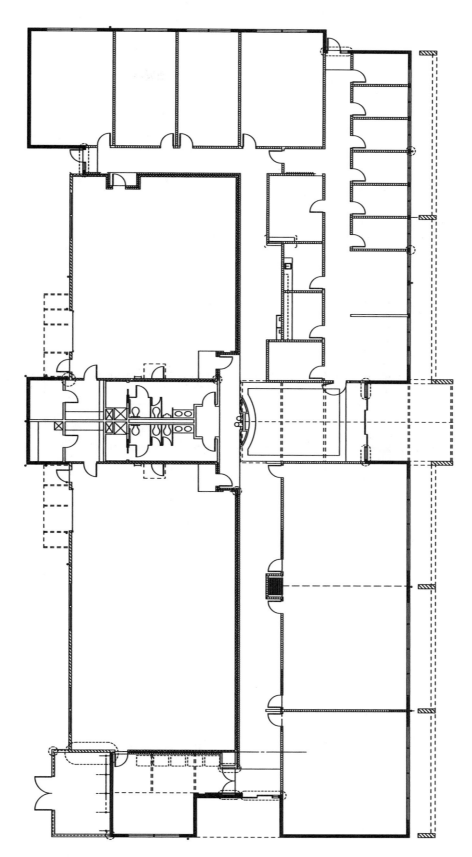

FIGURE 15-45 ■ The floor plan drawn showing walls, openings, cabinets and built-ins, and plumbing fixtures. *Courtesy Architects Barrentine, Bates & Lee, A.I.A.*

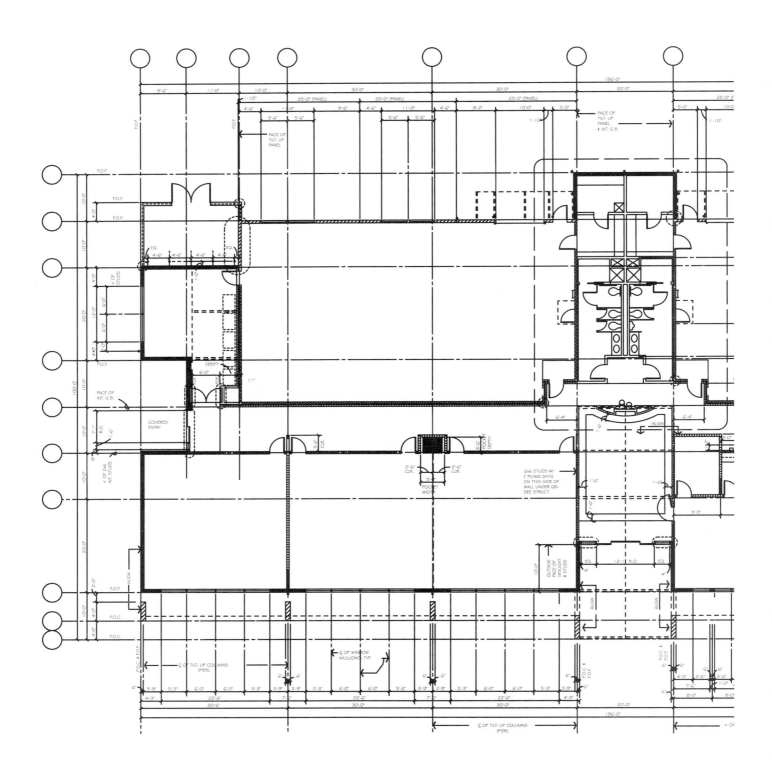

FIGURE 15-46 ■ A floor plan with completed dimensions. *Courtesy Architects Barrentine, Bates & Lee, A.I.A.*

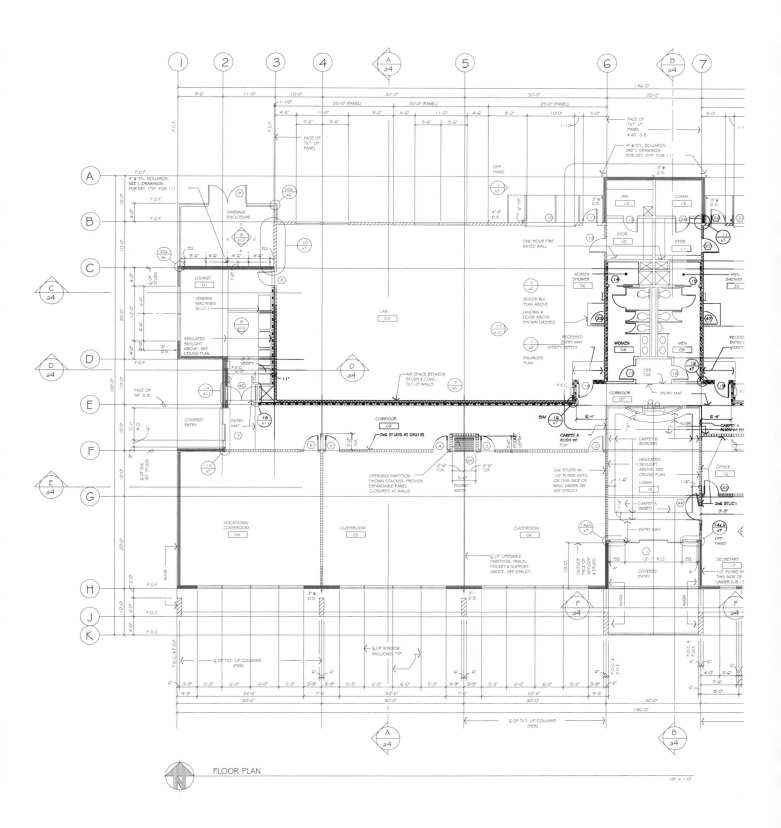

FLOOR PLAN

FIGURE 15-47 ■ A completed floor plan with notations and symbols. *Courtesy Architects Barrentine, Bates & Lee, A.I.A.*

SPACE PLANS

When large amounts of interior furnishing must be represented on the floor plan, the plan can often become unclear. Many firms represent interior furnishing on a separate plan called a *space plan*. This plan uses the walls and openings shown on the floor plan as a base, but does not show the other material typically associated with a floor plan. The space plan shows all specialty items such as interior built-ins, movable furnishings such as shelving, display cabinets, book racks, sales counters, display cases, and seating. Equipment specific to the occupancy such as dental chairs, turnstiles, or examination tables is also represented on space plans. Figure 15-48 shows an example of a space plan for a small theater.

REFLECTED CEILING PLANS

A *reflected ceiling plan* shows the ceiling of a specific level of a structure. The floor plan is created by the cutting plane passing through the walls five feet above the floor, allowing the viewer to see the floor. The ceiling plan is a mirror image of the ceiling, which is the source of the drawing title. Acoustical ceilings supported by wires are typically used on commercial structures to hide the structural material. These ceilings are often referred to as *T-bar* ceilings, because of the lightweight metal frames that support the ceiling tiles. A reflected ceiling plan shows the location, starting point, and type of material used to form the ceiling. The plan may also show the location of light panels and ceiling-mounted heat registers. These features may also be shown on a separate plan supplied by the mechanical engineer. Figure 15-49 shows an example of a reflected ceiling plan and examples of the symbol legend and notes that explain construction.

LAYERING

Because the floor plan is used for a base of so many different drawings, careful thought should be given to layering as the drawing file is being established. Walls, windows, doors, built-in cabinets, and plumbing symbols are represented on other drawings that use the floor plan as a drawing base. All other material must be placed on a layer that can be frozen so that it will not be displayed. Chapter 3 introduced the AIA formats for defining layers. A partial list of recommended layer names is provided at the end of this chapter and should be used to represent walls, doors, glazing, floor information, ceiling information, interior and exterior elevations, details, and reflected ceiling plan information.

Because of the wide variety of structures that will be drawn, each office usually establishes its own variation of the AIA system for naming layers. On complex structures, instead of having one layer for all walls, layers may be established for full-height, partial-height, curtain, and movable walls. Each of the eight major areas of the floor plan should be further divided as the floor plan grows more complicated. Added layers might include layers for furniture, area calculations, and occupancy information.

In addition to layer names that define information within the drawing, specific layers that describe the drawing should also be established. This would include layers to contain information that will be placed in the title block, dimensions, symbols, hatching symbols, schedules, and enlarged floor plans. The AIA has a complete list of modifiers available, which should be consulted as drawing standards are established.

LAYER GUIDELINES FOR FLOOR PLANS

A-FLOR	Floor information
A-FLOR-ANNO	Floor text
A-FLOR-DOOR	Doors
A-FLOR-EVTR	Elevator car and equipment
A-FLOR-EXST	Existing walls to remain
A-FLOR-EQPM	Special equipment
A-FLOR-FIRE	Fire walls
A-FLOR-FIXD	Fixed equipment
A-FLOR-FIXT	Floor information
A-FLOR-FUTR	Future walls
A-FLOR-GLAZ	Windows
A-FLOR-GRID	Floor and building grids
A-FLOR-HRAL	Handrails
A-FLOR-IDEN	Room numbers or names
A-FLOR-LEVL	Changes in floor levels such as ramps and pits
A-FLOR-NEWW	New walls
A-FLOR-NICN	Equipment not in contract
A-FLOR-OTLN	Floor and building outlines
A-FLOR-OVHD	Overhead items such as skylights
A-FLOR-PATT	Wall hatching and fill patterns
A-FLOR-PFIX	Plumbing fixtures
A-FLOR-REVS	Floor revisions
A-FLOR-RISR	Stair risers
A-FLOR-SPCL	Architectural specialties
A-FLOR- STRS	Stair treads
A-FLOR-TPTN	Toilet partitions

Status field modifiers similar to other drawing areas should be used throughout architectural drawings. See Chapter 3 and the AIA Layer Guidelines for additional layer names.

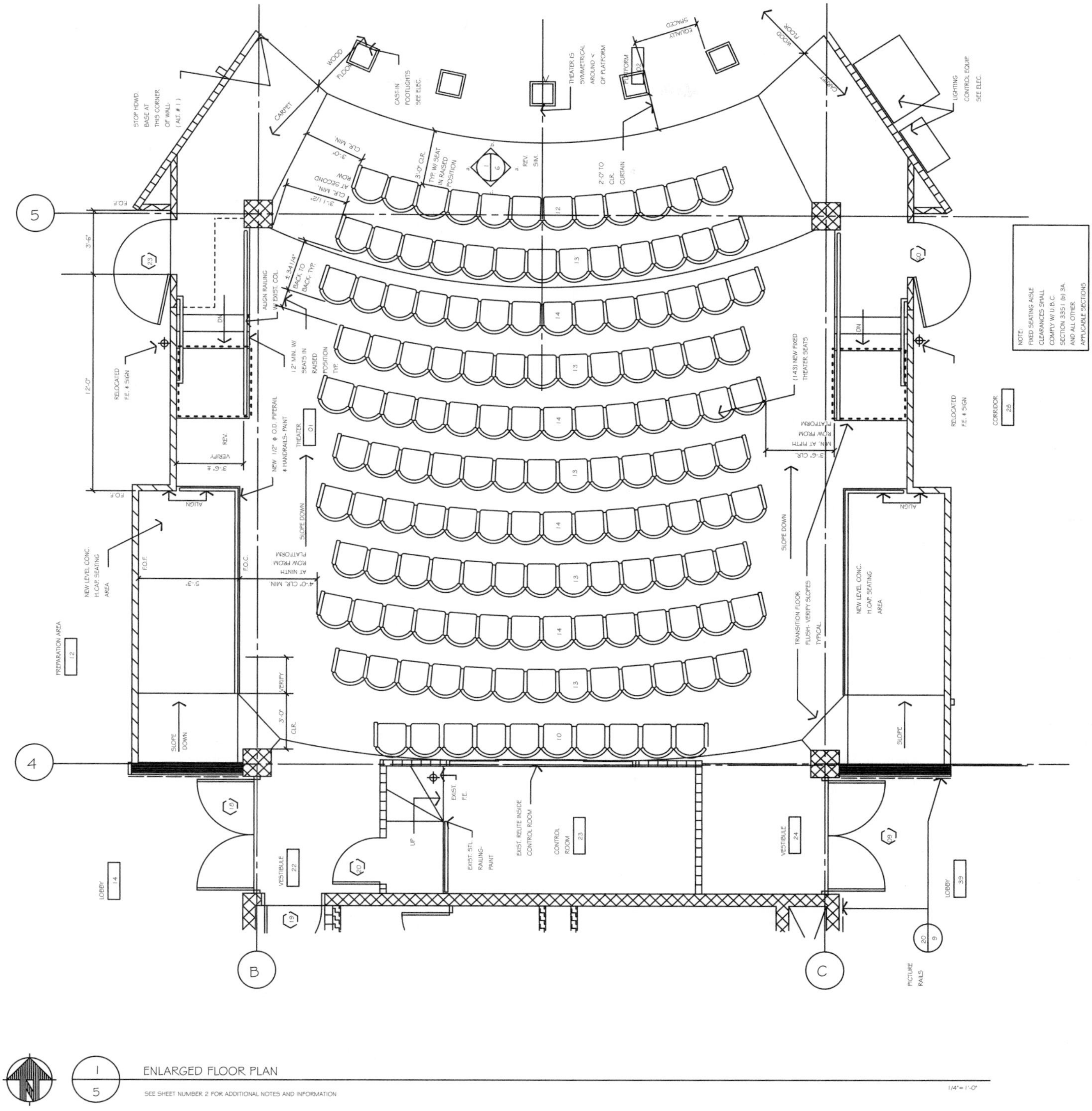

FIGURE 15-48 ■ A space plan showing interior furnishing can be created using the base of the floor plan. *Courtesy Architects Barrentine, Bates & Lee, A.I.A.*

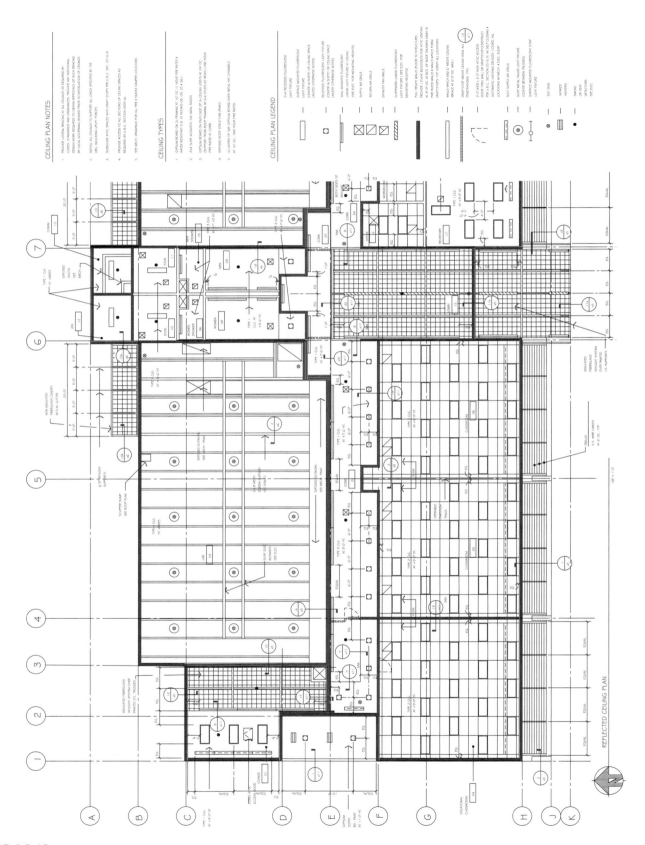

FIGURE 15-49 ■ A reflected ceiling plan can be created from the base layers of a floor plan. *Courtesy Architects Barrentine, Bates & Lee, A.I.A.*

Floor Plan Components, Symbols, and Development

CHAPTER QUIZ

Use a separate sheet of paper to answer the following questions.

Question 15-1 List three types of floor plans, explain their origin, and state which plan a drafter is most likely to work on.

Question 15-2 List five common drawing scales for a floor plan.

Question 15-3 List three common drawing scales appropriate for drawing an enlarged floor plan.

Question 15-4 A floor plan for a large office structure needs to be drawn but it will not fit on the specified vellum size. Without reducing the scale, how can the plan be completed?

Question 15-5 List two drawing symbols that must be added to the floor plan described in Question 15-4.

Question 15-6 List five categories of information that must be shown on the floor plan.

Question 15-7 List five categories of exterior dimensions.

Question 15-8 Describe two common types of notes found on a floor plan.

Question 15-9 List the dimension commands that would be useful for dimensioning a rectangular building.

Question 15-10 Describe the relationship between a floor plan and a reflected ceiling plan.

Question 15-11 Use graph paper and make a sketch of the floor plan of your drafting lab showing the dimensions of all walls and openings.

Question 15-12 Use graph paper and make a sketch of the floor plan of your drafting lab and show the layout of all furniture and equipment.

Question 15-13 Use graph paper and make a sketch of the floor plan of your drafting lab and show a reflected ceiling plan of your drafting lab.

Question 15-14 Contact architectural offices or contractors in your city and collect blueprints of five different structures. Compare differences in drawing techniques between these drawings.

Question 15-15 Visit five different construction sites, and with the permission of the superintendent, photograph the construction of wood, concrete block, concrete tilt-up, poured concrete, steel studs, and steel frame construction.

DRAWING PROBLEMS

Use the reference material from preceding chapters, local codes, and vendor catalogs to complete one of the following projects. Unless other instructions are given by your instructor, draw the floor plan that corresponds to the site plan that was drawn in Chapter 14. Skeletons of the plans can be accessed from http://www.delmar.com/resources/ocl.html. Use these drawings as a base to complete the assignment. Use appropriate symbols, linetypes, dimensioning methods, and notations to complete the drawing. Determine any unspecified sizes based on material or practical requirements. Unless specified, select a scale appropriate to plotting on "D" size material and determine LTSCALE and DIMVARS.

Note: *The sketches in this chapter are to be used as a guide only.* As you progress through the floor plan you will find that some portions of the drawings do not match things that have been drawn on the site plan. Each project has errors that will need to be solved. The errors are placed to force you to think in addition to draw. Most of the errors are so obvious that you will have no trouble finding them. If you think you have found an error, do not make changes to the drawings until you have discussed the problem and possible solutions with your engineer (your instructor).

Information is provided on drawings in Chapters 16 through 23 relating to each project that might be needed to make decisions regarding the floor plan. You will act as the project manager and will be required to make decisions about how to complete the project. Any information not provided must be researched and determined by you unless your instructor (the project architect and engineer) provides other instructions. When conflicting information is found, information from preceding chapters should take precedence. If conflicting information is found, use the following order of precedence:

1. Written changes given to you by the engineer (your instructor) as change orders

2. Verbal changes given to you by the engineer (your instructor)

3. The engineer's calculations (written instruction in the text)

4. Sketches provided by the engineer (sketches provided in each of the following chapters)

5. Lecture notes and sketches provided by your instructor

6. Sketches that you make to solve problems; verify alternatives with your instructor

Problem 15-1 Four-Unit Condominium—Wood or Metal Framing, Trussed Roof. Use the attached three drawings to complete a building floor plan for each level. The building is to contain four units that have aligning front walls. Each end unit is to have a bay window with center units to have flat glass. Place all required text in one unit, electrical information in another unit, and place all required unit dimensions in another unit. A separate unit is to be completed at a later time to be used for showing all framing material, which will be discussed in Chapter 21. Specify all door and window sizes using a schedule.

Problem 15-2 Retail Sales: Concrete Block, Wood Frame, Truss Roof. Use the attached drawings to complete the floor plan for the retail sales outlet. All exterior walls are $8 \times 8 \times 16$ grade A concrete block. Interior walls can be either light frame or steel stud construction. The south and east walls are to have a 4" brick veneer. The west wall should have an attractive block or pattern because it will be visible. The north wall will be hidden by an existing structure on adjoining property. The interior side of masonry walls, except those in the inventory and parts department, and all interior walls are to be covered with sheet rock. Panel the office and sales areas. All windows are to be 84" high storefront. Exterior doors to be $3' \times 7'$ double swinging tempered glass with a solar tint. Provide a sign over each pair of exterior doors to note that they are to remain unlocked during business hours. Use 32" wide doors throughout interior.

Problem 15-3 Concrete block warehouse. Material: concrete block, light-frame construction, panelized roof. Use the attached drawing to complete the floor plan for the concrete block warehouse to be used by a factory direct furniture company. Draw the building floor plan at the largest scale possible to fit on "D" size material. Provide pilasters as required by the framing plan. Provide a 1-hour fire wall framed with 2×6 studs between the display and storage area at grid 5. Install counters in the break room to allow for a stove, refrigerator, and double sink. Provide for a suspended ceiling in the display, sales, and break rooms 9' above the floor. Provide a unisex rest room in the storage area with its own water heater, lavatory, toilet, and urinal. Provide bathrooms near the sales /

display area that is open to the public. Provide a 3" diameter drain for each bath.

Provide a pair of glass doors to provide access from the exterior into the sales area, with as much glass as code allows on the south and west walls. Provide windows or skylights for the break room. Provide a pair of $3'$–$0'' \times 7'$–$0''$ steel doors to provide access from the sales area into the display area. Provide approximately 10' of storefront glass on the west and south walls of the display areas, (1)-$3'$–$0''$ exterior steel door and (1)-$8'$–$0'' \times 8'$–$0''$ overhead steel roll-up door on the west wall of the display area. Provide (2)-8' doors, (3)-$11' \times 11'$ overhead steel roll-up doors and (1)-3' door in the west wall of the storage area. Provide (1)-$11' \times 11'$ overhead steel roll-up door in the west wall of the display area. The three 8' doors are to be in a loading dock that is 48" below the storage/display finish floor. Parking on the site plan will need to be coordinated with the loading dock. Provide (1)-$8' \times 8'$ roll-up door and (1)-3' door to provide access from the display to the storage area.

Provide for 2 rows of $6 \times 6 \times 3/16''$ steel columns running north/south. Columns are to be at the midpoint of the display area, and at $42'$–$8''$ o.c. in the storage area. Provide one column on each side of the firewall for a total of 10 columns. Provide pilasters as per sketches contained in structural drawings.

Plan for a mezzanine floor system to be placed above the display area. Assume 12' from finish floor to finish floor. The west portion of the upper floor will be used as office space and will have a suspended ceiling at 8'. Provide windows that open to the south and west sides. Provide one unisex bathroom for staff only that will be above the bathroom located in the display area. The east portion of the upper floor will be used for lightweight storage with a minimum ceiling height of 9'. Research and specify an $8' \times 8'$ service elevator capable of lifting 4000 pounds that will not interfere with the roof framing. Submit copies of vender material showing requirements for the elevator to your instructor prior to completing the floor plan.

Problem 15-4 Tilt-up warehouse. Material: concrete tilt-up, truss roof. Use the attached drawing to complete the floor plan for the concrete tilt-up warehouse to be used by a cabinet manufacturer. Part of the structure will be used by the sales staff, for display, and for assembly and storage of units to be delivered to the job site. Although there will be no hazardous manufacturing on site, varnish and other types of stains will be applied to cabinets on site in the northeast portion of the storage area.

Draw a floor plan showing all walls, openings, and columns at a scale of $3/32'' = 1'$–$0''$. Draw an enlarged

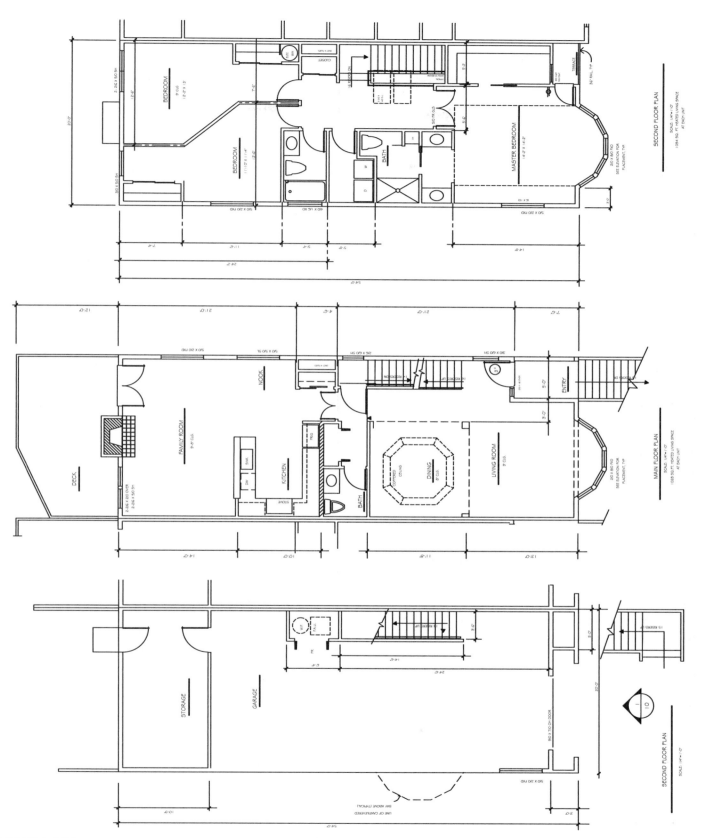

SECOND FLOOR PLAN

MAIN FLOOR PLAN

SECOND FLOOR PLAN

PROBLEM 15-1

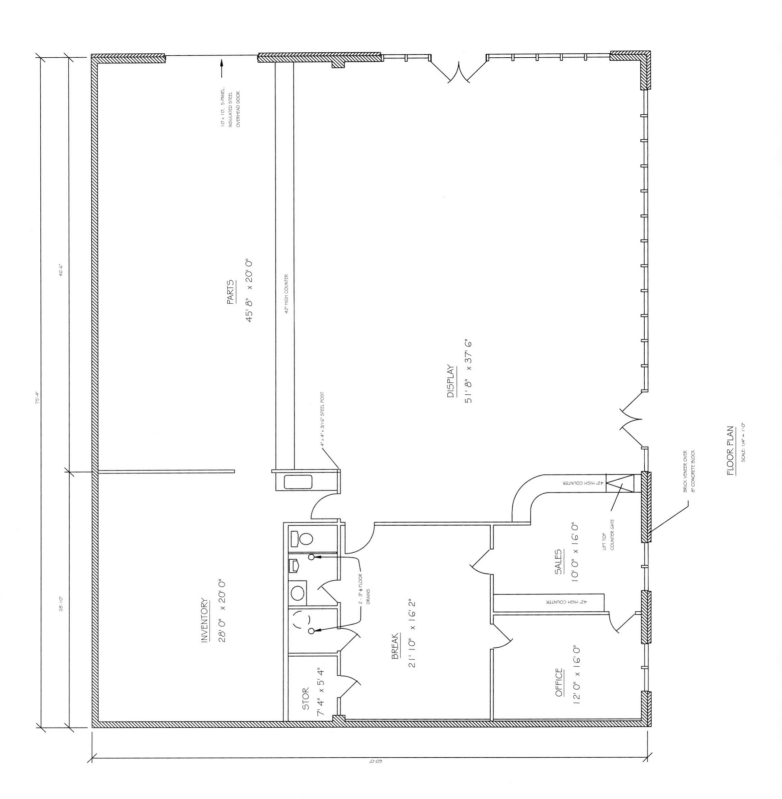

INVENTORY
28' 0" x 20' 0"

PARTS
45' 8" x 20' 0"

10' x 10', 5-PANEL
INSULATED STEEL
OVERHEAD DOOR

42" HIGH COUNTER

4" x 4" x 3/16" STEEL POST

DISPLAY
51' 8" x 37' 6"

STOR
7' 4" x 5' 4"

BREAK
21' 10" x 16' 2"

2 - 3" ⌀ FLOOR
DRAINS

SALES
10' 0" x 16' 0"

42" HIGH COUNTER

OFFICE
12' 0" x 16' 0"

42" HIGH COUNTER

LIFT TOP
COUNTER GATE

BRICK VENEER OVER
8" CONCRETE BLOCK

FLOOR PLAN
SCALE: 1/4" = 1'-0"

46'-6"

75'-4"

28'-10"

60'-0"

PROBLEM 15-2

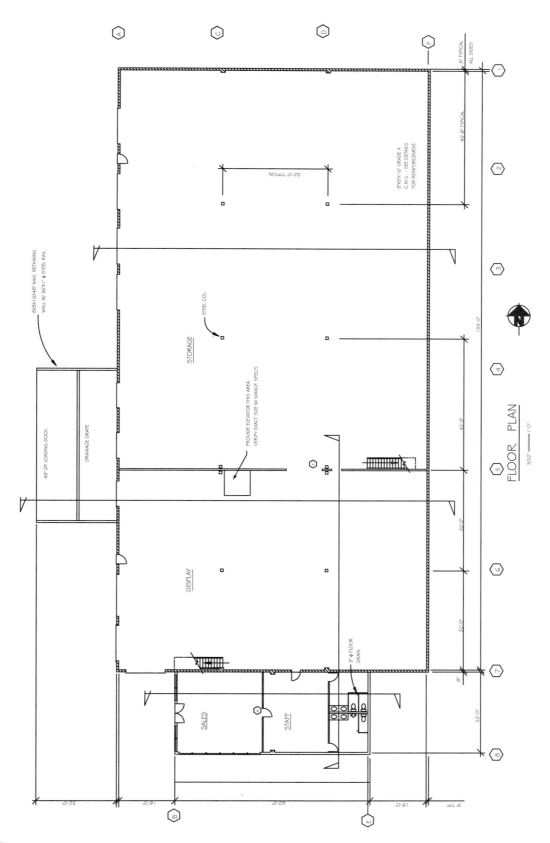

FLOOR PLAN
3/32" = 1'-0"

PROBLEM 15-3

floor plan of the office area at a scale of 1/4" = 1'–0". Assume all concrete walls will be 6" wide, all interior wood walls will be 2 × 4 stud walls. Use 2 × 6 studs for the firewall at grid F. Assume all grids to be 24'–0" wide. Dimension the location of all opening in the concrete walls on separate panel elevations completed in Chapter 22.

A 6 × 6 × 3/16 TS column will be located along grid line grid 3 at approximately 36' spacing. Exact placement will be

Column 1: 12' to the west of grid J.

Column 2: @ grid H.

Column 3: 12" to the east of grid F.

Column 4: 12" to the west of grid F.

Column 5: 24' to the west of grid.

Column 6: 24' to the west of grid F, 24' north of Grid 3.

Place a wall running e/w 24' from the south side of the storage area to divide the storage and display areas. Provide direct access from the storage areas to the break/office area.

Design a 48' × 70' office structure on the west end of the structure. Use this area for a sales area on the northeast side with a sales counter between the reception area placed in the northwest corner. Provide two offices on the west side, two bathrooms, and a break room in the balance of this area. Bathrooms should have easy access to the display/storage area and the break room and must be available for public access. Provide for a gas water heater. All heating and cooling units will be roof mount-

ed. Assume the ceiling area to be at 9'. Provide 7'–0" high storefront glass in the north and west sides of the display area. Provide for 24" minimum panel support beside each opening or limit openings to 10'–0" maximum at panel edge. Provide a pair of 3'–0" × 7'–0" glass entry doors from the exterior into the sales area. Storefront windows in each office would be nice but not required. The sales area is to have an 8'–0" suspended ceiling with 12'–0" from finish floor to finish floor for the mezzanine.

The lower storage area will be used for storage of all chemical materials. The area over the display will be used for storage of lightweight assembled units. Provide a freight elevator that will have a low impact on the slab to service the upper storage area capable of lifting 4000 pounds. Provide at least one stairway to this area. Verify with local codes if more are required. Provide (4)-24" × 48" sky lights per each grid.

Provide (3)-8'-0" × 8'-0" overhead doors in display area and (4)-8'-0" × 8'-0" overhead doors in the storage area at floor level. Provide an 8'-0" roll-up door, and a 3'-0" metal door in the tilt-up wall between the display and the sales areas. Provide a 3'-0" metal door between the shop and the break rooms. Provide an 8' metal roll-up door in the wall at grid F centered in bay and another door on the south section of wall F from storage area to storage area. Provide a 3' × 7' personnel door near each roll-up door in the storage areas. Provide a loading dock with a 36" high rail on the north side anywhere along the storage area. The three doors must be centered on the dock. The 3' door should not be in the loading dock.

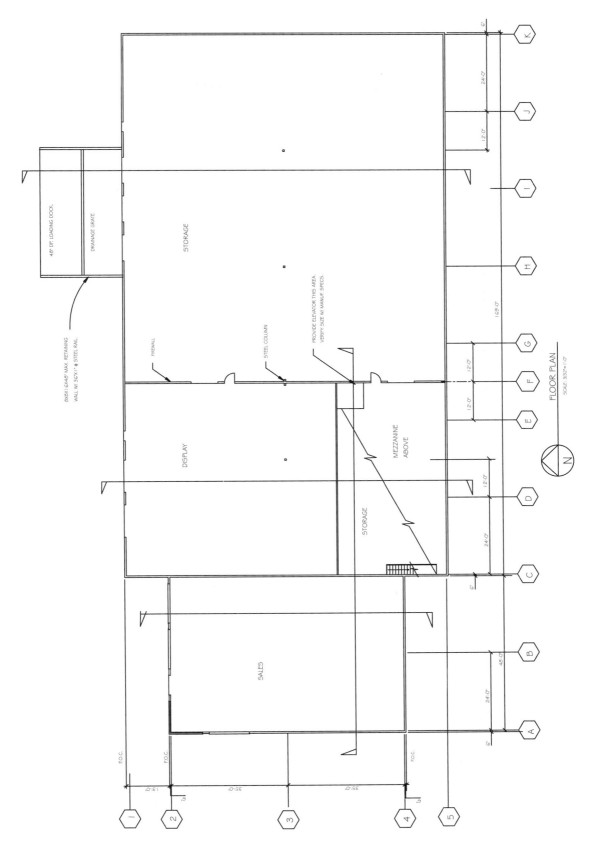

FLOOR PLAN
SCALE: 3/32"=1'-0"

PROBLEM 15-4

Orthographic Projection and Elevations

Elevations are the portions of the architectural drawings that show the exterior of the structure. In addition to showing the exterior shapes and finishes of a structure, these drawings also show vertical relationships of the material shown in the plan views. Figure 16-1 shows an example of an exterior elevation.

ELEVATIONS WITHIN THE DRAWING SET

The exterior elevations will initially be drawn by the architect as part of the preliminary designs. In conjunction with the floor and site plans, the elevations are initially developed as sketches to investigate the shapes, materials, and relationship of the structure to the site. The front and one side elevation are typically drawn, but other elevations may be provided for complex structures. Major materials used to protect the structure from the elements are represented and briefly noted. Shades and shadows, landscaping, vehicles, and people may be shown to add a sense of reality. Figure 16-2 shows an example of a preliminary elevation. Although a drafter may not draw the preliminary elevations, they will be useful later, when drawing the elevations that will be a part of the working drawings. Elevations are generally placed after the floor plans within the architectural drawings on sheets labeled A-2.

PRINCIPLES OF ORTHOGRAPHIC PROJECTION

Elevations are drawn using a drawing system called *orthographic projection*. This drawing system assumes a structure is placed within a glass box, so that each surface of the structure is parallel to a surface of the glass box. Each surface of the glass box is referred to as *picture plane*. Each surface of the structure parallel to picture plane is then presumed to be projected to the surface of the glass box. The box is then unfolded to produce the front, right side, rear, left side, top and bottom views. The top view is not shown with the elevations but will be used to form the roof plan, which is discussed in Chapter 17. Rather than having a bottom view of the structure, the floor and foundation plans are used to show the outline of the structure.

Required Elevations

An elevation of each side of the structure is required to accurately represent construction. Although each parallel surface of the structure is projected to a picture plane, the distance of surfaces from the picture plane is very difficult to represent. An elevation of each sur-

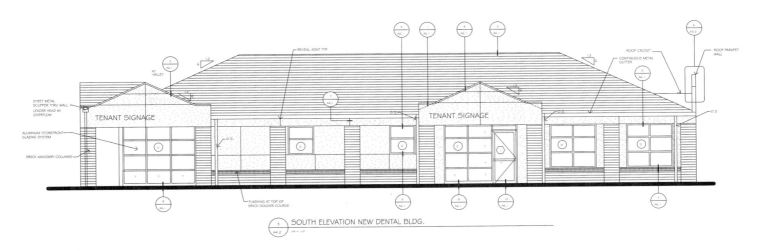

FIGURE 16-1 ■ Elevations are used to show the materials used to protect the skeleton from the elements, and to show vertical relationships. *Courtesy Scott R. Beck, Architect.*

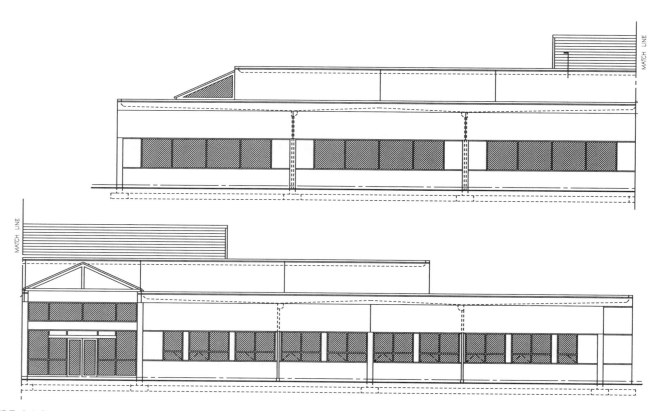

FIGURE 16-2 ■ A preliminary sketch is often given to a drafter to help in the creation of an elevation.

face is required to represent changes in shape and varied distance from the picture plane. With the comparison between two or more elevations, a structure can be clearly explained.

A minimum of four elevations are required to explain exterior materials for a structure. When a structure has an irregular shape such as a U shape, parts of the building are hidden in the basic four elevations. An elevation of each surface should be drawn, as shown in Figure 16-3. If a structure has walls that are not at 90° to each other, as shown in Figure 16-4, these walls will not be parallel to a picture plane and will be distorted when shown in an elevation. To eliminate this distortion, the structure is rotated within the "glass box" so that each surface of the structure is seen in true projection. Only the portion of the structure parallel to the viewing plane is shown, with the balance of the structure seen in another view. Match lines are used to relate each portion of the elevation to related parts. Figure 16-5 shows an example of an irregularly shaped structure with elevations drawn to eliminate distortion.

Elevation Scales

Elevations are typically drawn at the same scale as the floor plan. Matching the scales allows the elevations to be drawn by using the floor plan or plans to project horizontal distances for walls, openings and roof projections. Some offices provide elevations at a smaller scale to show the basic shape of the structure, with an enlarged set of elevations to provide details for construction. Figure 16-6 shows a small-scale elevation.

DRAWING PROJECTION

Exterior elevations can be easily constructed by using the floor, roof, and grading plans, as well as the preliminary sections, to project needed information. If the plan views are combined in one drawing file, required layers can be thawed for viewing to produce a base drawing to project the elevations form. A wblock of the required layers should be created so that the floor plan is not accidentally damaged. This wblock can be copied and rotated to allow each surface to be projected. Figure 16-7 shows an example of how elevations can be projected from a wblock using AutoCAD.

Drawing Major Shapes

With the wblock created to project walls and openings, lines should be established to represent each finish floor, the upper and lower limits of the windows and doors, the ceiling, roof, and the top of any rails, balconies, or walls. The elevation should be projected by drawing surfaces closest to the picture plane first and then drawing each receding surface. Drawing all surfaces prior to attempting to show any openings will aid in the projection. Figure 16-8 shows the projection for major shapes of an elevation.

Projecting Roof Shapes

The shape of the roof must be determined from preliminary elevations and sections. For structures that have shapes other than a flat

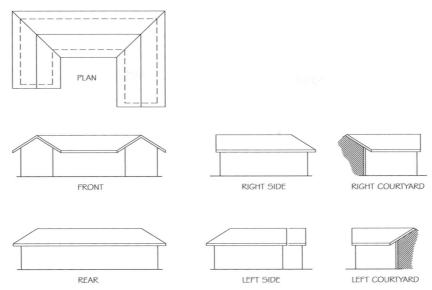

PLAN

FRONT   RIGHT SIDE   RIGHT COURTYARD

REAR   LEFT SIDE   LEFT COURTYARD

FIGURE 16-3 ■ An elevation should be drawn to show each side of a structure. Structures with irregular shapes will require an elevation of each hidden surface.

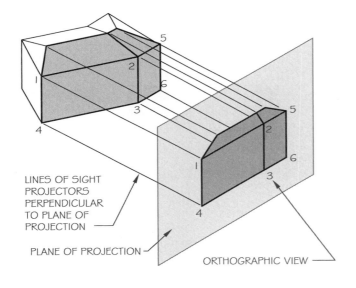

LINES OF SIGHT PROJECTORS PERPENDICULAR TO PLANE OF PROJECTION

PLANE OF PROJECTION

ORTHOGRAPHIC VIEW

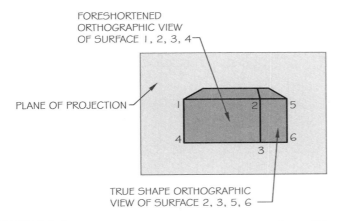

FORESHORTENED ORTHOGRAPHIC VIEW OF SURFACE 1, 2, 3, 4

PLANE OF PROJECTION

TRUE SHAPE ORTHOGRAPHIC VIEW OF SURFACE 2, 3, 5, 6

FIGURE 16-4 ■ Walls that are not parallel to the picture plane will be distorted.

roof, the roof plan must be drawn before the elevations can be completed. The overhangs of the roof should be included in the wblock used to project the elevations. The roof shape is established by projecting the limits of overhangs from the floor plan down to the elevations. The pitch of the roof and the height of bearing walls can be determined from preliminary sections. Figure 16-9 shows the projection of the roof to the base drawing.

Drawing Grades

As the elevations are started, major shapes of a structure can be established from the finished floor line. To complete the elevations, the structure must be referenced to the site by showing the finished grade elevations. Most architects use a bold line to represent the fin-

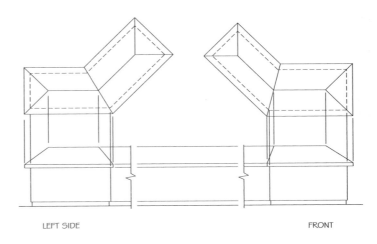

LEFT SIDE   FRONT

FIGURE 16-5 ■ When walls are not parallel to the picture plane, the elevation should be broken by a match line and an elevation in true orthographic projection should be drawn.

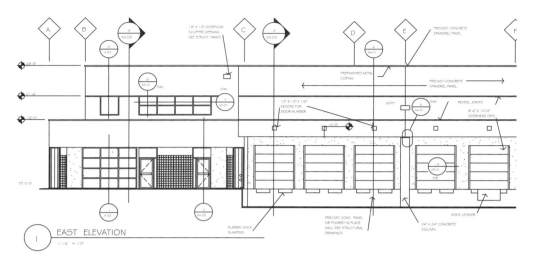

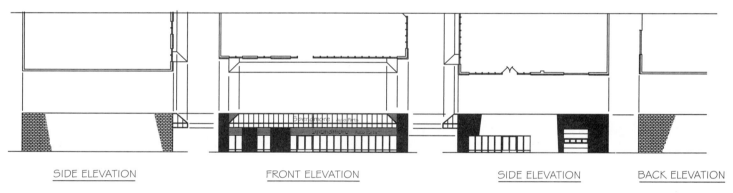

FIGURE 16-6 ■ Elevations drawn at small scale do not show all materials and depend on details or enlarged elevations to explain material placement.

SIDE ELEVATION   FRONT ELEVATION   SIDE ELEVATION   BACK ELEVATION

FIGURE 16-7 ■ With a block created from the floor plan, the elevations can be projected easily. Place elevations side by side to project horizontally. Once the elevations are completed, they can be moved into other positions for plotting.

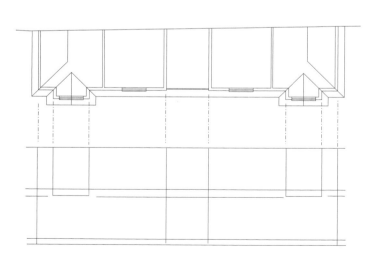

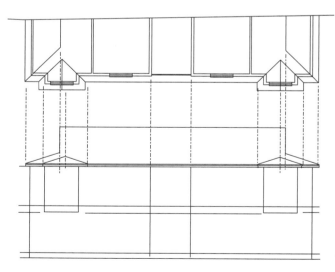

FIGURE 16-8 ■ The elevations can be started by projecting major shapes reflected on the floor plans.

FIGURE 16-9 ■ Once the major shapes have been outlined, the roof can be projected.

ished grade. Many firms place a small break in the ground line at each surface change and use a progressively smaller line to represent increasing distance of the surface to the picture plane. For relatively flat sites, the grade can be determined based on the type of foundation to be used. For structures with concrete slabs, the finished floor must be a minimum of 8" (200 mm) above the finished grade. For sloping sites, the ground needs to be established by using the grading plan. The grading plan shows the finished floor elevation and spot grades for each corner of the structure. By starting at the floor line, the drafter can measure down the required distance to establish the ground line specified on the grading plan. Once grades are determined, the outline of exterior footings can be drawn using a dashed-line pattern. Figure 16-10 shows a grading plan for a structure and the resulting ground line and footings on the elevation.

Drawing Windows, Doors, and Skylights

Once each major shape has been projected, doors and windows can be drawn matching the specifications from the floor plan or window schedule. Typically each window is inserted as a block from a reference library that has already been established. If you must create your own blocks, create a block of each type of window to be used and store the block on a disk in a directory labeled \PROTO\ELEV\WIN####. Text describing the size or type of window, such as sldg, fix, or awn can be inserted in place of the ####. The size can be altered by stretching the block once it has been inserted into the drawing and exploded. This results in slight distortions. Figure 16-11 shows an example of several different styles of window blocks for use on commercial drawings. The exact loca-

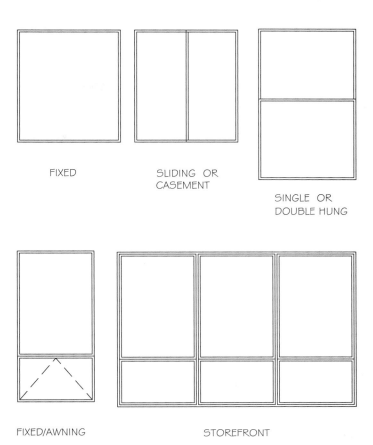

FIXED   SLIDING OR CASEMENT

SINGLE OR DOUBLE HUNG

FIXED/AWNING   STOREFRONT

FIGURE 16-11 ■ Common windows can be created and stored as a block.

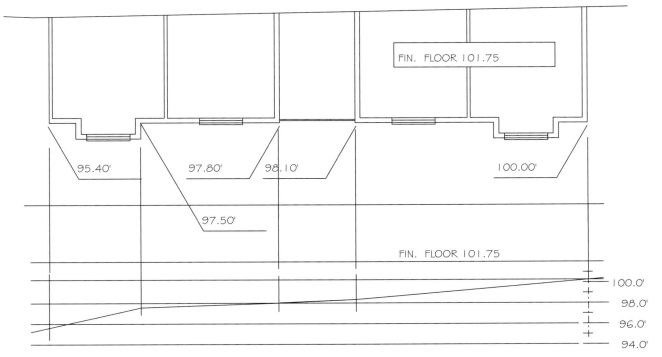

FIGURE 16-10 ■ When the site is not level, the finished floor must be projected. Elevations taken from the site or grading plan can be used to establish the finish grade by using the finished floor line as a base and offsetting the required distance for each point.

tion of the block can be determined by projecting the opening from the floor plan.

Door locations can be projected from the floor plan in the same manner as a window. Door styles should be confirmed with the door schedule. As door blocks are inserted, be sure that the sill of the door aligns with the finish floor. Figure 16-12 shows a sample of the wide variety of doors available from third-party vendors.

When an opening must be projected in an inclined wall or roof surface, the opening will not be seen in its true shape because it is not parallel to the picture plane. Inclined materials must be projected from an elevation or section where the inclined surface shows as a line rather than a plane. Figure 16-13 shows an example of projecting a skylight on an inclined roof.

Material Representation

The method of representing materials varies depending on the scale of the drawing, the type of material, and which elevation is being represented. As the scale of the drawing increases, so should the amount of detail to represent each material. The drafter should be careful not to spend too much time showing a material in elevation that will be completely detailed in a detail. The elevation should have enough detail to accurately represent materials and surfaces,

MULTI-FAMILY ENTRY DOORS

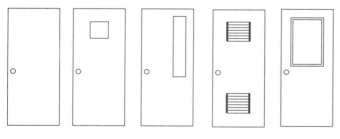

METAL ENTRANCE DOORS

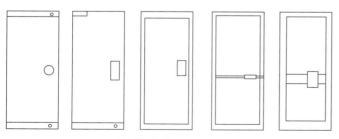

GLASS ENTRANCE DOORS

FIGURE 16-12 ■ Doors can be stored as blocks by type or material.

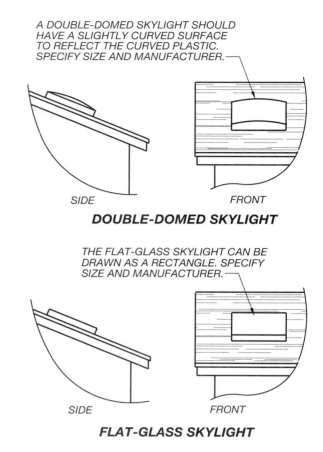

DOUBLE-DOMED SKYLIGHT

FLAT-GLASS SKYLIGHT

FIGURE 16-13 ■ Materials to be represented on an inclined surface must be projected from the surface where the material shows in its true size.

but not to the point of wasting time or disk space. Figure 16-14 shows methods of representing glass in elevations. The front elevation is generally the only elevation that has materials represented throughout the entire elevation. Other elevations show just enough of the material to accurately represent the material. Figure 16-15 shows an example of representing materials in elevations.

Using AutoCAD Hatch Patterns

Materials are often represented by using a hatch pattern created by AutoCAD or a third-party vendor. Common hatch patterns used to represent materials in elevations can be seen in the elevations presented throughout this chapter. Drafters typically encounter problems in scaling the hatch pattern and defining the extents of the pattern placement when working with the **HATCH** command. Inserting a hatch pattern is often a matter of trial and error in picking the scale pattern. For a pattern to be used accurately, it should fit the scale of the drawing. Adding a line of the desired length of the hatch pattern provides a reference for the pattern size. To insert a hatch pattern of CMUs similar to Figure 16-16, a 16" (400 mm) long line could be drawn as a reference point, and removed once the scale pattern is determined. Figure 16-17 shows an example of elevation with hatch patterns.

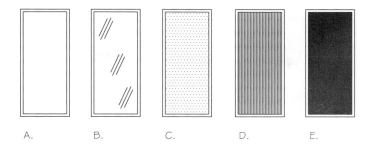

A.   B.   C.   D.   E.

FIGURE 16-14 ■ Representation of glass and other materials should be influenced by the drawing scale, and by which elevation is being created. Methods used to represent glass include (A) clear, (B) simple representation, (C) dot, (D) solid line, and (E) solid fill.

FIGURE 16-15 ■ Representing materials in elevation.

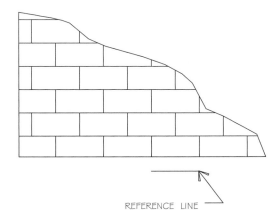

REFERENCE LINE

FIGURE 16-16 ■ A reference line should be temporally created to establish proper hatch pattern size.

Dimensioning

The use of dimensions varies for each office and depends on the complexity of the structure. Typically the height from floor to ceiling and floor to wall are indicated on the elevations. Other dimensions that may be specified on the elevations include the following:

■ Floor to ceiling for multilevel structures

■ Floor to rails at balconies or above ground decks

■ Window height and size

■ Height and width exterior finishes

■ Roof pitches

Figure 16-18 shows an example of a fully dimensioned elevation.

Elevations Symbols

The elevations often use symbols to keep the drawing uncluttered. Common symbols include the following:

■ Finished floor lines

■ Match lines

■ Grid line markers

■ Elevation symbols

■ Section markers

■ Detail markers

Floor and ceiling lines can be represented by a long-short-long pattern. If a horizontal siding is used, the floor line should

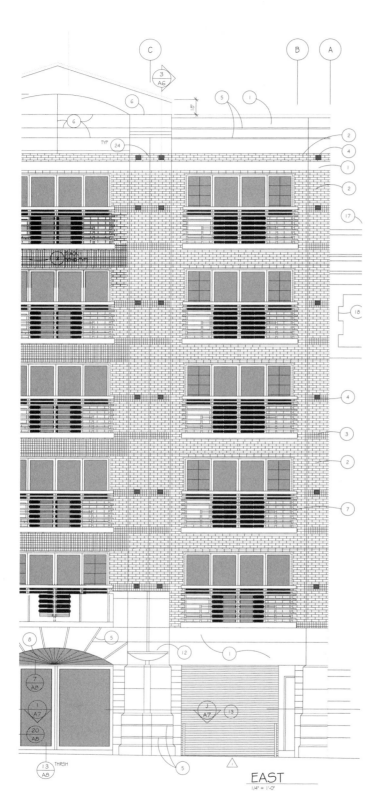

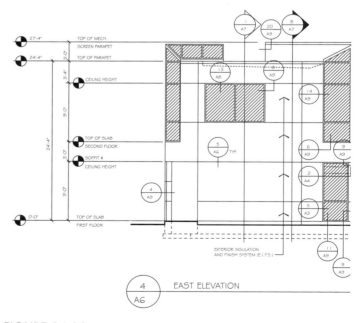

FIGURE 16-17 ■ Materials can be created in AutoCAD by using hatch patterns, inserting blocks, or by using the ARRAY and OFFSET commands. *Courtesy H.D.N. Architects A.I.A.*

FIGURE 16-18 ■ Dimensions should be placed and identified to clearly specify all vertical relationships. *Courtesy Architects Barrentine, Bates & Lee, A.I.A.*

be represented with a thick line so that it can be easily distinguished from the siding. If the elevation must be split because of length, the match line should be placed in the same location that was used on the floor plan. Grid lines should be placed on the elevation to match those used on the floor plan. Because the elevation is seen from the exterior of the structure, and the floor plan is seen when looking down on the structure (as shown in Figure 16-19), the grid lines will be opposite on the elevations to what they are on the floor plan.

Heights that are specified or dimensioned on the elevation are usually highlighted by an elevation symbol, as shown in Figure 16-20. Section markers like those used on the floor plan are placed on the elevations. The same symbol and location should be identical to that of the floor plan. Detail markers to represent exterior treatment, window or door construction, and special construction are also referenced on the elevation. Detail symbols are like those used on the floor plan. Figure 16-20 shows an example of the use of each of these symbols.

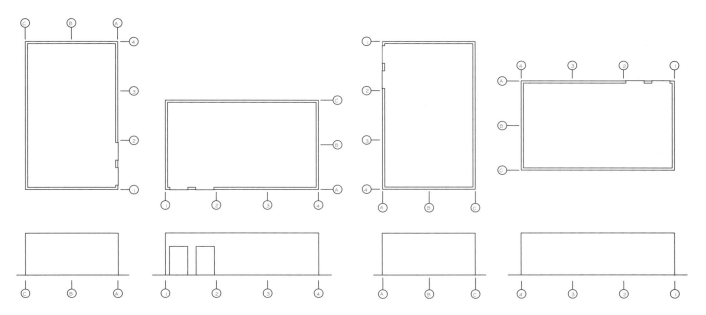

FIGURE 16-19 ■ Although most drafters understand the basics of orthographic projection, the elevations might be drawn backwards if the floor plan is not properly rotated

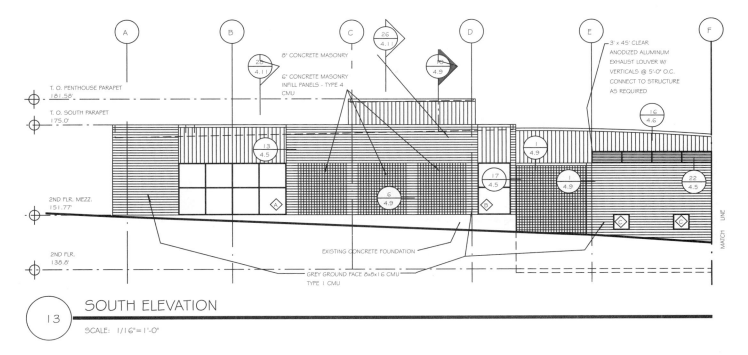

FIGURE 16-20 ■ Notes and dimensions can be included by the use of local notes. *Courtesy Architects Barrentine, Bates & Lee, A.I.A.*

Elevation Notations

Each material represented on the elevation should be specified by a note or detail. Notes can be either a local note referenced to the material by a leader line, or a general note. When general notes are used, a short reference should be specified on the elevation. Many offices use notes with a reference symbol, which can be placed in the elevation to maintain clarity in the drawing. Notes should be carefully worded so that conflict with specifications is not created. Figure 16-21 shows an elevation with key notes.

COMPLETING ELEVATIONS

The elements comprising elevations have been presented throughout this chapter. These components will be combined to complete the front elevation that matches the structure presented throughout Chapter 20.

1. Freeze all material on the floor plan except for layers that contain the walls, exterior openings, roof outlines, and grades. Make a wblock of this drawing and save it as ELEVBASE.

2. Close the current FLOOR file and open the ELEVBASE file

3. Establish layers for dimensions, text grids, outlines, roofs, doors, and windows with a prefix of AELEV

4. Make three additional copies of the floor wblock and rotate them for projection for each elevation similar to the layout presented in Figure 16-7

5. Project lines from the floor plans to represent the limits of each exterior wall

6. Draw a line to represent the finish floor line for each level

7. Establish lines to represent the tops of all walls, rails, and the bottoms of all footings

8. Use preliminary drawings to establish the roof slope and draw the required inclined roof pitches

The drawing should now resemble Figure 16-22.

9. Establish lines to represent the tops and bottom of all openings

10. Project the locations of all exterior openings from the floor plans

11. Use appropriate layers to insert blocks for all doors and windows

12. Use appropriate layers to provide hatch patterns to represent all exterior materials.

The elevation should now be completely drawn and should resemble Figure 16-23. Although only one elevation is shown, all elevations should be drawn as each stage is completed. The elevations can be completed by using the following steps.

13. Provide all grid and match lines and their notations

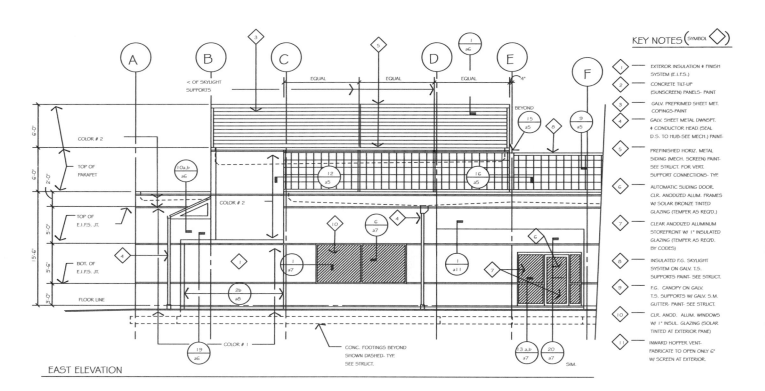

FIGURE 16-21 ■ Notes can be referenced on the drawing by a symbol, with the complete note placed beside the drawing to enhance clarity. *Courtesy Architects Barrentine, Bates & Lee, A.I.A.*

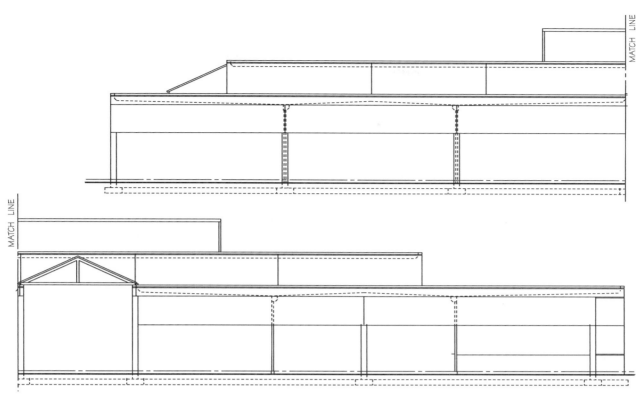

FIGURE 16-22 ■ The structure presented in Chapter 15 is used to project each of the major shapes. Although only one elevation is being presented, a drafter would be required to represent each surface. Figure 16-26 shows the entire structure. *Courtesy Architects Barrentine, Bates & Lee, A.I.A.*

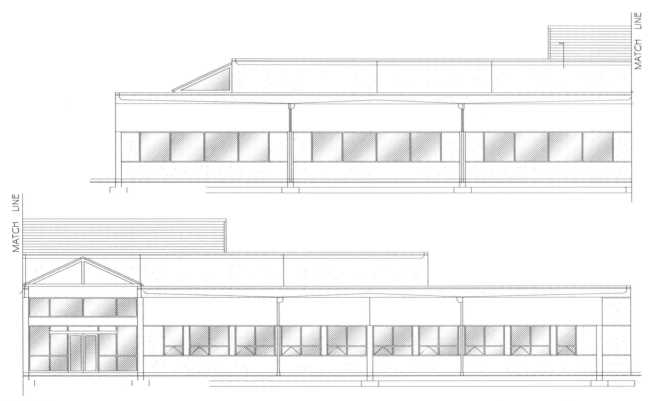

FIGURE 16-23 ■ Once major shapes have been created, windows, doors, and trim patterns can be represented. *Courtesy Architects Barrentine, Bates & Lee, A.I.A.*

14. Label all floor, rail, ceiling, and roof lines

15. Dimension all floors, rails, and window heights so that your drawing resembles Figure 16-24

16. Provide notations to specify all exterior materials

17. Provide all required general notes

18. Provide reference markers and notations for all required details. Detail markers are added as the details are drawn.

19. Provide a title and scale

The completed front elevations should now resemble Figure 16-25. The completed elevations for the entire structure are shown in Figure 16-26.

STRUCTURAL ELEVATIONS

Structures made of site-poured concrete, wood or steel frame, or tilt-up concrete panels often require structural elevations as well as the exterior elevations. Figure 16-27 shows an example of the structural elevation that corresponds to Figure 16-20. Various types of structural elevations and guidelines for completing them are covered in Chapter 22.

LAYERING

Material represented on elevations should be separated by layers. Titles that start with A-ELEV and are followed by an appropriate modifier are used by most offices. Modifiers that describe the contents or use vary for each office. Suggested AIA modifiers will be listed at the end of the chapter. Other common methods of naming layers include names based on the type of line to be used or the weight of the line when plotted. Layer names such as FINE, LIGHT, MED, HEAVY, or DASH, FLOOR or CONTINUOUS, describe the contents in general terms but may not be specific enough to adequately separate material for plotting.

LAYER GUIDELINES FOR EXTERIOR ELEVATIONS

Common Layer names based on standard AIA modifiers:

A-ELEV
A-ELEV-ANNO   Local and general notes
A-ELEV-CASE   Wall mounted casework
A-ELEV-DIM   Dimensions
A-ELEV-DRS   Doors
A-ELEV-SHD   Shade

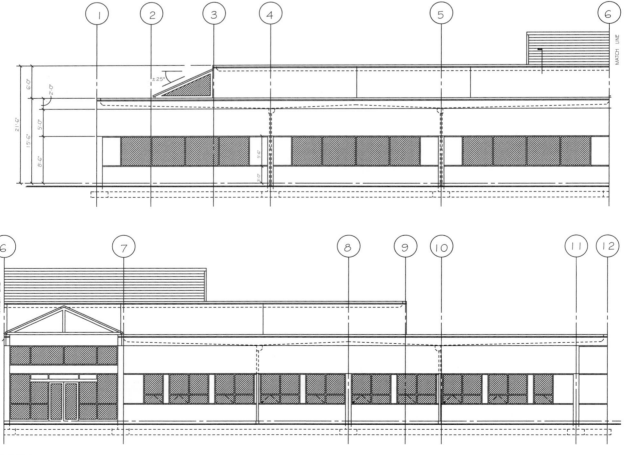

FIGURE 16-24 ■ With everything drawn, information about construction can now be placed. *Courtesy Architects Barrentine, Bates & Lee, A.I.A.*

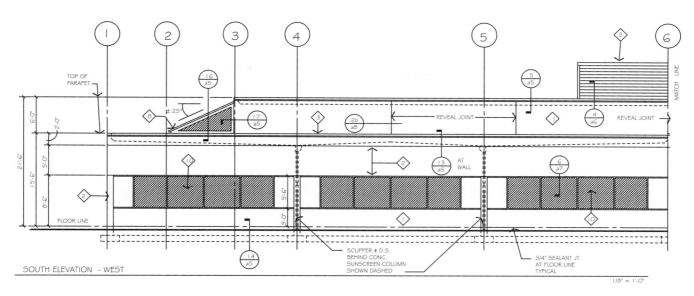

SOUTH ELEVATION - WEST

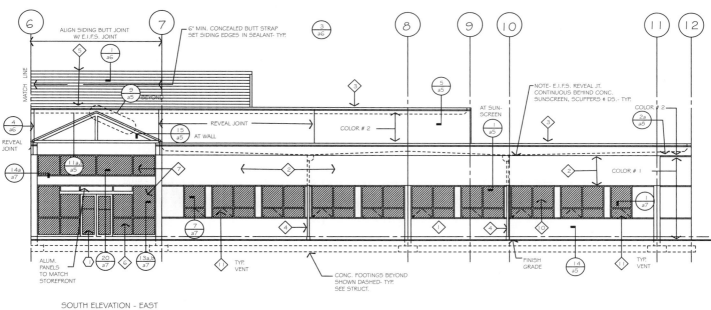

SOUTH ELEVATION - EAST

FIGURE 16-25 ■ The completed elevation. *Courtesy Architects Barrentine, Bates & Lee, A.I.A.*

A-ELEV-FNSH	Finishes, woodwork or trim	A-ELEV-RAIL	Railings
A-ELEV-FIX	Fixtures	A-ELEV-ROOF	Roof components
A-ELEV-FND	Footing and foundation lines	A-ELEV-SIGN	Signs which are attached to the structure
A-ELEV-GRID	Grid lines and markers		
A-ELEV-IDEN	Detail, section, and marker specifications	A-ELEV-SYMB	Symbols, detail, and section bubbles
A-ELEV-OUTL	Wall outlines for layout of elevations	A-ELEV-WALL	Wall lines and openings
A-ELEV-PATT	Textures and hatch patterns		
A-ELEV-PFIXT	Plumbing fixtures in elevations		

Status field modifiers similar to other drawing areas should be used throughout architectural drawings. See Chapter 3 and the AIA Layer Guidelines for additional layer names.

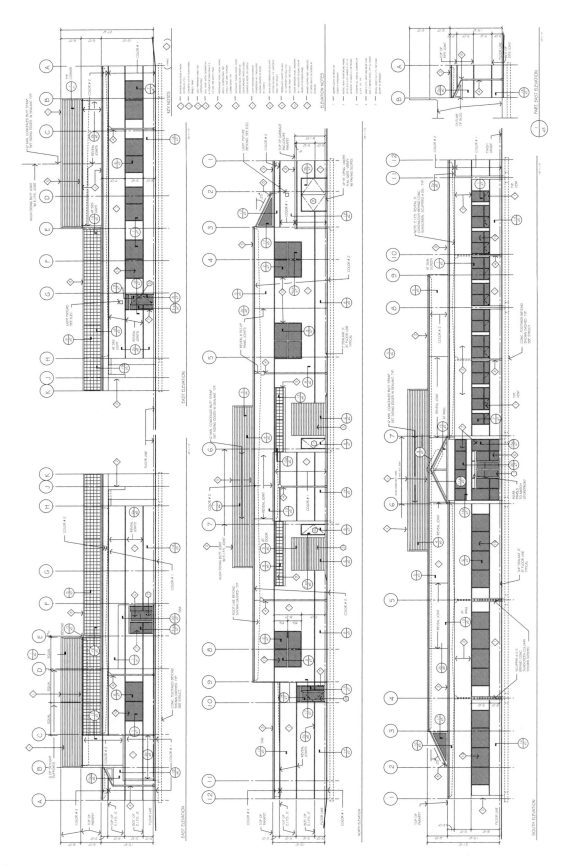

FIGURE 16-26 ■ The elevation page layout for the entire structure. *Courtesy Architects Barrentine, Bates & Lee, A.I.A.*

BUILDING SECTION / SOUTH WALL INTERIOR ELEVATION
3
SCALE 1/8"=1'-0"

FIGURE 16-27 ■ Structural elevations are drawn to show how major components of the skeleton relate to each other. Chapter 22 discusses the relationship of structural elevations to the other structural drawings. *Courtesy DiLoreto Architects, LLC.*

CHAPTER 16

Orthographic Projection and Elevations

CHAPTER QUIZ

Use a separate sheet of paper to answer the following questions.

Question 16-1 An elevation is too long to fit on the paper that is being used. How can this elevation be drawn?

Question 16-2 What process is used to draw elevations?

Question 16-3 List the drawings needed to draw exterior elevations.

Question 16-4 Describe how irregularly shaped structures should be handled for exterior elevations.

Question 16-5 Describe the differences in material representation for the front and other elevations.

Question 16-6 What materials are typically dimensioned on elevations?

Question 16-7 Describe symbols that are typically placed on elevations.

Question 16-8 List the complete layer names for the following materials: concrete footings, grid numbers and letters, window glazing, detail and section markers, and roof materials.

Question 16-9 A structure will have several sliding windows but each will have a different size. How can the window be easily inserted for each application?

Question 16-10 In drawing a structure on a sloping site, what should be used as base point?

DRAWING PROBLEMS

Use the reference material from preceding chapters, local codes, and vendor catalogs to complete one of the following projects. Unless other instructions are given by your instructor, draw the elevations that corresponds to the floor plan that was drawn in Chapter 15 (see pages {XREF 397–400 in previous edition}). Skeletons of the elevations for problems 16-3 through 16-6 can be accessed from http://www.delmar.com/resources/ocl.html. Use these drawings as a base to complete the assignment. Use appropriate symbols, linetypes, dimensioning methods, and notations to complete the drawing. Determine any unspecified sizes based on material or practical requirements. Unless specified, select a scale appropriate to plotting on "D" size material and determine LTSCALE and DIMVARS.

Note: *The sketches in this chapter are to be used as a guide only.* As you progress through the elevations you will find that some portions of the drawings do not match things that have been drawn on the floor plan. If you think you have found an error, do not make changes to the drawings until you have discussed the problem and possible solutions with your engineer (your instructor).

Information is provided on drawings in Chapters 16 through 23 relating to each project that might be needed to make decisions regarding the floor plan. You will act as the project manager and will be required to make decisions about how to complete the project. Any information not provided must be researched and determined by you unless your instructor (the project architect and engineer) provides other instructions. When conflicting information is found, information from preceding chapters should take precedence.

Problem 16-1 Draw blocks to represent a plain glass door, a pair of glass doors, and a solid metal personnel door. Save each block with an appropriate name.

Problem 16-2 Draw a block to represent a sliding, single-hung, and fixed window. Save each block with an appropriate name.

Problem 16-3 Using the site plan drawn in Chapter 14, the floor plan drawn in Chapter 15 , and the attached drawing of an exterior and an interior unit, draw the required elevations. Use 300# composition shingles, and horizontal siding with a 6" exposure over 1/2" wafer board and Tyvek. Use 1 × 3 corner trim and a 1 × 6 fascia. Provide notes to specify all exterior materials. Provide grid lines to match the floor plan.

Alter exterior materials and window sizes and shapes to provide an alternative front elevation using a Spanish style, and another front elevation providing a very traditional style.

12 ⌐12

LP LAP SIDING OVER 1/2" CX PLYWOOD 6" EXPOSURE (TYPICAL)

STUCCO OVER #15 FELT ⨥1/2" CX PLYWOOD

1X4 CEDAR TRIM (TYPICAL)

FRONT ELEVATION

SCALE 1/4" = 1'-0"

PROBLEM 16-3

Problem 16-4 Draw the required elevations to represent the retail sales outlet. In addition to showing the 8 × 8 × 16 grade A concrete blocks, show a brick veneer over south and east wall. Use 26 ga. metal roofing installed as per manufacturer's specifications.

Problem 16-5 Use the floor and site plan to draw each of the elevations for this structure. Use horizontal cedar siding with a 6" exposure over 1/2" wafer board at the roof awning. Cover the north and west office walls wall with

EIFS. Design a decorative pattern to enhance the north elevation of the warehouse. A sign displaying the company name and logo should also be prominent on the north wall.

Problem 16-6 Use the floor and site plan to draw each of the elevations for this structure. Use 26 ga. metal roofing for the projected roof. Use a combination of a grooved inset into each panel and a painted band on all walls to minimize the building height.

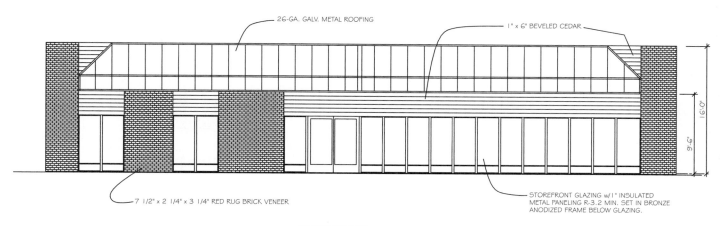

26-GA. GALV. METAL ROOFING

1" x 6" BEVELED CEDAR

7 1/2" x 2 1/4" x 3 1/4" RED RUG BRICK VENEER

STOREFRONT GLAZING w/1" INSULATED METAL PANELING R-3.2 MIN. SET IN BRONZE ANODIZED FRAME BELOW GLAZING.

9'-6"

16'-0"

FRONT ELEVATION   SOUTH
SCALE: 1/4" = 1'-0"

PROBLEM 16-4

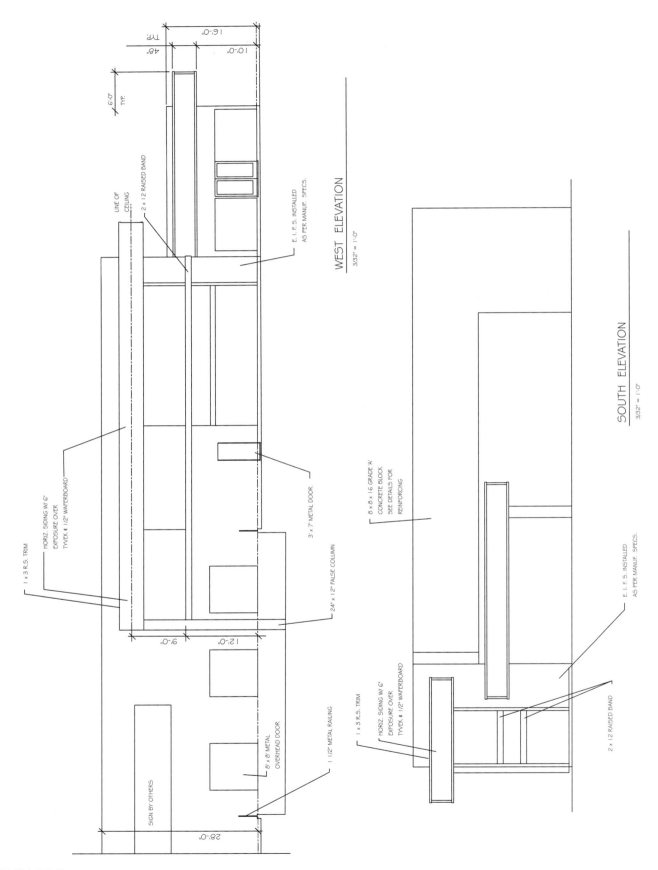

WEST ELEVATION
3/32" = 1'-0"

SOUTH ELEVATION
3/32" = 1'-0"

PROBLEM 16-5

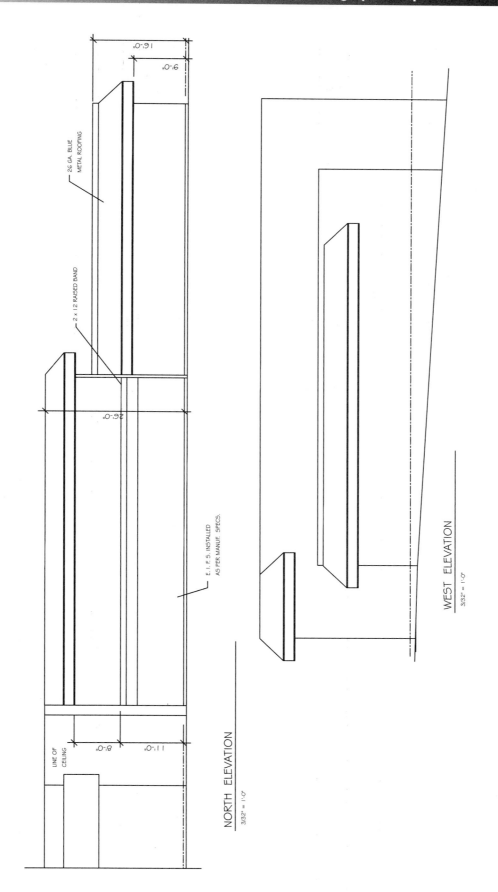

16'-0"

9'-0"

26 GA. BLUE
METAL ROOFING

2 x 12 RAISED BAND

26'-0"

E. I. F. S. INSTALLED
AS PER MANUF. SPECS.

LINE OF
CEILING

8'-0"

11'-0"

NORTH ELEVATION
3/32" = 1'-0"

WEST ELEVATION
3/32" = 1'-0"

PROBLEM 16-6

Roof Plan Components and Drawings

The roof plan is used to show the shape, the roofing materials, the direction of slope, and the drainage method of each roof surface. Although many commercial structures have a roof with a slight slope, each of the shapes traditionally associated with light-frame construction can also be used with commercial structures. This chapter will explore the components and methods of developing roof plans for both flat and inclined roofs.

The roof plan is the part of the architectural drawings developed to show a top view of the structure in orthographic projection. An example of a roof plan is shown in Figure 17-1. On high sloped roofs, the roof plan is developed with other preliminary drawings to help the clients visualize how the structure will look. On low-sloped roofs, a roof plan may not be started until the working drawings are started. On simple structures, the plan may be drawn at a small scale such as 1/8" = 1'–0" (1:100) or smaller. On more complicated structures, the roof plan is drawn at the same size as the floor plan, and divided into sections by match lines that match the floor plan and elevations. The construction of the roof is represented on a roof framing plan similar to Figure 17-2. The roof framing plan is part of the structural drawings which will be discussed in Chapter 21.

ROOF PITCH

Pitch is a ratio of the vertical rise to the horizontal run used to describe the angle of the roof slope. Pitch is described in units comparing horizontal run to the vertical rise. The slope may be measured in inches, millimeters, feet, or any other unit as long as the same unit is used for each measurement. The slope will be shown on the sections and elevations. The direction and intersections between roof surfaces must be shown on the roof plan, but the pitch is not represented. The angle equivalents for representing each pitch can be seen in Table 17-1.

Low-Sloped Roofs

Low-sloped roofs are divided into two categories by the IBC. The IBC requires a minimum slope of 1/4":12" (2% slope) for drainage. Manufacturers of roofing material typically consider roofs with a slope of less than a 2:12 (17% slope) pitch as a low-sloped roof. A second category of material can be used for roofs with slopes that range from 2 1/2:12 (21% slope) to 4:12 pitch (33% slope). Because of the size of many commercial structures, a flat, low-sloped roof is the only practical shape that can economically be used. The method of framing the roof will depend on the size of

TABLE 17-1 ■ Common Angles for Drawing Roof Pitches

Roof Pitch	Angle
1:12	4°30'
2:12	9°30'
3:12	14°0'
4:12	18°30'
5:12	22°30'
6:12	26°30'
7:12	30°0'
8:12	33°45'
9:12	37°0'
10:12	40°0'
11:12	42°30'
12:12	45°0'

the structure and the arrangement of space. For a structure composed of a single space, a single-slope shed roof similar to Figure 17-3 can be used. In locations that do not receive much rain, any water collected on the roof may be allowed to drain off the low edge of the roof directly to the ground. In moister regions, water will be diverted to one or more collection points and then removed from the roof by a series of drains. Water removal methods are discussed later in this chapter.

Many low-sloped roofs are surrounded by *parapet walls*, which are exterior walls that extend above the roof. Parapet walls are provided for low-sloped roofs to hide mechanical equipment from public view. Depending on the location of the structure relative to the property line, building codes require a minimum of a 30" (750 mm) high parapet wall regardless of the pitch. This is to protect the roof from a fire on adjoining property, or vice versa. The parapet wall location and height must be specified on the roof plan. Wall construction is specified in details similar to Figure 17-4, which must be referenced to the roof plan. The parapet wall creates a dam effect, trapping water on the low side of the roof. Lightweight metal called *flashing* is used to protect each wall/roof intersection from water seepage. To protect from water penetration when two different materials meet, flashing is also used on the tops of parapet walls and at each penetration of the roof. The type and gauge of the metal and the type of lap are specified on the roof details.

A *cant strip* is a small block of wood that is placed below the roofing material where the parapet wall intersects the roof. The cant strip provides support to the flashing and roofing. As shown in Figure 17-4, the cant strip is a piece of wood that is cut to provide an inclined surface to slope water away from the intersection.

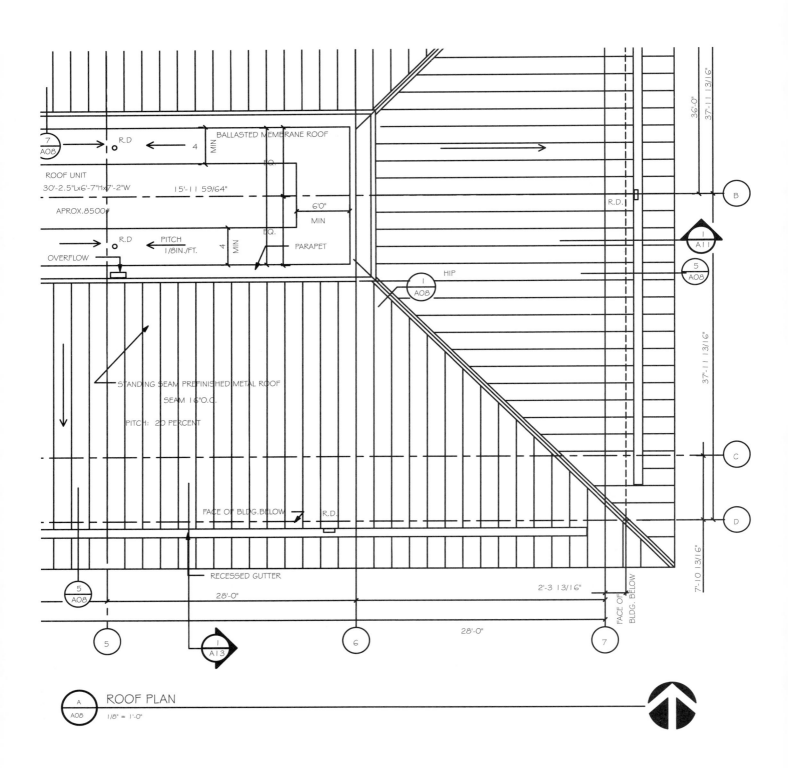

FIGURE 17-1 ■ A roof plan is drawn to show the contours of the top level of a structure and is considered the top view in orthographic projection. *Courtesy Peck, Smiley, Ettlin Architects.*

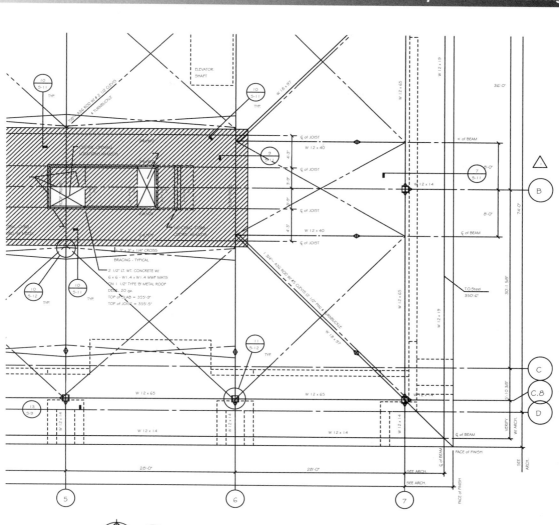

FIGURE 17-2 ■ A roof framing plan is used to show the structural materials used to construct the roof. *Courtesy Van Domelen/Looijenga/McGarrigle/Knauf Consulting Engineers.*

The cant strip is not always specified on the roof plan, but is generally referenced in details similar to Figure 17-4.

As the length of the structure increases, a single direction slope can still be used, but additional measures are required to distribute rainwater. As the distance between roof drains is increased, the cant strip is replaced by a cricket. A *cricket* or *saddle* is a small "fake roof" built to divert water. The cricket is built over the roof on its low side to divert water to roof drains. It can be constructed of plywood or sheet metal depending on the size. Figure 17-5 shows an example of a roof framed with crickets. Smaller crickets are also framed on the upper edge of any roof penetrations such as skylights and chimneys. An alternative to framing crickets is to alter the slope of the entire roof by using a framing system, typically referred to as a butterfly roof. With a single-sloped roof, framing members hang from a ledger that is attached to the exterior wall. The ledger is parallel to the floor. With a butterfly roof, the ledger is set on an incline, which provides additional slope to what is normally con-

sidered a flat roof. Figure 17-6 shows the outline of a butterfly roof in elevations. Figure 17-7 shows an example of a butterfly roof plan.

High-Sloped Roofs

High-sloped roofs are generally considered to be any roof with a slope greater than a 4:12 pitch (33% slope). Many of the roof shapes found in residential construction are also used in light construction and small commercial uses. The most common high-sloped roof shapes include mansard, gable, hip, Dutch hip, and gambrel roofs.

Mansard

A *mansard* roof is an inclined roof that is used to cover one or more full stories of a structure. A partial mansard can be used as a visual

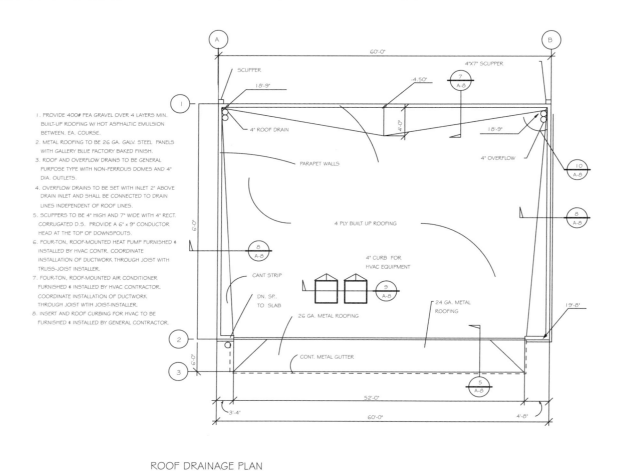

ROOF DRAINAGE PLAN
1/8" = 1' - 0"

FIGURE 17-3 ■ Simple structures often use a single-slope roof to direct water off the roof.

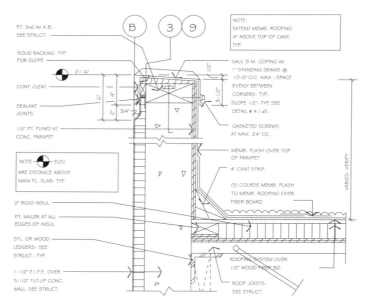

FIGURE 17-4 ■ A parapet wall that surrounds the roof must be referenced on the roof plan with construction methods showing the intersection in details, which are referenced to the roof plan. *Courtesy Architects Barrentine, Bates & Lee, A.I.A.*

screen, as a shading device, or as a decoration to break up a parapet wall. Figure 17-8 shows examples of full and partial mansard roofs in elevation. Figure 17-9 shows an example of representing a mansard on a roof plan. When shown on the roof plan, the mansard must be located by dimension and referenced to construction details.

Gable

A *gable* roof is a roof formed by two intersecting roof planes and can be found on both low- and high-sloped structures. When the roof is formed between equal-height walls, with equal-pitched roof, the intersection of the roof planes will be located at the midpoint between the supporting walls. The intersection of the roof planes is referred to as the *ridge* and represents the highest point of the roof. The ridge is parallel to the floor. The ridge location will be shown on the roof plan and located on the roof framing plan. Figure 17-10 shows an example of a gable roof.

Hip

A *hip roof* is formed by four intersecting roof planes. When a hip roof is formed on a square structure, the roof will intersect at a point. When a hip is framed on a rectangular-shaped structure, a

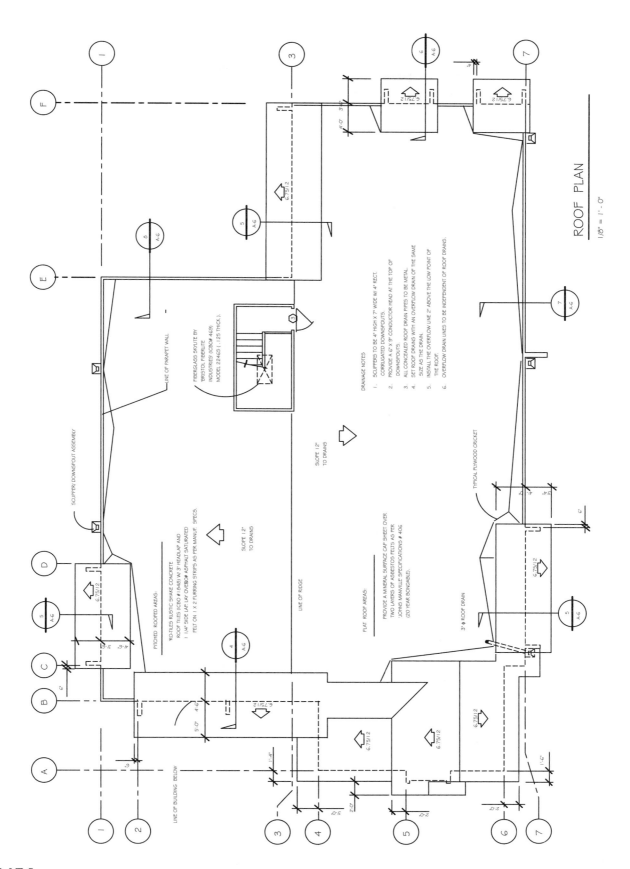

FIGURE 17-5 ■ A cricket is a fake roof built over the main roof, which is used to divert water to roof drains. *Courtesy StructureForm Masters, Inc.*

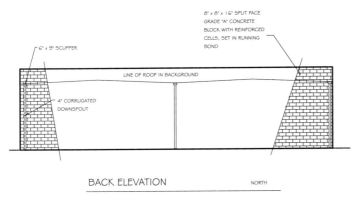

FIGURE 17-6 ■ The ledger that supports the roof-framing members at the wall can be inclined to divert water to roof drains. *Courtesy Wil Warner*

ridge will be formed halfway between the long walls of the structure. A *hip* is an inclined roof member that forms an external corner between two intersecting planes of a roof. A *valley* is an inclined roof member that forms an internal corner between two intersecting planes of a roof. Both hips and valleys are common to high-sloped roofs and are shown in Figure 17-11.

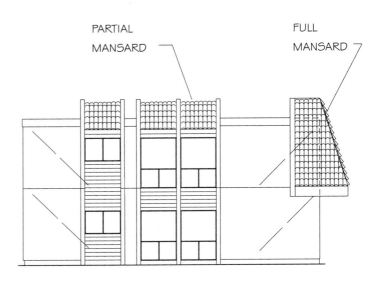

FIGURE 17-8 ■ A partial mansard roof can be used on low-slope roofs to hide mechanical equipment. A full mansard can be used to hide the height of a structure.

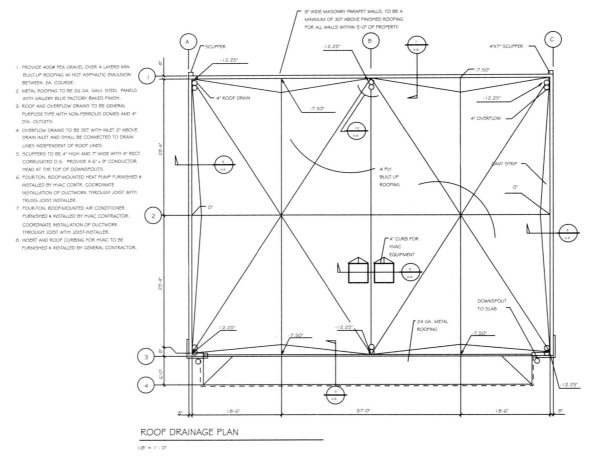

FIGURE 17-7 ■ A butterfly roof diverts water by altering the height of the ledger, causing high and low points. Water is diverted from the ridge to the edge, and along the edge to roof drains. *Courtesy Wil Warner.*

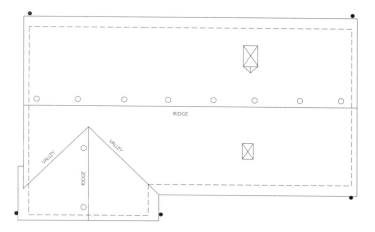

FIGURE 17-9 ■ The walls and sloped roofs of the mansard must be reflected on the roof plan.

Dutch Hip

A *Dutch hip*, which is sometimes referred to as a Dutch gable roof, is a combination of a gable and a hip roof. A wall is placed parallel to the short support wall at the roof level. This wall is used to form a break in the roof plan and is shown in Figure 17-12. The roof pitch and desired effect determines the location of the wall. Figure 17-13 shows a Dutch hip represented on a roof plan.

Gambrel

A *gambrel* roof is formed by two different roof levels. The upper portion is a gable roof, and the lower portion is a steep-shed roof. The ridge and the limits of each roof plane must be shown on the roof plan. Figure 17-14 shows a gambrel roof plan.

ROOF MATERIALS

The type of material used for weather protection must be specified on the roof plan. Common roofing materials include single-ply and built-up roofs, shingles, metal slate, and tile roofs. Many roof materials are also described by their weight per square. A *square* of roofing is equal to 100 sq ft (9.3 m²)

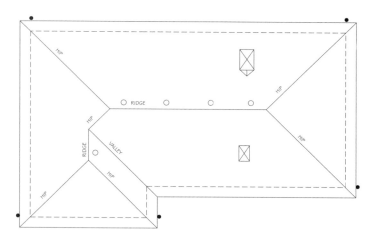

FIGURE 17-10 ■ The major shapes of a gable roof include the ridge, valley, and hips. The walls perpendicular to each ridge are referred to as gable end walls.

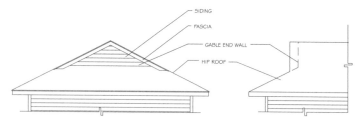

FIGURE 17-12 ■ A Dutch hip roof combines features of a hip and gable roof. The hip is broken before it reaches the ridge, and a gable wall is built at the roof level.

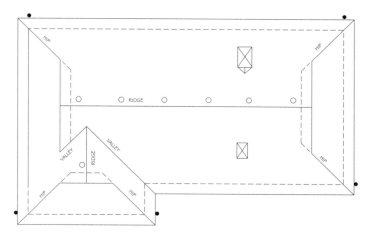

FIGURE 17-11 ■ The shapes of a hip roof include ridges, hips, and valleys. A hip roof provides shade on all sides of the structure.

FIGURE 17-13 ■ Representing a Dutch hip roof on a roof plan.

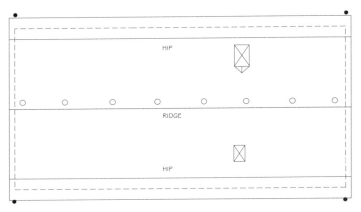

FIGURE 17-14 ■ Representing a gambrel roof on a roof plan.

Single-Ply Roofs

Single-ply roofs can be applied as a thin liquid or sheet made from Ethylene Propylene Diene Monomer (EPDM), which is an elastomeric or synthetic rubber material. Polyvinyl chloride (PVC), chlorosulfonated polyethylene (CSPE) and polymer-modified bitumens are also used. These materials are designed to be applied to roof decks ranging with a minimum slope 1/4:12 (2% slope). In liquid form, these materials can be applied directly to the roof decking with a roller or sprayer to conform to irregular-shaped roofs. Single-sheet roofs are rolled out and bonded together to form one large sheet. The sheet can be bonded to the roof deck by mechanical fasteners. Some applications are not attached to the roof deck and are held in place by gravel material that is placed over the roofing material to provide ballast. A typical specification for a single-ply roof would specify the material, the application method and the aggregate size. Many firms also specify a reference for the material to be installed according to the manufacturer's specification. Details are also typically supplied to show how roof intersections and sheet seams will be flashed similar to Figure 17-4.

Built-Up Roofs

Built-up roofs consist of two or more layers of bituminous-saturated roofing felt, cemented together with bitumen and surfaced with a cap sheet, mineral aggregate, or similar surfacing material. The bitumen material used to bond the layers together is usually tar or asphalt. In addition to the roofing felts and bitumens, a gravel surfacing material is used to protect the exposed surface from abrasions. Built-up roof systems can be applied over any type of roof deck and are suitable for low- or high-sloped roofs. Generally they are to be applied to roofs with a minimum slope of 1/4:12 (2% slope) but less than a 2:12 pitch 17% slope). Figure 17-15 shows the process for applying a built-up roof system.

Composition Shingles

Composition or asphalt shingles come in a variety of colors and patterns. Standard shingles come in a 12 × 36" (300 × 900 mm)

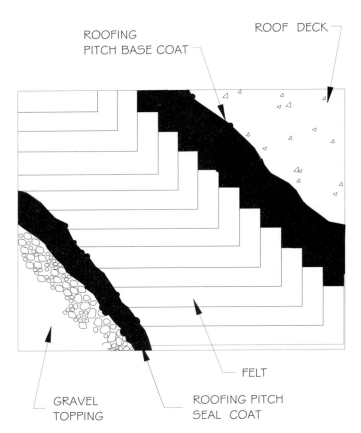

FIGURE 17-15 ■ The construction of a four-ply roof over a concrete roof deck.

shingle and are divided into three tabs, with a weight based on the square footage weight. Three-tab composition shingles have a weight of 235 lbs per square and are installed over 15# felt. Shingles divided into random widths and thickness with a weight of 300 lbs per square are also available. Composition shingles are designed for use on roofs with a minimum pitch of 2:12 (17% slope) or greater. Underlayment requirements vary depending on the roof pitch and the design wind speed.

Wood Shingles

Wood shingles are applied to roofs with a 3:12 pitch (25% slope) or greater over 1 × 4 or 1 × 6 (25 × 100 or 25 × 150) spaced sheathing and 15# felt. In areas of the country where the average daily temperature in January is 25°F (-4°C) or less, wood shingles must be applied over special underlayment. Wood shingles laid in cold climates must be installed over two layers of underlayment cemented together or of a self-adhering polymer modified bitumen sheet laid over plywood sheathing for the first 24" (610 mm) from the eave edge. An additional layer of 15# felt is placed between each row of shingles. Wood shakes are thicker and have a rougher texture than shingles. Wood shakes are applied in a similar manner as a shingle but must be applied to roofs with a 4:12 pitch (33% slope) or greater. Both wood shakes and shingles can be chemically treated for fire resistance. Shingles made from masonite are also available. Masonite shingles simulate wood shakes and provide a fire-resistant material.

Seamed Metal

A sheet metal roof made of copper, zinc alloy, or galvanized or coated stainless steel is a common alternative for a commercial roof when a 3:12 or greater pitch (25% slope) is used. A pitch as low as 1/4:12 (2% slope) can be used, depending on the type of seam used to join the metal sheets. Figure 17-16 shows common seam methods. When specified on the roof plan, the metal gauge, panel size, material, and seam pattern must be specified.

Corrugated Metal

Corrugated steel sheets can be used for either a siding or roofing material. Metal sheets in either rounded or angular bent patterns similar to Figure 17-17 are available and can be installed on roofs with a pitch of 3:12 (25% slope) or greater. In addition to steel, panels are also available in aluminum, galvanized steel, fiberglass and corrugated structural glass. When specified, the manufacturer, material, the panel size and weight, required lap, finish, and fastening method must be placed on the plans. Details of installation are also referenced to the roof plan.

Tile

Tile roofing can be clay, concrete, or metal units that overlap to form the weather protection. Tiles come in a variety of patterns, colors and shapes and are usually installed on roofs having a roof pitch of 3:12 or greater. Tile can be installed over either solid or spaced sheathing. Attachment methods will vary depending on the pitch and the maximum basic wind speed. When specified on the roof plan, the manufacturer, style, color, weight, and application method must be specified.

COMMON ROOF PLAN COMPONENTS

Although the plan of a low-sloped roof looks different than a high-sloped roof plan, many of the components are the same. Common

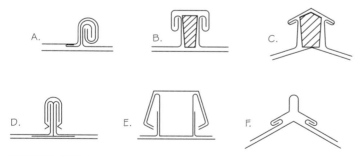

FIGURE 17-16 ■ Examples of metal seams include (A) double lock standing seam, (B) wood batten seam, (C) wood batten ridge seam, (D) prefabricated standing seam, (E) prefabricated batten seam, and (F) ridge seam.

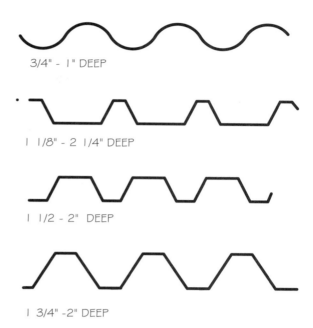

3/4" - 1" DEEP

1 1/8" - 2 1/4" DEEP

1 1/2 - 2" DEEP

1 3/4" -2" DEEP

FIGURE 17-17 ■ Standard corrugated and ribbed roofing pat terns.

components include changes in roof shape, roof openings, drainage methods, notations, and dimensions.

Changes in Shape

Changes in roof shape are shown on the roof plan with continuous object lines. A ridge, hip, and valley are the most common changes of roof shape.

Locating Ridges

The ridge between equal-pitched planes formed from equal-height walls can be located by locating the midpoint between the two supporting walls. When the walls are an unequal height, or a saw tooth roof is formed with unequal roof pitches, a section will need to be drawn so that the ridge location can be projected in the roof plan, as shown in Figure 17-18.

When two gabled structures intersect each other, the intersection of the ridges needs to be determined. With equal pitches over each portion of the structure, the portion that is widest will have the highest ridge. The ridge over the narrower portion of the structure will intersect the highest gable at a point where both roofs are of equal height. This point can be determined by drawing a line to represent the valley formed between the two planes. The valley will always be formed at an angle that is one-half the angle formed between the two intersecting walls. Figure 17-19 shows the intersection of two unequal width roofs and the merging of the ridge lines.

Representing Valleys

When one gable or hip roof intersects another gable roof, a valley is created. If the two intersecting roofs planes are at an equal pitch, the line representing the valley is drawn on the roof plan at an angle

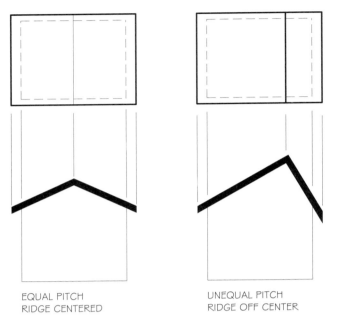

EQUAL PITCH
RIDGE CENTERED

UNEQUAL PITCH
RIDGE OFF CENTER

FIGURE 17-18 ■ When roof pitches are unequal, the location of the ridge on the roof plan must be projected from sections.

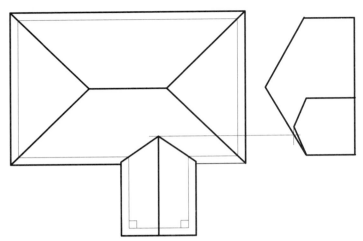

FIGURE 17-20 ■ When intersecting roofs are at different pitches, the ridge intersection and valleys must be projected from sections.

that is one-half the angle formed by the intersecting support walls. For perpendicular support walls, the valley is represented by a line drawn at a 45° on the roof plan, as shown in Figure 17-11. When the intersecting roof planes are at unequal pitches, sections must be drawn to determine where the valley will be created. Figure 17-20 shows an example of determining the intersection of two ridges with different heights and pitches. When the intersecting roof planes are not perpendicular to each other and are over portions of the structure with different widths, the ridges, valleys, and hips must be determined using sections, as seen in Figure 17-21.

The intersection of two gables that are supported by walls with different heights often must be represented on a roof plan. Using the roof pitch, the drafter must determine the distance the lower roof will require to rise to an equal height on the upper roof. Figure 17-22 shows the intersection of an entry portico roof framed with 12' (3700 mm) high support walls, the main roof framed with 10'

(3000 mm) high walls and each roof framed with a pitch of 6:12 (50% slope).

Roof Openings

Each penetration of the roof is usually shown on the roof plan. Major openings shown on the roof plan include skylights, chimneys, vent pipes, and openings for HVAC ducts.

Skylights

Individual skylights are shown on the roof plan by a rectangle that represents the size of the skylights. On inclined roofs, the true size of the skylight will be altered because of the roof angle. An *X* is sometimes placed in the skylight to accent its location. Figure 17-23 shows an example of how skylights are represented and specified on a roof plan. On plans with a large number of skylights, the size and type of the skylight can be shown in a schedule similar to Figure 17-24. Large skylight units need to be represented by a method that distinguishes the skylight from the roof. The hatch pattern used to represent the skylight varies for each office, but it must be represented in a symbol schedule attached to the roof plan. In addition to showing the location, the method of protecting the opening from water must also be shown. The waterproofing

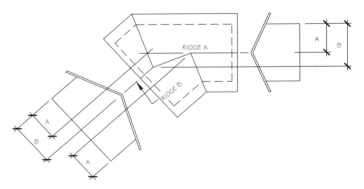

FIGURE 17-19 ■ The intersection of two roofs that are not perpendicular to each other. The hip and valley will be parallel to each other and will be formed at an angle that is equal to one-half the angle formed between the two intersecting support walls.

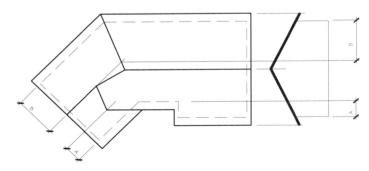

FIGURE 17-21 ■ Projecting the intersection of roofs of different widths.

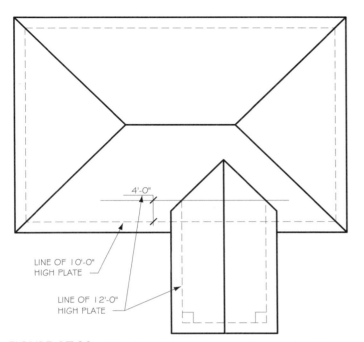

FIGURE 17-22 ■ The slope of the roof must be known when drawing the intersection of two different roofs supported on walls of differing heights. With a 6:12 pitch, a horizontal distance of 48" (1200 mm) is required to reach the starting height of the upper roof. By offsetting the 10' (3000 mm) wall 48" (1200 mm), the valley between the two roofs can be located from the point where the 12' (3600 mm) high wall intersects this 48" (1200 mm) offset.

method is shown in details similar to Figure 17-25, which are referenced on the roof plan.

Chimneys

A fireplace is common in many multifamily units and business occupancies such as restaurants and small offices. Chimneys are represented on the roof plan according to their type. Masonry and metal chimneys in wood chases are represented as a rectangle with an X through it. Metal chimneys are represented by a circle, which represents the required diameter of the chimney. Figure 17-11 shows a method of representing chimneys.

SKYLITE SCHEDULE

MARK.	LOCATION	PANEL DIMENSION	INSIDE CURB DIMENSION	FINISH DRYWALL DIMEN.
SL-1	CORRIDOR	5'-0" X 5'-0"	4'-8 1/2" X 4'-8 1/2"	4'-7 1/4" X 4'-7 1/4"
SL-2	PRODUCTION	5'-0" X 5'-0"	4'-8 1/2" X 4'-8 1/2"	----------------
SL-3	DROP-OFF	5'-0" X 16'-0"	4'-8 1/2" X 15'-8 1/2"	----------------
SL-4	RETAIL	5'-0" X 16'-0"	4'-8 1/2" X 15'-8 1/2"	----------------
SL-5	LUNCH ROOM	5'-0" X 12'-0"	4'-8 1/2" X 11'-8 1/2"	----------------
SL-6	OFFICES	4'-4" X 6'-4"	4'-1 1/4" X 6'-1 1/4"	4'-0" X 6'-0"
SL-7	OFFICES	5'-0" X 5'-0"	4'-8 1/2" X 4'-8 1/2"	4'-7 1/4" X 4'-7 1/4"
SL-8	CORRIDOR	5'-0" X 16'-0"	4'-8 1/2" X 15'-8 1/2"	4'-1 1/4" X 15'-7 1/4"

FIGURE 17-24 ■ On plans with a large number of skylights, a schedule can be used to help increase clarity. *Courtesy Michael & Kuhns Architects, P.C.*

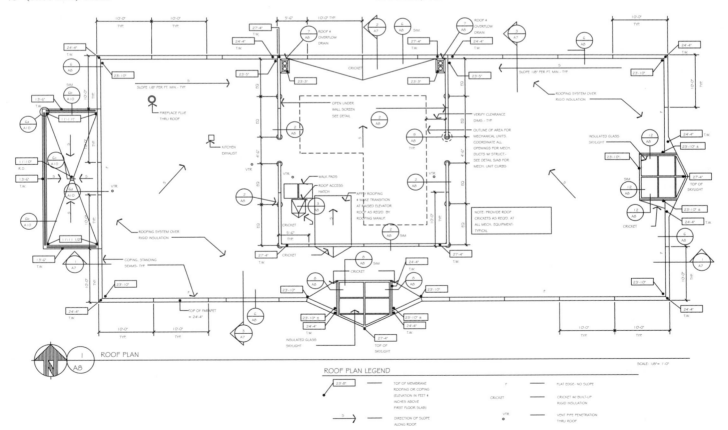

FIGURE 17-23 ■ A completed roof plan for a single-slope roof. *Courtesy Architects Barrentine, Bates & Lee, A.I.A.*

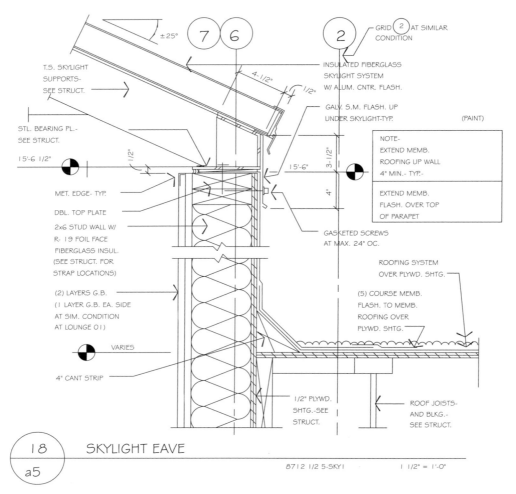

±25°

⑦ ⑥   ② GRID ② AT SIMILAR
                           CONDITION

T.S. SKYLIGHT
SUPPORTS-
SEE STRUCT.

4-1/2"   1/2"

INSULATED FIBERGLASS
SKYLIGHT SYSTEM
W/ ALUM. CNTR. FLASH.

STL. BEARING PL.-
SEE STRUCT.

GALV. S.M. FLASH. UP
UNDER SKYLIGHT-TYP.   (PAINT)

15'-6 1/2"   1/2"   3-1/2"

15'-6"

NOTE-
EXTEND MEMB.
ROOFING UP WALL
4" MIN.- TYP.-

MET. EDGE- TYP.

4"

EXTEND MEMB.
FLASH. OVER TOP
OF PARAPET

DBL. TOP PLATE

GASKETED SCREWS
AT MAX. 24" OC.

2x6 STUD WALL W/
R- 19 FOIL FACE
FIBERGLASS INSUL.
(SEE STRUCT. FOR
STRAP LOCATIONS)

ROOFING SYSTEM
OVER PLYWD. SHTG.

(2) LAYERS G.B.
(1 LAYER G.B. EA. SIDE
AT SIM. CONDITION
AT LOUNGE 01)

(5) COURSE MEMB.
FLASH. TO MEMB.
ROOFING OVER
PLYWD. SHTG.

VARIES

4" CANT STRIP

1/2" PLYWD.
SHTG.-SEE
STRUCT.

ROOF JOISTS-
AND BLKG.-
SEE STRUCT.

⎛ 18 ⎞   SKYLIGHT EAVE
⎝ a5 ⎠

8712 1/2 5-SKY1   1 1/2" = 1'-0"

FIGURE 17-25 ■ Details showing the construction and mounting of skylights must be referenced on the roof plan. *Courtesy Architects Barrentine, Bates & Lee, A.I.A.*

Plumbing and Mechanical Penetrations

Because of the risk of leaking, plumbing vents are shown on many low-pitched roofs. Circles or squares, depending on their shape, represent vents and exhaust flues. Their location is determined by their location on the floor plan and is generally not located by dimensions on the roof plan. Details such as Figure 17-26 are referenced to the roof plan to show waterproofing methods. Because openings for mechanical equipment must be shown on the roof framing drawing, they may or may not be shown on the roof plan. The location of equipment is generally shown and specified in relation to surrounding walls or screens on the roof plan. Both the screens and curbs for mounting the equipment should be referenced on the roof plan as shown in Figure 17-27.

Roof Drainage

Once the shape of the roof is indicated, the next priority is to show how water will be diverted from the surface. To indicate drainage, slope indicators, elevation markers, drains, and overflow drains must all be shown to indicate how water will be removed from the roof. Each is shown in Figure 17-23.

Slope Indicators

A *slope indicator* is an arrow that indicates the flow of water on the roof. Arrows are generally placed so that they point away from the ridge to the low points of the roof.

Elevation Markers

Once the slope has been indicated, the elevation of the roof must be shown. Elevations for the roof are represented with symbols similar to those used on the floor and site plans. Elevations for the roof are typically given from either the finished floor level or from the height of the ridge. The base point for the elevations should be indicated on the roof plan in either a legend or in general notes. Figures 17-23 and 17-28 show examples of elevations that are based on the finished floor. When elevations are given from the ridge, the height of the finished roof at the ridge is usually referred to as elevation *00*. All other points are expressed as heights relative

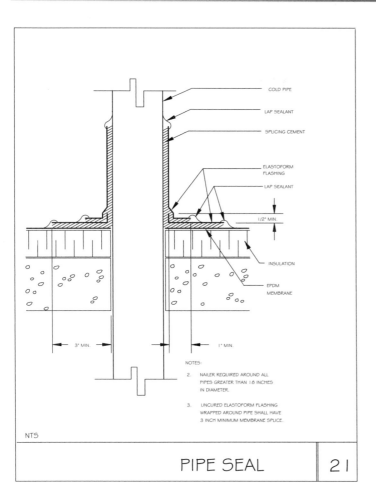

FIGURE 17-26 ■ Details showing the waterproofing of roof penetrations are an important addition to a roof plan. *Courtesy H.D.N. Architects A.I.A.*

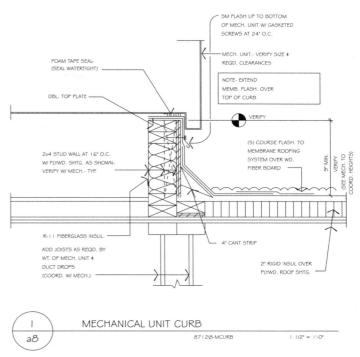

FIGURE 17-27 ■ Details showing the construction and water proofing of curbs for mechanical equipment must be referenced to the roof plan. *Courtesy Architects Barrentine, Bates & Lee, A.I.A.*

to the ridge. Points lower than the ridge will have negative numbers, and points above the ridge are represented by positive numbers. Heights are determined by multiplying the distance by the desired slope. For instance, with the minimum roof pitch, a ridge 30'–0" from a wall will require 7.5" (30' × .25") of fall to properly shed water. Figure 17-29 shows an example of elevations based on the ridge height.

Gutters

The roof system is designed to direct water from the ridge downward. For sloped roofs built in areas of the country that receive large amounts of rain, gutters are placed on the low edge of the roof to collect the water. The gutter has a slight slope to direct water to a downspout. The downspout transfers the water from the roof level to grade level. Depending on the area of the country and the rain to be dispersed, gutters may be connected to a dry well or to a storm sewer. The location of the gutter is specified on the roof plan but not drawn. The location, size and material of the downspouts should be indicated on the roof plan. Construction of the gutter and downspouts are shown in details similar to Figure 17-30.

Drains Overflow Drains and Scuppers

Low-sloped roofs are often enclosed by a parapet wall that traps water on the low side of the roof. *Drains* are placed in the roof to remove water from the roof wherever the water cannot run over the edge of the roof. A cricket is used to direct water to the drains in a method similar to that shown in Figure 17-5. Where drains are required, an overflow drain must also be provided to drain water from the roof in case the drain becomes blocked. The overflow drain must be the same size and be installed 2" (50 mm) above the roof drain. The drain directs water to a downspout, which in turn transports water to the waste disposal system. The overflow drain is connected to a separate drain that often connects to a hole in the wall. Water flowing from this drain indicates that the drainage system requires maintenance.

An opening in the parapet wall is an alternative to providing drains and overflow drains. A funnel-like collector called a *scupper* is placed on the outside of the hole to funnel water from the roof to a downspout. A scupper can be represented on the roof plan as shown in Figure 17-7 and would be shown in details similar to those shown in Figure 22–31.

Diverters

For structures built in areas of the country that receive only small amounts of rain, water is often allowed to flow over the edge of the roof directly to the ground. To keep water from dripping onto walkways at entries, a metal strip called a *diverter* is generally placed on the roof. The diverter is placed so that it will direct water on the

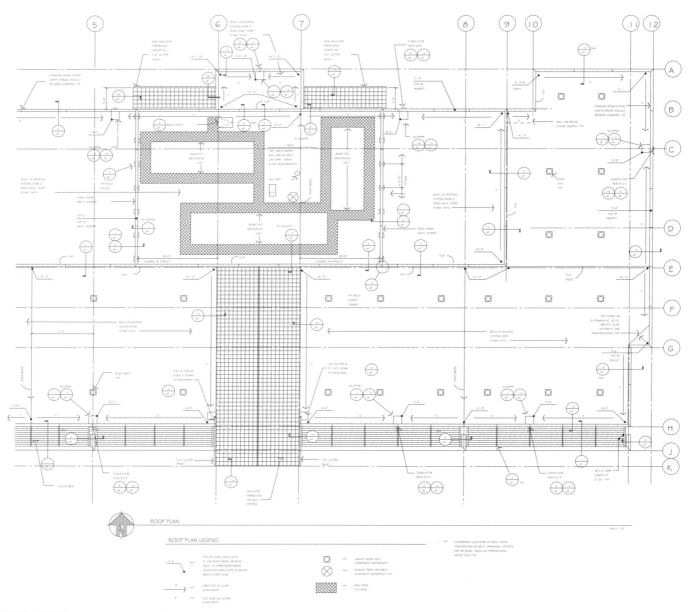

FIGURE 17-28 ■ A completed roof plan for a shed roof with multiple slopes for drainage. Roof elevations are based on their height above the finish floor level. *Courtesy Architects Barrentine, Bates & Lee, A.I.A.*

roof to an area where no one will be splashed. A bed of gravel is generally placed on the ground below the low point of the diverter to decrease the chance of erosion. A diverter can be represented on the roof plan as shown in Figure 17-32.

Notations

As with other drawings, notations must be placed on the roof plan to specify the material that has been drawn. Much of the material that will need to be specified has been presented throughout this chapter. Where possible, local notes can be used to identify each material. If space does not allow for the entire specification to be placed in or near the drawing, an abbreviated portion of the note

should be placed by the object, and the complete note should be provided in a listing of general notes. To add clarity to the drawing, keyed notes similar to Figure 17-33 can be used to reference general notes to the plan. Specifications should be provided to specify all changes in shape, roof material, equipment, and openings.

Dimensions

Although the roof plan is not used to construct the roof, dimensions should be provided on the roof plan to locate material that is located on the roof. Base dimensions should be provided to locate each grid, just as with the floor plan and elevations. Equipment should also be dimensioned. On sloped roofs, overhang

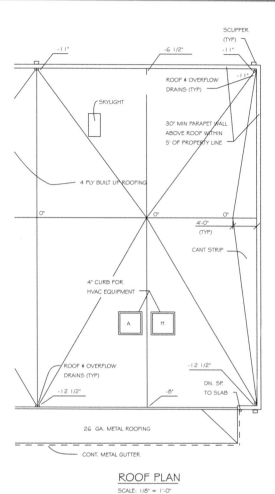

ROOF PLAN
SCALE: 1/8" = 1'-0"

FIGURE 17-29 ■ Roof elevations based on the height of the ridge of the roof. Heights shown on the front and rear walls represent a height that is below the ridge line.

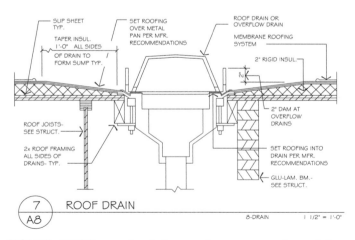

7 ROOF DRAIN
A8
8-DRAIN   1 1/2" = 1'-0"

FIGURE 17-30 ■ A detail showing the mounting of a roof drain. *Courtesy Architects Barrentine, Bates & Lee, A.I.A.*

dimensions should also be provided. Anything that is located on another plan is often not dimensioned on the roof plan. Figures 17-5, 17-7, 17-23 and 17-28 show examples of dimensions placed on roof plans.

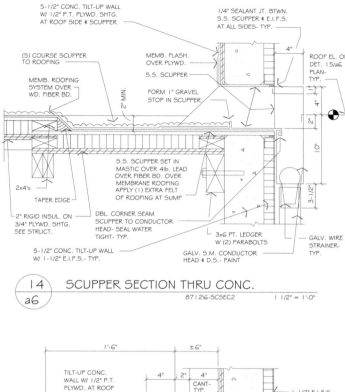

14 SCUPPER SECTION THRU CONC.
a6
8712\6-SCSEC2   1 1/2" = 1'-0"

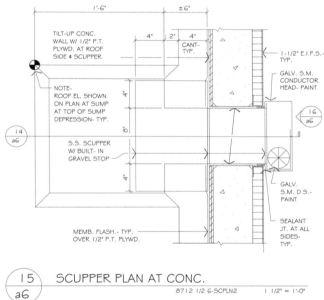

15 SCUPPER PLAN AT CONC.
a6
8712 1/2 6-SCPLN2   1 1/2" = 1'-0"

FIGURE 17-31 ■ Details showing the construction of a scupper through a concrete parapet wall. *Courtesy Architects Barrentine, Bates & Lee, A.I.A.*

COMPLETING A LOW-SLOPED ROOF PLAN

The convenience store shown in Chapter 15 will be used to show the process for drawing a low- sloped roof. The roof plan will be completed by using the architect's sketch shown in Figure 17-34.

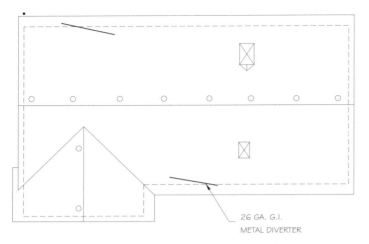

26 GA. G.I.
METAL DIVERTER

FIGURE 17-32 ■ In areas with minute amounts of rainfall, a metal diverter can be installed on the roof to drain water away from doorways.

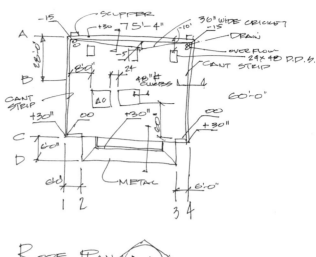

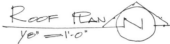

FIGURE 17-34 ■ The sketch of an architect or project designer is used by the drafter to develop a roof plan.

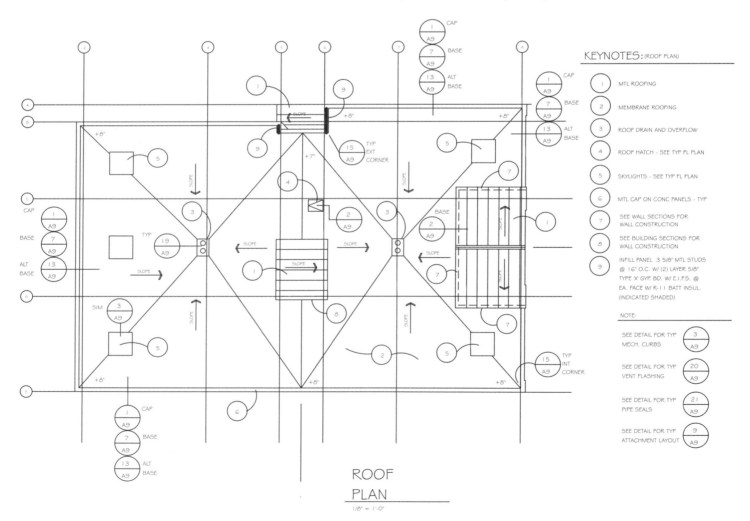

KEYNOTES: (ROOF PLAN)

1. MTL ROOFING
2. MEMBRANE ROOFING
3. ROOF DRAIN AND OVERFLOW
4. ROOF HATCH - SEE TYP FL PLAN
5. SKYLIGHTS - SEE TYP FL PLAN
6. MTL CAP ON CONC PANELS - TYP
7. SEE WALL SECTIONS FOR WALL CONSTRUCTION
8. SEE BUILDING SECTIONS FOR WALL CONSTRUCTION
9. INFILL PANEL 3 5/8" MTL STUDS @ 16" O.C. W/ (2) LAYER 5/8" TYPE 'X' GYP. BD. W/ E.I.F.S. @ EA. FACE W/ R-11 BATT INSUL. (INDICATED SHADED)

NOTE:

SEE DETAIL FOR TYP MECH. CURBS — 3/A9

SEE DETAIL FOR TYP VENT FLASHING — 20/A9

SEE DETAIL FOR TYP PIPE SEALS — 21/A9

SEE DETAIL FOR TYP ATTACHMENT LAYOUT — 9/A9

ROOF
PLAN
1/8" = 1'-0"

FIGURE 17-33 ■ To add clarity, keyed notes can be used. *Courtesy KPFF Consulting Engineers.*

1. Using the sketch, draw the exterior walls
2. Draw the outline of the mansard roof
3. Draw the drain, overflow drains, and scuppers based on the architect's specifications
4. Draw the cricket on the inside of the north parapet wall
5. Draw required 48" × 48" × 4" high curbs for HVAC equipment
6. Draw required skylights

With all necessary materials drawn, the roof plan will resemble Figure 17-35. The plan can be completed by using the following steps:

7. Draw all grid lines
8. Provide overall dimensions
9. Dimension the mansard
10. Dimension all openings
11. Dimension each skylight

The roof plan is now dimensioned and will resemble Figure 17-36. The drawing can be completed by adding the required notations:

12. Place the title and scale below the drawing.
13. Specify all materials that can be grouped as general notes.
14. Draw the north arrow
15. Provide identification text for each grid symbol
16. Locate and specify all elevations
17. Specify all changes in roof shape
18. Specify all drainage devices
19. Specify all roof-mounted equipment
20. Specify all roof openings
21. Draw detail markers for anticipated details

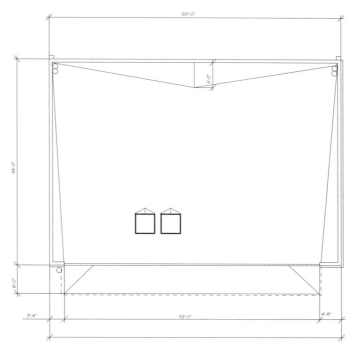

FIGURE 17-36 ■ Adding dimensions for each feature.

The completed roof plan is shown in Figure 17-3.

A similar structure with a ridge at grid B and a butterfly roof is shown in Figure 17-37. This structure can be drawn using steps similar to those used to draw the single-sloped roof. Once the walls and mansard have been drawn, the ridge should be located. Because the architect has specified that three sets of drains are to be

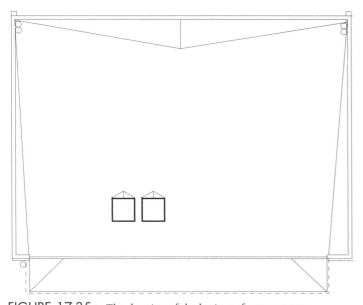

FIGURE 17-35 ■ The drawing of the basic roof components.

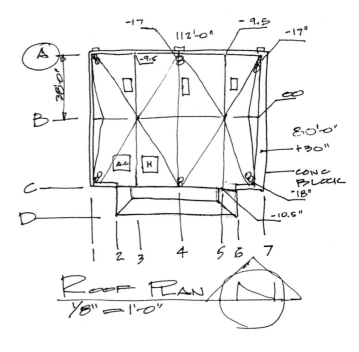

FIGURE 17-37 ■ The architect's sketch for a butterfly roof.

located on grid *A* and *C*, the roof can be divided into four quadrants to locate the butterfly.

The roof pitch for points 3 and 5 should be determined prior to determining the height at the drains. Using a slope of 0.25:12", the slope from B-3 to A-3 would require a 9 1/2" drop. The distance from A-3 to A-4 is 28 ft, which would require an additional drop of 7". Using the same materials that were used for the structure in Figure 17-4, the completed roof plan would resemble the structure of Figure 17-7.

COMPLETING A HIGH-SLOPED ROOF PLAN

A light-framed retail outlet will be used to show the process for drawing a high-sloped roof. The roof plan will be completed by using the architect's sketch, shown in Figure 17-38.

1. Using the sketch, draw the exterior walls. Assume the north and west walls to be 30" parapet walls for property line protection.
2. Draw all changes in roof shape
3. Draw the roof overhangs
4. Draw the drain, overflow drains, scuppers, and downspouts based on the architect's specifications
5. Draw the cricket on the inside of the north parapet wall
6. Draw required skylights

With all necessary materials drawn, the roof plan will resemble Figure 17-39. The plan can be completed by using the following steps:

7. Draw all grid lines
8. Provide overall dimensions
9. Dimension all overhangs

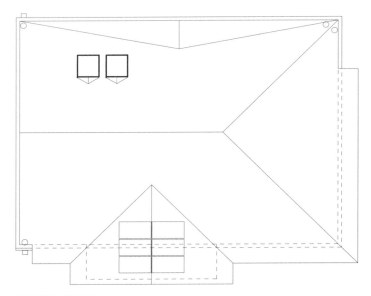

FIGURE 17-39 ■ The drawing of basic roof shapes and components.

10. Dimension all openings
11. Dimension the skylights
12. Place the title and scale below the drawing
13. Draw the north arrow
14. Provide identification text for each grid symbol
15. Locate and specify all elevations
16. Specify all changes in roof shape
17. Specify all drainage devices
18. Specify any roof-mounted equipment
19. Specify all roof openings
20. Draw detail markers for anticipated details

The completed roof plan is shown in Figure 17-40.

LAYER GUIDELINES FOR ROOF PLANS

Common Layer names based on standard AIA modifiers:

A-ROOF	Roof
A-ROOF-ANNO	Roof text
A-ROOF-DETL	Roof details
A-ROOF-DRAN	Roof drains
A-ROOF-ELEV	Roof elevations
A-ROOF-EQPM	Roof mounted equipment
A-ROOF-HRAL	Stair handrails, nosing, and guardrails
A-ROOF-LEVL	Roof level changes
A-ROOF-OTLN	Roof outline

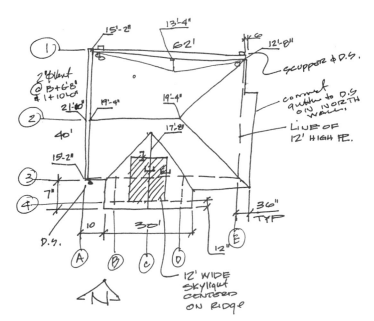

FIGURE 17-38 ■ The architect's sketch for a high-pitched gable roof.

DN. SP.
TO SLAB

4" x 7" SCUPPER

12'-8"

T.W. 15'-2"

13'-4"

12'-8"

PARAPET WALLS

CANT STRIP

4" MIN. CURB AND
LEVEL PLATFORM FOR
HVAC EQUIPMENT

24 GA. METAL
ROOFING

4" ROOF DRAIN

CONT. METAL GUTTER

3'-0"
TYP.

13'-0"

12'-0"

12'-0"

60'-0"

A   B   C   D   E

ROOF DRAINAGE PLAN

1/8" = 1' - 0"

1. METAL ROOFING TO BE 26 GA. GALV. STEEL ROOFING PANELS
   WITH GALLERY BLUE FACTORY BAKED FINISH.
2. ROOF AND OVERFLOW DRAINS TO BE GENERAL PURPOSE
   TYPE WITH NON-FERROUS DOMES AND 4" DIA. OUTLETS.
3. OVERFLOW DRAINS TO BE SET WITH INLET 2" ABOVE DRAIN
   INLET AND SHALL BE CONNECTED TO DRAIN LINES
   INDEPENDENT OF ROOF LINES.
4. SCUPPERS TO BE 4" HIGH AND 7" WIDE WITH 4" RECTANGULAR
   CORRUGATED DOWNSPOUTS. PROVIDE A 6" x 9" CONDUCTOR
   HEAD AT THE TOP OF DOWNSPOUTS.
5. FOUR-TON, ROOF-MOUNTED HEAT PUMP FURNISHED AND
   INSTALLED BY HVAC CONTRACTOR. COORDINATE INSTALLATION
   OF DUCTWORK THROUGH JOIST WITH TRUSS-JOIST INSTALLER.
6. FOUR-TON, ROOF-MOUNTED AIR CONDITIONER FURNISHED AND
   INSTALLED BY HVAC CONTRACTOR. COORDINATE INSTALLATION
   OF DUCTWORK THROUGH JOIST WTIH JOIST-INSTALLER.
7. INSERT AND ROOF CURBING FOR HVAC TO BE FURNISHED AND
   INSTALLED BY GENERAL CONTRACTOR.

FIGURE 17-40 ■ The completed high-pitched gable roof.

A-ROOF-PATT	Roof hatching patterns	A-ROOF-WALL	Walls at roof level
A-ROOFRISR	Roof stair risers		
A-ROOF-STRS	Roof ladders and stair treads		
A-ROOF-SYMB	Roof symbols		

Status field modifiers similar to other drawing areas should be used throughout architectural drawings. See Chapter 3 and the AIA Layer Guidelines for additional layer names.

CHAPTER 17

Roof Plan Components and Drawings

CHAPTER QUIZ

Use a separate sheet of paper to answer the following questions. Print the chapter title, the question number, and a short complete statement for each answer.

Question 17-1 Why are flat roofs used for a majority of commercial projects?

Question 17-2 List two reasons for using parapet walls.

Question 17-3 At what angle does the building code consider a roof a *steep roof*?

Question 17-4 What is the difference between a cricket and a cant strip?

Question 17-5 What is typically used to bind layers of a built-up roof together?

Question 17-6 List the weight per square foot for the two common types of composition shingles.

Question 17-7 Describe how the ridge of a saw-toothed roof is located.

Question 17-8 Describe the angle used to represent a valley between two intersecting portions of a structure.

Question 17-9 How should skylights be specified on a roof plan?

Question 17-10 List two methods of determining elevations on a roof plan.

DRAWING PROBLEMS

Use the reference material from preceding chapters, local codes, and vendor catalogs to complete one of the following projects. Unless other instructions are given by your instructor, draw the roof plan that corresponds to the floor plan that was drawn in Chapter 15. Skeletons of the roof plan for problems 17-1 through 17-4 can be accessed from http://www.delmar.com/resources/ocl.html. Use these drawings as a base to complete the assignment. Use appropriate symbols, linetypes, dimensioning methods, and notations to complete the drawing. Determine any unspecified sizes based on material or practical requirements. Unless specified, select a scale appropriate to plotting on "D" size material and determine LTSCALE and DIMVARS.

Information is provided on drawings in Chapters 14 through 23 relating to each project that might be needed to make decisions regarding the roof plan. You will act as the project manager and will be required to make decisions about how to complete the project. Any information not provided must be researched and determined by you unless your instructor (the project architect and engineer) provides other instructions. When conflicting information is found, information from preceding chapters should take precedence.

Minimum standards for all projects unless noted:

■ Create the necessary layers to keep the drawings stacked, but separated from the floor plan. Freeze all material not required by the roof plan.

■ Note that all downspouts are to be connected to a separate storm sewer connection separate from site drainage

■ Roof and overflow drains to be general purpose type with nonferrous domes and 4" drains. Overflow drains to be set 2" above drain inlet and shall be connected to drain lines independent of roof drains.

■ Scuppers to be 4" high and 7" wide with 4" rectangular corrugated downspouts. Provide a 6" × 9" conductor head at the top of each downspout.

■ Flat roofs to have 400 lb pea gravel over a 4-ply built-up roof

■ Inclined roofs to have 26-ga. metal roofing

Problem 17-1 Draw a roof plan for the four-unit complex. The roof is to be built at a pitch of 12:12 and will be covered with 300# composition shingles. For the two interi-

or units, provide a 4' minimum covered entry deck. The gable end walls are to have a 12" overhang. Exterior units will have a 24" overhang. All units will have a 6" overhang as it projects over lower units. Form a 24" high maximum cricket over the roof between units to divert water to scuppers and downspouts. Provide 5 scuppers and downspouts on each gable end wall. Provide a continuous gutter on the outer walls of each outer unit and connect to a downspout. Provide a 12:12 roof over each bay with a continuous gutter connected to a downspout located where it will not be seen.

Problem 17-2 Draw a roof plan to reflect a double shed roof with the ridge formed by a beam running east/west centered over the 4 × 4 steel column. Establish the ridge as 13'-0" above the finished floor. Provide a slope of 3/8:12 with all elevations listed from the finish floor. Place three scuppers and downspouts on the front and rear and establish cant strips to provide drainage to each roof drain. Provide continuous gutters on the mansard surrounding the main roof. Project the mansard 6' from the edge of the block wall. Provide a 4" curb for roof-mounted heating and air conditioning equipment. Curb to be installed by general contractor, based on sizes supplied by the HVAC supplier. Show (4)-24" × 48" double-domed skylights equally spaced over the parts and inventory areas and one centered over the break area.

Problem 17-3 Design a roof plan for the structure started in Problem 15-3. A ridge will run along the beam placed over the west row of columns (grid C) from 1 to 7. Use a double-shed system over the warehouse and provide a slope with a minimum of 1/4:12 to grids A and F. Provide four drains on each side of the warehouse. Use a single-shed roof over the office area sloping from grid 7 to 8 and a drain at B7.5 and E7.5. Establish the warehouse ridge so that it is 30" below the parapet wall but that minimum ceiling heights specified with the floor plan are maintained. Coordinate these plans with the framing plan and adjust the height accordingly. Provide (4)-24" × 48" skylights per each grid.

Problem 17-4 Serve as the project manager and design a butterfly roof plan for this structure. A ridge will run along grid 3 from A to C and from C to K with a minimum slope of 1/4:12 to grids 1, 2, 4 and 5. Provide three drains on each side of the warehouse and two on each side of the office area. Establish the ridge so that it is 30" below the parapet wall but that minimum ceiling heights specified with the floor plan are maintained. Coordinate these plans with the framing plan and adjust the height accordingly. Provide (4)- 24" × 48" skylights per each grid.

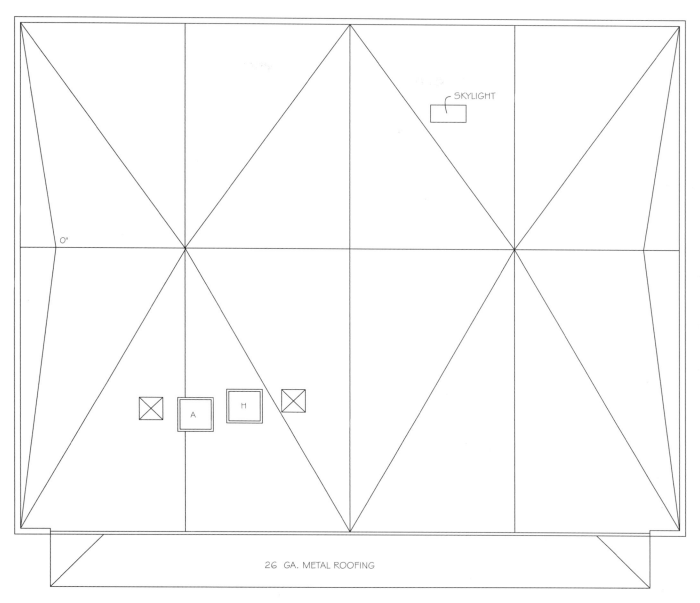

26 GA. METAL ROOFING

PROBLEM 17-2

## CHAPTER 18

Drawing Sections, Wall Sections, and Details

Building sections are drawn to show the vertical relationships of materials represented on each of the plan views and the exterior elevations. Exterior elevations show the material that will be used to shield the structure from the weather. Sections show the materials used to construct the walls, floors, ceilings, and roof components, and their vertical relationships to each other. In addition to considering the different types of sections, this chapter examines partial sections and details that supplement each type of section.

SECTION ORIGINATION

Building sections are the result of passing a viewing plane through a structure to reveal the construction methods being used. Material that is in front of the viewing plane cannot be seen. Material that is behind the viewing plane is projected to the plane and reproduced in the section. Another method of visualizing a section is to think of the viewing plane as a giant saw that slices vertically through a structure and divides it into two portions. One portion is removed to allow viewing of the portion that remains. Using the saw analogy, the viewing plane is referred to as a *cutting plane*. The location of the cutting plane is shown on the floor plan using a symbol similar to those shown in Figure 18-1. Notice that with each symbol, an arrow is included to indicate which portion of the structure is being viewed. Figure 18-2 shows an example of a floor plan with section markers. Figure 18-3 shows the section that is specified by the section marker 1/A12. Notice that material the cutting plane passes through on the floor plan is represented on the section. Material in the background such as doors, windows, and cabinets are also usually shown.

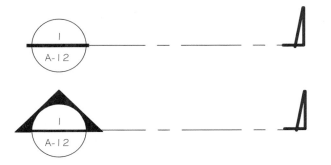

FIGURE 18-1 ■ The location of a section is shown on the floor plan by use of a cutting plane. The upper letter represents a specific section. The bottom number represents the page where the section can be found.

TYPES OF SECTIONS

Full sections, partial sections, and details may be used to represent a structure. Each has a specific use in representing how walls are constructed and how they intersect with floor and roof systems.

Full Sections

Full sections are the views that result from passing the cutting plane through the entire structure. Full sections are meant to give an overall view of a specific area of the structure and may be either longitudinal or transverse. A *longitudinal section* is produced by a cutting plane that is parallel to the long axis of the structure and is generally perpendicular to most structural materials used to frame the roof, ceiling, and floor systems. Because the framing members are perpendicular to the cutting plane, they are seen as if they had been cut. Figure 18-4 shows an example of a longitudinal section.

A *transverse section* is produced by a cutting plane that is parallel to the short axis of the structure, and it is often referred to as a cross section. The cutting plane for a transverse section is usually parallel to the materials used to frame the roof, ceiling, and floor systems and generally shows the shape of the structure better. Figure 18-5 shows a transverse section for a structure formed from precast concrete.

Partial Sections

A *partial section* is a section that does not go completely through the structure and can be used to show construction materials that are not seen in other sections. Figure 18-6 shows an example of a partial section that is referenced to the section shown in Figure 18-4. Partial sections are used to show only a specific area of the structure, while other sections are used to define the balance of the structure.

A partial section can also be used to show construction materials of a specific wall. When a section is used to show one specific wall, it is also referred to as a *wall section*. Wall sections can be referenced to the floor plan or to other sections. Figure 18-7 shows a wall section. Wall sections are drawn at a larger scale than a full section and are drawn to provide information on one specific type of wall.

Detailed Sections

Full and partial sections are often drawn at a scale that cannot adequately display the needed detail of all materials. *Detailed sections*

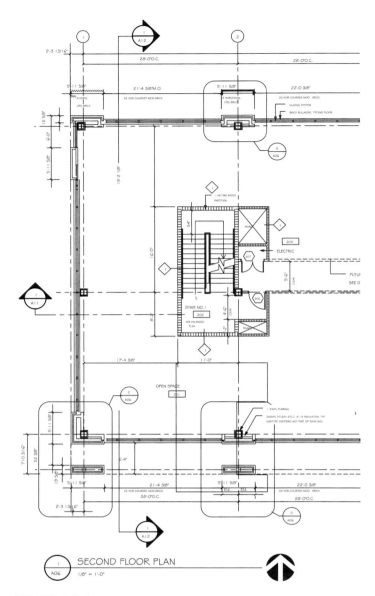

FIGURE 18-2 ■ Cutting plane representation on a floor plan. *Courtesy Peck, Smiley, Ettlin Architects.*

are drawn to provide clarity for a small area, such as the intersection of the floor system to a wall. Sections, wall sections, and details can be thought of as different stages of the **ZOOM** command in AutoCAD. With each zoom, a smaller area can be seen, but each material in the display increases in size. Details provide more detail and specific information about a smaller area than seen in other types of sections. Figure 18-8 shows an example of a detail that is referenced to the section shown in Figure 18-6.

Section/Elevations

Although not as common as other types of sections, sections can be combined with elevations to explain construction of a structure. Section/elevations can be either full or partial views. Figure 16-27 shows an example of a structural elevation that superimposes structural material over the exterior elevation. Figure 18-9 shows a structural section of the same area shown in Figure 16-27.

SECTION DEVELOPMENT

Sections are part of both the preliminary and architectural drawings.

Preliminary Drawings

A section is one of the preliminary drawings developed by the design team to represent the typical shape of the structure to the client. The architect generally develops the section as the design is being explored. A senior drafter may occasionally develop the section from very rough sketches developed by the architect. In the design stage, the section is not intended to accurately represent materials, but to show the shape of the structure, major areas of openings such as windows and skylights, and the vertical relationships of the structure. Walls, floors, and roof systems are represented by solid black lines, which represent thickness rather than specific construction methods. Materials in the background are represented by thin lines, or may not be shown at all. The distances between floor and ceiling, or overall heights, can be shown to represent heights critical to the design. Elevation symbols can be used in place of dimensions to specify heights. Figure 18-10 shows an example of a preliminary section.

To help the print reader relate to the section, room names are specified with large text, and important materials are represented with smaller text. A title and scale should be placed below the drawing. The actual size of the text will vary depending on the size and complexity of the structure. People, plants, automobiles, or other items that will help to identify the use are also placed in the section. Sun angles or shading can be placed to help define how the structure will relate to the site and environment.

Sections in the Architectural Drawings

A drafter will draw the sections for a structure. The preliminary section and other sketches developed by the design team, as well as a print of the floor plan with the desired cutting planes, are usually given to the drafter to indicate which sections should be determined. For experienced drafters, the architect will depend on the drafting team to determine the required sections. Junior drafters will typically not be involved with the initial drawing of the sections but will usually be introduced to sections by adding information to the drawing base from check prints.

Section Scales

The number of sections to be drawn is determined by the complexity of the structure. The size of each section is determined by the complexity and material used for construction. Full sections are generally drawn at the same size as the floor plan. Wall sections are usually drawn at a scale of 3/8" = 1'–0" (1:20) or larger. A scale of 1/2" = 1'–0" (1:20) or larger is used to represent materials shown in detail.

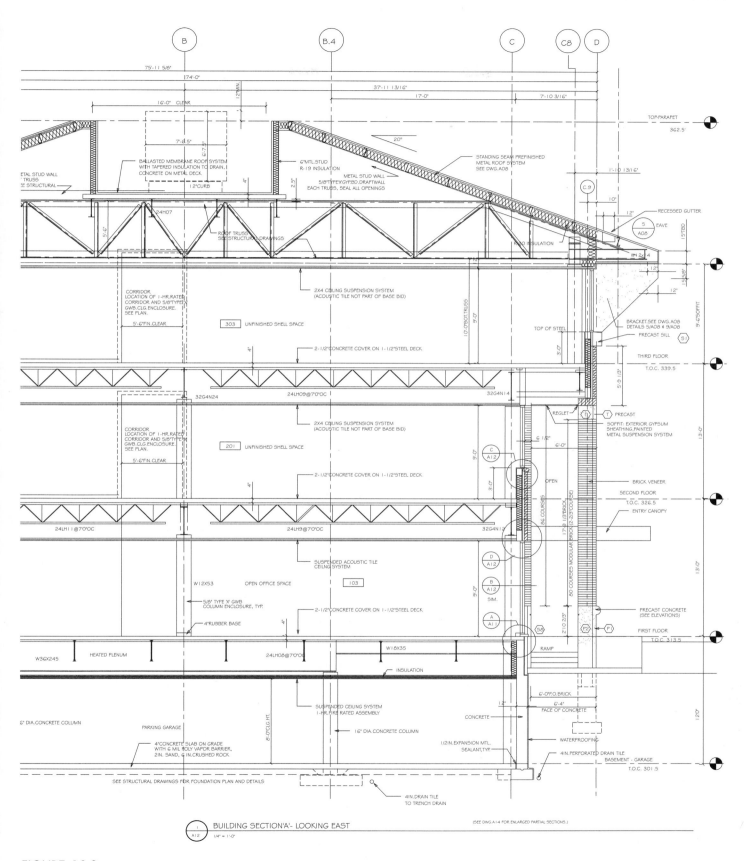

BUILDING SECTION 'A'- LOOKING EAST
1/4" = 1'-0"
(SEE DWG. A14 FOR ENLARGED PARTIAL SECTIONS.)

FIGURE 18-3 ■ The section represented on the floor plan in Figure 18-2 shows the major parts of the building skeleton. Details that explain construction are referenced to the section and the floor plan. *Courtesy Peck, Smiley, Ettlin Architects.*

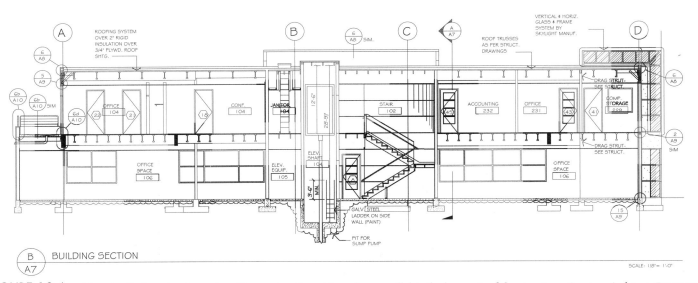

FIGURE 18-4 ■ A longitudinal section is produced by passing the cutting plane parallel to the long axis of the structure. *Courtesy Architects Barrentine, Bates & Lee, A.I.A.*

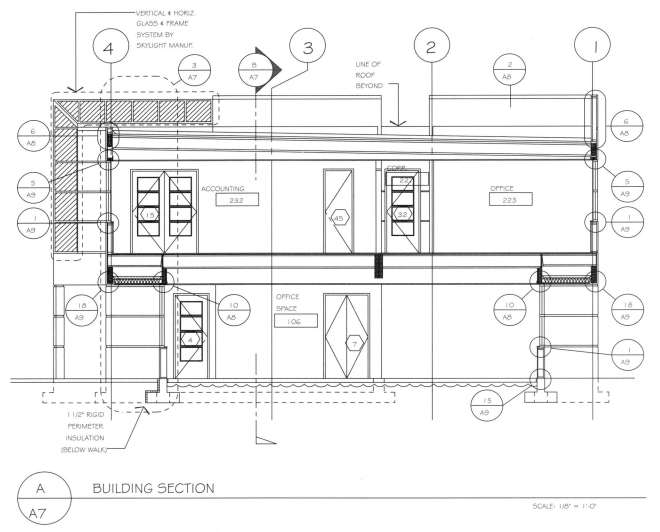

FIGURE 18-5 ■ A transverse section is produced by passing the cutting plane parallel to the short axis of the structure. *Courtesy Architects Barrentine, Bates & Lee, A.I.A.*

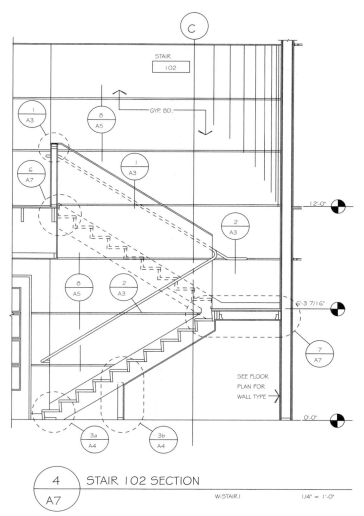

STAIR
102

1
A3

8
A5

6
A7

1
A3

2
A3

8
A5

2
A3

GYP. BD.

12'-0"

6'-3 7/16"

SEE FLOOR
PLAN FOR
WALL TYPE

7
A7

0'-0"

3a
A4

3b
A4

4

A7

STAIR 102 SECTION

W-STAIR1   1/4" = 1'-0"

FIGURE 18-6 ■ A partial section shows construction features of one specific area. This section is an enlargement of the stair shown in Figure 18-4. *Courtesy Architects Barrentine, Bates & Lee, A.I.A.*

Details should be drawn at a scale that will be sufficient to show the detail of all materials in the area being represented. Generally a scale of 3/4" = 1'–0" (1:10) or larger should be used. Common scales used for details and their closest metric counterpart include 1" = 1'–0", 1 1/2" = 1'–0", and 3" = 1'–0" (1:5). The scale used for wall sections and details is often determined by blocks, which are on file in the office library. Figure 18-11 shows an example of a party wall that is stored as a block and then inserted into the drawing set. Once inserted into the drawing base, the section can be edited to meet specific conditions of the job.

Material Representation

The method used to represent each material will vary depending on the scale that is used. Different methods are also used to represent materials that are continuous and are cut by the cutting plan, or that are intermittent and beyond the cutting plan. Although the method of representing materials may vary with each office, it is

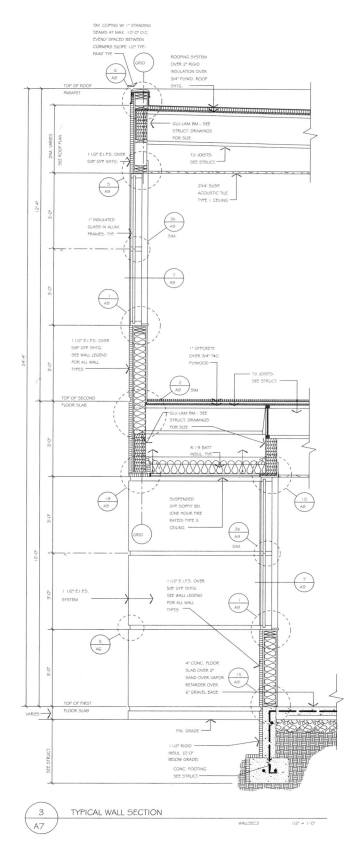

3

A7

TYPICAL WALL SECTION

WALLSEC3   1/2" = 1'-0"

FIGURE 18-7 ■ A wall section can be used to show typical construction for a project. *Courtesy Architects Barrentine, Bates & Lee, A.I.A.*

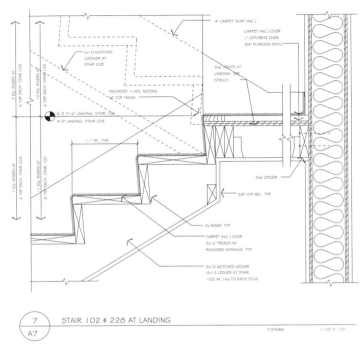

FIGURE 18-8 ■ Detail sections are drawn to provide clarity to small areas. See Figure 18-6. *Courtesy Architects Barrentine, Bates & Lee, A.I.A.*

critical that the drafter distinguish each material from other materials shown in the section to provide drawing clarity. Common materials shown in section are illustrated in Figure 18-12. It is also important not to spend more time than necessary detailing materials. If a product is delivered to the site ready to be installed, minimal attention to representing the product is required, but attention to installing the product is necessary. If an item in a detail must be constructed at the job site, the drawings must provide enough information for all the different trades depending on the drawings.

Wood, Timber, and Engineered Products

Figure 18-13 shows a section for a wood-framed medical facility. Notice that the studs beyond the cutting plane are represented by thin lines and the plates are represented by polylines. On small-scale sections, the lumber and timber products can be drawn using their nominal size. Thin materials such as plywood (at K in Figure 18-12), may have to be exaggerated so that they can be clearly represented. Trusses perpendicular to the cutting plane

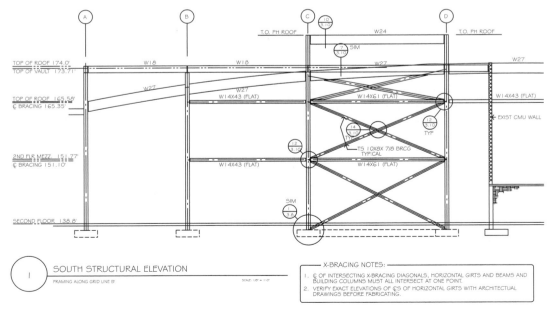

FIGURE 18-9 ■ Drawing techniques of sections and elevations can be combined to show the relationship of structural skeleton to other materials. See Figure 16-27 to compare the elevation for the same area. *Courtesy Charles J. Conlee P.E., Conlee Engineers, Inc.*

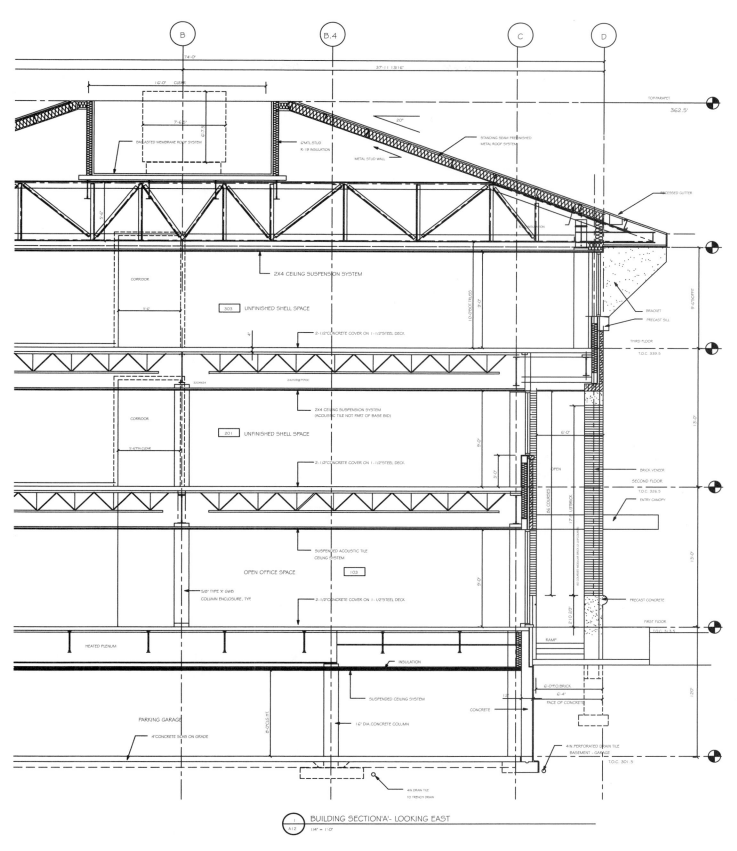

BUILDING SECTION 'A' - LOOKING EAST
1/4" = 1'-0"

FIGURE 18-10 ■ A preliminary section is one of the first drawings completed to show the client the vertical relationship of materials. Figure 18-3 shows this drawing in its completed form. *Courtesy Peck, Smiley, Ettlin Architects.*

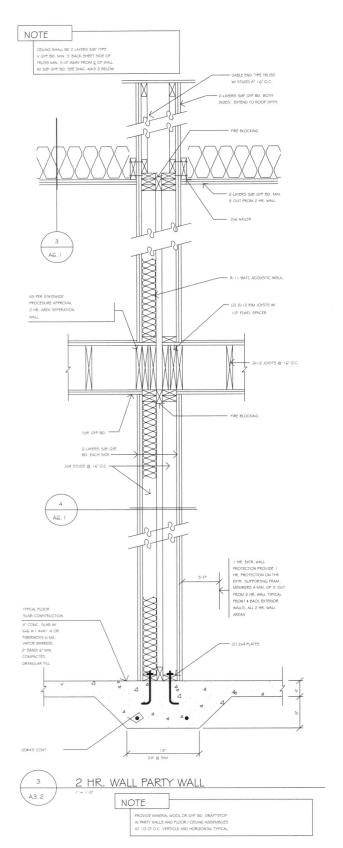

FIGURE 18-11 ■ Sections that are used repeatedly such as this 2-hour firewall can be stored as a wblock and inserted and edited to meet the demands of similar multifamily projects. *Courtesy Scott R. Beck, Architect.*

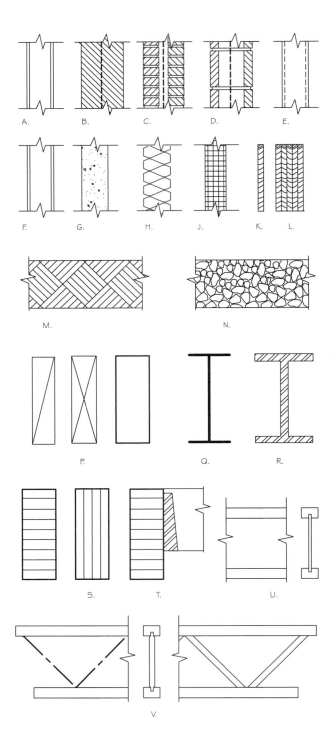

FIGURE 18-12 ■ Material representation in section views include (A) wood framed wall, (B) small scale masonry, (C) double wythe brick wall, (D) concrete masonry units, (E) steel tubes, (F) steel I or W shapes, (G) poured concrete walls, (H) batt insulation, (J) rigid insulation, (K) small-scale plywood, (L) large-scale plywood, (M) soil, (N) gravel, (P) wood and timber in end view (blocking and two methods of showing continuous members), (Q) small-scale steel shapes in end view, (R) large-scale steel shapes in end view, (S) laminated timbers in end view, (T) wood member supported by metal hanger in side view on a laminated member in end view, (U) solid-web trusses in side and end view, and (V) open-web trusses in side and end views.

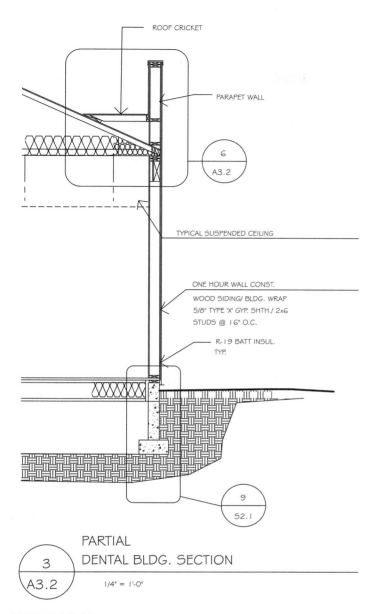

ROOF CRICKET

PARAPET WALL

6
A3.2

TYPICAL SUSPENDED CEILING

ONE HOUR WALL CONST.

WOOD SIDING/ BLDG. WRAP
5/8" TYPE 'X' GYP. SHTH./ 2x6
STUDS @ 16" O.C.

R-19 BATT INSUL.
TYP.

9
S2.1

3
A3.2

PARTIAL
DENTAL BLDG. SECTION

1/4" = 1'-0"

FIGURE 18-13 ■ A partial section of a wood framed structure. *Courtesy Scott R. Beck, Architect.*

can be represented by polylines showing the shapes of the truss, similar to those in Figure 18-4. When parallel to the cutting plane, trusses can be represented by a thin line, which represent the chords and webs, similar to Figure 18-3 or at V in Figure 18-12.

In partial sections and details, the actual size of lumber and timber should be represented. In addition to using a polyline to outline members shown in end views, several methods can be used to represent the material. Figure 18-12 at P and S shows common methods to represent materials such as plates, ledgers, and beams. Plywood, sheet rock, and other finishes are represented by hatch patterns similar to those shown at A, K, and L in Figure 18-12. See Chapters 6 and 7 for other methods of representing lumber and timber products.

Steel

The size of the drawing will affect how sectioned steel members are represented. At small scale, steel members are generally represented by a solid, thick polyline that represents the desired shape (at Q in Figure 18-12). As the scale increases, sectioned members can be shown by pairs of lines that represent the desired shape. In details, sectioned steel is represented by pairs of polylines with a hatch pattern consisting of pairs of parallel diagonal lines (at R in Figure 18-12). Steel columns, beams or trusses that are beyond the cutting plane are represented by thin lines that represent the nominal thickness. Steel trusses are represented in a manner similar to wood trusses. Figure 18-14 shows a wall section formed with steel products. See Chapter 8 for a complete description of steel products.

Unit Masonry

Methods of representing brick and masonry products vary as the scale of the drawing increases. In small-scale sections, units are typically hatched with diagonal lines and no attempt is made to represent cavities or individual units. As the size of the drawing increases, individual units are represented, as well as cavities within the unit and grouting between the units. Individual hatch patterns are used to differentiate between the masonry unit and the grout. Steel reinforcing can be represented by either a hidden or continuous polyline. Figure 18-12 at B, C, and D and Figure 18-15 show an example of a wall section, representing unit masonry and brick veneer. Figure 18-16 shows a detail representing masonry units in both section and elevation views.

Concrete

Poured and precast concrete members are represented by polylines representing the edges of the unit and a hatch pattern consisting of dots and small triangles (at G in Figure 18-12). On small-scale drawings, cavities in precast units are often omitted. Figure 18-17 shows an example of a wall section for a precast structure. Because of the complexity of concrete construction, the section depends on many details to show construction of each concrete member. Large-scale sections and details such as Figure 18-18 are usually part of the structural drawings and show individual cavities, grouting, and caulking materials.

Glass

When shown in section, glass is represented by a single line or pairs of lines depending on the drawing scale. In full and partial sections, glass is generally represented by thin lines with little attention given to intersections between the glazing and window frames. As the drawing scale is increased, the detail shown to the glass and the frame also increases. Figure 18-19 shows an example of a glazing detail.

Insulation

The type of insulation will dictate how it is drawn. Batt insulation is generally represented as shown at H in Figure 18-12. Depending on the complexity of the section, the insulation may be shown across its entire span, or it may be shown in only one portion of the

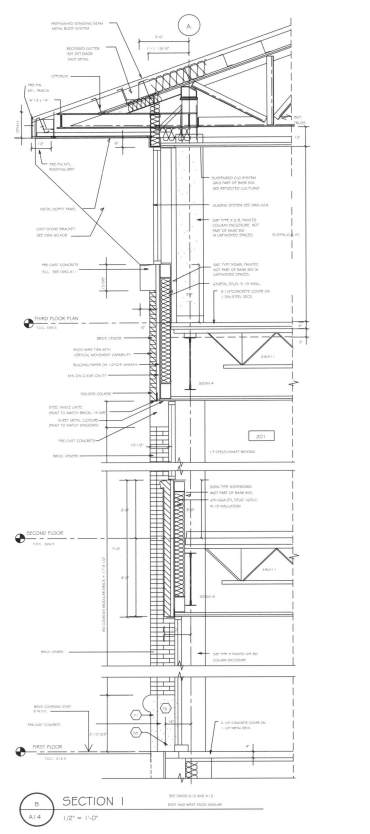

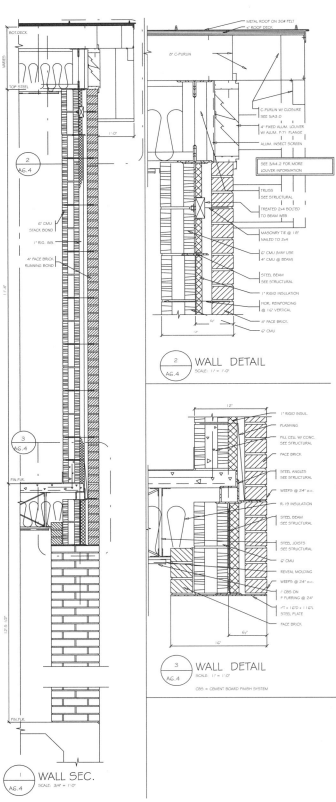

FIGURE 18-14 ■ Steel materials seen in section. *Courtesy Peck, Smiley, Ettlin Architects.*

FIGURE 18-15 ■ Unit masonry and brick seen in section. *Courtesy G. Williamson Archer A.I.A., Archer & Archer P.A.*

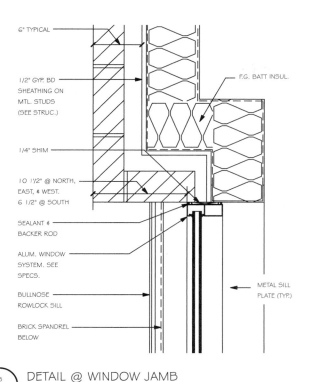

6" TYPICAL

1/2" GYP. BD
SHEATHING ON
MTL. STUDS
(SEE STRUC.)

F.G. BATT INSUL.

1/4" SHIM

10 1/2" @ NORTH,
EAST, ≠ WEST.
6 1/2" @ SOUTH

SEALANT ≠
BACKER ROD

ALUM. WINDOW
SYSTEM. SEE
SPECS.

BULLNOSE
ROWLOCK SILL

METAL SILL
PLATE (TYP.)

BRICK SPANDREL
BELOW

B / A12 DETAIL @ WINDOW JAMB   1" = 1'-0"

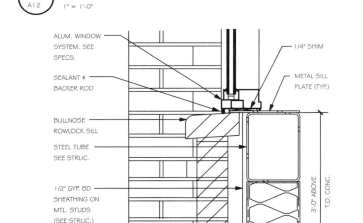

ALUM. WINDOW
SYSTEM. SEE
SPECS.

1/4" SHIM

METAL SILL
PLATE (TYP.)

SEALANT ≠
BACKER ROD

BULLNOSE
ROWLOCK SILL

STEEL TUBE
SEE STRUC.

1/2" GYP. BD
SHEATHING ON
MTL. STUDS
(SEE STRUC.)

3'-0" ABOVE T.O. CONC.

6" TYPICAL

F.G. BATT INSUL.

C / A12 DETAIL @ WINDOW SILL   1" = 1'-0"

FIGURE 18-16 ■ Representing brick in section and elevation in a detail. *Courtesy Peck, Smiley, Ettlin Architects.*

section. When only a portion of the insulation is drawn, notes must be placed, which clearly define the limits of the insulation. Rigid insulation can be represented as shown at J in Figure 18-12 and can be shown using the same considerations as batt insulation. As the scale increases, insulation should be shown throughout the entire detail.

Dimensions

Dimensions are an important element of full, partial, and wall sections. Both vertical and horizontal dimensions may be placed on sections, while partial sections and details generally show only vertical dimensions. On small-scale sections, the use of dimensions depends on the area being represented.

Vertical Dimensions

Vertical heights can be represented by using typical dimension methods or can be represented by elevation symbols from a known point. Each type of dimension is usually placed on the outside of the section. Figure 18-7 shows an example of each method. For wood frame structures, dimensions are generally given from the bottom of the sole plate to the top of the top plate. This dimension also provides the height from the top of the plywood floor to the bottom of the framing member used to frame the next level. A common alternative is to provide a height from the top of the floor sheathing to the top of the next level of floor sheathing. Other common vertical exterior dimensions include the following:

■ Steel decking: from the top of decking

■ Steel stud walls: from plate to top of channel

■ Structural steel: to top of steel member

■ Masonry units: to top of unit with distance and number of courses provided

■ Concrete slab: from top of slab or panel

The exact dimensions are generally provided for inexperienced drafters on a check print.

Once the major shapes of the structure have been defined, dimensions should be provided to define openings, floor changes, or protective devices. Openings are located by providing a height from the top of the floor decking or sheathing to the bottom of the header. Changes in floor height and the height of landings are dimensioned in a similar manner as changes in height between floor levels. Inclined floors can either be defined by a slope indicator, similar to the method of defining roof pitch, or they can be defined by a vertical dimension, which defines the total height difference. Other common interior dimensions that should be provided include height of railings, partial walls, balconies, planters, and decorative screens. When possible, interior dimensions should be grouped together.

Horizontal Dimensions

The use of horizontal dimensions on full sections varies greatly depending on each office. Horizontal dimensions are usually not placed on partial sections or details. When provided, horizontal dimensions are generally located from grid lines to the desired member. Exterior wood and concrete members should be referenced to their edge. Interior wood members are referenced to a centerline. Interior concrete members are referenced to an edge. Steel members are referenced to their centers. The distance for roof overhangs and balcony projections also may be placed on sections. Figure 18-3 shows the use of horizontal dimensions on a full section.

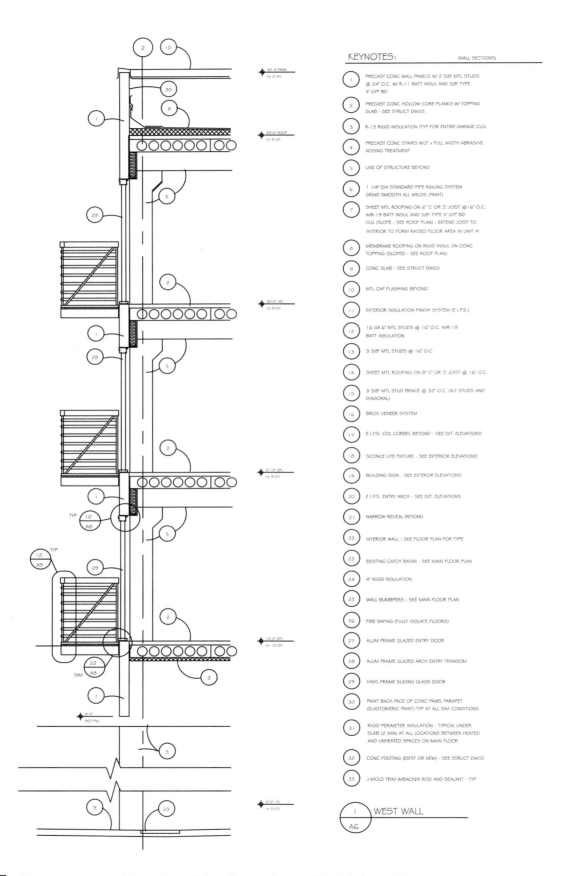

KEYNOTES: (WALL SECTIONS)

1   PRECAST CONC WALL PANELS W/ 2 5/8" MTL STUDS @ 24" O.C. W/ R-11 BATT INSUL AND 5/8" TYPE 'X' GYP BD

2   PRECAST CONC HOLLOW CORE PLANKS W/ TOPPING SLAB - SEE STRUCT DWGS.

3   R-15 RIGID INSULATION (TYP FOR ENTIRE GARAGE CLG)

4   PRECAST CONC STAIRS W/2" x FULL WIDTH ABRASIVE NOSING TREATMENT

5   LINE OF STRUCTURE BEYOND

6   1 1/4" DIA STANDARD PIPE RAILING SYSTEM GRIND SMOOTH ALL WELDS (PAINT)

7   SHEET MTL ROOFING ON 6' 'C' OR 'Z' JOIST @ 16" O.C. W/ R-19 BATT INSUL AND 5/8" TYPE 'X' GYP BD CLG (SLOPE - SEE ROOF PLAN) - EXTEND JOIST TO INTERIOR TO FORM RAISED FLOOR AREA IN UNIT 'A'

8   MEMBRANE ROOFING ON RIGID INSUL ON CONC TOPPING (SLOPED - SEE ROOF PLAN)

9   CONC SLAB - SEE STRUCT DWGS

10 MTL CAP FLASHING BEYOND

11 EXTERIOR INSULATION FINISH SYSTEM (E.I.F.S.)

12 16 GA 6" MTL STUDS @ 16" O.C. W/ R-19 BATT INSULATION

13 3 5/8" MTL STUDS @ 16" O.C.

14 SHEET MTL ROOFING ON 8' 'C' OR 'Z' JOIST @ 16" O.C.

15 3 5/8" MTL STUD BRACE @ 32" O.C. (ALT STUDS AND DIAGONAL)

16 BRICK VENEER SYSTEM

17 E.I.F.S. COL CORBEL BEYOND - SEE EXT. ELEVATIONS

18 SCONCE LITE FIXTURE - SEE EXTERIOR ELEVATIONS

19 BUILDING SIGN - SEE EXTERIOR ELEVATIONS

20 E.I.F.S. ENTRY ARCH - SEE EXT. ELEVATIONS

21 NARROW REVEAL BEYOND

22 INTERIOR WALL - SEE FLOOR PLAN FOR TYPE

23 EXISTING CATCH BASIN - SEE MAIN FLOOR PLAN

24 4" RIGID INSULATION

25 WALL BUMPERS - SEE MAIN FLOOR PLAN

26 FIRE SAFING (FULLY ISOLATE FLOORS)

27 ALUM FRAME GLAZED ENTRY DOOR

28 ALUM FRAME GLAZED ARCH ENTRY TRANSOM

29 VINYL FRAME SLIDING GLASS DOOR

30 PAINT BACK FACE OF CONC PANEL PARAPET (ELASTOMERIC PAINT) TYP AT ALL SIM CONDITIONS

31 RIGID PERIMETER INSULATION - TYPICAL UNDER SLAB (2 MIN) AT ALL LOCATIONS BETWEEN HEATED AND UNHEATED SPACES ON MAIN FLOOR

32 CONC FOOTING (EXIST OR NEW) - SEE STRUCT DWGS

33 J-MOLD TRIM W/BACKER ROD AND SEALANT - TYP

1   WEST WALL
A6

FIGURE 18-17 ■ Precast concrete panels in section with keyed notes. *Courtesy H.D.N. Architects A.I.A.*

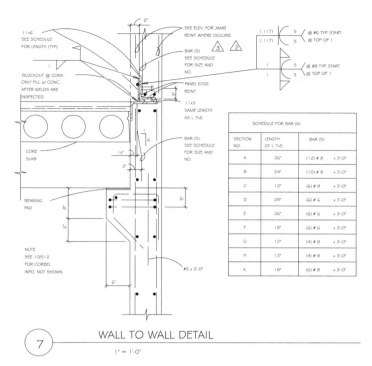

WALL TO WALL DETAIL

7 1" = 1'-0"

FIGURE 18-18 ▪ Details drawn by the structural engineer supplement the sections drawn by the architectural team. *Courtesy Van Domelen/Looijenga/McGarrigle/Knauf Consulting Engineers.*

Drawing Symbols

The sections use symbols that match those of the floor, roof, and elevation drawings to reference material. Symbols that might be found on the section include the following:

- Grid markers
- Elevation markers
- Section markers
- Detail markers
- Room names and numbers

Examples of each are shown in Figure 18-20. Grid markers should match in both style and reference symbol to those used on other drawings so the sections can be easily matched to other drawings. Elevations that are specified on the floor plan and elevation drawings should also be referenced on the section by use of a datum line or an elevation placed over a leader line. Examples of each can be seen throughout this chapter.

Each section is referenced to other drawings by a section marker, which defines the page the section is drawn on, and which section is being viewed. A reference such as 3 over A–7 would indicate that the section is drawing number 3 on page A–7. The smaller the scale used to draw the section, the more likely section and detail markers are used to reference other drawings to the section. Detail markers are especially prevalent on sections to provide enlarged views of intersections.

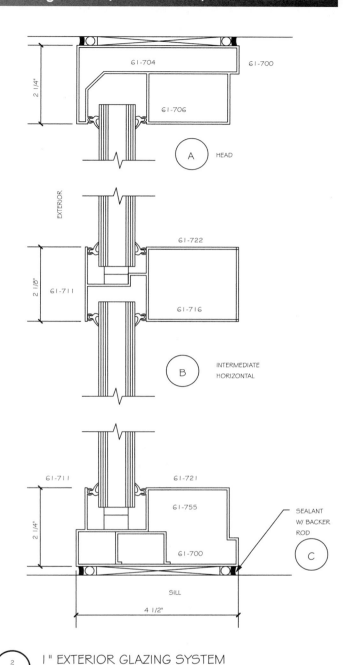

2   1" EXTERIOR GLAZING SYSTEM
A24   HALF SCALE

FIGURE 18-19 ▪ Glass is shown in details as intersections between glazing and other finishing materials are represented. *Courtesy Peck, Smiley, Ettlin Architects.*

Drawing Notations

Lettering on each type of section is used to specify material and explain special insulation procedures. As with other drawings, notes may be either placed as local or keyed notes. Most offices use local notes with a leader line that connects the note to the material. Local notes should be aligned to be parallel to the section to aid

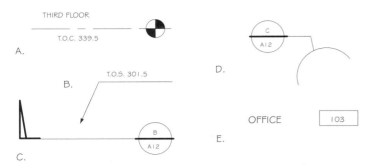

FIGURE 18-20 ■ Symbols used on sections and details are common to other architectural drawings and include (A) baseline elevation marker, (B) leader elevation marker, (C) section marker, (D) detail marker, and (E) room designation.

the print reader. On full sections, notes need to be placed neatly throughout the entire drawing. Wherever possible, notes should be placed on the exterior of the building. For wall sections, aligned notes can greatly add to drawing neatness.

The smaller the scale, the more generic the notes tend to be on a section. For instance, on a full section, roofing that might be specified as

MEMB. ROOFING SYSTEM OVER RIGID INSUL.
OVER PLYWD. SHTG.

would be referenced by complete notes for the roofing, insulation, and plywood in the roofing details. Related notes should be grouped together within the same area of a section. The contents of notes will also be affected by the project manual, which is discussed in Chapter 19.

Blocks

Careful planning can greatly reduce the drafter's job as sections are prepared. By planning and sketching the required sections, repetitive features can be drawn once and copied to other sections. Because of the similarities of many structures, common intersections can be drawn and stored as a wblock. These wblocks can be inserted into a drawing with all required notations, symbols and dimensions, greatly reducing drawing time. Complete sections can often be created by assembling and connecting wblocks, rather than drawing the section line by line. Figure 18-21 shows an example of a footing and roof detail for a concrete block structure. These details can be inserted into a drawing base and moved the proper distance to reflect the distance from floor to roofing as seen in Figure 18-22. Once a wall section has been created, the **MIRROR** command can be used to create the opposite wall, which can then be moved the appropriate distance to represent the total width of the building. Wblocks of details showing a truss to laminated beam and a column to footing were inserted at the center and connected to each other. The section can then be completed by stretching the slab from one detail to another, and by drawing the truss. The completed section, shown in Figure 18-23, required very little actual drawing because of the use of wblocks. For students, access to block libraries is limited. For that reason, the following explanation

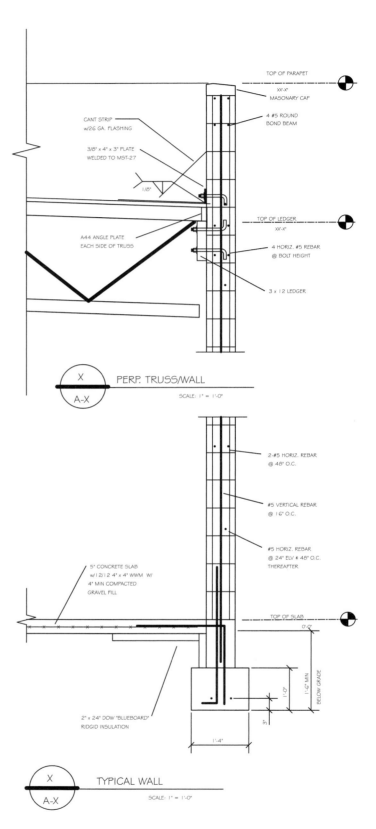

FIGURE 18-21 ■ Stock details such as these footing and truss details can be joined together to form a partial wall section. *Courtesy Wil Warner.*

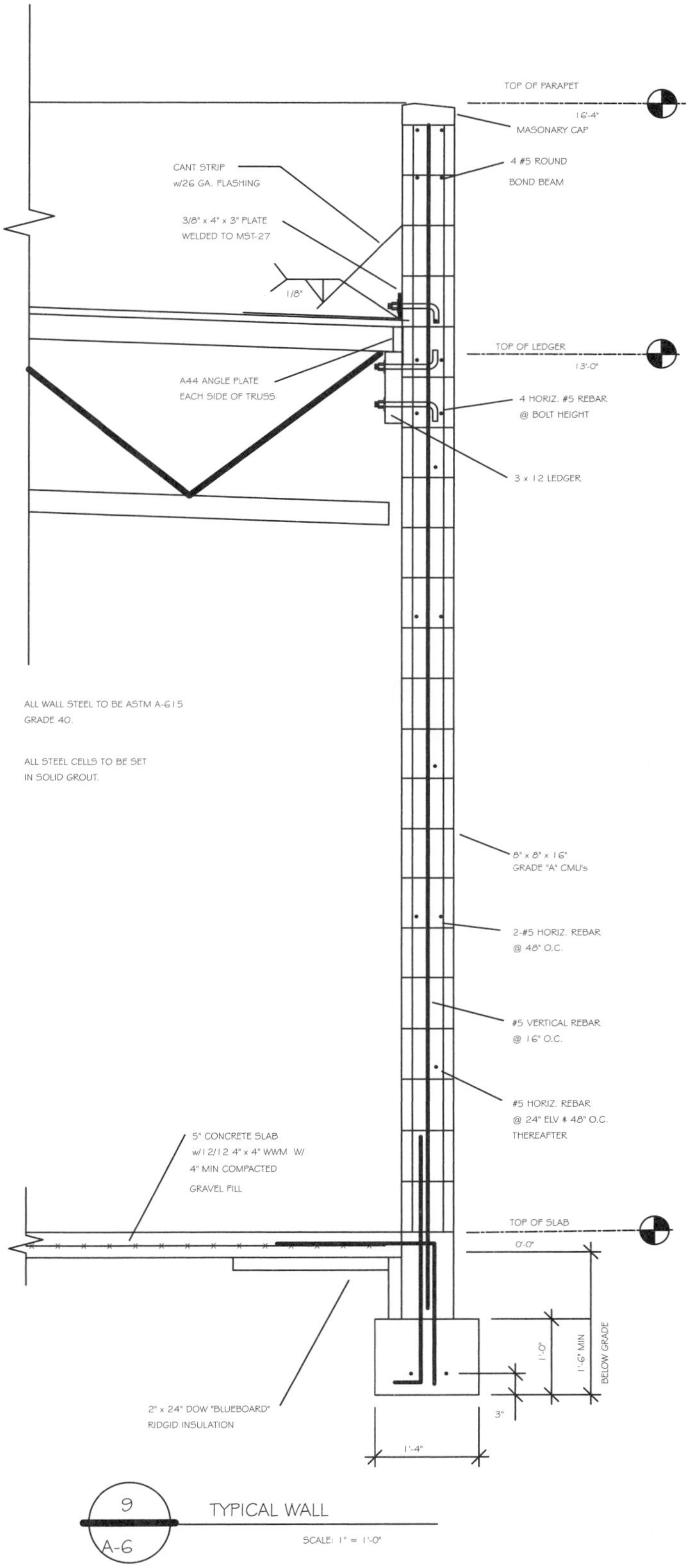
TOP OF PARAPET
16'-4"
MASONARY CAP
CANT STRIP
w/26 GA. FLASHING
4 #5 ROUND
BOND BEAM
3/8" x 4" x 3" PLATE
WELDED TO MST-27
1/8"
TOP OF LEDGER
13'-0"
A44 ANGLE PLATE
EACH SIDE OF TRUSS
4 HORIZ. #5 REBAR
@ BOLT HEIGHT
3 x 12 LEDGER
ALL WALL STEEL TO BE ASTM A-615
GRADE 40.
ALL STEEL CELLS TO BE SET
IN SOLID GROUT.
8" x 8" x 16"
GRADE "A" CMU's
2-#5 HORIZ. REBAR
@ 48" O.C.
#5 VERTICAL REBAR
@ 16" O.C.
#5 HORIZ. REBAR
@ 24" ELV & 48" O.C.
THEREAFTER
5" CONCRETE SLAB
w/12/12 4" x 4" WWM W/
4" MIN COMPACTED
GRAVEL FILL
TOP OF SLAB
0'-0"
1'-0"
1'-6" MIN
BELOW GRADE
3"
1'-4"
2" x 24" DOW "BLUEBOARD"
RIDGID INSULATION
9
A-6
TYPICAL WALL
SCALE: 1" = 1'-0"

FIGURE 18-22 ■ A wall section completed by joining stock details. *Courtesy Wil Warner.*

of drawing sections will assume no use of blocks.

COMPLETING A SECTION

A drafter will need a floor plan, foundation plan, grading plan, and roof plan, exterior elevations, and preliminary sections to complete the building sections. For relatively flat lots or if spot grades are indicated on the elevations, the grading plan is not required.

1. Place a cutting plan on the floor plan to determine the view to be created. The following guidelines will be used to draw section A-A4, shown in Figure 15-42.
2. Create a separate block of the floor plan, which can be used to project the section. *Do not use the floor plan.*
3. Establish lines to represent the finished floor, ceiling, and roof levels
4. Project lines representing each wall cut by the cutting plane from the floor plan into the drawing area
5. Project lines to represent each column and beam cut by the cutting plan from the floor plan into the drawing area
6. Using the foundation and grading plans, establish the foundation locations
7. Establish a line to represent the finish grade location
8. Using elevations on the roof or exterior elevations, establish the height of all exterior walls

The projection of the section should now resemble Figure 18-24. The base drawing can be completed by the following steps.

9. Trim all projection lines to form the outline of the section
10. Use the elevations on the roof plan to establish the pitch of the trusses
11. Use vendor catalogs to determine the depth of the trusses, and draw the roof trusses
12. Draw the finished roofing system above the trusses
13. Draw all beams and ledgers required to support the trusses

Figure 18-25 shows the development of the section. Once the base of the section is complete, detail can now be added to the drawing.

14. Draw any required ceilings
15. Show any interior walls
16. Show any false ceilings
17. Show all material used to complete the roof and floor systems

The entire section cut by the cutting plane has now been drawn and should resemble Figure 18-26. The following steps can be used to complete the drawing.

18. Show materials that lie beyond the cutting plane
19. Provide grid lines and specifications

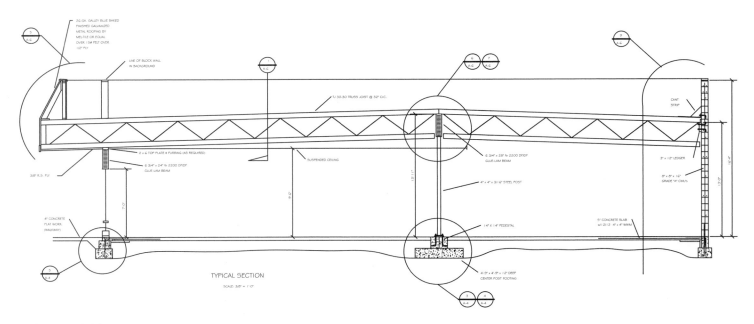

FIGURE 18-23 ■ ■ Full sections can be created by combining blocks. *Courtesy Wil Warner.*

20. Provide horizontal reference lines and elevations
21. Show section and detail markers for areas to be enlarged. Although the section and detail specifications are provided in this example, they can't usually be provided until the entire drawing is drawn.
22. Show all room names and numbers
23. Place all required notes based on the architect's specifications
24. Place a title and scale below the drawing

The completed section should now resemble Figure 18-5.

Completing A Wall Section

The layout of each wall section will vary depending on the material being represented. The steps used to lay out a full section can be used as a guideline to draw a partial section.

Completing A Detail

No one set of guidelines can be used to complete each detail. The drafter must be careful to distinguish between each material with varied line quality and hatch patterns. Generally several line weights will be required to complete a detail. Thin lines can be used to represent studs, interior finishes, and thin materials such as plywood. Thicker lines can be used to represent materials cut by the cutting plane. Outlines of concrete features are often drawn with a line slightly thicker than the line used to represent sectioned material. The thickest lines are generally used to represent the ground. Detailed hatch patterns should also be used to distinguish between materials. Examples of the varied line work and hatch patterns used on details are shown throughout this chapter.

LAYER GUIDELINES FOR SECTIONS & DETAILS

Separation of material by layer and color can greatly aid in the development of a section. A prefix of A-SECT is recommended by the AIA cad guidelines. Modifiers can be used to define materials. Many firms divide modifiers by materials such as CONC, CBLK, BRK, STL, WD, or PTWD.

A-DETL	Details
A-DETL-ANNO	Notes
A-DETL-DIMS	Dimensions
A-DETL-IDEN	Detail identification numbers
A-DETL-MBND	Material beyond the cutting plane
A-DETL-MCUT	Material cut by the cutting plane
A-DETL-PATT	Textures and hatch patterns
A-SECT	Sections
A-SECT-ANNO	Notes
A-SECT-DIMS	Dimensions
A-SECT-IDEN	Markers used to identify elevations, grids, etc.
A-SECT-MBND	Material beyond the cutting plane
A-SECT-MCUT	Material cut by the cutting plane
A-SECT-PATT	Hatch patterns used to represent various materials

Status field modifiers similar to other drawing areas should be used throughout structural drawings. See Chapter 3 and the *AIA Layer Guidelines* for additional layer names.

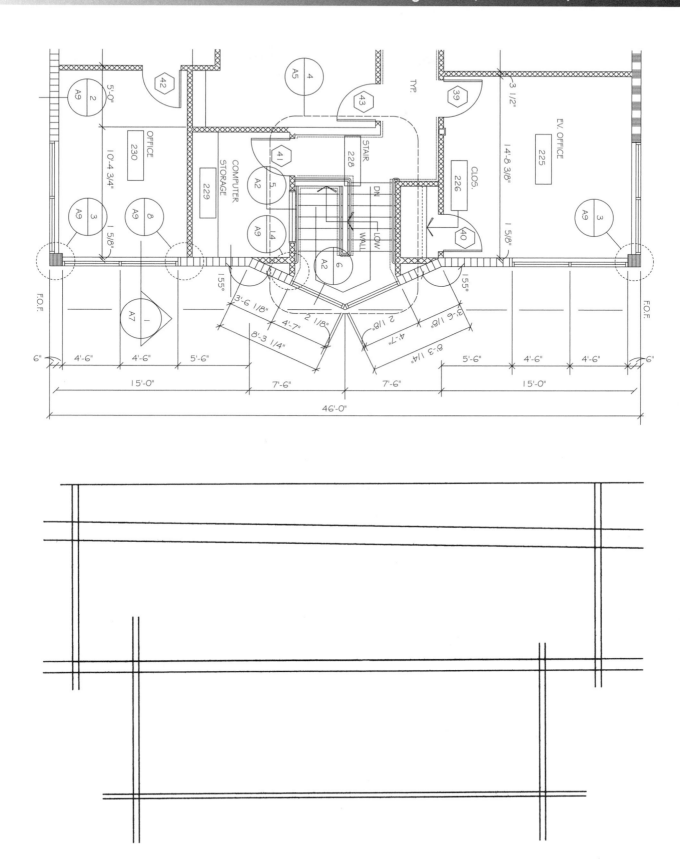

FIGURE 18-24 ■ Initial layout steps for projecting a section from a wblock of the floor plan. *Courtesy Architects Barrentine, Bates & Lee, A.I.A.*

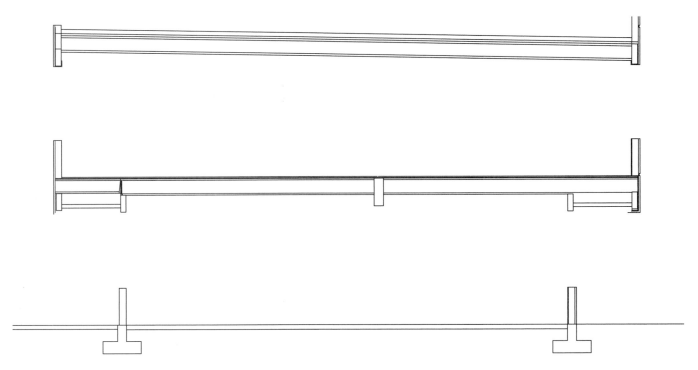

FIGURE 18-25 ■ Drawing the major shapes of the structure and projection of all ceilings and detailing of each floor and openings. *Courtesy Architects Barrentine, Bates & Lee, A.I.A.*

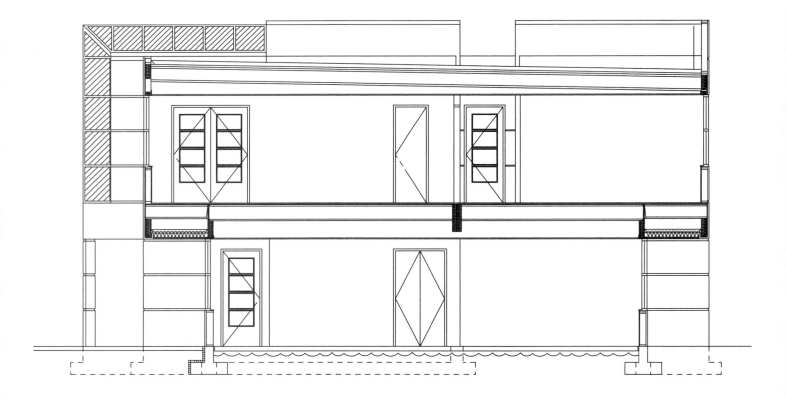

FIGURE 18-26 ■ The addition of materials that lie beyond the cutting plane. *Courtesy Architects Barrentine, Bates & Lee, A.I.A.*

CHAPTER

18 | Drawing Sections, Wall Sections and Details

CHAPTER QUIZ

Use a separate sheet of paper to answer the following questions. Print the chapter title, the question number, and a short complete statement for each answer.

Question 18-1 What drawings might be required to draw a full section?

Question 18-2 Explain the difference between a full and a wall section.

Question 18-3 List two types of full sections and explain the differences.

Question 18-4 What relates the section to the floor plan?

Question 18-5 Briefly describe two methods of completing a section.

Question 18-6 List guidelines for selecting the scale for a section, wall section, and a detail.

Question 18-7 Describe two common methods of describing vertical heights.

Question 18-8 List common symbols associated with sections.

Question 18-9 Explain variations in notes between a section and a detail.

Question 18-10 Explain the various roles a drafter will play in the development of sections.

DRAWING PROBLEMS

Use the reference material from preceding chapters, local codes, and vendor catalogs to complete one of the following projects. Unless other instructions are given by your instructor, draw the sections that corresponds to the floor plan that was drawn in Chapter 15. Skeletons for problems 18-1 through 18-4 can be accessed from http://www.delmar.com/resources/ocl.html. Use these drawings as a base to complete the assignment. Use appropriate symbols, linetypes, dimensioning methods, and notations to complete the drawing. Determine any unspecified sizes based on material or practical requirements. Unless specified, select a scale appropriate to plotting on "D" size material and determine LTSCALE and DIMVARS.

You will act as the project manager and will be required to make decisions about how to complete the project. By now you're far enough into this structure to realize that no one drawing problem contains all of the information

needed to complete the problem. Study each of the drawing problems related to the structure prior to starting a drawing problem. Any information not provided must be researched and determined by you unless your instructor (the project architect and engineer) provides other instructions. When conflicting information is found, information from preceding chapters should take precedence.

Although most of the information required to complete a project can be found by searching other related problems, some of the problems will require you to make a decision on how you would solve a particular problem. Make several sketches of possible solutions based on examples found throughout this text and submit them to your instructor prior to completing each problem with missing information.

Use the floor plans provided in Chapter 15 to complete the following sections. Decide on an appropriate scale to complete the drawing.

Place the necessary cutting plane and detail bubbles on the floor plan to indicate each drawing.

Note: Keep in mind as you complete these drawings that the finished drawing set may contain more sections and in some cases hundreds of details more than you are being asked to draw. Some of the required details will be introduced in succeeding chapters. Other details that would be required for the building department or the construction crew will not be drawn due to the limitations of space and time.

Problem 18-1 Using the floor plan drawn in Problem 15-1, draw a section cutting through each unit, showing the elevation change for each floor level at a scale of 1/4" = 1'-0". In addition to the full section draw a wall section of a party wall at a scale of 1/2" = 1'-0" and draw a detail of each retaining wall at a scale of 3/4" = 1'-0".

■ Dimension each major material and provide unit dimensions, floor-to-floor heights, and elevations

■ Use the partial section and details to completely specify required materials

■ The lower level is to have a 4" concrete slab and 8'-0" high walls; the main and upper floors are to have 9'-0" ceilings

■ Frame the roof using standard roof trusses. Use standard/scissor trusses to provide a vaulted ceiling over the master bedroom.

9'-0"

9'-0"

8'-0"

18"

BEDROOM   BEDROOM   CLOS

BEDROOM   BEDROOM   CLOS

FAMILY ROOM

KITCHEN

NOOK

GARAGE

FAMILY ROOM

KITCHEN

NOOK

GARAGE

2'-4"

2'-4"

PERIMETER FOOTING
TO BE STEPPED

36"   4'-0"   4'-0"

20'-0"

20'-0"

① / 9   SECTION A-A
         SCALE: 1/4" = 1'-0"

PROBLEM 18-1A

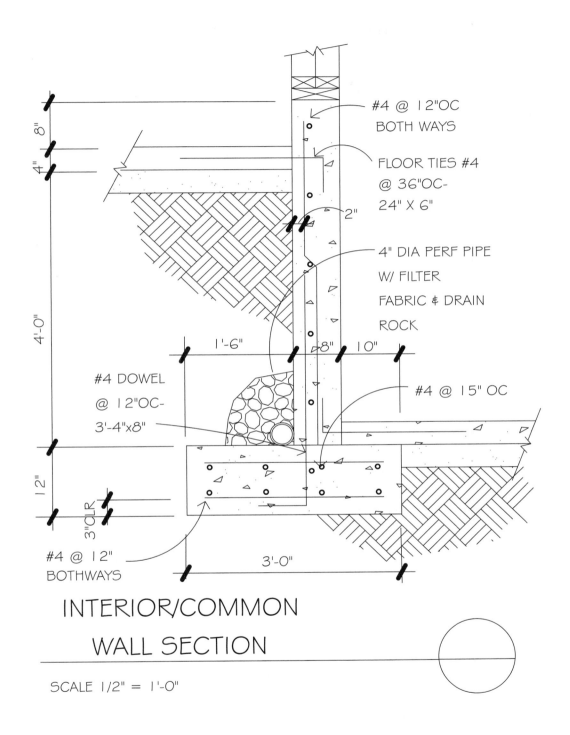

#4 @ 12"OC
BOTH WAYS

FLOOR TIES #4
@ 36"OC-
24" X 6"

4" DIA PERF PIPE
W/ FILTER
FABRIC & DRAIN
ROCK

2"

#4 @ 15" OC

#4 DOWEL
@ 12"OC-
3'-4"x8"

8"

10"

1'-6"

4'-0"

8"

4"

12"

3"CLR

#4 @ 12"
BOTHWAYS

3'-0"

INTERIOR/COMMON
WALL SECTION

SCALE 1/2" = 1'-0"

PROBLEM 18-1B

■ Use 2 × 6 studs at 16" o.c. for exterior walls and 2 × 4 studs @ 16" o.c. for interior walls

■ Use Figure 18-11 as a guide, and design a firewall that reflects the change in elevation for each unit based on local codes for your area. Use solid blocking in the upper wall of the party wall opposite the top plate for the lower wall.

■ Select and specify sawn floor joist, TJIs or open-web trusses suitable to span from party wall to party wall

■ Support the exterior wall at the upper end of the project on an 8" wide × 8' high concrete retaining wall supported on a 12 × 16" wide footing. Reinforce the wall with #5 bars at 18" o.c. E.W. Steel to be 2" clear of interior side of wall.

■ Support the party wall as per attached drawing

Problem 18-2a Use the floor plan created for Problem 15-2, the attached drawing, and Figure 18-23 as a guide to draw a full section of the structure at a scale of 3/8" = 1'–0". Run the cutting plane north/south through the cantilevered roof and the storefront glass of the display area. Reference and draw the indicated details. Specify the following heights:

Top of storefront glass	8'–0"
Suspended ceiling	9'–6"
Top of 2 × 6 stud wall	11'–0"
Top of ledger @ north wall	12'–0"
Top of masonry wall	16'–0" min.

Block walls to be 8 × 8 × 16 grade A concrete blocks with #5 vertical at 36" o.c. and #5 at 24" o.c. horizontally. Provide (2)-#5 bond beam at 48" o.c. Use (4)-#5 at top of wall. Solid grout all steel cells. Support walls as per the foundation drawings in Chapter 23. Provide (2)-#5 continuous rebar in the footing 2" clear of the bottom

and extend a #5 × 36" L from the footing to the wall with a 6" toe. Space footing steel to match the wall steel. Provide a slab as per foundation plan.

Use an 8 3/4" × 28 1/2" glu-lam for the ridge. Determine the location of the ridge based on the location of the steel column and pilaster on the floor plan. Although the location of the pilaster *must* remain modular, the centerline of the beam should be placed an even number of feet from each wall if possible. *Edit the dimensions on the floor plan as required.*

Support trusses on 3 × 12 DFPT ledger w/3/4" diameter A.B. at 24" o.c. staggered 3" up/down with 2" diameter washers. Use (4)-#5 rebar in 8 × 16 bond beam at ledger height. Use 3 × 4 blocking between trusses with an appropriate "A" clip from block to ledger. Use an MST strap capable of resisting 2700#. Place the strap over the top chord of trusses at 8'–0" o.c. Weld the strap to a 3" × 4" × 1/4" steel plate with 1/8" fillet weld topside only. Weld (2)-3/4" × 6" studs to the back side of the plate and weld with 1/8" fillet welds all around. Tie studs to wall steel with 1/8" weld.

At east and west nonbearing walls, use a 3 × 10 pressure-treated ledger, bolted with 5/8"Ø × 10" A.B. staggered 3" up/dn at 32" o.c. through 2" washers. Use (2)-#5 rebar in an 8 × 8" bond beam at ledger height. Use 4× ripped cant strip to support 26 ga. flashing at all wall/roof intersections. Use the attached drawings to complete the following details.

Problem 18-2b-1 Draw and label an enlarged wall section showing the roof /wall intersection and wall construction.

Problem 18-2b-2 Draw and label a detail showing the nonbearing wall/ truss intersection. Use a 3 × 10 pressure treated ledger, bolted with 5/8"Ø × 10" A.B. staggered 3" up/dn at 32" o.c. through 2" washers. Use (2)-#5 rebar

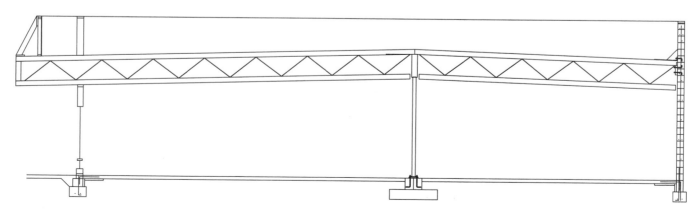

PROBLEM 18-2A

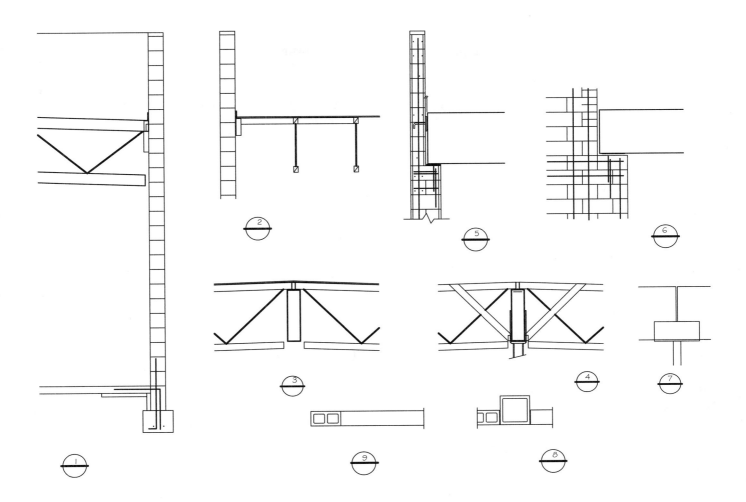

PROBLEM 18-2B

in an 8 × 8" bond beam at ledger height. Use 4× ripped cant strip to support 26-ga. flashing at all wall/roof intersections.

Problem 18-2b-3 Draw a detail to show the connection of the trusses to the ridge beam. Use an MST48 strap at every other truss. Place a solid block between trusses along the ridge and anchor to the ridge beam with an A35 anchor at each end of block, staggered each side.

Problem 18-2b-4 Draw a detail to show the truss to column bracing. Provide 2 × 4 braces @ each truss at each side of column set at a 45° maximum angle from vertical. Provide (2)-3 × 4 blocks between each brace at equal spacing along the brace and a 3 × 4 nailer where the brace intersects the trusses. Use an appropriate U-hanger to support the brace at the beam.

Problem 18-2b-5 Draw a detail showing the connection of the glu-lam beam to concrete block. Use a suitable glu-lam beam seat based on the load to be supported at each pilaster. Tie seat to pilaster steel with 1/8" welds. Bolt beam to seat as per manufacturer's specifications. At the upper end of the beam bolt a 3" × 4" × 3/8" steel plate to the wall. Provide (1)-3/4" × 6" stud welded with 1/8" fillet welded all around to back side of plate extending into the block wall and tied into the wall steel. Weld a MST 27 strap on each side of the beam to the front side of plate to support glu-lam. Weld straps to end plate with 1/8" fillet weld on each outer side of the strap. Place the centerline of straps at 3" from the top of the beam.

Tie the leg of the beam seat to a #5 × 24" rebar with 1/8" fillet weld both sides of bar to plate. Tie the 24" rebar into the steel used to reinforce the pilaster.

Problem 18-2b-6 Draw a detail showing the connection of the glu-lam beam to concrete block. Use a suitable glu-lam beam seat based on the load to be supported for the south and east walls at windows. Tie seat to wall steel with 1/8" welds. Bolt beam to seat as per manufacturer's specifications. At the upper end of the beam bolt a steel plate to the wall using detail 5 as a guide. Place the centerline of straps at 3" from the top of the beam.

Problem 18-2b-7 Draw a detail showing the connection of the ridge beam to the 4 × 4 × 5/16" steel column. Select a suitable CC column cap and weld to the column with 3/16" fillet weld all around. Specify bolts recommended by the manufacturer for the cap to beam connection. Provide a 3" × 27" × 5/16" steel strap with the centerline of (4)-3/4" M.B. 4" from the top of the beam, centered over the beam splice at the column.

Problem 18-2b-8 Draw a detail showing the plan view of the 16" × 16" pilaster construction. Reinforce with (8)-#5 vertical bars with #3 horizontal ties @ 16" o.c. Stagger laps 180°. Show a side view showing the termination of the glu-lam beam seat. Select a suitable beam seat for a 6 3/4" beam. Determine the loads to be supported by the pilaster and select a suitable beam seat.

Problem 18-2b-9 Draw a detail showing a plan view of reinforcement for the block wall at each opening. Use (6)-#5 verticals (2 per cell) with #3 ties at 16" o.c. horizontally. Show normal wall steel in other cells.

Problem 18-3 Use the floor plan from Problem 15–3 to complete the required drawings. Use the parameters outlined below and on the attached drawings for all drawings. All beams to be based on the framing drawings. Roofing to be 4-ply built-up asphaltic fiberglass class A roofing over 5/8" 42/20 interior APA plywood with exterior glue roof sheathing laid perpendicular to purlins. Stagger seams at each purlin. Nail with 10d common nails @ 4" o.c. at all panel edges and blocked areas, @ 6" o.c. all supported panel edges @ unblocked areas and 12" o.c. @ all intermediate supports. Use MST 27 straps @ 8'–0" o.c.

Walls to be 8 × 8 × 16" grade A concrete blocks reinforced with #5 verticals at 32" o.c. and #5 horizontal @ 24" o.c. with (2)-#5 @ 48" o.c. Provide 8 × 16 bond beam with (4)-#5 at each ledger and at top of walls.

Provide an 8" × 16" bond beam over openings in the walls up to 10' wide and support with (4)-#5 horizontal bars with #3 ties @ 24" o.c. Provide an 8" × 24" bond beam over openings in the walls greater than 10' wide and support with (6)-#5 horizontal bars with #3 ties @ 24" o.c. Support walls of office and warehouse as per the specifications in Chapter 23. All footings to extend 18" into natural grade.

Provide a 4" thick slab for the office area and a 5" thick slab in the warehouse.

Problem 18-3a Use the architectural and structural drawings in Chapter 15 through 23 to complete the required section through the office area.

Problem 18-3b Use attached drawing and the balance of the architectural and structural drawings to complete the required section through the warehouse and mezzanine area.

Problem 18-3c Use the drawings provided for Problem 18-3b and the balance of the architectural and structural drawings to complete the required section through the warehouse with no mezzanine.

Problem 18-3d Use the drawings provided for Problem 18-3b and the balance of the architectural and structural drawings to complete the required section through the office, warehouse and mezzanine.

Problem 18-3e Draw details to explain construction of each pilaster. Provide reinforcing similar to detail 18-2b-8 and 21-3b-1.

Problem 18-4 Use the floor plan from Problem 15–4 and the attached drawings to complete the required drawings. Use the parameters that follow for all drawings. All beams sizes to be based on the framing drawings in Chapter 21. Roofing to be 4-ply built-up asphaltic fiberglass class A roofing over 5/8" 42/20 interior APA plywood with exterior glue roof sheathing laid perpendicular to rafters. Stagger seams at each rafter. Nail with 10d common nails @ 4" o.c. at all panel edges and blocked areas, @ 6" o.c. all supported panel edges @ unblocked areas and 12" o.c. @ all intermediate supports. Use MST 27 straps @ 8'–0" o.c. Align straps at ridge with wall straps.

All concrete walls are to be 6" thick with steel reinforcement as per the panel elevations in chapter 22. Use trusses based on the sketches provide with the framing plans. Provide a concrete slab for the office and warehouse areas based on the foundation drawings.

Problem 18-4a Use the balance of the architectural and structural drawings to complete the required section through the office area.

Problem 18-4b Use drawing below and the balance of the architectural and structural drawings to complete the required partial section through the warehouse and mezzanine area.

Problem 18-4c Use the drawings provided for problem 24–4b and the balance of the architectural and structural drawings to complete the required partial section through the warehouse with no mezzanine.

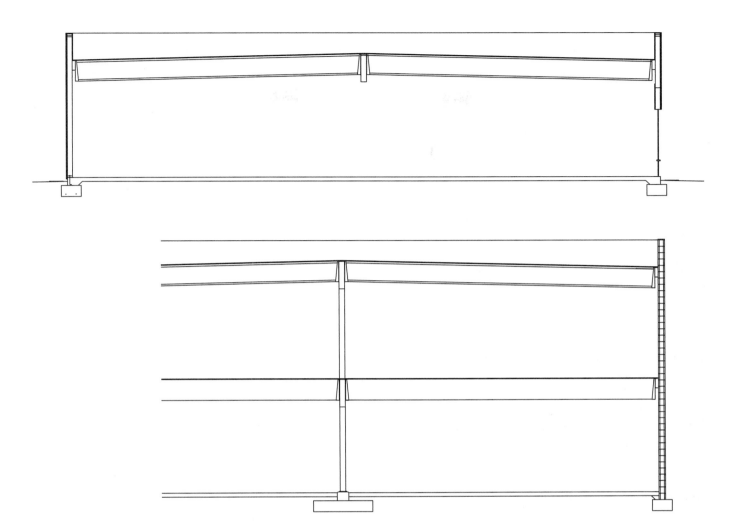

PROBLEM 18-3

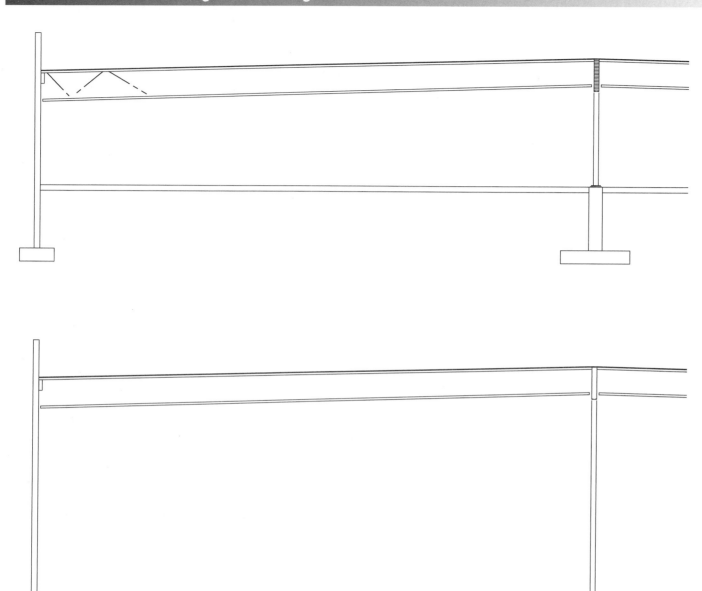

PROBLEM 18-4

## CHAPTER 19

Interior Elevations

Interior elevations are drawn to show the shape and finishes of the interior of a structure in a similar method that was used to draw the exterior elevations. Interior elevations similar to Figure 19-1 are drawn whenever a wall has features that are built into the wall. The complexity of the structure and office practice affect the method used to present materials in elevation. This chapter explains common methods of drawing interior elevations and referencing them to other architectural drawings.

ELEVATIONS WITHIN THE DRAWING SET

Except for a few features that are critical to the design of a structure, interior elevations are generally not a part of the preliminary elevations. Features such as an elaborate fireplace or intricate entry that is related to the design would be included in the preliminary design. Interior elevations are usually drawn after each of the other architectural drawings has been started. The floor plan, exterior elevations, and sections do not have to be complete before the inte-

rior elevations are started, but the basic arrangement of materials must be fairly well established before the interior elevations can be started. Once basic shapes have been defined and the architectural team is progressing with the working drawings, the design team can now plan the arrangements of interior features by making preliminary sketches. Once approved by the client, these sketches can be converted to drawings by the drafter.

Referencing Interior Elevations

Interior elevations are based on the position of materials that are shown on the floor plan. Elevations are referenced to the floor plan by the use of a title or a symbol. Titles such as **North Lobby Elevation** can be used to reference an elevation to a floor plan if only a few elevations need to be drawn. An elevation symbol should be placed on the floor plan where any possibility of confusion exists. Figure 19-2 shows an example of placing the symbol on the floor plan and Figure 19-3 shows the corresponding elevations. Although interior elevations are referenced to the floor plan, their location in the architectural drawings will vary greatly. Interior elevations can be placed on a sheet containing any other architectural drawing, as space allows. Every effort should be made to keep all interior elevations on the same sheet. Because they are referenced

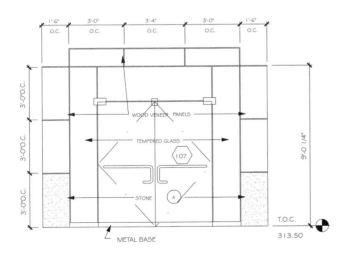

FIGURE 19-1 ■ Interior elevations are drawn whenever a wall has features that must be built or installed as part of the construction contract. *Courtesy Ned Peck, Peck, Smiley, Ettlin Architects.*

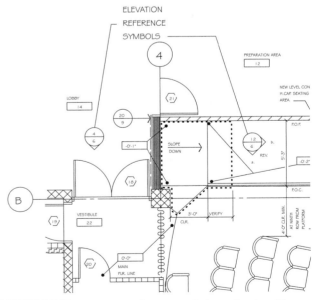

FIGURE 19-2 ■ Elevation reference symbols can be placed in a room to define the elevations to be drawn and their location. *Courtesy Architects Barrentine, Bates & Lee, A.I.A.*

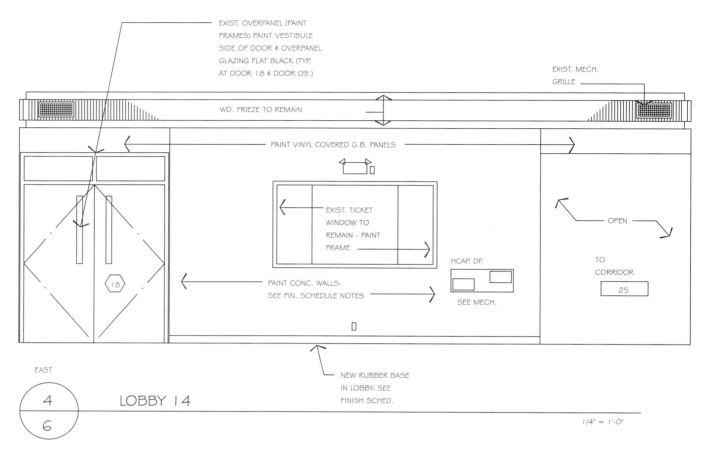

EXIST. OVERPANEL (PAINT FRAMES) PAINT VESTIBULE SIDE OF DOOR & OVERPANEL GLAZING FLAT BLACK (TYP. AT DOOR 18 & DOOR 09.)

EXIST. MECH. GRILLE

WD. FRIEZE TO REMAIN

PAINT VINYL COVERED G.B. PANELS

EXIST. TICKET WINDOW TO REMAIN - PAINT FRAME

OPEN

HCAP. DF.

TO CORRIDOR

25

PAINT CONC. WALLS- SEE FIN. SCHEDULE NOTES

SEE MECH.

18

NEW RUBBER BASE IN LOBBY- SEE FINISH SCHED.

EAST

4
6

LOBBY 14

1/4" = 1'-0"

FIGURE 19-3 ■ The elevation referenced to the floor plan shown in Figure 19-2. *Courtesy Architects Barrentine, Bates & Lee, A.I.A.*

to the floor plan and they are similar to the exterior elevations, interior elevations are often placed on one of these drawings. When a separate sheet is required to show the interior elevations, they are often placed immediately behind the floor plan in the drawing set. Regardless of where they are placed within the drawings, the location of interior elevations, like other drawings, should be clearly indicated in the table of contents, which is contained on the title page. See Chapter 2 for a review of title pages and drawing placement.

Drawing Projection

Interior elevations are similar to the exterior elevations in that they are drawn using orthographic projections based on a viewing plane. In theory, a viewing plane is placed on the floor plan parallel to each wall containing material that must be explained. Objects are then projected onto the plane to create a two-dimensional view of the wall. The location of the viewing plane is indicated on the floor plan by using the reference symbol. Objects parallel to the viewing plane are projected to the viewing plane just as with exterior elevations. In drawing the interior elevations, in addition to using the floor plan to determine the shape, the sections can be used to determine interior heights. Base cabinets are often 36" (900 mm) high, with upper cabinets typically placed 18", 24" or 30"

(450, 600, or 750 mm) above the lower counters. Because cabinet placement and sizes are so dependent on the equipment that will be installed in the cabinets, the drafter should rely on vender catalogs and drawing standards such as *Time Saver Standards* and *Architectural Graphics Standards*.

Types of Elevations

The material shown in an interior elevation will vary depending on the complexity of the structure. Generally elevations can be grouped into the categories of wall elevations and cabinet elevations.

Wall Elevations

Wall elevations can show an entire wall or a partial wall. All materials that are built into or mounted on the wall should be shown. This would include showing special finishes as well as items such as automated teller machines, fire extinguishers, drinking fountains, and phone stalls, similar to the elevation shown in Figure 19-4. The elevation becomes a source to reference details similar to those shown in Figure 19-5, which can be used to explain construction of the enclosures as well as mounting of equipment. Wall elevations are also used for defining the limits of interior finishes.

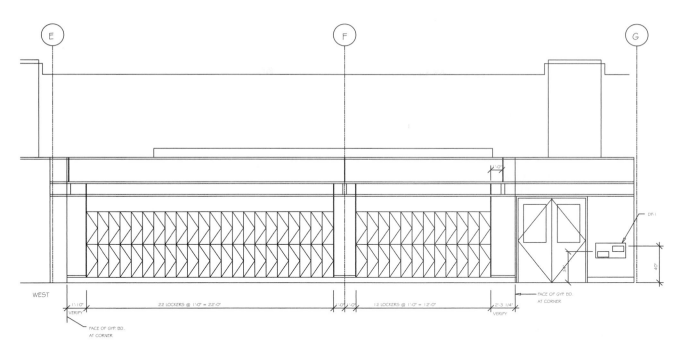

FIGURE 19-4 ■ An interior elevation is used to show the location of all special equipment. *Courtesy Michael & Kuhns Architects, P.C.*

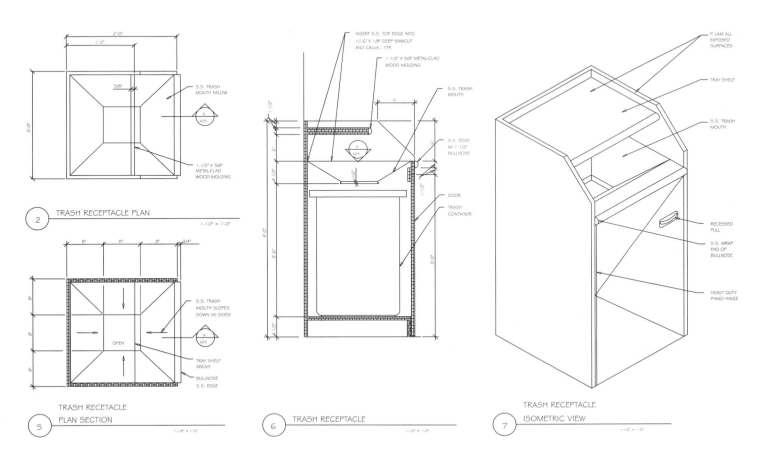

FIGURE 19-5 ■ Interior elevations are used to show construction of specialized cabinets. *Courtesy Michael & Kuhns Architects, P.C.*

Figure 19-6 shows an example of the lobby of a two-story structure and the various materials used to cover the structural frame.

Cabinet Elevations

Perhaps the most common use of interior elevations is to show the cabinets, appliances, grab bars, and plumbing fixtures used in bathrooms and kitchens. Nearly every structure drawn contains at least one bathroom that requires elevations. Figure 19-7 shows the elevation for a simple washroom and standard details that accompany the elevation to ensure minimum allowances to meet both the model code and ADA standards. Figure 19-8 shows the elevation of each wall of the bathroom shown in Figure 15–47.

For multifamily occupancies, a drafter working with the architectural team draws an elevation of each wall showing kitchen and bathroom cabinets and fixtures. Drawings must show each fixture and all door and drawer locations of the adjoining cabinets. Figure 19-9 shows the bath and kitchen elevations for a unit in a multilevel apartment complex. Elevations for each unit are provided. When a cooking area is used to serve the public, the supplier of the kitchen equipment generally develops the elevations for the kitchen equipment. Figure 19-10 shows the floor plan developed for a commercial kitchen. In addition to the cabinet drawings, many plans include drawings that show specialty equipment to be used in the kitchen. Drafters working for a consulting firm specializing in interior fixtures draw these drawings. Several of the specialty drawings can be found in Chapter 2. Figure 19-11 shows the interior elevations for this kitchen. Generally drawings for specialty equipment are placed separate from the interior drawings and are placed in a section of the drawings titled **Food Service (FS)**. The location of these drawings is listed in the table of contents.

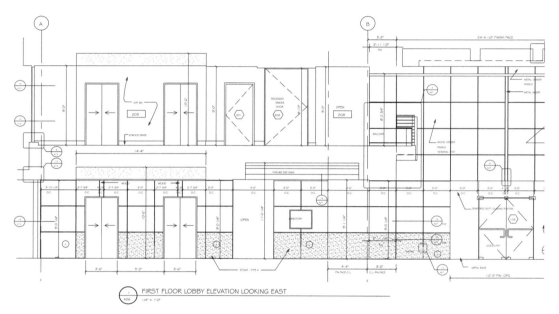

FIGURE 19-6 ■ Interior elevations are used to define the limits of interior finishes. *Courtesy Peck, Smiley, Ettlin Architects.*

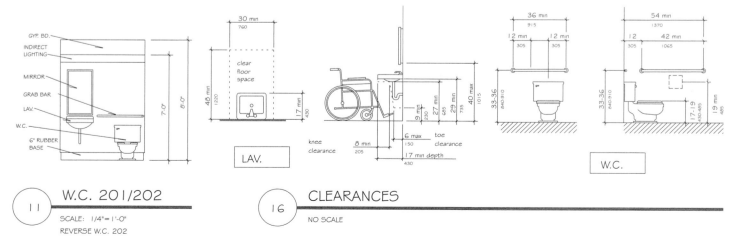

FIGURE 19-7 ■ Elevations are used to ensure that all minimum standards are met. *Courtesy DiLoreto Architects, LLC.*

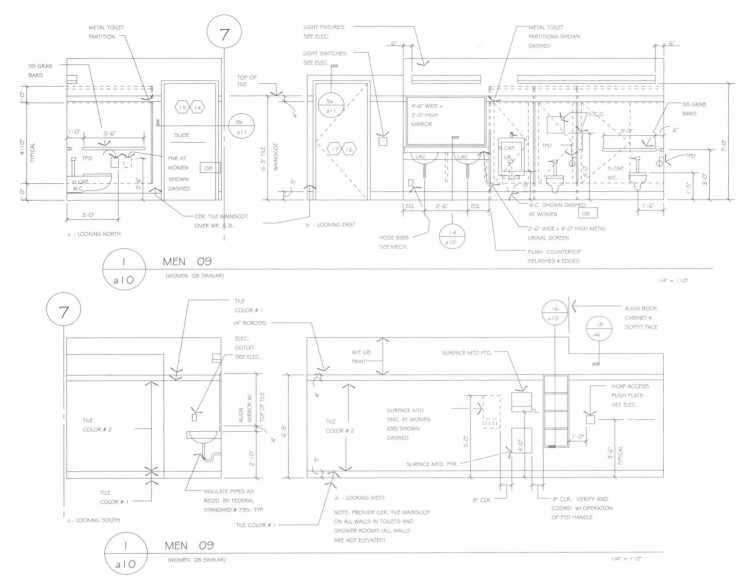

FIGURE 19-8 ■ Because of the large amount of fixtures, equipment and special finishes, an elevation for a bathroom can become quite complex. *Courtesy Architects Barrentine, Bates & Lee, A.I.A.*

Cabinet Sections

A third type of interior drawing that may be provided is an interior section. An interior section is used to show construction methods and materials of interior built-ins. Cabinet sections should be provided when the built-in is fabricated at the job site. When cabinets are built off site, the cabinet supplier often develops the drawings required for cabinet construction, and submits them to the architectural team for approval. When required, the drafter must be careful to clearly define each material that will be used. Figure 19-12 shows a cabinet section and common methods of representing thin materials.

Choosing A Drawing Scale

The complexity of the materials to be drawn determines the scale used to complete the cabinet elevations and details. A scale of 1/4" = 1'-0" (1:50) is often used for elevations, but both smaller and larger scales can be found on professional examples. In choosing the drawing scale, the drafter must be able to show sufficient detail to meet the demands of the drawings. If the drawings are to be used to represent pre-manufactured materials, a drawing using a small scale showing little detail can be used. If the drawings are used to construct the item, a scale sufficient to show all detail must be used. Scales for construction elevations are often drawn at a scale of 3/4" or 1/2" = 1'-0" (1:20). Details are usually drawn at a scale of 1" = 1'-0" or larger (1:10).

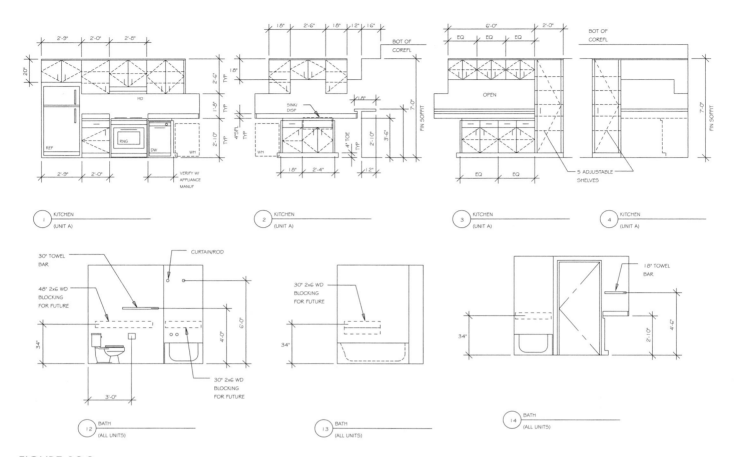

FIGURE 19-9 ■ The elevations for a multifamily complex resemble the elevations drawn for a single-family residence. *Courtesy H.D.N. Architects A.I.A.*

Material Representation

The symbols that are used to represent materials in sections and details are also used to represent material drawn in cabinet details. Continuous lines represent materials that are represented in elevation, with varied thickness. Thick lines usually represent the floor, walls, and ceiling, which define the view to be seen. On L-shaped cabinets, the line that represents the drawing limits will follow the contour of the cabinets when a cabinet is shown in end view. Figure 19-13 shows the outline of a room with L-shaped cabinets. Common features represented in the elevations, such as the range, dishwasher, and the refrigerator, are usually inserted as wblocks.

Cabinetry

The outline of cabinets and the lines representing doors, drawers, and exposed shelves are thin, continuous lines. A single thin, dashed line is used to represent shelves that are hidden behind doors. On larger scale drawings, pairs of continuous lines are used to represent shelves. Knobs and handles can be represented by a short polyline or a circle, to show the approximate style of hardware. Hinges are not usually shown, but door swings are represented by continuous or dashed lines in the shape of a V, drawn through the door. The point of the V represents the side of the door that is hinged. Figure 19-13 shows an elevation of cabinets with

door swings represented. When drawn in section, doors, drawers, and hardware can be represented as shown in Figure 19-12. The location of partitions for public toilet stalls should be represented on the elevations. Thin continuous lines can be used to represent walls that are perpendicular to the viewing plane. Doors can be omitted to show the location of toilets, or they can be represented by thin dashed lines to show the direction of swing. Figure 19-8 shows the location of premanufactured toilet partitions.

Plumbing Fixtures

Because of the repetitive nature of elevations, plumbing symbols should be created as wblocks for quick installation. Cabinet-mounted lavatories and sinks can be represented by hidden or continuous lines to define the outline of the unit. Fixtures built into a counter can be drawn using the same methods as for a cabinet-mounted unit. Wall-mounted and pedestal lavatories are drawn using thin continuous lines to represent the outline and some detailing of the unit. Figures 19-7, 19-8, 19-9, and 19-13 show examples of sinks and lavatories in elevation. Figure 19-14 shows a detail of a counter-mounted sink.

Thin lines are used to represent water closets and urinals, showing the basic components of the fixture. Blocks should be developed to show both a front and side view of each unit. Figure 19-15 shows common methods of representing water closets and urinals.

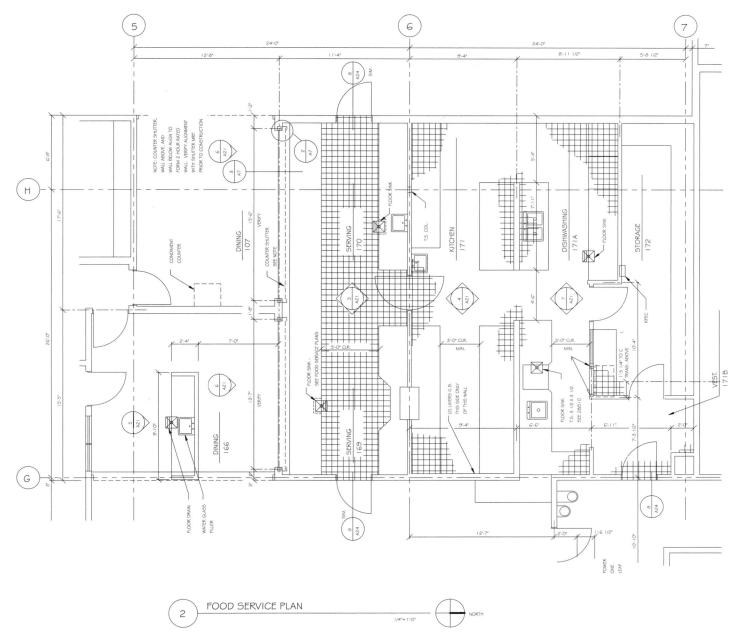

FOOD SERVICE PLAN

1/4" = 1'-0"

NORTH

FIGURE 19-10 ■ The kitchen for a commercial complex is far more detailed than that of a multifamily project and requires additional drawings from subcontractors. See Figure 2-5. *Courtesy Michael & Kuhns Architects, P.C.*

A thin continuous line should be used to represent the outlines of tubs and showers, with the interior shape represented by dashed lines. When the unit is perpendicular to the viewing plane, both the outline and the interior shape are shown with continuous lines. Figure 19-9 shows examples of tubs. The drawings should show the location of all grab bars near each unit.

Appliances

Kitchen appliances are represented by a box that outlines the shape or by a detailed representation. The detailed representation should

not be drawn to represent an exact model, but a generic representation. Figures 19-9 and 19-13 show cabinet elevations with a stove and a refrigerator. These appliances can be easily recognized but are not of a specific unit. In creating blocks, show doors or drawers that open and the general location of control features, but avoid spending time adding features that do not aid in installation.

Special Equipment

The occupancy of the structure will dictate what equipment is required. Equipment for fire protection, drinking fountains, tele-

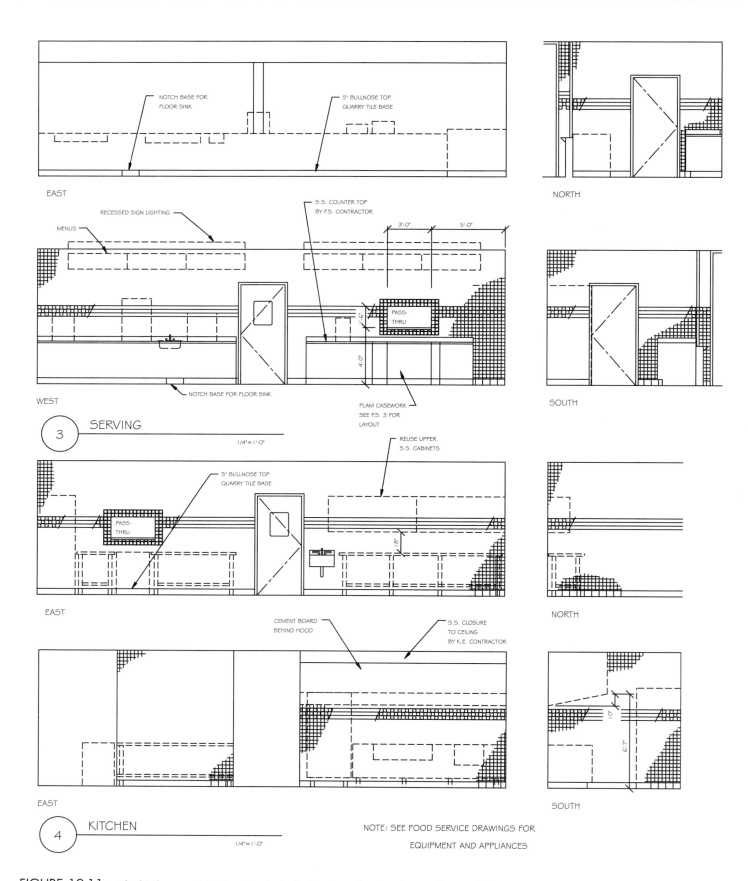

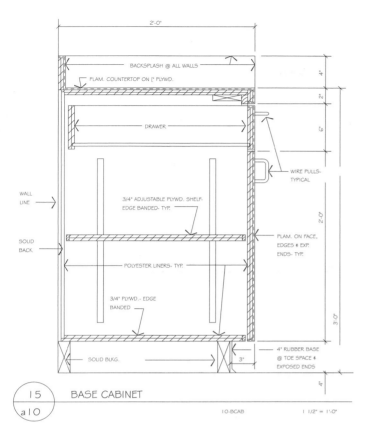

FIGURE 19-12 ■ Cabinet sections are used to show the construction of specialty equipment. *Courtesy Architects Barrentine, Bates & Lee, A.I.A.*

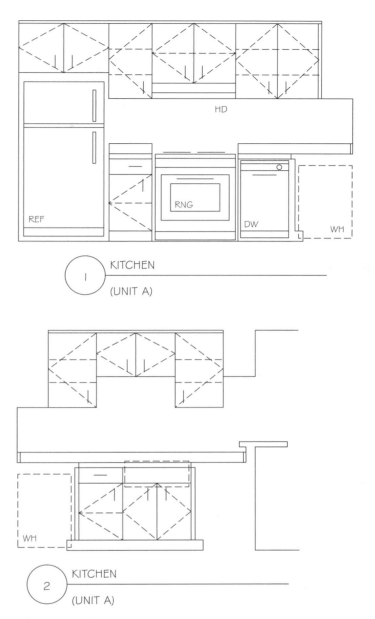

FIGURE 19-13 ■ Common components used in interior elevations can be stored as wblocks and inserted into drawings as needed. *Courtesy H.D.N. Architects A.I.A.*

phone stalls, mirrors, paper towel dispensers, access ladders, and building directories should all be represented on the elevations. Vendor catalogs can be an excellent source for obtaining information for drawing the needed feature. Many equipment suppliers will provide disk copies of their products, similar to the blocks shown in Figure 19-16.

Drawing Notations

The materials represented in elevation require both written and dimensional clarification. Notations must be added to each elevation to clearly define objects that have been drawn. As with sections, the larger the drawing scale, the more detailed the specifications should be. Information to be specified on the elevation is provided by the architect's preliminary sketches and check prints, and by consultation with *Time Saver Standards* or *Architectural Graphics Standards*. Vendor catalogs or *Sweet's Catalogs* should also be consulted to determine complete specifications. In addition to notations specifying each material, many cabinet elevations contain complete specifications to provide instructions for mounting and installation of each material to be used.

Vertical dimensions should be provided to locate the limits of all counter heights, cabinets, special equipment, and soffits that are not specified on the sections. Horizontal dimensions are used to locate special equipment from grid lines. Horizontal dimensions are also used to locate divisions in cabinets as shown in Figure 19-17. Cabinets and specialty items are generally referenced to a grid line or to a wall that is referenced to a grid. Although no numbers have been given for the dimensions shown in Figure 19-17, the cabinet maker will determine the exact size required to make four equally spaced units, based on the finished dimensions of the construction site. When the architect does not specify dimensions in the preliminary sketches, the drafter should provide dimensions based on common practice, building and ADA codes, and supplier recommendations.

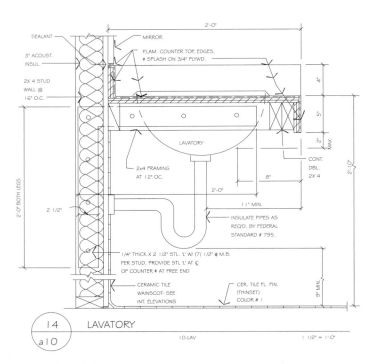

FIGURE 19-14 ■ Details are required to show how cabinets will be constructed and how fixtures will be installed. *Courtesy Architects Barrentine, Bates & Lee, A.I.A.*

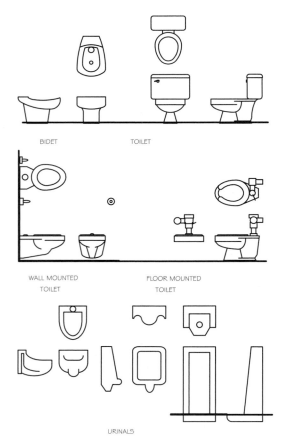

FIGURE 19-15 ■ Plumbing fixtures common to architectural drawings. *Courtesy Berol Rapidesign.*

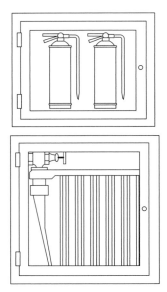

FIGURE 19-16 ■ Special equipment is often available in disk form from vendors or in libraries available from third-party suppliers. *Courtesy Berol Rapidesign.*

DRAWING SEQUENCE

Figure 19-18 shows a portion of the floor plan for two rooms from the educational complex shown in Figure 15–45. This unit is used for completing the cabinets for these rooms. Assume 36" base cabinets, with the bottom of the upper cabinets set at 4'–6" and the top at 7'–0" above the floor.

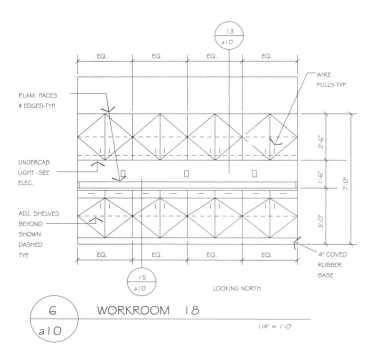

FIGURE 19-17 ■ Dimension must be supplied for all material or equipment that will be installed. *Courtesy Architects Barrentine, Bates & Lee, A.I.A.*

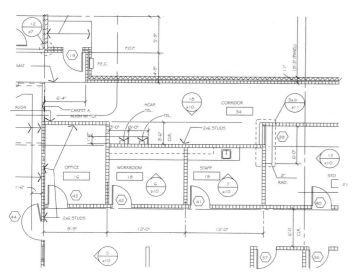

FIGURE 19-18 ■ The cabinets can be projected from a wblock of the floor plan, or drawn by following the dimensions on the plan.

1. Establish a viewing plane on the floor plan to indicate needed elevations
2. Use a block of the floor plan to project the outline of all walls that enclose the elevations
3. Establish grid lines on the elevations
4. Establish a line to represent the finish floor
5. Use the building sections to determine the heights of all ceilings and openings that are parallel to the viewing plane
6. Draw the outline of each cabinet shape

The initial layout should resemble Figure 19-19.

7. Insert blocks to represent each appliance
8. Draw doors and shelves in the upper cabinets

9. Draw doors and swings, drawers, toe kicks, and back splash required for the lower cabinets
10. Draw all equipment such as sinks, breadboards, and fixtures
11. Draw all required interior finishes

The cabinet elevations should now resemble Figure 19-20.

12. Draw and label each grid line
13. Provide horizontal dimensions to locate cabinets to each grid line
14. Place all horizontal dimensions, locating the edge of cabinets and door locations
15. Place all vertical dimensions, locating all the heights of all cabinets and equipment
16. Place notes to specify all materials and equipment
17. Reference any required details
18. Provide titles and drawing scales

The completed elevations are shown in Figure 19-21.

LAYER GUIDELINES FOR INTERIOR ELEVATIONS

When cabinet drawings are placed on a separate sheet, the drawing file is generally named I-ELEV-####. The owner's name or the job number should be substituted for ####. Cabinet elevations should be placed on layers with a prefix of A-ELEV if they are being placed in drawing files that contain other drawings. Modifiers such as APPL, CABS, CNTR, EQIP, PLMB or HDWR can be used to describe layers such as appliances, cabinets, counters, equipment, plumbing fixtures, and hardware. Cabinets that are placed in their own .dwg file can have a wider variety of file names. Many professional firms label interior elevations based on line type or usage. Layers titles such as FINE, HEAVY, or LIGHT can be used to define linetypes. Titles such as BORDER, TEXT, and DIMEN can be used to describe both the drawing contents and general drawing infor-

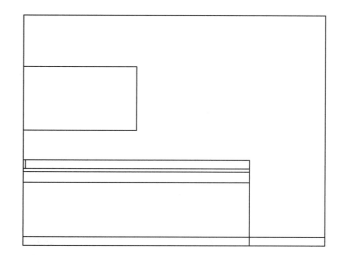

FIGURE 19-19 ■ The initial layout of interior elevations.

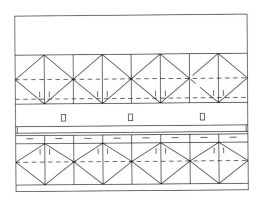

FIGURE 19-20 ■ Layout of the major cabinet components.

mation. Suggested layer names based on AIA standards include the following:

I-ELEV	Interior elevation
I-ELEV-CASE	Wall mounted casework
I-ELEV-FIXT	Miscellaneous fixtures
I-ELEV-FNSH	Finished and trim
I-ELEV-IDEN	Component identification numbers
I-ELEV-PATT	Hatch patterns
I-ELEV-PFIXT	Plumbing fixtures
I-ELEV-SIGN	Signage

Status field modifiers similar to other drawing areas should be used throughout architectural drawings. See Chapter 3 and the *AIA Layer Guidelines* for additional layer names.

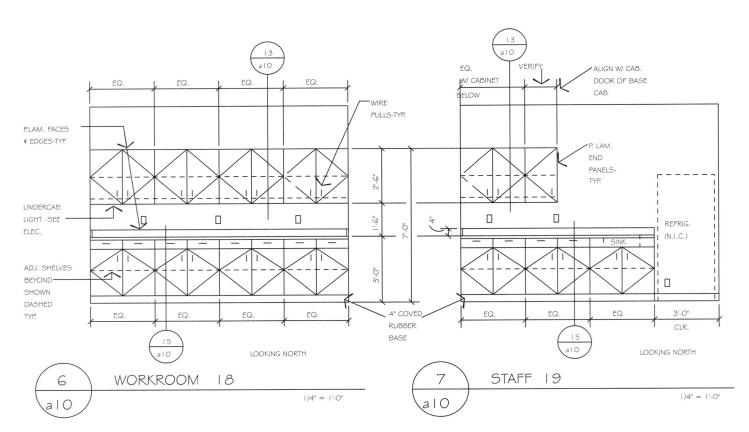

FIGURE 19-21 ■ The completed elevations for kitchen cabinet.

CHAPTER

19 *Interior Elevations*

CHAPTER QUIZ

Use a separate sheet of paper to answer the following questions. Print the chapter title, the question number, and a short complete statement for each answer.

Question 19-1 What drawings are required to draw the interior elevations?

Question 19-2 Describe placement of the interior elevations within the drawing set.

Question 19-3 List three common sources for cabinet and equipment specifications required for cabinet drawings.

Question 19-4 Describe two methods of referencing cabinet drawings to other drawings.

Question 19-5 Describe the selection of scales for cabinet drawings.

Question 19-6 To what are horizontal dimensions on cabinet elevations referenced?

Question 19-7 What is the common height of a base and upper cabinet?

Question 19-8 What is the goal of a cabinet section?

Question 19-9 List two methods of showing the direction of door swings.

Question 19-10 List three different types of interior drawings

DRAWING PROBLEMS

Use the reference material from preceding chapters and vendor catalogs to complete one of the following projects. Unless other instructions are given by your instructor, draw the elevations that correspond to the floor plan that was drawn in Chapter 15. Skeletons for the problems can be accessed from http://www.delmar.com/resources/ocl.html. Use these drawings as a base to complete the assignment. Use appropriate symbols, linetypes, dimensioning methods, and notations to complete the drawing. Determine any unspecified sizes based on material or practical requirements. Unless specified, select a scale appropriate to plotting on "D" size material and determine LTSCALE and DIMVARS.

You will act as the project manager and will be required to make decisions about how to complete the project. *Study each of the drawing problems related to the structure prior to starting a drawing problem. Any information not provided must be researched and determined by you unless your instructor (the project architect and engineer) provides other instructions.*

■ Place the necessary viewing planes on the floor plan to indicate each drawing

■ Use a 3" toe space unless noted

Problem 19-1a Use the floor plan from Problem 15-1 to draw the elevations for the kitchen. Use the following guidelines to complete the drawings:

■ Drop the ceiling to 8'–0" high in the kitchen and indicate the limits of the lowered ceiling on the floor plan

■ Draw the base cabinets with a 36" high ceramic tile counter with a full splash

■ Draw the bar behind the sink as 42" high and expand the width to 15" wide

■ Provide 18" clearance between the countertop and the upper cabinets

■ Amend the floor plan to show the upper cabinets to be over the stove and provide a range hood with 30" clearance to the cabinet

■ Provide a bank of drawers on the nook side of the sink

■ Provide an area for tray storage on the bathroom side of the stove with the balance of the base cabinets to have 4" deep drawers over doors

■ Provide a breadboard over a drawer between the refrigerator and the stove

Problem 19-1b Design three different elevations for the fireplace in the family room. Use entirely different styles and materials, and include the doors and windows on each side for each elevation.

Problem 19-1c Draw an elevation of the master bathroom cabinets including the linen closet. The linen closet is to have shelves above (2)-12" deep drawers. The vanity countertop is to be 32" high plastic laminate with a 4" back splash. Provide a mirror from the back splash to 6'–0" above the floor.

Problem 19-2 Use the floor plan created for Problem 15-2 to design and draw the elevation for an oak cabinet with shelves along the north wall of the office. The lower 36" is to have 4" deep drawers over doors. Shelves are to be on rollers. Part of the lower cabinet should include roll-out drawers (behind the doors) suitable for legal size filing. The upper portion should include shelves for books and a stereo system.

Problem 19-3 Use the floor plan in Problem 15-3 and design and draw an elevation of the cabinet in the break room.

Problem 19-4 Design a 24" deep storage cabinet from floor to ceiling along the north wall of the lower storage area. The 20 feet closest to the warehouse area will be used to store flammable materials and should be lockable. Ten feet of the cabinet should be divided into lockable storage units for employees. The balance of the area will be used for general storage and should have a base cabinet with enclosed shelves, a plastic laminate countertop at 36" high with a 4" splash, and upper shelves behind doors.

## CHAPTER 20

Ramp, Stair, and Elevator Drawings

Changes in floor elevation are accomplished by using a ramp, stairs, elevator, escalators, or a combination of each. Although the architect is responsible for the design, a drafter should be familiar with the code requirements and vendor drawings associated with each method of changing floor levels.

RAMPS

A ramp is an inclined surface that connects two different floor levels. Design and construction are regulated by both the building code and ADA requirements. (See Chapter 5 for a review of ADA requirements.) Chapter 15 introduced ramp representation on the floor plan. This chapter addresses code requirements, methods of representation on drawings, and common sections and details that a drafter is required to work with.

Code Requirements

Ramps are classified by IBC based on usage. Ramps that are used in an aisle for an exit from a structure must also meet the requirements for exits that were presented in Chapter 11. This chapter will focus on code requirements for ramps that are used to achieve a change in floor levels.

Minimum Ramp Sizes

The IBC requires ramps to have a minimum width equal to the required width for the corridor and a minimum clear width of 36" (900 mm) for the ramp and clear width between the handrails. Handrails are allowed to project into the required ramp width a maximum of 4 1/2" (114 mm) on each side provided the minimum clear width of 36" (900 mm) is maintained. Any decorative trim or other features may only project a maximum of 1 1/2" (40 mm) on each side. The rise of ramps is restricted to 30" (762 mm) total with a maximum slope of 1 vertical unit per each 8 horizontal units or a 12% slope (1:12 / 8% slope for means of egress). The cross slope of a ramp, measured perpendicular to the direction of travel, cannot be greater than 1 vertical unit per each 48 horizontal units (2% slope). If a door swings over an egress ramp, the door when opened in any position cannot reduce the width of the ramp to less than 42" (1070 mm). A minimum of 80" (2032 mm) headroom must be provided above a ramp.

Landings

Except in non-accessible Group R2 and R3 units, a landing with a length of 60" (1525 mm) measured in the direction of the ramp is required at the top and bottom, points of turning, doors, and the entrance and exit of all ramps. Group R2 and R3 units can have a landing with a minimum length of 36" (914 mm). The landing width must be as wide as the widest ramp that connects to the ramp. The slope of a landing cannot be greater than 1 vertical unit per each 48 horizontal units (2% slope).

Railings, Handrails, and Curbs

With the exception of ramps in aisles that serve seating, ramps with a rise greater than 6" (150 mm) must have a handrail on both sides of the ramp. The handrail must be between 34 and 38 inches (864 and 965 mm) in height. The handgrip portion of the rail must be between 1 1/4 and 2" (32 and 50 mm) wide and must be smooth with no sharp corners. Rails must project 1 1/2" (38 mm) clear from walls to provide hand space. A railing is required to be mounted below the handrail. The rail must be between 17 and 19 inches (432 to 483 mm) above the ramp or landing surface. A curb or barrier must be provided at the floor or ground surface of the ramp and landing. The curb must be constructed to prevent the passage of a 4" diameter sphere.

Ramp Representation

The width and length of a ramp can be represented on the floor plan using thin lines. The angle can be represented by listing either the slope or the elevation of each end, as seen in Figure 20-1. The slope and construction method must be represented by means of a section or detail that is referenced to the floor plan. Figure 20-2 shows an example of a ramp detail.

STAIRS

Stairs are the primary method of changing floor levels in a multi-level structure. To effectively work with the architect's stair drawings, the drafter must be familiar with common terms, layout methods, and code requirements.

Stair Terminology

The exact terminology for stair construction may vary in different parts of the country. Figure 20-3 shows each of the common components of a stair that must be represented by a drafter. The layout of a stair is dictated by the rise and run.

■ The *rise* is the vertical distance from one floor level to the next

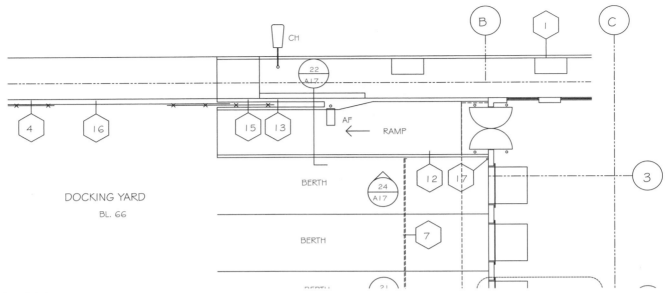

FIGURE 20-1 ■ A ramp is an inclined floor surface used to connect two different floor elevations that are often represented on the floor and site plans. *Courtesy Michael & Kuhns Architects, P.C.*

■ The *run* is the total length required to form the stairs

■ The total rise is comprised of individual *risers*, which are the vertical portion of a step. The height of a riser is measured from one step to the next.

■ The *tread* is the horizontal portion of a step. The length of a tread is measured from the face of the riser at the rear of the tread to the nosing at the front end of a tread.

■ The *nose* is the front edge of a step. Nosing is material that is added to the nose of the tread for decoration and to provide a slip-resistant surface. The type and design of nosing varies depending on the material used to build the stair. Nosings usually project approximately 1 1/8" (28 mm) past the riser.

■ A *stringer* provides support for the tread and riser. In some areas of the country, it is referred to as a *carriage* or *stair jack*. The stringer may be placed below or beside the risers.

Figure 20-4 shows the relationship of the stringer to the tread for various materials. The stringer terminates at a floor level or at an intermediate platform called a *landing*. Landings are inserted into long stair runs to provide users with a resting point, or as a means of changing the direction of the run. In multilevel structures, stair landings are often stacked above each other.

■ The *headroom* for stairs is the vertical distance above the stair measured from the tread nosing to any obstruction above the stair

To provide stability to the user, a handrail must extend 12" (300 mm) past the upper riser. The lower end of the rail must continue

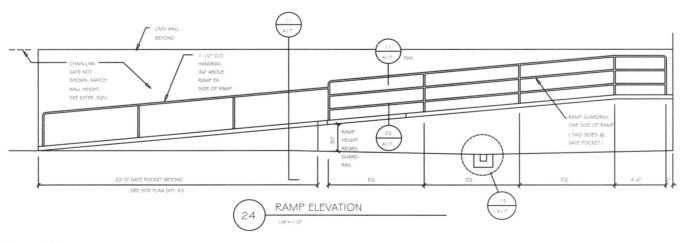

FIGURE 20-2 ■ Sections or details should be drawn to explain ramp angle, construction methods, and railings. *Courtesy Michael & Kuhns Architects, P.C.*

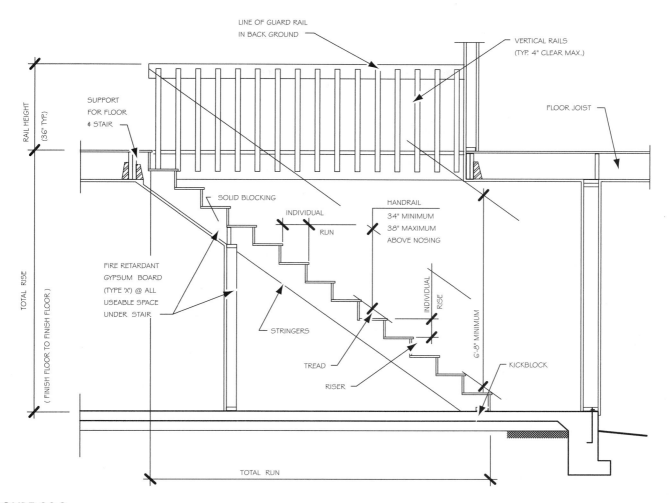

FIGURE 20-3 ■ Common stair terms.

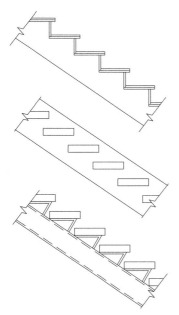

FIGURE 20-4 ■ Stair treads may be directly supported by the stringer, placed between the stringers, or be above the stringer and supported by metal brackets.

to slope for the depth of one tread beyond the bottom riser. The number of handrails required is based on the width of the stairs and is discussed later in this chapter. The handrail on the open side of a stair terminates at a newel post.

■ A *newel post* is a large, fancy post used to provide support at the end of a railing. The handrail is connected to the open side of the stair by vertical *baluster*.

■ A *guardrail* is the railing that is placed around stairs to protect people from falling from the floor into the stairwell

■ A *stairwell* is the vertical shaft where the stairs are to be placed. The design of the baluster connection to the stair and floor systems is of great importance for the safety of the stair users.

Depending on the height of the structure, the occupancy, and fire-resistive construction used, stairs may need to be separated from the balance of the structure by an enclosed stairwell. To control the spread of smoke and flames throughout a structure, access to stairwells is restricted by doors that must have a minimum 1-hour fire rating. Review Chapter 11 for fireproofing of stairwells.

Stair Types

Depending on the material used, stairs may either be open or enclosed. Stairs framed with wood, steel, or precast concrete risers on a steel stringer can be framed as open stairs with no riser. Each of these materials can also be used to form enclosed stairs. Stairways that serve as the only access to another floor cannot be open because of ADA restrictions. The method of framing is not represented in plan view but is shown in sections and details. In addition to the type of riser construction, the configuration of the run can also be altered to meet design needs.

■ Straight stair runs are the most economical to build. If space for the run is limited, a landing can be inserted into the run and the direction of the run can be altered. L- and U-shaped stairs are two common variations to a straight run.

■ An L-shaped stair contains a 90° bend in the total run. The change in direction occurs at a landing, which may be placed in any portion of the run. A U-shaped stair causes the user to make a 180° turn at a landing. The landing can be placed at any point of the run, but it is usually placed at the midpoint so that each run is of equal length.

■ A curved or circular stairwell can be used to enhance the design of an entry. *Curved stairways* are formed by using one or more radii to form the stair curve. Curved stairs have wedge-shaped treads.

■ A *spiral stair* is a stair that has a circular form in plan view with treads that radiate about a central support column. Spiral stairs can only be used within an individual unit or an apartment that serves an area less than 250 sq ft (23 m^2) and serving less than five occupants.

Figure 20-5 shows an example of how each type of stair is represented on a floor plan.

Minimum Code Design Requirements

Any stairs having two or more risers, (except stairs used exclusively to service mechanical equipment) are regulated by the IBC and the ADA. A drafter must be familiar with these codes so that drawings and details are accurate.

Required Sizes For Straight Run Stairs

The IBC specifies minimum sizes for the width, height, individual rise and run and the headroom for stair construction.

Required Width—Stairways serving an occupant load of 49 or less can be a minimum of 36" (900 mm) wide. Stairs serving 50 people or more must be a minimum of 44" (1120 mm) wide.

Required Riser Height—4" (102 mm) and 7" (178 mm) maximum measured vertically between the leading edges of adjacent treads. For dwelling units in Group R-2 and R-3 occupancies, the rise can be increased to 7 3/4" (197 mm).

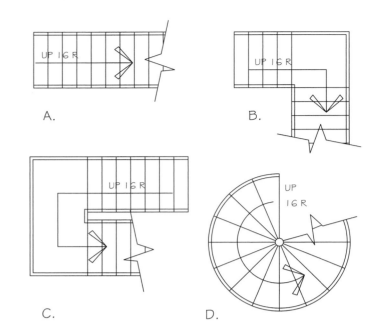

FIGURE 20-5 ■ Common stair types include (A) straight, (B) L-shaped with a landing at any point of the run, (C) U-shaped with a landing at any point of the run, and (D) spiral.

Maximum Stair Rise—A landing must be provided for every 12' (3600 mm) of vertical rise.

Required Tread Depth—11" (279 mm) minimum measured horizontally between the vertical planes of the foremost projection of adjacent treads and at right angles to the tread's leading edge.

Landing—A landing equal in length and width to the width of the stair must be provided at the top and bottom of a stairway. If a landing is provided in the stair span, the length of the landing measured in the direction of travel must be equal to the width of the stair run but is not required to exceed 48" (1219 mm).

Tread and Landing Slope—The walking surface of a tread or ramp can not be greater than 1 vertical unit per each 48 horizontal units (2% slope) in any direction.

Required Headroom—Stairways must be designed so that a minimum height of 80" (2032 mm) is provided. The minimum allowable headroom is to be measured vertically from a plane parallel and tangent to the stairway nosing to any obstruction.

In addition to the requirements for straight stairways, circular, spiral, and winding stairs must meet other code requirements.

Circular Stairs

Circular stairs can be used for an exit as long as the minimum tread run is 11" (279 mm) when measured at a point 12" (305 mm) from the narrow end of the tread (6" (152 mm) for dwelling units in Group R-2). The minimum tread depth shall not be less than 10" (254 mm) at the narrow end. The smallest radius used to layout the stairway curve cannot be less than twice the width of the stair. Curved stairs that do not have a uniform radius are considered *winders* by the building code.

Winders

Except for use within a dwelling unit, winders may not be used in a means of egress. When used, a required width for the run of 11" (279 mm) is provided at a point 12" (300 mm) from the narrowest portion of the stair. Winders also must be a minimum of 6" (150 mm) at the narrowest portion of the tread. Required tread sizes are shown in Figure 20-6.

Spiral Stairs

The tread of a spiral stair must provide a clear walking area of 26" (660mm) from the outer edge of the supporting column to the inner edge of the handrail. A minimum run of 7 1/2" (191 mm) must be provided at a point 12" (305 mm) from the narrowest portion of the tread. The maximum rise for spiral stairs cannot exceed 9 1/2" (241 mm) and a minimum headroom of 78" (1981 mm) must be provided. Figure 20-7 shows the minimum tread requirements for a spiral stair.

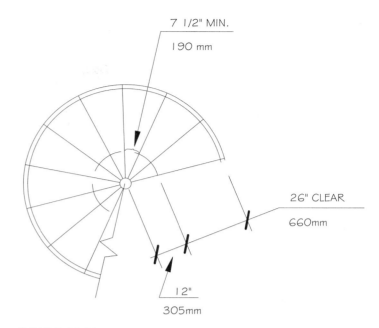

FIGURE 20-7 ■ Tread requirements for a spiral stair.

Rail Design

The placement of handrails and guardrails must be considered as stair drawings are completed.

Handrails

Handrails must be provided on each side of the stair run. An intermediate handrail must be provided for wide stairs so that no portion of the width required for egress is more than 30" (762 mm) of a handrail. Handrails must be between 34" and 38" (864 and 965 mm) high above the nosing of treads or landings. The guidelines used for placing handrails by ramps also apply for stairs.

Guardrails

Guardrails are required at non-enclosed floor areas that are 30" (760 mm) or more above the ground or another floor level. Guardrails are not required on the loading side of loading docks, on the auditorium side of stages or around service pits that are not accessible to the public. Guardrails must be 42" (1067 mm) high, except for rails serving individual apartment units that are allowed to be between 34" to 38" (864 to 964 mm) high. Intermediate railings in a guardrail must be designed so that a 4" (100 mm) diameter sphere cannot pass through any openings up to a height of 34" (864 mm). Openings in the railing above 34" (864 mm) to 42" (1067 mm) can be designed using an 8" sphere. Railings for industrial occupancies that are not open to the public can be designed so that a 21" (533 mm) sphere will not pass through the railing.

Lateral Load Design for Railings

Railings must be able to withstand the pressure of people pushing or pulling on the rail. The IBC requires rails serving an occupant load of less than 50 people and not used as an exit facility to be able

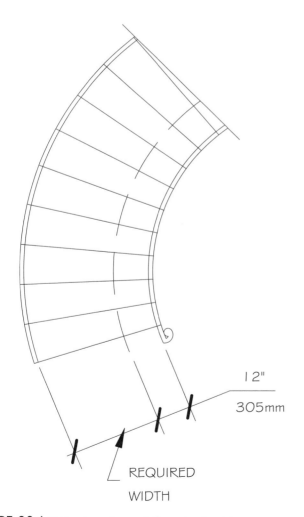

FIGURE 20-6 ■ Tread requirements for a circular stair.

to resist a lateral load of 20 psf (.958 kN/m²). Rails serving occupant loads of 50 or greater are required to resist 50 psf (2.395 Kn/m²). Rails must also be designed so that the rail can support 200 pounds (890 Kn) applied from any direction to any portion of the railing. These loads must be transferred down through balusters into the floor or stair system. Both the connectors used to attach balusters and the material they are connected to must be strong enough to resist the loads to be supported and to resist the tendency to twist as the loads are applied. Figure 20-8 shows an example of a detail used to show the connection of a railing to the floor system.

Determining Required Rise and Run

The total rise and run of a stair is based on the legal requirements for the rise and run of each tread. If the floor-to-floor height is 9'–6" and a rise of 7" is used, 17 risers will be required. This is determined by dividing 114" (the total height) by 7" (the maximum rise). Although the answer is 16.28, this must be rounded up to 17 risers. The actual riser height of 6.70" can be determined by dividing 114 inches by 17. The number of treads is always one less than the number of risers. This stair would require 17 risers and 16 treads. If a run of 11 inches is used, the total run would be 176 inches (16 × 11").

STAIR MATERIAL

The occupancy and type of construction will dictate the material used to form the stairs. Wood, timber, steel, concrete, or a combination of steel and concrete are used to form the stair.

Wood

Wood stairs are made by using a 2 × 12 or 2 × 14 (50 × 300 or 50 × 350) stringers. Treads can be formed by using 1" (25 mm) tread material or 3/4" (19.1 mm) plywood. Both the treads and the risers are covered by a non-slip surfacing material. If carpet is used as the finish material, the riser should be set on an angle. Figure 20-9 shows a detail of a stair made of wood. The stringer is supported by a metal hanger, which is connected to a header in the floor system. Where the stringer rests on the floor platform, a metal angle or a kicker block is used to keep the stringer from sliding across the floor. A block of 2× (50×) material is placed between the stringers at mid-span to minimize stringer vibrations and to reduce the spread of fire through the chase formed between the stringers. Figure 20-10 shows an example of a wood framed stair.

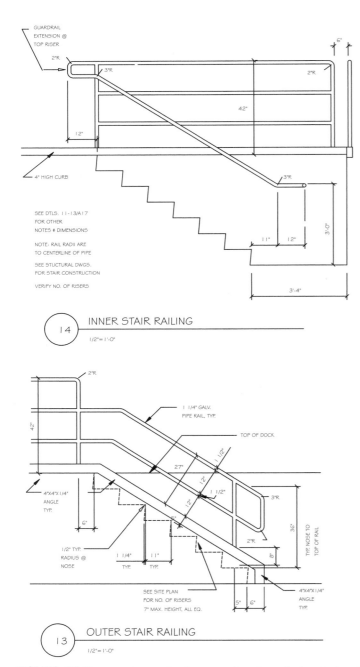

FIGURE 20-8 ■ Details must be drawn to show how the handrail and guardrail will be constructed. *Courtesy Scott R. Beck, Architect.*

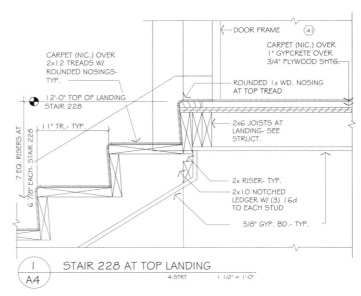

FIGURE 20-9 ■ Although typically drawn as a vertical riser in section, a riser covered with carpet is generally inclined and must be represented in detail. *Courtesy Architects Barrentine, Bates & Lee, A.I.A.*

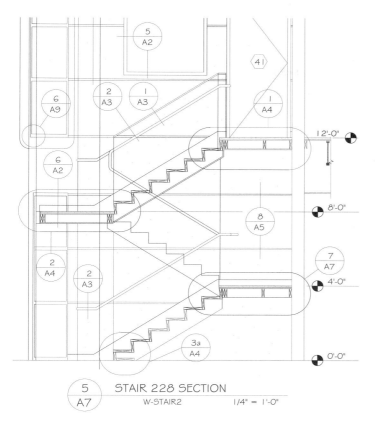

FIGURE 20-10 ■ A wood framed stair in section. *Courtesy Architects Barrentine, Bates & Lee, A.I.A.*

Timber

Timber can be used to form open stairs. A 3 × 12 (75 × 300) or larger is used for the stringer. The size of the stringer must be determined based on the span and load to be supported. Stringers are attached to the floor platform by use of a metal angle at each end of the stringer. Treads are generally made of 2" or 3" (50 or 75 mm) material and are attached to the stringer by use of a metal angle or let into the stringer. Figure 20-11 shows examples of each method of attachment. Figure 20-12 shows an example of a stair using timber construction.

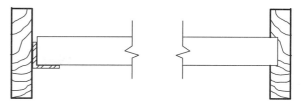

FIGURE 20-11 ■ Timber treads may be supported by a metal angle or let into the stringer.

Steel

Steel stairs are found in many types of commercial and light industrial construction because of the need for noncombustible construction. Steel stairs can be built similar to wood stairs, using a channel as the stringer and steel plates for the risers and treads. Stringers are attached to each floor system by the use of a steel angle. Handrails are generally made of 1 1/4" diameter metal pipe and are supported on metal uprights called *balusters*. Balusters are welded to the channel stringer. Figure 20-13 shows a portion of a steel stair. As shown in Figure 20-14, steel treads can be made in several configurations. For exterior applications, plates forming a grating or made from extruded metal are used to shed water. Metal is also used to form a pan to support concrete steps. The combination of a metal frame and concrete steps is quite common in many office structures. Figure 20-15 shows an example of a stair framed with steel and concrete. Figure 20-16 shows examples of details required to supplement the section.

Concrete

Because of cost, concrete stairs are used in the most restrictive fire rating situations. When used, the stairs are designed as an inclined one-way slab with an irregular upper surface. Concrete stairs can be cast after the wall and floor systems have been cast, or at the same time. Drawings for concrete stairs are generally done by drafters working for the engineering team. Figure 20-17 shows an example of a stair formed using concrete. Reinforcing details can be seen in Figure 20-18. Concrete stairs similar to Figure 20-19 are formed at grade.

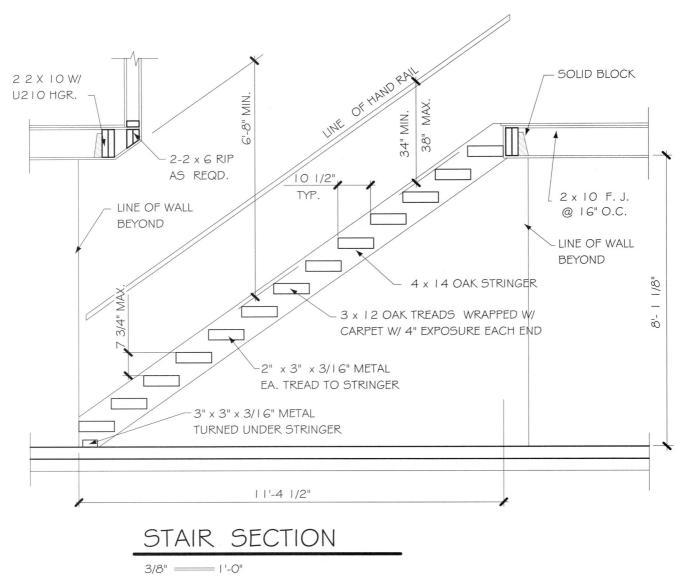

2 2 X 10 W/
U210 HGR.

SOLID BLOCK

6'-8" MIN.

LINE OF HAND RAIL

34" MIN.

38" MAX.

2-2 x 6 RIP
AS REQD.

LINE OF WALL
BEYOND

10 1/2"
TYP.

2 x 10 F. J.
@ 16" O.C.

LINE OF WALL
BEYOND

8'- 1 1/8"

7 3/4" MAX.

4 x 14 OAK STRINGER

3 x 12 OAK TREADS WRAPPED W/
CARPET W/ 4" EXPOSURE EACH END

2" x 3" x 3/16" METAL
EA. TREAD TO STRINGER

3" x 3" x 3/16" METAL
TURNED UNDER STRINGER

11'-4 1/2"

STAIR SECTION

3/8" ══════ 1'-0"

FIGURE 20-12 ■ A stair section for a timber stair in a Group R-3 occupancy.

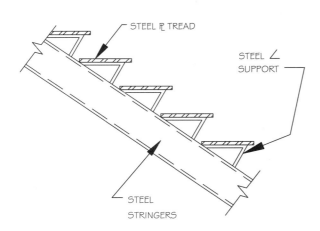

STEEL P TREAD

STEEL ∠
SUPPORT

STEEL
STRINGERS

FIGURE 20-13 ■ A steel plate stair.

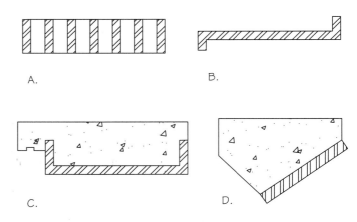

A.

B.

C.

D.

FIGURE 20-14 ■ Steel treads can be made from (A) bar grating, (B) steel plates with a textured surface, (C) concrete treads supported by a steel pan, and (D) precast concrete treads with a steel mounting plate.

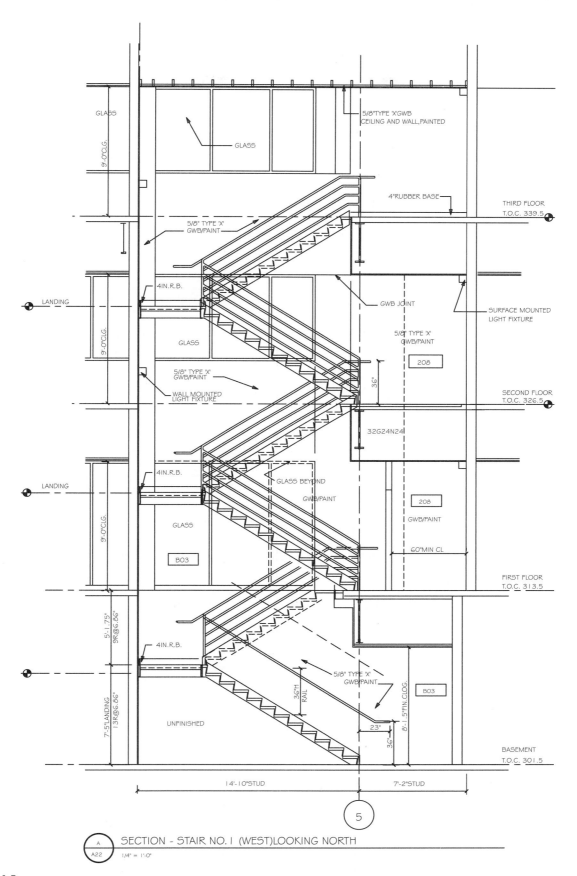

FIGURE 20-15 ■ A stair section for a multilevel steel stair. The plan view is shown in Figure 20-23. *Courtesy Peck, Smiley, Ettlin Architects.*

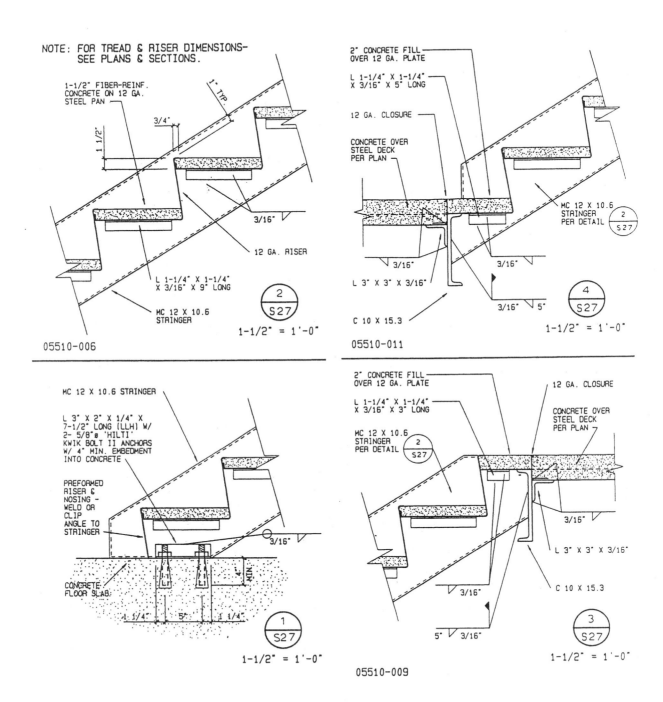

NOTE: FOR TREAD & RISER DIMENSIONS-
SEE PLANS & SECTIONS.

1-1/2" FIBER-REINF.
CONCRETE ON 12 GA.
STEEL PAN

3/4"

1 1/2"

1" TYP.

3/16"

12 GA. RISER

L 1-1/4" X 1-1/4"
X 3/16" X 9" LONG

MC 12 X 10.6
STRINGER

2
S27

1-1/2" = 1'-0"

05510-006

2" CONCRETE FILL
OVER 12 GA. PLATE

L 1-1/4" X 1-1/4"
X 3/16" X 5" LONG

12 GA. CLOSURE

CONCRETE OVER
STEEL DECK
PER PLAN

MC 12 X 10.6
STRINGER
PER DETAIL

2
S27

3/16"

3/16"

L 3" X 3" X 3/16"

3/16"   5"

C 10 X 15.3

4
S27

1-1/2" = 1'-0"

05510-011

MC 12 X 10.6 STRINGER

L 3" X 2" X 1/4" X
7-1/2" LONG (LLH) W/
2- 5/8"ø 'HILTI'
KWIK BOLT II ANCHORS
W/ 4" MIN. EMBEDMENT
INTO CONCRETE

PREFORMED
RISER &
NOSING -
WELD OR
CLIP
ANGLE TO
STRINGER

3/16"

CONCRETE
FLOOR SLAB

4" MIN.

1 1/4"   5"   1 1/4"

1
S27

1-1/2" = 1'-0"

2" CONCRETE FILL
OVER 12 GA. PLATE

12 GA. CLOSURE

L 1-1/4" X 1-1/4"
X 3/16" X 3" LONG

CONCRETE OVER
STEEL DECK
PER PLAN

MC 12 X 10.6
STRINGER
PER DETAIL

2
S27

3/16"

3/16"

L 3" X 3" X 3/16"

C 10 X 15.3

5"   3/16"

3
S27

1-1/2" = 1'-0"

05510-009

FIGURE 20-16 ■ Details for construction of a steel stair. *Courtesy Dean Smith, Kenneth D. Smith Architect and Associates, Inc.*

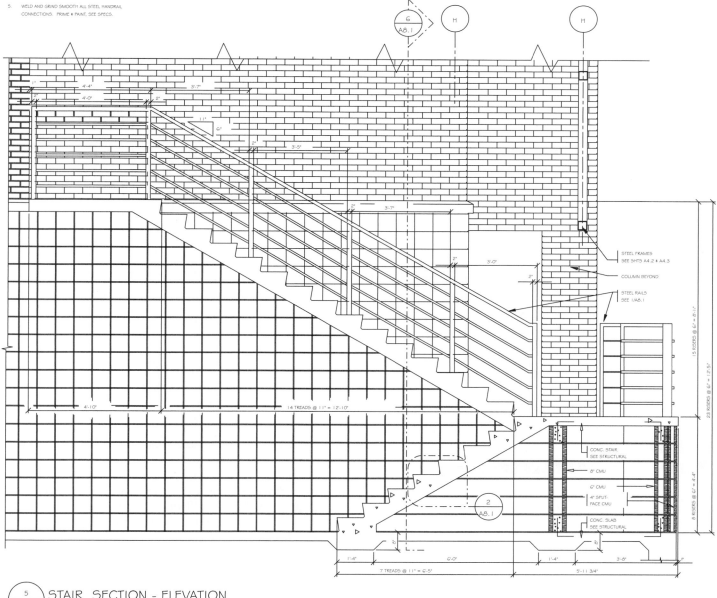

GENERAL NOTES:

1. PROVIDE WEEPS @ 24" o.c. AT CMU BELOW STAIR.

2. STEPS AND LANDINGS TO RECIEVE BROOM FINISH.

3. PAINT ALL STEEL HANDRAILS AND STEEL FRAMES.

4. PLACE STEEL RAIL SLEEVES BEFORE PLACING CONCRETE. IF CONCRETE IS LESS THAN 5" THICK, WELD SLEEVE TO STEEL DECK OR STRUCTURE.

5. WELD AND GRIND SMOOTH ALL STEEL HANDRAIL CONNECTIONS. PRIME & PAINT, SEE SPECS.

5 / A8.1 STAIR SECTION - ELEVATION
SCALE: 1/2" = 1'-0"

FIGURE 20-17 ■ Sections showing the construction of a concrete stairway. *Courtesy G. Williamson Archer A.I.A., Archer & Archer P.A..*

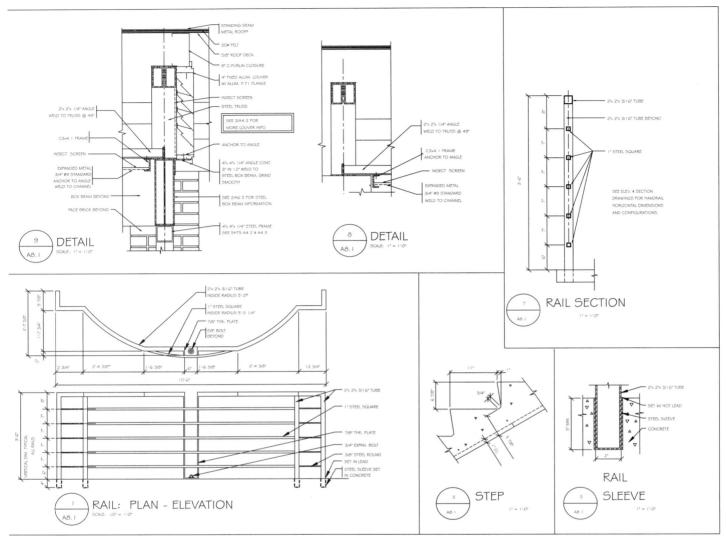

FIGURE 20-18 ■ The details for the stair shown in Figure 20-17. *Courtesy G. Williamson Archer A.I.A., Archer & Archer P.A.*

DRAWING REPRESENTATION

The drafter will generally be required to work with stairs in plan views, sections, and details.

Plan Views

Stairs and the line work used to represent them were first introduced in Chapter 15. The stairway must be shown on the floor plan before details and sections can be started. The sections must be planned however, before the floor plan can be drawn. Using steps described earlier in this chapter, the number of risers and treads must be determined. When drawing a stair that goes to another level on the same floor, the entire run of the stair is shown. Figure 20-20 shows an example of a stair going up to an intermediate floor level. For a stair that extends between two full floors, the lower portion of the stair is shown on the main floor plan, and the upper end is shown on the second floor plan, as shown in Figure 20-21.

In planning the layout of the stairs on the floor plan, the required width, tread length, and number of risers must be known. Using a height of 9'–6" from floor to floor, it is determined that 17 risers and 16 treads are required with an overall run of 176". The layout of the stair on the floor plan can be completed by using the following steps.

1. Draw a line to represent the width of the stair

2. Start at the upper floor landing, and place a line to represent the ending point. The length of the run can be determined as illustrated in Figure 20-22.

3. Use either **OFFSET** or **ARRAY** to show enough treads to establish the pattern. The run should be terminated with a short break line at the end of the run.

4. Represent the handrail by a line offset into the stairs approximately 1". There is no need to accurately repre-

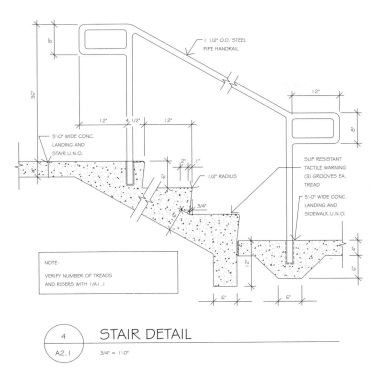

FIGURE 20-19 ■ A concrete stairway formed at grade. *Courtesy Scott R. Beck, Architect.*

sent the actual rail offset on the floor plan because it will be detailed in the stair details.

5. Label the rails, floor above, the direction of travel, and the number of risers per flight

Figure 20-21 shows the completed lower floor plan. A similar process can be used to draw the upper portion of the run on the upper floor plan.

For structures with U-shaped or other shaped stairs, the layout process is similar. Remember, on multilevel stairs the highest and

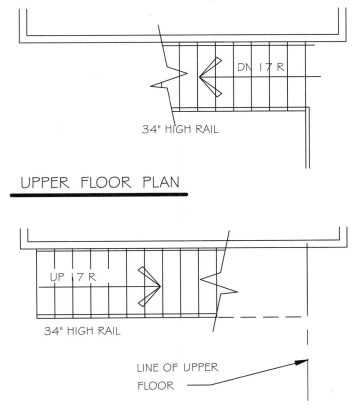

UPPER FLOOR PLAN

MAIN FLOOR PLAN

FIGURE 20-21 ■ When a stair serves two different floors, only one end of the run is shown on each floor plan.

lowest floor levels show only one run. Middle floors show runs going both up and down. Figure 20-23 shows how multilevel stairs would be represented. Notice that the stair going up is always on the same side of the stair and the upper run is opposite the lowest run.

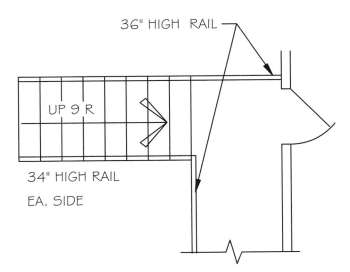

FIGURE 20-20 ■ When a stair serves two different levels of the same floor, both ends of the stair run are shown in the plan view.

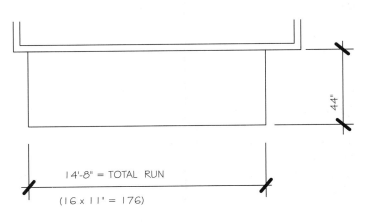

FIGURE 20-22 ■ The stair plan is started by determining and representing the limits and width of the stair.

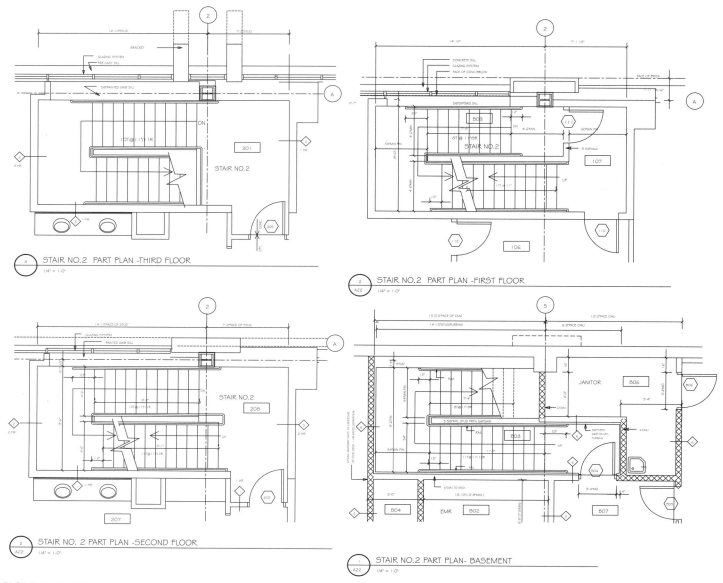

FIGURE 20-23 ■ When a stair serves more than two levels, the drafter must coordinate each run so that it does not interfere with the headroom of other stairs in the run. *Courtesy Peck, Smiley, Ettlin Architects.*

Drawing Stair Sections

Stair sections are typically drawn at a scale of 1/4" = 1'–0", 3/8" = 1'–0" or 1/2" = 1'–0". The space available, the detail to be shown, and the amount of related details to be used affect the selection of the scale. If the stair is to be pre-manufactured, only drawings required to attach the stair are supplied. The section for the stair shown in Figure 20-21 can be completed by using the following steps. Straight steel, concrete, and combination stairways are drawn using similar methods.

1. Draw a line to represent each floor level

2. Establish a line to represent the edge of the upper floor

3. Establish a line to represent the total run

4. Draw lines to represent the required number of treads and risers

5. Draw a line to represent where each tread and riser will be placed

6. Draw a line through the nose of the lowest tread to the nose of the highest tread

7. Offset the line just drawn to represent the bottom side of the stringer

The drawing should now resemble Figure 20-24.

8. Use the line drawn in step 7 to determine where the stringer intersects the floor framing. Use this point to locate the stair support. In this example, (2)-2 × 12

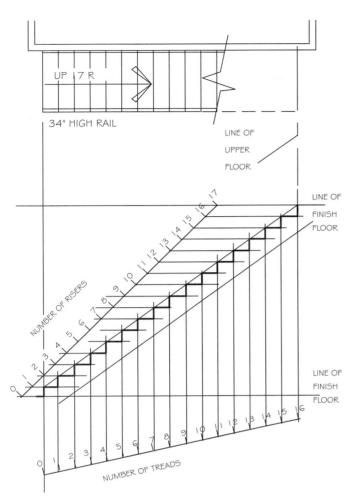

FIGURE 20-24 ■ A stair section is started by defining the limits and representing each finished floor level. The distance from floor to floor is divided by the number of risers. The distance representing the run is divided by the number of treads.

floor joists are used with metal hangers to the floor joist and the stringer.

9. Using the grid that was established in Step 4, draw a tread and a riser similar to Figure 20-25

10. Using the COPY command with a running OSNAP of INT, copy the tread and riser to reproduce the required number of treads

11. Eliminate all layout lines

12. Provide a 2× kick block at the toe of the stair and blocking at the mid-span of the stringer

13. Show required finish material on the bottom side of the stringer and floor system so that the stair resembles Figure 20-26

14. Locate required guard and handrails

15. Provide dimensions to represent the following sizes:
    ■ Total rise and run
    ■ Individual rise and runs

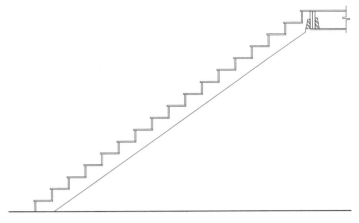

FIGURE 20-25 ■ Once a grid for the stair has been determined, individual treads and risers can be drawn.

    ■ Required headroom
    ■ Handrail and guardrail heights

16. Use leaders to label each size and component used to make the stair. Material represented in a detail does not have to be completely specified.

Notice the lack of notes in each of the professional examples throughout this chapter and the dependence on details for a clear explanation of construction.

17. Place a title and scale below the drawing

18. Place required detail tags to locate related details to the section

The completed stair section should resemble Figure 20-27.

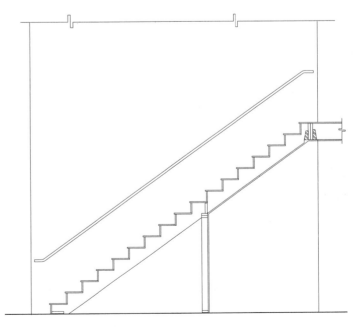

FIGURE 20-26 ■ The layout is completed by adding rails, support walls, and required supports.

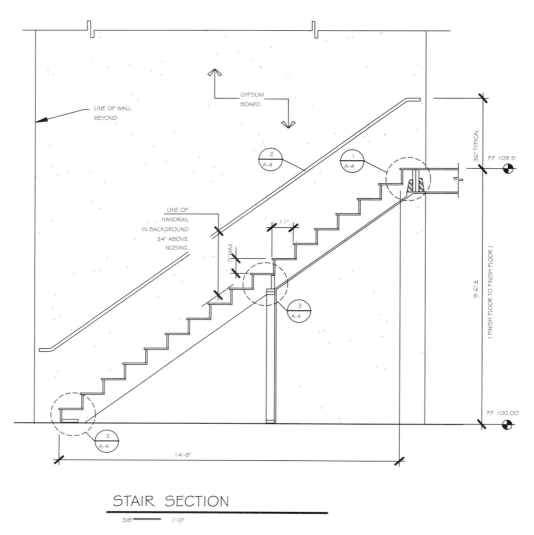

STAIR SECTION

3/8"＝ 1'-0"

FIGURE 20-27 ■ A completed stair section. The absence of notation is allowed because of the use of four details to further explain construction.

Drawing Stair Details

Stair details should be drawn at a scale appropriate to the detail that must be shown. Scales ranging from 3/4" = 1'–0" to 3 "= 1'–0" (1:20 to 1:5) are typically used for stair details. Material typically shown in detail include

■ Stringer-to-upper-floor-system connection

■ Stringer-to-lower-floor-system connection

■ Baluster-to-stringer connection

■ Stringer-to-tread connection

Figure 20-28 shows each of these details for the stair drawn in Figure 20-27. Each is a stock detail edited to meet the specific needs of this stair. Because of the scale used for details, it is important to use varied line weight and appropriate hatch patterns to accent each material used.

ELEVATOR DRAWINGS

The elevator manufacturer provides the bulk of the drawings required to install required elevators. The responsibility of drafters working with the architectural team is to accurately represent the shaft size and location on each plan view based on the design of the architect. Figure 20-29 shows how an elevator can be represented on enlarged floor plans. Sections and details similar to Figure 20-30 also need to be drawn to show vertical heights between floors around the elevator shaft, the depth of the elevator pit, and the height of the elevator loft. The *elevator pit* is the area below the lowest floor required to provide clearance for the elevator and the shaft mechanism. The *elevator loft* is provided at the upper end of the shaft for the same purpose. The drafter needs to work very closely with specifications supplied by the manufacturer to detail required sizes for the elevator. Occasionally, elevations of the car are drawn similar to those in Figure 20-31, so that the interior of the elevator matches the building interior.

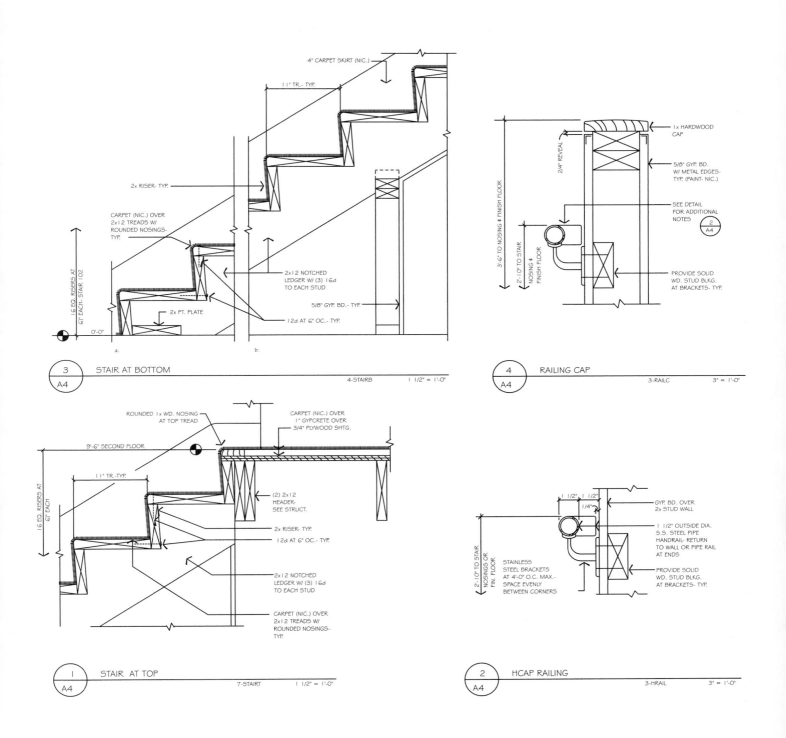

4" CARPET SKIRT (NIC.)

11" TR.- TYP.

2x RISER- TYP.

CARPET (NIC.) OVER
2x12 TREADS W/
ROUNDED NOSINGS-
TYP.

2x12 NOTCHED
LEDGER W/ (3) 16d
TO EACH STUD

5/8" GYP. BD.- TYP.

12d AT 6" OC.- TYP.

2x PT. PLATE

16 EQ. RISERS AT
6" EACH- STAIR 102

0'-0"

a.

b.

③/A4 STAIR AT BOTTOM

4-STAIRB   1 1/2" = 1'-0"

1x HARDWOOD
CAP

5/8" GYP. BD.
W/ METAL EDGES-
TYP. (PAINT- NIC.)

SEE DETAIL
FOR ADDITIONAL
NOTES ②/A4

PROVIDE SOLID
WD. STUD BLKG.
AT BRACKETS- TYP.

2/4" REVEAL

3'-6" TO NOSING ¢ FINISH FLOOR

2'-0" TO STAIR
NOSING ¢
FINISH FLOOR

④/A4 RAILING CAP

3-RAILC   3" = 1'-0"

ROUNDED 1x WD. NOSING
AT TOP TREAD

CARPET (NIC.) OVER
1" GYPCRETE OVER
3/4" PLYWOOD SHTG.

9'-6" SECOND FLOOR

11" TR.-TYP.

(2) 2x12
HEADER-
SEE STRUCT.

2x RISER- TYP.

12d AT 6" OC.- TYP.

2x12 NOTCHED
LEDGER W/ (3) 16d
TO EACH STUD

CARPET (NIC.) OVER
2x12 TREADS W/
ROUNDED NOSINGS-
TYP.

16 EQ. RISERS AT
6" EACH

①/A4 STAIR AT TOP

7-STAIRT   1 1/2" = 1'-0"

GYP. BD. OVER
2x STUD WALL

1 1/2" OUTSIDE DIA.
S.S. STEEL PIPE
HANDRAIL- RETURN
TO WALL OR PIPE RAIL
AT ENDS

PROVIDE SOLID
WD. STUD BLKG.
AT BRACKETS- TYP.

1 1/2" 1 1/2"
1/4"

STAINLESS
STEEL BRACKETS
AT 4'-0" O.C. MAX.-
SPACE EVENLY
BETWEEN CORNERS

2'-10" TO STAIR
NOSINGS OR
FIN. FLOOR

②/A4 HCAP RAILING

3-HRAIL   3" = 1'-0"

FIGURE 20-28 ■ Details stored as wblocks can be used to explain construction of the stair shown in Figure 20-27. *Courtesy Architects Barrentine, Bates & Lee, A.I.A.*

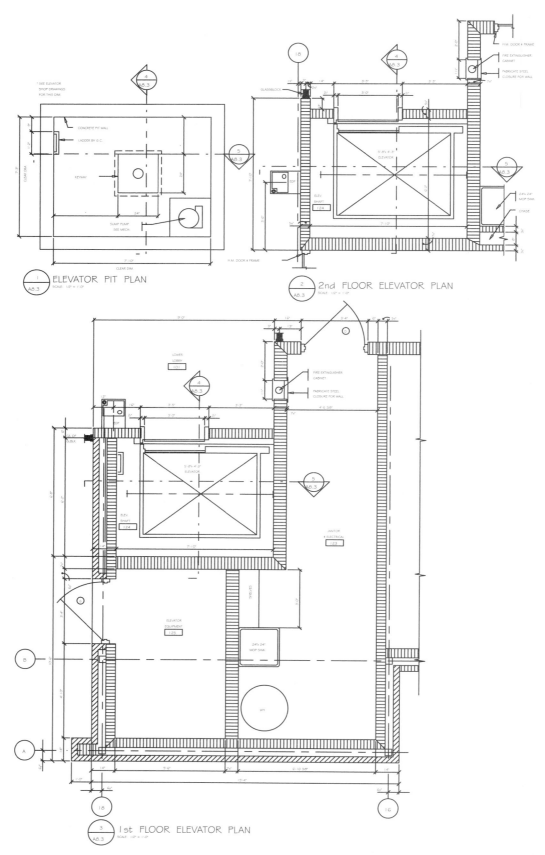

FIGURE 20-29 ■ The shaft for an elevator must be coordinated with the supplier and carefully detailed on each floor. *Courtesy G. Williamson Archer A.I.A., Archer & Archer P.A.*

NOTES:
1. HOISTWAY ENCLOSURES MUST HAVE SUBSTANTIALLY FLUSH SURFACES ON HOISTWAY SIDE EXCEPT ON SIDES WHERE LANDING OCCURS. ANY SETBACKS, PROJECTIONS, OR BEAMS OF MORE THAN 2" MUST BE BEVELED ON TOP SIDE NOT LESS THAN 75° FROM HORIZONTAL.

2. LEAVE OUT FRONT WALL OF HOISTWAY FULL WIDTH BY 7'4" HIGH, WHERE OPENINGS OCCUR OR LEAVE ROUGH OPENING 1'0" WIDER AND 4" HIGHER SO THAT ENTRANCES CAN BE SET IN PROPER RELATIONSHIP TO GUIDE RAILS. ONCE ENTRANCES ARE SET, GENERAL CONTRACTOR IS TO COMPLETE FRONT WALLS AND GROUT FRAMES AND SILLS.

3. ANY SLEEVES, CUT OUTS, CHASES, AND RECESSES REQUIRED BY ELEVATOR CONTRACTOR SHALL BE PROVIDED BY GENERAL CONTRACTOR IN LOCATIONS DESIGNATED ON ELEVATOR LAYOUT DRAWINGS. GENERAL CONTRACTOR TO FILL VOIDS AFTER ELEVATOR EQUIPMENT IS IN PLACE.

4. PROVIDE VINYL COMPOSITION TILE FOR ELEVATOR CAR FLOOR.

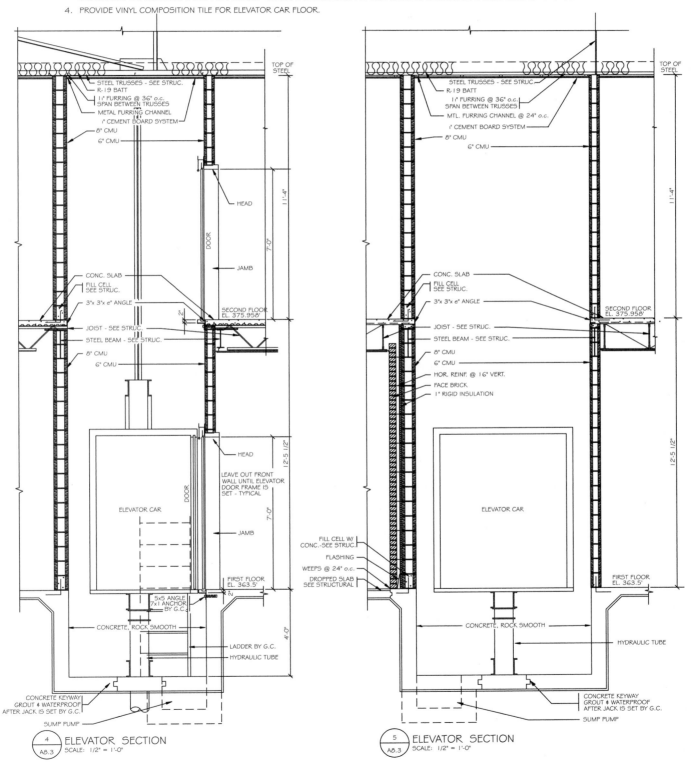

FIGURE 20-30 ■ The sections necessary to supplement the stair shown in Figure 20-29. *Courtesy G. Williamson Archer A.I.A., Archer & Archer P.A.*

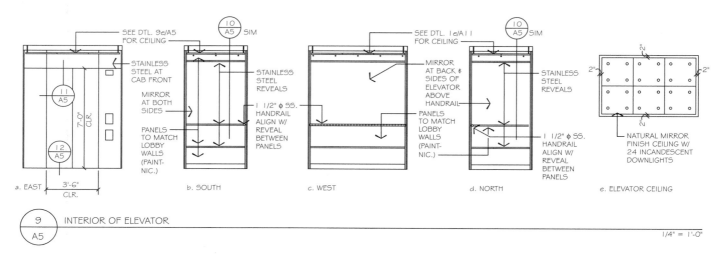

a. EAST   b. SOUTH   c. WEST   d. NORTH   e. ELEVATOR CEILING

9 / A5 INTERIOR OF ELEVATOR   1/4" = 1'-0"

FIGURE 20-31 ■ The architect may control the interior of the elevator design by the use of interior elevations. *Courtesy Architects Barrentine, Bates & Lee, A.I.A.*

ESCALATOR DRAWINGS

The manufacturer of the escalator provides the bulk of the drawings required to install escalators. The responsibility of drafters working with the architectural team is to accurately represent the width, length, and location on each plan view based on the design of the architect.

CHAPTER

20 *Ramp, Stair, and Elevator Drawings*

CHAPTER QUIZ

Use a separate sheet of paper to answer the following questions. Print the chapter title, the question number, and a short complete statement for each answer.

Question 20-1 What restrictions govern stairs and ramps?

Question 20-2 When are ramps required to have a handrail?

Question 20-3 Can a handrail be mounted to a bracket that will extend three inches from the support wall?

Question 20-4 What is the maximum riser height allowed by the IBC for a stairway serving an office? An upper floor of an apartment?

Question 20-5 When is a guardrail required?

Question 20-6 What is the required spacing of the balusters of a guardrail?

Question 20-7 What is the minimum width required for a spiral stair tread?

Question 20-8 Determine the total run and the number of risers and treads required for an office structure with 11'-0" from floor to floor.

Question 20-9 List five materials commonly used for treads.

Question 20-10 Describe two methods of attaching a wood tread to its support.

Question 20-11 Why are the drafter's responsibilities limited when specifying an elevator?

Question 20-12 List details that typically accompany a stair section.

Question 20-13 What elevator drawings is a drafter usually required to work on?

Question 20-14 What information must be specified for an escalator?

Question 20-15 What is the minimum information that should be labeled on a stairway on a floor plan?

DRAWING PROBLEMS

Use the reference material from preceding chapters, local codes, and vendor catalogs to complete one of the following projects. Unless other instructions are given by your instructor, draw the stair sections that correspond to the floor plan that was drawn in Chapter 15. Skeletons for problems 20-4 and 20-5 can be accessed from http://www.delmar.com/resources/ocl.html. Use these drawings as a base to complete the assignment. Use appropriate symbols, linetypes, dimensioning methods, and notations to complete the drawing. Determine any unspecified sizes based on material or practical requirements. Unless specified, select a scale appropriate to plotting on "D" size material and determine LTSCALE and DIMVARS.

Although most of the information required to complete a project can be found by searching other related problems, some of the problems will require you to make a decision on how you would solve a particular problem. Make sketches of possible solutions and submit them to your instructor prior to completing each problem.

■ Place the necessary cutting plane on the floor plan to indicate each drawing.

■ All heights listed can be assumed to be from finish floor to finish floor.

■ Save each drawing as a wblock with a name that accurately reflects the contents.

Problem 20-1 Draw a ramp that will connect two different floor levels in a warehouse that are not open to the public. The ramp will be a total of 30" high. Determine the required length. Assume a 5" thick concrete slab, and thicken the intersection of the top and bottom of the ramp to 10" × 20". Reinforce the slab with (3)-#5ø bars continuous throughout the ramp, 2" up from the bottom of the slab. Use (2)-#5 bars at the top and bottom of the ramp (at the thickened areas) 1 1/2" clear of the footing bottom. Use # 5 @ 24" o.c. for the length of the ramp.

Problem 20-2 Draw a U-shaped stair to be used in an office structure with 10'–0" from floor to floor. Determine the total required run for each flight and the rise and run for each tread. Use steel stringers, rails, and balusters, and a metal pan to support each riser. Use the drawings in this chapter to determine material sizes unless your instructor provides other instructions.

Problem 20-3 Draw a U-shaped stair to be used in an office structure with 12'–0" from floor to floor. Determine the total required run for each flight and rise and run for each tread. Use open wood treads with laminated stringers. Use the examples in this chapter for selecting materials.

Problem 20-4 Draw a section for a stair to be used in a town house unit of an apartment complex. The lower floor will be a concrete slab and the upper floor will be framed with 2 × 12 f.j. with 3/4" plywood and 1 1/2" concrete deck. Assume a total rise of 9'–0". Determine the required number of treads, risers and the total run. Frame the stair using (3)-2× material for the stringers. Select appropriate material for threads and risers.

Problem 20-5 Draw the required stairs for the structure that was started in Chapter 15. If suitable, use one of the blocks created in Problem 1-3 as a base drawing and edit as required. Provide details to show stringer-to-floor, tread-to-stringer, and stringer-to-baluster connections.

SECTION 4

Preparing Structural Drawings

Drawing Framing Plans

In Chapter 2 you were introduced to the drawings that comprise a set of construction drawings. The drawings used to build the frame of a structure are the structural drawings. Understanding the major concepts contained on these drawings and how the drawings are integrated into the entire set of construction drawings are a must for drafting a set of plans.

STRUCTURAL DRAWINGS

The same types of drawings used to present the architectural information can be used to present structural information. These include plan views, elevations, sections, details, schedules, and written specifications. Structural plan views include framing plans of the roof level and each floor level. This chapter discusses each of the framing plans, and subsequent chapters introduce the remaining structural drawings.

The area of the country you are in, the type of building to be erected, and the occupancy of the structure dictate what materials are to be used. Common materials used to form the shell of the structure that are represented on a framing plan include sawn lumber, heavy timber, poured concrete, and concrete block. With the exception of concrete block, each of these materials can be used to frame the roof and floor systems represented on the framing plan. Typically, several of these materials may be incorporated into the framework and reflected in the framing plan.

FRAMING PLANS

Framing plans are drawn at a scale that matches the scale used for other plan views. If drawn using CAD, the base layers for the floor plan are used to provide the basis for a framing plan. Figure 15-45 shows the base layers of the floor plan that is displayed when starting a framing plan. The drawing can be started by making copy of the floor plan prior to adding information specific to the floor plan, to serve as a base for the framing plan. Material can then be attached to the base drawing using the **INSERT** or **XREF** command. If a large complex is being drawn, the plan may be divided into zones and placed on two or more sheets. Zones must always match those used for the architectural drawings.

Framing plans are drawn for each level of the structure. For a one-level structure, the material used to frame the roof system is shown on a plan that resembles the roof plan. Because the framing for the first level floor is shown on the foundation, no framing plan is required. For a multilevel structure, a plan is provided for each level. A three-level structure requires plans for the lower, sec-

ond, and upper levels. Plans are arranged within the drawing set from the ground level to the roof, reflecting how the structure will be built. The lowest level of the structure is represented on the foundation plan, followed by succeeding floor levels. Framing plans are drawn working from the top of the structure down to the ground, so that the location for beams and supports is better understood.

The main goal of the framing plan is to represent the location and size of framing members such as beams, joists, posts, and columns, which resist the stresses applied to the structure. These are the major elements that comprise the skeleton of a structure and transfer the weight of vertical loads to the supporting foundation. Framing plans are also used to show the materials that are used to resist the horizontal stress from forces caused by wind, flooding, and seismic activity. Information regarding nailing, bolting, and welding to resist these stresses is also found on the framing plan. The plans are drawn by drafters from sketches prepared by the engineering team. Figure 21-1 shows a portion of a framing plan.

Using CAD Skills for Framing Plans

The framing plan should be used to specify the locations and the materials used to resist the forces of rotation, uplift, and shear. These features can easily be placed using AutoCAD commands such as **LAYER, ARRAY**, **COPY**, **DIM**, **TEXT**, **WBLOCK**, and **ATTRIBUTE** and features such as DesignCenter.

Layers

The complexity of the structure dictates how the framing plan will be arranged. On a simple structure such as a retail sales outlet, the framing plan could be included with other plan views and separated by different layers. With proper planning, information can be divided into layers to define each plan view contained in the drawing file. If the architectural and structural drawings are kept in the same file, the drafter must carefully separate each layer using names based on the AIA guidelines. Recommended names are presented at the end of the chapter. Prefixes such as S-FRAM can be used to define material on the framing plan, with modifiers such as BEAM, JOIS, and TEXT used to further define materials stored on the framing plan.

Larger structures are also more likely to involve several different firms to complete the plans. This will require each firm to have a copy of the base floor plan file to add their material. A separate framing plan can be created by making a wblock of the walls and

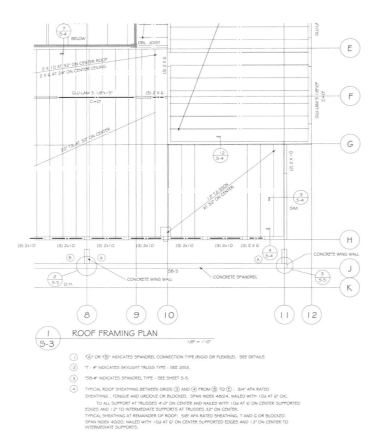

FIGURE 21-1 ■ A plan is drawn of each level of a structure to show all materials needed to construct the skeleton. *Courtesy Van Domelen/Looijenga/McGarrigle/Knauf Consulting Engineers.*

columns of the floor plan. Information specific to the framing plan can now be added to this drawing, and the drawing file is then stored as a new drawing file.

By using the original floor plan walls and column layers as an external reference (**XREF**) for the framing plan, changes can be made to the floor plan, and the framing plan will be updated automatically.

Inserting BLOCKS and Assigning Attributes

Once the method of creating the framing plan has been determined, information specific to the framing plan can be added. Repetitive information should be created as a wblock or block and inserted or referenced into the drawing base. Items such as grid, detail, and section markers should be created as a block and inserted with **ATTRIBUTE** to control page and detail numbers. An alternative to inserting common features is to store them in a template on appropriate layers and thaw them as needed.

Common Features Shown on a Framing Plan

Regardless of the material used, framing plans have many common features. Because of the large amount of information that needs to be placed on the framing plan, it is important that the drafter develop the framing plan in a logical order.

Bearing Walls and Support Columns

Using the walls and columns drawn for the floor plan is the most efficient way to draw the framing plan. This can be done by either using the **WBLOCK** or **XREF** method. All items shown on the floor plan that are not directly related to forming or supporting the structure should be removed from the drawing to be used to create the framing plan. Figure 21-2 shows the base floor plan used to draw the framing plan for an apartment complex. The right unit shows the information required for the floor plan and the left unit the framing plan for the structure.

Locations for Each Beam

Once the drawing base has been prepared, support beams should be drawn. Start with major beams and work to intermediate beams and then to purlins. As beams are labeled, text is typically placed parallel to the member using the methods presented in Section II and Figure 21-2. Beams can also be specified using a schedule as shown in Figure 21-3. Notes to specify columns are placed on an angle to help distinguish the post or column from other building materials. Beams and related material should be placed on layers separate from the base material with a layer name such as S-FRAM-BEAM.

Dimensions

The floor plan contains all of the dimensions needed to locate walls. The framing plan typically contains very few dimensions other than grid dimensions to help the print reader gain an overall sense of the building size without having to go to the architectural drawings. Some offices provide no dimensions on the framing plan except those needed to locate structural material. Figure 21-4 shows an example of dimensions on a framing plan. Methods for dimensioning structural members vary based on the material to be used. These methods are explored later in this chapter. Regardless of the material used, dimensions locating structural material should be referenced to a wall, column, or grid that is already located on the floor plan.

Thought should also be given to other drawings that require dimensioning. The overall and grid dimensions on the floor plan will be the same for the framing plan. The majority of dimensions on the lowest framing level are often the same as the dimensions on the foundation plan. Placing the dimensions on several layers can aid in dimensioning the overall project. A layer such as S-FRAM-DIMN-EXT can be used to contain all exterior dimensions that will be needed on all drawings to be dimensioned. A layer such as S-FRAM-DIMN-INT1 could be used for all dimensions that are specific to the first level.

Local Notes

Notes on the framing plan should be placed using the guidelines presented in Chapter 20 for placing notes on a floor plan. Many items on the framing plans will be specified but not drawn. This includes items such as shear panels, hold down anchors, or metal ties, which can be noted as shown in Figure 21-5. Types of notes

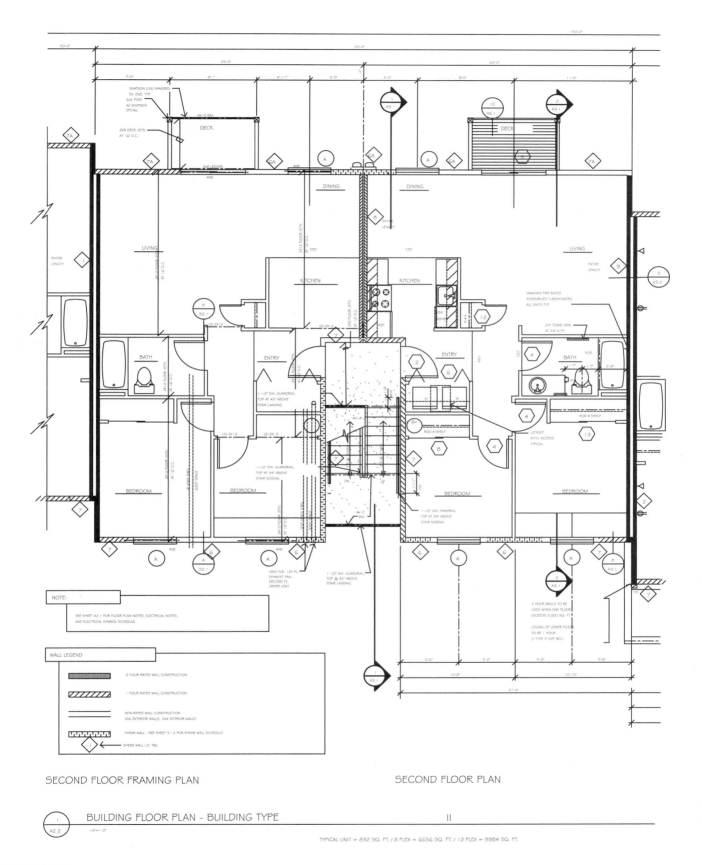

SECOND FLOOR FRAMING PLAN

SECOND FLOOR PLAN

BUILDING FLOOR PLAN - BUILDING TYPE   II

TYPICAL UNIT = 832 SQ. FT. / 8 PLEX = 6656 SQ. FT. / 12 PLEX = 9984 SQ. FT.

FIGURE 21-2 ■ Framing information can be placed in one unit so that the floor plan remains uncluttered. *Courtesy Scott R. Beck, Architect.*

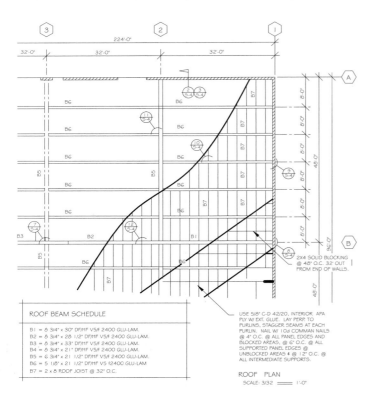

ROOF BEAM SCHEDULE

B1 = 8 3/4" x 30" DF/HF V5/f 2400 GLU-LAM.
B2 = 8 3/4" x 28 1/2" DF/HF V5/f 2400 GLU-LAM.
B3 = 8 3/4" x 33" DF/HF V5/f 2400 GLU-LAM.
B4 = 8 3/4" x 21" DF/HF V5/f 2400 GLU-LAM.
B5 = 6 3/4" x 21 1/2" DF/HF V5/f 2400 GLU-LAM.
B6 = 5 1/8" x 21 1/2" DF/HF V5 f2400 GLU-LAM.
B7 = 2 x 8 ROOF JOIST @ 32" O.C.

USE 5/8" C-D 42/20, INTERIOR, APA
PLY W/ EXT. GLUE. LAY PERP. TO
PURLINS. STAGGER SEAMS AT EACH
PURLIN. NAIL W/ 10d COMMAN NAILS
@ 4" O.C. @ ALL PANEL EDGES AND
BLOCKED AREAS, 6" O.C. @ ALL
SUPPORTED PANEL EDGES @
UNBLOCKED AREAS 4 @ 12" O.C. @
ALL INTERMEDIATE SUPPORTS.

2X4 SOLID BLOCKING
@ 48" O.C. 32' OUT
FROM END OF WALLS.

ROOF PLAN
SCALE: 3/32 = 1'-0"

FIGURE 21-3 ■ The specification for a beam can be written parallel to the beam it describes or placed in a schedule.

specific to each material will be explored later in this chapter. A layer name such as S-FRAM-NOTE can be used to keep framing text separate from other material.

Drawing Tags

In addition to showing major materials, framing plans are used as a reference map to coordinate the structural elevations, sections, and details. Tags relating the elevations to the framing plan resemble those introduced in Chapter 15. Detail markers are placed on the framing plan to show the location of details that relate to the framing plan, just as they were on the floor plan. Tags should be placed on a layer separate from other tags, with a title such as S-FRAM-SYMB. The section and detail tags can be inserted before the detail or section is actually drawn, but the attributes for page and detail number cannot be defined until the entire job is near completion. Detail and section tags numbering is explained in Chapter 2.

General Notes

Lengthy notes that specify materials can be placed as a general note to keep the framing plan uncluttered. When this is done, an abbreviated version of the note should be placed on the drawing, and referenced to the full note in the framing notes. An example of this form of notation would be

Abbreviated note:

1/2" ply roof sheath. See note 1.

General note:

All roof sheathing to be 1/2" APA 32/16 interior grade with exterior glue. Lay face grain perpendicular to joist and stagger all joints. Use 10d common nails @ 6" o.c. @ boundary and edges. Use 10d common nails @ 10" o.c. @ field unless noted.

General notes should be set on layers with prefixes specific to the building level being represented. A layer name such as S-FRAM-TEXT can be used to store general notes.

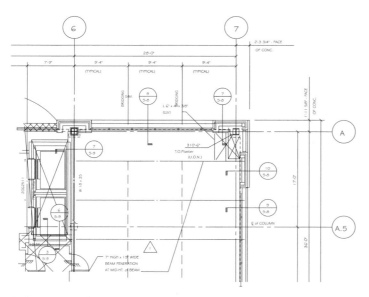

FIGURE 21-4 ■ Dimensions are typically not placed on the framing plan except to describe grid locations and beams or columns that are not shown on the floor plan. *Courtesy Van Domelen/Looijenga/McGarrigle/ Knauf Consulting Engineers.*

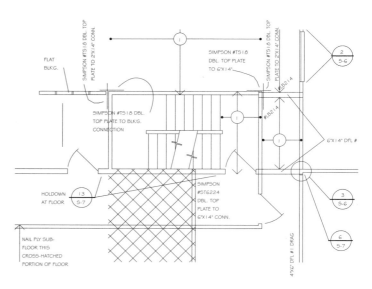

FIGURE 21-5 ■ Notes are required to explain all materials shown on the framing plan. Notes are also used to specify materials such as shear panels, which are not drawn but must be referenced.

Representing Wood and Timber on Framing Plans

Before specifics can be given to describe the drawing process for a framing plan, the drafter must consider the material used to form the roof or floor system. The type of material used influences what is shown on the framing plan. Chapters 6 and 7 introduced the common uses for wood and timber in construction. Major materials represented on a wood framing plan are studs, post, sawn and glu-lam beams, joists, engineered joists, trusses, and plywood. Shear walls, diaphragms, and drag struts (see Web site) must be specified on framing plans with wood members. Figure 21-6 shows an example of a framing plan for a wood framed office structure.

Walls

Stud walls and wood posts appear on the framing plan just as they do on a floor plan. Walls that have special construction such as an extra base or top plate or plywood panels for resisting shear need to be noted and detailed based on the engineer's calculations. Figure 21-6 shows the locations of several different shear walls. Shear walls and metal hangers can be specified by local notes and explained in a detail similar to Figure 21-7.

Beams

Sawn and laminated beams are represented by thin dashed lines as shown in Figure 21-8. Beams are located by dimensions from the edge of an exterior wood or masonry wall and from the center of wood interior walls to the center of the beam. When a beam cantilevers past a supporting post, the end of the beam should be dimensioned from the end to the center of the supporting post. The locations of main support columns and beams should be dimensioned from exterior wall or grid lines. The drafter's job is to provide dimensions that define the locations of each beam and column based on the design of the engineer. Drafters working for the lumber fabricator or truss manufacturer use the information provided on the framing plan to produce drawings that indicate exactly how the prefabricated material will be constructed, as well as precise measurements for how it will be cut.

Joists and Trusses

Joists or trusses can each be represented on framing plans using two techniques. Figure 21-9 shows a symbol that can be used on simple structures to locate the direction and member to be used. The joist symbol can be created as a block with attributes, which can be altered for each application. Many offices show all of the required framing members except where other information cannot be clearly displayed.

Figure 21-10 shows a typical method of representing framing members. Details that are required to show how forces will be transferred from the roof or floor system through walls to other areas of the structure must be referenced on the framing plan. Structures framed with wood typically require details showing beam-to-beam, beam-to-wall, joist-to-beam, and joist-to-wall con-

nections. These can be referenced as shown in Figure 21-10, to represent a cutting plane or a view.

Structural Steel Framing Systems

Steel framing can be either a moment frame as shown in Figure 21-11 or a rigid frame as shown in Figure 21-12. Each system is introduced in Section II. Each system has its own unique material to be shown on a framing plan.

Moment Frames

Figure 21-13 shows an example of a framing plan using structural steel to form the major support system. Common steel materials were first introduced in Chapter 8. The most common shapes to support major loads are typically M-, S- and W-shaped steel beams. Rectangular and circular steel columns as well as W-shapes are used as vertical supports. Steel angles, tees, and channels are often used to support intermediate loads. Steel cables are used to resist lateral, wind, and seismic loads.

Just as with structures framed with wood, framing plans reflecting steel-framed structures are formed using a base drawing that is similar to the floor plan. Drafters prepare the plan using the engineer's sketches and calculations, as well as reference manuals published by AISC such as *Manual of Steel Construction* and *Structural Steel Detailing*. The engineer provides sketches and specifications for the selection and location of materials, as would be provided for wood-framed structures. Drafters working for the steel fabricator produce drawings providing exact measurements of each piece of steel to be fabricated.

Columns and Beams

On the framing plan, steel columns are typically represented by a polyline and dimensioned from center to center of columns. Exterior columns are dimensioned from the face of the exterior shell to the center of the column. Beams may be represented by centerlines, solid lines, or hidden lines and are located using centerline dimensions from one member to the next. Figure 21-14 shows how steel columns and beams can be specified and dimensioned.

Steel framing members are specified using methods similar to those used with wood members. On complicated plans, framing members are labeled using tables to help the print reader understand the location and quantity required for construction. Steel columns should be represented in a table separate from beams to provide better clarity. Beams of different materials should also be kept in separate beam tables. Depending on the complexity of the structure, a separate plan can be provided for vertical supports and horizontal steel to provide clarity. Figure 21-15 shows a piling plan for the hillside office structure shown in Figure 21-6. Figure 21-16 shows the beam plan for the lower floor of the same structure. Steel framing often requires the elevation of a specific member to be shown on the plan. The height above a specific point, or surface, such as a finish floor level, is noted near the beam or listed in the table specifying the member size.

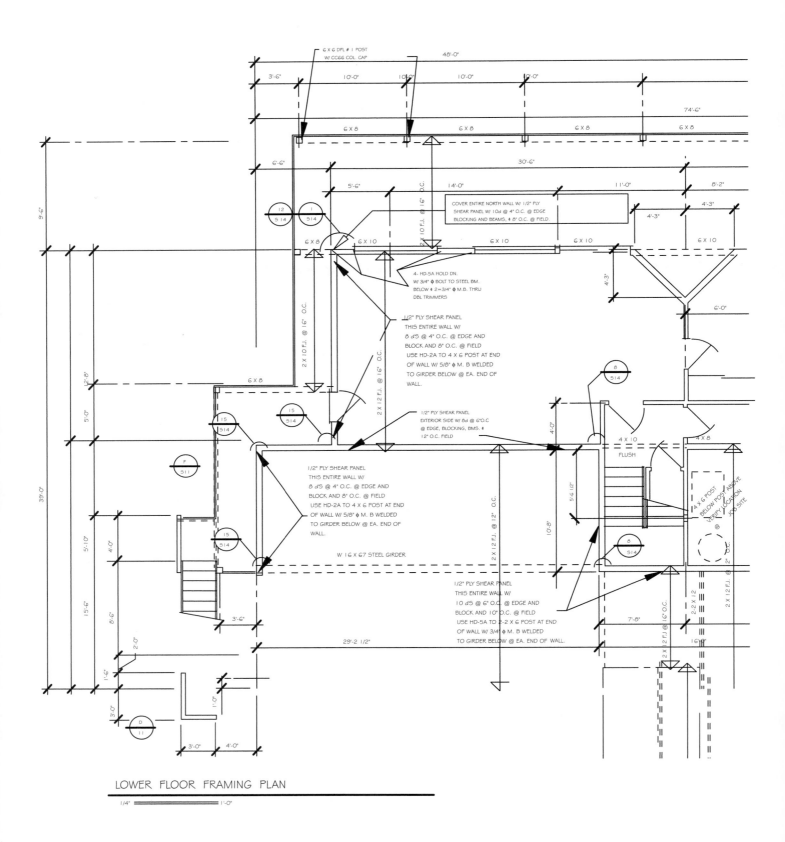

LOWER FLOOR FRAMING PLAN

1/4" = 1'-0"

FIGURE 21-6 ■ The framing plan for a hillside wood-framed office structure. Because the back side of the office is 16' off the ground, walls must be stiffened to resist wracking.

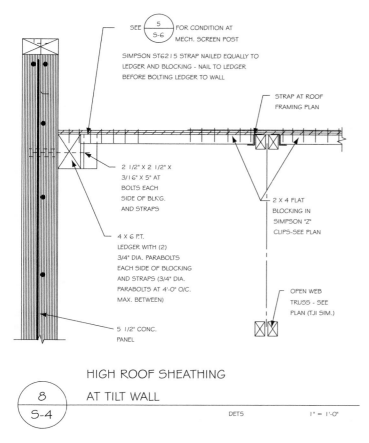

SEE $\frac{5}{S-6}$ FOR CONDITION AT
MECH. SCREEN POST

SIMPSON ST6215 STRAP NAILED EQUALLY TO
LEDGER AND BLOCKING - NAIL TO LEDGER
BEFORE BOLTING LEDGER TO WALL

STRAP AT ROOF
FRAMING PLAN

2 1/2" X 2 1/2" X
3/16" X 5" AT
BOLTS EACH
SIDE OF BLK'G.
AND STRAPS

2 X 4 FLAT
BLOCKING IN
SIMPSON "Z"
CLIPS-SEE PLAN

4 X 6 P.T.
LEDGER WITH (2)
3/4" DIA. PARABOLTS
EACH SIDE OF BLOCKING
AND STRAPS (3/4" DIA.
PARABOLTS AT 4'-0" O/C.
MAX. BETWEEN)

OPEN WEB
TRUSS - SEE
PLAN (TJI SIM.)

5 1/2" CONC.
PANEL

HIGH ROOF SHEATHING

$\frac{8}{S-4}$ AT TILT WALL

DET5   1" = 1'-0"

FIGURE 21-7 ■ Drawing showing that intersections between building systems must be detailed to explain construction methods.

6 3/4" x 45 GLU-LAM BM.

5 1/8" X 28 1/2" GLU-LAM BM.

FIGURE 21-8 ■ A wide variation in presentation methods is used to represent beams. Two common methods include a single line or double lines representing the width of the beam.

22" TJL @ 32" O.C.

FIGURE 21-9 ■ On simple plans, one symbol can be used to represent the span and spacing of repetitive members.

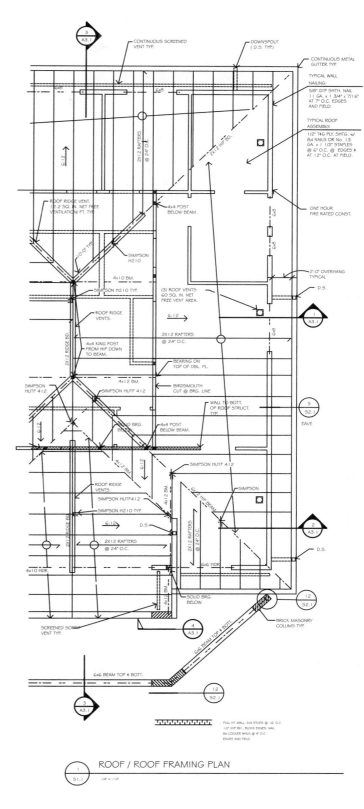

ROOF / ROOF FRAMING PLAN

FIGURE 21-10 ■ Most structural drawings show the majority of repetitive members to avoid confusion during the construction process. *Courtesy Scott R. Beck, Architect.*

FIGURE 21-11 ■ Steel is often used to form the skeleton using connections referred to as a moment frame. *Courtesy Mike Jefferis.*

FIGURE 21-12 ■ Many single-story steel industrial buildings are built using a framing system referred to as rigid frame. *Courtesy Mike Jefferis.*

Decking

Steel decking is often used to form a diaphragm in horizontal surfaces. Poured concrete floors placed over the steel deck can also be used to provide rigidity. Steel cables and turnbuckles, steel rods, or steel tubing are used between major supports to resist sheer and rotational stresses. Figure 21-17 shows an example of how the reinforcing steel is specified on a framing plan.

Rigid Frames

Rigid framing or prefab methods were first introduced in Chapter 8. Members that comprise the frame are typically W-shapes, but S- and M-shapes are also used, depending on the size and spacing of the frame. Horizontal members that span between the frame are primarily channels, but angles and tees are also used. Figure 21-18 shows the framing plan for an industrial building framed with a rigid steel frame. Frame and intermediate members are located based on centerline locations. Support between members is usually developed by the use of steel cables. Framing elevations similar to Figure 21-19 are drawn to show the locations of members used to form the shell supports. Chapter 22 explains how the elevations are drawn. A structure framed using rigid frame methods is drafted using methods similar to those used to draw a steel-framed structure.

Precast Concrete Framing Systems

Precast concrete structures similar to those in Figure 21-20 offer exceptional strength and resistance to seismic stresses as well as a high degree of fire safety. Concrete is also widely used because it can be cast into almost any shape. Concrete structures require drawings to represent the walls, beams, and columns for each specific level. Depending on the complexity of the project, concrete framing plans can be divided into column, beam, wall, and floor and roof plans. Drafters receive the information they need to draw a concrete framing plan from the architectural drawings and the engineer's sketches and calculations. Walls are located to their edges, and columns are located to their center in a method similar to that used in steel structure.

Concrete Tilt-Up Plans

Forming walls with precast concrete panels, similar to those shown in Figure 21-21, is a common method of construction. The tilt-up plan must show the location of each panel as well as the members that are used to form and support floor or roof framing. The panel plan is used to reference the location of each panel and resembles Figure 21-22. Panels are dimensioned from center of joint to center of joint, and from edge to edge. Notice in Figure 21-22 that panel 21 has a length of 19'–11 1/2" from center to center, with a distance of 1/2" required at each end between panels. On simple structures, this information can be placed on the floor or roof plans without the use of a separate plan, with specific information for the construction of each panel shown in a panel elevation similar to Figure 21-23.

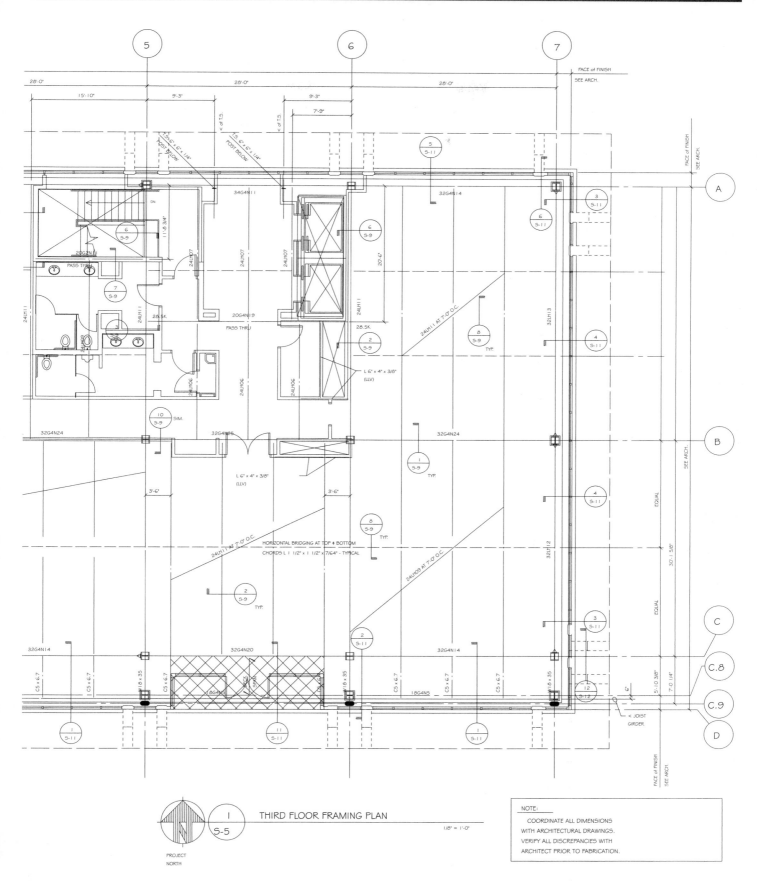

THIRD FLOOR FRAMING PLAN

1/8" = 1'-0"

NOTE:
COORDINATE ALL DIMENSIONS
WITH ARCHITECTURAL DRAWINGS.
VERIFY ALL DISCREPANCIES WITH
ARCHITECT PRIOR TO FABRICATION.

FIGURE 21-13 ■ The framing plan for a steel structure. *Courtesy Van Domelen/Looijenga/McGarrigle/Knauf Consulting Engineers.*

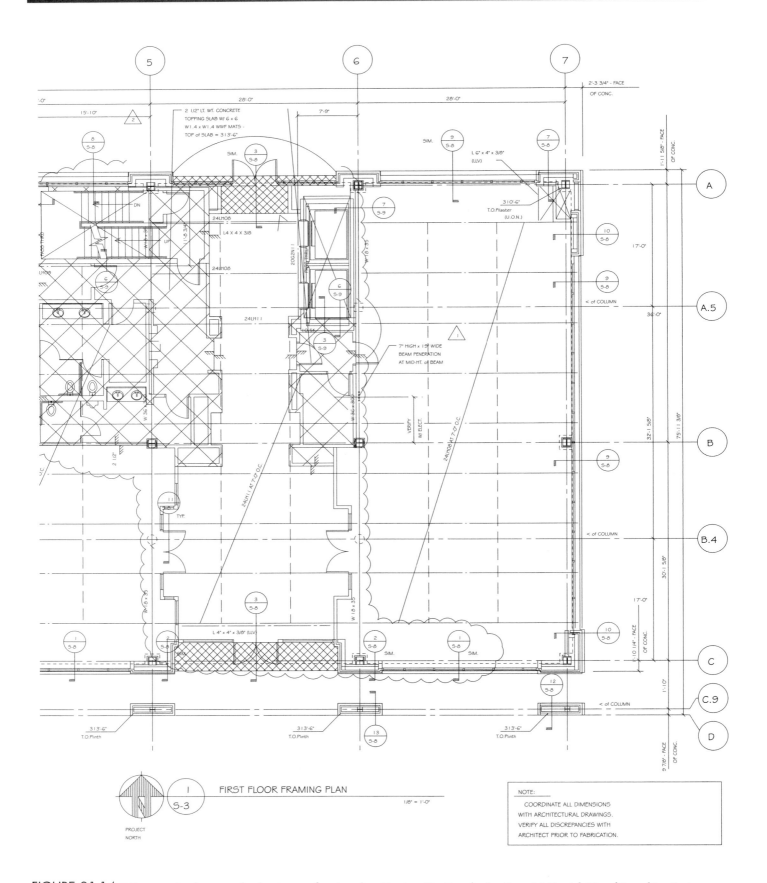

FIGURE 21-14 ■ Representing beams and columns on a framing plan. *Courtesy Van Domelen/Looijenga/McGarrigle/Knauf Consulting Engineers.*

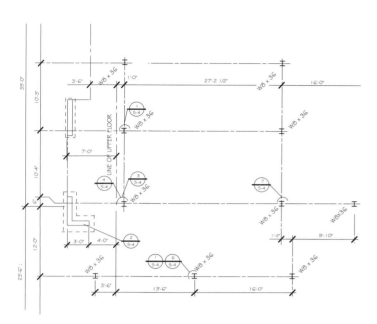

FIGURE 21-15 ■ A portion of the piling plan for the hillside office shown in Figure 21-6.

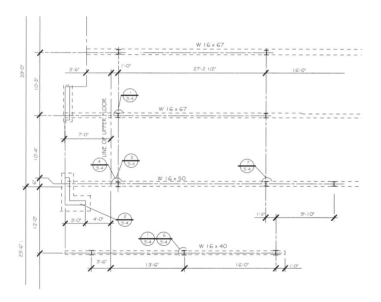

FIGURE 21-16 ■ Because a different crew sets the beams from the crew that drives the pilings, each is placed on a separate plan.

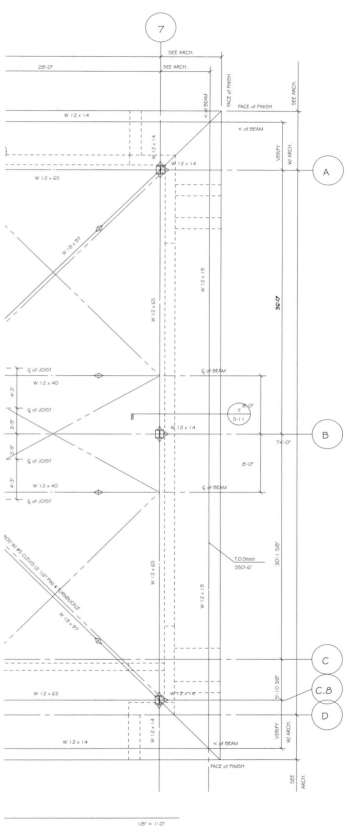

FIGURE 21-17 ■ Representing reinforcing steel on the framing plan.
Courtesy Van Domelen/Looijenga/McGarrigle/Knauf Consulting Engineers.

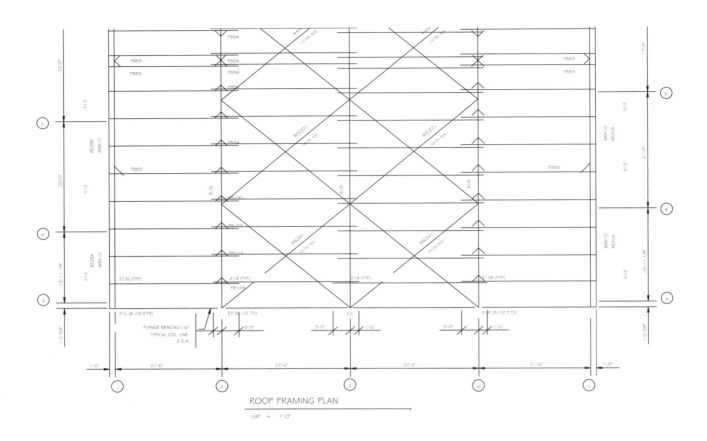

ROOF FRAMING PLAN

1/4" = 1'-0"

FIGURE 21-18 ■ The roof plan for a rigid frame structure.

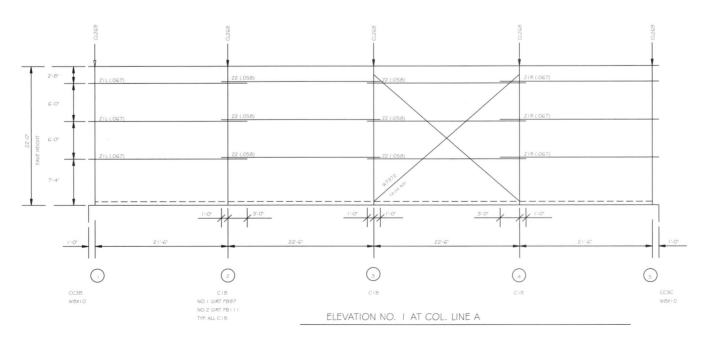

ELEVATION NO. 1 AT COL. LINE A

FIGURE 21-19 ■ The elevation for a steel structure shows where each member of the skeleton will be placed.

FIGURE 21-20 ■ Precast and poured-in-place concrete offers exceptional strength and resistance to seismic forces and a high degree of fire safety. *Courtesy Janice Jefferis.*

FIGURE 21-21 ■ Precast concrete panels are poured on the ground and then lifted into positions once they have cured. Drawings must show the materials used to construct each panel as well as the location of each panel. *Courtesy Janice Jefferis.*

DRAWING A FRAMING PLAN

Rarely is a drafter given the task of drawing a framing plan from its inception. Remember that a drafter is a person who draws the ideas of another in a clear, logical manner. As a new employee in a company, you may be working with a marked-up print, adding information to the framing plan. Experienced drafters with an understanding of how the company is organized usually work from the engineer's calculations to complete the framing plan.

The use of an engineer's calculations is introduced in Chapter 2. Typically the information that the engineer requires to be placed on the plan is highlighted so that the drafter can easily understand the calculations. It is the drafter's responsibility to place everything on the drawings that is highlighted in the calculations.

Roof Framing Plans

The shape of the roof effects how the drawing is started. Framing plans for a steep-pitched roof can be drawn using the roof plan as a base. Roof plans for structures with a low-sloped roof can use the floor plan for the drawing base.

Drawing Low-Sloped Roof Framing Plans

The structure drawn in Figures 15-45 and 17-33 will be used throughout this section.

1. Use the floor plan base drawing to create the roof framing plan. Freeze all material except the outline of exterior walls, interior bearing walls, and openings in each. The base plan should now resemble Figure 21-24.

2. Thaw or draw all grid lines and markers

3. Draw and specify all beams

4. Draw and specify all openings

5. Draw and specify the outline of all roof projections

6. Provide dimensions to locate all beams, openings and overhangs not dimensioned on the floor or roof plan

The roof plan should now resemble Figure 21-25.

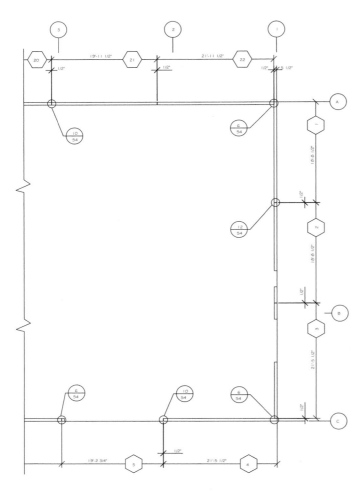

FIGURE 21-22 ■ A panel plan can be used to locate where each panel fits into a structure. Some offices reference elevations to the floor plan. *Courtesy StructureForm Masters, Inc.*

7. Locate and specify all trusses

8. Place all local and general notes specified by the engineer

9. Place all detail and section markers

10. Place a title and scale below the drawing

The completed drawing should now resemble Figure 21-26.

Drawing High-Sloped Roof Framing Plans

The plans for a high-sloped roof can be completed using methods similar to those used for a low-sloped roof. The steel structure shown in Chapter 8 and Figures 17-1 and 21-13 will be used as the example.

1. Use the floor plan base drawing to create the roof framing plan. Freeze all material except the outline of exterior walls, interior bearing walls, and columns. This particular plan includes a dashed line to represent supports indicated on the architectural drawings that will support the overhang. (See Figure 21-27).

2. Thaw or draw all grid lines and markers

3. Draw and specify the outline of all overhangs

4. Draw and specify all changes in roof shape such as ridges, hips, and valleys

The base plan should now resemble Figure 21-28.

5. Draw and specify all primary beams and their elevations

6. Draw and specify all openings

7. Locate and specify all trusses (See Figure 21-29.)

8. Locate and specify all bracing and cross bracing

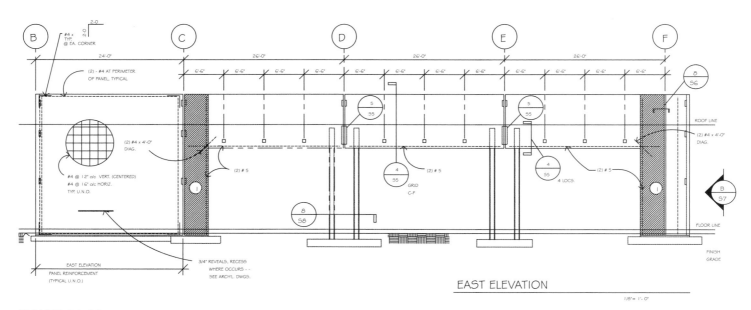

EAST ELEVATION

FIGURE 21-23 ■ Many offices reference panel elevations by name rather than by number. *Courtesy Charles J. Conlee P.E., Conlee Engineers, Inc.*

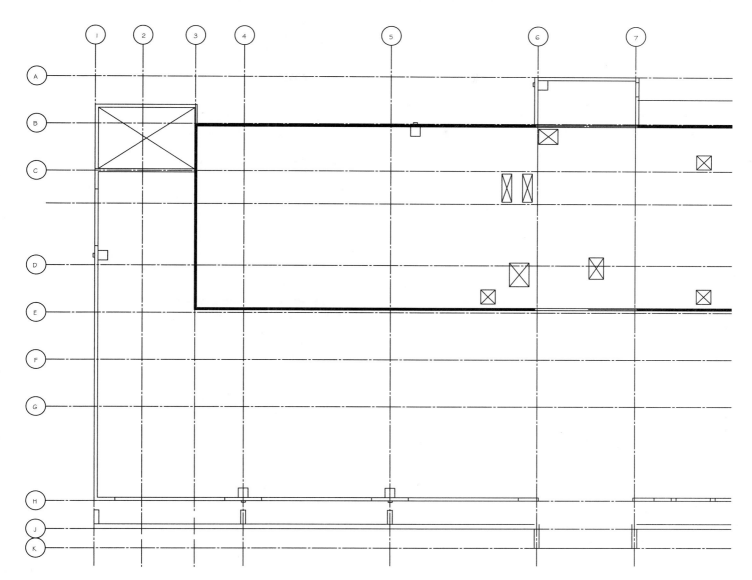

FIGURE 21-24 ■ The roof plan can be started by using the base plan used to develop the floor plan. *Courtesy Architects Barrentine, Bates & Lee, A.I.A.*

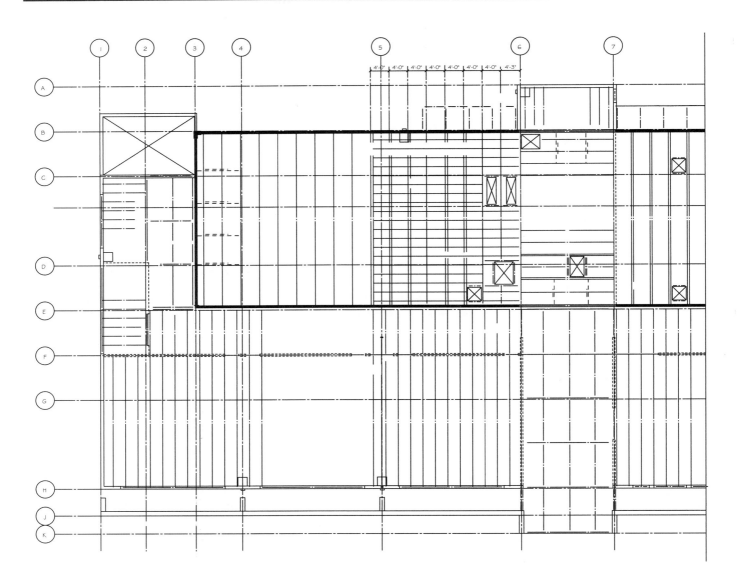

FIGURE 21-25 ■ Using the base plan supplied by the architectural team, the structural team adds beams and trusses. *Courtesy Van Domelen/Looijenga/McGarrigle/Knauf Consulting Engineers.*

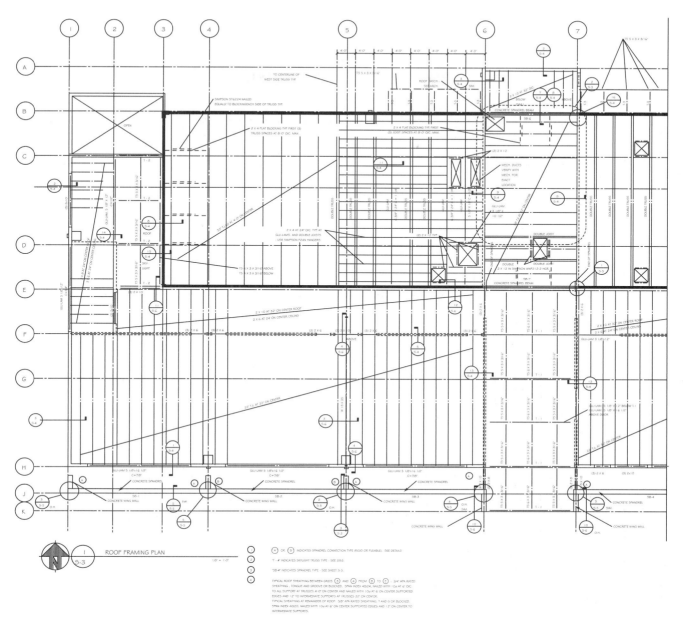

FIGURE 21-26 ■ The completed roof plan with all notes and specifications. *Courtesy Van Domelen/Looijenga/McGarrigle/Knauf Consulting Engineers.*

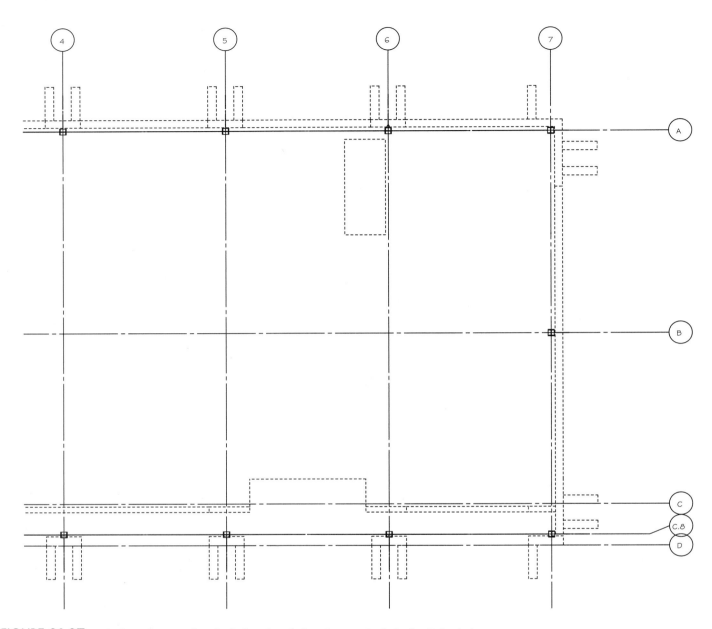

FIGURE 21-27 ■ The base drawings for a high-sloped roof plan. *Courtesy Peck, Smiley, Ettlin Architects.*

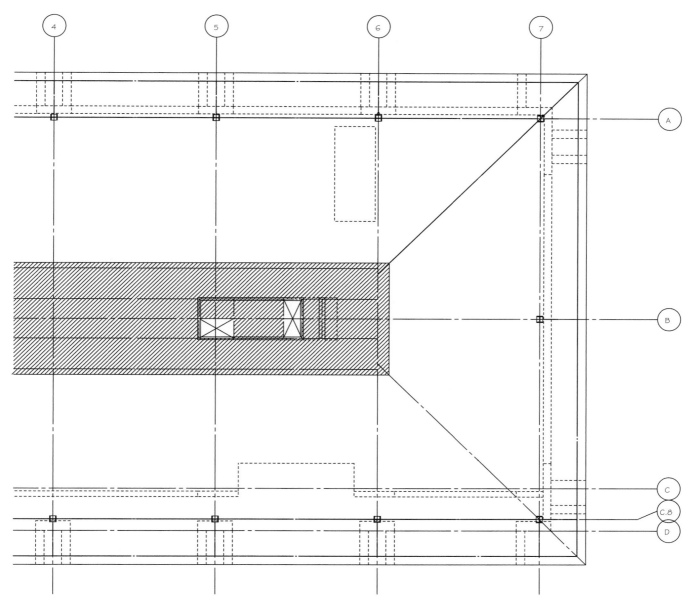

FIGURE 21-28 ■ The layout of major changes in shape of the roof. *Courtesy Van Domelen/Looijenga/McGarrigle/Knauf Consulting Engineers.*

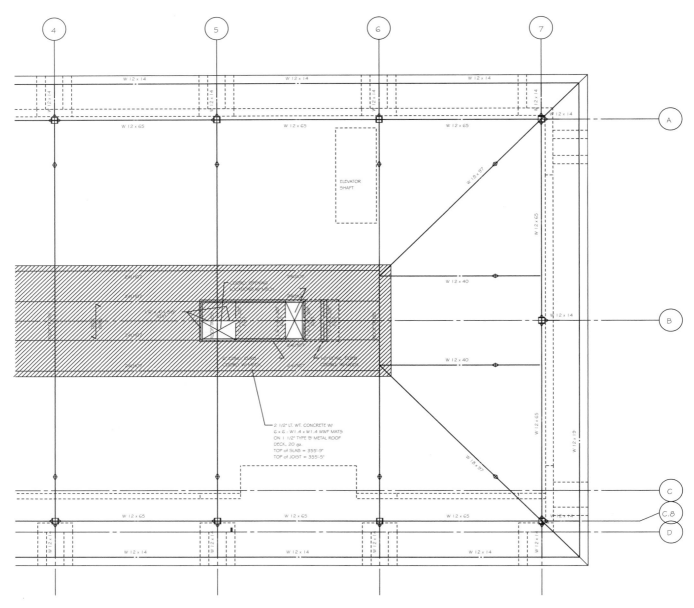

FIGURE 21-29 ■ The specification of major structural supports.

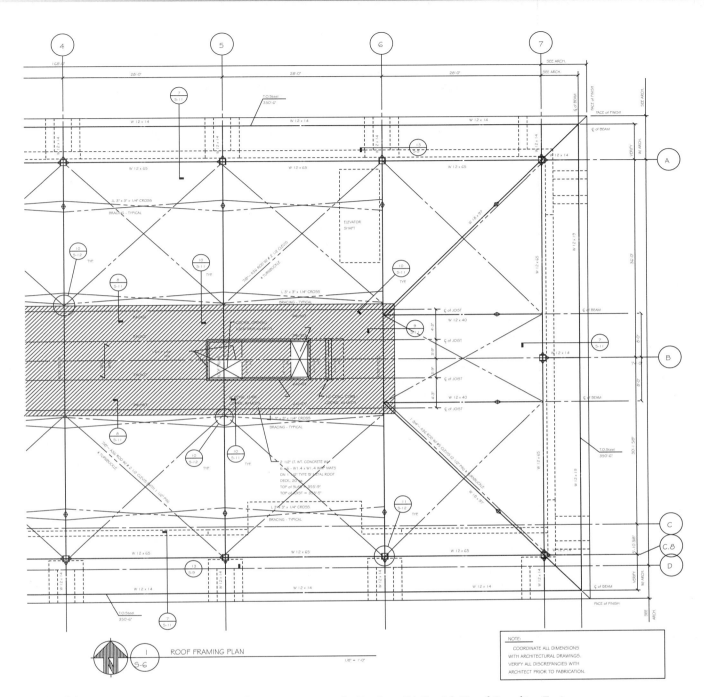

FIGURE 21-30 ■ The completed high-sloped roof. *Courtesy Van Domelen/Looijenga/McGarrigle/Knauf Consulting Engineers.*

9. Provide dimensions to locate all beams, openings and overhangs not dimensioned on the floor or roof plan

10. Place all local and general notes specified by the engineer

11. Place all detail and section markers

12. Place a title and scale below the drawing

The completed drawing should now resemble Figure 21-30.

Floor Framing Plans

A framing plan is started using methods similar to those that were used to draw the roof framing plan. The plan shown in Figure 15-3 will be used as the example.

1. Use the floor plan base drawing to create the framing plan. Freeze all material except the outline of exterior walls, interior bearing walls, and openings in each. Thaw or draw all grid lines and markers.

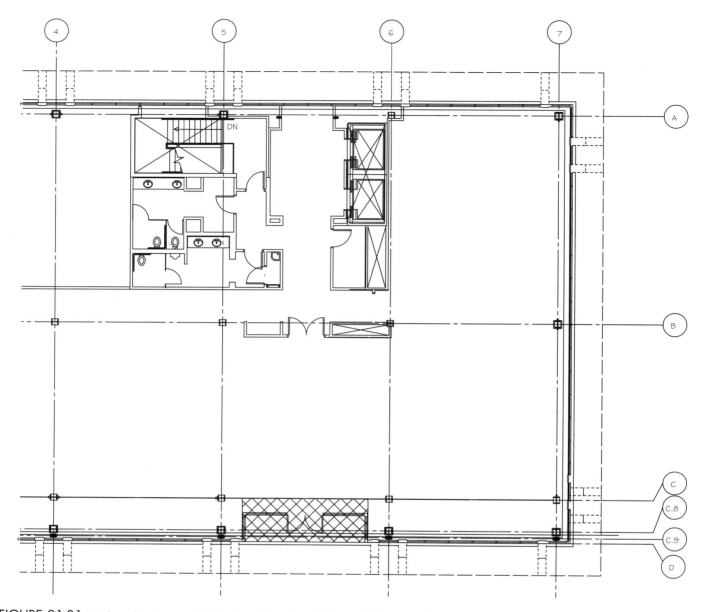

FIGURE 21-31 ■ A base drawing supplied by the architect for the layout of a framing plan. *Courtesy Peck, Smiley, Ettlin Architects.*

The base plan should now resemble Figure 21-31.

1. Draw and specify interior columns

2. Draw the centerline locations for the primary and intermediate beams

3. Draw and specify the outline of all projections of upper floors (Figure 21-32.)

4. Label each beam and provide elevation specifications

5. Dimension the locations of each wall, column, beam, opening, and overhang not dimensioned on the floor plan

6. Place detail markers at each beam-to-beam, beam-to-wall, or column connection

7. Draw and label each joist or truss span for each area of the structure

8. Place detail markers at each joist-to-beam and joist-to-wall connection

9. Locate common section markers

10. Place all local and general notes specified by the engineer

11. Place a title and scale below the drawing.

The completed drawing should now resemble Figure 21-33.

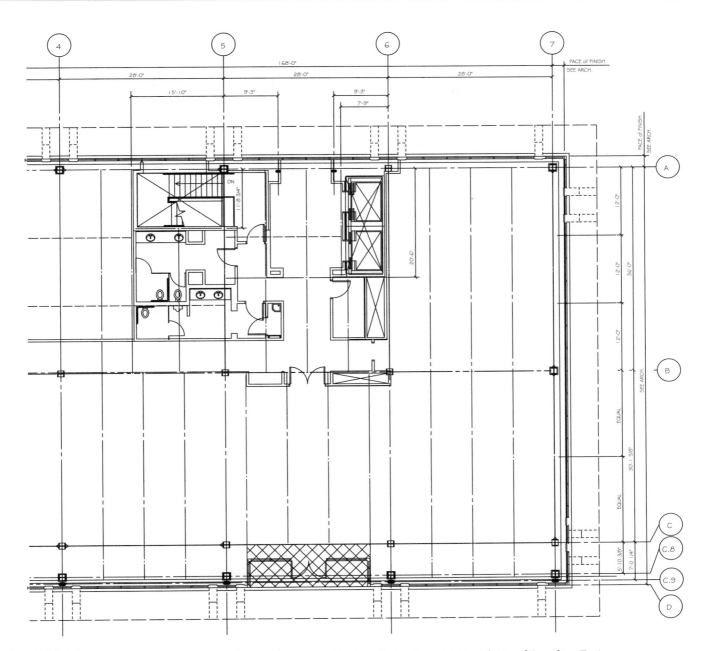

FIGURE 21-32 ■ The layout of major structural materials. *Courtesy Van Domelen/Looijenga/McGarrigle/Knauf Consulting Engineers.*

Drawing a Framing Plan for Concrete

The drawing procedure for a concrete frame is similar to that used for a steel framing plan. Steps for completing the drawing include the following:

1. Use the floor plan base drawing to create the framing plan. Freeze all material except the outline of exterior walls, interior bearing walls, grids, and openings in each. The base plan should resemble Figure 21-34.

2. Draw and specify interior columns

3. Draw the centerline locations for each beam not located at a grid line

4. Draw major primary support beams and then draw intermediate beams

5. Label each beam and provide elevation specifications.

6. Locate panel indicators

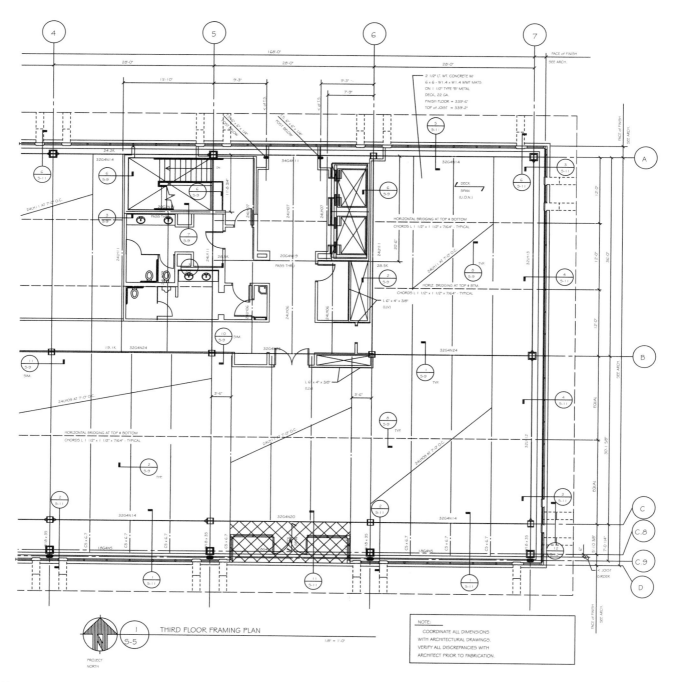

THIRD FLOOR FRAMING PLAN

NOTE:
COORDINATE ALL DIMENSIONS
WITH ARCHITECTURAL DRAWINGS.
VERIFY ALL DISCREPANCIES WITH
ARCHITECT PRIOR TO FABRICATION.

PROJECT NORTH

FIGURE 21-33 ■ The completed framing plan. *Courtesy Van Domelen/Looijenga/McGarrigle/Knauf Consulting Engineers.*

7. Place detail markers at each beam-to-beam and beam-to-wall or column connection

8. Dimension the locations of each wall, column, and beam

9. Locate elevation and section markers

10. Provide all general notes and grid designations

11. Place a title and scale below the drawing

The completed drawing should now resemble Figure 21-35.

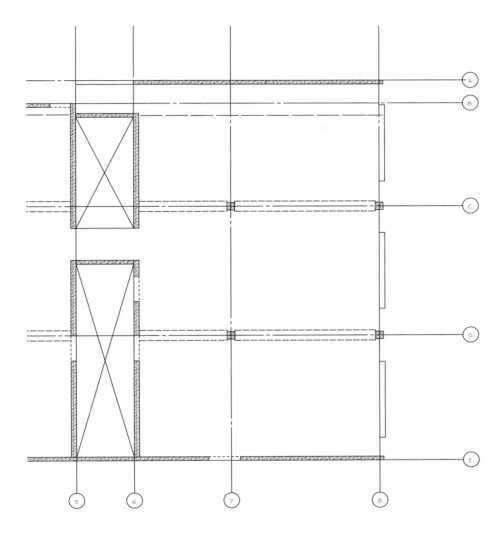

FIGURE 21-34 ■ The base floor plan for a precast concrete structure. *Courtesy H.D.N. Architects A.I.A.*

LAYER GUIDELINES FOR FRAMING PLANS

S-ABLT	Anchor bolts
S-ANNO	Framing text
S-ANNO-NOTE	General notes
S-ANNO-LEGN	Legends and schedules
S-BEAM	Beams
S-COLS	Column
S-DECK	Structural floor or roof decking
S-ELEV	Elevations
S-GRID	Column grids

S-GRID-DIMS	Column grid dimensions
S-GRID-EXTR	Exterior column grids
S-GRID-IDEN	Column grid tags
S-GRID-INTR	Interior column grids
S-IDEN	Identification tags
S-JOIS	Joists
S-SYMB	Symbols
S-PATT	Hatch patterns
S-WALL	Bearing walls

Status field modifiers similar to other drawing areas should be used throughout structural drawings. See Chapter 3 and the AIA Layer Guidelines for additional layer names.

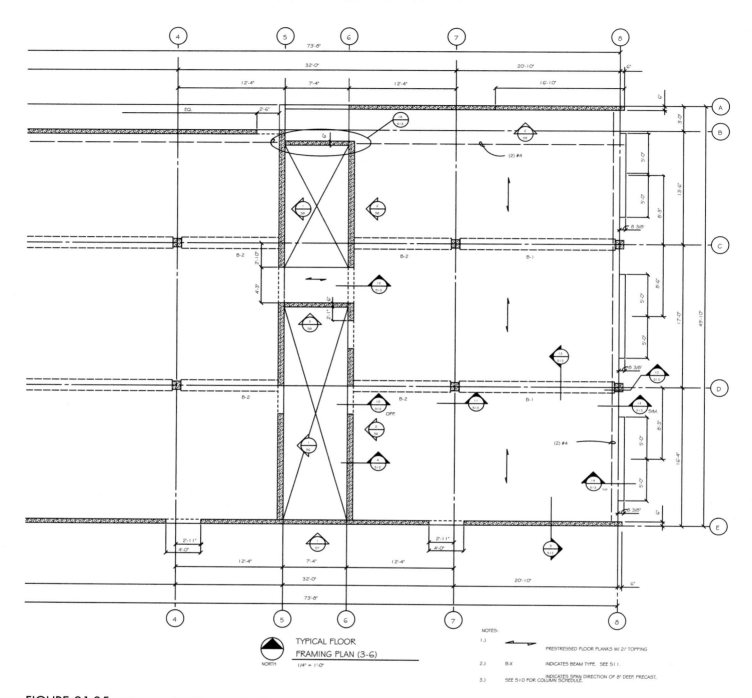

TYPICAL FLOOR
FRAMING PLAN (3-6)
NORTH   1/4" = 1'-0"

NOTES:
1.) ⟷ PRESTRESSED FLOOR PLANKS W/ 2½" TOPPING

2.) B-X   INDICATES BEAM TYPE. SEE S11.

3.) SEE S10 FOR COLUMN SCHEDULE.
     INDICATES SPAN DIRECTION OF 8" DEEP, PRECAST,

FIGURE 21-35 ■ The completed framing plan for a precast concrete structure. *Courtesy KPFF Consulting Engineers.*

CHAPTER 21

Drawing Framing Plans

CHAPTER QUIZ

Answer the following questions on a separate sheet of paper. Print the chapter title, the question number, and a short complete statement for each question.

Question 21-1 List the common scales used to draw a framing plan.

Question 21-2 Describe the major differences between a steel frame and a steel rigid frame structure.

Question 21-3 What is the main goal of a framing plan?

Question 21-4 What is a diaphragm, and how does it affect a framing plan?

Question 21-5 What type of details will typically be referenced on a framing plan for a heavy timber structure?

Question 21-6 What are two common uses for tables on a framing plan for a steel framed structure?

Question 21-7 How is the elevation of a steel beam typically referenced on a framing plan?

Question 21-8 How are precast concrete panels typically attached to the foundation?

Question 21-9 A concrete component is listed as a 155. What type of structural member would it be?

Question 21-10 List possible layer titles that could be used to divide information on a framing plan.

DRAWING PROBLEMS

Use the reference material from preceding chapters, local codes, and vendor catalogs to complete one of the following projects. Unless other instructions are given by your instructor, draw the framing plan that corresponds to the floor plan that was drawn in Chapter 15. Skeletons of the plans and details can be accessed from http://www.delmar.com/resources/ocl.html. Use these drawings as a base to complete the assignment.

■ Use appropriate symbols, linetypes, dimensioning methods, and notations to complete the drawing

■ Create the needed layers to keep major groups of information separated

■ Provide complete dimensions to locate all walls, columns, and beams not represented on the floor plan

■ Determine any unspecified sizes based on material or practical requirements

■ Place a detail reference bubble on the framing plan to represent only the details that you or your team will draw

■ Unless specified, select a scale appropriate to plotting on "D" size material and determine LTSCALE and DIMVARS

Note: *The sketches in this chapter are to be used as a guide only.* You will find that some portions of the drawings do not match things that have been drawn on other portions of the project. Each project has errors that will need to be solved. If you think you have found an error, do not make changes to the drawings until you have discussed the problem and possible solutions with your engineer (your instructor).

Information is provided on drawings in Chapters 16 through 23 relating to each project that might be needed to make decisions regarding the framing plan. You will act as the project manager and will be required to make decisions about how to complete the project. Any information not provided must be researched and determined by you unless your instructor (the project architect and engineer) provides other instructions. When conflicting information is found, information from preceding chapters should take precedence. If conflicting information is found, use the order of precedence introduced in chapter 14.

Unless noted, all shear panels to be 1/2" plywood with 8d nails @ 4" o.c. @ edge and 8d nails @ 8" o.c. in field. Use (2)-2× studs or 4× post at each end of shear panel and an appropriate Simpson Co. strap or tie to the lower floor or foundation.

Problem 21-1 Draw the framing plan for each level of the condominium started in Problem 15-1 One unit should contain all framing information for the entire floor level.

■ Frame the upper level with standard roof trusses

■ Use combination standard/scissor trusses to form a vaulted ceiling over the master bedroom

- Select and specify appropriate solid-web truss joists to span the width of the unit for each floor system
- Use (2)-2 × 12 hdrs. for all openings unless noted
- Use 2 × 8 joists to frame each deck and support with a 5 1/8" × 13 1/2" glu-lam beam at the outer edge of each deck. Support each bm. on a 2 × 6 wall btwn. units.
- Use a 5 1/8" × 13 1/2" glu-lam for the hdr. over the garage door

Use the framing elevations from Chapter 22 (page {XREF 583 in previous edition}) as a guide and specify all shear walls and metal straps on each framing plan as well as the elevations and foundation plan.

- The front and rear walls are to be full height shear walls for all units
- The wall between the kitchen/bathroom on the main floor is to also be a shear wall
- Place a 5 1/8" × 10 1/2" glu-lam beam in the garage ceil. below the shear wall for all units

Problem 21-2 Use the attached drawings to complete the roof framing plan for the retail sales outlet started in Problem 15-2.

- Use 8 3/4" × 28 1/2" glu-lam beams for the center span with a splice over the steel column

- Select appropriate GLB steel beam seats for the beam at each pilaster and detail each connection
- Use a 3 × 4 × 3/8" steel plate with an MST27 strap at the top of ea. bm. Attach the MST straps as per the attached sketch 21-2-1.
- Use the attached sketch 21-2-2 as a guide to detail the support for the beam to steel column

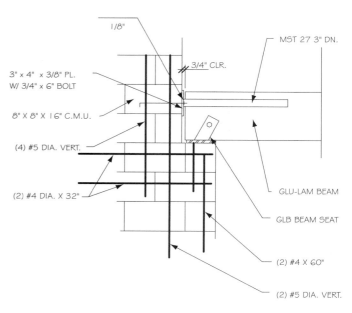

PROBLEM 21-2-1

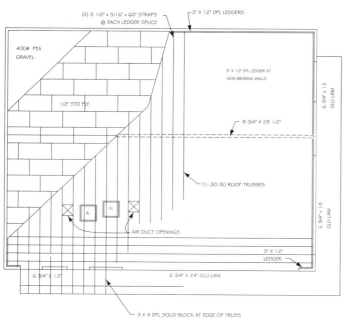

PROBLEM 21-2

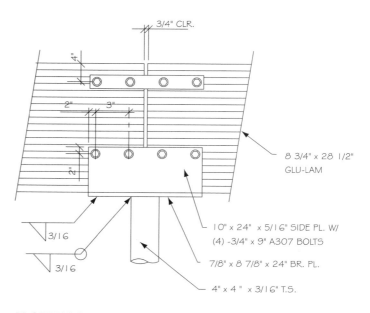

PROBLEM 21-2-2

■ Select, specify, and detail suitable GLB seats for each of the following beams and attach the top of the beam to the wall similar to ridge beam

■ Use 6 3/4" × 12" beams over the two smaller windows on the south and northeast side, and a 6 3/4" × 24" beam over the opening on the south end of the east side

■ Use a 6 3/4" × 18" beam over the windows on the south end of east wall

Specify each ledger on the framing plan as per specifications given for Problems 18-2b-1 through 18-2b-9. Use vendor catalogs to determine open-web trusses for each span. Provide 3 × 4 solid blocking between each truss at each bearing point and attach blocks to ledger w/ A34 each end to ledger. Use H3 hurricane ties to each side of each truss. Use 1/2" APA group 1 interior ply with exterior glue roof sheathing.

Problem 21-3a Use the attached sketch to draw the roof framing plan for the furniture store started in Problem 15-3.

Roof Beams: All beams to be DF/HF, V-5, Fb 2400 unless noted. Listed spans refer to distance from the column to the indicate cantilever.

Beam 1: 6'–4" into span 2-3. Use 6 3/4" × 34 1/2"

Beam 2: Use 6 3/4" × 27"

Beam 3: 6'–4" into grid 3. Same as beam 1

Beam 4: Same as Beam 2

Beam 5: +8' Same as Beam 1

Beam 6: Grid B+ 30'. Use 6 3/4" × 31 1/2"

Beam 7: Use TJI60/30 truss @ 8'–0" o.c.

Beam 8: Use 2 × 6 purlins @ 24" o.c.

Problem 21-3b Use the attached drawings to detail the required roof connections. Use MST36 straps to tie trusses to wall and MST27 straps to tie purlins to ledger. Use 8'–0" spacing for each. Choose a scale suitable for clearly showing the required information. Show skylights referenced in Chapter 15. Submit preliminary drawings to your instructor prior to completing the drawings. Assume all welds to be 3/16" fillet unless noted.

Problem 21-3b-1 Beams to wall. (Place lower portion of this detail with the balance of the mezzanine details.) Support upper bm. on 16" × 16" pilaster, and lower bm. on 16" × 24" pilaster. Detail upper portion of pilaster to match problem 18-2b-8. Detail the lower portion of the pilaster with (10)-#5 vertical bars, with #3 horizontal ties @ 16" o.c. Provide 3/4" clr. for ea. bm. to block wall. Provide fire-cut to glu-lams and 6x sld. blk. to wall. Connect to wall with MST 27 ea. side of bm, 3"

dn. Weld straps to 4 × 6 7/8 × 3/8" steel pl. w/ 1/8" fillet weld. Bolt plate to wall w/ 1/2"Ø × 6" stud, and weld stud to pl. w/ 1/8" fillet weld, all around. Provide (4)-5Ø bond beam behind plate & where wall changes width. Use Simpson Co. GLB beam seat. Bolt as per manuf. recommendations. Tie bolts to (3)-#5 horiz. ties @ 8" o.c.

Problem 21-3b-2 Beam to beam w/ saddle. See framing plan for range of beam sizes. Support bm. W/ 6 7/8 × 7 1/4 × 7/8" bearing plates @ top and btm. Dap top pl. into bm. Provide 5/16 × 3" h side pl. ea. side at top and btm. W/ (3)-3/4"Ø bolts @ 3" o.c. and 1 1/2" from strap end. Set bolts 4 1/2" u/d, with first bolt 8" from bm. splice. Attach straps to side pl. w/ 5/16" weld.

Problem 21-3b-3 Beam to beam @ fire wall. Frame wall w/ 2 × 6 studs @ 16"o.c. with sld. Blk. @ 10' max. o.c. Provide 3 × 6 top pl. w/ 2 × 6 × 1 dp. counter bore for 5/8"Ø × 6" bolts. with 2"Ø wash. thru (2)-2 × 6 top pl. Provide 3 × 4 cont. nailer ea. side of top pl, w/ mtl. Hgrs. for blk. Provide 2 × 4 sld. blk. @ 24" o.c. btwn. trusses, 60" out from wall min. and cover with 5/8" Type X gyp. bd.

Support lam. bms. on ea. side of wall w/ appropriate CCO connector to 6 × 6 T.S. col. Center ea. col. 12" from center of stud wall. Provide 3/4" clr. of gyp. Bd. Use 3/8" fillet for CCO to cols. Provide an MST48 strap, 6" dn. from top of bms. centered on wall.

Problem 21-3b-4 Beam to column. Support ea. glu-lam on 6 7/8 × 20 × 7/8 bearing plate welded to T.S. col. W/ 3/8" weld. Provide (2)-9 × 20 × 5/16" side pl. w/ (2)-bolts to ea. bm. Place bolts 2" dn. 2" in from upper edge of side plate and 3" o.c. Provide std. washers.

Problem 21-3b-5 I-joist to beam. Select a suitable HIT joist hanger and specify the required connection for TJI 60-30 trusses to each laminated beam. Provide an MST48 @ 8'-0" o.c. from truss/truss over beam top.

Problem 21-3b-6 I-joist / ledger—perpendicular. Provide 3 × 12 d.f.p.t. ledger w/ 3/4"Ø bolts @ 32" o.c. staggered 3" u/dn. Select suitable hangers for the rafters and purlins (HIT w/ web stiffner). Reinforce wall w/ 8 × 16 bond beam @ ledger. Provide 3 × 4h × 5/8" plate w/ 3/4"Ø × 4 1/8" stud, attach w/ 1/8" fillet weld. Weld MST36 to plate w/ 1/8" fillet weld @ 8'-0" O.C

Problem 21-3b-7 Purlin/ledger–parallel. Provide 3 × 12 d.f.p.t. ledger w/ 3/4"Ø bolts @ 60" o.c. staggered 3" u/dn. Select suitable U hangers for the rafters. Reinforce wall @ ledger w/ 8 × 16 bond beam. Provide 3 × 4h × 5/8" plate w/ 3/4Ø × 4 1/8" stud, attach w/ 1/8" fillet weld. Weld MST27 to plate w/ 1/8" fillet weld @ 8'-0" o.c.

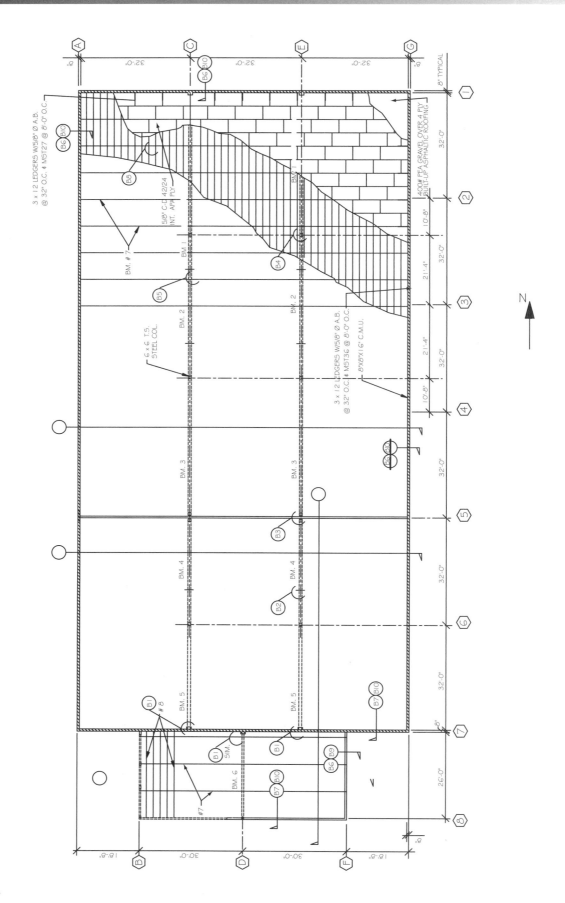

PROBLEM 21-3A

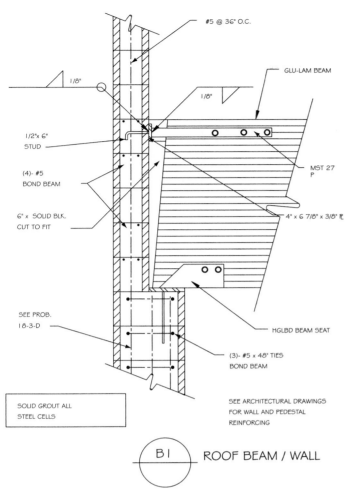

#5 @ 36" O.C.

1/8"

GLU-LAM BEAM

1/8"

1/2"x 6"
STUD

(4)- #5
BOND BEAM

6" x SOLID BLK.
CUT TO FIT

MST 27
P

4" x 6 7/8" x 3/8" ℞

SEE PROB.
18-3-D

HGLBD BEAM SEAT

(3)- #5 x 48" TIES
BOND BEAM

SOLID GROUT ALL
STEEL CELLS

SEE ARCHITECTURAL DRAWINGS
FOR WALL AND PEDESTAL
REINFORCING

B1   ROOF BEAM / WALL

PROBLEM 21-3B-1

16" x 16" PEDESTAL
UP TO ROOF BEAM

1/8"

(3)- #5 Ø @ 18' O.C.

(4)- #4 Ø x 36"
BOND BEAM

GLU-LAM BEAM

1/8"

(2)- 1/2"~ x 5"
STL. ROD
@ 8" O.C.

MST 27, W/ (3) 1/2"
BOLTS EA. SIDE

6" x SOLID BLK.
CUT TO FIT

4" x 6 7/8" x 3/8" ℞

SOLID GROUT ALL
STEEL CELLS

HGLBD W/
2X 3/4" Ø BOLTS

(3)- # 5 TIES

SEE ARCHITECTURAL DRAWINGS
FOR WALL AND PEDESTAL
REINFORCING

B1-a   BEAM / WALL

PROBLEM 21-3B-1A

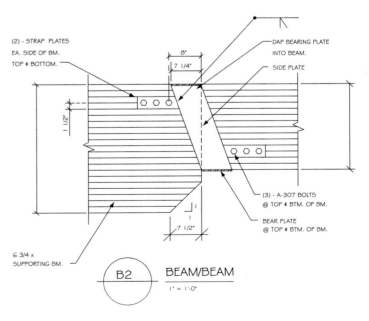

(2) - STRAP PLATES
EA. SIDE OF BM.
TOP & BOTTOM.

8"

7 1/4"

DAP BEARING PLATE
INTO BEAM.

SIDE PLATE

1 1/2"

(3) - A-307 BOLTS
@ TOP & BTM. OF BM.

BEAR PLATE
@ TOP & BTM. OF BM.

6 3/4 x
SUPPORTING BM.

7 1/2"

B2   BEAM/BEAM
1" = 1'-0"

PROBLEM 21-3B-2

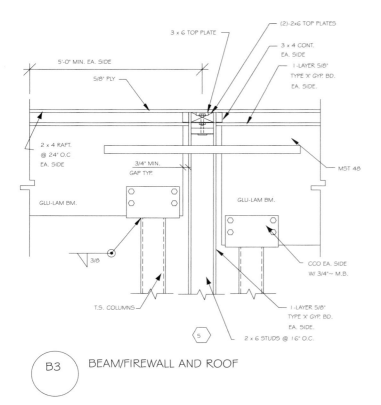

3 x 6 TOP PLATE
(2)-2x6 TOP PLATES
3 x 4 CONT. EA. SIDE
1-LAYER 5/8" TYPE 'X' GYP. BD. EA. SIDE.
5'-0" MIN. EA. SIDE
5/8" PLY
2 x 4 RAFT. @ 24" O.C. EA. SIDE
3/4" MIN. GAP TYP.
MST 48
GLU-LAM BM.
GLU-LAM BM.
3/8
T.S. COLUMNS
CCO EA. SIDE W/ 3/4"~ M.B.
1-LAYER 5/8" TYPE 'X' GYP. BD. EA. SIDE.
5
2 x 6 STUDS @ 16" O.C.

B3   BEAM/FIREWALL AND ROOF

PROBLEM 21-3B-3

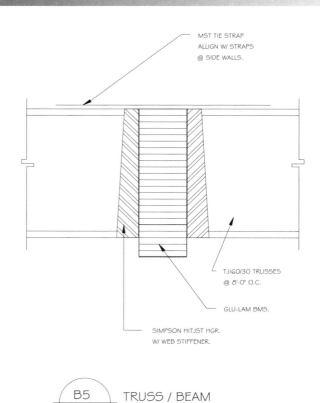

MST TIE STRAP ALLIGN W/ STRAPS @ SIDE WALLS.

TJI60/30 TRUSSES @ 8'-0" O.C.
GLU-LAM BMS.
SIMPSON HITJST HGR. W/ WEB STIFFENER.

B5   TRUSS / BEAM
3/4" ===== 1'-0"

PROBLEM 21-3B-5

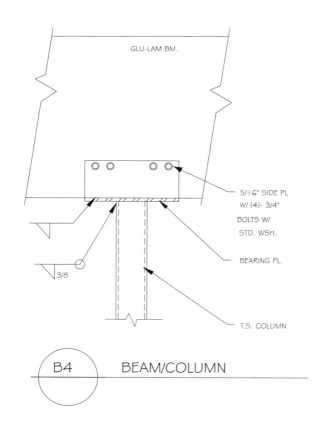

GLU-LAM BM.
5/16" SIDE PL W/ (4)- 3/4" BOLTS W/ STD. WSH.
BEARING PL.
3/8
T.S. COLUMN

B4   BEAM/COLUMN

PROBLEM 21-3B-4

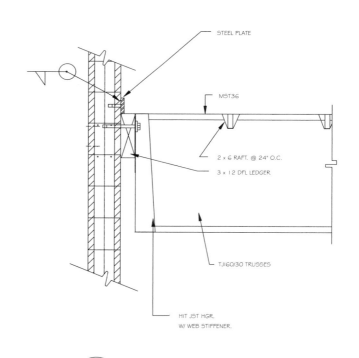

STEEL PLATE
MST36
2 x 6 RAFT. @ 24" O.C.
3 x 12 DFL LEDGER
TJI60/30 TRUSSES
HIT JST HGR. W/ WEB STIFFENER.

B6   TRUSS/WALL @ 'A & G'

PROBLEM 21-3B-6

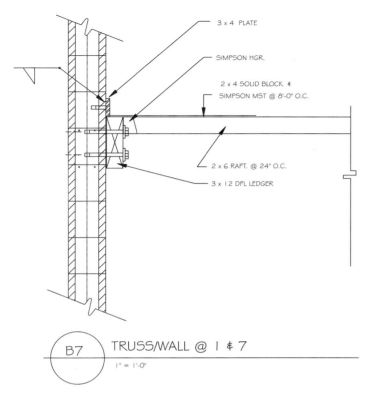

3 x 4 PLATE

SIMPSON HGR.

2 x 4 SOLID BLOCK &
SIMPSON MST @ 8'-0" O.C.

2 x 6 RAFT. @ 24" O.C.

3 x 12 DFL LEDGER

B7 TRUSS/WALL @ 1 & 7
1" = 1'-0"

PROBLEM 21-3B-7

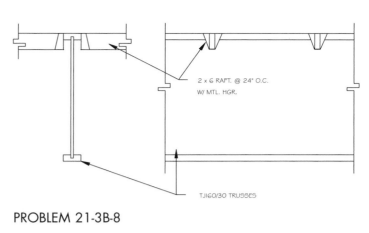

2 x 6 RAFT. @ 24" O.C.
W/ MTL. HGR.

TJI60/30 TRUSSES

PROBLEM 21-3B-8

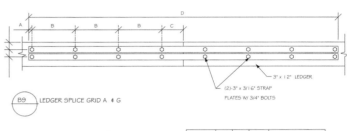

3" x 12" LEDGER

(2)-3" x 3/16" STRAP
PLATES W/ 3/4" BOLTS

B9 LEDGER SPLICE GRID A & G

	A	B	C	BOLTING
ROOF				
FLOOR				

PROBLEM 21-3B-9

Problem 21-3b-8 Purlin / I-joist. Determine the load to be supported by each purlin and select a suitable PF hanger.

Problem 21-3b-9 Ledger splice—bearing wall. See roof/floor ledger notes. Establish a detail to show floor and roof usage.

Problem 21-3b-10 Ledger splice—nonbearing wall. See roof /floor ledger notes. Establish a detail to show flor and roof usage.

Roof Ledgers: Use 3 × 12 d.f.p.t. ledger with 3/4"Ø anchor bolts at 32" o.c. at grids A and F. Staggered bolts 3" u/d. Space at 60" o.c. at grids 1 and 7. At splices in ledger at grids A and G use (2)-3" × 144" × 3/16" straps with (8)-3/4" bolts 2" from ends and 20" o.c. and 10" from splice. Place straps 3" down from top of ledger and 3 1/2" from center to center.

At splices at walls 1 and 7 use (3)-3/16" × 3" × 24" straps at 3 1/2" o.c. with 4 bolts per straps placed 2" from strap end, and 6"o.c.

Problem 21-3c Use the attached drawings as a base to develop the framing plan for the mezzanine level started in Problem 15-3. Develop a beam schedule for framing members. Use DF/HF, V-5, Fb 2400 beams unless noted. Listed spans refer to grids on framing plan.

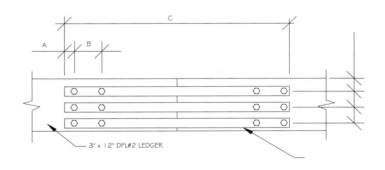

3" x 12" DFL#2 LEDGER

B10 LEDGER SPLICE-GRID 1 & 7

	A	B	C	BOLTING
ROOF				
FLOOR				

PROBLEM 21-3B-10

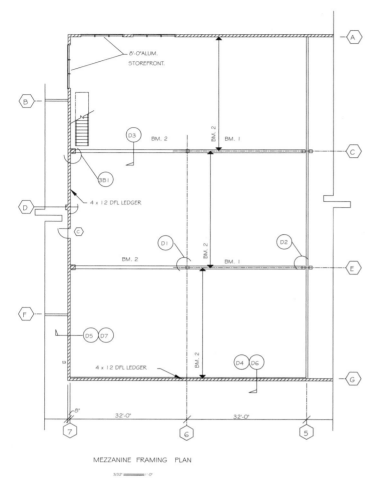

MEZZANINE FRAMING PLAN

3/32" = 1'-0"

PROBLEM 21-3C

Problem 21-3c Mezzanine Floor Framing. Use 1 1/8" plywood floor sheathing.

■ Beam 1. Span 32'. Use 6 3/4" × 37 1/2".

■ Beam 2. Span 32'. Use TJI60P/30" I-joist for 32" o.c.

Floor Ledgers. Use 4 × 12 d.f.p.t. ledger with 3/4"Ø anchor bolts at 24" o.c. at grids A and F. Staggered bolts 3" u/d. Space at 48" o.c. at grid 7. At splices in ledger at grids A and F use (2)-144" long × 3" straps with (10)-3/4"Ø bolts 1 1/2" from ends and 15" o.c. and 10 1/2" min. from splice. Place straps 3" down from top of ledger and 3 1/2" from center to center. At splices at wall 7 use (3)-3/16" × 3" × 36" straps at 3 1/2" o.c. with 4 bolts per straps placed 1 1/2" from strap end, and 6"o.c.

Problem 21-3d Use the attached drawings as a base and draw details showing the connections for the floor framing plan. Choose a scale suitable for clearly showing the required information. Submit preliminary drawings to your instructor prior to completing the drawings. Required floor connections:

Problem 21-3d-1 Beam to column. Support the glu-lam with a 3/4 × 15 × 6 7/8" bearing plate with 7/16 × 9 × 15 side plates ea. side, and 7/16 × 15 × 6 7/8" end plate. Weld side plates to base and end plate to 6 × 6 × 3/16" col. w/ 3/8" fillet welds ea. side of plate. Bolt with (2)-3/4" bolts 2" dn, and 2" from edge of side plate and 3" o.c. Provide an 8 × 8 × 3/8" gusset with 1/4" fillet weld ea. side to CCO/T.S.

Provide an 8 × 3 × 3/8" strap @ ea. side of beam w/ (1) 3/4" Ø bolt thru beam, 3" dn from top of bm. Weld side of pl. to TS w/ 3/8" fillet weld @ ea. Set bolts 2" from end of strap.

Problem 21-3d-2 Beam to firewall. Use detail 21-3d-1 as a guide for weld, plates, straps, gusset, and bolting sizes. Connect the glu-lam to the T.S. col. w/ an appropriate CCO col. cap. w/ a 7 1/8" × 8 1/4" × 3/8" end plate. Complete fire wall to match detail 21-3b-3.

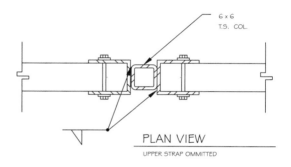

PLAN VIEW

UPPER STRAP OMMITTED

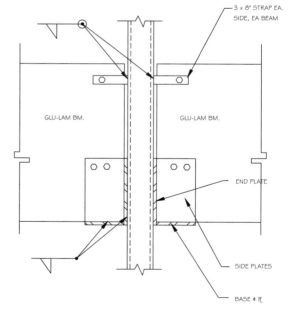

3d1 BEAM / BEAM / COL.

PROBLEM 21-3D-1

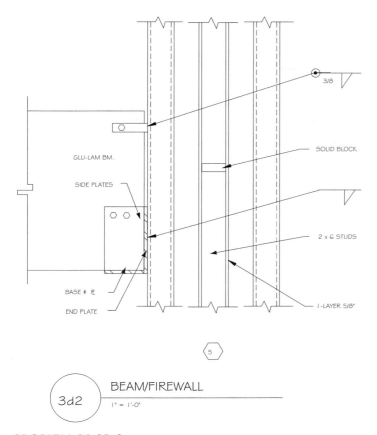

BEAM/FIREWALL
1" = 1'-0"

3d2

PROBLEM 21-3D-2

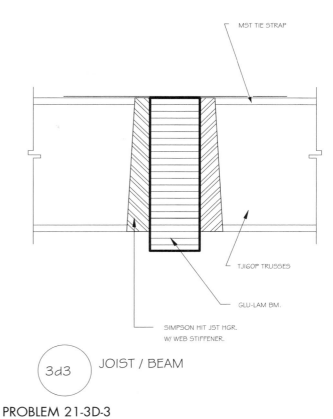

JOIST / BEAM

3d3

PROBLEM 21-3D-3

Problem 21-3d-3 I-joist to beam. Use appropriate HIT hgr, to connect TJI60/30 I-joists to glu-lam.

Problem 21-3d-4 I-joist to ledger—perpendicular. See framing plan to specify ledger size and bolting. Use appropriate HIT hgr, to connect TJI60/30 I-joists to ledger. Connect to wall similar to detail 21-3b-6.

Problem 21-3d-5 I-joist to ledger—parallel. See sections for steel locations. See framing plan to specify ledger size and bolting. Use appropriate HIT hgr, to connect blocking btwn. ledger and TJI. Connect to wall similar to detail 21-3b-7.

Problem 21-3d-6 Ledger splice—bearing wall. Use 4 × 12 ledger with bolting as per detail 21-3b-9.

Problem 21-3d-7 Ledger splice—non-bearing wall. Use 4 × 12 ledger with bolting as per detail 21-3b-10.

Problem 21-4a Use the attached sketch to draw the roof framing plan for the structure started in Problem 15-4. Show sky lights referenced in Chapter 15. Submit preliminary drawings to your instructor prior to completing the drawings. Assume all welds to be 3/16" fillet unless noted.

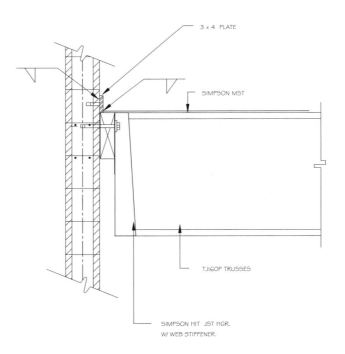

TRUSS/WALL @ 'A & G'

3d4

PROBLEM 21-3D-4

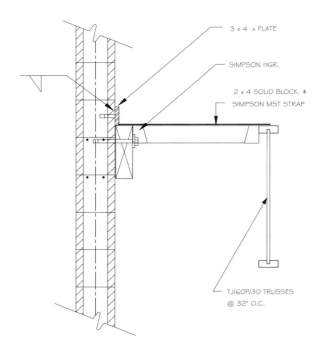

3d5   TRUSS/WALL @ 5 & 7
        1" = 1'-0"

PROBLEM 21-3D-5

Develop a beam schedule for the roof framing. Use

Roof: All beams to be DF/HF, V-5, Fb 2400 unless noted. Listed spans assume 36' column spacing plus cantilevers. Listed spans refer to distance from the column to the indicate cantilever.

Beam 1:   8'–0" east of grid I. Use 6 3/4" × 31 1/2"

Beam 2:   8'–0" east of grid H. Use 6 3/4" × 21"

Beam 3:   8'–0" east of H to F. Same as beam 1

Beam 4:   12'–0" east of D to F. Use 6 3/4" × 24"

Beam 5:   16'–0" east of D to C. Same as beam 1

Beam 6:   48'–0" clear span. Use 6 3/4" × 34 1/2"

Problem 21-4b Use the attached sketches to design and detail the following roof connections. Use the appropriate sketch from Problem 21-3 as a guide for *unlisted* information. Choose a scale suitable for clearly showing the required information. Submit preliminary drawings to your instructor prior to completing the drawings. Required connections:

Problem 21-4b-1 Beam to wall. Grids 3C & 3K (3A & 3C similar @ sales). Provide 1/4" min. clr. From beam to wall & support glu-lam on 6 7/8 × 10 × 3/4" base pl. Provide 10 × 10 × 16h × 5/16" angle ea. side of beam, weld to base pl. w/ 5/16" fillet. Provide (3)-3/4"Ø × 9 bolts thru beam. Provide (6)-3/4"Ø × 4 1/2" taper bolts to 6" tilt-up wall ea. side in 2 columns, 5" o.c. Place beam bolts 2" min. from pl. edge and 3"o.c. Place taper bolts 3" dn from plate edge.

Provide a 8 × 4 × 1/2" × 3h angle each side of beam 4 1/2" dn. from top of beams. Place (1)-3/4 Ø bolt in 13/16 × 1 7/8" slotted hole in 8" leg thru bm. to opposite angle. Bolt to wall w/ 3/4"Ø × 4 1/2 taper bolts.

Problem 21-4b-1a Beam to col. grid 3. Support glu-lam on 6 7/8 × 22 × 7/8" base pl. Provide 10h × 22 × 5/16" side pl. ea. side of beam, weld to base pl. w/ 1/4" fillet. Weld base pl. to col w/ 1/4" fillet. Provide (4)-3/4"Ø bolts thru beam. Place bolts 1 1/2" min. from pl. edge and 6 1/4"o.c.

Problem 21-4b-2 Beam to beam w/ saddle. Support beam on 7/8 × 6 7/8 × 8" long top & base pl. Dap top pl/ into bm. Attach w/ 5/16 × 8" side pl. ea. side of bm. Provide 3 × 5/16" side straps each side, top & btm. w/ (3)-3/4"Ø × 8 bolts thru bm. @ 3" o.c. 1 1/2" from end of strap. Bolts to be 8" min. from bm. splice & 4" up & 3" dn. Attach straps to side pl. w/ 5/16" × 3" fillet & back weld.

Problem 21-4b-3 Beam to beam @ fire wall. Grid 3F. Use appropriate CCO col. caps to connect glu-lam to 6 × 6 T.S. Use 3/8" fillet weld, all around. Provide 3/4" min. clr. Beam to gyp. bd. Center col. 12"o.c. from center of 2 × 6 fire wall w/ 5/8" type X gyp. bd. ea. side. Provide 5/8" type X gyp. bd. on btm. side of top chord for 60" min ea. side. Provide 2 × 4 solid blk. @ 24" o.c. btwn. top chords to support block.

Provide 3 × 6 top plate over (2)-2 × 6 pls. match bolts per detail 21-3b-4.

Problem 21-4b-4 Truss to beam, Grid 3. In addition to truss clips supplied by manuf. provide 2 × 4 solid blk. btwn TJ60 trusses and anchor to glu-lam w/ A35 anchors ea. end of blk. Provide MST48 @ ea.48" o.c. Align with straps at wall.

Problem 21-4b-5 Truss to beam @ column. Grid 3D+12', 3H, & 3J+12'. Provide 2 × 4 brace @ 32" o.c. @ 45° for 2 truss spaces min. ea. side of col. Attach to bm. w/ U24 hgr. 2" up from btm of bm. & blk. upper end w/ 3 × 4 × 10' min. cont. nailer. w/ (2)-16d ea. truss. Provide 3 × 4 sld. blk. equally spaced along brace & nail to brace w/ (5)-16d ea. Provide 2× blocking btwn. trusses per detail 21-4b-4. Provide MST 60 @ ea. truss over brace.

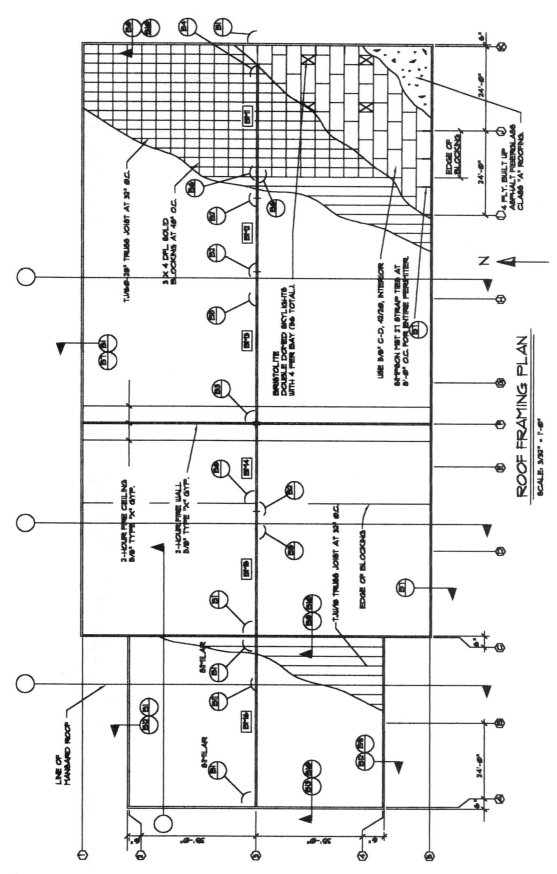

ROOF FRAMING PLAN
SCALE: 3/32" = 1'-0"

PROBLEM 21-4A

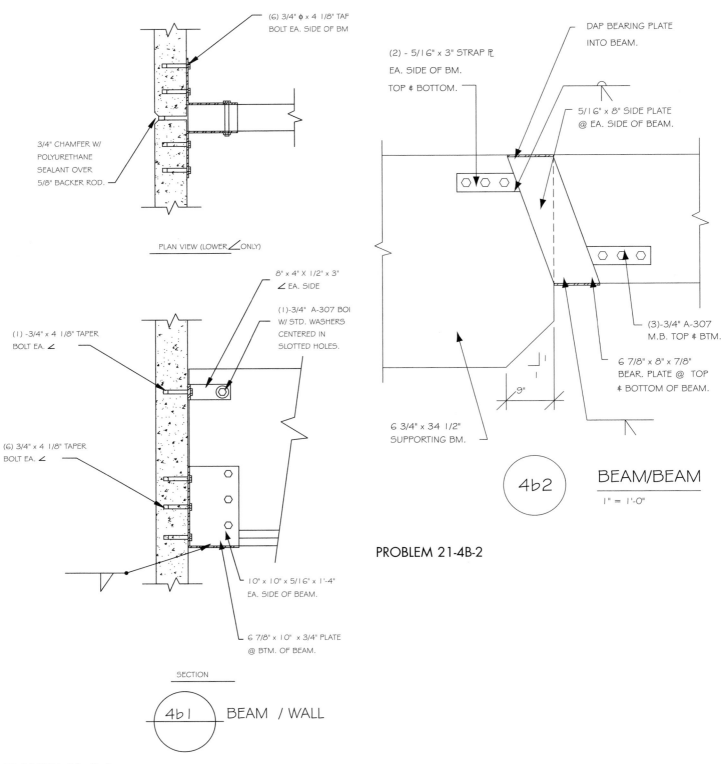

(6) 3/4" φ x 4 1/8" TAF
BOLT EA. SIDE OF BM

(2) - 5/16" x 3" STRAP PL
EA. SIDE OF BM.
TOP & BOTTOM.

DAP BEARING PLATE
INTO BEAM.

5/16" x 8" SIDE PLATE
@ EA. SIDE OF BEAM.

3/4" CHAMFER W/
POLYURETHANE
SEALANT OVER
5/8" BACKER ROD.

PLAN VIEW (LOWER ∠ ONLY)

(3)-3/4" A-307
M.B. TOP & BTM.

6 7/8" x 8" x 7/8"
BEAR. PLATE @ TOP
& BOTTOM OF BEAM.

8" x 4" X 1/2" x 3"
∠ EA. SIDE

(1)-3/4" A-307 BOI
W/ STD. WASHERS
CENTERED IN
SLOTTED HOLES.

(1) -3/4" x 4 1/8" TAPER
BOLT EA. ∠

9"

6 3/4" x 34 1/2"
SUPPORTING BM.

(6) 3/4" x 4 1/8" TAPER
BOLT EA. ∠

4b2   BEAM/BEAM

1" = 1'-0"

PROBLEM 21-4B-2

10" x 10" x 5/16" x 1'-4"
EA. SIDE OF BEAM.

6 7/8" x 10" x 3/4" PLATE
@ BTM. OF BEAM.

SECTION

4b1   BEAM / WALL

PROBLEM 21-4B-1

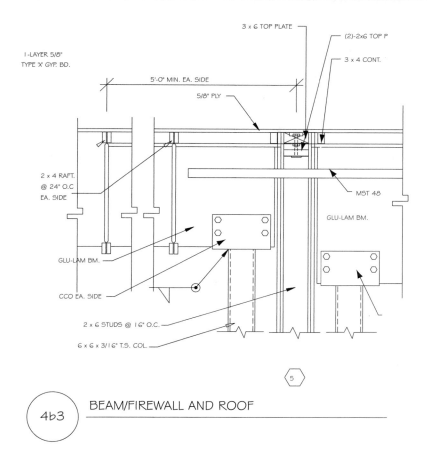

1-LAYER 5/8"
TYPE 'X' GYP. BD.

3 x 6 TOP PLATE

(2)-2x6 TOP P

3 x 4 CONT.

5'-0" MIN. EA. SIDE

5/8" PLY

2 x 4 RAFT.
@ 24" O.C.
EA. SIDE

MST 48

GLU-LAM BM.

GLU-LAM BM.

CCO EA. SIDE

2 x 6 STUDS @ 16" O.C.

6 x 6 x 3/16" T.S. COL.

5

4b3 BEAM/FIREWALL AND ROOF

PROBLEM 21-4B-3

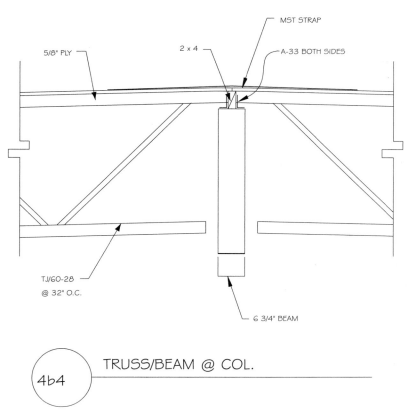

MST STRAP

5/8" PLY

2 x 4

A-33 BOTH SIDES

TJ/60-28
@ 32" O.C.

6 3/4" BEAM

4b4 TRUSS/BEAM @ COL.

PROBLEM 21-4B-4

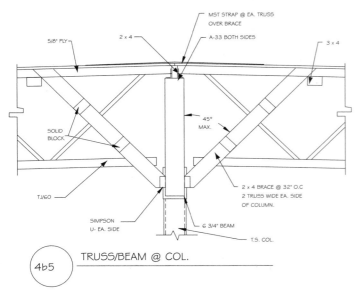

4b5 TRUSS/BEAM @ COL.

PROBLEM 21-4B-5

Problem 21-4b-6 Truss to beam @ diaphragm edge. As edge of roof blocking (36' min.) provide (2) Simpson HD-2 anchors 8" min. ea side of splice w/ (2)-5/8" × 4 1/2 bolts thru truss, & 5/8 Ø × 16 bolt across splice. Provide solid blk. btwn trusses similar to detail 21-4b-4. Maintain 6" min from anchor to end of blk.

Problem 21-4b-7 Truss to ledger—perpendicular. Grid 1 & 5. Provide 3 × 12 d.f.p.t. ledger w/ 3/4"Ø bolts @ 32" o.c. staggered 3" u/dn. Provide 3 × 4h × 5/8" plate w/ 3/4"Ø × 4 parabolt. Weld MST36 to plate w/ 1/8" fillet weld @ 8'–0" O.C Provide 2× blocking btwn. trusses per detail 21-4b-4.

Problem 21-4b-8 Truss to ledger—parallel. Grid A, C (both sides) & K. Provide 3 × 12 d.f.p.t. ledger w/ 3/4"Ø bolts @ 60" o.c. staggered 3" u/dn. Select suitable U hangers for the solid blk. Provide 3 × 4h × 5/8" plate w/ 3/4"Ø × 4 parabolt. Weld MST27 to plate w/ 1/8" fillet weld @ 8'–0" O.C

Problem 21-4b-9 Ledger splice—bearing wall. Grid 1, 2, 4 & 5. See roof ledger notes. Establish a detail to show floor and roof usage. See detail 27-3b-9

Problem 21-4b-10 Ledger splice—nonbearing wall. Grid A, C (both sides) & K. See roof/floor ledger notes. Establish a detail to show flor and roof usage. See detail 27-3b-10.

Problem 21-4b-11 Truss/beam—TJWs. Grid 3. Provide 2 × 4 solid blk. laid flat btwn TJW18 trusses. Provide MST48 @ ea. 48"o.c. Align with straps at wall.

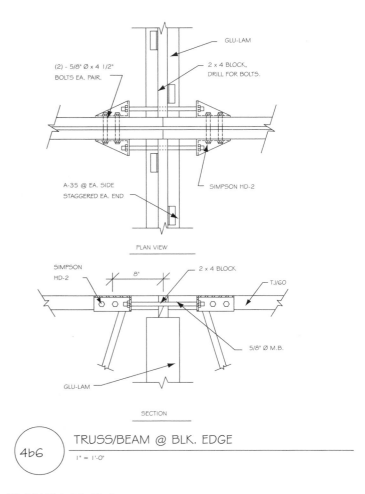

PLAN VIEW

SECTION

4b6 TRUSS/BEAM @ BLK. EDGE
1" = 1'-0"

PROBLEM 21-4B-6

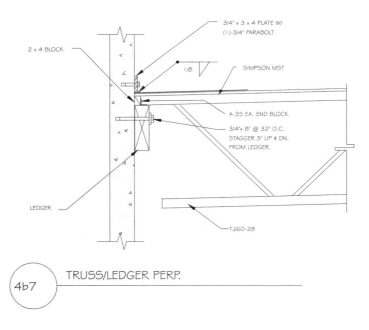

4b7 TRUSS/LEDGER PERP.

PROBLEM 21-4B-7

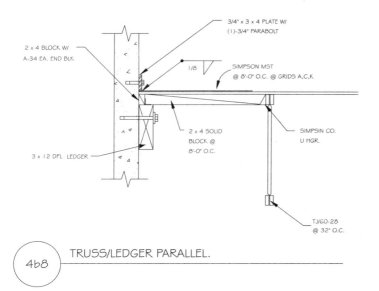

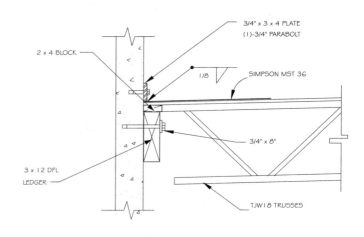

4b8 — TRUSS/LEDGER PARALLEL.

PROBLEM 21-4B-8

4b12 — TRUSS/LEDGER PERP.

PROBLEM 21-4B-12

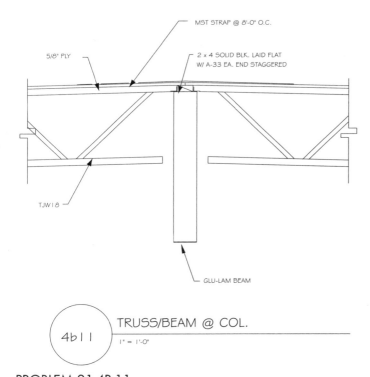

4b11 — TRUSS/BEAM @ COL.
1" = 1'-0"

PROBLEM 21-4B-11

Problem 21-4b-12 Truss/ledger—TJWs—perpendicular. Grids 2 & 4. Provide 3 × 12 d.f.p.t. ledger w/ 3/4"Ø bolts @ 32" o.c. staggered 3" u/dn. Provide 3 × 4h × 3/4" plate w/ 3/4 Ø × 4 parabolt. Weld MST36 to plate w/ 1/8" fillet weld @ 8'–0" O.C Provide 2 × 4 blocking laid flat btwn. trusses.

Problem 21-4b-13 Truss/ledger—TJWs—parallel. Grid A & C. Provide 3 × 12 d.f.p.t. ledger w/ 3/4"Ø bolts @ 60" o.c. staggered 3" u/dn. Select suitable U hangers for the solid blk. Provide 3 × 4h × 3/4" plate w/ 3/4"Ø × 4 parabolt. Weld MST27 to plate w/ 1/8" fillet weld @ 8'–0" O.C

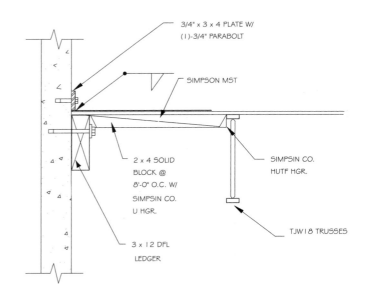

4b13 — TRUSS/LEDGER PARALLEL.

PROBLEM 21-4B-13

Roof Ledgers: Use 3 × 12 d.f.p.t. ledger with anchor bolts at 32" o.c. at grids 1, 2, 4, and 5. Staggered bolts 3" u/d. Space at 60" o.c. at grids A, C, and K. At splices in ledger at grids 1, 2, 4, and 5 use (2)-3"× 1/4" (length based on detail) straps with 3/4"Ø bolts 1 1/2" from ends and 20" o.c. Place straps 3" down from top of ledger and 3 1/2" from center to center. At splices at walls A, C, and K use (3)-MST48 straps at 3 1/2" o.c. with (8)-1/2"Ø bolts. Provide 2× blocking between trusses and connect to ledger with Simpson Company A-35 angles at each end of block.

Roof blocking: Provide 3 × 4 solid blocking between trusses at each end of the warehouse structure. Blocks to extend a minimum of 36' (and modular based on truss locations) from each end wall and to be placed at 48" o.c. Bond last truss of blocking pattern w/ (2)-Simpson HD-2 hold down anchors.

Problem 21-4c Use the attached drawings and complete the framing plan for the mezzanine level started in Problem 15-4. Develop a beam schedule for framing members. Use DF/HF, V-5, *Fb 2400* beams unless noted. Listed spans refer to grids on framing plan.

Mezzanine Floor Framing: Use 1 1/8" floor sheathing.

Beam 1: Span 36'. Use 6 3/4 × 37 1/2"

Beam 2: Span 24'. Use 6 3/4 × 30"

Beam 3: Span 24'. Use TJL22 @ 24" o.c.

Beam 4: Span 16'. Use TJL22 @ 32" o.c.

Beam 5: Span 32'. Use TJL30 @ 24" o.c.

Problem 21-4d Choose a scale suitable for clearly showing the required information. Use the appropriate drawing from Problem 21-3 as a guide for unlisted information. Submit preliminary drawings to your instructor prior to completing the drawings.

Problem 21-4d-1 Beam to column—floor. Provide 1/4" clr. btwn. bm and col. Support the glu-lam with a 3/4 × 9 × 6 7/8" bearing plate with 7/16 × 9 × 15" side plates ea. side, and 7/16 × 15 × 6 7/8" end plate. Weld side plates to base and end plate to 6 × 6 col. w/ 3/8" fillet welds ea. side of plate. Bolt with (4) 3/4" bolts 1 1/2" dn, and 1 1/2" from edge of side plate and 3" o.c. Provide an 9 × 9 × 3/8" gusset with 1/4" fillet weld ea. side to CCO/T.S.

Provide an 8 × 3 × 6 7/8" U × 7/16" strap w/ (1)-3/4"Ø bolt thru beam, 3" dn from top of bm. Weld side of pl. to TS w/ 3/16" fillet weld @ ea. Set bolts 1 1/2" from end of strap.

Problem 21-4d-2 Beam to wall. See detail 21-4b-1 Copy and edit as required.

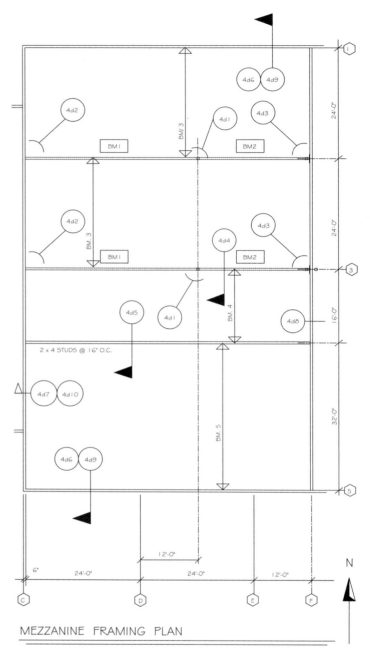

MEZZANINE FRAMING PLAN

PROBLEM 21-4C

Problem 21-4d-3 Beam to firewall. See detail 21-3d-2. Weld side plates to base and end plate to 6 × 6 col. w/ 3/8" fillet welds ea. side of plate. Bolt with (2)-3/4" bolts 2" dn, and 2" from edge of side plate and 3" o.c. Provide an 8 × 8 × 3/8" gusset with 1/4" fillet weld ea. side to CCO/T.S.

Provide an 8 × 3 × 3/8" strap @ ea. side of beam w/ (1)-3/4"Ø bolt thru beam, 3" dn from top of bm. Weld side of pl. to TS w/ 3/8" fillet weld @ ea. Set bolts 2" from end of strap.

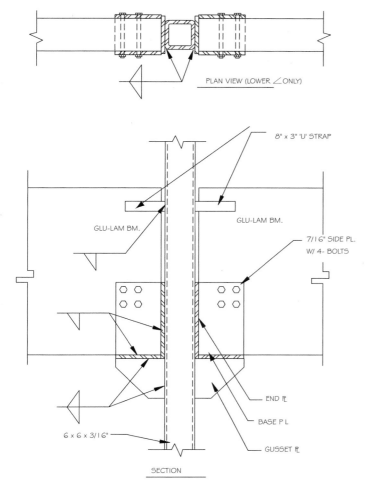

PLAN VIEW (LOWER ∠ ONLY)

8" x 3" 'U' STRAP

GLU-LAM BM.

GLU-LAM BM.

7/16" SIDE PL.
W/ 4- BOLTS

END PL

BASE PL

6 x 6 x 3/16"

GUSSET PL

SECTION

4d1   BEAM/COLUMN

PROBLEM 21-4D-1

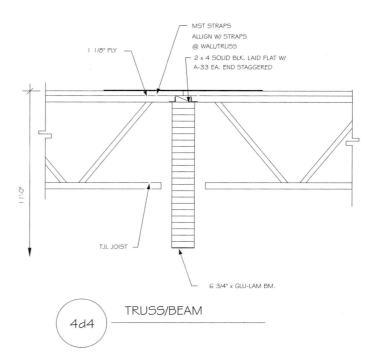

MST STRAPS
ALLIGN W/ STRAPS
@ WALL/TRUSS

1 1/8" PLY

2 x 4 SOLID BLK. LAID FLAT W/
A-33 EA. END STAGGERED

1'-0"

TJL JOIST

6 3/4" x GLU-LAM BM.

4d4   TRUSS/BEAM

PROBLEM 21-4D-4

Problem 21-4d-4 Truss to beam. Grid 3C to F. Provide 2 × 4 solid blk. laid flat btwn TJL trusses. Provide MST36 @ ea. 96"o.c. Align w/ straps at ledger.

Problem 21-4d-5 Truss to wood wall. Grid 3+16' btwn. C to F. Extend 1/2" sheet rock to 10' high & show suspended ceiling on display side. Provide 2 × 4 solid blk. laid flat btwn TJL trusses over (2)-2 × 4 top plates. Provide MST36 @ ea. 96"o.c. Align w/ straps at ledger.

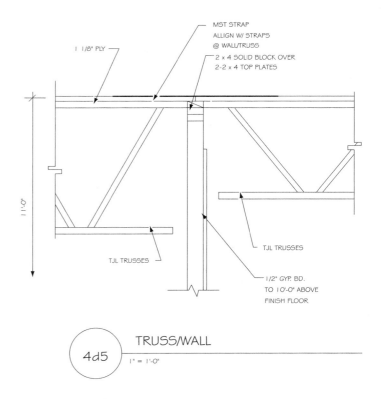

MST STRAP
ALLIGN W/ STRAPS
@ WALL/TRUSS

1 1/8" PLY

2 x 4 SOLID BLOCK OVER
2-2 x 4 TOP PLATES

1'-0"

TJL TRUSSES

TJL TRUSSES

1/2" GYP. BD.
TO 10'-0" ABOVE
FINISH FLOOR

4d5   TRUSS/WALL

1" = 1'-0"

PROBLEM 21-4D-5

Problem 21-4d-6 Truss to ledger—perpendicular. Grids 1 & 5. Provide 4 × 12 d.f.p.t. ledger w/ 3/4"Ø bolts @ 24" o.c. staggered 3" u/dn. Provide 3 × 4h × 3/4" plate w/ 3/4"Ø × 4 parabolt. Weld MST36 to plate w/ 1/8" fillet weld @ 8'–0" o.c. Provide 2 × 4 blocking laid flat btwn. trusses.

Problem 21-4d-7 Truss to ledger—parallel. Grid C. (similar to detail 21-4b-13). Provide 4 × 12 d.f.p.t. ledger w/ 3/4"Ø bolts @ 48" o.c. staggered 3" u/dn. Provide 3 × 4h × 3/4" plate w/ 3/4"Ø × 4 parabolt. Weld MST36 to plate w/ 1/8" fillet weld @ 8'–0" o.c. Provide 2 × 4 blocking btwn truss @ 8'–0" o.c. and select suitable U hangers for the blk. Provide 3 × 4h × 5/8" plate w/ 3/4"Ø × 4 parabolt. Weld MST27 to plate w/ 1/8" fillet weld @ 8'–0" o.c.

Problem 21-4d-8 Truss to firewall. Provide 2 × 12 d.f.l. ledger over 5/8"gyp. bd. Provide solid blk. in wall behind ledger.

Problem 21-4d-9 Ledger splice—bearing wall. See floor ledger notes. Establish a detail to show floor and roof usage. See detail 21-4b-12.

Problem 21-4d-10 Ledger splice—nonbearing wall. See floor ledger notes. Establish a detail to show floor and roof usage. See detail 21-4b-13.

Floor Ledgers: Use 4 × 12 d.f.p.t. ledger with 3/4"Ø anchor bolts at 24" o.c. at grids 1 and 5. Staggered bolts 3" u/d. Space at 48" o.c. at grid C. At splices in ledger at grids 1 and 5 use (2)-144" long × 3" straps with 3/4" bolts 1 1/2" from ends and 20" o.c. Place straps 3" down from top of ledger and 3 1/2" from center to center. Use (3)-MST48 straps at 3 1/2" o.c. at splices at wall C with (8) bolts per strap.

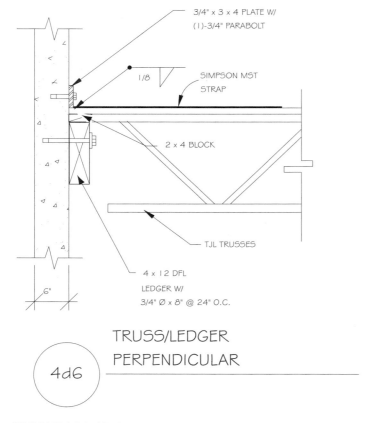

TRUSS/LEDGER
PERPENDICULAR

4d6

PROBLEM 21-4D-6

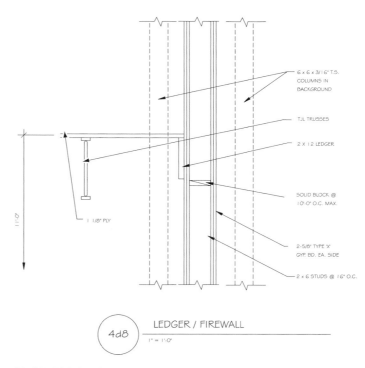

LEDGER / FIREWALL

4d8

PROBLEM 21-4D-8

CHAPTER 22

Drawing Structural Elevations and Sections

Chapters 15 and 18 introduced the drawing of elevations and sections used in the architectural drawings. Similar drawings can also be used in the structural drawings to describe structural materials. Structural elevations are generally associated with structural steel and concrete tilt-up construction. Structural sections can be used with any type of building material to show complicated framing intersections.

STRUCTURAL STEEL ELEVATIONS

Three common types of elevations are used to explain structural steel. Elevations can be used to show the vertical relationship of major framing members shown on the framing plans without showing any architectural information. Figure 22-1 shows an example of the steel skeleton for a warehouse. This type of eleva-

tion provides a vertical reference for materials shown along one specific grid line on the framing plans. Several elevations can be created depending on the complexity of the structure. Details similar to Figure 22-2 are referenced to the elevation to explain construction.

Elevations can also be used to show how the steel framework relates to the architectural members of the structure. Figure 22-3 shows an example of this type of elevation. Notice that this drawing shows similarities to both a structural elevation and an architectural section. Details for structural information are referenced to the elevation.

A third type of steel elevation is used to show major steel components similar to the truss shown in Figure 22-4. This type of elevation is used to fabricate a component off site. The drawing shows major structural components as well as connection details similar to Figure 22-5.

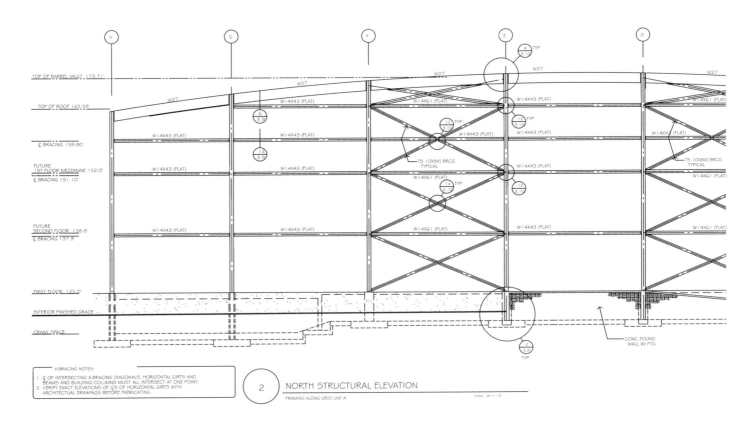

FIGURE 22-1 ■ Structural elevations show the shape of the skeleton. *Courtesy Charles J. Conlee, P.E., Conlee Engineers, Inc.*

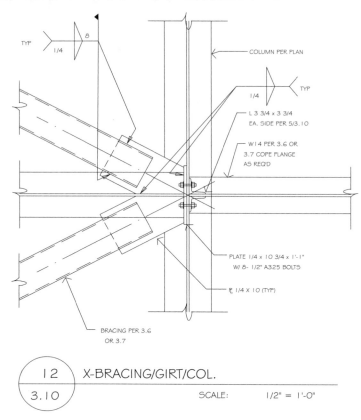

TYP $\frac{}{1/4}$ 8

COLUMN PER PLAN

$\frac{}{1/4}$ TYP

L 3 3/4 x 3 3/4
EA. SIDE PER 5/3.10

W14 PER 3.6 OR
3.7 COPE FLANGE
AS REQ'D

PLATE 1/4 x 10 3/4 x 1'-1"
W/ 8- 1/2" A325 BOLTS

℄ 1/4 X 10 (TYP)

BRACING PER 3.6
OR 3.7

12	X-BRACING/GIRT/COL.
3.10	

SCALE: 1/2" = 1'-0"

FIGURE 22-2 ■ Details are keyed to structural sections to explain connections. *Courtesy Charles J. Conlee, P.E., Conlee Engineers, Inc.*

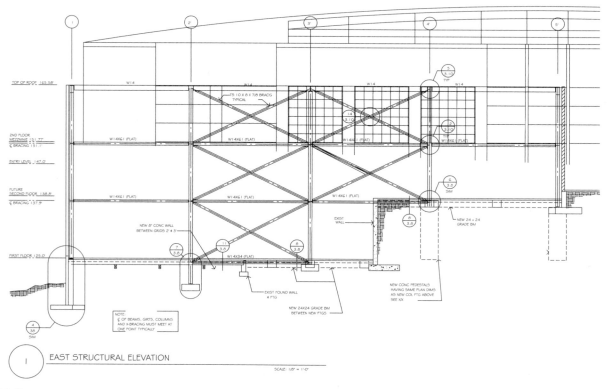

EAST STRUCTURAL ELEVATION

SCALE: 1/8" = 1'-0"

FIGURE 22-3 ■ Structural elevations can be used to show how the framework relates to the architectural members. *Courtesy Charles J. Conlee, P.E., Conlee Engineers, Inc.*

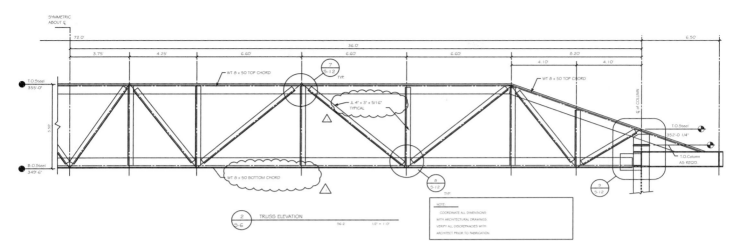

FIGURE 22-4 ■ Structural elevations can be used to show how a structural system is constructed. *Courtesy Van Domelen/Looijenga/McGarrigle/Knauf Consulting Engineers.*

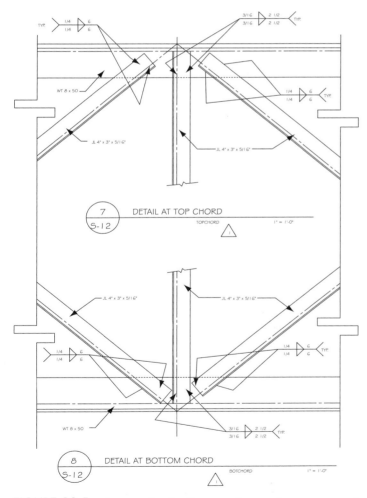

FIGURE 22-5 ■ Construction details are used to supplement structural elevations. *Courtesy Van Domelen/Looijenga/McGarrigle/Knauf Consulting Engineers.*

POURED CONCRETE ELEVATIONS

The elevations for poured concrete are similar to those used for structural steel. Drawings show the location of structural beams and columns along a specific grid in the structure, as shown in Figure 22-6. In addition to describing distances between members, the elevation also serves as a reference map for listing beam sizes and for connection details similar to Figure 22-7. Elevations can also be more specific and show only specific types of construction, similar to the column elevations shown in Figure 22-8, as well as the size, shape, and reinforcing of specific members.

CONCRETE TILT-UP ELEVATIONS

Elevations for a tilt-up structure are used to show the size, shape, opening locations, and reinforcing for each panel of a structure. Elevations are drawn for tilt-up construction using two common formats. The format used depends on the size and complexity of the structure. Each method is usually referenced on the title page by the name *panel elevation*. One method of showing the construction of panels is to show an entire face of a structure similar to the architectural elevations. A scale of 1/8" or 1/4" = 1'–0" is often used to draw panel elevations. Structural elevations similar to Figure 22-9 show the components required to construct the panels. The panel elevations are drawn as if you are standing inside the building, looking out. Elevations are drawn showing the inside surface, because what will become the exterior surface is against the slab that is used to form the panels. Dimensions showing the size of all panels and the size and locations of all openings should be provided. Because of the small scale, reinforcing steel patterns are referenced by note but not shown on these drawings. Individual steel

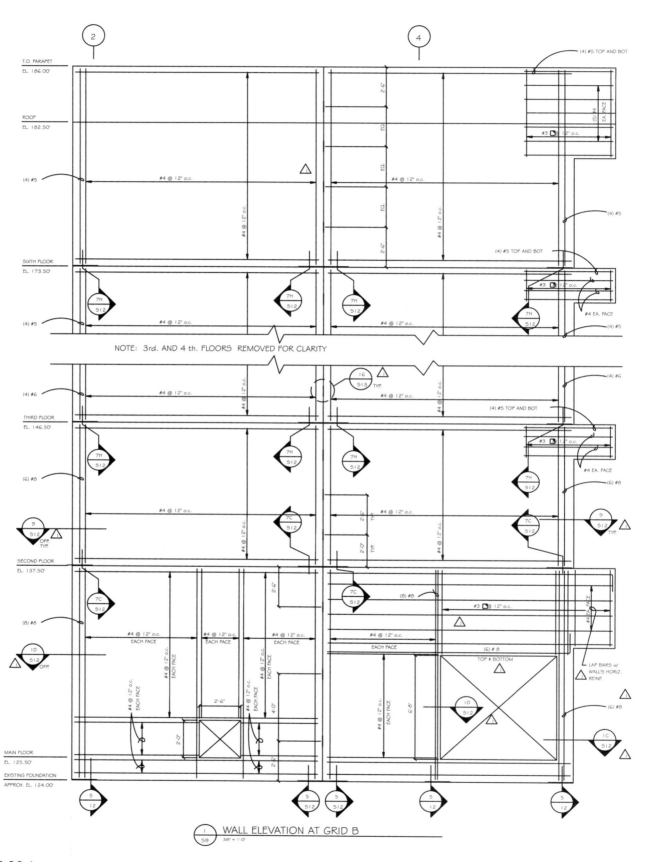

FIGURE 22-6 ■ Elevations for poured concrete show the location of each column and beam. *Courtesy KPFF Consulting Engineers.*

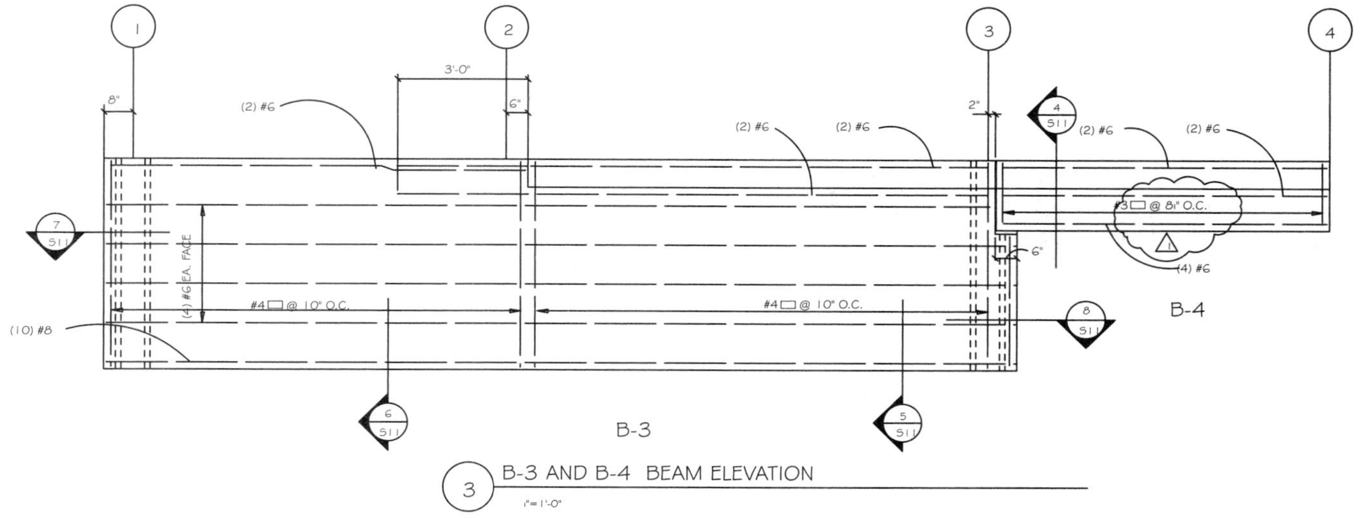

FIGURE 22-7 ■ Details must be drawn to show how each column and beam will be constructed. *Courtesy KPFF Consulting Engineers.*

required for openings is shown in schedules or details similar to Figure 22-10 and referenced to the elevations.

An alternative to drawing all of the panels along one grid line as a whole unit is to draw each panel as an individual panel elevation, similar to the panel seen in Figure 22-11. Using this method, a large-scale elevation is drawn of each panel. The scale depends on the size of the panel and the complexity of the reinforcing pattern to be represented. One elevation can be used to describe several panels if they are exactly the same in every respect. If any variation exists, an elevation of each panel must be drawn. A detail is typically included within each panel elevation to show and specify typical reinforcing steel. Special steel required to reinforce each opening is drawn and specified on each panel elevation. Panels can be referenced to other drawings by referring to the grids located at each end of the panel. Panels can also be referenced to the framing plan by using an elevation reference symbol. This method is similar to the way interior drawings are referenced to the floor plan (see Chapter 19).

Details similar to Figure 22-12 are referenced to each type of panel elevations to show typical corner and joint details. In addition to showing whole panels, elevation of smaller panels, called spandrels are also usually provided. *Spandrels* are concrete panels that are not full height, located over or between areas of glazing. When spandrels are required for a project, an elevation should be provided to specify reinforcing steel. Details similar to Figure 22-13 are typically required to show both reinforcing steel within the panel and restraining steel at connections.

STRUCTURAL SECTIONS

Sections can be drawn by the engineering team to show any number of material intersections. When drawn, the section would be created using the same guidelines used to create architectural sections. Although referred to as sections on the project title page,

structural sections are usually details of a specific area rather than of an entire wall, or a portion of the structure. The structural drawings detail how each component is to be constructed. The architectural drawings are used to show how each component relates to others in size and location. The same guidelines used to construct architectural details should be used to draft structural details. See Chapter 3 for a review of linetypes and hatch patterns typically used with sections and details.

Drawing Scales

The scale used to create sections and details depends on the intent of the detail. The senior drafter typically provides a sketch and indicates the required scale for a drafter new to the firm to follow. When determining the size for a detail, the smallest component to be represented should determine the drawing scale. Use a scale that allows the smallest member to be shown clearly. Common scales used for structural details and the closest metric counterpart include the following:

In/ft Scale	Metric
3/4" = 1'-0"	1:20
1" = 1'-0"	1:10
1 1/2" = 1'-0"	(no common scale)
3" = 1'-0"	1:5

When several details are to be drawn that relate to the same component, each should be drawn at the same scale if space permits.

When scales smaller than 3/4" = 1'-0" (1:20) are used, structural materials are sometimes drawn at their nominal size of 6 × 12 (150 × 300) rather than their actual size 5 1/2" × 11 1/2" (140 × 292). Structural details drawn at a scale of 3/4" (1:20) and larger should always represent materials with their actual sizes. Review Section II for various material sizes. Although the use of comput-

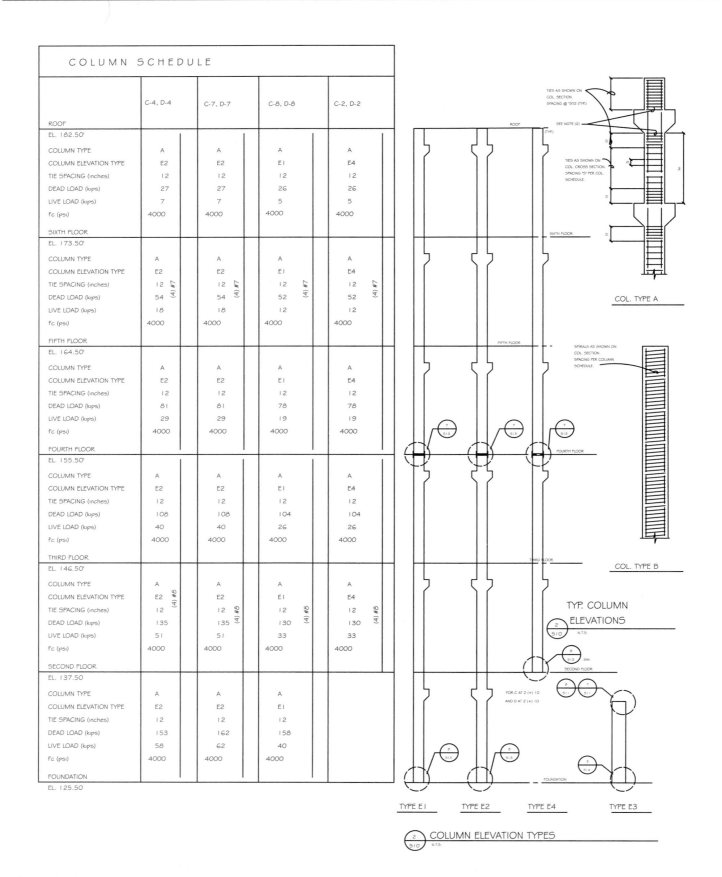

FIGURE 22-8 ■ Schedules and elevations are typically used to explain column and beam construction. *Courtesy KPFF Consulting Engineers.*

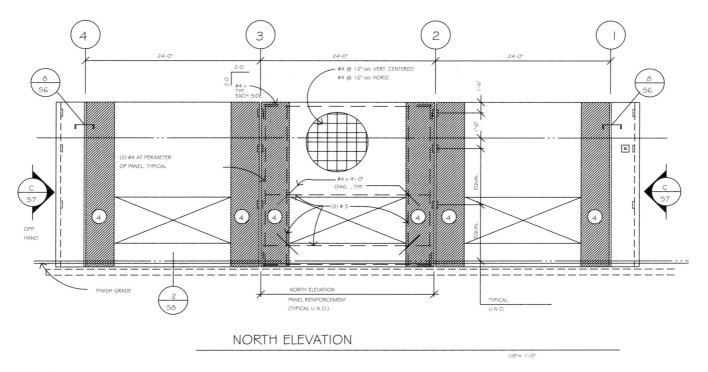

NORTH ELEVATION

1/8"= 1'-0"

FIGURE 22-9 ■ Panel elevations can be drawn of an entire face of a structure. *Courtesy Bill Berry, Berry-Nordling Engineers, Inc.*

TILT- UP JAMB REINFORCEMENT

MARK	VERT. REINF. *	DETAIL
1	#4 @ 6" CENTERED	2 / 56
2	#4 @ 9" CENTERED	2 / 56
3	#4 @ 9" CENTERED	3 / 56
4	#4 @ 12" INSIDE FACE / #4 @ 8" OUTSIDE FACE	4 / 56
5	# 5 @ 8" INSIDE FACE / # 5 @ 8" OUTSIDE FACE	5 / 56
6	# 5 @ 8" INSIDE FACE / # 5 @ 8" OUTSIDE FACE	6 / 56
7	(3) - #5 EACH FACE	7 / 56

* IN ADDITION PROVIDE (2) # 5 VERTS. AT EDGE OF
ALL OPENINGS, (2) #4 VERTS. AT PANEL PERIMETER -
SEE PANEL ELEVATIONALS. (TYP. U.N.O.)

 * SEE 1 / 56 FOR TYPICAL VERTICAL REINFORCEMENT LOCATION WITHIN PANEL

FIGURE 22-10 ■ Steel required to reinforce openings can be repre-sented in a schedule to keep the drawings uncluttered. *Courtesy Bill Berry, Berry-Nordling Engineers, Inc.*

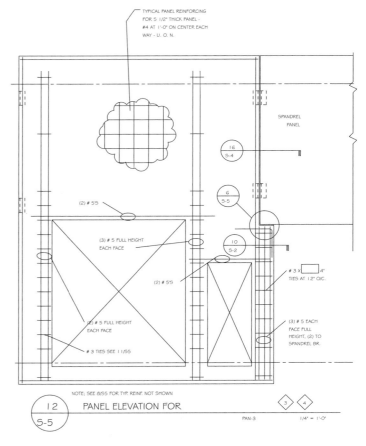

PANEL ELEVATION FOR
12 / S-5
PAN-3   1/4" = 1'-0"

FIGURE 22-11 ■ An individual elevation can be drawn for each panel. *Courtesy Van Domelen/Looijenga/McGarrigle/Knauf Consulting Engineers.*

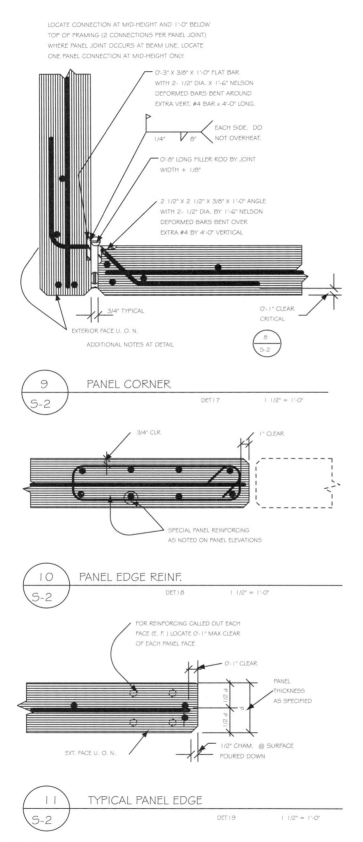

LOCATE CONNECTION AT MID-HEIGHT AND 1'-0" BELOW
TOP OF FRAMING (2 CONNECTIONS PER PANEL JOINT)
WHERE PANEL JOINT OCCURS AT BEAM LINE, LOCATE
ONE PANEL CONNECTION AT MID-HEIGHT ONLY.

0'-3" X 3/8" X 1'-0" FLAT BAR
WITH 2- 1/2" DIA. X 1'-6" NELSON
DEFORMED BARS BENT AROUND
EXTRA VERT. #4 BAR x 4'-0" LONG.

EACH SIDE. DO
NOT OVERHEAT.

0'-8" LONG FILLER ROD BY JOINT
WIDTH + 1/8"

2 1/2" X 2 1/2" X 3/8" X 1'-0" ANGLE
WITH 2- 1/2" DIA. BY 1'-6" NELSON
DEFORMED BARS BENT OVER
EXTRA #4 BY 4'-0" VERTICAL

3/4" TYPICAL

EXTERIOR FACE U. O. N.

ADDITIONAL NOTES AT DETAIL

0'-1" CLEAR
CRITICAL

9 / S-2   PANEL CORNER

DET17   1 1/2" = 1'-0"

3/4" CLR   1" CLEAR

SPECIAL PANEL REINFORCING
AS NOTED ON PANEL ELEVATIONS

10 / S-2   PANEL EDGE REINF.

DET18   1 1/2" = 1'-0"

FOR REINFORCING CALLED OUT EACH
FACE (E. F.) LOCATE 0'-1" MAX CLEAR
OF EACH PANEL FACE

0'-1" CLEAR

PANEL
THICKNESS
AS SPECIFIED

EXT. FACE U. O. N.

1/2" CHAM. @ SURFACE
POURED DOWN

11 / S-2   TYPICAL PANEL EDGE

DET19   1 1/2" = 1'-0"

FIGURE 22-12 ■ No matter which method is used to represent the panels, details must be drawn to represent each panel intersection. *Courtesy Van Domelen/Looijenga/McGarrigle/Knauf Consulting Engineers.*

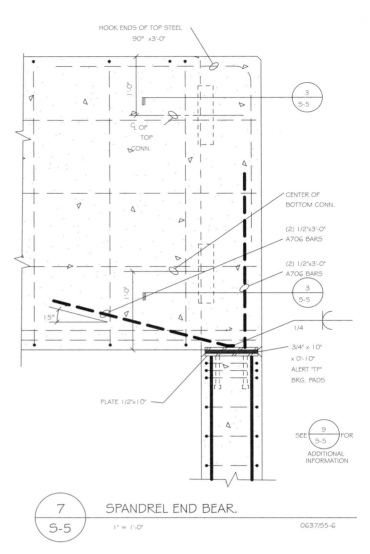

HOOK ENDS OF TOP STEEL
90° x3'-0"

CL OF
TOP
CONN.

CENTER OF
BOTTOM CONN.

(2) 1/2"x3'-0"
A706 BARS

(2) 1/2"x3'-0"
A706 BARS

3 / S-5

3/4" x 10"
x 0'-10"
ALERT "TF"
BRG. PADS

PLATE 1/2"x10"

SEE **9 / S-5** FOR
ADDITIONAL
INFORMATION

7 / S-5   SPANDREL END BEAR.

1" = 1'-0"   0637/55-6

FIGURE 22-13 ■ Reinforcing steel details are required to be drawn for individual panels. *Courtesy Van Domelen/Looijenga/McGarrigle/Knauf Consulting Engineers.*

ers enables drafters a very high degree of drawing accuracy, some components of a detail may be drawn out of scale for drawing clarity. In small-scale drawings, thin materials such as plywood, sheet rock, or steel plates may be represented thicker than they really are to add clarity. Care should be taken as components are enlarged so that their size remains consistent when compared to other material. Figure 22-14 shows examples of enlarging materials for clarity.

Drawing Placement

Structural sections and details should be drawn as close as possible to where they are referenced on the drawing. The drafter needs to place the details so that they conform to the job layout established by the engineer. Whenever possible, details should be drawn so that common features such as walls, floors, or roof lines are aligned. This is especially true if details are not placed in individual detail boxes. Aligning the top of the slab in each detail adds clarity to the

A detail should generally be started by drawing the major component to be shown first. Figure 22-15 shows the layout for a beam-column-foundation connection with the three major components represented. The detail can be started by representing the foundation first, because it will be used to support the column. Once the major shapes have been defined, specific details such as straps and bolt holes can be added, as shown in Figure 22-16. The

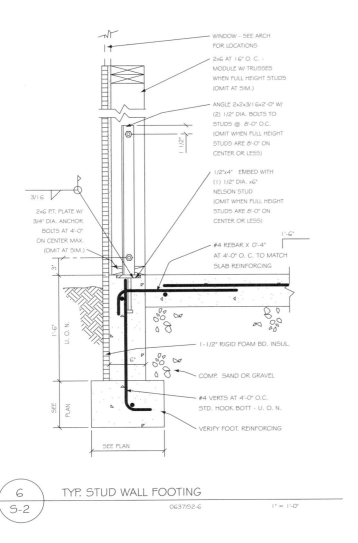

WINDOW - SEE ARCH
FOR LOCATIONS

2x6 AT 16" O. C. -
MODULE W/ TRUSSES
WHEN FULL HEIGHT STUDS
(OMIT AT SIM.)

ANGLE 2x2x3/16x2'-0" W/
(2) 1/2" DIA. BOLTS TO
STUDS @ 8'-0" O.C.
(OMIT WHEN FULL HEIGHT
STUDS ARE 8'-0" ON
CENTER OR LESS)

1/2"x4" EMBED WITH
(1) 1/2" DIA. x6"
NELSON STUD
(OMIT WHEN FULL HEIGHT
STUDS ARE 8'-0" ON
CENTER OR LESS)

#4 REBAR X 0'-4"
AT 4'-0" O. C. TO MATCH
SLAB REINFORCING

1-1/2" RIGID FOAM BD. INSUL.

COMP. SAND OR GRAVEL

#4 VERTS AT 4'-0" O.C.
STD. HOOK BOTT - U. O. N.

VERIFY FOOT. REINFORCING

3/16

2x6 P.T. PLATE W/
3/4" DIA. ANCHOR
BOLTS AT 4'-0"
ON CENTER MAX.
(OMIT AT SIM.)

6 / S-2   TYP. STUD WALL FOOTING

0637/52-6   1" = 1'-0"

FIGURE 22-14 ■ Small materials such as bolts or plywood are often enlarged so that they can be clearly seen. *Courtesy Van Domelen/ Looijenga/McGarrigle/Knauf Consulting Engineers.*

total page layout. When individual detail boxes are used, an attempt should still be made to group details so that similar items can be aligned. Although similar features cannot always be aligned, the box outlines provide an order and drawing clarity.

Drawing Process

Since no two details are alike, one step-by-step process cannot be used for their layout. Every drafter should have a logical method of attacking drawing layout, however. A common method of layout would be as follows:

1. Draw the materials to be represented
2. Detail each material
3. Dimension required components
4. Place all required notes

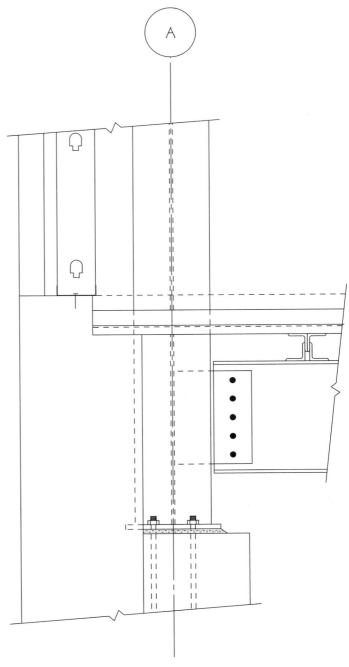

FIGURE 22-15 ■ Details are started by drawing the outlines of major construction materials.

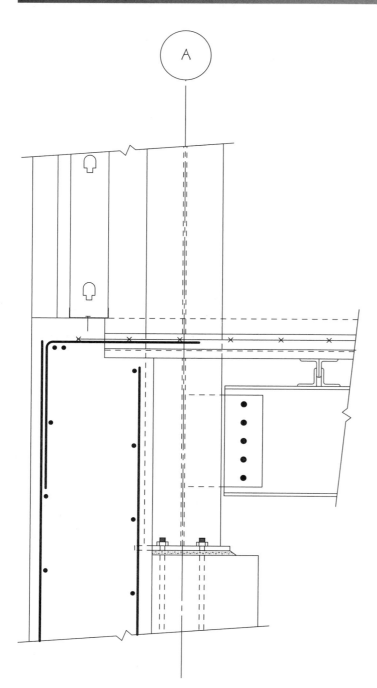

FIGURE 22-16 ■ Once major shapes have been defined, specific details should be added.

FIGURE 22-17 ■ Line quality and hatch patterns are added to add drawing quality.

drawing process can be completed by adding varied line weights and hatch patterns to represent each material. See Figure 22-17.

With all material drawn, dimensions should be added. Any feature that is to be fabricated should be dimensioned. Any feature that is pre-manufactured does not need to be dimensioned. Place dimensions so that smaller dimensions do not cross over larger dimensions. On large-scale drawings, dimensions can be placed within clear spaces of the detail. Never place dimensions in the component if clarity will be compromised.

Notes should be placed to reference all material represented in the detail. Care should be taken not to repeat information specified in the architectural drawings or within the specifications. Generally sizes should be specified in dimension form rather than as a note. When a drawing is scanned, numerals on dimension lines are easier to visualize than numbers placed within fields of text. Text should be placed using the guidelines given in Chapters 2 and 15. Information presented in a note should be placed using the following format:

Size, component name, information required to fabricate or install.

Although the information cannot be completed at this time, a detail title, scale, and reference bubble should also be placed as the text is being provided. Figure 22-18 shows the completed detail.

Representing Materials

The drafter's main consideration when drawing sections or details is to clearly distinguish between materials. Section II and Chapters 16 and 18 give examples of how construction materials are represented in elevation and section views. The amount of detail to be drawn is determined by the scale that will be used to plot the drawing. As drawing scale increases, the complexity of material representation should also increase. As the detail zooms in on a smaller area, drawing representation should also increase.

Figure 22-19 shows an example of a foundation detail showing six different materials. Using standard hatch patterns, the soil, gravel fill, concrete, steel reinforcing, grout, and structural steel can each be quickly identified. Even though a continuous line is used to represent the shape of the structural steel, concrete, and reinforcing, each can be easily distinguished because of varied line widths and the use of hatch patterns. Care should be taken to keep the size of the hatch pattern from becoming ineffective. Using too large a pattern will not provide the repetition needed to distinguish the pattern. Too small a pattern will have the same effect as using the **FILL** command.

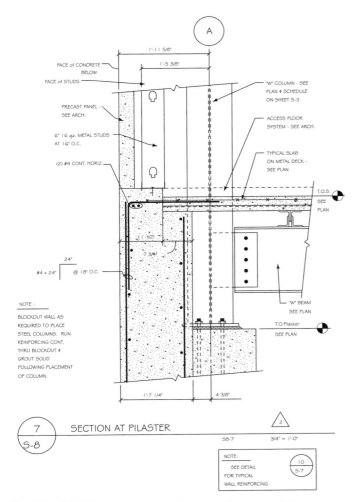

FIGURE 22-18 ■ The completed detail with notes and dimensions.

LAYER GUIDELINES FOR STRUCTURAL SECTIONS & DETAILS

S-DETL	Details
S-DETL-ANNO	Notes
S-DETL-DIMS	Dimensions
S-DETL-IDEN	Detail identification numbers
S-DETL-MBND	Material beyond the cutting plane
S-DETL-MCUT	Material cut by the cutting plane
S-DETL-PATT	Textures and hatch patterns
S-SECT	Sections
S-SECT-ANNO	Notes
S-SECT-DIMS	Dimensions
S-SECT-IDEN	Markers used to identify elevations, grids, etc.
S-SECT-MBND	Material beyond the cutting plane
S-SECT-MCUT	Material cut by the cutting plane
S-SECT-PATT	Hatch patterns used to represent various materials

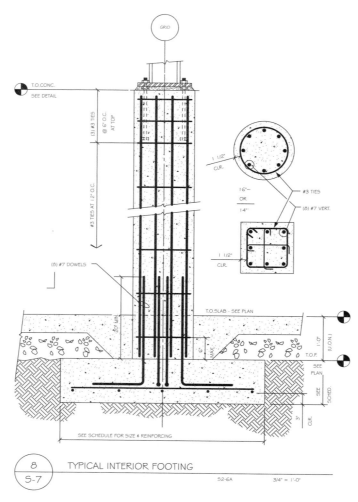

GRID

T.O.CONC.
SEE DETAIL

(3) #3 TIES
@ 6" O.C.
AT TOP

#3 TIES AT 12" O.C.

1 1/2"
CLR.

16"—
OR
14"

#3 TIES

(8) #7 VERT.

1 1/2"
CLR.

(8) #7 DOWELS

T.O.SLAB - SEE PLAN

24" MIN

T.O.F.

1'-0" (U.O.N.)

SEE
PLAN

SEE
SCHED.

3" CLR.

SEE SCHEDULE FOR SIZE & REINFORCING

8
S-7
TYPICAL INTERIOR FOOTING
S2-6A   3/4" = 1'-0"

FIGURE 22-19 ■ A key to good detailing is to use varied line weight, line types, and hatch patterns to distinguish between materials.

CHAPTER 22
Drawing Structural Elevations and Sections

CHAPTER QUIZ

Answer the questions on a separate sheet of paper. Print the chapter title, the question number, and a short complete statement for each answer.

Question 22-1 What types of drawings typically require structural elevations?

Question 22-2 What is the purpose of structural steel elevations?

Question 22-3 What is the advantage of drawing individual panel elevations for concrete tilt-up construction?

Question 22-4 What is a spandrel?

Question 22-5 List common scales typically used with structural details and describe how the scale affects the drawing method.

Question 22-6 How can a drafter decide if something should be dimensioned?

Question 22-7 What should be drawn first as a detail is being laid out?

Question 22-8 Rewrite the following note in its proper format:

Install bolts through slotted holes using 2" dia. washers each side of beam. Machine bolts to be 3/4" dia. × 10" A325-N high strength bolts.

DRAWING PROBLEMS

Use the reference material from preceding chapters, local codes, and vendor catalogs to complete one of the following projects. Unless other instructions are given by your instructor, complete the project (22-1, 22-2, & 22-4) that corresponds to the floor plan that was drawn in Chapter 15. Skeletons for problems 22-1 through 22-4 can be accessed from http://www.delmar.com/resources/ocl.html. Use these drawings as a base to complete the assignment. Use appropriate symbols, linetypes, dimensioning methods, and notations to complete the drawing. Determine any unspecified sizes based on material or practical requirements. Unless specified, select a scale appropriate to plotting on "D" size material and determine LTSCALE and DIMVARS.

You will act as the project manager and will be required to make decisions about how to complete the project. Study each of the drawing problems related to the structure prior to starting a drawing problem. Any information not provided must be researched and determined by you unless your instructor (the project architect and engineer) provides other instructions. When conflicting information is found, information from preceding chapters should take precedence. Make several sketches of possible solutions based on examples found throughout this text and submit them to your instructor prior to completing each problem with missing information.

Place the necessary cutting plane and detail bubbles on the framing plan to indicate each drawing.

Note: Keep in mind as you complete these drawings that the finished drawing set may contain more sections and in some cases hundreds of details more than you are being asked to draw. Other details that would be required for the building department or the construction crew will not be drawn due to the limitations of space and time.

Problem 22-1 Use the drawing as a guide and complete a framing elevation for the front, rear, and interior shear walls of the condominium started in Project 15-1.

Front wall—upper floor: On exterior face of wall use 1/2" c-d exterior plywood Struct II or better with 8d @ 4" o.c. at all edges and 8d @ 6" o.c. at field.

Front wall—middle and lower floor: Same plywood as above, applied to interior and exterior surfaces of wall.

Front straps—vertical: Upper to middle floor—use MST48 over double top plate with (48)-10d at exterior face of each opening. Middle to lower floor—MST60 over double top plate with (60)-10d to interior and exterior side of wall. Lower to foundation—Use HD-7A to each vertical support at each side of each opening.

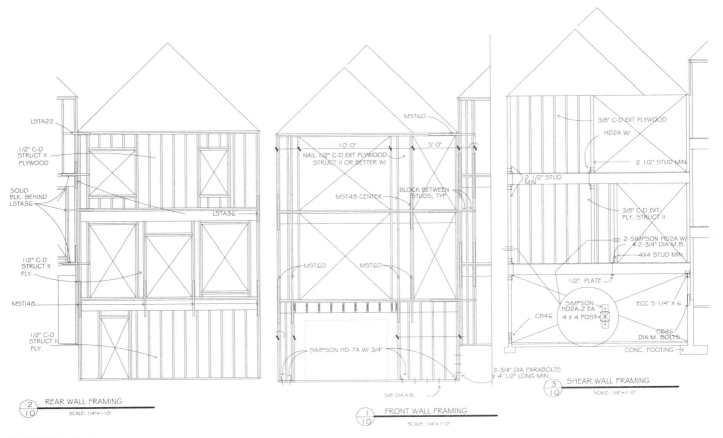

PROBLEM 22-1

Attach to foundation with 3/4" A.B. with 12" embedment, through 3× dfpt plate. Bolt to vertical supports as per manufacturer's specifications. Use 5/8" × 12" A.B. at 16" o.c. for plate to foundation. Use (3)-3/4" parabolts × 4 1/2" long at end of shear wall into 4' high retaining wall.

Front straps—horizontal: Block between studs for all horizontal straps as plates from one unit are tied into the wall system of the next unit. Use MST60 with 10d nails at each plate level.

Rear wall—upper floor: Use plywood as per front wall with nailing at 6" and 12" spacings.

Rear wall—middle floor: Use plywood as per upper floor applied to interior and exterior face of wall at each unit.

Rear wall—lower floor: Use plywood as per upper floor applied to exterior face only.

Rear straps—vertical: Upper to middle floor—Use LSTA36 with (24)-10d nails. Middle to lower floor—use MSTI48 w/ (48)-10d nails.

Rear Straps—horizontal: Use LSTA22 at top plate, and LSTA36 at all other plates this wall.

Interior wall—upper floor: Use 3/8" ply with nails at 4/6" spacing one side only. Use 3× stud at each end of shear wall. Use HD2A with 5/8" bolts as per manufacturer's specifications to each end post to post located in wall below.

Interior wall—middle floor: Use plywood as per wall above applied to each side of wall. Use 4× studs each end of shear wall. Place 4× stud below the post in wall above. Use HD2A with 5/8" bolts as per manufacturer's specifications. Use a 1/2" × 4 × 9 plate on bottom side of glu-lam beam for bolt attachment.

Interior wall-lower floor: Place 5 1/2" × 10 1/2" glu-lam beam below shear wall above. Support on 4 × 6 post with EPC cap and appropriate bolts at each end. Attach post to foundation with CB column base.

Problem 22-2 Use the drawings provided to detail the truss-to-beam and truss-to-ledger connections for the structure started in Chapter 15-2. Refer to Chapter 18 for information regarding the ledgers and use Figure 18-21 as guide.

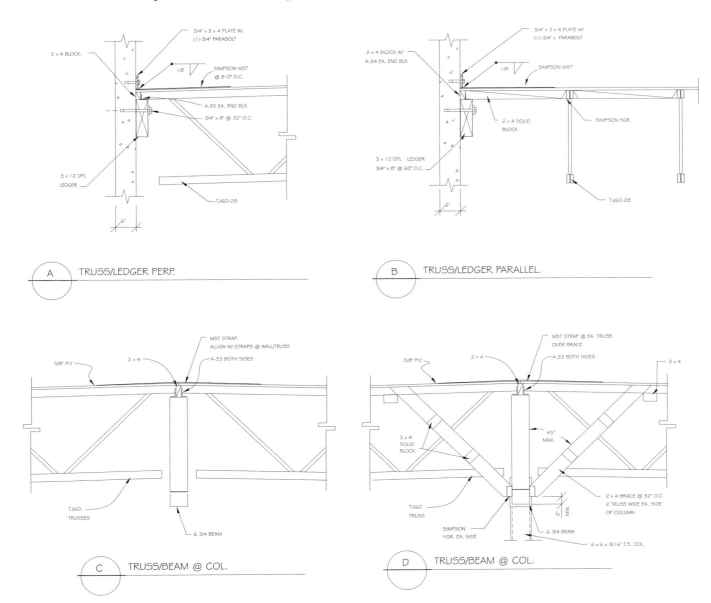

A TRUSS/LEDGER PERP.

B TRUSS/LEDGER PARALLEL.

C TRUSS/BEAM @ COL.

D TRUSS/BEAM @ COL.

PROBLEM 22-2

Problem 22-3 Use the drawings provided as a guide to complete the details required to explain construction of the shear wall referenced in Problems 22-3a to 22-3e. For each detail, provide a plan view of each plate to locate bolting patterns.

Problem 22-3a Draw an elevation of the steel bracing system used in a wall using a scale of 1/4" = 1'–0". Assume a distance of 12'–0" from center to center of columns, and 10'–0" from floor to floor.

Problem 22-3b Support the W18 × 50 beam on a 7 1/2" × 9 × 1/2" steel cap plate bolted to beam with 2 machine bolts. Weld plate to a TS 5 × 5 × 3/16" steel tube column with 3/16" fillet all around weld. Hang a W16 × 26 from the beam with 6 × 9 × 5/16" steel connection plate welded to the W18 × 50 with 1/4" fillet with 2" return at the top and bottom. Bolt the plate to the W16 × 26 beam with (3)-3/4"Ø machine bolts through plate and web. Bolts to be 1 1/2" from each edge and 3" o.c. Set top bolt 5" down from the top flange. Provide a note to indicate that all bolts are to be A325-N high strength bolts and require special inspection.

Problem 22-3c Use a 5 × 5 × 3/16" steel tube column welded to the base plate with 5/16" fillet welds all around at field. The plate is to be a 12 × 12 × 7/8" steel base plate on 1" dry pack with (4)-1 1/8" × 18" long anchor bolts into footing. Base plate bolts to be 15/16" minimum from the edge of plate. Below the base plate weld an 1 1/2" × 4 × 12" long shear plate with 1/2" × 12" fillet weld each side. Set shear plate in 3 1/2" × 5 1/2" deep ×18" long keyway filled with dry pack.

Provide a 5 × 5 × 1/4 " steel tube brace between each steel column. Weld the brace to the base and the column with a 3/8" fillet weld all around. Provide a 14" × 10" × 3/8" steel gusset plate. Extend the gusset plate 1/2" past the backside of the column. Weld the plate to the column with a 1/4" × 9" fillet weld on the back and side of the column. Weld the plate to the brace with a 3/8 × 3 1/2" fillet weld on each side.

Provide 22-3d Use a TS 5 × 5 × 1/4" column to support a W16 × 40 beam bolted to a 6 1/2 × 24 × 1 1/4" steel cap plate with (2)-5/8" diameter machine bolts. Weld the beam to the plate with a 3/16" fillet weld all around. Butt the W16 × 26 beam into each side and support on the cap plate. Bolt each beam with (4)-A325N high-strength bolts to plate at 3" o.c. 1 1/2" from the end of the beam. Weld the plate to column with 5/16" fillet weld all around. Provide a 14 × 10 × 3/8" gusset plate and weld the plate to the column and base plate with 1/4" × 9" fillet weld on each side. Weld the TS 5 × 5 × 1/4" steel tube brace to the column with 3/16" fillet and 1/4" × 6"

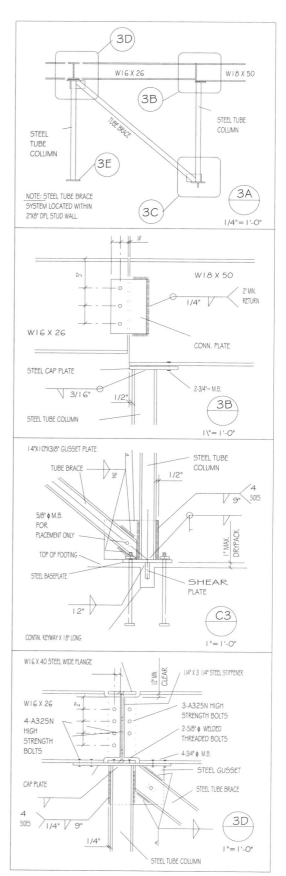

PROBLEMS 22-3A TO 22-3D

fillet on each side of the gusset.

Problem 22-3e Use a 10 × 10 × 7/8" steel base plate with four 18" anchor bolts into foundation. Set bolts at 19/16" minimum from the edge of the plate. Weld the column to the plate with 1/4" fillet weld all around. Set the plate on 1" drypack.

Problem 22-3f Use the drawings provided to draw the support for the steel beam. Support a W16 × 26 beam on a 5 1/2" × 7 × 1" steel cap plate bolted to the beam with two machine bolts. Weld the plate to the TS 3 1/2 × 3 1/2 × 1/4" steel tube column with a 3/16" fillet weld all around. Hang the W12 × 22 from the beam with (2)-1/2" × 3 1/2" × 1/4" steel angles × 5 1/2" long with two machine bolts through the 3 1/2" legs and W12 × 22. Weld the angles to the W16 × 26 with 5/16" fillet weld.

Problem 22-3g Use the detail provided to reference the beam and column. Support the W12 × 22 on a 5 1/2" × 7 × 3/4" steel cap plate welded to a 3" diameter steel pipe column. Bolt the beam to the plate with two machine bolts 1 1/2" from the edges. Provide a 1/4" steel stiffener plate welded all around. Weld a 5 × 9 × 7 gauge bucket connector to the end plate with a 3/16" fillet weld all around to support (2)-2 × 14. Set the 2 × 14 flush with a 3× plate connected to the steel beam.

Problem 22-4a Draw a panel elevation for the structure drawn in Problem 15-4. See the foundation and site plans to determine the depth of the foundation below the finish floor. Break out a section of the panel to display grade 60 #5 vertical bars at 10" o.c. Use grade 40 #4 horizontal bars at 12" o.c. Place steel centered in the wall and extend the steel to with 11/2" of all panel edges. Specify that all panels are to be 6" thick, 4000 psi concrete with an exposed finish.

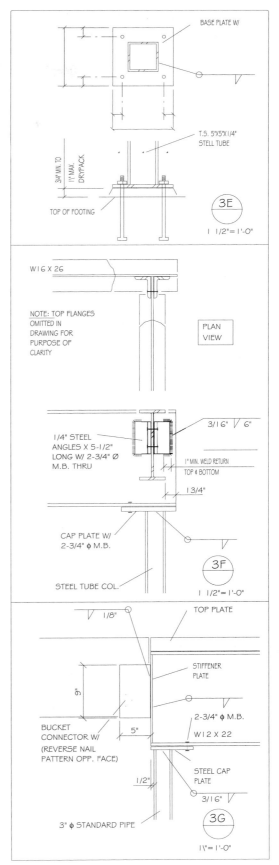

PROBLEMS 22-3E TO 22-3G

Problem 22-4b Use the attached drawing, information in Problem 22-4a, and the following information to complete a panel elevation for panel 1-E-D. Provide dimensions for the entire panel as well as for each opening. In addition to the normal steel use (2)-grade 40, # 5 × 48" diagonal steel bars at each corner of each door 2" clear of opening and 8" o.c. In addition, for 3'–0" × 7'–0" door use (2)-grade 60, #6 @ 6" o.c. vertically for full height of panel. Provide (2)-horizontal grade 40, #5 × 7' long rebar at top and bottom of door at 8" o.c.

For the 8'–0" door, use (2)-grade 40, #5 @ 8" o.c. vertical bars on each side of door, extending full panel height. Use (3)-grade 60, #4, @12" o.c. horizontal bars above and below door. Extend horizontal steel 24" each side of door. All door reinforcing steel to be 1" clear of exterior face.

Problem 22-4c Using the guidelines for the panels on Problem 22-3a or 22-3b, design and draw the panel elevations for the panels located on grids 1, 2, 4, 5, and A, C, and K for this structure. Show and dimension all openings. Verify foundation locations with information provided in Chapter 23.

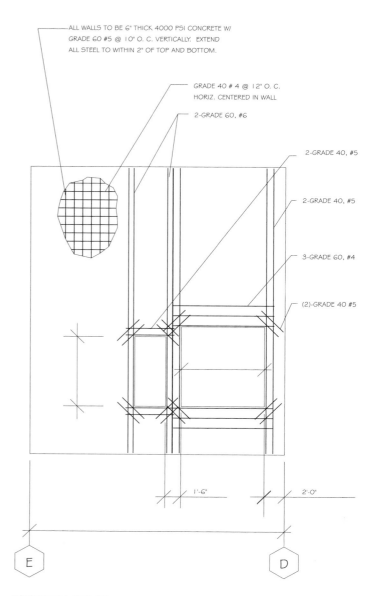

PROBLEM 22-4B

CHAPTER 23

Foundation Systems and Components

All structures are required to have a foundation to resist gravity and lateral loads. The foundation must also be designed to resist forces from floodwaters, seismic, wind, freezing, and seismic activity. The loads from the structure must be spread throughout the soil that surrounds the foundation. Common methods of spreading the loads of the structure into the soil, common types of foundations, and concrete and wood floor systems are introduced in this chapter, as well as methods of drawing the plans to show the structural system to be used.

SOIL CONSIDERATIONS

Before the foundation drawings are considered, the type of soil at the construction site must be determined. A soils engineer tests the soil for its strength, the affect of moisture on the soil-bearing capacity, compressibility, liquefaction, and expansiveness. Soil samples are taken from holes bored in representative areas of the site. Using the findings of the soils engineer, the structural engineer can design a foundation to resist the loads of the proposed structure.

Soil Texture

The texture of the soil influences each aspect of foundation design. Soil *texture* refers to the granular composition of the soil and influences the bearing capacity. The bearing capacity of soil depends on its composition and the moisture content. The IBC lists five basic classifications for soil. Soils and their allowable bearing pressure include the following:

Crystalline bedrock	12,000 psf (574.8 kPa)
Sedimentary and foliated rock	4,000 psf (191.6Kpa)
Sandy gravel and gravel	3,000 psf (143.7 kPa)
Sand and silty sand	2,000 psf (95.8 kPa)
Clay and sandy clay	1,500 psf (47.9 kPa)

Structures built on low-bearing capacity soils require footings that extend into stable soil, or footings that are spread over a wide area. Both options will be discussed later in this chapter.

Settlement

Settlement of soil will occur as the loads of the structure are placed on the footing. *Settlement* is caused by reduction of space between soil grains. Granular soils have fewer void areas and suffer only slight compaction as loads are applied. Clay-based soil has a larger

percentage of voids than sandy soil and compacts more under loads. Clay soils tend to compact over long periods of time because any moisture that is trapped during construction cannot move easily through the clay. A properly designed foundation takes into account the type and depth of soil, so that the effects of settlement are not seen throughout the structure. The foundation must be designed so that settlement is evenly distributed throughout the foundation. This is accomplished by designing the structure so that the loads are distributed into uniform types of soil.

Moisture Content

The amount of water the foundation will be exposed to, as well as the permeability of the soil, must also be considered in the design of the foundation. As the soil absorbs water it expands, causing the foundation to heave. The amount of rain and the type of soil can also cause the foundation to heave. The evaporation rate of soil is also of great importance in the design of commercial foundations, because many are formed by the use of an on-grade concrete slab. Unless the site has been properly prepared, the slab can trap large amounts of moisture. In areas that receive little rain, the contrast between the dry soil under the slab and damp soil beside the foundation provides a tendency for heaving to occur at the edge of the slab. If the soil beneath the slab experiences a change of moisture content after the slab is poured, the center of the slab can heave. The heaving can be resisted by reinforcing provided in the foundation and throughout the floor slab, and by proper drainage.

Surface and ground water must be properly diverted from the foundation to maintain the soil's ability to support the building loads. Proper drainage also minimizes water leaking into the crawl space or basement, reducing mildew or rotting. The IBC requires the finish grade to slope away from the foundation at a minimum slope of one vertical unit per 20 horizontal units (5% slope) for a distance of 10' (3048 mm). For arid climates, the IBC allows the slope to be reduced to 1 vertical unit per 48 horizontal units.

Gravel or coarse-grain soils can be placed beside the foundation to increase percolation. Because of the void spaces between grains, course-grain soils are more permeable and drain better than fine-grain soils. A drain is often required to be placed beside the foundation at the base of a gravel bed in damp climates to facilitate drainage of water from the foundation. As the amount of water is reduced in the soil surrounding the foundation, the lateral loads imposed on the foundation are reduced. The weight of the soil becomes increasingly important as the height of the foundation wall is increased. Foundation walls enclosing basements should be waterproofed. Asphaltic emulsion is often used to prevent water penetration into the basement. Floor slabs placed below grade are required to be placed over a vapor barrier.

Freezing

Because water expands as it freezes, the depth of the frost level must be considered in the design of the foundation. Expansion and shrinking of the soil causes heaving in the foundation. As the concrete expands with the soil, the foundation can crack. As the soil thaws, water not absorbed by the soil can cause the soil to lose much of its bearing capacity, causing further cracking of the foundation. In addition to the geographic location, the soil type also affects the influences of freezing. Based on the recommendations of the soils engineer and requirements of the building department, the engineer provides detailed information to the drafter to ensure that the foundation extends below the depth of the deepest expected frost penetration. Specifications are typically included to ensure that concrete is not poured under severe weather conditions, so that excess moisture is not trapped under the foundation.

Soil Elevation

The height of the soil above a known point is referred to as its *elevation*. Construction sites often include soil that has been brought to or moved onto the site. Terms that the drafter is required to specify on plans include natural grade, existing grade, finished grade, excavation, fill, backfill and compacted fill. Each is shown in Figure 23-1.

The height of soil in its undisturbed condition is called the *natural grade*. The use of the term natural grade assumes that the soil has never been disturbed. The depth of the foundation is generally based on the depth into the natural grade. The *existing grade* is the result of adding or removing soil from the site prior to the start of construction. Soil that was moved to the site as access roads were placed is an example of soil that forms the existing grade. *Finished grade* is used to define the height of the soil after all work at the site has been completed. The finished grade may be higher or lower than the natural grade, depending on the design of the project.

The removal of soil from a construction site is known as *excavation*. Figure 23-2 shows a site that has been excavated, and the forming of the foundation. Because the foundation must be placed below the maximum penetration level of frost, soil must be moved to provide a stable base for the foundation. Soil is also removed from the site to alter the contour of the finished grade. A bank created by excavation is referred to as a *cut bank*. The engineer provides specifications controlling the slope of banks adjacent to the structure.

Soil that is placed over the natural grade is called *fill material*. Once the foundation system has been formed and the floor system is in place, soil is placed in the footing trench to create the finished grade. This fill material, which is placed against the foundation, is called *backfill*. Fill can be either uncompacted or compacted. *Uncompacted fill* is soil that is brought to a site to raise the elevation of the natural grade. Although the fill may be leveled after it is placed, no special care is given to compact the material.

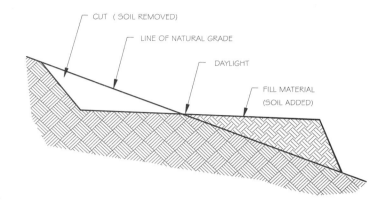

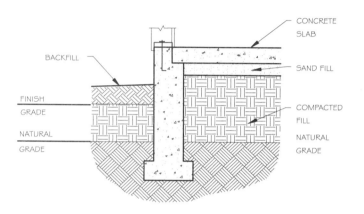

FIGURE 23-1 ■ Terms associated with the placement of soil include natural grade, cut and fill material, and the daylight point. Fill can be further described as uncompacted and compacted.

FIGURE 23-2 ■ Once a site has been excavated to the desired elevation, foundation forms and reinforcing steel can be placed. *Courtesy Sara Jefferis.*

Uncompacted fill material is unusable for supporting the loads from a structure. Where uncompacted fill is found at the building site, the foundation must be placed so that it is bearing on the natural grade. *Compacted fill* is soil that is added to the site in a very systematic method to ensure its load-bearing capacity.

Soil Compaction

Fill material can be compacted to increase its bearing capacity. Compaction is typically accomplished by vibrating, tamping, rolling, or by adding a temporary weight. Three major ways to compact soil include the following:

Static force—A heavy roller presses soil particles together

Impact forces—A ramming shoe repeatedly strikes the ground at high speed

Vibration—High-frequency vibration is applied to soil through a steel plate

Each of these methods reduces the number of air voids contained between grains of soil. Proper compaction lessens the effect of settling and increases the stability of the soil, which will in turn increase the load-bearing capacity. The effects of frost damage are minimized in compacted soil because penetration of water into voids in the soil is minimized. A soils or geotechnical engineer specifies the requirements for soils excavation and compaction. The drafter's job is to place the specification of the engineer clearly on the plans.

TYPES OF FOUNDATIONS

The structure to be supported, the type of soil, and the contour of the site affect the type of foundation to be used. Common methods of forming the foundation include spread footings, pilings, and piers. The type of foundation used affects the type of floor system used. The drafter represents both the foundation and floor system on the foundation drawings.

Continuous or Spread Footings

The most common type of foundation used in light commercial structures is a continuous or spread footing. This type of footing consists of a footing and a stem wall. Each is typically made of poured concrete, although concrete blocks can be used for the stem wall. The outline of the footing can be represented by dashed lines, and the width and depth of the footing should be dimensioned. Figure 23-3 shows examples of each component as it would be represented in details, sections and on a foundation plan. A *footing* is the base of the foundation system and is used to disperse loads over the soil. Figure 23-4 shows common sizes for foundation components based on the IBC.

Footing Reinforcement

When the footing is placed near expansive soil, the bottom of the footing will bend, placing the footing in tension. Steel is placed in the bottom of the footing to resist the forces of tension in concrete. This reinforcing steel or rebar is usually not shown on the founda-

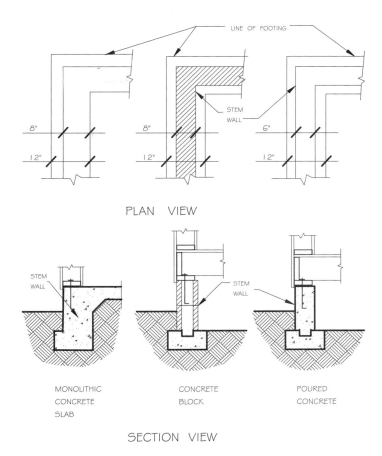

PLAN VIEW

SECTION VIEW

FIGURE 23-3 ■ Representation of common foundations in plan and section view.

tion plan but is specified in a note that gives the size, quantity, and spacing of the steel. Only rebar that is #5 diameter or larger is considered as reinforcing, with the use of smaller steel considered as plain (non-reinforced) concrete. See Chapter 9 for a review of reinforcing steel sizes and numbers. Steel rebar may be smooth, but most uses of steel require deformations to help the concrete bind to the steel.

Stem Walls

The stem wall is the vertical portion of the foundation that brings the foundation up above the finished grade. The stem wall can be represented by continuous lines and is usually centered on the footing. Most municipalities require the wall to extend 6" (150 mm) above the finished grade. The material used to construct the foundation and the wall affects how they are joined together. Common stem wall materials include poured concrete and concrete blocks. When both are made of concrete, they can be formed separately or in one pour. Figure 23-5 shows common methods of forming concrete foundations. The exact shape used depends on the location of the footing to the property line, and the type of floor system to be used.

If the footing and stem wall are formed separately, a *keyway* is formed in the footing by placing a 2 × 4 in the wet concrete. Once

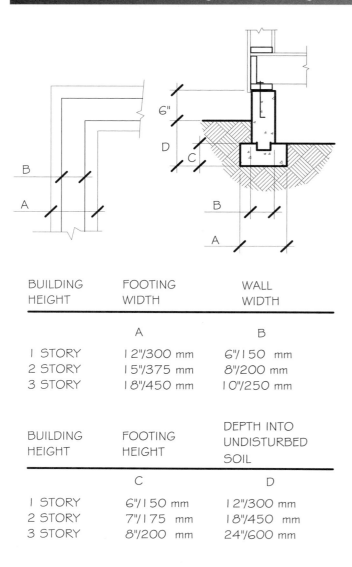

BUILDING HEIGHT	FOOTING WIDTH	WALL WIDTH
	A	B
1 STORY	12"/300 mm	6"/150 mm
2 STORY	15"/375 mm	8"/200 mm
3 STORY	18"/450 mm	10"/250 mm

BUILDING HEIGHT	FOOTING HEIGHT	DEPTH INTO UNDISTURBED SOIL
	C	D
1 STORY	6"/150 mm	12"/300 mm
2 STORY	7"/175 mm	18"/450 mm
3 STORY	8"/200 mm	24"/600 mm

FIGURE 23-4 ■ Minimum required footing and stem wall sizes based on the International Building Code.

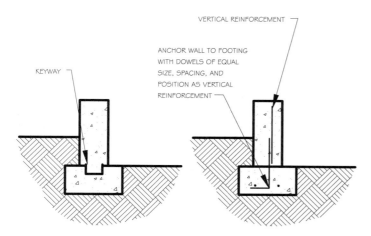

FIGURE 23-5 ■ A keyway or reinforcing steel can be used to tie the stem wall to the foundation to resist lateral loads.

the concrete has set, the 2 × 4 is removed, leaving an indentation in the footing. When the stem wall is poured, the concrete fills in the keyway. The interlocking nature formed by the keyway helps the wall resist lateral pressure from the soil.

An L-shaped piece of steel is normally placed in the footing to increase the bond with the wall. The seismic zone, the loads to be supported, and the height of the wall determine whether reinforcing is to be used. When the engineer requires steel to be shown, the drafter is typically required to dimension the size, spacing, and placement of the steel in details. Although the steel from the footing to the wall is not shown on the foundation plan, it is often specified in a note referenced to the stem wall. The short leg of the rebar is referred to as the *toe* and is usually placed below a continuous piece of steel in the footing. The extension of the footing steel into the stem wall must also be specified on the footing details. The distance that the stem wall steel overlaps the extended footing steel is referred to as the *lap*. Lap is often critical in the resistance of lateral pressure. Figure 23-6 shows an example of a foundation detail.

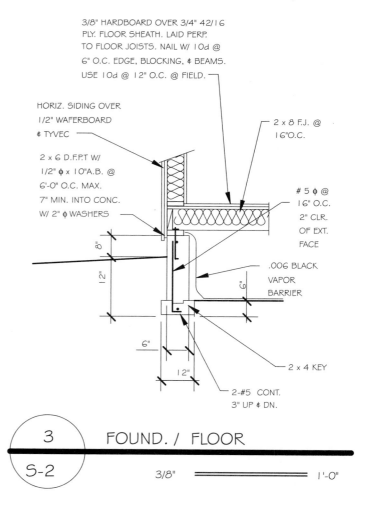

FIGURE 23-6 ■ Details drawn by drafters working with the engineering team show foundation construction. Each detail is referenced to the foundation drawings and sections.

Stepped Walls and Foundations

When the construction site is not level, the foundation must be stepped with the height of each portion of the stem wall adjusted to provide a level base for the floor system. The top of the stem wall will also change heights at each change in floor elevation. Figure 23-7 shows an example of a stepped footing. Most codes limit the height of each step to 24" (600 mm) and require a minimum of a 32" (810 mm) run. If wood studs are used to frame between the stem wall and the floor system, a minimum length of 14" (360 mm) is required.

Veneer Support

If a masonry or stone veneer is to be used, the footing width must be altered to support the veneer. The thickened footing width needs to be represented and specified on the foundation plan as well as in a detail showing construction. Figure 23-8 shows common methods of supporting the veneer and how it would be represented on a foundation plan.

Anchor Bolts

The type of floor system to be used affects how the top of the stem wall is constructed. An anchor bolt is placed at the top of the stem wall while it is wet to bond a wood floor system to the stem wall. If concrete blocks are used to form the stem wall, the cells containing the reinforcing steel and the anchor bolt are filled with solid grout. (See Chapter 10). An anchor bolt is set in the edge of a con-

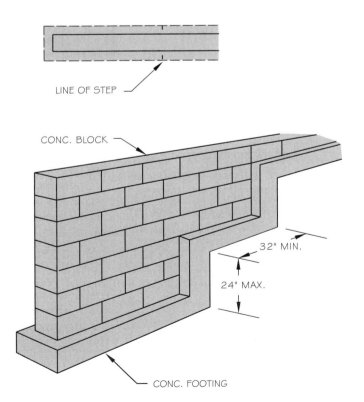

FIGURE 23-7 ■ The footing can be stepped on sloping lots to reduce the material required to build the stem wall.

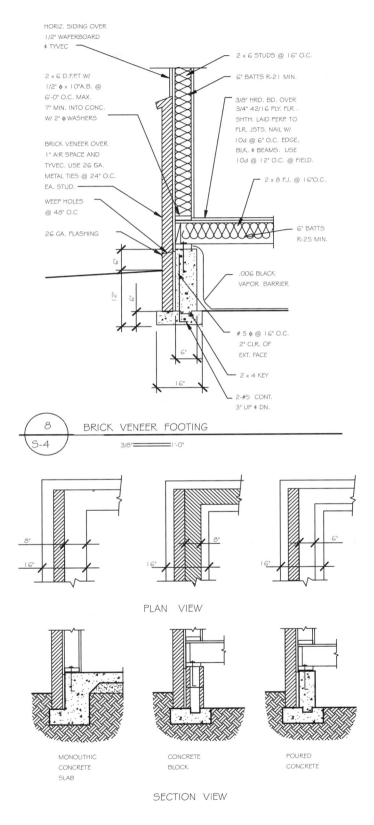

FIGURE 23-8 ■ If brick veneer is to be used, the support must be represented on the foundation plan and sections.

crete slab to bond the wall system to the concrete. Anchor bolt spacing is determined by the engineer based on the amount of lateral, seismic, and uplift loads that must be resisted. Minimum sizes of bolts allowed by code include a diameter of 1/2" (13 mm), a length of 10" (250 mm) with a 7" (180 mm) penetration into the concrete. Bolts are to be placed through a pressure treated mudsill and held in place by a nut resting on a 2" (50 mm) diameter washer. Spacing is not to exceed 6'–0" (1800 mm). In many parts of the country, the mudsill must be protected from termites by use of a metal guard placed between the plate and the top of the stem wall. The anchor bolts, termite shield, and mudsill are not drawn on the foundation plan, but are specified by a note similar to Figure 23-9. Bolts are represented in details similar to Figure 23-10. When a concrete slab is to be placed over compacted fill, steel is often placed in the stem wall, which will extend into the slab. This steel must be shown on both the foundation plan views and the details of the wall/slab intersection.

Metal Connectors

Metal connectors are often used to connect the wall or floor system to the foundation system, to transfer loads caused from wind and

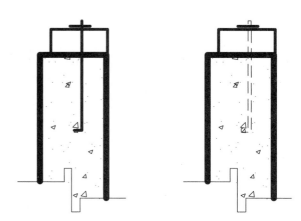

FIGURE 23-10 ■ Bolt representation in details.

seismic forces. Three common metal anchors represented in foundation drawings include column connectors, hold down anchors and metal angles. Each is shown in Figure 23-11. The engineer determines the size and location of any metal connectors, and the drafter is required to represent these on the foundation plan and details. Figure 23-12 shows common methods of representing metal connectors.

Beam Support

If the stem wall is to support a wood floor system, a method of supporting floor beams must be provided. A cavity called a *beam pocket* can be built into the wall to provide support for beams. A 3" (75 mm) bearing surface must be provided for a wood beam resting on concrete with a 1/2" (13 mm) air space on all sides of the beam. Beams resting in a beam pocket must be wrapped with 55# felt to protect the beam from moisture in the concrete. A metal hanger can be used in place of a beam pocket. Figure 23-13 shows examples of each method of beam support.

Foundation Access

When the stem wall is to support a wood floor system, a method of access must be provided to the space below the floor system, which is called the *crawl space*. Access can be provided from inside the structure through the floor, or through the stem wall. A 30" × 18" (750 × 450 mm) minimum access hole is required if access is through the stem wall. When access is to be through the floor, a 22" × 30" (560 × 750 mm) access panel must be provided and shown on the floor plan. The access should be placed out of traffic routes; for example in an area such as a closet. The drafter should verify that floor beams shown on the foundation plan will not interfere with the access hole.

Vents must be installed in the stem wall to allow air to circulate in the crawl space. A vent is required within 3'–0" (900 mm) of each foundation corner and at approximately 10'–0" (3000 mm) spacings. Consideration should be given to door placement so that vents are not to be placed where doors or steps will be located. Vents should also not be placed below a beam pocket. Figure 23-14 shows how the crawl access and vents can be represented and specified on a foundation plan.

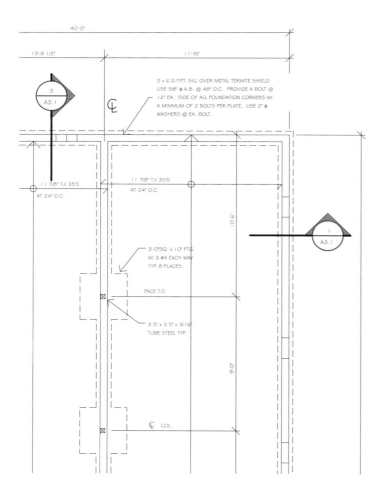

FIGURE 23-9 ■ Anchor bolts, termite shields, and mudsills are not drawn on the foundation plan but must be specified in a note. *Courtesy Scott R. Beck, Architect.*

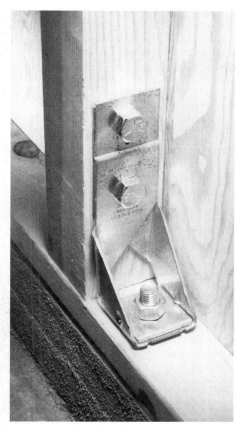

FIGURE 23-11 ■ Metal anchors and straps are often shown on the foundation plan. They help transfer loads from the skeleton into the foundation. Three of the most common include (left) column base, (middle) hold down anchor, and (right) framing straps. *Courtesy Simpson Strong-Tie Co., Inc.*

Stem Wall Insulation

In colder regions of the country, 2" (50 mm) rigid insulation is used to insulate the stem wall. Insulation can be placed on either side of the stem wall. If the insulation is placed on the exterior side of the wall, heat from the structure will be retained in the concrete. When placed on the exterior side of the stem wall, the insulation should extend down past the bottom of the foundation. Exposed insulation must be protected from punctures, which can usually be done by placing a protective covering such as 1/2" (13mm) concrete board over the insulation. Figure 23-15 shows a common method of placing insulation on the exterior side of the wall.

Alternative Wall Materials

Many companies have developed alternative products for forming the stem wall. Blocks made of expanded polystyrene form (EPS) or other lightweight materials that can be stacked into the desired position and fit together with interlocking teeth. EPS block forms can be assembled in a much shorter time than traditional form work and remain in place to become part of the finished wall. Reinforcing steel can be set inside the block forms in patterns similar to traditional block walls. Once the forms are assembled, concrete can be pumped into the forms using any of the common methods of pouring. The finished wall has an R-value of between R-22 and R-35 depending on the manufacturer. Figure 23-16 shows a foundation being built using EPS forms.

Masonry Wall Support

Masonry and precast concrete construction eliminate the need for a stem wall because the wall is supported directly on the footing. The reinforcing for the masonry wall is connected to the steel that extends from the footing. Figure 23-17 shows wall details for brick and concrete block walls. Precast wall panels are fastened directly to the footing, and soil is back-filled directly against the panel. Figure 23-18 shows a footing prepared for a precast panel. When this type of construction is used, a separate slab plan and foundation plan are generally drawn. Each is introduced later in this chapter.

Retaining Walls

A *retaining* or basement wall is typically made of 8" (200 mm) poured concrete or concrete block and is assumed to be anchored at the top and bottom by a floor system. The width will vary depending on the loads to be resisted and the height of the wall. Because of the height, the lateral pressure tends to bend the wall inward, placing the soil side of the wall in compression and the side away from the soil in tension. (See Figure 23-19.) To resist the tension, the engineer specifies reinforcing based on the height of the wall, the soil pressure, and the seismic zone. Figure 23-20 shows an example of a retaining wall detail. The wall will appear on a foundation plan just as any other stem wall. The detail would be

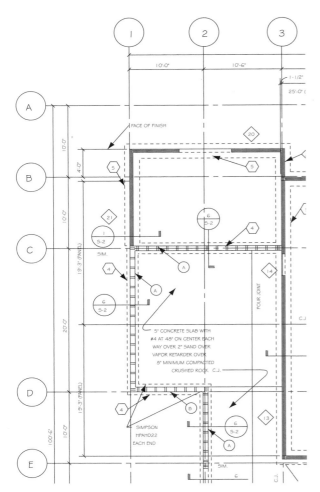

FIGURE 23-12 ■ Metal anchors and ties must be specified on the foundation plan. *Courtesy Architects Barrentine, Bates & Lee, A.I.A.*

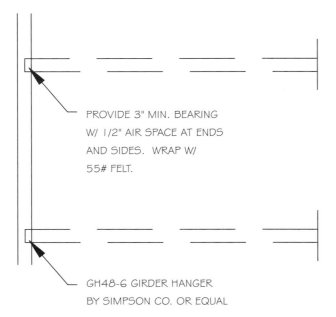

PROVIDE 3" MIN. BEARING
W/ 1/2" AIR SPACE AT ENDS
AND SIDES. WRAP W/
55# FELT.

GH48-6 GIRDER HANGER
BY SIMPSON CO. OR EQUAL

FIGURE 23-13 ■ A girder can be supported by the use of a metal hanger or by a recess placed in the stem wall.

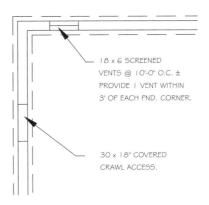

18 x 6 SCREENED
VENTS @ 10'-0" O.C. ±
PROVIDE 1 VENT WITHIN
3' OF EACH FND. CORNER.

30 x 18" COVERED
CRAWL ACCESS.

FIGURE 23-14 ■ Foundation access and vents must be represented on the foundation plan if a wood floor system is to be used.

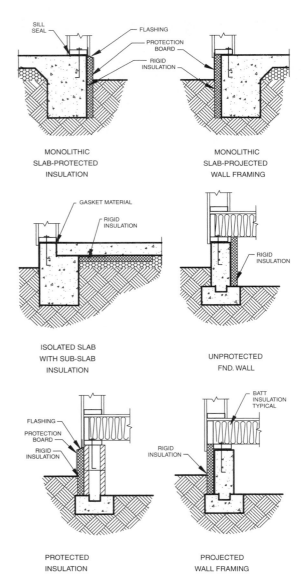

FIGURE 23-15 ■ The location of foundation/floor insulation varies depending on the type of floor framing system to be used. When placed on the exterior side of the stem wall, the insulation must be protected from punctures.

FIGURE 23-16 ■ Forms are available for constructing the stem wall. Forms remain in place and insulate the structure. *Courtesy American Polysteel Forms.*

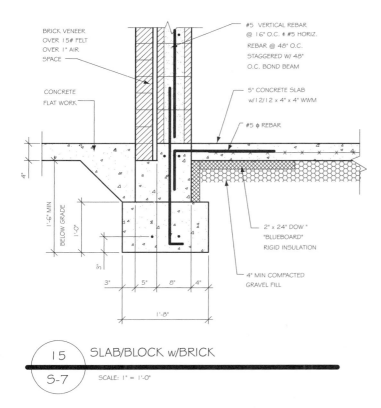

BRICK VENEER
OVER 15# FELT
OVER 1" AIR
SPACE

CONCRETE
FLAT WORK

#5 VERTICAL REBAR
@ 16" O.C. & #5 HORIZ.
REBAR @ 48" O.C.
STAGGERED W/ 48"
O.C. BOND BEAM

5" CONCRETE SLAB
w/12/12 x 4" x 4" WWM

#5 φ REBAR

2" x 24" DOW "
"BLUEBOARD"
RIGID INSULATION

4" MIN COMPACTED
GRAVEL FILL

4"

1'-6" MIN
BELOW GRADE
1'-0"

3"

3" 5" 8" 4"

1'-8"

(15 / S-7) SLAB/BLOCK w/BRICK

SCALE: 1" = 1'-0"

FIGURE 23-17 ■ Varied line quality and hatch patterns should be used to represent masonry products and reinforcement in foundation details. *Courtesy Wil Warner.*

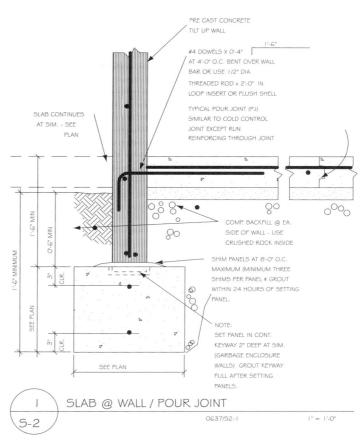

PRE CAST CONCRETE
TILT UP WALL

#4 DOWELS X 0'-4"
AT 4'-0" O.C. BENT OVER WALL
BAR OR USE 1/2" DIA
THREADED ROD x 2'-0" IN
LOOP INSERT OR FLUSH SHELL

TYPICAL POUR JOINT (PJ)
SIMILAR TO COLD CONTROL
JOINT EXCEPT RUN
REINFORCING THROUGH JOINT

1'-6"

SLAB CONTINUES
AT SIM. - SEE
PLAN

COMP. BACKFILL @ EA.
SIDE OF WALL - USE
CRUSHED ROCK INSIDE

SHIM PANELS AT 8'-0" O.C.
MAXIMUM (MINIMUM THREE
SHIMS PER PANEL & GROUT
WITHIN 24 HOURS OF SETTING
PANEL.

NOTE:
SET PANEL IN CONT.
KEYWAY 2" DEEP AT SIM.
(GARBAGE ENCLOSURE
WALLS) GROUT KEYWAY
FULL AFTER SETTING
PANELS.

1'-6" MINIMUM
1'-6" MIN
0'-6" MIN
SEE PLAN
3" CLR.
3" CLR.

SEE PLAN

(1 / S-2) SLAB @ WALL / POUR JOINT

0637/92-1   1" = 1'-0"

FIGURE 23-18 ■ Representing the intersection of a precast concrete panel with the foundation. *Courtesy Van Domelen/Looijenga/McGarrigle/Knauf Consulting Engineers.*

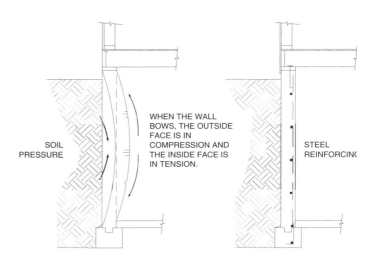

SOIL
PRESSURE

WHEN THE WALL
BOWS, THE OUTSIDE
FACE IS IN
COMPRESSION AND
THE INSIDE FACE IS
IN TENSION.

STEEL
REINFORCING

FIGURE 23-19 ■ Soil pressure bows a wall inward when the wall is anchored to a floor at the top and bottom. Because concrete has poor tensile strength, steel must be placed near the side in tension to help resist the soil load.

referenced to the foundation plan. Some method of waterproofing the wall is also generally noted on the foundation plan. Typically, walls are covered with at least two layers of bituminous waterproofing material or by fiber-reinforced asphaltic mastic.

Anchorage

Anchor bolts are placed in the retaining wall by a method similar to that used for stem walls. Bolt spacings are generally much closer for a retaining wall because of the greater loads to be resisted. Because the floor systems are used to resist the soil loads, a metal angle is typically used to securely bond the floor joist to the top of the wall. The anchor bolts and angles are specified in note form on the foundation and shown in details. When the floor joists are parallel to the retaining wall, blocks are typically placed between floor joists near the wall edge to transfer lateral pressure on the wall to the floor system. This blocking would be drawn and noted on the foundation plan and represented in a detail similar to Figure 23-21.

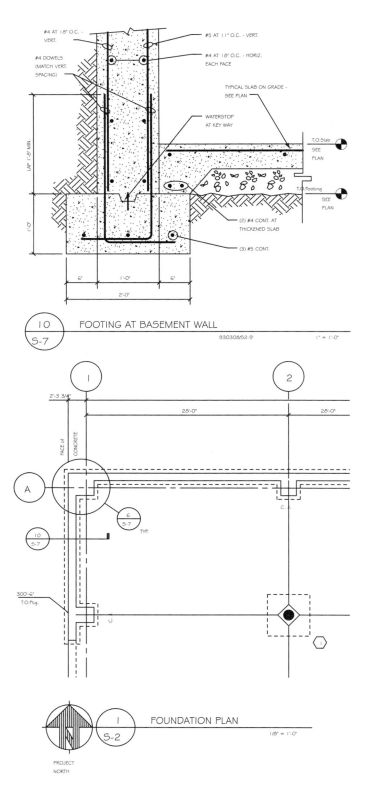

FIGURE 23-20 ■ A retaining wall is represented on the foundation plan using the same methods used to describe a stem wall. Details must be referenced to the plan to show wall construction. *Courtesy Van Domelen/Looijenga/McGarrigle/Knauf Consulting Engineers.*

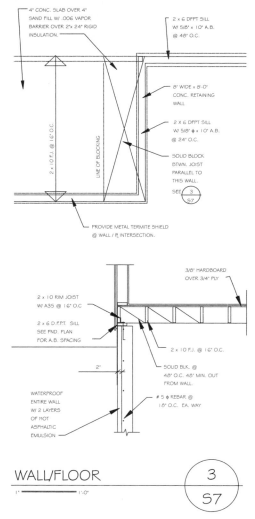

FIGURE 23-21 ■ If a retaining wall is to be placed parallel to wood floor joists, solid blocking is required between the joists to stiffen the floor system. Required blocking should be specified on the foundation plan and details.

Footings

The width of the footing for a one-story retaining wall is typically 16" (400 mm) wide. The exact size is determined by the engineer based on the loads to be supported and the soil-bearing pressure. The footing width should be drawn and dimensioned using the same methods used to describe other foundation footings. To reduce soil pressure, a drain is placed in a gravel bed at the base of the retaining wall to remove water from the soil. (See Figure 23-20.) The gravel allows water to percolate down to the drain and be diverted away from the foundation. The drain can be shown on the foundation plan, as in Figure 23-22, or it may be represented in a note and not shown.

Cantilevered Retaining Walls

When a structure is built on a sloping site, the retaining wall may not be required to extend to first floor level. (See Figure 23-23.) A wall that is not supported at both ends by a floor system is referred to as a *cantilevered wall*. Although less soil is retained than with a full height wall (as shown in Figure 23-24) the wall has a greater tendency to bend and create a hinge point. To resist the soil pressure, a wider footing must be used. The width of the footing is based on the height of the wall and is determined by the engineer. Figure 23-25 shows two methods of placing the footing in rela-

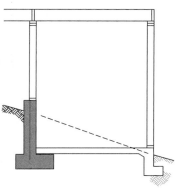

FIGURE 23-23 ■ On sloping sites, the retaining wall may not be required to span the full height between floor levels.

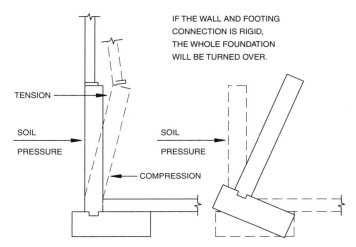

IF THE WALL AND FOOTING CONNECTION IS RIGID, THE WHOLE FOUNDATION WILL BE TURNED OVER.

TENSION

SOIL PRESSURE

SOIL PRESSURE

COMPRESSION

FIGURE 23-24 ■ When the top of a retaining wall is not anchored to the floor system, lateral pressure causes the wall to rotate inward.

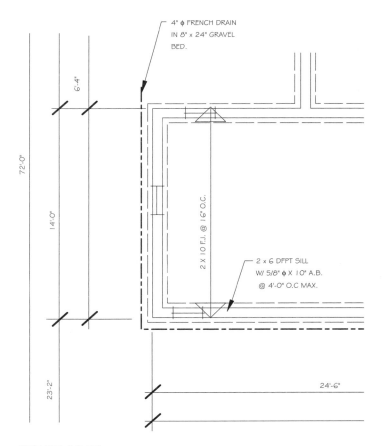

4" φ FRENCH DRAIN IN 8" x 24" GRAVEL BED.

2 X 10 F.J. @ 16" O.C.

2 x 6 DFPT SILL W/ 5/8" φ X 10" A.B. @ 4'-0" O.C MAX.

6'-4"

72'-0"

14'-0"

23'-2"

24'-6"

FIGURE 23-22 ■ A foundation drain may be represented on the foundation plan, or it may be referenced in general notes.

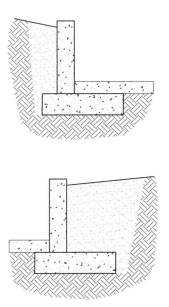

FIGURE 23-25 ■ Lateral pressure resisted by the wall is transferred to the foundation. The foundation in the upper example transfers loads to the soil below the wall. The foundation in the lower example resists the lateral loads by the weight of the soil above the foundation.

tionship to the wall. Depending on the height of the wall, or the angle of repose of the soil being retained, a key may be placed on the bottom of the footing. As shown in Figure 23-26, the *key* is the rectangular or square portion of concrete placed on the bottom side of the footing to resist lateral pressure. The width of the wall and the footing must be represented and specified on the foundation. The key is not drawn or referenced on the foundation plan. A detail similar to Figure 23-27 must be referenced on the foundation plan to show the reinforcing, footing and wall size, and key size and location.

Interior Supports

Stem walls and footings are used to support the exterior portion of the structure. Interior load-bearing walls over concrete floors are supported on continuous footings. Bearing walls resting on a wood floor system or columns or post on a concrete floor system are supported by a beam supported on repetitively spaced spot concrete footings called *piers*. Piers and continuous interior footings should be drawn with the same linetype used to draw exterior footings. Figure 23-28 shows interior piers for concrete and wood floor systems. Piers are also used to provide increased bearing values to exterior footing. (See Figure 23-29.) Piers can be formed by using preformed or framed forms, or by excavation. When placed under a stem wall, piers are usually square or rectangular. When used to support interior loads on a floor system, piers are often round. Piers that are too close together to form their full shape are often poured as a rectangle. The pier depth is usually the same depth required for the exterior footing, with the diameter based on the load to be supported and the soil-bearing pressure. Piers are usually required to be recessed into the soil to resist lateral movement. In areas subject to seismic activity, a piece of rebar or a metal post base may be required to bond the support post to the pier. Blocking or a metal strap can also be provided to add stability to the post-to-beam connection. Figure 23-30 shows piers in section and plan view.

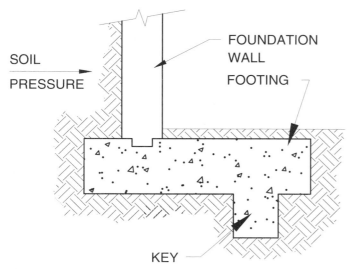

FIGURE 23-26 ■ When lateral pressure is too great for the footing to withstand, the surface area of the footing can be enlarged by adding a key.

Grade Beams

A *grade beam* can be used in place of a footing to provide added support for a foundation placed over unstable soil. An example of a grade beam is shown in Figure 23-31. The grade beam is similar to a beam that supports loads over an opening in the structure. The grade beam is placed below the stem wall and spans between stable supports, which may be concrete footings resting on stable soil or pilings. The depth and reinforcing required for a grade beam are determined by the engineer and are based on the loads to be supported. A grade beam resembles a footing when drawn on a foundation plan. Steel reinforcing may be specified by notes and referenced to details rather than on the foundation plan.

Pedestals

A *pedestal* is a poured concrete column that is built on top of a footing to extend through fill material up to a concrete floor level. (See Figure 23-32.) Pedestals are often used to provide support for post or columns so that loads on the vertical member are not transferred into the floor system. Figure 23-33 shows an example of a pedestal detail. Figure 23-34 shows how a pedestal and a pedestal schedule are represented on a foundation plan.

Pilings

A piling is a type of foundation system that uses beams placed between columns to support loads of a structure. The columns may extend into the natural grade or be supported on other material that extends into stable soil. Piling foundations are typically used

■ On steep hillside sites where it may not be feasible to use traditional excavating equipment

■ Where the loads imposed from the structure exceed the bearing capacity of the soil

■ On sites subject to flooding or other natural forces that might cause large amounts of soil to be removed

Coastal property and sites near other bodies of water subject to flooding often use a piling foundation to keep the useable space above the flood plane level.

When a structure is to be built above grade using a piling system, a support beam is typically placed under or near each bearing wall. Beams are supported on a grid of vertical supports that extends down to a stable strata of rock or dense soil. Beams can be steel, sawn or laminated wood, or prestressed concrete. Vertical supports may be concrete columns, steel tubes or beams, wood columns, or a combination of each material. When pilings are used for work at or below grade, concrete and steel are typically used to support the structure. Pilings can be several hundred feet in length. The engineer may require the depth to be based on the soil resistance measured as the pile is driven, or a minimum depth to solid rock based on the soils report. The desired depth is placed in the foundation notes by the drafter.

For shallow pilings, a hole can be bored and poured concrete with steel reinforcing can be used. Figure 23-35 shows an example

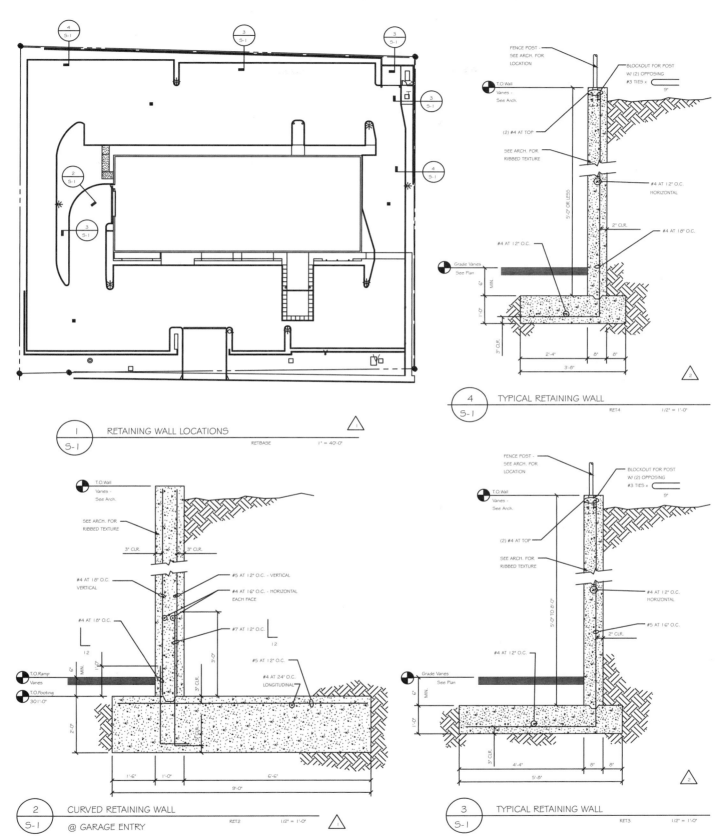

FIGURE 23-27 ■ Details are generally referenced to the foundation plan. When a project contains a wide variety of wall types, a simplified site plan can be placed near the foundation plan to reference details. *Courtesy Van Domelen/Looijenga/McGarrigle/Knauf Consulting Engineers.*

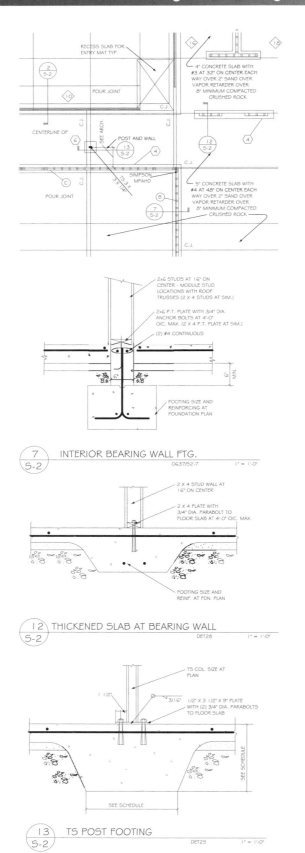

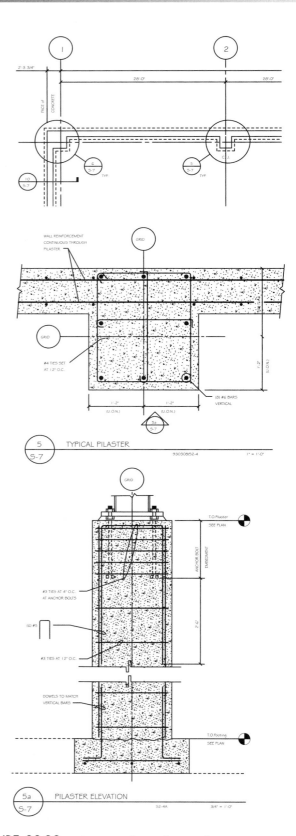

FIGURE 23-28 ■ Piers and continuous footings should be drawn with the same linetypes used to represent exterior footings. *Courtesy Architects Barrentine, Bates & Lee, A.I.A.*

FIGURE 23-29 ■ A pier can be used to reinforce a stem or retaining wall when loads to be supported exceed the bearing capacity of the foundation. *Courtesy Van Domelen/Looijenga/McGarrigle/Knauf Consulting Engineers.*

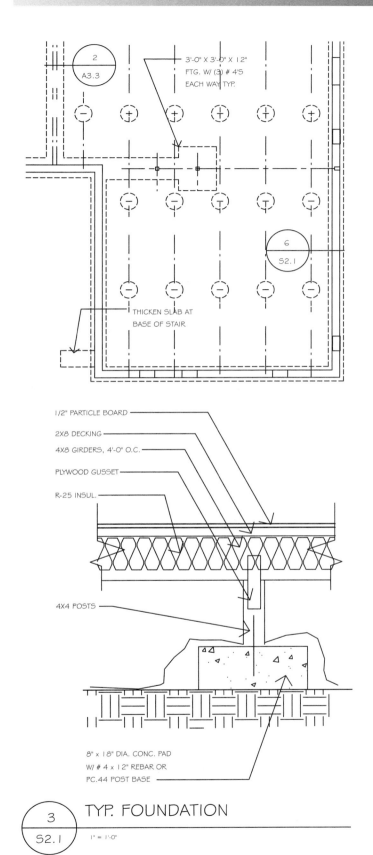

THICKEN SLAB AT
BASE OF STAIR

1/2" PARTICLE BOARD

2X8 DECKING

4X8 GIRDERS, 4'-0" O.C.

PLYWOOD GUSSET

R-25 INSUL.

4X4 POSTS

8" x 18" DIA. CONC. PAD
W/ # 4 x 12" REBAR OR
PC.44 POST BASE

TYP. FOUNDATION

3
S2.1   1" = 1'-0"

FIGURE 23-30 ■ Representing interior piers on the foundation plan and in detail. *Courtesy Scott R. Beck, Architect.*

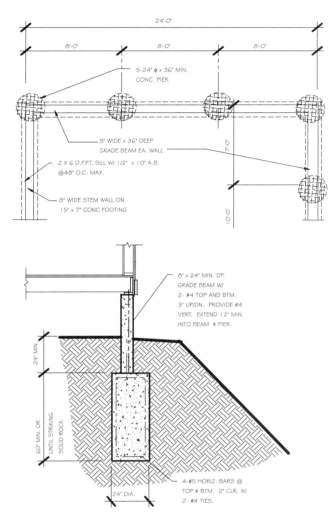

FIGURE 23-31 ■ In unstable soil, the stem wall can be reinforced and treated as a beam spanning between piers. *Courtesy Tereasa Jefferis.*

FIGURE 23-32 ■ The pedestal supporting the steel column extends from the footing, through the fill material that will be placed to support the floor. Notice the steel that extends from the pedestal into the future floor slab. *Courtesy Brian Sweeney.*

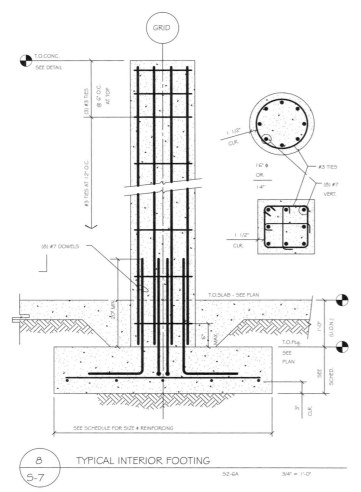

8 / S-7   TYPICAL INTERIOR FOOTING

S2-6A   3/4" = 1'-0"

FIGURE 23-33 ■ Details must be drawn to show the construction of a pedestal and referenced on the foundation plan. *Courtesy Van Domelen/Looijenga/McGarrigle/Knauf Consulting Engineers.*

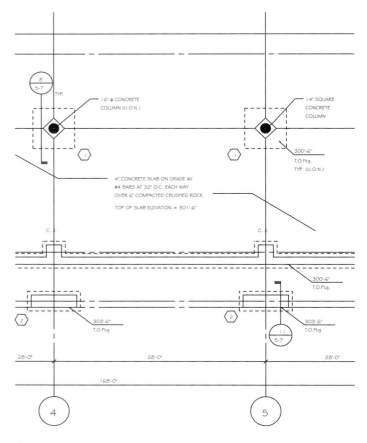

COLUMN AND FOOTING SCHEDULE			
	FOOTING SIZE	REINFORCING	NOTES
1	7'-6" x 7'-6" x 18" DEEP	(7) #6 BARS EACH WAY	
2	2'-8" x 7'-0" x 18" DEEP	(7) #5 BARS SHORT DIRECTION	
		(3) #5 BARS LONG DIRECTION	
3			

DESIGN SOIL PRESSURE: 5,000 p.s.f.

SEE SOILS INVESTIGATION BY: WRIGHT / DEACON & ASSOCIATES, INC., 7-28-98

1 / S-2   FOUNDATION PLAN   1/8" = 1'-0"

PROJECT NORTH

FIGURE 23-34 ■ The pedestal shown in Figure 23-33 can be seen in plan view with the supporting footing shown with dashed lines. When a wide variety of pedestals are to be used, a schedule can be used to specify reinforcing patterns and sizes. *Courtesy Van Domelen/Looijenga/ McGarrigle/Knauf Consulting Engineers.*

of a detail for a poured concrete piling. If the vertical support is required to extend deeper than 10' (3000 mm), a pressure-treated timber or steel column can be driven into the soil. Timber pilings that are above the water table must be pressure treated and are typically coated with creosote to extend their lifetime.

Steel Pilings

As the depth of the piling increases, or where layered soil conditions prevent wood from being used, steel H-sections can be used for sub-grade pilings. Steel is typically called for when the design requires the piling to be driven into solid rock. The pilings in Figure 23-36 were driven into stable soil and then capped with a base plate. From this platform, steel columns were welded to the plate to provide a support system for the structure. Figure 23-37 shows the components of a piling plan for a hillside office structure. Notice, in addition to the vertical columns and horizontal beams, diagonal steel cables are placed between the columns to resist lateral and rotational forces. In addition to resisting gravity loads, piling foundations must be able to resist forces from uplift,

lateral force, and rotation. Figure 23-38 shows a detail of a steel piling foundation that is used to support steel columns above grade. Notice that the engineer has specified a system that uses braces approximately parallel to the ground to stabilize the top of the pilings from lateral loads. These braces are not shown on the foundation plan but are specified on the elevations, sections, and details.

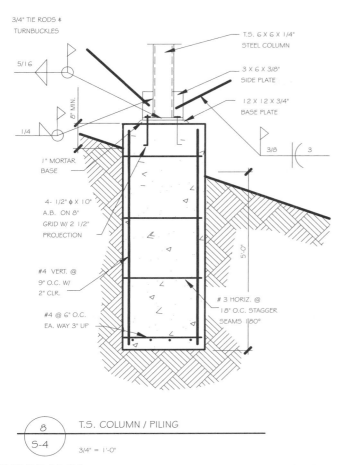

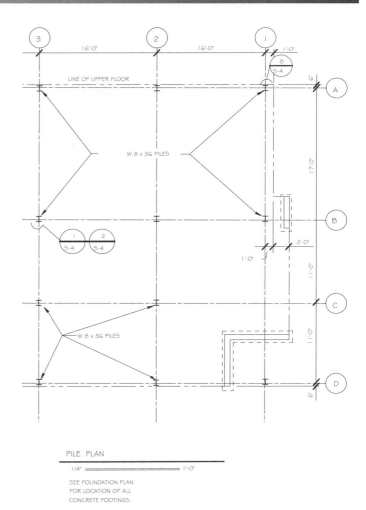

FIGURE 23-35 ■ Reinforced concrete pilings can be used for shallow pilings to provide support for the foundation.

FIGURE 23-37 ■ Representing pilings in plan view. Depending on the complexity of the structure, pilings may be shown on the foundation plan or on a separate piling plan.

FIGURE 23-36 ■ Steel pilings are often used to provide a stable base for a structure. A steel plate is welded to the top of the piling to provide support to a column or beam. The rectangular tube is used to span between pilings at the hinge point between the piling and support column. *Courtesy Donna Sweeney.*

Concrete Pilings

Pilings formed of precast and poured concrete are used for many structures as loads and heights are increased, to reduce forces causing uplift and overturning. Concrete pilings are also used in sandy soils to increase the soil friction of the pile. Concrete pilings offer excellent resistance to uplift. Reinforced precast concrete piles can be driven into the soil just like wood or steel pilings. Pilings can also be pretensioned, just as a vertical column to add to the strength of the piling. Figure 23-39 shows examples of common pile shapes.

Cast-in-Place Concrete Pilings

A steel shell can be used to place cast-in-place concrete pilings. The shell is driven into the ground to the required depth and concrete is poured into the shell. Shell lengths up to 40' (12,200 mm) can be driven with other sections welded or threaded together as the pile is driven. Reinforcing can be wired into position in the shell and the concrete placed in the pilings. Pilings can also be created by forming the hole by using a steel tube and then withdrawing the tube. The concrete conforms to the shape of the soil and increases

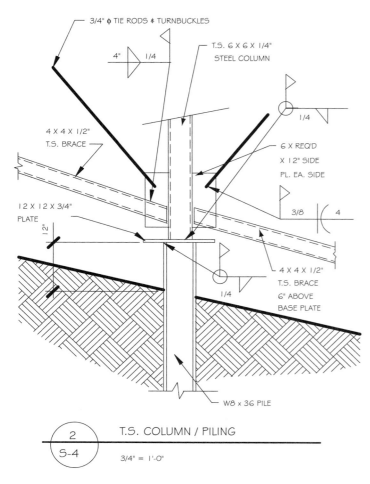

FIGURE 23-38 ■ Details showing the connections of the piling, support plates, and reinforcing must be referenced in the plan view.

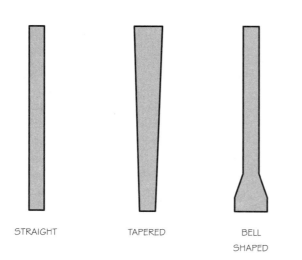

FIGURE 23-39 ■ Common shapes for deep depth cast-in-place concrete pilings.

the bonding strength over a steel encased concrete pile. The method used to place the concrete can also greatly affect the strength of the piling. If the steel casing is removed, concrete can be pumped under pressure, which will force the soil around the concrete to compress. As the soil compresses, the concrete forms an enlarged or bulbous tip at the end of the piling, which increase both the bearing and withdrawal strength.

TYPES OF FLOOR SYSTEMS

The most common floor system for commercial structures is an on-grade slab. Other types of floor systems used with commercial and light industrial construction include pretensioned slabs, slabs built below grade level, and slabs built on pilings. As building loads and height increase, the depth of the foundation must also increase. Typically, a piling foundation is used to provide stability. Multifamily applications and small construction projects in damp areas often use wood floor systems using joist or post-and-beam framing techniques supported on a concrete foundation.

On-Grade Slabs

A concrete slab provides a firm floor system with little or no maintenance and generally requires less material and labor than wood floor systems. The floor slab can be poured as an extension of foundation wall and footing as shown in Figure 23-40 in what is known as *monolithic construction*. Other common methods of pouring slabs are shown in Figure 23-41. The IBC requires slabs to be a

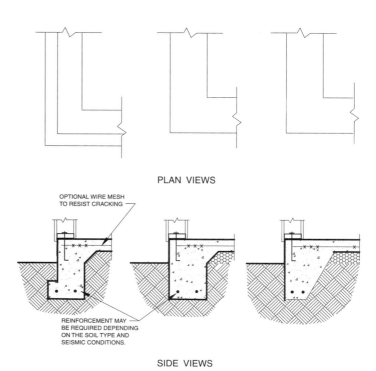

FIGURE 23-40 ■ Common methods of monolithic construction for on-grade concrete floor systems.

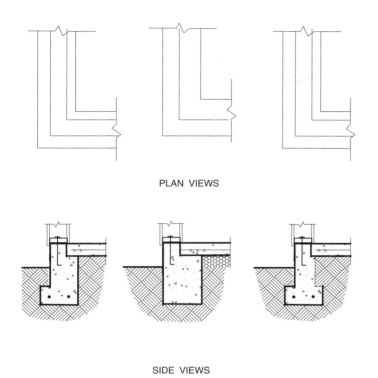

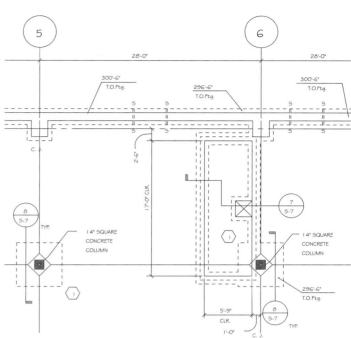

FIGURE 23-41 ■ Common methods of construction for on-grade concrete floor systems using two or more pours.

FIGURE 23-42 ■ Representing changes of floor level on a foundation plan. *Courtesy Van Domelen/Looijenga/McGarrigle/Knauf Consulting Engineers.*

minimum of 3 1/2" (90 mm), but design factors for commercial and industrial structures often require slabs of 5" to 6" (125 to 150 mm) in thickness. Keep in mind that plain (unreinforced) concrete slabs are only used as a floor system and not to transmit vertical loads. If load-bearing walls or columns are to be supported by the slab, reinforcing must be placed in the slab, a footing placed beneath the slab, or both may be required depending on the loads to be supported. Figure 23-28 shows common methods of supporting floor slabs at bearing walls. If a column is to be supported, piers can be placed below the slab as seen in Figure 23-29. The edge of the slab is represented on the foundation drawings by a continuous line.

Changes in Floor Elevation

The floor level is often required to change elevation to meet the design needs. A stem wall is formed between the two floor levels and should match the required width for an exterior stem wall. The lower slab is typically thickened to a depth of 8" (200 mm) to support the change in elevation. Figure 23-42 shows how a lowered slab can be represented on the foundation plan. Figure 23-43 shows common methods of changing floor elevations with a concrete slab, which would be shown in details referenced to the foundation plan. The step often occurs at what will be the edge of a wall when the framing plan is complete. Great care must be taken to coordinate the dimensions of a floor plan with those of the foundation plan, so that walls will match the foundation step.

Slab Joints

Joints in concrete were introduced in Chapter 10. Joints must be shown on the foundation because concrete tends to shrink approximately 0.66" per 100' (16.7 mm/30 500 mm) as the moisture in the mix hydrates and the concrete hardens. Concrete also expands and shrinks throughout its life based on the temperature and the moisture in the supporting soil. This can cause the floor slab to crack. To help control possible cracking, three types of joints can be placed in the slab: control, construction, and isolation. Each can be seen in Figure 23-44.

As the slab contracts during the initial drying process, tensile stress is created in the lower surface of the slab as it rubs against the soil. The friction between the slab and the soil causes cracking, which can be controlled by *control* or contraction joints. These joints do not prevent cracking, but they control where the cracks will develop in the slab. Control joints are usually one-third of the slab depth. Because the slab has been weakened, any cracking due to stress will happen along the joint. The American Concrete Institute (ACI) suggests control joints be placed a distance in feet equal to about 2 1/2 times the slab depth. For a 5" (125 mm) slab, joints would be placed at approximately 12.5' (3800 mm) intervals. The location of control joints should be represented by a line and specified by a note referenced to the joint. The spacing and method of placement can be specified in the general notes placed with the foundation plan or in the specifications within the project manual.

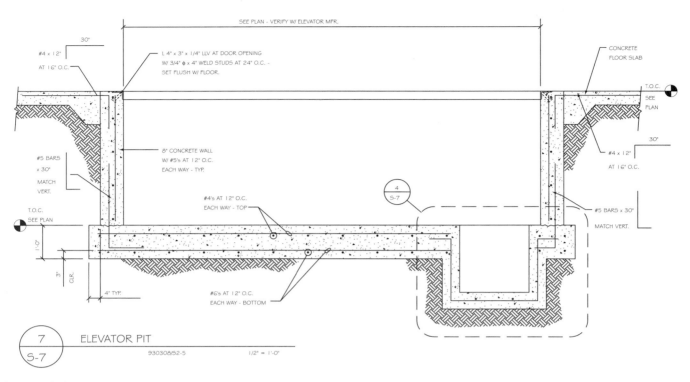

FIGURE 23-43 ■ Changes in floor level must be shown in detail and referenced to the foundation plan. *Courtesy Van Domelen/Looijenga/McGarrigle/Knauf Consulting Engineers.*

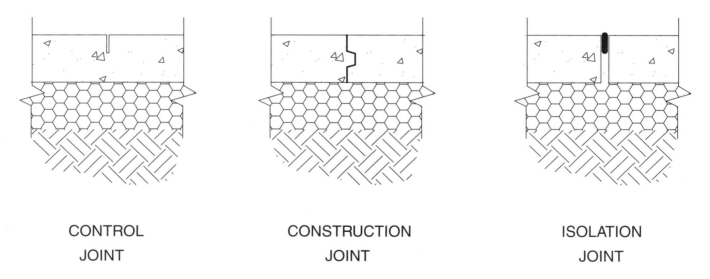

FIGURE 23-44 ■ Common slab joints specified on a foundation plan.

Isolation or *expansion joints* are used to separate a slab from adjacent structural members so that stress cannot be transferred into the slab and cause cracking. These joints also allow for expansion of the slab caused by moisture or temperature. Isolation joints are typically between 1/4" to 1/2" (6 to 13 mm) wide and should be drawn and specified on the foundation plan using methods similar to those used to represent control joints.

When slab construction must be interrupted, a *construction* or *keyed joint* is made to provide support between the two slabs. The key is a beveled edge that is about one-fifth of the slab thickness in depth, and one-tenth of the slab thickness in width. The method used to form the joint must be specified with information about other joint types. Because the crew placing the concrete determines the location, these joints are not shown on the drawings.

Slab Preparation and Placement

The slab may be placed above, below, or at grade level. Slabs placed above grade were considered in Chapter 10. Slabs built below grade are used in basement or subterranean construction. When used at grade level, the slab is placed 8" (200 mm) above grade level unless it is protected by a retaining wall. Approximately 8" to 12" (200 to 300 mm) of topsoil is removed to provide a stable, level, building site. Excavation usually extends several feet beyond the building footprint to allow for the operation of excavating equipment needed to trench for the footings. Once forms for the footings have been set, compacted fill material can be placed to support the slab. The slab is placed on a 4" (100 mm) minimum base of compacted sand or gravel fill to provide a level base for the pouring of the slab. The fill is not drawn on the foundation plan, but the material and required depth is specified on the foundation plan.

Slab Reinforcement

When the slab is placed on more than 4" (100 mm) of uncompacted fill, welded wire fabric similar to that shown in Figure 23-45 should be specified to help the slab resist cracking. The engineer will specify the spacings and sizes of welded wire fabric (WWF) to be identified on the foundation plan. Steel reinforcing bars can be added to a floor slab to prevent bending of the slab due to expansive soil. While mesh is placed in a slab to limit cracking, steel reinforcement is placed in a grid pattern near the surface of the concrete in tension to resist bending. (See Figure 23-46.) The amount of concrete placed around the steel is referred to as *coverage*. Proper coverage strengthens the bond between the steel and concrete and also protects the steel from corrosion or fire. If steel is required to reinforce a concrete slab, footing, or pier, the engineer will determine the size, spacing, coverage, and grade of bars to be used. Generally, steel reinforcement is not shown on the foundation plan but is specified by a note.

FIGURE 23-45 ■ Welded wire fabric can be used in concrete slabs to help resist cracking.

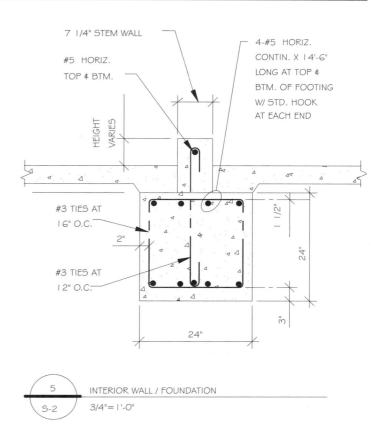

FIGURE 23-46 ■ Slab reinforcement should be specified on the foundation plan and shown in details.

Moisture Protection

The slab is required to be placed over 6 mil polyethylene sheet plastic to protect the floor from ground moisture. When a plastic vapor barrier is to be placed over gravel, sand should be specified to cover the gravel fill to avoid tearing the vapor barrier. The vapor barrier is not drawn but is specified with a note on the foundation plan.

Slab Insulation

Depending on the risk of freezing, some municipalities require the concrete slab to be insulated to prevent heat loss. The insulation can be placed under the slab if it is not placed on the outside of the stem wall. When placed under the slab, a 2" × 24" (50 × 600 mm) minimum rigid insulation material is usually required to insulate the slab. An isolation joint is typically provided between the stem wall and the slab to prevent heat loss through the stem wall. Figure 23-15 shows common methods of insulating the intersection of the floor and stem wall. Insulation is not shown on the foundation plan but is represented by a note and specified in details referenced to the foundation plan.

Plumbing and Heating Requirements

Figure 23-47 shows plumbing installation, which must be placed under the slab prior to the pouring of the slab. Information supplied by consulting engineers needs to be incorporated into the

FIGURE 23-47 ■ HVAC and plumbing lines that are required to be placed prior to the foundation must be indicated on a plan. The information is often provided by drafters working for consulting engineers. *Courtesy Margarita Miller.*

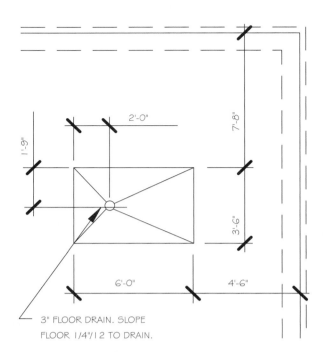

FIGURE 23-48 ■ Changes in the foundation caused by HVAC ducts and plumbing lines such as this floor drain should be indicated on the foundation plan.

foundation plan by the drafter to show required HVAC, electrical, and plumbing. Figure 23-48 shows a foundation plan referencing plumbing drains. Although piping runs are often not shown, terminations such as floor drains are shown, specified, and located on the foundation drawings.

Radon Protection

Structures built in areas of the country identified by the Environmental Protection Agency (EPA) as having high radon levels need to be protected from the cancer-causing gas. A common method of reducing the buildup of radon can be achieved by placing a 4" (100 mm) PVC vent pipe in the gravel fill covered with 6 mil polyethylene sheathing. Any joints, cracks, or penetrations in the floor slab must be caulked. The vent is run under the slab until it can be routed up through the framing system to an exhaust point that is a minimum of 10' (3000 mm) from other openings in the structure and 12" (300 mm) above the roof. The piping should be indicated on the foundation drawings and coordinated with other

drawings to show electrical wiring for installation of a fan located in the vent stack, and a system failure warning device.

Post-tensioned Concrete Reinforcement

The methods of reinforcement mentioned thus far are used when it is assumed that the slab will be poured over stable soil. Post-tensioning of slabs extends the range of where slabs can be placed. This method of construction was developed for reinforcement of slabs that were poured at ground level and then lifted into place for multilevel structures. Adopted for use on at-grade slabs, post-tensioning allows concrete slabs to be poured over expansive soil.

For design purposes, a concrete slab is considered to be a wide, shallow beam that is supported by a concrete foundation at the edge. While a beam may sag at the center due to loads and gravity, a concrete slab can either sag or bow depending on soil conditions such as excessive moisture differential, freezing, or expansive soil. Because concrete is very poor at resisting stress from tension, the slab must be reinforced with a material like steel, which has a high tensile strength. Steel tendons with anchors can be extended through the slab as it is poured. Usually between 3 and 10 days after the concrete has been poured, hydraulic jacks stretch these tendons. The tendon force is transferred to the concrete slab through anchorage devices at the ends of the tendons. This process creates an internal compressive force throughout the slab, increasing the ability of the slab to resist cracking and heaving. Post-tensioning usually allows for a thinner slab than normally would be required to span over expansive conditions, the elimination of other slab reinforcing, and the elimination of most slab joints. Two common methods of post-tensioning used for slabs are the flat slab and the ribbed slab method.

Post-tensioned Flat Slabs

The *flat slab* method of post-tensioning uses steel tendons ranging in size from 3/8" to 1/2" (9.5 to 13 mm) in diameter. Maximum spacings recommended by the Post-Tensioning Institute (PTI) for tendons include the following:

Tendon Diameter	Spacing
3/8" (9.5 mm) diameter	5'-0" (1500 mm) spacing.
7/16" (11 mm) diameter	6'-0" (1800 mm) spacing.
1/2" (13 mm) diameter	9'-0" (2700 mm) spacing.

The exact spacing and size of tendons must be determined by the engineer, based on the loads to be supported and the strength and conditions of the soil. When required, the tendons can be represented on the foundation plan as seen in Figure 23-49. Details also need to be provided to indicate how the tendons will be anchored, as well as showing the exact locations of the tendons. Figure 23-50 shows an example of a tendon detail that would be referenced to the foundation drawings. In addition to representing and specifying the steel throughout the floor system, the drafter needs to specify the engineer's requirements for the strength of the concrete at 28 days, the period when the concrete is to be stressed, as well as what strength the concrete should achieve before stressing. This information is generally placed in the specifications in a project manual.

Post-tensioned Ribbed Slabs

A second method of post-tensioning an on-grade floor slab is with the use of concrete ribs or beams placed below the slab. These

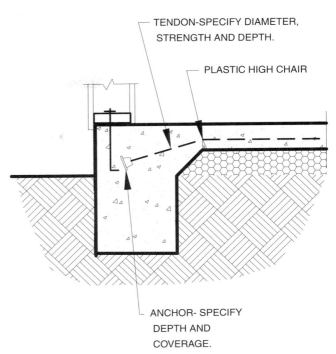

TENDON-SPECIFY DIAMETER, STRENGTH AND DEPTH.

PLASTIC HIGH CHAIR

ANCHOR- SPECIFY DEPTH AND COVERAGE.

FIGURE 23-50 ■ Details showing tendon locations must be referenced to the slab plan.

beams reduce the span of the slab over the soil and provide increased support. The width, depth, and spacing are determined by the engineer, based on the strength and condition of the soil and the size of the slab. Figure 23-51 shows an example of a beam detail that a drafter would be required to draw and reference on the foundation plan to show the reinforcing specified by the engineer. Figure 23-52 shows an example of how these beams could be shown on the foundation plan.

Wood Floor Systems Over a Crawl Space

Floor systems built over a crawl space are usually made using western platform or post-and-beam construction methods. Wood framing methods were introduced in Chapter 6 and 7 as wood construction was considered. This chapter considers only how wood and timber are drawn on the foundation plan.

Western Platform Framing

Floor joists are used to span between the stem walls in western platform construction. (See Figure 23-53.) The material size and spacing must be represented on the foundation plan. Figure 23-54 shows methods of representing floor joists on a foundation plan. The floor joists rest on top of the mudsill, which is supported by the stem wall. (See Figure 23-55.) Figure 23-56 shows other common methods of supporting floor joists.

When the distance between stem walls is too great for the floor joists to span, a girder is used to support the floor joists. Unless the floor joists are designed to support concentrated loads, a girder must also be placed under interior bearing walls that are not supported by stem walls. A beam pocket or hanger supports the

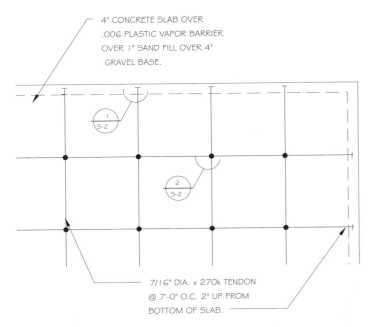

4" CONCRETE SLAB OVER .006 PLASTIC VAPOR BARRIER OVER 1" SAND FILL OVER 4" GRAVEL BASE.

$\frac{1}{5-2}$

$\frac{2}{5-2}$

7/16" DIA. x 270k TENDON @ 7'-0" O.C. 2" UP FROM BOTTOM OF SLAB.

FIGURE 23-49 ■ Post-tensioned concrete slabs can be used when the slab must be placed over unstable soil.

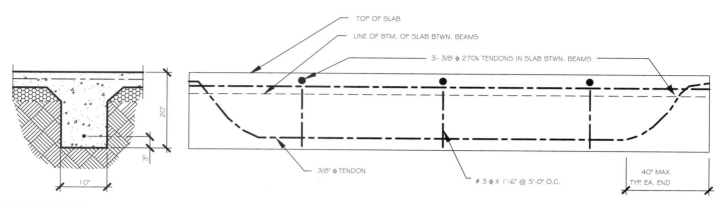

FIGURE 23-51 ■ A rib can be placed below post-tensioned slab to provide increased slab support.

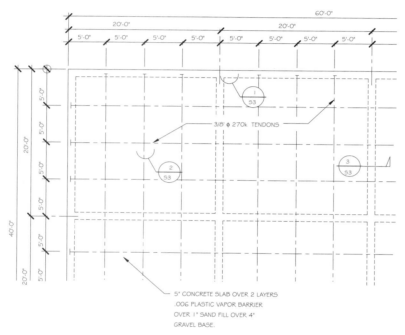

FIGURE 23-52 ■ Support ribs for a post-tensioned slab must be shown and located on the foundation plan.

FIGURE 23-53 ■ Western platform construction methods are used on many light commercial structures because of the system's many economical features.

girder at the stem wall and, by a post and pier, provides support as it spans across the foundation. In areas where seismic activity must be planned for, the girder must be attached to the post, and the post must be attached to the supporting concrete pier. Figure 23-54 shows the representation of girders, post, and piers on the foundation plan. Connections from girder to post and post to pier are not shown on the foundation plan but are specified by notes referenced to the pier.

Floor sheathing is nailed to the floor joists to provide the rough floor. The engineer sizes the subfloor depending on the spacing of the floor joists and the floor loads that must be supported. The floor sheathing is not represented on the foundation plan but is specified in a general note.

Figure 23-57 shows some common methods used to frame changes in floor height. If a stem wall is required, it must be represented on the foundation plan just as an exterior stem wall and footing would be represented. If the change is made over a girder,

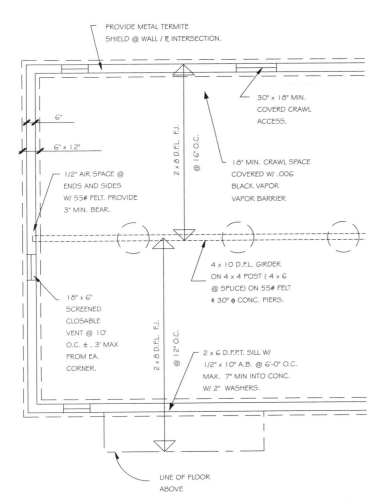

FIGURE 23-54 ■ Representing materials of a western platform floor system on the foundation plan. Common members that should be shown include the stem walls, footings, girders, piers, floor joists, floor overhangs, vents, crawl access, and metal anchors or straps.

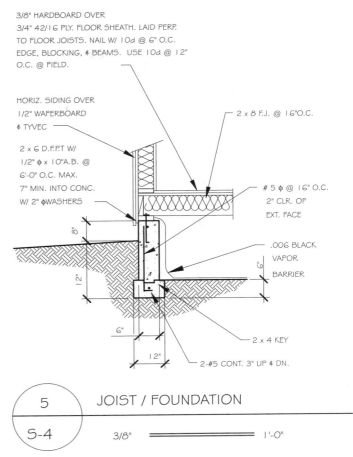

FIGURE 23-55 ■ A detail showing the intersection of the floor system to the foundation must be referenced to the foundation plan.

a linetype different from the line used to represent the girder should be used to represent the floor edge. Figure 23-58 shows methods of representing elevation changes on the foundation plan.

Post and Beam

Post-and-beam construction is used on small commercial and multifamily projects in areas of the country where cold or dampness is a consideration. Rather than having floor joists span between the stem walls, a series of beams is used to support the subfloor, as shown in Figure 23-59. Beams are typically 4 × 8 (100 × 200 mm) spaced at 48" (1200 mm) on center with supports placed at 8'–0" (2400 mm) along the beam. This pattern is based on the use of 4 × 8 (1200 × 2400 mm) sheets of plywood but the exact girder type, size, and spacing are affected by the loads to be supported. A post-and-beam floor system is well suited to projects that have very few interior bearing walls. As seen in Figure 23-60, additional girders must be placed under interior bearing walls that do not align with the normal placing of girders.

Once the mudsill is bolted to the foundation wall, girders can be placed so that the top of the girder is level with the top of the mudsill. Girders are supported just as with western platform construction. Plywood with a thickness of 1 1/8" (28.5 mm) and an APA rating of STURD-I-FLOOR 2-4-1 with and exposure rating of EXP-1 is generally used for the subfloor. Two-inch (50 mm) thick material such as 2 × 6 (50 × 150) T & G boards laid perpendicular to the beams can also be used for the subfloor. The strength and quality of the floor is greatly increased when the subfloor is glued to the support beams, eliminating squeaks, bounce, and nail popping.

Party Wall and Fire Wall Support

Regardless of the material used to form the floor system, care must be taken with special walls such as fire and party walls to ensure the wall will function as it is designed. Because unprotected material cannot pass through a fire wall, extra support at a fire wall may be required to be shown on the foundation plan to support beam terminations at a fire wall. Figure 23-61 shows construction of a 2-hour wall for a warehouse.

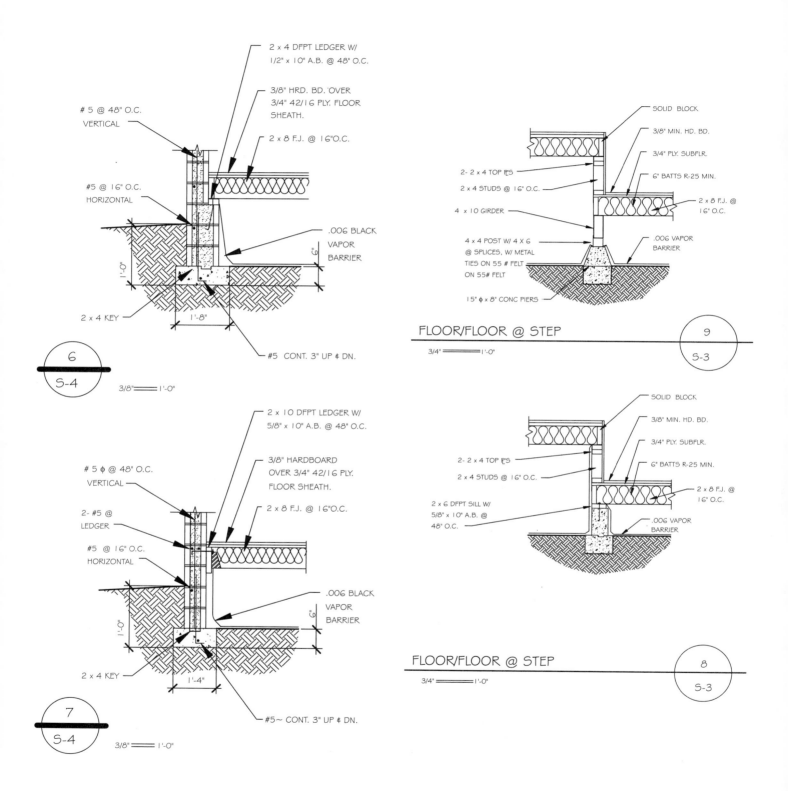

5 @ 48" O.C.
VERTICAL

2 x 4 DFPT LEDGER W/
1/2" x 10" A.B. @ 48" O.C.

3/8" HRD. BD. 'OVER
3/4" 42/16 PLY. FLOOR
SHEATH.

2 x 8 F.J. @ 16"O.C.

#5 @ 16" O.C.
HORIZONTAL

.006 BLACK
VAPOR
BARRIER

1'-0"

6"

2 x 4 KEY

1'-8"

#5 CONT. 3" UP & DN.

6
S-4

3/8" ═══ 1'-0"

SOLID BLOCK

3/8" MIN. HD. BD.

3/4" PLY. SUBFLR.

2- 2 x 4 TOP P'S

6" BATTS R-25 MIN.

2 x 4 STUDS @ 16" O.C.

2 x 8 F.J. @ 16" O.C.

4 x 10 GIRDER

4 x 4 POST W/ 4 X 6
@ SPLICES, W/ METAL
TIES ON 55 # FELT
ON 55# FELT

.006 VAPOR
BARRIER

15" φ x 8" CONC PIERS

FLOOR/FLOOR @ STEP

3/4" ═══ 1'-0"

9
S-3

5 φ @ 48" O.C.
VERTICAL

2 x 10 DFPT LEDGER W/
5/8" x 10" A.B. @ 48" O.C.

3/8" HARDBOARD
OVER 3/4" 42/16 PLY.
FLOOR SHEATH.

2 x 8 F.J. @ 16"O.C.

2- #5 @
LEDGER

#5 @ 16" O.C.
HORIZONTAL

.006 BLACK
VAPOR
BARRIER

1'-0"

6"

2 x 4 KEY

1'-4"

#5~ CONT. 3" UP & DN.

7
S-4

3/8" ═══ 1'-0"

SOLID BLOCK

3/8" MIN. HD. BD.

3/4" PLY. SUBFLR.

2- 2 x 4 TOP P'S

6" BATTS R-25 MIN.

2 x 4 STUDS @ 16" O.C.

2 x 8 F.J. @ 16" O.C.

2 x 6 DFPT SILL W/
5/8" x 10" A.B. @
48" O.C.

.006 VAPOR
BARRIER

FLOOR/FLOOR @ STEP

3/4" ═══ 1'-0"

8
S-3

FIGURE 23-56 ■ Alternative methods of supporting floor joists for masonry walls.

FIGURE 23-57 ■ Common methods of framing changes in floor elevation using western platform construction methods.

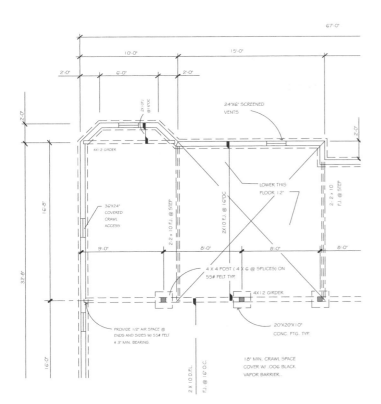

FIGURE 23-58 ■ Representing changes in floor elevation on the foundation plan with a joist floor system.

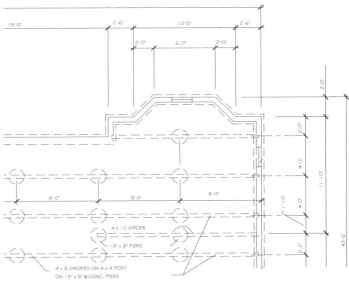

FIGURE 23-60 ■ Representing a post-and-beam floor system on the foundation plan. Members that should be shown include the stem walls, footings, girders, piers, vents, crawl access, and metal anchors or straps.

FIGURE 23-59 ■ A post-and-beam floor system is often used for multifamily construction projects in damp climates. *Courtesy Floyd Miller.*

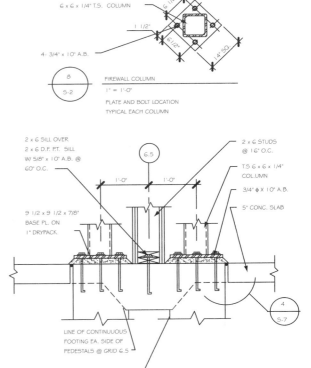

FIGURE 23-61 ■ Firewalls shown on the floor plan often require special construction that must be referenced to the foundation plan.

A party wall typically requires complete separation for adjacent units to help control vibrations. For a wood floor, this would require girders to support each portion of the wall. Figure 23-62 shows two options for framing the intersection of a party wall.

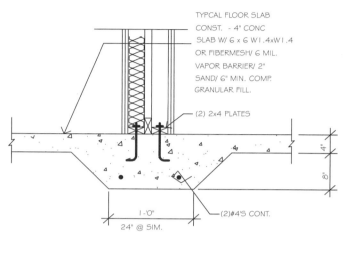

TYPICAL FLOOR SLAB
CONST. - 4" CONC
SLAB W/ 6 x 6 W1.4xW1.4
OR FIBERMESH/ 6 MIL.
VAPOR BARRIER/ 2"
SAND/ 6" MIN. COMP.
GRANULAR FILL.

(2) 2x4 PLATES

1'-0"
24" @ SIM.

(2)#4'S CONT.

4"

8"

2 HR. WALL
PARTY WALL
1" = 1'-0"

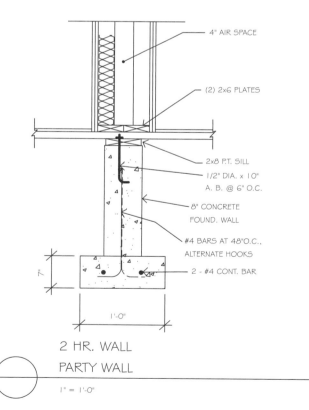

4" AIR SPACE

(2) 2x6 PLATES

2x8 P.T. SILL
1/2" DIA. x 10"
A. B. @ 6" O.C.

8" CONCRETE
FOUND. WALL

#4 BARS AT 48"O.C.,
ALTERNATE HOOKS

2 - #4 CONT. BAR

7"

1'-0"

2 HR. WALL
PARTY WALL
1" = 1'-0"

FIGURE 23-62 ■ A party wall is a wall framed between two separate tenants. Intersections of party walls and the floor system must be referenced to the foundation plan. *Courtesy Scott R. Beck, Architect.*

FOUNDATION DRAWINGS

Two types of drawings can be used to represent the information used to construct the floor and foundation systems. A *slab-on-grade plan* is usually drawn when the walls of the structure extend from the footing. Masonry and precast concrete panel construction are examples of structures where the wall would rest directly on the footing, eliminating the need for a stem wall. In each case, the walls of the structure determine the limits of the floor. The location of walls that form the perimeter of the structure are placed on the floor plan. The slab-on-grade plan is used to show the shape of the floor slab. Interior walls are shown on the floor plan and not on the slab plan. The slab plan is used to show only items necessary to construct the floor. Figure 23-63 shows an example of a slab-on-grade plan. The process for completing this plan is given later in this chapter.

A foundation plan is used to supplement the slab-on-grade plan. Only material used to represent the actual foundation is shown. Because the stem walls (the concrete panels) are shown on the slab-on-grade plan, they are not shown on the foundation plan. The foundation plan is used to show the size, shape, location, and elevation of all footings. Because no other level is represented, the outline of the footings and piers is drawn using continuous lines. Figure 23-64 shows the foundation plan that would accompany the slab plan shown in Figure 23-63. The process for completing this type of foundation plan is given later in this chapter.

When a monolithic slab/foundation is poured, it is represented in one foundation plan, similar to Figure 23-65. Western platform and post-and-beam floor systems are also represented on the foundation plan rather than having two separate plans. Material above grade is generally drawn using continuous lines, and material below grade such as footings and piers are usually drawn using dashed lines. Information to complete each type of foundation is presented in this chapter. Before considering differences, it is important to consider similarities between each type of plan.

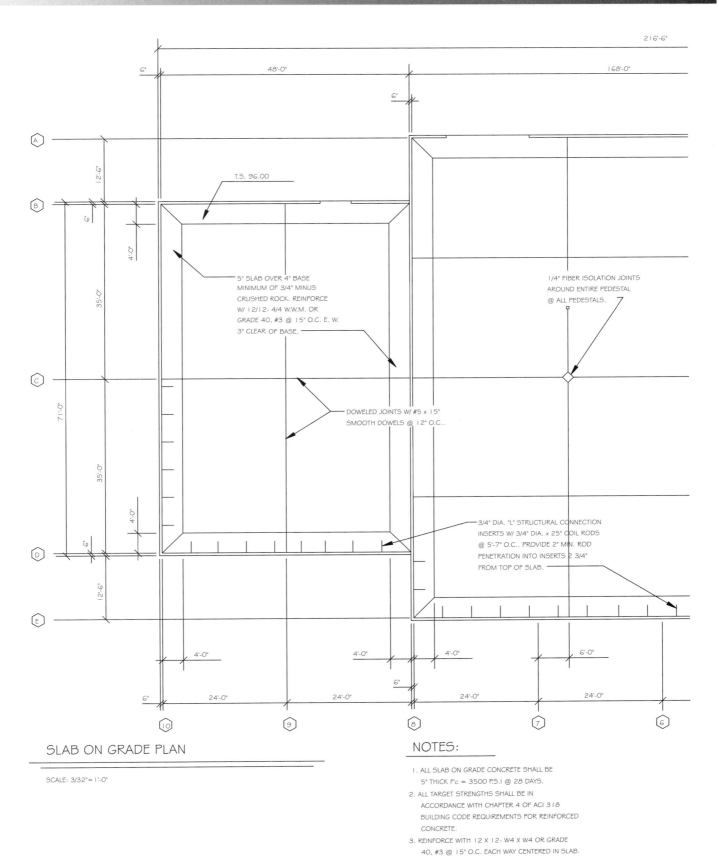

216'-6"

6" 48'-0" 168'-0"

6"

Ⓐ

12'-6"

Ⓑ

6"

4'-0"

35'-0"

T.S. 96.00

5" SLAB OVER 4" BASE
MINIMUM OF 3/4" MINUS
CRUSHED ROCK. REINFORCE
W/ 12/12- 4/4 W.W.M. OR
GRADE 40, #3 @ 15" O.C. E. W.
3" CLEAR OF BASE.

1/4" FIBER ISOLATION JOINTS
AROUND ENTIRE PEDESTAL
@ ALL PEDESTALS.

71'-0"

Ⓒ

DOWELED JOINTS W/ #5 x 15"
SMOOTH DOWELS @ 12" O.C..

35'-0"

4'-0"

3/4" DIA. "L" STRUCTURAL CONNECTION
INSERTS W/ 3/4" DIA. x 25" COIL RODS
@ 5'-7" O.C.. PROVIDE 2" MIN. ROD
PENETRATION INTO INSERTS 2 3/4"
FROM TOP OF SLAB.

Ⓓ

6"

12'-6"

Ⓔ

4'-0" 4'-0" 4'-0" 6'-0"

6"

6" 24'-0" 24'-0" 24'-0" 24'-0"

⑩ ⑨ ⑧ ⑦ ⑥

SLAB ON GRADE PLAN

SCALE: 3/32"= 1'-0"

NOTES:

1. ALL SLAB ON GRADE CONCRETE SHALL BE
   5" THICK F'c = 3500 P.S.I @ 28 DAYS.
2. ALL TARGET STRENGTHS SHALL BE IN
   ACCORDANCE WITH CHAPTER 4 OF ACI 318
   BUILDING CODE REQUIREMENTS FOR REINFORCED
   CONCRETE.
3. REINFORCE WITH 12 X 12- W4 X W4 OR GRADE
   40, #3 @ 15" O.C. EACH WAY CENTERED IN SLAB.

FIGURE 23-63 ■ The placement of slabs and slab joints are often separated from the foundation plan to provide clarity. A slab-on-grade plan shows only an on-grade concrete floor system.

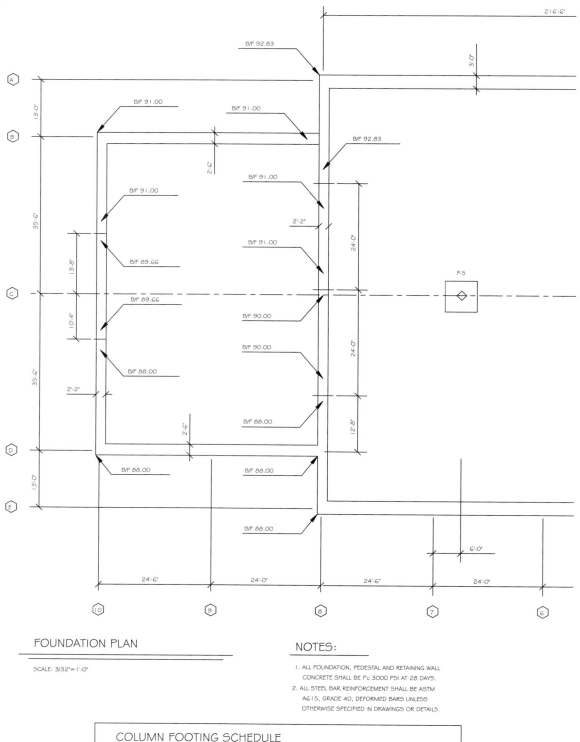

216'-6"

B/F 92.83

3'-0"

Ⓐ

13'-0"

B/F 91.00   B/F 91.00

Ⓑ   B/F 92.83

2'-6"

B/F 91.00

35'-6"   2'-2"   24'-0"

B/F 91.00

B/F 91.00

13'-8"   B/F 89.66

F-5

Ⓒ   B/F 89.66

10'-4"   B/F 90.00

B/F 88.00   B/F 90.00   24'-0"

35'-6"

2'-2"   B/F 88.00

2'-6"   12'-8"

Ⓓ

B/F 88.00   B/F 88.00

13'-0"

Ⓔ

B/F 88.00

6'-0"

24'-6"   24'-0"   24'-6"   24'-0"

⑩   ⑨   ⑧   ⑦   ⑥

FOUNDATION PLAN

SCALE: 3/32" = 1'-0"

NOTES:

1. ALL FOUNDATION, PEDESTAL AND RETAINING WALL
   CONCRETE SHALL BE f'c 3000 PSI AT 28 DAYS.

2. ALL STEEL BAR REINFORCEMENT SHALL BE ASTM
   A615, GRADE 40, DEFORMED BARS UNLESS
   OTHERWISE SPECIFIED IN DRAWINGS OR DETAILS.

COLUMN FOOTING SCHEDULE			
FOOTING	PEDESTAL HEIGHT	SIZE	BTM. OF FTG.
F-1	6.39	7'-3" x 7'-3" x 14"	88.44
F-2	5.96	7'-6" x 7'-6" x 14"	88.87
F-3	5.53	5'-0" x 5'-0" x 14"	89.30
F-4	5.53	4'-3" x 4'-3" x 14"	89.30
F-5	5.18	7'-0" x 7'-0" x 14"	89.65

FIGURE 23-64 ■ When a slab-on-grade plan is provided, the foundation plan shows only the footings. This foundation plan is for the structure shown in Figure 23-63.

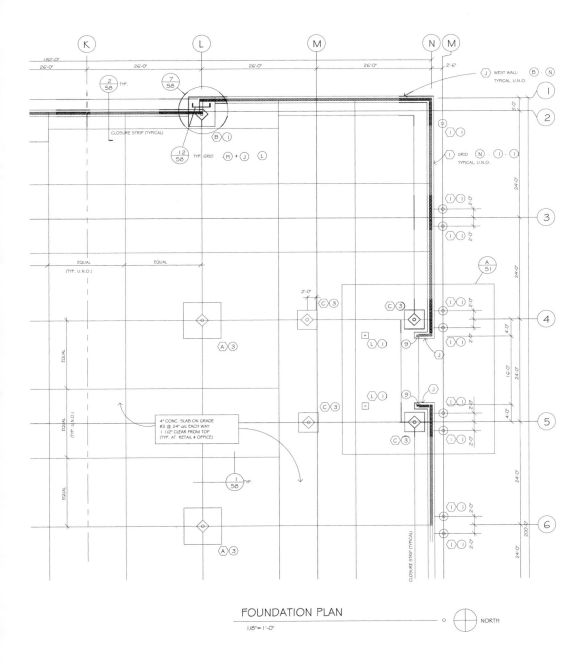

FOUNDATION PLAN

1/8" = 1'-0"

NORTH

FIGURE 23-65 ■ A foundation plan combining the slab and foundation plan. *Courtesy Bill Berry of Berry-Nordling Engineers, Inc. and Thomas J. Kuhns, A.I.A., Michael & Kuhns Architects, P. C.*

Symbols

Grid markers, elevation symbols, and a north arrow are the symbols typically associated with foundation drawings. The grid and north arrow can be thawed from other drawings. Elevation symbols should match symbols used for other drawings. Elevation symbols often include an elevation on both the top and bottom of the line, as shown in Figure 23-66. When more than one elevation is given per symbol, a schedule should be included with the drawing to define each label. Elevation labels used on foundation drawings include the following:

BOF	Bottom of footing
BOW	Bottom of wall
FF	Finish floor
TOF	Top of footing
TOG	Top of grate
TOW	Top of wall

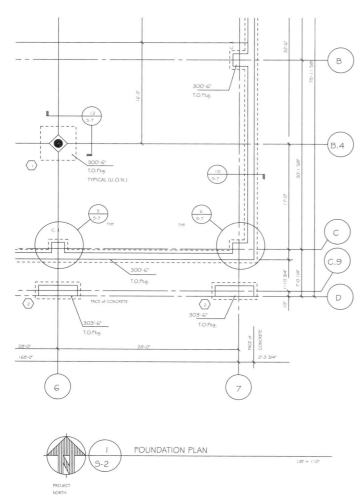

FIGURE 23-66 ■ Representing changes in elevation on the foundation plan may be referenced to the top of slab (T.O.S) top of footing (T.O.F.) or bottom of footing (B.O.F.). *Courtesy Van Domelen/Looijenga/McGarrigle/Knauf Consulting Engineers.*

Schedules

Just as with other drawings, schedules are an effective method of presenting foundation information. Separate schedules are often used to represent footings, piers, pilasters, beams, and repetitive framing members. Reference symbols that clearly represent the material to be specified should be used. Continuous footings could be represented by a symbol such as F-1 or F-6, with piers represented by P-1 or P-8. Many schedules include listings for the symbol, the quantity, size, and required reinforcing. Reinforcing can be divided into separate columns for upper and lower mats, as well as required hairpin size and spacing. Figure 23-67 shows an example of schedules for continuous and spot footings. Spot footing schedules often include additional listings for elevation, pedestal height, and pedestal reinforcing.

Notes

Foundation drawings require the use of local and general notes to specify all materials. Most notes should be placed using the same consideration that was used to place notes on other framing plans. More so than with other drawings, specifications controlling work quality and material are usually placed with the foundation plan. Figure 23-68 shows an example of the notes typically associated with a foundation plan.

Dimensioning

The method of dimensioning is as important as the lines and symbols used to represent the foundation material. Line quality for dimension and extensions should match those used for the floor or

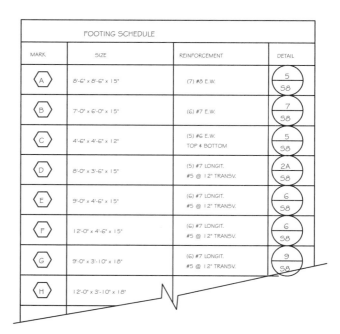

FIGURE 23-67 ■ Schedules can be used to keep the foundation plan uncluttered. *Courtesy Van Domelen/Looijenga/ McGarrigle/Knauf Consulting Engineers.*

FOUNDATION NOTES:

1. DESIGN SOIL PRESSURE 3500 PSF AT SOIL AND 5000 PSF AT ROCK PER GEOTECHNICAL REPORT PREPARED BY WRIGHT/DEACON AND ASSOCIATES, DATED JULY 28, 1998.

2. ALL FOOTINGS TO BEAR ON FIRM, UNDISTURBED SOIL OR APPROVED COMPACTED FILL MINIMUM 24" BELOW ADJACENT FINAL GRADE. NOTIFY VLMK BEFORE PROCEEDING IF ANY UNUSUAL CONDITIONS ARE ENCOUNTERED IN THE FOOTING EXCAVATIONS.

3. DO NOT EXCAVATE CLOSER THAN 2:1 SLOPE BELOW FOOTING EXCAVATIONS.

4. CLEAN ALL FOOTING EXCAVATIONS OF LOOSE MATERIAL BY HAND.

5. EXCAVATIONS MAY BE MADE UNDER CONTINUOUS FOOTINGS FOR PIPES. BACKFILL TO BE APPROVED BY VLMK.

CONCRETE

1. STRENGTH: AVERAGE CONCRETE STRENGTH AS DETERMINED BY JOB CAST LAB CURED CYLINDER TO BE 3000 PSI AT 28 DAYS PLUS INCREASE DEPENDING ON THE PLANTS STANDARD DEVIATION AS SPECIFIED IN ACI 318-83.

2. MINIMUM MIX REQUIREMENTS:

LOCATION	COMPRESSIVE STRENGTH (PSI)	SLUMP (A)	MIN. CEMENT CONTENT	ADMIXTURES	WATER/ CEMENT RATIO
FOOTINGS	4000	0-5"	5	NONE	0.55
SLABS ON GRADE (INTERIOR)	3500	2-4"	5 1/2	WRA (B)	0.45
SLABS ON GRADE (EXTERIOR)	3000	2-4"	5 1/2	WRA, AE (C)	0.50
WALLS AND COLUMNS	4000	2-4"	5 1/2	WRA/PT20 (D)	0.50
MISC. CONC.	3000	0-5"	5	WRA	0.55
LT.WT. SLAB ON METAL DECK	3500	0-5"	5	WRA	0.45

A. SLUMP EXCEEDING SPECIFIED LIMITS SHALL NOT BE INCORPORATED IN THE PROJECT EXCEPT BY WRITTEN APPROVAL FROM ENGINEER.

B. WRA = WATER REDUCING AGENT

C. AE = AIR ENTRAINMENT

D. POZZOTEC 20 REQUIRED FOR CONCRETE PLACED BELOW 40 DEG.

F. CALCIUM CHLORIDE IS NOT TO BE USED ON THIS PROJECT.

3. PLACE AND CURE ALL CONCRETE PER ACI CODES AND STANDARDS.

4. PROVIDE CONTROL JOINTS IN ALL INTERIOR SLABS ON GRADE AS SHOWN ON PLANS.

5. SLEEVES, PIPES OR CONDUITS OF ALUMINUM SHALL NOT BE EMBEDDED IN STRUCTURAL CONCRETE UNLESS EFFECTIVELY COATED.

FIGURE 23-68 ■ General notes can be used to organize information related to the foundation plan. Notes are often divided into separate divisions including concrete, reinforcing steel, soil, and framing lumber. *Courtesy Van Domelen/Looijenga/McGarrigle/Knauf Consulting Engineers.*

framing plans. Jogs in the foundation are dimensioned using the same methods used on the floor plan. Most of the dimensions for major shapes will be exactly the same as the corresponding dimensions on the floor plan. The foundation plan can be drawn in the same drawing file as the floor plan. This allows the overall dimensions for the floor plan to be displayed on the foundation plan if different layers are used to place the dimensions. Layers such as A-BASEDIM, A-FLOORDIM, and S-FNDDIMEN help differentiate between dimensions for the floor and foundation.

A different method is used to dimension the interior walls of a foundation than is used on a floor plan. Foundation walls are dimensioned from face to face rather than face to center, as on a floor plan. Footing widths are dimensioned from center to center. Each type of dimension is shown in Figure 23-69.

DRAWING FOUNDATION PLANS

The foundation plan should be drawn at the same scale and in the same orientation as the floor and framing plans. The plan can best be drawn by thawing the floor plan layers, which show the exterior walls and the interior load-bearing walls.

Concrete Slab

A concrete slab can be completed by using the following steps.

1. Thaw the layers containing the exterior walls and interior bearing walls

2. Using a running OSNAP of INT, draw the outline of the slab to match the outline of the exterior walls

3. Draw a centerline to represent all interior bearing walls

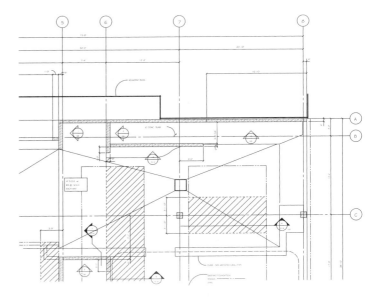

FIGURE 23-69 ■ Dimensions for concrete and masonry walls should be given from face of wall to face of wall. Interior footings are represented to their centers. *Courtesy KPFF Consulting Engineers.*

4. Block out for any openings in the stem wall such as those required for garage doors

5. Draw a support ledge if brick veneer is to be used

6. Draw the outline of the exterior footings

7. Draw the outline of all interior continuous footings

The base of the foundation plan has now been drawn and should now resemble Figure 23-70.

8. Mark the center of all posts and columns, and draw pedestals or the spot footings to support them.

9. Thaw all grid lines and markers

10. Draw all control joints and changes in floor level

11. Draw all metal connectors

12. Draw any plumbing work

13. Draw any HVAC duct work

14. Draw any required electrical conduit runs and slab penetrations

The foundation plan should now resemble Figure 23- 71. Use the following steps to place the needed dimensions.

15. Thaw the base dimension layer to provide overall dimensions for each side of the structure

16. Thaw dimensions used on the floor plan or provide dimensions to locate each jog in the foundation

17. Dimension the location of all interior continuous footings

18. Dimension the size and location of any changes in slab elevation

19. Dimension all metal connectors

20. Dimension the location to the centers of all spot piers

21. Dimension the location of any required piping, duct work, or electrical conduit

The major features of the drawing have now been dimensioned. The foundation plan should now resemble Figure 23-72. The drawing can be completed by adding all required text and detail markers:

22. Describe the concrete slab, fill material, vapor barrier, and insulation

23. Describe all footing and stem wall sizes

24. Describe all required mudsill, ledgers, and anchor bolts

25. Describe all metal straps, anchors, and connectors

26. Describe all required changes in elevation

27. Describe all footings and pedestals

28. Place any general notes specified by the engineer

29. Place a title and scale below the drawing

The completed foundation plan should now resemble Figure 23-73. Similar procedures could be used to draw a post-tensioned concrete slab foundation.

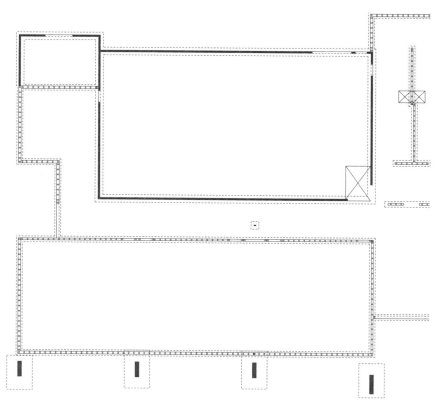

FIGURE 23-70 ■ The base drawing for a concrete slab foundation plan should show the support for each bearing wall and column. *Courtesy Van Domelen/Looijenga/McGarrigle/Knauf Consulting Engineers.*

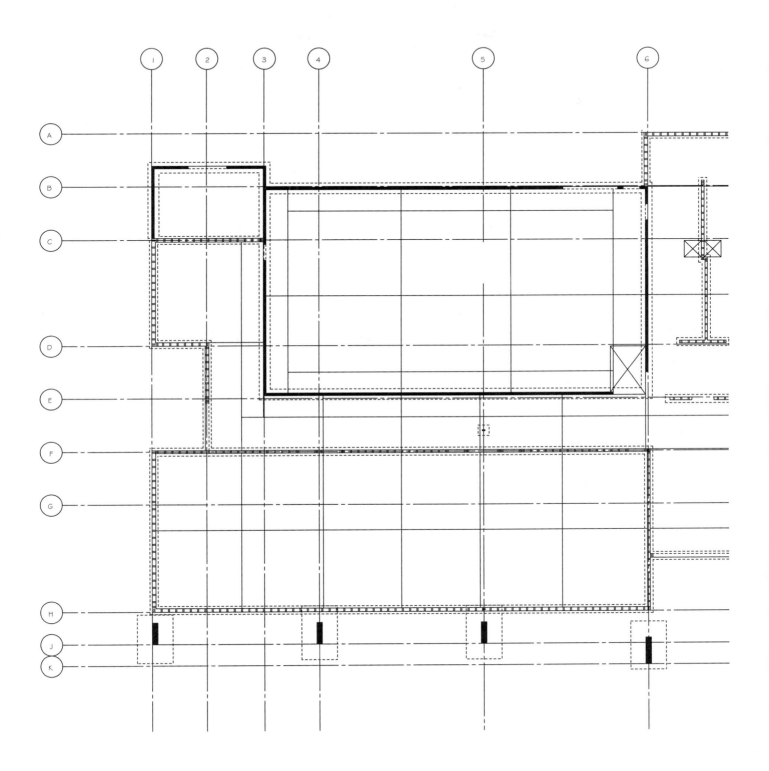

FIGURE 23-71 ■ The foundation plan with grids and slab joints added. *Courtesy Van Domelen/Looijenga/McGarrigle/Knauf Consulting Engineers.*

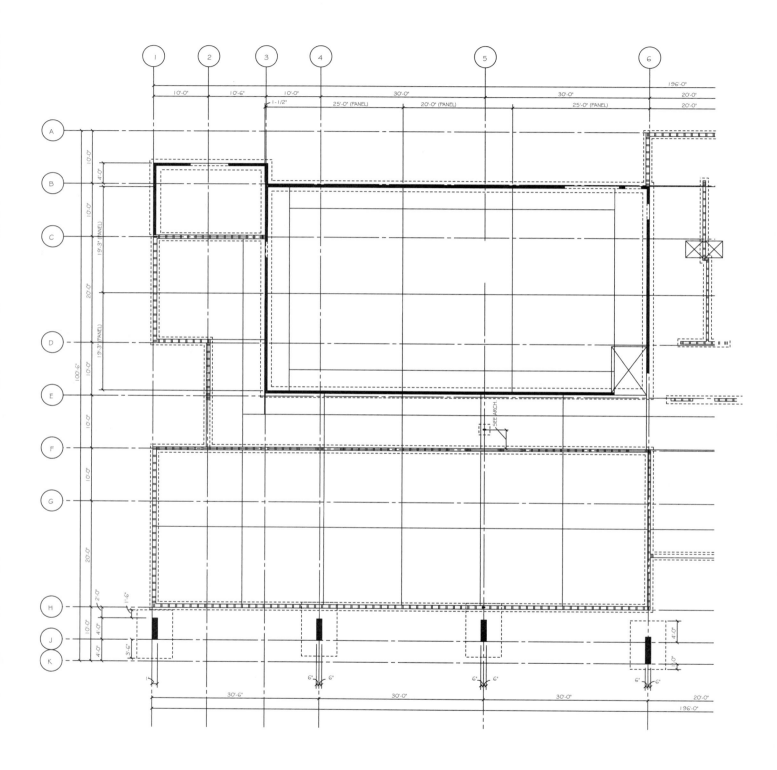

FIGURE 23-72 ■ Placement of dimensions to locate interior and exterior features. *Courtesy Van Domelen/Looijenga/McGarrigle/Knauf Consulting Engineers.*

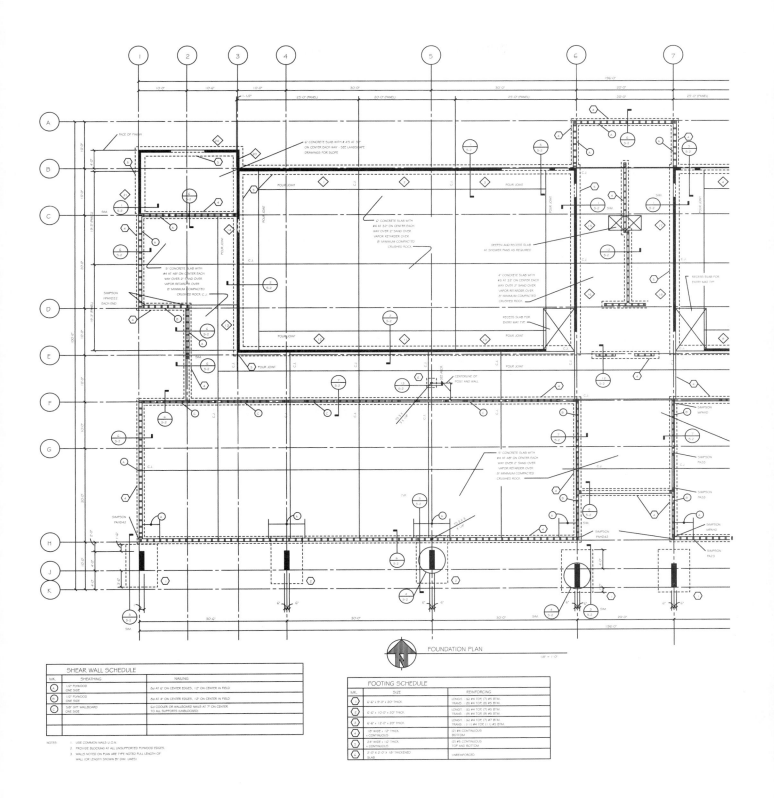

FIGURE 23-73 ■ The completed foundation plan with local and general notes, detail markers, and schedules. *Courtesy Van Domelen/Looijenga/McGarrigle/Knauf Consulting Engineers.*

Floor Joists

A foundation plan for a structure framed with floor joists can be completed by using the following steps.

1. Thaw the layers containing the exterior walls and interior bearing walls

2. With OSNAP set to ON, draw the outline of the stem wall to match the outline of the exterior walls

3. Use the **OFFSET** command to reproduce the interior side of the stem walls. Use **TRIM** and **FILLET** to clean up each intersecting corner.

4. Draw a centerline to represent all interior bearing walls

5. Use the engineer's specifications or determine the sizes for all girders to be placed under load-bearing walls

6. Block-out openings in the stem wall such as those required for garage doors

7. Draw a support ledge if brick veneer is to be used

8. Draw the outline of the exterior footings

9. Draw all interior piers

The base of the foundation plan has now been drawn and should now resemble Figure 23-74.

10. Thaw all grid lines and markers created for other plan views

11. Mark the center of all exterior posts or columns, and draw the spot footings for each

12. Draw floor joist indicators

13. Draw all changes in floor level

14. Draw all metal connectors

15. Draw all vents

16. Draw a crawl access

The foundation plan should now resemble Figure 23-75. Use the following steps to place the needed dimensions.

17. Thaw the base dimension layer to provide overall dimensions for each side of the structure

18. Thaw dimensions used on the floor plan or provide dimensions to locate each jog in the foundation

19. Dimension all openings in the stem wall

20. Dimension the location of all girders.

21. Dimension the location of all interior piers

22. Dimension the size and location of any changes in slab elevation

23. Dimension the location of all metal connectors

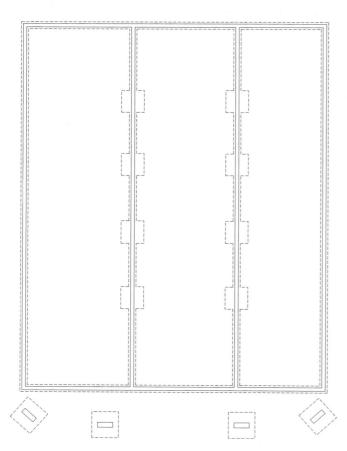

FIGURE 23-74 ■ The base plan for a joist floor system. *Courtesy Scott R. Beck, Architect.*

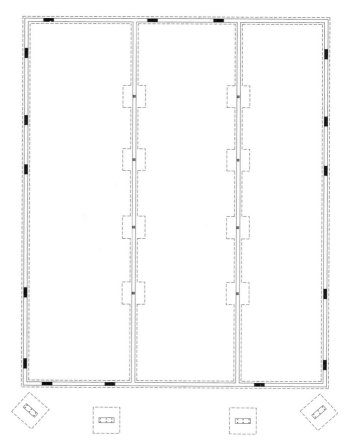

FIGURE 23-75 ■ Adding required columns, anchors, vents, and access. *Courtesy Scott R. Beck, Architect.*

The major features of the drawing have now been dimensioned. The foundation plan should now resemble Figure 23-76. The drawing can be completed by adding all required text.

24. Place notes, detail markers, and dimensions to describe all footing and stem wall sizes

25. Place notes and detail markers to describe all required mudsill, ledgers, and anchor bolts

26. Place notes, detail markers, and dimensions to locate and describe all metal straps, anchors, and connectors

27. Place notes to describe all vents, wall openings, and the crawl access

28. Describe all girders, post, and piers

29. Place notes to describe all joist type, size, and spacing

30. Place notes to describe all required changes in elevation

31. Place notes to describe the crawl space covering

32. Place notes to describe all special blocking

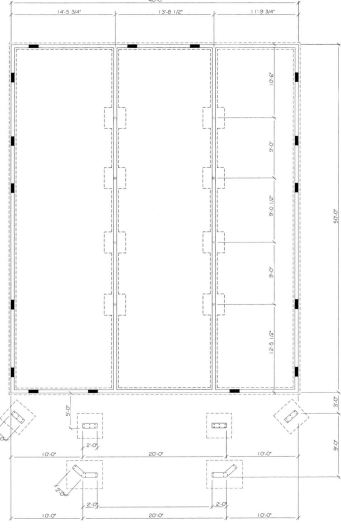

FIGURE 23-76 ■ Placement of interior and exterior dimensions.
Courtesy Scott R. Beck, Architect.

33. Place any general notes specified by the engineer

34. Place a title and scale below the drawing

The completed foundation plan should now resemble Figure 23-77. A similar procedure could be used to draw a structure supported on pilings instead of piers.

Post and Beam

A foundation plan for a structure framed with a post-and-beam floor system can be completed by using the following steps.

1. Thaw the layers containing the exterior walls and interior bearing walls

2. Use **OSNAP** to draw the outline of the stem wall to match the outline of the exterior walls

3. Use the **OFFSET** command to reproduce the interior side of the stem walls. Use **TRIM** and **FILLET** to clean up each intersecting corner.

4. Draw a centerline to represent all interior bearing walls

5. Use the engineer's specifications or determine the sizes for all girders to be placed under load-bearing walls

6. Layout all girders required for the floor support

7. Block out for any openings in the stem wall such as those required for garage doors

8. Draw a support ledge if brick veneer is to be used

9. Draw the outline of the exterior footings

10. Draw all interior piers for all girders

The base of the foundation plan has now been drawn and should now resemble Figure 23-78.

11. Thaw all grid lines and markers

12. Mark the center of all exterior posts or columns, and draw the spot footing for each

13. Draw all changes in floor level

14. Draw all metal connectors

15. Draw all vents

16. Draw a crawl access

The foundation plan should now resemble Figure 23-79. Use the following steps to place the needed dimensions.

17. Thaw the base dimension layer to provide overall dimensions for each side of the structure

18. Thaw dimensions used on the floor plan or provide dimensions to locate each jog in the foundation

19. Dimension all openings in the stem wall

20. Dimension the location of all girders

21. Dimension the location of all interior piers

22. Dimension the size and location of any changes in slab elevation

23. Dimension the location of all metal connectors

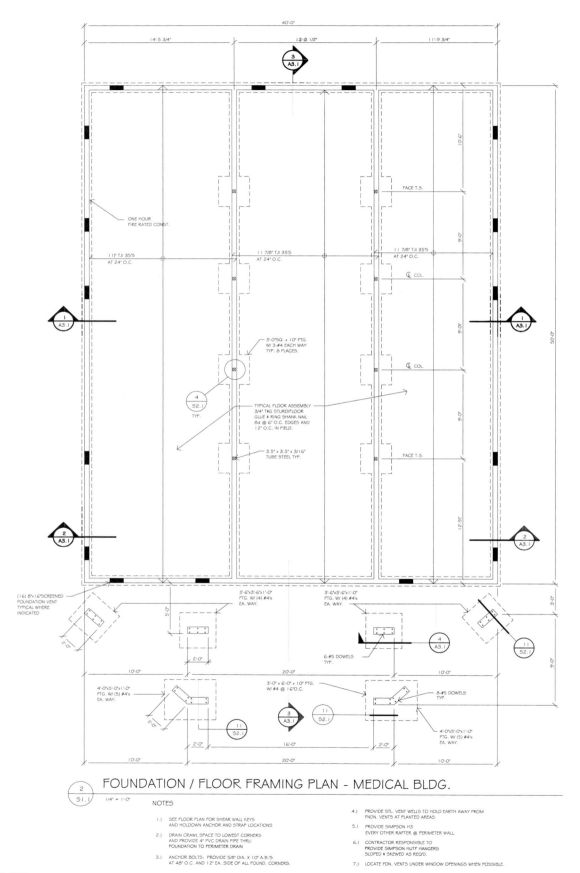

FIGURE 23-77 ■ The completed foundation plan. *Courtesy Scott R. Beck, Architect.*

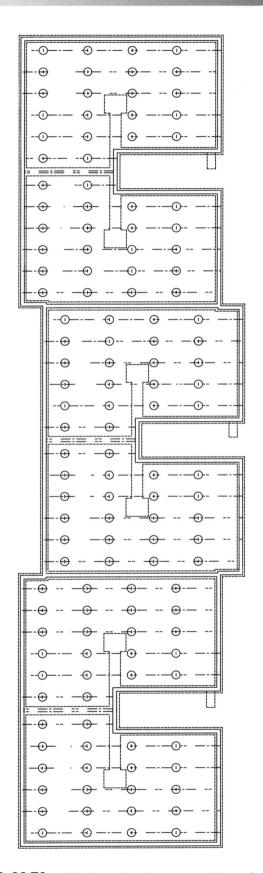

FIGURE 23-78 ■ The base plan for a post-and-beam foundation.
Courtesy Scott R. Beck, Architect.

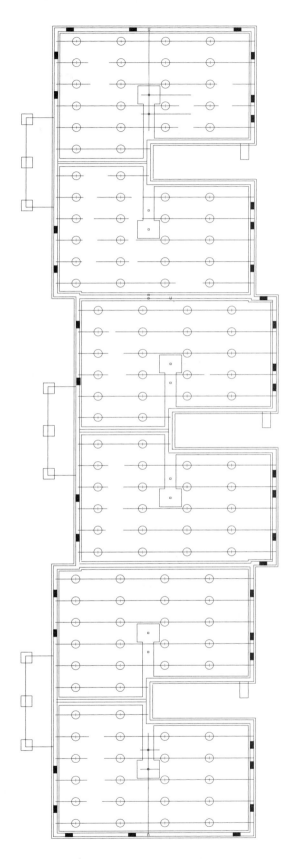

FIGURE 23-79 ■ Adding required columns, anchors, vents, and access.
Courtesy Scott R. Beck, Architect.

The major features of the drawing have now been dimensioned. The foundation plan should now resemble Figure 23-80. The drawing can be completed by adding all required text and detail markers:

24. Describe all footing and stem wall sizes

25. Describe all required mudsill, ledgers, and anchor bolts

26. Place notes and dimensions to locate and describe all metal straps, anchors, and connectors

27. Place notes to describe all vents, wall openings, and the crawl access

28. Describe all girders, post, and piers

29. Describe all required changes in elevation

30. Place notes to describe the crawl space covering

31. Place notes to describe all special blocking

32. Place any general notes specified by the engineer

33. Place a title and scale below the drawing

The completed foundation plan should now resemble Figure 23-81.

LAYER GUIDELINES FOR FOUNDATION PLANS

As with other drawings, the layers used to draw the foundation drawings should be carefully considered. Typically a prefix of S-FND or S-SLAB can be used to separate the drawing from other drawings that may be contained in the drawing file. Layers such as GRIDS and BASEDIM created for other drawings can be thawed and used with the foundation drawings. Common layer names based on AIA guidelines include the following:

Layer	Description
S-FNDN	Foundation
S-FNDN-ABLT	Anchor bolts
S-FNDN-ANNO	Local text
S-FNDN-ANNO-NOTE	General notes
S-FNDN-ANNO-LEGN	Legends and schedules
S-FNDN-BEAM	Beams
S-FNDN-DECK	Structural floor or roof decking
S-FNDN-JOIS	Joists
S-FNDN-PATT	Hatch patterns
S-FNDN-RBAR	Reinforcing
S-FNDN-SYMB	Symbols
S-GRID	Column grids
S-GRID-DIMS	Column grid dimensions
S-GRID-EXTR	Exterior column grids
S-IDEN	Identification tags
S-SLAB	Slab plan
S-SLAB-EDGE	Edge of slab
S-SLAB-JOIN	Slab control joints
S-SLAB-RBAR	Slab reinforcing

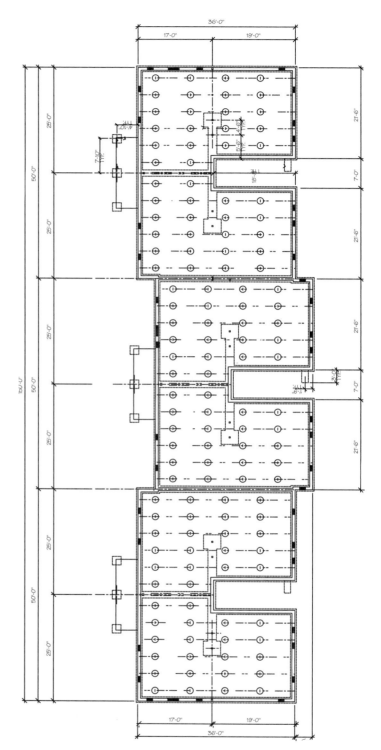

FIGURE 23-80 ■ Placement of interior and exterior dimensions. *Courtesy Scott R. Beck, Architect.*

Status field modifiers similar to other drawing areas should be used throughout structural drawings. See Chapter 3 and the *AIA Layer Guidelines* for additional layer names.

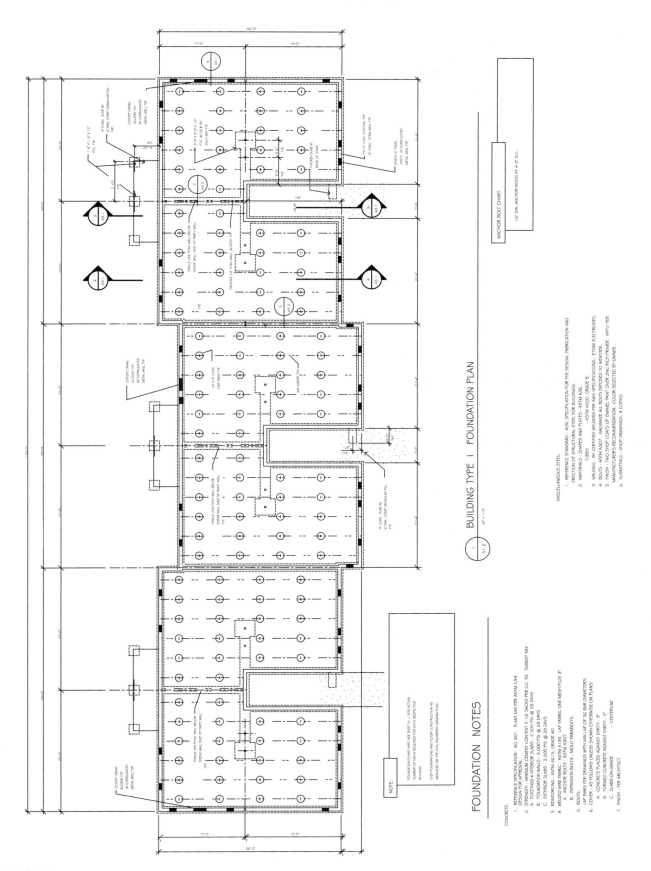

FIGURE 23-81 ■ The completed foundation plan. *Courtesy Scott R. Beck, Architect.*

CHAPTER QUIZ

Answer the following questions on a separate sheet of paper.

Question 23-1 Describe the difference between fill, backfill, uncompacted, and compacted fill.

Question 23-2 List four classifications used to describe soil texture, and describe the bearing capacity of each.

Question 23-3 What is the main cause of settlement?

Question 23-4 What problems does excess moisture in the soil cause?

Question 23-5 What type of soil is most susceptible to the effects of freezing?

Question 23-6 What are the names and sizes of the two components that comprise a spread footing for a two-story structure?

Question 23-7 What is the minimum size of reinforcing steel recognized by most codes?

Question 23-8 What is the required distance a stem wall must extend above the finish grade?

Question 23-9 List two methods of resisting lateral pressure applied to a stem wall.

CHAPTER 23 *Foundation Systems and Components*

Question 23-10 Describe the requirements for stepping a footing.

Question 23-11 How is masonry veneer typically supported on an exterior footing?

Question 23-12 Describe a method of transferring lateral loads to the floor system when joists are parallel to the retaining wall.

Question 23-13 Why is the spacing for anchor bolts usually different for stem walls and retaining walls?

Question 23-14 What is the minimum bearing required for a girder resting on concrete?

Question 23-15 List two methods of attaching a wood floor to a masonry wall.

Question 23-16 How is a cantilevered retaining wall different from a retaining wall?

Question 23-17 List two methods of providing support for a concrete slab floor.

Question 23-18 What is the purpose of a grade beam?

Question 23-19 Why are interior piers for wood floors required to be recessed in the excavated grade?

Question 23-20 Describe the difference between a piling and a pedestal.

Question 23-21 What two methods are used to determine the depth a piling is to be driven?

Question 23-22 What is the advantage of having a bulbous piling?

Question 23-23 What type of joint would be provided if enough time to pour a slab were not available?

Question 23-24 What is coverage and why is it important?

Question 23-25 List and describe two methods of post-tensioning a concrete slab.

DRAWING PROBLEMS

Draw the following details in generic form for future use in a library. Use a scale suitable for showing materials typically associated with the detail. Required details can then be amended for use with a specific foundation plan Skeletons of the plans and details can be accessed from http://www.delmar.com/resources/ocl.html. Use these drawings as a base to complete the assignment.

■ Use appropriate symbols, linetypes, dimensioning methods, and notations to complete the drawing

■ Create the needed layers to keep major groups of information separated

■ Unless specified, select a scale appropriate to display the required material and determine LTSCALE and DIM-VARS

Problem 23-1 Draw a footing suitable for a two-story office building with a concrete slab and 2 × 6 stud walls.

Problem 23-2 Draw a foundation suitable for a one-level concrete block structure with a concrete slab floor system.

Problem 23-3 Draw a foundation suitable for a one-level concrete block structure with brick veneer and a concrete slab floor system.

Problem 23-4 Draw a foundation suitable for a one-level masonry structure with a wood floor system supported by a ledger.

Problem 23-5 Draw a detail to support a two-level bearing wall with a post-and-beam floor for the lower system with beams parallel to the stem wall.

Problem 23-6 Draw a detail to represent a two-level post-and-beam floor with beams perpendicular to the stem wall.

Problem 23-7 Draw a detail to represent a non-bearing party wall resting on a concrete slab. Use 2 × 4 studs at 16" o.c. with 1" clear between walls.

Problem 23-8 Draw a detail to represent a nonbearing party wall resting on parallel wood beams. Use 2 × 4 studs at 16" o.c. with 1" clear between walls. Support each wall on 4× beams on wood posts on an 18" wide footing. Place beams parallel to the party wall.

Problem 23-9 Draw a retaining wall for an 8'–0" high wall supporting 12" deep floor joists resting on a 2 × 6 DFPT sill. Show details representing floor joists parallel to the wall.

Problem 23-10 Draw a 24" × 14" deep footing to support a 6" wide precast concrete panel. Show a 3 × 3 × 9 slot at 60"o.c. for panel supports in the footing filled with non-shrink grout. Show the wall resting on 1" non-shrink grout. Use the Internet, *Sweet's Catalogs*, or local vendor material to specify a structural connector that is suitable for concrete panels.

Use the reference material from preceding chapters, local codes, and vendor catalogs to complete one of the following projects. Unless other instructions are given by your instructor, draw the foundation plan that corresponds to the floor plan that was drawn in Chapter 15. Skeletons of the plans and details can be accessed from http://www.delmar.com/resources/ocl.html. Use these drawings as a base to complete the assignment.

■ Use appropriate symbols, linetypes, dimensioning methods, and notations to complete the drawing

■ Create the needed layers to keep major groups of information separated

■ Provide complete dimensions to locate all walls, footings, piers, anchors, and other structural material

■ Determine any unspecified sizes based on material or practical requirements

■ Place a detail reference bubble on the slab or foundation plan to represent only the details that you or your team will draw

■ Unless specified, select a scale appropriate to plotting on "D" size material and determine LTSCALE and DIMVARS.

Note: *The sketches in this chapter are to be used as a guide only.* You will find that some portions of the drawings do not match things that have been drawn on other portions of the project. Each project has errors that will need to be solved. If you think you have found an error, do not make changes to the drawings until you have discussed the problem and possible solutions with your engineer (your instructor).

Information is provided on drawings in Chapters 15 through 22 relating to each project that might be needed to make decisions regarding the slab or foundation plan. You will act as the project manager and will be required to make decisions about how to complete the project. Any information not provided must be researched and determined by you unless your instructor (the project architect and engineer) provides other instructions. When conflicting information is found, information from preceding chapters should take precedence. If conflicting information is found, use the order of precedence introduced in chapter 14.

Problem 23-11 Draw a foundation plan showing a 5" concrete slab floor for each unit of the four-unit condominium started in Problem 15-1. Place the slab over 4" of crushed rock or compacted sand and suitable slab insulation for your area. Use 3500 psi concrete for the garage floor and slope the garage floor to the 9'–0" door openings. Use 5/8" anchor bolts embedded 10" at 48" o.c. for all walls unless noted. Use 16" spacings on the south walls. Specify and locate all hold-down anchors and straps that were specified on the framing plans.

Use a 16" wide × 12" deep footing under the 8' high retaining wall at the east wall of the project. Show an 8" wide retaining wall in the northeast corner of each unit. Represent a stem wall and footing under the wall between the garage and storage area. Use a similar footing under the wall supporting the stairs. Use a 24" × 24"

× 12" deep pier at each corner of each deck with a suitable column base for a 6 × 6 post.

Problem 23-12 Draw the foundation plan showing a 5" concrete slab floor for the structure started in Problem 15-2. Use 12 × 12/4" × 4" WWM 2" up from the bottom of the slab. Place the slab over 4" crushed rock or compacted sand. Use 3000 psi concrete for the floor slab and 3500 psi for footings. Specify an expansion joint that is 48" in from each exterior wall, suitable slab insulation for your area, and a soil-bearing value of 2500 psf.

Use a 16" wide × 12" deep footing under the 8" non-bearing CMU perimeter walls and a 20" × 12" deep footing under the bearing walls. Use a continuous 12" × 12" footing under the slab at all wall openings. Use a 36" sq × 12" deep footing to support the left pilaster, and a 45" sq. × 12" footing to support the right pilaster. Use a 4'–9" sq. × 15" deep pier under the steel column. Support the column on a 14" square pilaster set at 45° to the footing. Provide a suitable joint to isolate the slab from the pilaster.

Use 21" sq × 12" deep piers on each side of the small opening on the south wall, and 30" square piers on each side of the large openings on the south wall. Specify that all footings are to extend a minimum of 18" into the natural grade.

Problem 23-13a Use the drawing below to complete the slab-on-grade plan showing a 5" concrete floor slab (4" @ office) for the structure started in Problem 15-3. Show all openings in block walls as per the floor plan. Use 12 × 12/ 4" × 4" WWM 2" up from the bottom of the slab. Place the slab over 4" crushed rock or compacted sand. Use 3500 psi concrete for the floor slab. Specify target strengths for concrete to be in accordance with ACI 318 requirements for reinforced concrete. Specify steel reinforcement to be ASTM A615 grade 40 smooth rebar at all joints.

Support each column on a 14" square pedestal, set at 45° to grid C/D. Provide a 1/4" fiber joint to isolate the slab from the pedestal. Provide suitable joints between each pedestal and between the pedestals and outer walls. Provide expansion joints 48" in from each exterior wall and one expansion joint @ grid D of the sales area. Provide a 3" diameter floor drain and a sloped floor for each bathroom with a 1/8"/12" slope.

Problem 23-13b Use the attached drawing to complete a foundation plan for the structure started in Problem 15-3. Show all openings in block walls as per floor plan. Use 3500 psi concrete for the footings and pedestals with each extending 24" minimum into the natural grade. Specify target strengths for concrete to be in accordance with ACI 318 requirements for reinforced concrete.

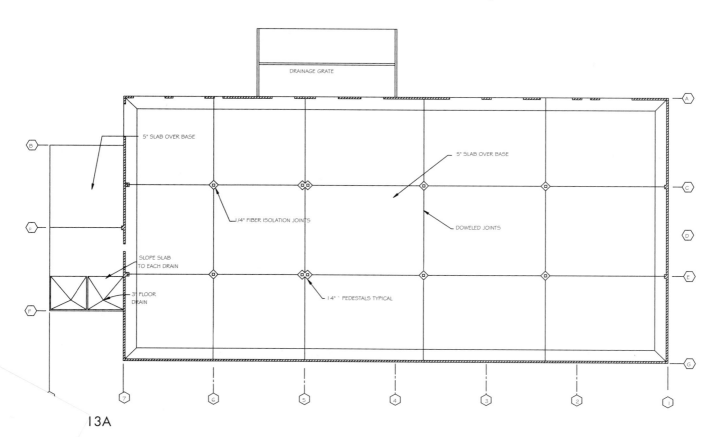

13A

Specify steel reinforcement to be ASTM A615 grade 40 deformed bars unless noted 1 1/2" clear of each surface. Draw and specify the following footings:

    a. 24" × 12" dp. w/ 3/4" dia. × 10" A.B. @ 48" o.c.

    b. 16" × 8" dp. w/ 5/8" dia.v10" A.B. @ 48" o.c.

    c. 42" × 15" dp. w/ (4)-#5 cont.

    d. 36" × 15" dp. w/ (4)-#5 cont.

    e. 24" × 15" dp. w/ (4)-#5 cont.

    f. 16" × 12" dp. w/ (2)-#5 cont.

    g. Thicken slab to 15" × 12" dp.

At the loading dock, support the 4' maximum high retaining wall on a 14" deep footing that tapers from 54" to 30". Reinforce footing w/ #5 rebar @ 10" o.c. ea. way/top & btm. 2" dn, and 3" up. Reinforce wall w/ #5 @ 16" o.c. both ways. Provide #5 × 16" (into wall) × varies into wall and ftg. Solid grout all steel cells. See the attached detail and complete per area requirements.

Draw and specify the following piers.

    1. 6'–0" × 6'–0" × 18" dp. (5)-#8 E.W. top /btm.

    2. 1'–6" × 4'–0" × 12" dp. unreinforced

    3. 8'–4" × 8'–4" × 18" dp. (7)-#8 E.W./top/btm.

    4. 6'–6" × 5'–6" × 18" dp. (6)-#6 E.W./top/btm.

    5. 4'–9" × 4'–9" × 14" dp. (6)-#5 E.W./btm. only

    6. 9'–4" × 9'–4" × 18" dp. (8)-#8 E.W./top/btm.

    7. 7'–3" × 7'–3" × 18" dp. (6)-#8 E.W./top/btm.

    8. 3'–3" × 3'–3" × 14" dp. (5)-#5 E.W./btm. only

    9. 6'–9" × 6'–9" × 14" dp. (6)-#8 E/W./top/btm.

  10. 4'–6" × 4'–6" × 14" dp. (5)-#6 E/W./btm. only (provide for each end of bm over office area)

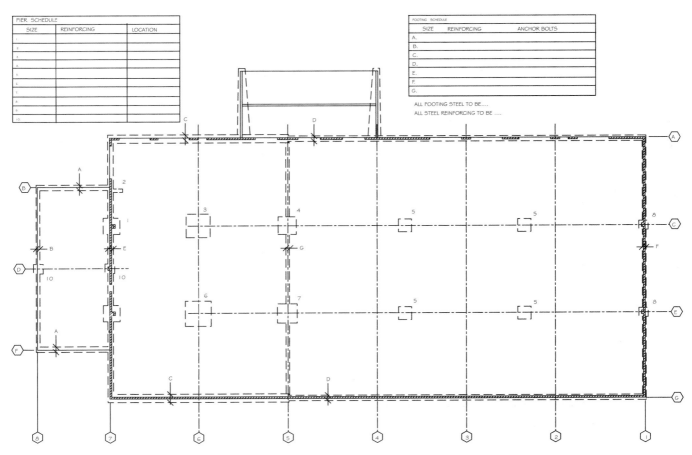

PROBLEM 23-13B

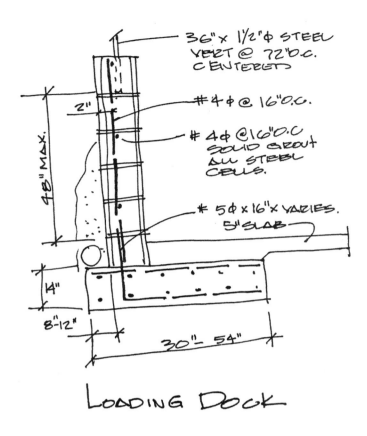

LOADING DOCK

PROBLEM 23-13C

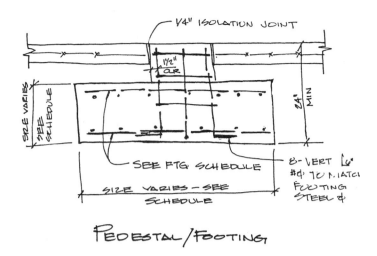

PEDESTAL/FOOTING

PROBLEM 23-13D

Problem 23-14a Use the attached drawing to complete the slab-on-grade plan showing a 5" concrete floor slab (4" in office /sales area) for the structure started in Problem 15-4. Show all openings in concrete walls as per the floor plan and panel elevations. Use 12 × 12/4" × 4" WWM 2" up from the bottom of the slab. Place the slab over 4" crushed rock or compacted sand. Use 3500 psi concrete for the floor slab. Specify target strengths for concrete to be in accordance with ACI 318 requirements for reinforced concrete. Specify steel reinforcement to be ASTM A615 grade 40 smooth rebar at all joints.

Support each column on a 14" square pedestal set at 45° to grid 3. Provide a 1/4" fiber isolation joint to isolate the slab from the pedestal. Provide suitable joints between each pedestal and between the pedestals and outer walls. Provide joints at grids 3, b, running N/S at each column from grids 1 to 5, and joints running E/W @ grid 3+24' and 5+24'. Provide an expansion joint that is 48" in from each exterior wall. Provide a 3" diameter floor drain and a sloped floor for each bathroom with a 1/8"/12" slope.

Provide 3/4" diameter L structural connection inserts w/ 3/4" dia. × 25" steel inserts @ 60" o.c. in all portions of the slab where footings are 6'–0"or deeper below the finish floor. See foundation plan for ftg. Elevations.

Insert details 10-2, 10-4, 10-5, and 10-6 on or near the slab plan and reference each to the plan. Edit as required.

Problem 23-14b Use the attached drawing to complete the foundation plan for structure started in Problem 15-4. Use 3500 psi concrete for the footings and pedestals with each extending 18" minimum into the natural grade using the bottom of footing elevations. Specify target strengths for concrete to be in accordance with ACI 318 building code requirements for reinforced concrete. Specify steel reinforcement to be ASTM A615 grade 40 deformed bars unless noted, 2" clear of each surface. Draw and specify the following footing:

    a. 24" × 14" dp. with (3)-#5

    b. 20" × 14" dp. with (2)-#5

    c. 42" × 14" dp. with (5)-#6

    d. 36" × 14" dp. with (4)-#6

    e. 34" × 14" dp. with (3)-#5

    f. 22" × 14" dp. with (2)-#5

    g. Thicken slab to 15" × 12" deep from top of slab with (2)-#4

    h. 15" × 18" deep from top of slab with (2)-#5

At the loading dock, support the 4' maximum high retaining wall on a 14" deep footing that tapers from 54" to 30". See detail in Problem 23-13c and adjust for poured concrete.

Draw and specify the following piers.

    1. 7'–9" × 7'–9" × 18" dp. (5)-#8 E.W. / top /btm.

    2. 9'–6" × 9'–6" × 18" dp. (8)-#8 E.W./ top/btm.

    3. 6'–6" × 6'–6" × 14" dp. (7)-#6 E.W./ top/btm.

    4. 3'–9" × 3'–9" × 14" dp. (4)-#6 E.W./ btm. only

    5. 5'–3" × 5'–3" × 14" dp. (6)-#5 E.W./ btm. only

Note: Locate pier #1 under the column in the north portion of the display area.

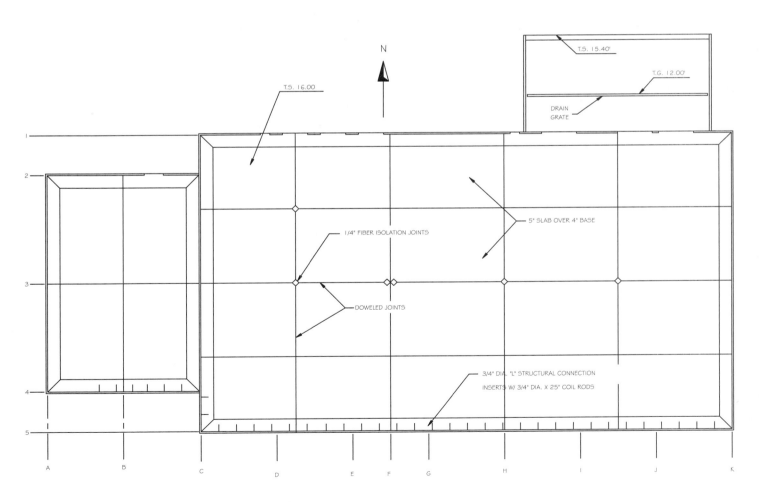

PROBLEM 23-14A

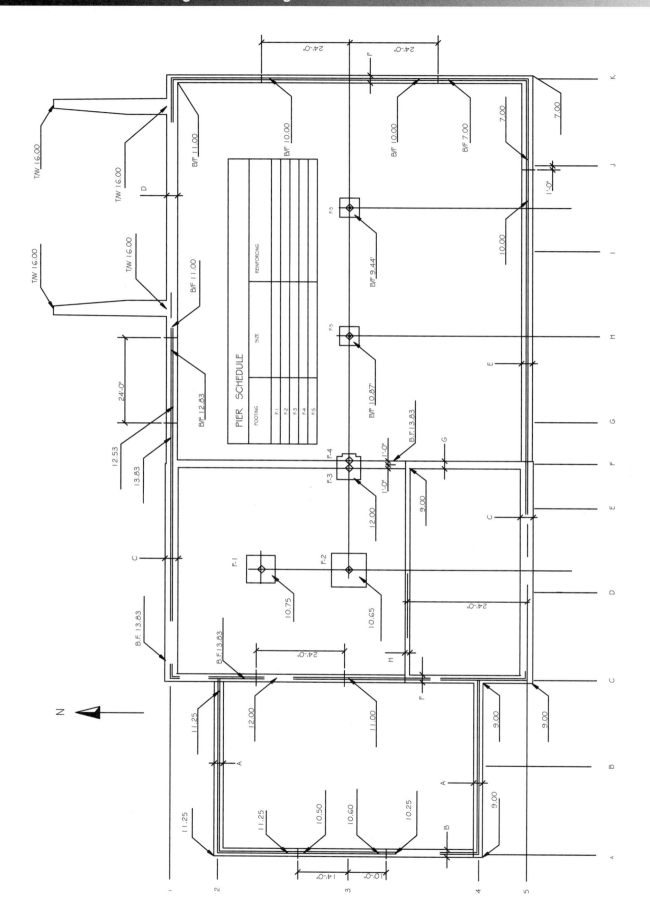

PROBLEM 23-14B

Index